W0018643

Springer Series in Statistics

Springer Series in Statistics

Alho/Spencer: Statistical Demography and Forecasting
Andersen/Borgan/Gill/Keiding: Statistical Models Based on Counting Processes
Atkinson/Riani: Robust Diagnostic Regression Analysis
Atkinson/Riani/Ceriloi: Exploring Multivariate Data with the Forward Search
Berger: Statistical Decision Theory and Bayesian Analysis, 2nd edition
Borg/Groenen: Modern Multidimensional Scaling: Theory and Applications, 2nd edition
Brockwell/Davis: Time Series: Theory and Methods, 2nd edition
Bucklew: Introduction to Rare Event Simulation
Cappé/Moulines/Rydén: Inference in Hidden Markov Models
Chan/Tong: Chaos: A Statistical Perspective
Chen/Shao/Ibrahim: Monte Carlo Methods in Bayesian Computation
Coles: An Introduction to Statistical Modeling of Extreme Values
Devroye/Lugosi: Combinatorial Methods in Density Estimation
Diggle/Ribeiro: Model-based Geostatistics
Dudoit/Van der Laan: Multiple Testing Procedures with Applications to Genomics
Efromovich: Nonparametric Curve Estimation: Methods, Theory, and Applications
Eggermont/LaRiccia: Maximum Penalized Likelihood Estimation, Volume I: Density Estimation
Fahrmeir/Tutz: Multivariate Statistical Modeling Based on Generalized Linear Models, 2nd edition
Fan/Yao: Nonlinear Time Series: Nonparametric and Parametric Methods
Ferraty/Vieu: Nonparametric Functional Data Analysis: Theory and Practice
Ferreira/Lee: Multiscale Modeling: A Bayesian Perspective
Fienberg/Hoaglin: Selected Papers of Frederick Mosteller
Frühwirth-Schnatter: Finite Mixture and Markov Switching Models
Ghosh/Ramamoorthi: Bayesian Nonparametrics
Glaz/Naus/Wallenstein: Scan Statistics
Good: Permutation Tests: Parametric and Bootstrap Tests of Hypotheses, 3rd edition
Gouriéroux: ARCH Models and Financial Applications
Gu: Smoothing Spline ANOVA Models
Györfi/Kohler/Krzyżak/Walk: A Distribution-Free Theory of Nonparametric Regression
Haberman: Advanced Statistics, Volume I: Description of Populations
Hall: The Bootstrap and Edgeworth Expansion
Härdle: Smoothing Techniques: With Implementation in S
Harrell: Regression Modeling Strategies: With Applications to Linear Models, Logistic Regression, and Survival Analysis
Hart: Nonparametric Smoothing and Lack-of-Fit Tests
Hastie/Tibshirani/Friedman: The Elements of Statistical Learning: Data Mining, Inference, and Prediction
Hedayat/Sloane/Stufken: Orthogonal Arrays: Theory and Applications
Heyde: Quasi-Likelihood and its Application: A General Approach to Optimal Parameter Estimation
Huet/Bouvier/Poursat/Jolivet: Statistical Tools for Nonlinear Regression: A Practical Guide with S-PLUS and R Examples, 2nd edition
Iacus: Simulation and Inference for Stochastic Differential Equations
Ibrahim/Chen/Sinha: Bayesian Survival Analysis
Jiang: Linear and Generalized Linear Mixed Models and Their Applications
Jolliffe: Principal Component Analysis, 2nd edition
Konishi/Kitagawa: Information Criteria and Statistical Modeling
Knottnerus: Sample Survey Theory: Some Pythagorean Perspectives
Kosorok: Introduction to Empirical Processes and Semiparametric Inference
Küchler/Sørensen: Exponential Families of Stochastic Processes
Kutoyants: Statistical Inference for Ergodic Diffusion Processes
Lahiri: Resampling Methods for Dependent Data
Lavallée: Indirect Sampling
Le Cam: Asymptotic Methods in Statistical Decision Theory
Le Cam/Yang: Asymptotics in Statistics: Some Basic Concepts, 2nd edition

(continued after index)

Carl N. Morris
Robert Tibshirani
Editors

The Science of Bradley Efron

Selected Papers

Carl N. Morris
Department of Statistics
Harvard University
Cambridge, MA 02138
USA
morris@stat.harvard.edu

Robert Tibshirani
Department of Statistics
Stanford University
Stanford, CA 94305
USA
rob.tibshirani@gmail.com

ISBN: 978-0-387-75691-2 e-ISBN: 978-0-387-75692-9
DOI: 10.1007/978-0-387-75692-9

Library of Congress Control Number: 20079401521

Printed on acid-free paper.

9 8 7 6 5 4 3 2 1

springer.com

Foreword

Nature didn't design human beings to be statisticians, and in fact our minds are more naturally attuned to spotting the saber-toothed tiger than seeing the jungle he springs from. Yet scientific discovery in practice is often more jungle than tiger. Those of us who devote our scientific lives to the deep and satisfying subject of statistical inference usually do so in the face of a certain under-appreciation from the public, and also (though less so these days) from the wider scientific world.

With this in mind, it feels very nice to be over-appreciated for a while, even at the expense of weathering a 70th birthday. (Are we certain that some terrible chronological error hasn't been made?) Carl Morris and Rob Tibshirani, the two colleagues I've worked most closely with, both fit my ideal profile of the statistician as a mathematical scientist working seamlessly across wide areas of theory and application. They seem to have chosen the papers here in the same catholic spirit, and then cajoled an all-star cast of statistical savants to comment on them.

I am deeply grateful to Rob, Carl, and the savants for this retrospective view of what I like to think of as "my best work so far". Good colleagues are an irreplaceable ingredient of good science, and here I am in better-than-good company. Statistics is the field of the future, now a future approaching closely; and of course I hope this look at the past also has something to say, both in the papers and the commentaries, about what might be happening next.

Bradley Efron
Stanford, CA
August 2007

Preface to this Special Edition

by Carl Morris and Rob Tibshirani

A Brief Biography and Career Profile

This volume honors Bradley Efron on the occasion of his 70th birthday. Brad is widely admired for his past and continuing genius, evidenced in his pioneering contributions to statistical theory, applications, and computation. Statisticians everywhere have benefited enormously from Brad's bullishness on the future of our subject and from his constant championing of statistics as crucial to scientific thinking. As an ambassador of statistics who is widely recognized by other fields and who reaches out as an administrator, a writer and a speaker, Brad has worked tirelessly to raise the visibility and appeal of statistics to other disciplines. In the end, however, it is Brad's humanity, his humility, and his generosity that especially inspire our enthusiasm for this project.

Bradley Efron was born in May, 1938, in St. Paul, Minnesota to Miles Efron and Esther Kaufmann, Jewish-Russian immigrants. The Efrons have always been an intellectual family. Brad's older brother, Arthur, is a retired Professor of English, while his younger brothers, twins Don and Ron, are psychiatric social workers. All three still are journalists and/or writers. Brad's only child is his son Miles, whose mother is Gael, a painter and Brad's former wife. Miles, now with his own Ph.D., serves on the University of Texas School of Information faculty. Some of Miles' current research interests connect with statistics, including devising statistics for information retrieval, machine learning with large bodies of text, information theory, and developing probabilistic frameworks for studying information in games. In 1987, Brad met Donna Spiker. Now a developmental psychologist at SRI in Palo Alto, Donna holds a Ph.D. in child development from the University of Minnesota and manages research programs investigating early childhood programs for children with disabilities and other special needs. Donna and Brad have been close companions for the past 20 years.

Brad's passion for scientific thinking and his natural attraction to vital research areas has led to a remarkable career, as evidenced by his many distinctions and honors. These include the MacArthur "genius" award, his elections to the National Academy of Sciences and to the American Academy of Arts and Sciences, Stanford's School of Humanities and Sciences endowed Max H. Stein Professorship, honorary doctorates from the University of Chicago as well as universities in Norway and Spain, and named-lectureship invitations from statistics societies around the world. He has been President of the IMS and of the ASA, Theory and Methods Editor for *JASA*, and he is the founding editor of the new *Annals of Applied Statistics*.

As we write this in the summer of 2007, Brad has just been named one of eight US scientists to receive the 2005 National Medal of Science, the sixth statistician to be so honored since the award's 1959 inception, with Jerzy Neyman, John Tukey, Joseph Doob, Sam Karlin and C. R. Rao his predecessors. On July 27, 2007, the President in a White House ceremony awarded this, the nation's highest scientific honor. The National Medal of Science citation acknowledges Brad with these words:

> For his contributions to theoretical and applied statistics, especially the bootstrap sampling technique; for his extraordinary geometric insight into nonlinear statistical problems; and for applications in medicine, physics, and astronomy.

Characteristically, Brad has accepted each new honor not only for himself but also to advance statistics as integral to scientific thinking. Each of his honors has served and made more visible the whole field of statistical science, as will this latest one, too.

Over the years, Brad has had an enormous impact at Stanford University. He has been Chair of the Statistics Department three times, Chair of the Stanford Faculty Senate, Chair of the Mathematical and Computational Science program for 25 years, Associate Dean of the School of Humanities and Sciences, and twice Chair of the Stanford University Advisory Board. He also was at the forefront of planning the new Sequoia Hall, a beautiful home for the Statistics and Math/CompSci departments, and a desirable home-away-from-home for dozens of visiting scholars and researchers every year.

John Tukey said, "The best thing about being a statistician is that you get to play in everyone's backyard." And it's a sentiment that holds true for many of us. Brad has thrived in that playground, statistical theory and scientific applications being two sides of the same coin for him. Brad has always been drawn to building new models for data, such as those concerning astrophysics red-shift measurements and biostatistics (two disciplines that Brad has labeled "the odd couple" because their censored data follow parallel models), and those from biology's microarrays, as well as data and models from many other fields, from Shakespeare's vocabulary to sports performance.

Brad's wide and deep theoretical contributions include areas he has named, such as the bootstrap, and biased coin designs. He even put the "curve" in curved exponential families. He has been a pioneer in Stein estimation and empirical Bayes, computer-intensive methods, the geometry of exponential families, survival analysis, decision theory, statistical foundations, Fisher information, and likelihood. You will find some of his best work on these topics in the papers reprinted in this book.

We are privileged to have been Brad's two primary co-authors: Carl in the first half of his career, and Rob in the period since 1980. We have chosen to write separately about our respective portions of Brad's career in order to include our personal perspectives, but of course that makes these recollections both selective and subjective. Naturally, Brad's many other friends and colleagues would emphasize other aspects of his work, and some of that emerges in the memories recorded by his former Ph.D. students, and in the introductions to his papers newly-written for this volume by many of those same friends and colleagues.

For those who want to know more of Efron's own point of view, two short interviews with Brad reveal much about how he thinks, and where he thinks the profession is going. One article appeared in the 2003 *Statistical Science* "Silver Anniversary of the Bootstrap" issue, and is titled "Bradley Efron: A Conversation with Good Friends," with Susan Holmes, Carl Morris and Rob Tibshirani. A second interview came out in 1995 and can be found online at the Irwin/McGraw-Hill Website (www.mhhe.com/business/opsci/bstat/efron.mhtml). In this piece, Brad comments on trends in statistics and on advances in the field based on improving uses of the computer.

At the end of the interview, when asked about his outside interests, Brad concluded by saying:

> I love statistics. I live right at the Stanford campus. I go in all the time. I don't work hard every day, though. I work every day. In the evenings I like to go to movies with my girlfriend, and I go to almost all movies. I'm interested in astronomy and science. I claim that I'm an amateur scientist as well as a professional scientist. And I love science. I think it's the greatest thing people ever thought of. So I guess my hobby is science, too. I like to sit around and talk . . . Movies, stat, and science.

Brad's Academic History 1956–1980, by Carl Morris

My friendship with Brad Efron began in 1956 when we were freshmen at The California Institute of Technology in Pasadena. In those years, Caltech's 600 undergraduates were all male, and we were all science or engineering majors. The extraordinary scientific environment there exposed us to faculty like Linus Pauling and Richard Feynman, who were regular lecturers for large undergraduate classes. Statistics, however, hardly existed at Caltech at the time, partly because it was a smaller and newer field then, and partly because Caltech faculty would remind us of Einstein's dictum: "God does not play dice with the universe." I still joke that I understood the sample mean before arriving at Caltech, and graduated without understanding the standard deviation: and it's utterly true.

Despite Caltech's deficit in probability and statistics, the scientific and empirical training there provided us with a deep background in mathematics and scientific thinking that made statistics a natural next step for surprisingly many. In fact, some of our former Caltech contemporaries are among today's academic leaders in statistics and probability. Undergraduates from our own student residence, Ricketts House, over which Brad presided in our senior year, include professors John Walsh, Mike Perlman and Gary Lorden, while Peter Bickel, Larry Brown, Chuck Stone and others resided in other houses nearby.

Brad majored in mathematics, but his main passion was for the parts of mathematics that empower scientific thinking. Brad developed an interest in statistics in his senior year and so requested a special reading course from Harald Cramér's book, *Mathematical Methods of Statistics* (1951). We can thank Cramér and that wonderful text for hooking Brad onto statistics.

Besides being an academic star at Caltech, Brad had various social and extracurricular interests that foreshadowed his leadership in our field. With a natural wit, beyond that suggested by his left-handedness, he wrote a humor column at Caltech, and did so with a flair that led him to flirt with humor writing as a profession. His wit is evident in many of his research papers, and even more so in other, lighter articles, such as his President's Corner columns for the ASA newsletter; the April 2004 column is reprinted as part of this retrospective. Even after settling on Ph.D. study in mathematics, Brad chose Stanford partly for the opportunity to write for *The Chaparral*, Stanford's well-known humor magazine.

Upon his 1960 arrival at Stanford, Brad joined the *Chaparral* staff and soon became its editor. Shortly after, he led The Chappy to publish a parody of *Playboy Magazine*. In those staid times, some articles in that now-famous issue offended members of the Stanford administration and of the Bay Area clergy. Brad was denounced from the pulpit of the Catholic Church for his authorship of one piece that Stanford's administration described as "a burlesque of the Incarnation or Virgin Birth." Local newspapers seized on that juicy topic, and Brad says he was more famous for those next few weeks than he's been ever since. Stanford came under heavy pressure to dismiss him, but with characteristic self-assurance, Brad fought back. Partly in recognition of his obvious intellectual potential, influential members of Stanford's faculty and administration, including statisticians and mathematicians Al Bowker, Herb Solomon and Halsey Royden, defended Brad and eventually arranged for a suspension that allowed his return in good standing.

On returning to Stanford in the Fall of 1961, Brad switched to the Statistics Department, which I also joined in 1962, after studying elsewhere in mathematics. For the next three years Brad, along with the rest of Stanford's statistics students, was exposed to an extraordinary environment. On the regular faculty were Herman Chernoff, Kai-Lai Chung, Vernon Johns, Sam Karlin, Jerry Lieberman, Rupert Miller, Lincoln Moses, Ingram Olkin, Manny Parzen, Herb Solomon and Charles Stein. Stanford's tradition of joint appointments kept Statistics intimately connected with faculty from various University departments. Our fellow Ph.D. students included Alvin Baranchik, Norm Breslow, Gerry Chase, Tom Cover, Joe Eaton, Jean-Paul Gimon, Mel Hinich, Paul Holland, Myles Hollander, Steve Portnoy, Sheldon Ross, Steve Samuels, Galen Shorack, Burt Singer, Marshall Sylvan, Howard Taylor, Grace Wahba, Ted Wallenius and James Zidek.

The alternating Berkeley–Stanford statistics colloquia, which met four or six times a year and was always heavily attended, gave students from both departments many opportunities to meet, dine, and talk with the great statisticians on those two faculties. At those gatherings, students regularly saw and talked with Berkeley's luminaries like David Blackwell, Lucien LeCam, Erich Lehmann, Jerzy Neyman, and Betty Scott, and also got to know Berkeley's Ph.D. students, including Peter Bickel, Kjell Doksum, Bruce Hoadley, John Rolph, and Stephen Stigler. The two schools together attracted famous seminar speakers, plus post-docs, summer and sabbatical visitors from all over the world. So Brad, like all of us, met and learned from the world's statistics leaders, and the world's statistics leaders knew of Brad's exceptional talents even before his graduation.

Brad's Ph.D. dissertation, called "Problems in Probability of a Geometric Nature" and officially advised by Rupert Miller and Herb Solomon, combined three related papers of his own choosing. One of those papers is reprinted in this book and commented on by Tom Cover, Brad's colleague and then a fellow Stanford Ph.D. student in electrical engineering. Brad's dissertation emphasized his omnipresent gift for geometrical thinking, and geometric ideas are in the titles of at least six of his papers.

After his 1964 Ph.D. degree, Brad accepted the Statistics Department's offer to continue on their faculty. Two years later his appointment expanded to the joint position he still holds, split between Statistics and the Biostatistics group in Stanford's School of Medicine. He has spent his entire academic career in those posts, with collateral assignments in Stanford's higher administration. Aside from sabbaticals taken at Harvard, Imperial College and Berkeley, Brad has lived at Stanford continuously since 1960.

At 31, in 1969, Brad became *JASA*'s Theory and Methods Editor. I vividly recall how he managed that time-consuming responsibility. On weekdays, his research, teaching and personal life continued as always. However, Saturdays from 1969 to 1972 were his *JASA* days. Brad would arrive at his office about 9 a.m. and work sequentially through a foot-high mound of submitted and reviewed manuscripts piled on the left side of his desk. During the day, that stack would shrink as papers were reviewed and correspondence was drafted, with each completed file then placed on the right-hand side of his desk. By dinner-time the pile, slightly expanded, would rest entirely on the right side, ready for his editorial assistant's attention on Monday, and Brad could return to his other work for another week.

Ever since his 1962 summer apprenticeship, Brad has been a regular consultant at RAND in Santa Monica, California, the original US public policy think tank. Brad's visits increased during my RAND years (1967–1978), and statistics began to flourish when John Rolph formed the RAND Statistics Group. Brad's involvement enhanced the Group's visibility and attractiveness, with his Ph.D. student Naihua Duan being among our early hires. Special among Brad's many contributions while I was there was applying his statistical wisdom to the design and analysis of RAND's famous national public policy Health Insurance Experiment. His key advice given and taken during the HIE design stage was to keep the number of insurance plans (treatments) at a minimum, and to give some HIE families extra-large participation bonuses to enable a test for whether or not families earmarked those bonuses for health care. At the end of each visit, Brad would leave with a fresh supply of RAND's gigantic paper tablets, dubbed the "Dead Sea Scrolls" by Rob. Brad always joked that these tablets, which he still relies upon to outline his papers, were his real compensation.

Especially important to me was how Brad's RAND visits supported our decade-long collaboration on Stein's shrinkage estimation and empirical Bayes. I remember our excitement one evening in 1968 outside Stanford's stat department, when we settled on this as the topic we would work on together, chosen both for its statistical promise and also for our great admiration of Stein. In those days, Stein's minimax shrinkage estimators were still too mysterious for statisticians, ourselves included, to trust for real applications. We hoped to demystify them, initially via empirical Bayes and multi-level models, and then to see if extensions could be devised for wider application.

Part of Brad's genius is revealed in his compelling and clear writing, which usually proceeds from the specific to the general. His papers often start with a deftly-chosen data set. Two such data sets greatly aided our joint work, and these data show up in the two Efron–Morris papers reprinted in this volume.

To illustrate the applicability of Stein's shrinkage estimators, we obtained the early-season (1970) batting averages of 18 baseball players, each sampled after batting 45 times. (Since Stein's estimators require homogeneous variances, shrinkage must involve estimates with the same sample sizes.) The James–Stein shrinkage estimator was sensible for these data because baseball players' batting averages were generally understood to exhibit regression-toward-the-mean patterns. What sealed the deal, after the 18 "true values" became available (these being the players' batting averages over the remainder of the season), was seeing that the shrinkage estimates for these data were a whopping three times more accurate than using the players' separate averages.

Our second example concerned estimating toxoplasmosis prevalence rates in 36 areas of El Salvador, from Brad's work at the Stanford School of Medicine. As is typical of most applications, the variances of prevalence estimates differed greatly. Accounting for that inequality required generalizing Stein's shrinkage approach via a two-level model. That led us to develop sensible estimators from a framework that data analysts now understand and apply regularly.

Our intensive co-authorship succumbed to the combination of my move from California in 1978 with Brad's discovery of the bootstrap while preparing for his 1977 IMS Rietz Lecture, which was aimed at explaining Tukey's jackknife. Our last joint paper in that 1970s series was "Stein's paradox in statistics," which appeared in the May 1977 issue of *Scientific American* (and is reprinted in this book). Individually, we both have continued to develop this topic since then. Microarray data, which demands making simultaneous statistical inferences, has especially inspired Brad to return repeatedly to empirical Bayes and multi-level models.

Brad, for over a half-century now you have been the quintessential brilliant, generous, and fun colleague and friend. Working with you has made so many of us feel special, as legions of others have been similarly blessed to discover.

Brad's Academic History After 1981, by Rob Tibshirani

I came to Stanford as a graduate student in 1981, after completing undergraduate and master's degrees at the Universities of Waterloo and Toronto, respectively. Two of my early teachers spoke in glowing terms about Brad, before I had met him. I had heard (second-hand) that Jack Kalbfleisch was in awe of Brad's work on the efficiency of the Cox estimator, a problem that Jack had worked on himself. And David Andrews, whose opinion I respect greatly, told me in his usual colorful style that "You can learn more from Brad in an hour that you can from most people in a year." After getting to know Brad, I realized that David's claim needed only slight modification. Technical conversations with Brad tended to be very intense but brief – usually it was not more than 20 minutes before his mind wandered over to his funny, sarcastic take on the political news of the day, or the latest (often dirty) joke he had heard from Ed George. But that 20-minute window of opportunity was priceless.

Stanford Statistics in 1981 had a wonderful faculty and student body. I took courses and learned from Ted Anderson, Byron Brown, Persi Diaconis, Jim Fill, Jerry Friedman, Vernon Johns, Rupert Miller, Lincoln Moses, David Siegmund, Charles Stein, Werner Stuetzle, Paul Switzer, and the visiting Andreas Buja. Ph.D. students around that time included Hani Doss, Trevor Hastie, Tim Hesterberg, Fred Huffer, Steve Lawley, Art Owen, Mark Segal, Michael Stein and Terry Therneau.

During my first year, I got to know Brad. He had (and still has) very methodical working habits. Every morning at 9 a.m. he would sit at his desk, making illegible scribbles on an enormous pad, one that looked like a modern version of the Dead Sea Scrolls. He would do this every day, including Saturday, and it was understood that you shouldn't disturb Brad before noon on any day unless it was really important.

I asked Brad to be my Ph.D. supervisor. He agreed, but made it clear that he was not a good supervisor. Unlike his own advisor, Rupert Miller (whom he admired greatly), Brad said that he was not good at posing appropriate problems for a thesis. They tended to be too easy – something he could solve in a few days, or too hard – something he couldn't solve in a few years. So I came up with my own thesis topic (Local Likelihood), and pretty much did the work on my own (with help from my classmate, Trevor Hastie). However, my experience as Brad's student was very enriching in so many other ways. We had frequent discussions about the bootstrap, at a time when he was actively working in that area. Like few others, I got a first-hand glimpse of the mind of a great scientist. I remember specifically the months he spent working on his "Better bootstrap confidence intervals" paper, in which he proposes the BCa interval. He had told me that his new paper would be a follow-up to his 1982 *Annals* paper on transformation theory. So I studied that paper closely, before I received his handwritten draft of the new manuscript. It was nearly illegible (as usual), but a brilliant piece of work. It took me the longest time to understand the clever constructions and arguments, and I have never understood how he could have thought of them. It reminds me of the quote by Mark Kac, about Richard Feynman:

> There are two kinds of geniuses: the "ordinary" and the "magicians." An ordinary genius is a fellow whom you and I would be just as good as, if we were only many times better. There is no mystery as to how his mind works. Once we understand what they've done, we feel certain that we, too, could have done it. It is different with the magicians. Even after we understand what they have done it is completely dark. Richard Feynman is a magician of the highest calibre.

This is how I feel about Brad. He is a magician. When he covers a research problem, it is from the ground up. He states (and often redefines) a problem, and then analyzes it with great mathematical dexterity and practical sensibility. And he is frustratingly thorough: many times I have read one of his papers, coming up with an idea or extension of my own along the way. Then I would turn the page and find my idea (and five others) discussed in detail. When he is done, Brad does not leave any low-hanging fruit: he usually cuts down the whole tree!

Between 1984 and 1998, I was at the University of Toronto. I visited Stanford fairly often and Brad and I became collaborators. We wrote our bootstrap book together in 1993. Along the way, and after I returned to Stanford as a Professor in 1998, I noticed that Brad's work had become more focused on applied problems. He told me that in the old days he would sit down and think of a topic to work on, while now almost all of his papers arose from a real data-analysis problem he had encountered. Brad seemed amazed that in the old days he ever could have worked "in a vacuum."

I have witnessed many more doses of Brad's magic. A special one was in the Least Angle Regression work with Brad, Trevor Hastie, Iain Johnstone and myself. It goes back to when we were writing our book, *The Elements of Statistical Learning*: we had noticed a close relationship between forward stagewise regression and the lasso, but could not nail down the details. I consulted with a number of Stanford experts on optimization, and they too could not figure it out. So in our book, we showed some numerical evidence of the relationship but left it as vague conjecture. Later, Brad started reading our newly-published book, and I suggested he might look at that section. He sat down and pretty much single-handedly solved the problem. Along the way, he developed a new algorithm, "least angle regression," which is interesting in its own right, and sheds great statistical insight on the lasso. In this work, Brad shows his great mathematical power – not the twentieth century, abstract kind of math, but the old-fashioned kind: geometric insight and analysis. He had hit another home run.

Brad has been a teacher, mentor, colleague, and great friend of mine now for over 25 years. He is a difficult collaborator (by his own admission), stubborn and single-minded, often preferring to think things through on his own. Perhaps that is why Carl and I have been only frequent collaborators. But the results are well worth the effort.

Brad loves the science of Statistics like no one else I have ever known. And he carries his greatness in his pocket, not on his lapel. He is never afraid to ask the simplest question at a seminar, inspiring me and countless other students and professors to do the same. He has been the leader of Statistics at Stanford for the past 30 years, intellectually and by the way he conducts himself. He would never, for example, push for a faculty candidate without the clear support of the other faculty. He always tries to put the Department's welfare before his own. When the guy at the top acts this way, it is hard for others not to follow. He has been incredibly supportive of me and my career, and for that I am extremely grateful.

Conclusion

In putting together this volume, we have tried to choose a collection of Brad's best work. Because of the quantity and quality of his accomplishments, this was not an easy task. There are many more papers that we would have liked to have included. We have asked distinguished colleagues in the field to comment on these papers, and we are delighted with the results. Their introductions show their admiration for Brad's legacy, and for Brad as a person. He is most deserving of this special edition, and we hope it does him justice.

Acknowledgements

We are grateful to the many contributors who made this volume possible. In the Stanford Statistics Department, Cindy Kirby's excellent ideas and considerable effort in assembling this volume were crucial. Holger Hoefling at Stanford carefully assembled the Index. John Kimmel of Springer-Verlag provided invaluable editorial expertise and leadership. And of course we especially thank Brad Efron for his inspiration.

Carl Morris
Harvard University

Robert Tibshirani
Stanford University

Comments from Former Students

Bradley Efron has had 18 Ph.D. students through 2007, all at Stanford's Department of Statistics, including Rob Tibshirani, Ph.D. in 1985. We contacted as many as we could to ask for their comments and here are their remembrances of Brad. [The remaining two are Stephen Peters, Class of 1983, and Abhinanda Sarkar, Class of 1996.]

Norman Breslow, Class of 1967

I've always felt very proud and privileged to have been Brad's first student. His mentorship and later research contributions were truly inspirational. What I remember most was Brad's emphasis on building statistical research around a meaningful scientific problem and of introducing that problem early in the research article as motivation for, rather than later as illustration of, the statistical methodology. Watching Brad's innovative approaches (and those of Lincoln Moses and Rupert Miller) to problems posed by Stanford Medical Center faculty provided invaluable training in the art of consulting, and a clear object lesson as to what statistics was fundamentally all about. All interactions with Brad were of course enlivened by his sometimes irreverent but always hilarious brand of humour. For example, while we were both in London he aptly described one of my colleagues there, who had somewhat unruly hair, as a man who looked like he was "sprouting theorems".

Jim Ware, Class of 1969

It is an honor to join the chorus of congratulations to Brad on the publication of this volume of collected works. I began my doctoral studies in 1964, graduated in 1969, and was Brad's second student. It was a privilege to be associated with the Department during those turbulent times. The faculty was uniformly distinguished and the student body was filled with future stars. Sequoia Hall provided a welcoming and collegial environment for research and conversation.

Though the Department had many distinguished senior faculty, we students welcomed the appointments of Brad and Paul Switzer. They were both superb teachers, receptive to graduate students, and great role models. We were particularly excited about Brad's work with Carl Morris on empirical Bayes estimation. Throughout my career, I have looked to Brad's work for insights about statistics and its applications to health. It is a pleasure to look back over all he has accomplished with so many colleagues and students.

Louis Gordon, Class of 1972

Brad once said to me "Programming is the only human endeavor for which 'clever' is a pejorative." It's a precept to which my thoughts frequently return, happily in the observance, all too sadly in the breach.

Stanley Shapiro, Class of 1972

It was a privilege to be [in] one of the first handfuls of students that Brad supervised, and I'm sure at times it was a handful for him. Those were exciting but turbulent times, and it's hard to imagine that 40 years have elapsed since I first encountered Brad on the side steps of Sequoia Hall, shortly after he had returned from a year at Harvard. It will be a treat to see what the next decades have in store for and from him.

Gary Simon, Class of 1972

There must be a wealth of stories about Brad Efron, so I'll just put in one little added note. This goes back to the time at which I was a graduate student, about 1970. In a group session, which might have been the regular biostatistics seminar, the topic of difficult computing came up. Brad set out this hypothetical: "Imagine that you had computers so powerful that you could get any calculations you wanted. What would you compute?" I was startled that someone would ask for something so far-out. Shouldn't we be thinking of what we can do for today's problem with today's technology? Brad didn't want to limit himself to the possible of today. It must have been this optimistic bravery that made him push forward to invent the bootstrap. And now we do not only bootstrap, but also MCMC and data mining. This little comment was a great insight to his fearlessly inventive mind.

Terence O'Neill, Class of 1976

Brad's approach to statistics made an immediate impression on me as a new graduate student from Australia. Many statisticians that I had encountered until then defined themselves by their area of research: "I'm time series," or "My area is survey sampling." Brad on the other hand seemed to go wherever his incredibly-varied interests took him, and no matter where he went, he worked deeply with the application and used or invented appropriate, sophisticated statistical methods. He always made statistics fun, engaging and important. I have tried to follow Brad's example to this day.

Arthur Peterson, Jr., Class of 1976

As one of Brad's Ph.D. thesis students from the mid-70s in old Sequoia Hall, I was privileged to witness three of Brad's special qualities. First, of course, there is Brad's inquisitive intellect and brilliance in topics both statistical and nonstatistical. Also, Brad's much admired human qualities: his genuine personableness toward all, encouraging spirit (quite helpful to me), and special love for his family. And then, there is his own practical style. Brad loved his unassuming, smallish office on his beloved Stanford campus, probably in part because it came equipped with a large artist's pad on his desk, where his mind created and his left hand scrawled his pioneering work. Congratulations, Brad, on your lifetime of achievement so far, and Happy Birthday.

Ronald Thisted, Class of 1977

When I was a graduate student, Brad invited me to work with him on estimating the size of Shakespeare's vocabulary. That work was stimulating and rewarding, and it became my model both for mentorship and for fostering good scientific collaboration. Brad takes the underlying problem seriously, and his statistical models always reflect important features of the problem that are often more easily ignored. He thinks broadly and he is generous with ideas. And he treats collaborators (no matter how junior) as equal partners, respecting and making good use of what they might know that he does not. The end result is both a satisfying collaboration and, often, scientific work that is much better than any single person could have produced.

Naihua Duan, Class of 1980

For more than 30 years, Brad provided statistical consultations regularly at RAND, a stronghold of Stanford alumni, including myself. Brad helped RAND Statistics Group grow into one of the leading applied statistics agencies. Brad's appreciation of practical applied problems, such as the consultations at RAND, probably serves as an important source of motivation for his rich and diverse statistical research.

Gail Gong, Class of 1982

I'd like to thank Brad for recognizing great talent when he saw it. But seriously folks, let me tell you a true story. My tennis partner, who I just stumbled onto on the public tennis courts in Davis several years ago, is married to a biologist. This biologist thinks I am a hot statistician because I did my thesis under the famous Bradley Efron! My memories of graduate school days include Brad walking down the hallways of Sequoia wearing a great big smile, Brad scribbling away on that giant pad of paper, Brad exuding excitement about statistics. Thank you, Brad, for giving me the bootstrap, and for sharing with me your love for statistics.

Terry Therneau, Class of 1984

One of my memories of Brad is his statement that, "By the end of the thesis, the student should know more about the topic than the advisor." This is reflected in the guidance he gave, which was designed to let you take charge of the direction of your thesis, with a helping hand as necessary. (But what a great helping hand!) It was a fantastic way to work and learn. One of the nicer things that have happened to me was being named a Fellow the year that Brad was president of the ASA: receiving the award and handshake from my one-time advisor was special, and a chance on my part to acknowledge the role that his training had in my success.

Timothy Hesterberg, Class of 1988

Brad would bike to work on an old black bicycle: not flashy, just utilitarian. He didn't wear a helmet, which strikes me as unnecessary endangerment of a unique global resource. His home was a little cottage on the Stanford campus, around the corner from the German house where I lived. He had great geometric intuition, and used to develop his work on huge scratchpads, divided into neat little boxes.

Alan Gous, Class of 1998

It was a real privilege to study under Brad. In discussions with him about some or other particular problem I was attempting to solve, he was always patient and generous, and I came away with insights into the whole business of how to think about and do research, which greatly impressed me then, and still influence me now. The particular problem may have remained, but that is another of Brad's trademarks as an advisor: you have to solve those yourself!

S. C. Samuel Kou, Class of 2001

Brad once said, "Never meet your heroes." I have to disagree, for I not only met my hero, but was fortunate enough to be his student. The four years at Stanford were some of my most cherished ones; Brad taught me not only about statistics but also how to approach problems. I remember in the summer of my first year, I began to do some reading with Brad. Marveling at his papers one after another, I asked Brad, "How did these big problems come to you?" He answered, "I had no plan to do 'big things'. I just work on whatever is fun." Brad went on to explain, "It is always fun to work on things you are really interested in." To a first-year graduate student, who was expecting some "secret recipe", this was very surprising. Years later, in my own research career, I gradually realize how true Brad's words are. Brad taught me many lessons like this, and those continue to guide my career and life. As a Chinese proverb aptly describes, "Once a teacher, forever a father."

Armin Schwartzman, Class of 2006

The first time I asked Brad to be my adviser, he recommended against it. He said I would not get much help from him. He only agreed to it a year later, after I had started my own project. Then I understood. He wanted me to learn how to be an independent researcher. And when we worked together, despite his previous claim, he helped me a lot. His comments would usually be brief, but extremely insightful and would send me off in the right direction. Brad turned out to be exactly the kind of mentor I was looking for: somebody with experience who would have vision of the field, while at the same time letting me keep my own initiative; knowledgeable, but at the same time personable; always interested in my well-being and that of my budding family. For this and more, we thank you, Brad.

Brit Turnbull, Class of 2007

Just as Brad gave statistics the Bootstrap for data, he gave me bootstraps for research, guidance that will last the rest of my career. Brad is extremely generous and supportive, both as a friend and as an advisor, and I always look forward to seeing him. His joking around makes meetings fun and keeps everyone from taking things too seriously.

Contents

Publications by Bradley Efron

1. Optimum evasion versus systematic search (1964). *SIAM J.* **12**, 450–457.
2. Increasing properties of Polya frequency functions (1965). *Ann. Math. Statist.* **36**, 272–279.
3. Note on the paper "Decision procedures for finite decision problems under complete ignorance" (1965). *Ann. Math. Statist.* **36**, 691–697.
4. The convex hull of a random set of points (1965). *Biometrika* **52**, 331–343.
5. The two-sample problem with censored data (1966). *Proceedings of the Fifth Berkeley Symposium on Mathematical Statistics and Probability*, University of California Press.
6. Geometrical probability and random points on a hypersphere (with T. Cover) (1967). *Ann. Math. Statist.* **38**, 213–220.
7. The power of the likelihood ratio test (1967). *Ann. Math. Statist.* **38**, 802–806.
8. Large deviations theory in exponential families (with D. Truax) (1968). *Ann. Math. Statist.* **39**, 1402–1424.
9. Student's t-test under symmetry conditions (1969). *JASA* **64**, 1278–1302.
10. Hotelling's T^2 test under symmetry conditions (with M. L. Eaton) (1970). *JASA* **65**, 702–711.
11. Some remarks on the inference and decision models of statistics (1970). *Ann. Math. Statist.* **41**, 1034–1058.
12. Studies on toxoplasmosis in El Salvador: Prevalence and incidence of toxoplasmosis as measured by the Sabin–Feldman dye test (1970). J. S. Remington, B. Efron, E. Cavanaugh, H. J. Simon and A. Trejos. *Trans. Roy. Soc. Trop. Med. Hyg.* **64**, 252–267.
13. Does an observed sequence of numbers follow a simple rule? (Another look at Bode's law) (1971). *JASA* **66**, 552–559.
14. Spurious appearance of mosaicism in three generations in one family with a $3/B$ translocation. II. Statistical model of the chromosomal abnormality (with R. G. Miller, Jr. and B. Wm. Brown, Jr.) (1971). Department of Statistics Technical Report 27, Stanford University.
15. Forcing a sequential experiment to be balanced (1971). *Biometrika* **58**, 403–417. Also appeared *International Symposium on Hodgkin's Disease* NCI Monograph **36**, 571–572.
16. Limiting the risk of Bayes and empirical Bayes estimators – Part I: The Bayes case (with C. Morris) (1971). *JASA* **66**, 807–815.
17. Limiting the risk of Bayes and empirical Bayes estimators – Part II: The empirical Bayes case (with C. Morris) (1972). *JASA* **67**, 130–139.
18. Empirical Bayes on vector observations: An extension of Stein's method (with C. Morris) (1972). *Biometrika* **59**, 335–347.
19. Improving the usual estimator of a normal covariance matrix (with C. Stein and C. Morris) (1972). Department of Statistics Technical Report 37, Stanford University.
20. Stein's estimation rule and its competitors – An empirical Bayes approach (with C. Morris) (1973). *JASA* **68**, 117–130.
21. Combining possibly related estimation problems (with C. Morris) (1973). *JRSS-B* **35**, 379–421 with discussion.
22. Defining the curvature of a statistical problem (with applications to second order efficiency) (1975). *Ann. Statist.* **3**, 1189–1242 with discussion and Reply.
23. Data analysis using Stein's estimator and its generalizations (with C. Morris) (1975). *JASA* **70**, 311–319.

24. The efficiency of logistic regression compared to normal discriminant analysis (1975). *JASA* **70**, 892–898.
25. The possible prognostic usefulness of assessing serum proteins and cholesterol in malignancy (with F. Chao and P. Wolf) (1975). *Cancer* **35**, 1223–1229.
26. Biased versus unbiased estimation (1975). *Advan. Math.* **16**, 259–277. Reprinted in *Surveys in Applied Mathematics* (1976). Academic Press, New York.
27. Families of minimax estimators of the mean of a multivariate normal distribution (with C. Morris) (1976). *Ann. Statist.* **4**, 11–21.
28. Multivariate empirical Bayes and estimation of covariance matrices (with C. Morris) (1976). *Ann. Statist.* **4**, 22–32.
29. Dealing with many problems simultaneously (with C. Antoniak) (1976). *On the History of Statistics and Probability* (D. Owen, Ed.). Marcel Dekker, New York.
30. Estimating the number of unseen species: How many words did Shakespeare know? (with R. Thisted) (1976). *Biometrika* **63**, 435–447.
31. The efficiency of Cox's likelihood function for censored data (1977). *JASA* **72**, 557–565.
32. Stein's paradox in statistics (with C. Morris) (1977). *Sci. Am.* **236**, 119–127.
33. The geometry of exponential families (1978). *Ann. Statist.* **6**, 362–376.
34. Regression and ANOVA with zero–one data: Measures of residual variation (1978). *JASA* **73**, 113–121.
35. Controversies in the foundations of statistics (1978). *Amer. Math. Mon.* **85**, 231–246.
36. How broad is the class of normal scale mixtures? (with R. Olshen) (1978). *Ann. Statist.* **6**, 1159–1164.
37. Assessing the accuracy of the MLE: Observed versus expected Fisher information (with D. V. Hinkley) (1978). *Biometrika* **65**, 457–487 with comments and Reply.
38. Bootstrap methods: Another look at the jackknife (1979). *Ann. Statist.* **7**, 1–26.
39. Computers and the theory of statistics: Thinking the unthinkable (1979). *SIAM Rev.* **21**, 460–480.
40. A distance theorem for exponential families (1980). *Pol. J. Prob. Math. Stat.* **1**, 95–98.
41. *Biostatistics Casebook* (with R. G. Miller, Jr., B. Wm. Brown, Jr. and L. Moses) (1980). John Wiley & Sons, New York.
42. Randomizing and balancing a complicated sequential experiment (1980). *Biostatistics Casebook, Part I*, 19–30. John Wiley & Sons, New York.
43. Which of two measurements is better? (1980). *Biostatistics Casebook, Part II*, 153–170. John Wiley & Sons, New York.
44. The jackknife estimate of variance (with C. Stein) (1981). *Ann. Stat.* **9**, 586–596.
45. Censored data and the bootstrap (1981). *JASA* **76**, 312–319.
46. Nonparametric estimates of standard error: The jackknife, the bootstrap, and other methods (1981). *Biometrika* **68**, 589–599.
47. Nonparametric standard errors and confidence intervals (1981). *Can. J. Stat.* **9**, 139–172 with discussion.
48. Computer and statistical theory (with G. Gong) (1981). Keynote address, *Proceedings of the 13th Conference on the Interface between Computers and Statistics*.
49. The jackknife, the bootstrap, and other resampling plans (1982). *SIAM CBMS–NSF Monogr.* **38**. Also published by J. W. Arrowsmith Ltd., Philadelphia.
50. Transformation theory: How normal is a family of distributions? (1982). *Ann. Statist.* **10**, 323–339.
51. Maximum likelihood and decision theory (1982). *Ann. Statist.* **10**, 340–356.
52. A leisurely look at the bootstrap, the jackknife, and cross validation (with G. Gong) (1983). *Amer. Statist.* **37**, 36–48.
53. Estimating the error rate of a prediction rule: Improvement on cross-validation (1983). *JASA* **78**, 316–331.

54. Computer intensive methods in statistics (1983). *Lisboa 200th Anniversary Volume*, Lisbon Academy of Sciences, 173–181.
55. Computer intensive methods in statistics (with P. Diaconis) (1983). *Sci. Am.* **248**, 116–130.
56. Comparing non-nested linear models (1984). *JASA* **79**, 791–803.
57. The art of learning from experience (with G. Kolata) (1984). *Science* **225**, 156–158.
58. Testing for independence in a two-way table: New interpretations of the chi-square statistic (with P. Diaconis) (1985). *Ann. Statist.* **13**, 845–913 with discussion and Rejoinder.
59. Bootstrap confidence intervals for a class of parametric problems (1985). *Biometrika* **72**, 45–58.
60. The bootstrap method for assessing statistical accuracy (with R. Tibshirani) (1985). *Behaviormetrika* **17**, 1–35. Also appeared as: Bootstrap methods for standard errors, confidence intervals, and other measures of statistical accuracy (with R. Tibshirani) (1986). *Statist. Sci.* **1**, 54–77 with comment and Rejoinder.
61. Why isn't everyone a Bayesian? (1986). *Amer. Statist.* **40**, 1–11.
62. How biased is the apparent error rate of a prediction rule? (1986). *JASA* **81**, 461–470.
63. Double exponential families and their use in generalized linear regression (1986). *JASA* **81**, 709–721.
64. Better bootstrap confidence intervals (1987). *JASA* **82**, 171–200 with discussion and Rejoinder.
65. Did Shakespeare write a newly discovered poem? (with R. Thisted) (1987). *Biometrika* **74**, 445–455.
66. Probabilistic–geometric theorems arising from the analysis of contingency tables (with P. Diaconis) (1987). *Contributions to the Theory and Application of Statistics, A Volume in Honor of Herbert Solomon*, 103–125. Academic Press, New York.
67. Computer-intensive methods in statistical regression (1988). *SIAM Rev.* **30**, 421–449.
68. Logistic regression, survival analysis, and the Kaplan–Meier curve (1988). *JASA* **83**, 414–425.
69. Bootstrap confidence intervals: Good or bad? (1988). *Psychol. Bull.* **104**, 293–296.
70. Three examples of computer-intensive statistical inference (1988). *Sankhya 50th Anniversary Volume* **50**, 338–362. Presented at the *20th Interface Symposium on Computing Science and Statistics* and *COMPSTAT88*.
71. Application of the bootstrap statistical method to the tau-decay-mode problem (with K. G. Hayes and M. L. Perl) (1989). *Phys. Rev. D* **39**, 274–279.
72. Prognostic indicators of laparotomy findings in clinical stage I–II supradiaphragmatic Hodgkin's disease (1989). M. H. Leibenhaut, R. T. Hoppe, B. Efron, J. Halpern, T. Nelsen and S. A. Rosenberg. *J. Clin. Oncol.* **7**, 81–91.
73. Fisher's information in terms of the hazard rate (with I. Johnstone) (1990). *Ann. Statist.* **18**, 38–62.
74. More efficient bootstrap computations (1990). *JASA* **85**, 79–89.
75. Regression percentiles using asymmetric squared error loss (1991). *Stat. Sinica* **1**, 93–125.
76. Compliance as an explanatory variable in clinical trials (with D. Feldman) (1991). *JASA* **86**, 9–26 with comments and Rejoinder.
77. Statistical data analysis in the computer age (with R. Tibshirani) (1991). *Science* **253**, 390–395.
78. Poisson overdispersion estimates based on the method of asymmetric maximum likelihood (1992). *JASA* **87**, 98–107.
79. Jackknife-after-bootstrap standard errors and influence functions (1992). *JRSS-B* **54**, 83–127 with discussion and Rejoinder.
80. More accurate confidence intervals in exponential families (with T. DiCiccio) (1992). *Biometrika* **79**, 231–245.
81. Six questions raised by the bootstrap (1992). In *Exploring the Limits of Bootstrap* (Raoul LePage and Lynne Billard, Eds.). John Wiley & Sons, New York.
82. A simple test of independence for truncated data with applications to redshift surveys (with V. Petrosian) (1992). *Astrophys. J.* **399**, 345–352.
83. Bayes and likelihood calculations from confidence intervals (1993). *Biometrika* **80**, 3–26.

84. *An Introduction to the Bootstrap* (with R. Tibshirani) (1993). Chapman and Hall, New York.
85. Multivariate analysis in the computer age (1993). In *Multivariate Analysis: Future Directions 2* (C. M. Cuadras and C. R. Rao, Eds.). Elsevier Science Publishers, Amsterdam.
86. Survival analysis of the gamma-ray burst data (with V. Petrosian) (1994). *JASA* **89**, 452–462.
87. Missing data, imputation, and the bootstrap (1994). *JASA* **89**, 463–479 with comment and Rejoinder.
88. On the correlation of angular position with time of occurrence of gamma-ray bursts (with V. Petrosian) (1995). *Astrophys. J. Lett.* **441**, L37–L38.
89. Testing isotropy versus clustering of gamma-ray bursts (with V. Petrosian) (1995). *Astrophys. J.* **449**, 216–223.
90. Computer-intensive statistical methods (with R. Tibshirani) (1995). Invited article for *Encyclopedia of Statistical Sciences*. Wiley-Interscience, New York.
91. Empirical Bayes methods for combining likelihoods (1996). *JASA* **91**, 538–565 with comments and Rejoinder.
92. Using specially designed exponential families for density estimation (with R. Tibshirani) (1996). *Ann. Statist.* **24**, 2431–2461.
93. Bootstrap confidence intervals (with T. DiCiccio) (1996). *Statist. Sci.* **11**, 189–212.
94. Bootstrap confidence levels for phylogenetic trees (with E. Halloran and S. Holmes) (1996). *Proc. Nat. Acad. Sci.* **93**, 13429–13434.
95. Improvements on cross-validation: The .632+ bootstrap method (with R. Tibshirani) (1997). *JASA* **92**, 548–560.
96. The length heuristic for simultaneous hypothesis tests (1997). *Biometrika* **84**, 143–157.
97. Risk for retinitis in patients with AIDS can be assessed by quantitation of threshold levels of cytomegalovirus DNA burden in blood (1997). L. Rasmussen, D. Zipeto, R. A. Wolitz, A. Dowling, B. Efron and T. C. Merigan. *J. Infect. Dis.* **176**, 1146–1155.
98. A predictive morphometric model for the obstructive sleep apnea syndrome (1997). C. A. Kushida, B. Efron and C. Guilleminault. *Ann. Intern. Med.* **127**, 581–587.
99. Circadian rhythms and enhanced athletic performance in the National Football League (1997). R. S. Smith, C. Guilleminault and B. Efron. *Sleep* **20**, 362–365.
100. The problem of regions (with R. Tibshirani) (1998). *Ann. Statist.* **26**, 1687–1718.
101. R. A. Fisher in the 21st century (1998). *Statist. Sci.* **13**, 95–122 with comments and Rejoinder.
102. Nonparametric methods for doubly truncated data (with V. Petrosian) (1999). *JASA* **94**, 824–834.
103. HIV-1 Genotypic resistance patterns predict response to saquinavir-ritonavir therapy in patients in whom previous protease inhibitor therapy had failed (1999). B. Efron, A. R. Zolopa, R. W. Shafer, et al. *Ann. Intern. Med.* **131**, 813–821.
104. The bootstrap and modern statistics (2000). *JASA* **95**, 1293–1296.
105. Selection criteria for scatterplot smoothers (2001). *Ann. Statist.* **29**, 470–504.
106. Infectious complications among 620 consecutive heart transplant patients at Stanford University Medical Center (2001). J. G. Montoya, L. F. Giraldo, B. Efron, et al. *CID* **33**, 629–640.
107. Empirical Bayes analysis of a microarray experiment (with R. Tibshirani, J. D. Storey and V. Tusher) (2001). *JASA* **96**, 1151–1160.
108. Scales of evidence for model selection: Fisher versus Jeffreys (with A. Gous) (2001). In *IMS Lecture Notes – Monograph Series: Model Selection* (P. Lahiri, Ed.) **38**, 208-256 with discussion and Rejoinder.
109. Empirical Bayes methods and false discovery rates for microarrays (with R. Tibshirani) (2002). *Genet. Epidemiol.* **23**, 70–86.
110. The two-way proportional hazards model (2002). *JRSS-B* **64**, 899–909.
111. Smoothers and the C_p, GML, and EE criteria: A geometric approach (with S. C. Kou) (2002). *JASA* **97**, 766–782.

112. Pre-validation and inference in microarrays (with R. Tibshirani) (2002). *Stat. Appl. Genet. Mol. Biol.* **1**, 1–8.
113. Robbins, empirical Bayes, and microarrays (2003). *Ann. Statist.* **31**, 366–378.
114. Bayesians, frequentists, and physicists (2003). *Proc. PHYSTAT2003*, 17–28.
115. Second thoughts on the bootstrap (2003). *Statist. Sci.* **18**, 135–140.
116. Bradley Efron: A conversation with good friends (with Susan Holmes, Carl Morris and Rob Tibshirani) (2003). *Statist. Sci.* **18**, 268–281.
117. The statistical century (2003). In *Stochastic Musings: Perspectives from the Pioneers of the Late 20th Century* (John Panaretos, Ed.), 29–44. Lawrence Erlbaum Associates, New Jersey.
118. Least angle regression (with T. Hastie, I. Johnstone and R. Tibshirani) (2004). *Ann. Statist.* **32**, 407–499 with discussion and Rejoinder.
119. Large-scale simultaneous hypothesis testing: The choice of a null hypothesis (2004). *JASA* **99**, 96–104.
120. The estimation of prediction error: Covariance penalties and cross-validation (2004). *JASA* **99**, 619–642 with discussion.
121. Impact of HIV-1 Subtype and antiretroviral therapy on protease and reverse transcriptase genotype: Results of a global collaboration (2005). R. Kantor, D.A. Katzenstein, B. Efron, et al. *PLoS Med.* **2**, 325–337.
122. Signature patterns of gene expression in mouse atherosclerosis and their correlation to human coronary disease (2005). R. Tabibiazar, R. A. Wagner, et al. *Physiol. Genomics* **22**, 213–226.
123. The 'miss rate' for the analysis of gene expression data (2005). J. Taylor, R. Tibshirani and B. Efron. *Biostatistics* **6**, 111–117.
124. Modern science and the Bayesian–frequentist controversy (2005). *Papers of the XII Congress of the Portuguese Statistical Society*, 9–20.
125. Minimum volume confidence regions for a multivariate normal mean vector (2006). *JRSS-B* **68**, Part 4, 1–16.
126. On testing the significance of sets of genes (with R. Tibshirani) (2007). *Ann. Appl. Statist.* **1**, 107–129.
127. Correlation and large-scale simultaneous significance testing (2007). *JASA* **102**, 93–103.
128. Doing thousands of hypothesis tests at the same time (2007). *METRON* **65**, 3–21.
129. Size, power, and false discovery rates (2007). *Ann. Statist.* **35**, 1351–1377.
130. Microarrays, empirical Bayes, and the two-groups model (2007). To appear *Statist. Sci.*. Available at: http://www-stat.stanford.edu/~brad/papers/twogroups.pdf.
131. Simultaneous inference: When should hypothesis testing problems be combined? (2007). To appear *Ann. Appl. Statist.* Available at: http://www-stat.stanford.edu/~brad/papers/combinationpaper.pdf.

1

The Convex Hull of a Random Set of Points

Introduction by Tom Cover
Stanford University

Brad Efron's interest in geometrical probability is shown in this paper and his paper on random points on a hypersphere, as well as his later papers on curved exponential families, in which he developed a natural parameter space through considerations of differential geometry. The curved family papers influenced Amari and Csiszar in their work on the geometry of information.

In the convex hull paper, Efron extends the work of Renyi and Sulanke on the expected number of vertices, edges and area of the convex hull of a random collection of points. I recall, during his writing of this paper (we were graduate students at the time), that Efron was particularly delighted with a trick for finding the number of extreme points. The idea is to find the number of edges. Two points form an edge if the remaining $n - 2$ points all fall on one side or the other of the line, a relatively simple probability to express. Thus the expected number of vertices is equal to the expected number of edges.

A similar observation then holds in three dimensions. If n points are independent identically distributed according to some density, all the faces of the convex hull will be triangles. Then Euler's formula $F - E + V = 2$ and some book-keeping gives the number of vertices in terms of the number of faces, and the expected number of faces is easy to express.

In the setup of this paper, the number of extreme points grows very slowly with n but grows without bound. However in the paper by Efron and Cover, "Geometrical probability and random points on a hypersphere" (*Annals of Mathematical Statistics* **38**, 213–220, 1967), where the points are distributed on the surface of a hypersphere, and one conditions on the convex cone being nonempty, a distribution-free result is obtained in which the expected number of extreme points tends to a finite limit as n tends to infinity. These two papers to some extent anticipate the well-known results of Stephen Smale in which he investigates why the simplex algorithm requires so few steps in practice. Apparently, a randomly-generated constraint set is much simpler than the number of constraints would suggest, thus reducing the number of edges traversed by the simplex algorithm. In particular, the expected cross-section of the nonempty intersection of n random half spaces in d-space tends to a d-cube as n tends to infinity.

Biometrika (1965), **52**, 3 *and* 4, *p.* 331
Printed in Great Britain

The convex hull of a random set of points†

By BRADLEY EFRON
Stanford University, California

SUMMARY

Various expectations concerning the convex hull of N independently and identically distributed random points in the plane or in space are evaluated. Integral expressions are given for the expected area, expected perimeter, expected probability content and expected number of sides. These integrals are shown to be particularly simple when the underlying distribution is normal or uniform over a disk or sphere.

1. Introduction

In two recent papers Renyi & Sulanki (1963, 1964) have given expressions for the expected area, perimeter, and number of vertices of the convex hull of N independently and identically selected random points in the plane. In these papers, limit theorems for asymptotically large N receive the greatest attention. Here the emphasis will be on the development of convenient formulae for fixed values of N. Some new results for random convex hulls in the plane are derived, such as the expected probability content, and also various expectations concerned with random convex hulls in three and higher dimensions. Special attention is given to the case of normally distributed points and also to the case of points drawn uniformly from an ellipse or an ellipsoid.

Historically, calculating the expected probability content of three random points in two dimensions is known as 'Sylvester's problem', and has been solved explicitly for many different distributions by Deltheil (1926, p. 42). The corresponding problem of four points in three dimensions is connected with the name of Hostinsky (1925). A discussion of Sylvester's problem is given in Kendall & Moran's monograph, *Geometrical Probability* (1963).

2. The expected number of vertices, faces, and edges

Let $\mathbf{W}_1, \mathbf{W}_2, \ldots, \mathbf{W}_N$ be N independent random points, each selected according to a probability distribution Γ in the plane (or in three-dimensional space). We will always assume that Γ is absolutely continuous with respect to Lebesgue measure in the plane (or in space), with density function $g(\mathbf{w})$. The convex hull of the points $\mathbf{W}_1, \mathbf{W}_2, \ldots, \mathbf{W}_N$ will be denoted by H_N, with number of vertices V_N in the two-dimensional case, and number of vertices, faces, and edges V_N, F_N and E_N in the three-dimensional case.

Renyi & Sulanke calculate the expected value of V_N by the following elegant argument, which will be used, with variations, repeatedly in the sequel. Given $\mathbf{W}_1 = \mathbf{w}_1, \mathbf{W}_2 = \mathbf{w}_2$, the line $L(\mathbf{w}_1, \mathbf{w}_2)$ through $\mathbf{w}_1$ and $\mathbf{w}_2$ divides the plane into two regions, whose probability content under Γ we denote by Γ_{12} and $1-\Gamma_{12}$, respectively. The probability that the remaining $N-2$ points all lie on the same side of $L(\mathbf{w}_1, \mathbf{w}_2)$ is then

$$\Gamma_{12}^{N-2} + (1-\Gamma_{12})^{N-2}, \qquad (2{\cdot}1)$$

† This research was sponsored by the Rand Corporation, Santa Monica, California and the Office of Naval Research.

and this is just the conditional probability that the segment $(\mathbf{W}_1 \mathbf{W}_2)$ is an extreme edge of H_N, given $\mathbf{W}_1 = \mathbf{w}_1$ and $\mathbf{W}_2 = \mathbf{w}_2$. The unconditional probability of the first two points defining an edge of H_N is thus

$$\iint (\Gamma_{12}^{N-2} + (1-\Gamma_{12})^{N-2})\, g(\mathbf{w}_1)\, g(\mathbf{w}_2)\, d\mathbf{w}_1\, d\mathbf{w}_2, \tag{2·2}$$

where the integral is over the entire plane in both points. Taking all $\binom{N}{2}$ pairs of points, the expected number of edges (and vertices) of H_N is, by symmetry,

$$EV_N = \binom{N}{2} \iint [\Gamma_{12}^{N-2} + (1-\Gamma_{12})^{N-2}]\, g(\mathbf{w}_1)\, g(\mathbf{w}_2)\, d\mathbf{w}_1\, d\mathbf{w}_2. \tag{2·3}$$

The evaluation of the integral above is simplified by changing from the rectangular co-ordinates $\mathbf{w}_1 = (x_1, y_1)$, $\mathbf{w}_2 = (x_2, y_2)$, to the four co-ordinates (p, θ, t_1, t_2) defined as follows: the perpendicular from $L(\mathbf{w}_1, \mathbf{w}_2)$ to the origin has (signed) length p and direction $\theta\,(-\infty < p < \infty, 0 \leqslant \theta < \pi)$, while the points $\mathbf{w}_1$ and $\mathbf{w}_2$ have co-ordinates t_1 and t_2, respectively, on the line $L(\mathbf{w}_1, \mathbf{w}_2)$, the origin being taken at the foot of the perpendicular. The Jacobian of this transformation is $|t_2 - t_1|$ (see Santalo, 1953, p. 16), and the integral (2·3) becomes

$$EV_N = \binom{N}{2} \int_{-\infty}^{\infty} \int_0^{\pi} \gamma_N(p, \theta) \left[\int_{-\infty}^{\infty} \int_{-\infty}^{\infty} |t_2 - t_1|\, g(\mathbf{w}_1)\, g(\mathbf{w}_2)\, dt_1\, dt_2 \right] d\theta\, dp, \tag{2·4}$$

where $\mathbf{w}_i = \mathbf{w}_i(p, \theta, t_i)\ (i = 1, 2)$, and

$$\gamma_N(p, \theta) = \Gamma_{12}^{N-2} + (1-\Gamma_{12})^{N-2}, \tag{2·5}$$

a quantity which obviously depends only on p and θ.

Now define

$$\tilde{g}_2(p, \theta) \equiv \left[\int_{-\infty}^{\infty} g(\mathbf{w}(p, \theta, t))\, dt \right]^2. \tag{2·6}$$

Multiplying and dividing (2·5) by $\tilde{g}(p, \theta)$, we have

$$EV_N = \binom{N}{2} \int_{-\infty}^{\infty} \int_0^{\pi} \gamma_N(p, \theta) \left[\frac{\int_{-\infty}^{\infty} \int_{-\infty}^{\infty} |t_2 - t_1|\, g(\mathbf{w}_1)\, g(\mathbf{w}_2)\, dt_1\, dt_2}{\int_{-\infty}^{\infty} \int_{-\infty}^{\infty} g(\mathbf{w}_1)\, g(\mathbf{w}_2)\, dt_1\, dt_2} \right] \tilde{g}_2(p, \theta)\, d\theta\, dp, \tag{2·7}$$

and the term in brackets can be recognized as a conditional expectation: let $\mathbf{n}(\theta)$ be the unit vector $(\cos\theta, \sin\theta)$, where θ is now considered fixed, and define

$$Q_i = \mathbf{W}_1 . \mathbf{n}(\theta) \quad (i = 1, 2). \tag{2·8}$$

Then, remembering that $\mathbf{W}_1$ and $\mathbf{W}_2$ are selected independently according to Γ, the bracketed term equals

$$E(|\mathbf{W}_2 - \mathbf{W}_1|\, |Q_1 = p, Q_2 = p), \tag{2·9}$$

a quantity which we will denote, for convenience, $\tilde{E}(L_{12}|p, \theta)$, L_{12} representing the length $|\mathbf{W}_2 - \mathbf{W}_1|$. Thus we have

$$EV_N = \binom{N}{2} \int_{-\infty}^{\infty} \int_0^{\pi} \gamma_N(p, \theta)\, \tilde{E}(L_{12}|p, \theta)\, \tilde{g}_2(p, \theta)\, d\theta\, dp. \tag{2·10}$$

A case of particular interest is that where the distribution Γ is circularly symmetric about the origin (i.e. invariant under rotations). In such a case let $G_2(\cdot)$ denote the marginal cumulative distribution function of Γ in any direction, say along the x-axis, and let $g_2(\cdot)$ be the derivative of G_2. Then

$$\left.\begin{aligned}\gamma_N(p,\theta) &= G_2^{N-2}(p)+(1-G_2(p))^{N-2},\\ \tilde{g}_2(p,\theta) &= g_2^2(p),\end{aligned}\right\} \tag{2·11}$$

while
$$\tilde{E}(L_{12}|p,\theta) = E(|Y_2-Y_1|\,|X_1=p, X_2=p), \tag{2·12}$$

(X_1, Y_1) and (X_2, Y_2) being selected independently according to Γ. The circular symmetry further implies that

$$G_2(-p) = 1-G_2(p),\quad E(|Y_2-Y_1|\,|X_1=p, X_2=p) = E(|Y_2-Y_1|\,|X_1=-p, X_2=-p),$$

and
$$g_2(p) = g_2(-p).$$

Therefore,
$$EV_N = 2\pi\binom{N}{2}\int_{-\infty}^{\infty} E(|Y_2-Y_1|\,|X_1=p, X_2=p)\,G_2^{N-2}(p)\,g_2^2(p)\,dp \tag{2·13}$$

in the circularly symmetric case.

If the points $\mathbf{W}_1, \mathbf{W}_2, \ldots, \mathbf{W}_N$ are drawn from a probability distribution Γ over three-dimensional space, absolute continuity implies that with probability one every face of the convex hull H_N will be a triangle. The relation between the number of edges and the number of faces is then

$$E_N = \tfrac{3}{2}F_N, \tag{2·14}$$

which, coupled with the Euler relation

$$V_N = E_N - F_N + 2, \tag{2·15}$$

implies
$$V_N = \tfrac{1}{2}F_N + 2 \tag{2·16}$$

with probability one.

The calculation of EF_N is almost identical with that for the expected number of edges in the two-dimensional case. The probability that the first three random points define a face of H_N is just the probability that the remaining $N-3$ points lie on the same side of the plane $P(\mathbf{w}_1, \mathbf{w}_2, \mathbf{w}_3)$ through $\mathbf{w}_1$, $\mathbf{w}_2$ and $\mathbf{w}_3$,

$$\iiint (\Gamma_{123}^{N-3} + (1-\Gamma_{123})^{N-3})\,g(\mathbf{w}_1)\,g(\mathbf{w}_2)\,g(\mathbf{w}_3)\,d\mathbf{w}_1\,d\mathbf{w}_2\,d\mathbf{w}_3. \tag{2·17}$$

Here Γ_{123} and $1-\Gamma_{123}$ are the probability contents of the two regions determined by $P(\mathbf{w}_1, \mathbf{w}_2, \mathbf{w}_3)$, and the integral is over all of three-space in each of the three points. By symmetry

$$EF_N = \binom{N}{3}\iiint (\Gamma_{123}^{N-3} + (1-\Gamma_{123})^{N-3})\,g(\mathbf{w}_1)\,g(\mathbf{w}_2)\,g(\mathbf{w}_3)\,d\mathbf{w}_1\,d\mathbf{w}_2\,d\mathbf{w}_3, \tag{2·18}$$

and
$$EE_N = \tfrac{3}{2}EF_N,\quad EV_N = \tfrac{1}{2}EF_N + 2 \tag{2·19}$$

by the relationships (2·14) and (2·16).

The natural co-ordinate system for evaluating the integral (2·18) is described as follows: the perpendicular from the plane $P(\mathbf{w}_1, \mathbf{w}_2, \mathbf{w}_3)$ to the origin has (signed) length p and direction (θ, ϕ), where θ is the polar angle and ϕ is the angle of longitude, $-\infty < p < \infty$, $0 \leqslant \theta \leqslant \frac{1}{2}\pi$, $0 \leqslant \phi < 2\pi$. In the plane $P(\mathbf{w}_1, \mathbf{w}_2, \mathbf{w}_3)$ let $\mathbf{u}_1 = (s_1, t_1)$, $\mathbf{u}_2 = (s_2, t_2)$, $\mathbf{u}_3 = (s_3, t_3)$ be the co-ordinates of $\mathbf{w}_1, \mathbf{w}_2, \mathbf{w}_3$, respectively, relative to a rectangular co-ordinate system (S, T) fixed rigidly in $P(\mathbf{w}_1, \mathbf{w}_2, \mathbf{w}_3)$. (More explicitly let R_θ be a positive rotation of angle θ about the y-axis and R_ϕ a positive rotation of angle ϕ about the z-axis. The transformation $(x^*, y^*, z^*)' = R_\phi R_\theta(x, y, z)'$ takes the plane $z = p$ into the plane (p, θ, ϕ). We now define the

(S, T) axes as the image of the (X, Y) axes in the plane $z = p$ under the transformation $R_\phi R_\theta$. Notice that the point $(s = 0, t = 0)$ is the foot of the perpendicular to the origin.) The Jacobian of the transformation from the nine co-ordinates $\mathbf{w}_i = (x_i, y_i, z_i)$ $(i = 1, 2, 3)$, to the nine co-ordinates $(p, \theta, \phi, \mathbf{u}_1, \mathbf{u}_2, \mathbf{u}_3)$ is evaluated as

$$J = 2A_{123}\sin\theta, \tag{2·20}$$

where A_{123} is the area (always taken positive) of the triangle determined by $\mathbf{w}_1$, $\mathbf{w}_2$ and $\mathbf{w}_3$.

In the new system of co-ordinates, (2·18) becomes

$$EF_N = 2\binom{N}{3}\int_{-\infty}^{\infty}\int_0^{\frac{1}{2}\pi}\int_0^{2\pi}\delta_N(p,\theta,\phi)\left[\iiint A_{123}\,g(\mathbf{w}_1)\,g(\mathbf{w}_2)\,g(\mathbf{w}_3)\,d\mathbf{u}_1\,d\mathbf{u}_2\,d\mathbf{u}_3\right]\sin\theta\,d\phi\,d\theta\,dp, \tag{2·21}$$

the bracketed integral being over the entire plane in each of the three points $\mathbf{u}_1, \mathbf{u}_2, \mathbf{u}_3$, and

$$\delta_N(p,\theta,\phi) = \Gamma_{123}^{N-3} + (1-\Gamma_{123})^{N-3}. \tag{2·22}$$

If the bracketed term is divided by

$$\tilde{g}_3(p,\theta,\phi) \equiv \left[\int g(\mathbf{w}(p,\theta,\phi,\mathbf{u}))\,d\mathbf{u}\right]^3, \tag{2·23}$$

the integral being taken over the entire plane, it becomes a conditional expectation

$$\frac{\iiint A_{123}\,g(\mathbf{w}_1)\,g(\mathbf{w}_2)\,g(\mathbf{w}_3)\,d\mathbf{u}_1\,d\mathbf{u}_2\,d\mathbf{u}_3}{\iiint g(\mathbf{w}_1)\,g(\mathbf{w}_2)\,g(\mathbf{w}_3)\,d\mathbf{u}_1\,d\mathbf{u}_2\,d\mathbf{u}_3}. \tag{2·24}$$

Let $\mathbf{n}(\theta, \phi)$ be the unit vector $(\sin\theta\cos\phi, \sin\theta\sin\phi, \cos\theta)$, and define

$$Q_i = \mathbf{W}_1.\mathbf{n}(\theta,\phi) \quad (i = 1, 2, 3). \tag{2·25}$$

Then, if the $\mathbf{W}_i$ are chosen independently according to Γ, expression (2·24) equals

$$E(A_{123}|Q_1 = p, Q_2 = p, Q_3 = p), \tag{2·26}$$

where θ and ϕ are considered fixed. This will be denoted $\tilde{E}(A_{123}|p,\theta,\phi)$, and so (2·21) becomes

$$EF_N = 2\binom{N}{3}\int_{-\infty}^{\infty}\int_0^{\frac{1}{2}\pi}\int_0^{2\pi}\delta_N(p,\theta,\phi)\,\tilde{E}(A_{123}|p,\theta,\phi)\,\tilde{g}_3(p,\theta,\phi)\sin\theta\,d\phi\,d\theta\,dp. \tag{2·27}$$

In the case where Γ is spherically symmetric about the origin, let $G_3(\cdot)$ be the cumulative marginal along the x-axis, and $g_3(\cdot)$ its derivative. Proceeding as before,

$$EF_N = 8\pi\binom{N}{3}\int_{-\infty}^{\infty} E(A_{123}|X_1 = p, X_2 = p, X_3 = p)\,G_3^{N-3}(p)\,g_3^3(p)\,dp, \tag{2·28}$$

where X_i is the first co-ordinate of $\mathbf{W}_i$ $(i = 1, 2, 3)$.

3. The expected probability content

Let $C_N = \Gamma(H_N)$ denote the probability content of the random convex hull in either two or three dimensions. Regardless of the dimension, the probability that none of the last r random points selected is an extreme point (vertex) of H_N is EC_{N-r}^r, $(C_n \equiv 0$ for $n \leqslant 0)$, since this is just the probability that the last r points lie in the convex hull of the first $N - r$.

Let I_i be the indicator random variable, i.e.

$$I_i = \begin{cases} 0 & \text{if } \mathbf{W}_i \text{ is an extreme point of } H_N \\ 1 & \text{otherwise} \end{cases} \quad (i = 1, 2, \ldots, N), \tag{3·1}$$

and let $U_N \equiv N - V_N$, the number of non-extreme points. Then

$$\begin{aligned} EU_N^k &= E(I_1 + I_2 + \ldots + I_N)^k \\ &= \Sigma P(\mathbf{W}_{i_1}, \mathbf{W}_{i_2}, \ldots, \mathbf{W}_{i_k} \text{ not vertices of } H_N), \end{aligned} \tag{3·2}$$

the sum being taken over all k-tuples $(i_1, i_2, \ldots, i_k)$ from the integers $1, 2, \ldots, N$. Making use of the argument above and symmetry in the indices, we have

$$EU_N^k = \sum_{r=1}^{k} n(N, k, r)\, EC_{N-r}^r, \tag{3·3}$$

where $n(N, k, r)$ is the number of k-tuples from $1, 2, \ldots, N$ having exactly r different entries,

$$n(N, k, r) = \binom{N}{r} \sum_{m=1}^{r} (-1)^{r-m} \binom{r}{m} m^k. \tag{3·4}$$

The cases $k = 1$ and $k = 2$ yield

$$EU_N = NEC_{N-1}, \quad EU_N^2 = NEC_{N-1} + N(N-1)\, EC_{N-2}^2, \tag{3·5}$$

and

$$\operatorname{var} U_N = N^2(EC_{N-2}^2 - (EC_{N-1})^2) + N(EC_{N-1} - EC_{N-2}^2). \tag{3·6}$$

If we define the probability non-content D_N as $D_N = 1 - C_N = 1 - \Gamma(H_N)$, then from (3·5) we have

$$ED_N = \frac{EV_{N+1}}{N+1}, \tag{3·7}$$

and $ED_N = 1 - EC_N$ can be calculated from the formulas of § 2. In the important case where the probability distribution Γ is uniform over a convex domain k, the expected area (or volume) of the convex hull is just EC_N times the area (or volume) of k.

4. The expected perimeter

A simpler argument than those above yields the expected perimeter EP_N of H_N in the two-dimensional situation. Let a line in the plane be parametrized by (p, θ), the (signed) length and direction of its normal to the origin, and let J_N be the area of the set of points (p, θ) in the infinite strip $-\infty < p < \infty$, $0 \leqslant \theta < \pi$, which correspond to lines intersecting H_N. It is well known (Kendall & Moran, 1963, p. 58; Santalo, 1953, p. 13) that $P_N = J_N$; hence

$$EP_N = EJ_N. \tag{4·1}$$

The probability that the line (p, θ) intersects H_N is just the probability that not all N random points lie on the same side of this line, and equals $1 - \gamma_{N+2}(p, \theta)$ by definition (2·5). Then

$$EP_N = EJ_N = \int_{-\infty}^{\infty} \int_0^{\pi} [1 - \gamma_{N+2}(p, \theta)]\, d\theta\, dp. \tag{4·2}$$

(This last equality, which is the continuous analog of 'the expectation of a sum is the sum of the expectations', is derived rigorously by Robbins (1944, 1947).)

Unfortunately, the preceding line of reasoning cannot be extended to three dimensions, while an argument similar to that used in § 2 can. Define

$$\hat{L}_{ij} = \begin{cases} L_{ij}(\equiv |\mathbf{W}_j - \mathbf{W}_i|) & \text{if the segment } (\mathbf{W}_i \mathbf{W}_j) \\ & \text{is a side of } H_N, \\ & 0 \text{ otherwise,} \end{cases} \tag{4·3}$$

so that

$$P_N = \sum_{1 \leqslant i < j \leqslant N} \hat{L}_{ij}, \tag{4·4}$$

and, by symmetry,

$$EP_N = \binom{N}{2} E\hat{L}_{12}. \tag{4·5}$$

Making use of (2·1), we have

$$EP_N = \binom{N}{2} \iint [\Gamma_{12}^{N-2} + (1-\Gamma_{12})^{N-2}] L_{12}\, g(\mathbf{w}_1)\, g(\mathbf{w}_2)\, d\mathbf{w}_1\, d\mathbf{w}_2, \tag{4·6}$$

which becomes, in the (p, θ, t_1, t_2) co-ordinates,

$$EP_N = \binom{N}{2} \int_{-\infty}^{\infty} \int_0^{\pi} \gamma_N(p,\theta) \left[\int_{-\infty}^{\infty} \int_{-\infty}^{\infty} (t_2 - t_1)^2 g(\mathbf{w}_1)\, g(\mathbf{w}_2)\, dt_1\, dt_2 \right] d\theta\, dp. \tag{4·7}$$

Repeating the argument following (7) gives

$$EP_N = \binom{N}{2} \int_{-\infty}^{\infty} \int_0^{\pi} \gamma_N(p,\theta)\, \tilde{E}[L_{12}^2 | p, \theta]\, \tilde{g}(p,\theta)\, dp_1\, d\theta_1, \tag{4·8}$$

where $\tilde{g}(p,\theta)$ is given by (2·6), and

$$\tilde{E}(L_{12}^2 | p, \theta) \equiv E(L_{12}^2 | Q_1 = p, Q_2 = p) \tag{4·9}$$

as in (2·8) and (2·9). (Partial integration verifies the identity of (4·2) and (4·8).)

In the case where Γ is circularly symmetric we have

$$EP_N = 2\pi \binom{N}{2} \int_{-\infty}^{\infty} E[(Y_2 - Y_1)^2 | X_1 = p, X_2 = p]\, G_2^{N-2}(p)\, g_2^2(p)\, dp, \tag{4·10}$$

where, as before, $G_2(\cdot)$ is the marginal, in any direction, of Γ, $g_2(\cdot) = G_2'(\cdot)$, and (X_1, Y_1), (X_2, Y_2) are independently selected according to Γ. This may also be written

$$EP_N = 4\pi \binom{N}{2} \int_{-\infty}^{\infty} \sigma_p^2(Y)\, G_2^{N-2}(p)\, g_2^2(p)\, dp, \tag{4·11}$$

$\sigma_p^2(Y)$ being the conditional variance of Y, given $X = p$, when (X, Y) is selected according to Γ.

By the 'total perimeter' T_N of a three-dimensional random convex hull, we shall mean the total length of all the line segments bounding the faces of H_N. As remarked previously, each face is a triangle, and each line bounds two faces. Repeating the argument above,

$$ET_N = \binom{N}{3} \iiint (\Gamma_{123}^{N-3} + (1-\Gamma_{123})^{N-3}) \frac{L_{123}}{2} g(\mathbf{w}_1)\, g(\mathbf{w}_2)\, g(\mathbf{w}_3)\, d\mathbf{w}_1\, d\mathbf{w}_2\, d\mathbf{w}_3, \tag{4·12}$$

where Γ_{123} is defined as before in (2·17), L_{123} equals the perimeter of the triangle defined by $\mathbf{w}_1$, $\mathbf{w}_2$ and $\mathbf{w}_3$, and A_{123} is the area.

Changing to the $(p, \theta, \phi, \mathbf{u}_1, \mathbf{u}_2, \mathbf{u}_3)$ co-ordinates,

$$ET_N = \binom{N}{3} \int_{-\infty}^{\infty} \int_0^{\frac{1}{2}\pi} \int_0^{2\pi} \delta_N(p,\theta,\phi)\, \tilde{E}(L_{123} A_{123} | p, \theta, \phi)\, \tilde{g}_3(p,\theta,\phi) \sin\theta\, d\phi\, d\theta\, dp, \tag{4·13}$$

δ_N and $\tilde{g}_3$ being given by (2·22) and (2·23), and

$$\tilde{E}(L_{123}A_{123}|p,\theta,\phi) = E(L_{123}A_{123}|Q_1 = p, Q_2 = p, Q_3 = p) \tag{4·14}$$

as in (2·26). Notice that $L_{123}A_{123} = (L_{12}+L_{13}+L_{23})A_{123}$, and so by symmetry,

$$E(L_{123}A_{123}|Q_1 = p, Q_2 = p, Q_3 = p) = 3E(L_{12}A_{123}|Q_1 = p, Q_2 = p, Q_3 = p), \tag{4·15}$$

yielding

$$ET_N = 3\binom{N}{3}\int_{-\infty}^{\infty}\int_0^{\frac{1}{2}\pi}\int_0^{2\pi}\delta_N(p,\theta,\phi)\,\tilde{E}(L_{12}A_{123}|p,\theta,\phi)\,\tilde{g}_3(p,\theta,\phi)\sin\theta\,d\phi\,d\theta\,dp. \tag{4·16}$$

In the case where Γ is spherically symmetric about the origin, (4·16) reduces to

$$ET_N = 12\pi\binom{N}{3}\int_{-\infty}^{\infty}E(L_{12}A_{123}|X_1 = p, X_2 = p, X_3 = p)\,G_3^{N-3}(p)\,g_3^3(p)\,dp, \tag{4·17}$$

G_3 and g_3 being respectively the marginal cumulative and density of Γ, and X_i the first component of the random point $\mathbf{W}_i$ $(i = 1, 2, 3)$.

(Note in the case $N = 3$ that the given expressions for ET_N must be doubled, since each side is counted only once.)

5. The expected area and surface area

The expected surface area ES_N of a random convex hull in three dimensions can be calculated by the same technique used to find the expected number of vertices and the expected total perimeter: going directly to the $(p, \theta, \phi, \mathbf{u}_1, \mathbf{u}_2, \mathbf{u}_3)$-co-ordinate system, the formula is

$$\begin{aligned} ES_N &= 2\binom{N}{3}\int_{-\infty}^{\infty}\int_0^{\frac{1}{2}\pi}\int_0^{2\pi}\delta_N(p,\theta,\phi)\left[\iiint A_{123}^2\,g(\mathbf{w}_1)\,g(\mathbf{w}_2)\,g(\mathbf{w}_3)\,d\mathbf{u}_1\,d\mathbf{u}_2\,d\mathbf{u}_3\right]\sin\theta\,d\phi\,d\theta\,dp \\ &= 2\binom{N}{3}\int_{-\infty}^{\infty}\int_0^{\frac{1}{2}\pi}\int_0^{2\pi}\delta_N(p,\theta,\phi)\,\tilde{E}(A_{123}^2|p,\theta,\phi)\,\tilde{g}_3(p,\theta,\phi)\sin\theta\,d\phi\,d\theta\,dp, \end{aligned} \tag{5·1}$$

where δ_N and $\tilde{g}_3$ are given by (2·22) and (2·23), and

$$\tilde{E}(A_{123}^2|p,\theta,\phi) = E(A_{123}^2|Q_1 = p, Q_2 = p, Q_3 = p). \tag{5·2}$$

In the case where Γ is spherically symmetric this becomes

$$ES_N = 8\pi\binom{N}{3}\int_{-\infty}^{\infty}E(A_{123}^2|X_1 = p, X_2 = p, X_3 = p)\,G_3^{N-3}(p)\,g_3^3(p)\,dp, \tag{5·3}$$

with the definitions the same as those following (4·17).

(Note that in the case $N = 3$, the right-hand sides of both (5·1) and (5·3) must be multiplied by 2 because both sides of the triangle $(\mathbf{W}_1\mathbf{W}_2\mathbf{W}_3)$ contribute to the surface area.)

Consider projecting H_N into the plane perpendicular to the vector

$$\mathbf{n}(\theta,\phi) = (\sin\theta\cos\phi, \sin\theta\sin\phi, \cos\theta).$$

The set so obtained, $H_N(\theta,\phi)$, is the convex hull of the projected points

$$\mathbf{W}_1(\theta,\phi), \mathbf{W}_2(\theta,\phi), \ldots, \mathbf{W}_N(\theta,\phi),$$

and we will denote the area by $A_N(\theta,\phi)$. The points $\mathbf{W}_i(\theta,\phi)$ are independent and identically distributed according to the projected distribution $\Gamma(\theta,\phi)$, which is the marginal distribution in the plane perpendicular to $\mathbf{n}(\theta,\phi)$.

It is well known (Kendall & Moran, 1963, p. 74) that

$$S_N = 4\int_0^{\frac{1}{2}\pi}\int_0^{2\pi} A_N(\theta,\phi)\frac{\sin\theta\, d\phi\, d\theta}{4\pi}, \tag{5·4}$$

and therefore

$$ES_N = 4\int_0^{\frac{1}{2}\pi}\int_0^{2\pi} EA_N(\theta,\phi)\frac{\sin\theta\, d\phi\, d\theta}{4\pi}. \tag{5·5}$$

If now Γ is assumed to be spherically symmetric, then $EA_N(\theta,\phi)$ will not depend on θ or ϕ, yielding

$$\begin{aligned} EA_N &= \tfrac{1}{4}ES_N \\ &= 2\pi\tbinom{N}{3}\int_{-\infty}^{\infty} E(A_{123}^2|X_1 = p, X_2 = p, X_3 = p)\, G_3^{N-3}(p)\, g_3^3(p)\, dp. \end{aligned} \tag{5·6}$$

(As before, the right-hand side of this equation must be multiplied by 2 when $N = 3$.)

Equation (58) applies only to those distributions such as the circular normal which are the two-dimensional projections of spherically symmetric distributions in 3-space. It has already been commented that EA_N can be obtained from (10) and (35) in the case where Γ is uniform over a convex region k. In Renyi & Sulanke (1964), a simple general expression for EA_N is developed. In our notation, this expression is

$$EA_N = 2\pi\tbinom{N}{2}\int_{-\infty}^{\infty} p\sigma_p^2(Y)\, G_2^{N-2}(p)\, g_2^2(p)\, dp$$

is the case of a circularly symmetric distribution, the definitions here being the same as those following (4·11).

Another procedure, which is more complicated but yields more information, is the following: first consider a distribution $\hat{\Gamma}$ on the *circumference* of the unit circle, with probability density $f_1(\theta)$ $(0 \leqslant \theta < 2\pi)$. Define $F(\theta) = \int_\theta^{\theta+\pi} f_1(\theta)\, d\theta$. If N points $\hat{W}_1, \hat{W}_2, \ldots, \hat{W}_N$, are selected independently according to $\hat{\Gamma}$, the probability P that there will exist a semi-circle containing them all is

$$P = N\int_0^{2\pi} F^{N-1}(\theta) f_1(\theta)\, d\theta. \tag{5·7}$$

(Since $\int_0^{2\pi} F^{N-1}(\theta) f_1(\theta)\, d\theta$ is the probability that $\hat{W}_2, \ldots, \hat{W}_N$ all lie in the semi-circle whose clockwise extreme is $\hat{W}_1$.) Another expression for this probability is

$$P = N\int_0^{2\pi} F^{N-1}(\theta) f_1(\theta+\pi)\, d\theta. \tag{5·8}$$

Adding these two expressions gives a third,

$$P = N\int_0^{\pi} \beta(\theta) f_2(\theta)\, d\theta, \tag{5·9}$$

where

$$\beta(\theta) = F^{N-1}(\theta) + F^{N-1}(\theta+\pi) \tag{5·10}$$

and

$$f_2(\theta) = \tfrac{1}{2}\{f_1(\theta) + f_1(\theta+\pi)\}. \tag{5·11}$$

Returning to the distribution Γ, let $\mathbf{w}$ be any fixed point in the plane, and let $\hat{\Gamma}$ be the marginal distribution induced by projection Γ along rays onto the unit circle centred at $\mathbf{w}$. Then if $\mathbf{w}$ has the co-ordinate t on the line (p,θ) the angular density $f_2(\theta)$ used in expression (5·9) will be given by

$$f_2(\theta) = \frac{1}{2}\int_{-\infty}^{\infty} |s-t|\, g(p,\theta,s)\, ds \tag{5·12}$$

(as before $g(p,\theta,s)$ is the density $g(w)$ of Γ at the point w with co-ordinate s on the line (p,θ)).

From (59), the probability that $\mathbf{w}$ is *not* in the convex hull of N points selected independently according to Γ is easily seen to be

$$P(\mathbf{w}) = N\int_0^\pi \gamma_{N+1}(p,\theta) f_2(\theta)\, d\theta$$

$$= \tfrac{1}{2}N\int_0^\pi \int_{-\infty}^{\infty} \gamma_{N+1}(p,\theta)\, |s-t|\, g(p,\theta,s)\, ds\, d\theta, \tag{5·13}$$

$\gamma_N(p,\theta)$ being defined by equation (5).

The integral (with respect to Lebesgue measure) of $P(\mathbf{w})$ over any set A in the plane is the expected area of the intersection of A with the complement of the convex hull H_N (this is another application of Robbins' theorem (1944)). In particular let $A(R)$ be the circle of radius R centred at the origin, and denote by $B_N(R)$ the area of $A(R) \cap H_N^C$. Then, letting $q = \sqrt{(R^2-p^2)}$ and interchanging integrations in the (p,θ,t,s) co-ordinates,

$$EB_N(R) = \iint_{|\mathbf{w}|\leqslant R} P(\mathbf{w})\, d\mathbf{w}$$

$$= \tfrac{1}{2}N\int_0^\pi \int_{-R}^{R} \int_{-\infty}^{\infty} \int_{-q}^{q} \gamma_{N+1}(p,\theta)\, g(p,\theta,s)\, |s-t|\, dt\, ds\, dp\, d\theta. \tag{5·14}$$

Define
$$I_p(s) = \int_{-q}^{q} |t-s|\, dt = \begin{matrix} q^2+s^2 & \text{for} & |s| \leqslant q \\ 2q|s| & \text{for} & |s| > q. \end{matrix} \tag{5·15}$$

Using the tilda notation,

$$EB_N(R) = \tfrac{1}{2}N\int_0^\pi \int_{-R}^{R} \gamma_{N+1}(p,\theta)\, \tilde{E}(I_p|p,\theta)\, \tilde{g}_1(p,\theta)\, dp\, d\theta, \tag{5·16}$$

where
$$\tilde{g}_1(p,\theta) = \int_{-\infty}^{\infty} g(p,\theta,s)\, ds \tag{5·17}$$

and
$$E(\tilde{I}_p|p,\theta) = \frac{\int_{-\infty}^{\infty} g(p,\theta,s)\, I_p(s)\, ds}{\int_{-\infty}^{\infty} g(p,\theta,s)\, ds}. \tag{5·18}$$

In the case where Γ is circularly symmetric with marginal *cdf* and density G_2 and g_2, this reduces to

$$EB_N(R) = \pi N\int_{-R}^{R} E(I_p(Y)|X = p)\, G_2^{N-1}(p)\, g_2(p)\, dp, \tag{5·19}$$

the conditional expectation being with respect to a random point $\mathbf{W} = (X, Y)$ selected according to Γ. An integration by parts yields an equivalent expression,

$$EB_N(R) = -\pi\int_{-R}^{R} G_2^N(p)\frac{d}{dp}E(I_p(Y)|X = p)\, dp. \tag{5·20}$$

6. Linear transformations

A linear transformation T on two- or three-dimensional space takes the convex hull H_N of the points $\mathbf{W}_i$ $(i=1,2,\ldots,N)$, into the convex hull H_N^* of the transformed points $\mathbf{W}_i^* = T\mathbf{W}_i$ $(i=1,2,\ldots,N)$. It follows that the expected number of vertices, edges, and faces and the expected probability content under the transformed distribution $\Gamma^* = T\Gamma$ are the same as the corresponding quantities under Γ.

Consider the transformation in 2-space $T(x,y)' = (\alpha x, y)'$ $(\alpha > 0)$. In this case the area satisfies $A_N^* = \alpha A_N$ for every realization of H_N, and hence $EA_N^* = \alpha EA_N$. The length

$L_{ij} = |\mathbf{W}_j - \mathbf{W}_i|$ goes into $L^*_{ij} = L_{ij}\sqrt{(\alpha^2\cos^2\theta + \sin^2\theta)}$, where θ is the angle the line $L(\mathbf{W}_i, \mathbf{W}_j)$ makes with the x-axis. If Γ is circularly symmetric, θ will be uniformly distributed between 0 and π, and independent of L_{ij}. It follows that

$$EL^*_{ij} = EL_{ij}\left[\frac{1}{\pi}\int_0^\pi \sqrt{(\alpha^2\cos^2\theta + \sin^2\theta)}\,d\theta\right], \tag{6·1}$$

with the corresponding relation holding between $E\hat{L}^*_{ij}$ and $E\hat{L}_{ij}$ under definition (4·3). The term in brackets is an elliptic integral, and can be evaluated in terms of the elliptic function of the second kind (Dwight, 1957, p. 171, formula 774·1), i.e.

$$E_2(k) = \frac{\pi}{2}\left(1 - \frac{1}{2^2}k^2 - \frac{1^2}{2^2}\frac{3}{4^2}k^4 - \frac{1^2}{2^2}\frac{3^2}{4^2}\frac{5}{6^2}k^6\ldots\right). \tag{6·2}$$

Equation (4·4) then gives the relation between the perimeters

$$\left.\begin{aligned} EP^*_N &= \frac{2}{\pi}E_2\{\sqrt{(1-\alpha^2)}\}\,EP_N \qquad \text{for} \quad \alpha \leqslant 1,\\ EP^*_N &= \frac{2\alpha}{\pi}E_2\left\{\sqrt{\left(1-\frac{1}{\alpha^2}\right)}\right\}EP_N \quad \text{for} \quad \alpha > 1,\end{aligned}\right\} \tag{6·3}$$

in the case where Γ is circularly symmetric.

Transformations such as the one above enable us to reduce all normal distributions to the case of the circular (or spherical) normal, and all uniform distributions over an ellipse (or ellipsoid) to the uniform distribution over the interior of the unit circle (or sphere).

7. The normal distribution and the uniform distribution over a circle or a sphere

Let Γ now be the circular normal distribution in two dimensions;

$$g(x,y) = \frac{1}{2\pi}\exp -\tfrac{1}{2}(x^2+y^2), \tag{7·1}$$

or the spherical normal distribution in three dimensions,

$$g(x,y,z) = \frac{1}{(2\pi)^{\frac{3}{2}}}\exp -\tfrac{1}{2}(x^2+y^2+z^2). \tag{7·2}$$

The formulas developed in the previous sections are particularly simple in these cases, since the conditional expectations $E(L_{12}|X_1 = p, X_2 = p)$, $E(A_{123}|X_1 = p, X_2 = p, X_3 = p)$, etc., do not depend on p. We have†

$$\left.\begin{aligned} E(L_{12}|X_1 = p, X_2 = p) &= \frac{2}{\sqrt{\pi}},\\ E(L^2_{12}|X_1 = p, X_2 = p) &= 2,\\ E(A_{123}|X_1 = p, X_2 = p, X_3 = p) &= \frac{\sqrt{3}}{2},\\ E(A^2_{123}|X_1 = p, X_2 = p, X_3 = p) &= \frac{3}{2},\\ E(L_{123}A_{123}|X_1 = p, X_2 = p, X_3 = p) &= 2\sqrt{\frac{3}{\pi}}.\end{aligned}\right\} \tag{7·3}$$

† The first two of these relations are trivial, while the last three are best computed by expressing two of the three points (y_i, z_i) $(i = 1, 2, 3)$, in the (p, θ, t_1, t_2)-co-ordinates introduced following equation (2·3).

Denote by $\phi(x) = (2\pi)^{-\frac{1}{2}} \exp(-\frac{1}{2}x^2)$ the density of the univariate normal, and let $\Phi(x)$ be the cumulative distribution function

$$\Phi(x) = \int_{-\infty}^{\infty} \phi(y)\, dy. \tag{7·4}$$

In terms of these standard functions the various expectations are, for $N > 3$,

Two dimensions

$$\left.\begin{aligned}
&\text{vertices:} && EV_N = 4\sqrt{\pi}\tbinom{N}{2}\int_{-\infty}^{\infty} \Phi^{N-2}(p)\,\phi^2(p)\,dp,\\
&\text{probability content:} && EC_N = 1 - 2\sqrt{\pi}\tbinom{N}{1}\int_{-\infty}^{\infty} \Phi^{N-1}(p)\,\phi^2(p)\,dp,\\
&\text{perimeter:} && EP_N = 4\pi\tbinom{N}{2}\int_{-\infty}^{\infty} \Phi^{N-2}(p)\,\phi^2(p)\,dp,\\
&\text{area:} && EA_N = 3\pi\tbinom{N}{3}\int_{-\infty}^{\infty} \Phi^{N-3}(p)\,\phi^3(p)\,dp.
\end{aligned}\right\} \tag{7·5}$$

and

Three dimensions

$$\left.\begin{aligned}
&\text{vertices, faces and edges:} \begin{cases} EF_N = 4\sqrt{(3\pi)}\tbinom{N}{3}\int_{-\infty}^{\infty} \Phi^{N-3}(p)\,\phi^3(p)\,dp,\\ EE_N = \frac{3}{2}EF_N,\ EV_N = \frac{1}{2}EF_N + 2,\end{cases}\\
&\text{probability content:}\quad EC_N = 1 - \frac{2\pi}{\sqrt{3}}\tbinom{N}{2}\int_{-\infty}^{\infty} \Phi^{N-2}(p)\,\phi^3(p)\,dp - \frac{2}{N+1},\\
&\text{total perimeter:}\quad ET_n = 24\sqrt{(3\pi)}\tbinom{N}{3}\int_{-\infty}^{\infty} \Phi^{N-3}(p)\,\phi^3(p)\,dp,\\
&\text{surface area:}\quad ES_N = 12\pi\tbinom{N}{3}\int_{-\infty}^{\infty} \Phi^{N-3}(p)\,\phi^3(p)\,dp.
\end{aligned}\right\} \tag{7·6}$$

In the case $N = 3$, the expressions given for the area, total perimeter, and surface area should be doubled.

We now consider the distribution Γ_2 which is uniform over the interior of a circle of unit radius, and the three-dimensional distribution Γ_3 which is uniform over the interior of the unit sphere. The marginal distribution along the x-axis of Γ_2 has density function

$$\lambda_2(p) = \left\{\begin{array}{ll} \frac{2}{\pi}\sqrt{(1-p^2)} & (-1 \leqslant p \leqslant 1),\\ 0 & \text{elsewhere,}\end{array}\right\} \tag{7·7}$$

and cumulative distribution function $\Lambda_2(p) \equiv \int_{-1}^{p} \lambda_2(x)\,dx$ given by

$$\Lambda_2(p) = \left\{\begin{array}{ll} \frac{1}{2} + \frac{1}{\pi}(p\sqrt{(1-p^2)} + \sin^{-1} p) & \text{for}\quad 0 \leqslant p \leqslant 1,\\ 1 & \text{for}\quad 1 \leqslant p,\\ 1 - \Lambda_2(-p) & \text{for}\quad p < 0,\end{array}\right\} \tag{7·8}$$

while the corresponding univariate marginal density and cumulative for Γ_3 are

$$\lambda_3(p) = \begin{cases} \tfrac{3}{4}(1-p^2) & (-1 \leqslant p \leqslant 1), \\ 0 & \text{elsewhere}, \end{cases} \tag{7·9}$$

and

$$\Lambda_3(p) = \begin{cases} \tfrac{1}{2}+\tfrac{3}{4}p-\tfrac{1}{4}p^3 & (-1 \leqslant p \leqslant 1), \\ 0 & (p < -1), \\ 1 & (p > +1). \end{cases} \tag{7·10}$$

It is directly verified that

$$\left.\begin{aligned} &E(L_{12}|X_1 = p, X_2 = p) = \frac{\pi}{3}\lambda_2(p), \\ &E(L_{12}^2|X_1 = p, X_2 = p) = \frac{\pi^2}{6}\lambda_2^2(p), \\ &E(A_{123}|X_1 = p, X_2 = p, X_3 = p) = \tfrac{4}{3}c_1\lambda_3(p), \\ &E(A_{123}^2|X_1 = p, X_2 = p, X_3 = p) = \tfrac{16}{9}c_2\lambda_3^2(p), \\ &E(L_{12}A_{123}|X_1 = p, X_2 = p, X_3 = p) = (\tfrac{4}{3})^{\frac{3}{2}}c_3\lambda_3^{\frac{3}{2}}(p), \end{aligned}\right\} \tag{7.11}$$

where

$$\left.\begin{aligned} c_1 &= \frac{105}{144\pi} = 0{\cdot}2321, \\ c_2 &= \frac{3}{32} = 0{\cdot}09375, \\ c_3 &= \frac{16}{5\pi^2}\left(\frac{608}{945}+\frac{32}{245}\right) = 0{\cdot}2509. \end{aligned}\right\} \tag{7·12}$$

(The constants c_1, c_2 and c_3 are, respectively, $E(A_{123})$, $E(A_{123}^2)$, and $E(L_{12}A_{123})$ for three independent random points from Γ_2.)

Substituting into the expectation formulae gives the following relationships:

vertices:
$$EV_N = \frac{2\pi^2}{3}\binom{N}{2}\int_{-1}^{1} \Lambda_2^{N-2}(p)\,\lambda_2^3(p)\,dp,$$

probability content and area:
$$\begin{cases} EC_N = 1-\dfrac{\pi^2}{3}\dbinom{N}{1}\displaystyle\int_{-1}^{1} \Lambda_2^{N-1}(p)\,\lambda_2^3(p)\,dp, \\ EA_N = \pi EC_N, \end{cases}$$

perimeter:
$$EP_N = \frac{\pi^3}{3}\binom{N}{2}\int_{-1}^{1} \Lambda_2^{N-2}(p)\,\lambda_2^4(p)\,dp, \tag{7·13}$$

and

Γ_3:

vertices, faces and edges:
$$\begin{cases} EF_N = \tfrac{32}{3}\pi c_1\dbinom{N}{3}\displaystyle\int_{-1}^{1} \Lambda_3^{N-3}(p)\,\lambda_3^4(p)\,dp, \\ EE_N = \tfrac{3}{2}EF_N,\ EV_N = \tfrac{1}{2}EF_N+2, \end{cases}$$

probability content and volume:
$$\begin{cases} EC_N = 1-\tfrac{16}{9}\pi c_1\dbinom{N}{2}\displaystyle\int_{-1}^{1} \Lambda_3^{N-2}(p)\,\lambda_3^4(p)\,dp - \dfrac{2}{N+1}, \\ E\,\mathrm{vol}_N = \tfrac{4}{3}\pi EC_N, \end{cases}$$

total perimeter:
$$ET_N = 12(\tfrac{4}{3})^{\frac{3}{2}}\pi c_3\binom{N}{3}\int_{-1}^{1} \Lambda_3^{N-3}(p)\,\lambda_3^{\frac{9}{2}}(p)\,dp,$$

surface area:
$$ES_N = \tfrac{128}{9}\pi c_2\binom{N}{3}\int_{-1}^{1} \Lambda_3^{N-3}(p)\,\lambda_3^5(p)\,dp. \tag{7·14}$$

The formulae for the total perimeter and the surface area should be doubled in the case $N = 3$.

8. Normal vectors in higher dimensions

If $\mathbf{W}_1, \mathbf{W}_2, \ldots, \mathbf{W}_N$ are independent normal vectors in n dimensions, $N > n$ each with mean zero and covariance matrix the identity, the expected volume of the convex hull H_N can be computed by an argument analogous to that proceeding equation (5·6)

$$E\,\mathrm{vol}_N\,(n \text{ dimensions}) = 2\frac{n+1}{n}\frac{\pi^{\frac{1}{2}n}}{\Gamma(\frac{1}{2}n)}\binom{N}{n+1}\int_{-\infty}^{\infty}\Phi^{N-n-1}(p)\,\phi^{n+1}(p)\,dp$$

(this expression must be doubled when $N = n+1$.)

The integral on the right has been computed by Ruben (1954). In terms of the function $\mathbf{u}$ tabled on his pages 222–223,

$$\int_{-\infty}^{\infty}\phi^{\alpha}(p)\,\Phi^{\beta}(p)\,dp = (2\pi)^{-\frac{1}{2}(\alpha-1)}\,\alpha^{-\frac{1}{2}}\mathbf{u}_{\beta}(\alpha+1).$$

It should be noted that this expression for the expected volume reduces in the one-dimensional case to the well-known formula for the expected range of N independent normal points with unit standard deviation

$$ER_N = 4\binom{N}{2}\int_{-\infty}^{\infty}\phi^{N-2}(p)\,\phi^{2}(p)\,dp.$$

As noted previously, these integrals can be expressed in many different forms by the systematic use of partial integration.

References

Deltheil, R. (1926). Probabilities geometriques, (E. Borel, Ed.) *Traité du calcul des Probabilities*, vol. 22, p. 42.

Dwight, H. B. (1957). *Tables of Integrals and Other Mathematical Data.* New York: MacMillan Company.

Hostinsky, B. (1925). *Sur les probabilities geometrique. Pub. Fac. Sci. Univ. Masaryk.*

Kendall, M. G. & Moran, P. A. P. (1963). *Geometrical Probability.* Griffin's Statistical Monographs and Courses, no. 10 (1963, 1964).

Renyi, A. & Sulanke, R. (1963, 1964). Uber die convexe hulle von is zufallig gewahlten punkten, I and II. *Z. Wahr.* **2**, 75–84; **3**, 138–48.

Robbins, H. E. (1944). The measure of a random set. *Ann. Math. Statist.* **15**, 70–4.

Robbins, H. E. (1947). Acknowledgment of priority. *Ann. Math. Statist.* **18**, 297.

Ruben, H. (1954). On the moments of order statistics in samples from normal populations. *Biometrika*, **41**, 200–27.

Santalo, L. A. (1953). Introduction to integral geometry. *Actualities Scientifiques et Industrielles*, no. 1198. Paris: Hermann.

2

Forcing a Sequential Experiment to be Balanced

Introduction by Herman Chernoff
Harvard University

One of the most important contributions to Statistics in the twentieth century was the development of Experimental Design with the use of randomization. A recent book by David Salsburg, entitled *The Lady Tasting Tea: How Statistics Revolutionized Science in the Twentieth Century*, begins with that classical experiment of R.A. Fisher. But suppose that the random design had chosen to add milk after tea for the first four of the eight cups. Not only would Bayesians, for whom randomization raises a problem, be troubled, but Fisher would probably have had serious concerns.

The use of randomly selected treatments in clinical trials was considered the gold standard, but raised issues, some of which were not seriously addressed until Efron wrote the paper "Forcing a sequential experiment to be balanced."

The object is to reduce imbalances in small-sized experiments, involving treatments and controls that arrive sequentially, while avoiding various forms of experimental bias characteristic of perfectly balanced plans. Complete randomization has the advantages of freedom from selection bias and from accidental bias due to the presence of nuisance factors. It also provides an inferential basis for calculating appropriate P-values. But in small samples it can lead to serious imbalances, degrading the power of the test. Efron offers a relatively simple alternative plan which gives the experimenter a slight ability for guessing which treatment will be next. This plan, *biased coin design* or BCD(p), consists of selecting the next subject to be treatment or control with probability $p > 1/2$ if there is an excess of subjects in the other group. If the trial is well balanced at the time, the next subject is chosen to be treatment with probability 0.5. Efron indicates a preference for $p = 2/3$ where $r = p/(1-p) = 2$. The experimenter knows which choice is more likely to be next, but his knowledge is limited by chance.

The tendency for lack of balance using BCD(2/3) is much reduced from that of complete randomization, but the plan suffers from some tendencies to bias. These are evaluated and compared with those of Randomized Block Design, RB(b), which consists of complete randomization within a block of $2b$ subjects of which b must be selected to be treatments.

Efron measures selection bias of a sequential design by the expected number of correct guesses about treatment choice that the experimenter can make if he guesses optimally. For complete randomization, this measure is $n/2 = b$ for an experiment of length $n = 2b$ or $1/2$ per subject. For BCD(2/3), the *excess selection bias* exceeding b approaches $b(r-1)/2r$ or 1/8 per subject. For RB(b), the excess selection bias is slightly greater for $b < 8$ and less for $b > 9$.

Efron indicates why the maximum vulnerability to accidental bias may be measured by the largest eigenvalue of the covariance matrix of $\mathbf{T} = (T_1, T_2, \ldots, T_n)$ where T_i is $+1$ or -1 depending on whether the ith subject is a treatment or control. This measure is 1 for complete randomization and it is $1 + 1/(b-1)$ for RB(b). This gives $10/9$ for $b = 5$. For BCD(2) it is also approximately $10/9$, and this approximation is conjectured to be more generally $1 + (2p-1)^2$. The argument used suggests that under some circumstances the accidental bias, as opposed to the maximum possible bias, can be less than that for complete randomization.

References

Salsburg, D. (2001). *The Lady Tasting Tea: How Statistics Revolutionized Science in the Twentieth Century.* W.H. Freeman, New York.

Biometrika (1971), 58, 3, p. 403
With 2 text-figures
Printed in Great Britain

Forcing a sequential experiment to be balanced

By BRADLEY EFRON
Stanford University

SUMMARY

Subjects arrive sequentially at an experimental site and must be assigned immediately to treatment or control groups. In order to avoid biasing the results of the experiment it is customary to make the assignments by independent flips of a fair coin, but in small-sized experiments this may result in a severe imbalance between the numbers of treatments and controls. This paper discusses a new method of assigning the subjects which tends to balance the experiment, but at the same time is not over vulnerable to various common forms of experimental bias.

1. The problem

The random assignment of subjects to treatment and control groups is a canon of good experimental design. In many situations, the subjects are not available to the experimenter all at one time but rather arrive sequentially, as is often the case in medical experimentation. Complete randomization is achieved by the flip of a coin, that is by assigning each subject randomly to the treatment or control group with equal probability, independently of the assignment of the other subjects.

Complete randomization has several attractive properties, which are listed below, but suffers from the disadvantage that in experiments which are limited to a small number of subjects, the final distribution of treatments and controls can be very unbalanced. For example, Table 1 shows the distribution of treatments and controls in an investigation of a new treatment for Hodgkin's disease; the data are given by courtesy of Dr Henry Kaplan, Department of Radiology, Stanford University. The patients were divided into four age categories since age was thought to be a possible factor influencing the response. It obviously would be preferable to have the youngest age-group more equitably divided, as would certainly have been done if all 29 patients could have been assigned at once.

Table 1. *Distribution of treatments and controls in four age categories, Hodgkin's disease investigation*

	Age			
	10–19	20–29	30–39	40–49
Treatments	7	5	4	1
Controls	2	5	3	2

It is simple to devise methods of sequential experimentation which force approximately equal numbers of treatments and controls. For example, the completely nonrandom systematic design *TCTCTC*..., '*T*' for treatment, '*C*' for control, followed separately in each category of subjects guarantees that the imbalance will never exceed one in any category. A variant is the Student sandwich plan *TCCTTCCTT*.... However, these plans

27-2

can easily bias the results of the experiment, as shown below, and are not serious competitors to complete randomization. A controversy between Student and Fisher arose on a closely related point in the 1930's; see Student (1937), Barbacki & Fisher (1936), Yates (1939) and Greenberg (1951).

The problem then is to compromise between a perfectly balanced experiment and the advantages of complete randomization. These advantages are:

(1) *Freedom from selection bias.* If the experimenter knows for certain that the next assignment will be a treatment, or a control, he may consciously or unconsciously bias the experiment by such decisions as who is or is not a suitable experimental subject, in which category the subject belongs, etc. It is obvious that complete randomization eliminates selection bias, and that the systematic designs maximize it. Blackwell & Hodges (1957) coined the term 'selection bias'. They advocate using complete randomization but continuing the experiment until there is a certain minimum number of both treatments and controls. In practice this can be very difficult advice to follow, particularly in an experiment such as the Hodgkin's disease investigation where there are many categories of subjects. Selection bias is not a factor in blind experiments where admission to the study and related decisions are made by someone ignorant of the past assignment of treatments and controls. However, even in this situation complete randomization enjoys the two advantages which follow.

(2) *Freedom from accidental bias.* Known or unknown to the experimenter, there may be nuisance factors systematically affecting the experimental units. Typical examples are time trends, sex-linked differences, differing experimental conditions, etc. Complete randomization tends to balance out such factors (see § 5) and thus protect the significance level of the usual hypothesis tests. The systematic designs mentioned above are quite vulnerable to accidental bias.

(3) *Randomization as a basis for inference.* Probability statements, such as the obtained significance level of the experiment, can be based entirely on the randomness induced by the complete randomization between treatments and controls. This eliminates the need for probability assumptions on the responses of the individual experimental units and guarantees the validity of the stated significance level.

Advantage (3) is closely related to but not identical with advantage (2).

One compromise between complete randomization and balanced systematic designs that is used in practice is the permuted block design. This design divides the experiment into blocks of even length, say $2b$, and within each block randomly assigns b units to treatment and b units to control, all $\binom{2b}{b}$ combinations being equally likely. Permuted blocks can be quite effective in eliminating unbalanced designs but they suffer from the disadvantage that at certain points in the experiment the experimenter knows for certain whether the next subject will be assigned as a treatment or as a control. For example, if $b = 5$ the probability is $\frac{1}{6}$ that the experimenter will know for certain the assignment of units 8, 9 and 10, and $\frac{4}{9}$ that he will know for certain the assignment of units 9 and 10.

The biased coin designs introduced in § 2 are motivated by the desire to achieve balanced experiments without ever giving the experimenter a high probability of guessing the assignment of the next unit. Comparisons are made between these designs and permuted blocks in the later sections.

2. Biased coin designs

Suppose that at a certain stage in the experiment a new subject arrives and is noted to be in a category which has had $\tilde{D}$ more treatments than controls previously assigned to it. We assign the new subject as follows:

$$\begin{array}{ll} \text{If } \tilde{D} > 0, & \text{assign treatment with probability } q \text{ and control with probability } p. \\ \text{If } \tilde{D} = 0, & \text{assign treatment with probability } \tfrac{1}{2} \text{ and control with probability } \tfrac{1}{2}. \\ \text{If } \tilde{D} < 0, & \text{assign treatment with probability } p \text{ and control with probability } q. \end{array} \quad (2{\cdot}1)$$

Here $p \geqslant q, p+q = 1$, so that the assignment rule tends to balance the number of treatments and controls, the tendency being weakest if $p = \frac{1}{2}$, complete randomization, and strongest, if $p = 1$, permuted block design with $b = 1$.

We will call the rule described in (2·1) the *biased coin design* with bias p, abbreviated *BCD*(p) for convenient reference. The rule is meant to be applied separately within each category of the subjects, and so for our purposes we can think of each category as being a separate experiment in which we are trying to balance treatments and controls. In what follows we drop all reference to separate categories.

The value $p = \frac{2}{3}$, which is the author's personal favourite, will be seen to yield generally good designs and will be featured in the numerical computations.

Section 3 discusses the balancing properties of the biased coin designs. Selection bias, accidental bias and randomization as a basis for inference are discussed in §§4, 5 and 6. All proofs are deferred until § 7.

A note on methodology. Complete randomization is usually implemented by means of a deck of envelopes which the statistician gives the experimenter. After each patient is admitted to the experiment, an envelope is opened which directs him into either the treatment or control group, these choices having been made by the statistician using a random device or random number table. To implement a permuted block design the deck must be ordered in blocks of the proper size, and a separate deck provided for each category of patient. Usually the statistician makes the assignments by means of a table of random permutations.

A biased coin design for a multicategory experiment can be implemented using a single unordered deck of envelopes, each of which contains instructions covering the three cases $\tilde{D} > 0, \tilde{D} = 0$ and $\tilde{D} < 0$ described in (2·1). It is probably best to have the three instructions in separate envelopes within the envelope so that the experimenter does not see the ones he does not use. The assignments can be made with a simple random device, such as a die if $p = \frac{2}{3}$, or a random number table.

3. Balancing properties of the biased coin designs

Define D_n to be the absolute difference between the number of treatments and number of controls after n assignments have been made, $D_0 = 0$. Under *BCD*(p), the D_n form a Markov chain with states $0, 1, 2, \ldots$ and transition probabilities

$$\begin{aligned} \operatorname{pr}(D_{n+1} = j+1 \mid D_n = j) &= q \quad (j \geqslant 1), \\ \operatorname{pr}(D_{n+1} = j-1 \mid D_n = j) &= p \quad (j \geqslant 1), \end{aligned} \quad (3{\cdot}1)$$

$$\operatorname{pr}(D_{n+1} = 1 \mid D_n = 0) = 1.$$

This is a random walk with a reflecting barrier at the origin (Cox & Miller, 1965, p. 41) and has stationary probabilities π_j given by

$$\pi_0 = \frac{r-1}{2r}, \quad \pi_j = \frac{r-1}{2r}\frac{r+1}{r^j} \quad (j \geqslant 1), \tag{3·2}$$

where $r = p/q$. The first few values of the π_j are given for $r = 2$, 3 and 4 in Table 2.

Table 2. *Values of the first few stationary probabilities*

	π_0	π_1	π_2	π_3	π_4	π_5
$r = 2$ ($p = \frac{2}{3}$)	$\frac{1}{4}$	$\frac{3}{8}$	$\frac{3}{16}$	$\frac{3}{32}$	$\frac{3}{64}$	$\frac{3}{128}$
$r = 3$ ($p = \frac{3}{4}$)	$\frac{1}{3}$	$\frac{4}{9}$	$\frac{4}{27}$	$\frac{4}{81}$	$\frac{4}{243}$	$\frac{4}{729}$
$r = 4$ ($p = \frac{4}{5}$)	$\frac{3}{8}$	$\frac{15}{32}$	$\frac{15}{124}$	$\frac{15}{512}$	$\frac{15}{2048}$	$\frac{15}{8192}$

Since the D_n can take on only odd or even values as n is odd or even, the Markov chain has period 2, and the limiting probabilities should be doubled accordingly,

$$\lim_{m\to\infty} \mathrm{pr}(D_{2m} = 0) = 2\pi_0 = (r-1)/r, \quad \lim_{m\to\infty} \mathrm{pr}(D_{2m+1} = 1) = 2\pi_1 = (r^2-1)/r^2, \tag{3·3}$$

etc. Thus, with $p = \frac{2}{3}$, the experiment has asymptotic probability $\frac{1}{2}$ of being exactly balanced for n even, and asymptotic probability $\frac{3}{4}$ of being as close as possible to balanced for n odd.

The experiment starts with $D_0 = 0$ so it is natural to expect the limiting distribution of D_n to be approached from below. This is made precise as follows.

THEOREM 1. *If $h(j)$ is a nondecreasing function of j $(j = 0, 1, \ldots)$ and D_n is the absolute difference between the numbers of treatments and controls after n assignments have been made, then $E\{h(D_{n+2})\} \geqslant E\{h(D_n)\}$ for every value of n.*

Taking $h(0) = 0$, $h(j) = 1$ for $j > 0$, in the theorem shows that $\mathrm{pr}(D_{2m} = 0)$ is a decreasing function of m, and taking $h(0) = 0$, $h(1) = 0$ and $h(j) = 1$ $(j > 1)$, shows that $\mathrm{pr}(D_{2m+1} = 1)$ is a decreasing function of m. We can write down the total bonus probability of having $D_{2m} = 0$ or $D_{2m+1} = 1$ explicitly as

THEOREM 2. *The following relations hold:*

$$\sum_{m=1}^{\infty} \{\mathrm{pr}(D_{2m} = 0) - 2\pi(0)\} = \frac{1}{r(r-1)},$$

$$\sum_{m=1}^{\infty} \{\mathrm{pr}(D_{2m+1} = 1) - 2\pi(1)\} = \frac{2}{r^2(r-1)}.$$

Notice that we are not including the trivial cases $D_0 = 0$ and $D_1 = 1$.

Table 3 shows the actual distribution of D_n $(n = 2, \ldots, 10)$ for the case $r = 2$. For $r = 2$ both sums in Theorem 2 equal 0·5, and it can be seen from the table that 0·391 of the even-n bonus and 0·336 of the odd-n bonus have occurred by $n = 10$.

A comparison of $BCD(p)$ with $p = \frac{2}{3}$ and the permuted block design with $b = 5$ is given in Table 4, where the two plans are seen to behave rather similarly for the crucial small values of n. Comparisons of the probabilities of some early extreme imbalances, $D_4 = 4$, $D_5 = 3$, etc., reinforce this impression, though the different natures of the two rules make such comparisons difficult.

It seems obvious that increasing the value of p, or equivalently of r, should shrink the distribution of D_n toward 0. We conclude this section with a theorem to that effect.

Table 3. *Percentage probabilities* 100 $(D_n = j)$ *for* $r = 2$

$j \backslash n$	2	3	4	5	6	7	8	9	10
0	66·7	—	59·3	—	56·0	—	54·1	—	53·0
1	—	88·9	—	84·0	—	81·2	—	79·5	—
2	33·3	—	37·0	—	37·9	—	38·0	—	38·0
3	—	11·1	—	14·8	—	16·5	—	17·3	—
4	—	—	3·7	—	5·8	—	7·0	—	7·7
5	—	—	—	1·2	—	2·2	—	2·9	—
6	—	—	—	—	0·4	—	0·8	—	1·2
7	—	—	—	—	—	0·1	—	0·3	—
8	—	—	—	—	—	—	0·0	—	0·1
9	—	—	—	—	—	—	—	0·0	—
10	—	—	—	—	—	—	—	—	0·0

Table 4. *Percentage probabilities that experiment is exactly balanced at stage n, within* 1 *if n is odd*

	n								
	2	3	4	5	6	7	8	9	10
$BCD(\frac{2}{3})$	66·7	88·9	59·3	84·0	56·0	81·2	54·1	79·5	53·0
Permuted blocks $b = 5$	55·6	83·3	47·6	79·3	47·6	83·3	55·5	100·0	100·0

THEOREM 3. *If* $h(j)$ *is a nondecreasing function of* j $(j = 0, 1, \ldots)$, *then* $E\{h(D_n)\}$ *is a non-increasing function of* r *for* $r \geq 1$ *and any value of* n.

4. SELECTION BIAS

A natural measure of the selection bias of a sequential design is the expected number of correct guesses the experimenter can make if he guesses optimally. Every guessing strategy is equally useless against complete randomization, yielding an expected $\frac{1}{2}n$ correct guesses in n trials. It is intuitively clear, and proved by Blackwell & Hodges (1957), that the best strategy against a permuted block design is to guess treatment or control on the basis of which has so far occurred least often in the block. Blackwell and Hodges show that this results in an expected $2^{2b} \Big/ \binom{2b}{b} - 1$ correct guesses per block of length $2b$ against the corresponding permuted block design. We can think of this as an excess selection bias of

$$2^{2b} \Big/ \binom{2b}{b} - (b+1) \tag{4·1}$$

expected correct guesses per block for the permuted block design compared to complete randomization.

The best guessing strategy against a biased coin design is to guess treatment or control on the basis of which has so far occurred least often in the experiment, with no preferred guess if there is a tie. The probability of guessing correctly on trial n is

$$\tfrac{1}{2} \operatorname{pr}(D_{n-1} = 0) + p \operatorname{pr}(D_{n-1} > 0), \tag{4·2}$$

which asymptotically approaches

$$\tfrac{1}{2}\pi_0 + p(1-\pi_0) = \tfrac{1}{2} + \frac{r-1}{4r}. \tag{4·3}$$

The excess selection bias of $BCD(p)$ in $2b$ trials, ignoring initial effects, is therefore

$$\frac{r-1}{2r} b. \tag{4·4}$$

Figure 1 compares (4·1) with (4·4) as a function of b. The case $r = 2$ and $p = \frac{2}{3}$ is seen to yield the same excess selection bias as a permuted block design with b between 8 and 9, that is with block length between 16 and 18. For $r = 2$ the excess correct guesses per trial are asymptotically $(r-1)/(4r) = \frac{1}{8}$.

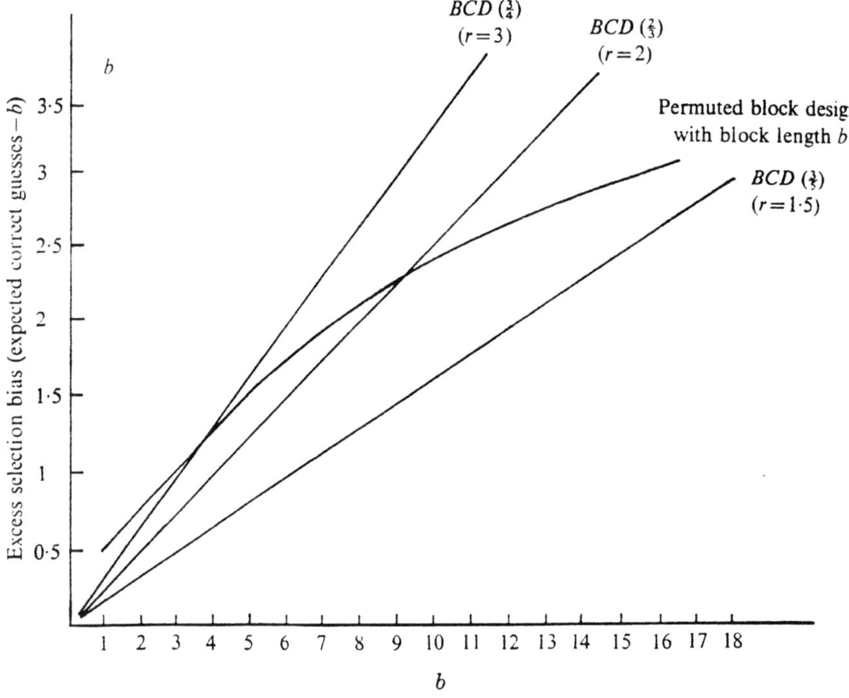

Fig. 1. Excess selection bias in $2b$ trials. In a long sequence of assignments $BCD(\frac{2}{3})$ would have approximately the same selection bias as the permuted block design with $2b = 18$.

5. Accidental bias

Let $T_k = +1$ or -1 as the kth experimental unit is assigned to the treatment or control group, and let $\mathbf{T} = (T_1, \ldots, T_N)$ be the vector of assignments after some fixed number N of trials. All of the random designs we have mentioned have $E(\mathbf{T}) = \mathbf{0}$. In this section we assess the vulnerability of a design to accidental bias by the magnitude of the largest eigenvalue of the covariance matrix $\mathbf{\Sigma_T}$ of $\mathbf{T}$. This choice is motivated in the following way.

Suppose the responses y_k of the experimental units are determined by the linear model

$$y_k = \mu + \alpha t_k + \beta z_k + \epsilon_k \quad (k = 1, \ldots, N), \tag{5·1}$$

where $t_k = +1$ or -1 as unit k is in the treatment or control group, z_k is the measurement of some nuisance factor, for example, age of subject on unit k, and the ϵ_k are independent $\mathcal{N}(0, \sigma^2)$ random variables. In vector notation, $\mathbf{y} = \mu\mathbf{e} + \alpha\mathbf{t} + \beta\mathbf{z} + \boldsymbol{\epsilon}$, where $\mathbf{e}' = (1, \ldots, 1)$, $\mathbf{t}' = (t_1, \ldots, t_N)$, etc. There is no loss of generality in this model in assuming that

$$\mathbf{z}'\mathbf{e} = 0, \quad \|\mathbf{z}\|^2 = 1. \tag{5·2}$$

The nuisance factor $\mathbf{z}$ can harmfully affect our inferences about α in two ways: if we test $\alpha = 0$ using Student's t in the usual way, without allowing for the covariate $\mathbf{z}$, there will be a spurious noncentrality parameter of magnitude $(\beta^2/\sigma^2)(\mathbf{z}'\mathbf{t})^2$ in the numerator of Student's statistic when the null hypothesis is true. If the null hypothesis is false, and if we do allow for the factor $\mathbf{z}$ in our analysis, then the noncentrality parameter for testing $\alpha = 0$ is

$$\frac{N\alpha^2}{\sigma^2}\left[1 - \frac{1}{N}\left\{\left(\frac{\mathbf{e}'}{\sqrt{N}}\mathbf{t}\right)^2 + (\mathbf{z}'\mathbf{t})^2\right\}\right] \tag{5·3}$$

and we see that the loss of noncentrality due to the nuisance factor is $(\alpha^2/\sigma^2)(\mathbf{z}'\mathbf{t})^2$.

In both cases it is ideal to have $\mathbf{t}$ orthogonal to $\mathbf{z}$, and we are penalized proportionately to $(\mathbf{z}'\mathbf{t})^2$ as we depart from this ideal. It is also good to have $\mathbf{t}$ orthogonal to $\mathbf{e}$, i.e. to balance the experiment, which is what we are trying to do in this paper.

Now if $\mathbf{t}$ is the realization of the random vector $\mathbf{T}$, with mean vector $\mathbf{0}$ and covariance matrix $\boldsymbol{\Sigma}_\mathbf{T}$, then for a fixed vector $\mathbf{z}$

$$E\{(\mathbf{z}'\mathbf{T})^2\} = \mathbf{z}'\boldsymbol{\Sigma}_\mathbf{T}\mathbf{z}. \tag{5·4}$$

The least favourable vector $\mathbf{z}$ is that vector satisfying (5·2) which maximizes (5·4), and we see that

$$E\{(\mathbf{z}'\mathbf{T})^2\} \leqslant \text{largest eigenvalue of } \boldsymbol{\Sigma}_\mathbf{T}, \tag{5·5}$$

with equality if the corresponding eigenvector is orthogonal to $\mathbf{e}$. This will turn out to be the case for all the designs we are studying.

For example, under complete randomization $\boldsymbol{\Sigma}_\mathbf{T} = \mathbf{I}$ and $E\{(\mathbf{z}'\mathbf{T})^2\} = 1$ for every $\mathbf{z}$. Notice that 1 is the smallest possible value for the maximum eigenvalue of $\boldsymbol{\Sigma}_\mathbf{T}$ since $E(T_k) = 0$ implies var $(T_k) = 1$ and hence tr $(\boldsymbol{\Sigma}_\mathbf{T}) = n$ = sum of eigenvalues of $\boldsymbol{\Sigma}_\mathbf{T}$. For the permuted block design with block length $2b$, we have $\boldsymbol{\Sigma}_\mathbf{T} = \{1 + 1/(2b-1)\}\mathbf{I} - 1/(2b-1)\,\mathbf{ee}'$ for N satisfying $2 \leqslant N \leqslant 2b$. This matrix has largest eigenvalue $1 + 1/(2b-1)$, attained for any vector $\mathbf{z}$ orthogonal to $\mathbf{e}$. This same result holds for $N > 2b$. If we use permuted blocks with $b = 5$, we therefore increase the maximum vulnerability to accidental bias from 1 to $1\frac{1}{9}$.

We now compute the maximum eigenvalue for the biased coin design. A much simpler, and in some ways more informative, answer is possible if we analyze the process $T_1, \ldots, T_N$ as if it were derived from a stationary process. Specifically, the results obtained below pertain to the sequence $T_{h+1}, \ldots, T_{h+N}$ with both h and N approaching infinity. It then turns out that the maximum value of the spectral density of the process is the maximum eigenvalue we are looking for.

The limiting autocovariance function

$$\rho_k \equiv \lim_{h\to\infty} E(T_h T_{h+k}) \tag{5·6}$$

has a closed form expression. Define $B_k(i)$ as the cumulative distribution function of a binomial random variable with parameters k and p,

$$B_k(i) = \sum_{i'=0}^{i} \binom{k}{i'} p^{i'} q^{k-i'}. \tag{5·7}$$

THEOREM 4. *The limiting autocovariances ρ_k are given by*

$$\rho_1 = -\phi(r)\frac{1}{2(r-1)}, \quad \rho_{k+1} - \rho_k = \phi(r)\frac{1}{k}\left\{\sum_{i=0}^{[\frac{1}{2}(k-2)]} B_k(i) + \delta_k B_k([\tfrac{1}{2}k])\right\},$$

where

$$\phi(r) = \frac{(r-1)^3}{r(r+1)},$$

$\delta_k = 0$ *if k is even and* $\delta_k = 1$ *if k is odd. The square brackets indicate the greatest integer function.*

COROLLARY 1. *For k even and positive* $\rho_{k+1} - \rho_k = \rho_{k+2} - \rho_{k+1}$.

COROLLARY 2. *For $k \geqslant 1$, ρ_k is negative and $\rho_{k+1} - \rho_k$ is positive and decreasing.*

COROLLARY 3. *Also*

$$\sum_{k=1}^{\infty} \rho_k = -0·5, \quad \sum_{k=1}^{\infty} (-1)^k \rho_k = \frac{1}{2}\left(\frac{r-1}{r+1}\right)^2.$$

Table 5 gives the first 11 values of ρ_k for $r = 2$, 3 and 4.

Table 5. *Values of the autocorrelation function* ρ_k

$k \backslash r$	2	3	4
0	1·000	1·000	1·000
1	−0·0833	−0·1667	−0·2250
2	−0·0556	−0·0833	−0·0900
3	−0·0463	−0·0625	−0·0630
4	−0·0370	−0·0417	−0·0360
5	−0·0319	−0·0325	−0·0263
6	−0·0267	−0·0234	−0·0166
7	−0·0234	−0·0187	−0·0124
8	−0·0201	−0·0140	−0·0082
9	−0·0178	−0·0113	−0·0062
10	−0·0155	−0·0087	−0·0042

Since $\Sigma |\rho_k| < \infty$, the process has an absolutely continuous spectrum with spectral density

$$f(\omega) \equiv \sum_{k=-\infty}^{\infty} \rho_k e^{-i\omega k} = 1 + 2\sum_{k=1}^{\infty} \rho_k \cos(\omega k); \tag{5·8}$$

this is a factor of 2π times the usual definition (Cox & Miller 1965, p. 315). Corollary 3 can be restated in terms of $f(\omega)$ as

$$f(0) = 0, \quad f(\pi) = 1 + \left(\frac{r-1}{r+1}\right)^2 = 1 + (p-q)^2. \tag{5·9}$$

Since $f(-\omega) = f(\omega)$ the abscissa is plotted only from $\omega = 0$ to $\omega = \pi$ in Fig. 2. We see that the spectral density increases monotonically as ω goes from 0 to π and so obtains its maximum value at π. By (5·9) this maximum is equal to $1 + (p-q)^2$. Therefore $BCD(\frac{2}{3})$ has maximum vulnerability to accidental bias of amount $1+(\frac{2}{3}-\frac{1}{3})^2 = 1\frac{1}{9}$, the same as the permuted block design with $b = 5$.

Although $f(\omega)$ was monotone for every value of r numerically investigated, the author was not able to prove the result in general, and so it remains as a conjecture that $1+(p-q)^2$ is the maximum value for $BCD(p)$.

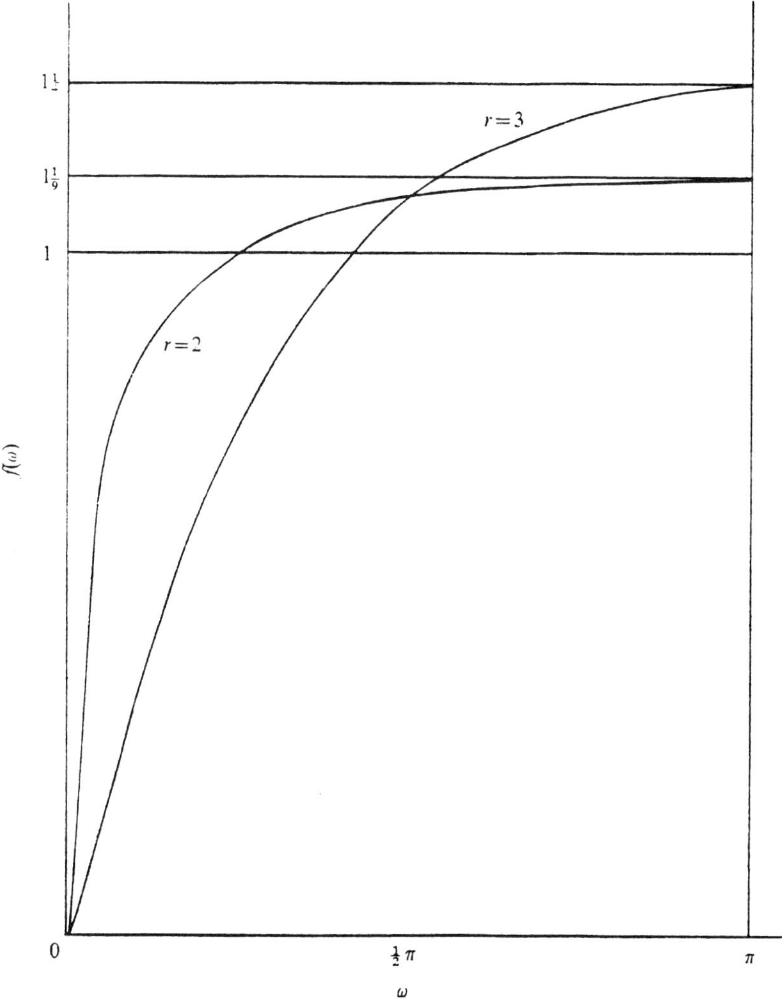

Fig. 2. Spectral density of biased coin assignments, showing maximum value at $\omega = \pi$. The spectral density for complete randomization is the line $f(\omega) = 1$.

It is easy to see that the maximum of the spectral density corresponds to the maximum eigenvalue definition given before. Let $z_1, \ldots, z_N$ be any sequence satisfying (5·2) and define

$$l_{\mathbf{z}}(\omega) = \sum_{k=1}^{N} z_k e^{-i\omega k}, \quad L_{\mathbf{z}}(\omega) = \frac{1}{2\pi}|l_{\mathbf{z}}(\omega)|^2. \tag{5·10}$$

Then by Parseval's lemma $L_{\mathbf{z}}(\omega)$ is a probability density on $[-\pi, \pi]$,

$$\int_{-\pi}^{\pi} L_{\mathbf{z}}(\omega)\, d\omega = 1,$$

and by the standard theory of linear transforms (Cox & Miller, 1965, p. 321)

$$E\{(\mathbf{z}'\mathbf{T})^2\} = \int_{-\pi}^{\pi} f(\omega)\, L_{\mathbf{z}}(\omega)\, d\omega. \tag{5·11}$$

We see that the integral is bounded by the maximum value of $f(\omega)$, a bad $\mathbf{z}$ being one with $L_{\mathbf{z}}(\omega)$ concentrated near the maximizing value of ω. In the cases above the least favourable $\mathbf{z}$ is asymptotically proportional to the sequence $+1, -1, +1, -1, \ldots$.

The definition of vulnerability to accidental bias we have given is of a minimax nature and so favours complete randomization. Figure 2 and (5·11) show that for some values of $\mathbf{z}$, namely those concentrated at low frequencies, the biased coin designs are superior to complete randomization.

6. Randomization as a basis for inference

Suppose that we have a fixed sample size experiment for comparing m treatment measurements $x_1, \ldots, x_m$ with n control measurements $y_1, \ldots, y_n$ $(m+n = N)$. Let $U_1, \ldots, U_N$ be the experimental units and let c represent the collection of m indices corresponding to treatments, say $c \in \mathscr{C}$, where $\mathscr{C}$ is the set of all choices of m integers from $1, \ldots, N$.

Whatever test we are using for the hypothesis of no difference between treatment and control responses will have a significance level α_c depending on the choice of $c \in \mathscr{C}$. The value of α_c will also depend, of course, on the particular experimental conditions, but these are considered fixed in this discussion. Notice that we are assuming that the test statistic is invariant under permutations of the x values and permutations of the y values separately, which is the usual case. If we use a level α permutation test, such as the conditional t test, Wilcoxon's test, etc., then

$$\frac{1}{\binom{N}{m}} \sum_{c \in C} \alpha_c = \alpha, \tag{6·1}$$

which is just another way of saying that for each set of N experimental responses the test rejects for $\alpha \binom{N}{m}$ of the $\binom{N}{m}$ possible assignments of these responses to the treatment and control groups.

Under complete randomization, if we condition on the number of treatments which have occurred, then c is in fact randomly selected from $\mathscr{C}$ with equal probability for all $\binom{N}{m}$ members, and (6·1) allows one to claim an exact significance level of α for the test without any probability model on the responses of the experimental units. This argument is distinct from that of § 5, which said that under a reasonable probability model complete randomization, and also $BCD(p)$, were likely to yield a c with α_c nearly equal to the nominal α level. In practice the two points of view conflict only if the randomly selected c happens to look particularly nonrandom, hence prone to accidental bias; for example, if it selects most of the treatments early in the experiment.

The biased coin designs do not give the same conditional distribution of c and so (6·1) does not apply directly. Theoretically we could redefine the rejection region of any permutation test to give level α with respect to the distribution of c under $BCD(p)$ but this is hard work and probably unnecessary as the following asymptotic argument shows.

Assume we are using $BCD(p)$ and that N is large. We can assume $m = n = \frac{1}{2}N$ without affecting the rough calculation which follows. Let $w_1, \ldots, w_N$ be the observed responses on units $1, \ldots, N$, with the common mean subtracted off so that $\bar{w} = 0$, and let

$$\sigma_w^2 = \sum_{k=1}^{N} w_k^2/N.$$

Suppose our test rejects if the observed value of $|\bar{x} - \bar{y}|$ is among the $\alpha\binom{N}{\frac{1}{2}N}$ largest of the $\binom{N}{\frac{1}{2}N}$ possible values of $|\bar{x} - \bar{y}|$ given w. This is the conditional t test for the treatment and control measurements having identical distributions. If n is large, this test is actually carried out by a normal approximation: if c is chosen with equal probability from among the elements of $\mathscr{C}$ then $\bar{x} - \bar{y}$ is approximately $\mathscr{N}\left(0, \frac{4}{N}\sigma_w^2\right)$.

If c has been selected by $BCD(p)$ then, conditioned on the observed value of w, $\bar{x} - \bar{y}$ has mean 0 and variance approximately

$$\frac{4}{N}\sigma_w^2 \int_{-\pi}^{\pi} f(\omega)\, L_{\mathbf{z}}(\omega)\, d\omega, \tag{6·2}$$

where $z_k = w_k/(\sigma_w \sqrt{N})$, as in (5·10) and (5·11). For $p = \frac{2}{3}$ the permutation standard deviation of $\bar{x} - \bar{y}$ therefore cannot exceed $\sqrt{(1\frac{1}{9})} = 1{\cdot}055$ times its value under complete randomization, so we will not make a seriously anticonservative error using the latter value. The standard deviation rather than the variance is the crucial factor for the significance test. On the other hand, the integral in (6·2) can be arbitrarily small, as commented before, so the complete randomization variance can conceivably be very conservative from this point of view. In most cases we would expect the w sequence to be noisy, that is to have $L_{\mathbf{z}}(\omega)$ spread rather evenly over $[-\pi, \pi]$, in which case the two answers tend to coincide since

$$\frac{1}{2\pi}\int_{-\pi}^{\pi} f(\omega)\, d\omega = 1.$$

The situation is similar for the permuted block designs. If $N = 2bI$ then the variance of $\bar{x} - \bar{y}$, conditioned on w, is

$$\frac{4}{N}\sigma_w^2 \left(1 + \frac{1}{2b-1}\right)\left(1 - \sum_{i=1}^{I} \bar{w}_i^2/I\sigma_w^2\right), \tag{6·3}$$

where $\bar{w}_i$ is the average of the w_k within the ith block. Again the factor multiplying $(4/N)\sigma_w^2$ can range from a high of

$$1 + \frac{1}{2b-1},$$

which equals $1\frac{1}{9}$ if $b = 5$, down to zero.

We have ignored the question of asymptotic normality in this discussion for which reasonable conditions on the w_k are necessary (Serfling, 1968).

7. PROOFS

Define $\pi_n(j) = \text{pr}(D_n = j)$ and $E_k\{h(D_n)\} = E\{h(D_n)|D_0 = k\}$, the expected value of $h(D_n)$ given that we start the Markov chain (3·1) with $D_0 = k$.

LEMMA 0. *For $j \geqslant 2$ and $n \geqslant 2$,*

$$\pi_n(0) = p\pi_{n-1}(1), \quad \pi_n(1) = \pi_{n-1}(0) + p\pi_{n-1}(2),$$
$$\pi_n(j) = q\pi_{n-1}(j-1) + p\pi_{n-1}(j+1).$$

LEMMA 1. *For any function h and $n \geqslant 2$,*

$$E\{h(D_n)\} = p\{h(0)\pi_{n-1}(1) + h(1)\pi_{n-1}(0)\} + q\sum_{j=0}^{\infty} h(j+1)\pi_{n-1}(j) + p\sum_{j=2}^{\infty} h(j-1)\pi_{n-1}(j).$$

LEMMA 2. *If $h(j)$ is a nondecreasing function of j then*

$$E_{k+2}\{h(D_n)\} \geqslant E_k\{h(D_n)\} \quad (k \geqslant 0,\ n \geqslant 1).$$

Lemma 0 follows immediately from the definition (3·1) of the Markov chain. Lemma 1 is the expression

$$E\{h(D_n)\} = \sum_{j=0}^{\infty} \pi_n(j) h(j)$$

with the $\pi_n(j)$ expressed in terms of the $\pi_{n-1}(j)$ via Lemma 1. Lemma 2 is obviously true for $n = 1$, and assume by induction it is true for the case $n-1$. Then, making use of (3·1) again,

$$E_{k+2}\{h(D_n)\} = pE_{k+1}\{h(D_{n-1})\} + qE_{k+3}\{h(D_{n-1})\} \geqslant E_{k+1}\{h(D_{n-1})\} \tag{7·1}$$

by the induction hypothesis, and likewise, if $k > 0$,

$$E_k\{h(D_n)\} = pE_{k-1}\{h(D_{n-1})\} + qE_{k+1}\{h(D_{n-1})\} \leqslant E_{k+1}\{h(D_{n-1})\}, \tag{7·2}$$

while for $k = 0$

$$E_k\{h(D_n)\} = E_{k+1}\{h(D_{n-1})\}. \tag{7·3}$$

Combining (7·1) with (7·2), or (7·3), gives Lemma 2.

Proof of Theorem 1. By Lemma 2,

$$E_0\{h(D_{n+2})\} = pE_0\{h(D_n)\} + qE_2\{h(D_n)\} \geqslant E_0\{h(D_n)\}.$$

Proof of Theorem 2. Let

$$E_n = E(D_n) = p\pi_{n-1}(0) + q\sum_{j=0}^{\infty}(j+1)\pi_{n-1}(j) + p\sum_{j=2}^{\infty}(j-1)\pi_{n-1}(j)$$

by Lemma 1, which simplifies to

$$E_n - E_{n-1} = 1 - 2p + 2p\pi_{n-1}(0) = \left\{\pi_{n-1}(0) - \frac{r-1}{2r}\right\}\frac{2r}{r+1}. \tag{7·4}$$

Writing $E_{2n} = (E_{2n} - E_{2n-1}) + (E_{2n-1} - E_{2n-2}) + \ldots + E_1$ and remembering that $\pi_j(0) = 0$ for j odd, we have

$$E_{2n} = \frac{2r}{r+1}\sum_{j=0}^{n-1}\left\{\pi_{2j}(0) - \frac{r-1}{r}\right\}. \tag{7·5}$$

But

$$\lim_{l\to\infty} E_{2l} = \sum_{j=1}^{\infty} \frac{(r-1)(r+1)}{r}(2j)(1/r)^{2j}$$

by (3·2) and (3·3), which is evaluated as $2r/(r^2-1)$, the interchange of limit and expectation being easily justified from Theorem 3·1 and standard theorems; see Loève (1963, p. 183). Passing to the limit in (7·5) gives the first half of Theorem 2 upon subtraction of the term for $j=0$. The second half is proved in the same way starting from the function

$$h(0)=h(1)=0, \quad h(j)=j-1 \ (j \geqslant 1).$$

Proof of Theorem 3. The theorem is trivially true for $n=1$, and assume as an induction hypothesis that it is true for the case $n-1$.

In the proof for case n we can assume that $h(0)=h(1)=0$. For example, if n is odd we can subtract $h(1)$ from every value $h(j)$ without affecting the validity of the result, and then we can assume $h(0)=0$ since $\pi_n(0)=0$, a similar trick working for n even. Under these conditions Lemma 1 becomes

$$E\{h(D_n)\}=qE\{h'(D_{n-1})\}+pE\{h''(D_{n-1})\}, \tag{7·6}$$

where $h'(j)$ takes values $0, h(2), h(3), \ldots$ and $h''(j)$ takes values $0, 0, 0, h(2), h(3), \ldots$ $(j=0,1,\ldots)$.

By the induction hypothesis $E\{h'(D_{n-1})\}$ and $E\{h''(D_{n-1})\}$ are nonincreasing functions of r. Also $E\{h'(D_{n-1})\} \geqslant E\{h''(D_{n-1})\}$ since $h'(j) \geqslant h''(j)$ for all j. Since q is a decreasing function of r and p an increasing function of r, the theorem now follows from (7·6).

Corollaries 1 and 2 are easily derived from Theorem 4 by calculating that

$$\frac{d}{dp}\frac{1}{j}\left\{\sum_{i=0}^{[\frac{1}{2}(j-2)]} B_j(i)+\delta_j B_j([\tfrac{1}{2}j])\right\}=-B_{2[\frac{1}{2}j]-1}([\tfrac{1}{2}j]-1) \tag{7·7}$$

for $j \geqslant 2$ odd or even. For $j=1$ the derivative equals $-\frac{1}{2}$. Letting $\Delta_j(p)$ represent the quantity being differentiated in (7·7), this says that $d\Delta_j(p)/dp=d\Delta_{j+1}(p)/dp$ for j even and greater than or equal to 2. Since $\Delta_j(1)=0$ for all j by the definition of $B_j(i)$, Corollary 1 follows by integration of this equality from 1 to p. It is easily shown from (7·7) that $0>-d\Delta_{j+2}(p)/dp>-d\Delta_j(p)/dp$ for $\frac{1}{2}<p<1$ and $j \geqslant 1$, and Corollary 2 also follows by integration.

The second part of Corollary 3 follows from Corollary 1 by writing

$$\begin{aligned}\sum_{k=1}^{\infty}(-1)^k\rho_k &= -\tfrac{1}{2}\{\rho_1-(\rho_2-\rho_1)+(\rho_3-\rho_2)-(\rho_4-\rho_3)+\ldots\}\\ &= -\tfrac{1}{2}\{\rho_1-(\rho_2-\rho_1)\},\end{aligned} \tag{7·8}$$

and evaluating $\rho_1=-\phi_r/\{2(r+1)\}$ and $\rho_2-\rho_1=\phi_r/\{2(r+1)\}$ from Theorem 4. The first part of Corollary 3 can be obtained by direct summation in Theorem 4, but it is simpler to note that

$$1+2\sum_{k=1}^{\infty}\rho_k=\lim_{N\to\infty}\operatorname{var}\left(\sum_{k=1}^{N}T_k/\surd N\right)=0, \tag{7·9}$$

the last equality following from

$$\operatorname{var}\left(\sum_{k=1}^{N}T_k/\surd N\right)=\operatorname{var}\{(D_N-D_0)/\surd N\} \leqslant 4\operatorname{var}(D_0)/N. \tag{7·10}$$

Finally, for the proof of Theorem 4 itself we need two definitions:

$$e_{kj} \equiv E_k(T_j) \equiv E(T_j|D_0=k, \text{ control in excess}) \quad (k \geqslant 0, j>1), \tag{7·11}$$

$$b_{kj}=\operatorname{pr}\left\{\sum_{i=1}^{l}Y_i>-k \ (l=0,1,\ldots,j-1)\right\} \quad (k \geqslant 0, j \geqslant 1), \tag{7·12}$$

where the Y_i are independently and identically distributed random variables taking values $+1$ and -1 with probabilities q and p respectively. That is b_{kj} is the probability that a random walk with negative drift starts from the origin and stays above the level $-k$ at least until step j. By definition $b_{0j} = 0$ for all j.

LEMMA 4. $e_{kj} = (q-p)\, b_{kj}$.

Proof. For $k > 0, j > 1, e_{0,j} = 0, e_{k,1} = q-p$, and $e_{kj} = qe_{k+1,j-1} + pe_{k-1,j-1}$. An elementary calculation shows that the right hand side satisfies the same relationships.

LEMMA 5. *If* $\frac{1}{2}(j-k)$ *is a nonnegative integer,*

$$b_{k,j} - b_{k,j+1} = (k/j)\left\{\binom{j}{\frac{1}{2}(j-k)} p^{\frac{1}{2}(j+k)} q^{\frac{1}{2}(j-k)}\right\},$$

and equals zero otherwise.

Proof. By (7·12),

$$\begin{aligned} b_{kj} - b_{k,j+1} &= \mathrm{pr}\left\{\sum_{i=1}^{l} Y_i > -k \quad (l = 0, 1, \ldots, j-1),\quad \sum_{i=1}^{j} Y_i = -k\right\} \\ &= \mathrm{pr}\left(\sum_{i=1}^{j} Y_i = -k\right) \mathrm{pr}\left\{\sum_{i=1}^{l} Y_i > -k \quad (l = 0, 1, \ldots, j-1) \,\middle|\, \sum_{i=1}^{j} Y_i = -k\right\}. \end{aligned}$$

Unless $\frac{1}{2}(j-k)$ is a nonnegative integer the first factor is zero, while if it is a nonnegative integer it equals $\binom{j}{\frac{1}{2}(j-k)} p^{\frac{1}{2}(j+k)} q^{\frac{1}{2}(j-k)}$. The second factor then equals k/j by the ballot theorem (Feller, 1968, p. 66).

LEMMA 6.

$$\rho_j = (p-q) \sum_{k=1}^{\infty} \pi_k e_{kj} = -\left(\frac{r-1}{r+1}\right)^2 \sum_{k=1}^{\infty} \pi_k b_{kj}.$$

Proof. The second statement follows from the first by Lemma 4. We have

$$\rho_j = E(T_1 T_{j+1}) = \sum_{k=0}^{\infty} \pi_k E(T_1 T_{j+1} | D_1 = k),$$

where we have taken advantage of the symmetry of the process about zero. But

$$E(T_1 T_{j+1} | D_1 = k) = E(T_1 | D_1 = k)\, E(T_{j+1} | D_1 = k) = e_{jk} E(T_1 | D_1 = k)$$

by the Markov property of the chain

$$\tilde{D}_n = \tilde{D}_0 + \sum_{j=1}^{n} T_j.$$

Here $\tilde{D}_0$ is given the stationary distribution $\tilde{\pi}_0 = \pi_0$, $\tilde{\pi}_k = \frac{1}{2}\pi_{|k|}$ $(k \neq 0)$ which makes the entire chain stationary. Finally, $E(T_1 | D_1 = k) = 0$ if $k = 0$, and equals $p-q$ if $k > 0$.

Theorem 4 follows by substituting the values of b_{kj} implicitly given in Lemma 5 into Lemma 6.

This work was done under the auspices of a National Science Foundation Grant.

References

BARBACKI, S. & FISHER, R. A. (1936). A test of the supposed precision of systematic arrangements. *Ann. Eugen.* **7**, 189.

BLACKWELL, D. & HODGES, J. L. (1957). Design for the control of selection bias. *Ann. Math. Statist.* **28**, 449–60.

COX, D. R. & MILLER, H. D. (1965). *The Theory of Stochastic Processes.* New York: Wiley.

FELLER, W. (1968). *An Introduction to Probability Theory and Its Applications.* **1**, 3rd edition. New York: Wiley.

GREENBERG, B. G. (1951). Why randomize? *Biometrics* **7**, 309–22.

LOÈVE, M. (1963). *Probability Theory*, 3rd edition. Princeton: Van Nostrand.

SERFLING, R. J. (1968). Contributions to central limit theory for dependent variables. *Ann. Math. Statist.* **39**, 1158–75.

STUDENT. (1937). Comparison between balanced and random arrangements of field plots. *Biometrika* **29**, 363–79.

YATES, F. (1939). The comparative advantages of systematic and randomized arrangements in the design of agricultural and biological experiments. *Biometrika* **30**, 441–64.

[*Received November* 1970. *Revised April* 1971]

Some key words: Randomization in experimental design; Balance; Two treatment sequential experiments; Permutation tests.

3

Defining the Curvature of a Statistical Problem (with Applications to Second-Order Efficiency)

Introduction by Robert E. Kass
Carnegie Mellon University

and Paul W. Vos
East Carolina University

Re-examining Efron's 1975 paper from the vantage point of more than 30 years of subsequent history, we see it not only as pathbreaking, but as representing a landmark in the development of statistical theory. The paper generated unusual excitement: the notion of statistical curvature added new insights into old problems, and suggested novel methods of attacking them; it even raised the possibility that invariants, and geometrical arguments generally, might solve deep and vexing problems of statistical inference.

The paper certainly spawned quite a bit of research, much of it summarized in several books (Amari (1990), Murray and Rice (1993), Kass and Vos (1997), Marriott and Salmon (2000), Amari and Nagaoka (2000)), and its influence is especially important if we see it as the first of a trilogy: Efron (1975, 1978) and Efron and Hinkley (1978). On the other hand, the "vexing problems" mainly involved conditionality, and it now seems impossible to create a general and compelling theory of inference without some arbitrariness entering as soon as one considers terms that "correct" for inferences based on asymptotic Normality. What, then, are the enduring contributions here? We see several. The paper introduced the idea of statistical curvature, identified its role in Fisher's notion of information loss, and provided a rigorous approach to the issue. It also focused attention on curved exponential families as useful settings for understanding parametric statistical methods. Furthermore, the discussants added unusually important insights, partly by sketching a path toward multivariate generalization through differential geometry, and partly by making further connections with problems of statistical inference.

3.1. Information Loss

We continue to see information loss as the most basic concept elucidated by Efron's geometrical considerations. Information loss was Fisher's fundamental quantification of departure from sufficiency: he had claimed (but never demonstrated) that the MLE minimized the loss of information among efficient estimators. Recall that according to the Koopman–Darmois theorem, under regularity conditions, the families of continuous distributions with fixed support that admit finite-dimensional sufficient statistics of i.i.d. sequences are precisely the exponential families. It is thus intuitive that (for such regular families) departures from sufficiency, i.e., information loss, should correspond to deviations from exponentiality. The remarkable reality is that the correspondence takes a beautifully simple form. The most transparent case, especially for the untrained eye, occurs for a one-parameter subfamily of a two-dimensional exponential family.[1] There, the relative information loss, in Fisher's sense, from using statistics T in place of the whole sample is

$$\lim_{n\to\infty} \frac{ni(\theta) - i^T(\theta)}{i(\theta)} = \gamma^2 + \frac{1}{2}\beta^2, \tag{3.1}$$

where $ni(\theta)$ is the Fisher information in the whole sample, $i^T(\theta)$ is the Fisher information calculated from the distribution of T, γ is the statistical curvature of the family and β is the mixture curvature of the "ancillary family" associated with the estimator T. When the estimator T is the MLE, β vanishes; this supports Fisher's claim. [Stronger support for the superiority of the MLE over other estimators is provided by the dual geometries described by Efron (1978) and more fully developed by Amari (1990). The key geometrical properties are the Pythagorean relationship of the Kullback–Leibler divergence in exponential families and the fact that the MLE is the unique orthogonal projection satisfying this relationship.] Efron derived the two-term expression for the information loss as his equation (10.25), discussed the geometrical interpretation of the first term, and noted that the second term is zero for the MLE. He defined γ to be the curvature of the curve in the natural parameter space that describes the subfamily, with the inner product determined by Fisher information replacing the usual Euclidean inner product. The definition of β is exactly analogous to that of γ, with the mean value parameter space used instead of the natural parameter space, but Efron did not recognize this – it came as a result of the commentary by Dawid and subsequent work by Amari – and so did not identify the mixture curvature. He did stress the role of the ancillary family associated with the estimator T (see his Remark 3 of Sect. 9 and his reply to the discussants, p. 1240).

Prior to Efron's paper, Rao (1961) had introduced the definitions of efficiency and second-order efficiency that were intended to classify estimators just as Fisher's definitions did, but using more tractable expressions. This led to the same measure of minimum information loss used by Fisher, corresponding to γ^2 in (3.1). Rao (1962) computed the information loss in the case of multinomial distribution for several different methods of estimation. Rao (1963) then went on to provide a decision-theoretic definition of second-order efficiency of an estimator T, measuring it according to the magnitude of the second-order term in the asymptotic expansion of the bias-corrected version of T. Efron's analysis clarified the relationship between Fisher's definition and Rao's first definition. Efron then provided a decomposition of the second-order variance term in which the right-hand side of (3.1) appeared, together with a parameterization-dependent third term. The extension to the multiparameter case was outlined in the discussion by Reeds. In his contribution to the discussion, Rao listed some topics that would require further investigation. One of these topics was the question of what can be said about loss functions more general than quadratic loss. Recent work by Eguchi and Yanagimoto (2007) consider geometric adjustments to the maximum likelihood estimator that reduce loss incurred when using Kullback–Leibler and other risk functions.

3.2. Curved Exponential Families

An analytically and conceptually important first step of Efron's analysis was to begin by considering smooth subfamilies of regular exponential families, which he called "curved exponential families." Analytically, this made possible rigorous derivations of results, and for this reason such families were analyzed concurrently by Ghosh and Subramanyam (1974). Efron recognized the information

[1] A portion of the text in this section is taken, with permission, from the Introduction by R.E. Kass to the volume Amari et al. (1987).

loss result required regularity conditions. He also produced an ingenious counterexample when the conditions do not hold: he gave a prescription for a curved exponential family for which γ does not vanish and yet the MLE loses no information. The idea was to construct a model where each ancillary submanifold contains exactly one sample point. Efron's solution involved the trinomial distribution and the simplest curve, a circle, where a circle is understood as points equidistant in terms of Kullback–Leibler divergence rather than Euclidean distance. However, Kumagai and Inagaki (1996) showed that Efron's example was actually vacuous, meaning that there was no family that satisfied his prescription. In Kass and Vos (1997, p.72) we modified Efron's counterexample so that it produced families that exist: the modification was simply to move the center of the "circle" into the expectation parameter space of the trinomial. The inadequacy of the original counterexample was not an indication of carelessness; rather, it illustrates how the Euclidean geometry that works so well for Normal theory, falls short in more general settings. In the Euclidean geometry that describes Normal models, one can, to a large extent, safely identify the parameter space with a tangent space. In a general exponential family this will not work. The dual geometries that followed from this paper clearly distinguish between these ideas, and this made our modification to the counterexample intuitive.

Conceptually, Efron's introduction of curved exponential families allowed specification of the ancillary families associated with an estimator T: for exponential families with continuous sample space the ancillary family associated with T at t is the set of points y in the sample space – equivalently, the set of points μ in the mean value parameter space for the full exponential family – for which $T(y) = t$. The terminology and subsequent detailed analysis is due to Amari but, as noted above, the importance of the ancillary family, at once emphasized and obscured by Fisher, was apparent from Efron's presentation.

When a full exponential family is given an inner product defined by Fisher information, the curvature of a curved exponential subfamily is the quantity γ appearing in (3.1). Furthermore, curved exponential families become non-curved one-dimensional exponential families exactly when the curvature γ vanishes, thereby establishing statistical curvature as a measure of departure from being an exponential family. Efron used regularity conditions that made a curved exponential family into the geometrical quantity known as an immersed submanifold. To insure that the topology of the submanifold is homomorphic to that inherited from the larger manifold, submanifolds are typically imbedded rather than immersed. This is a technical distinction; all of Efron's examples involved imbedded submanifolds. In our treatment (Kass and Vos 1997), we took curved exponential families to be imbedded submanifolds because they are precisely the subfamilies that are invariant to smooth reparameterization (see our Theorem 2.3.1).

The geometry of curved exponential families turns out to be much richer than was apparent in Efron's treatment, but much of the richness began to be sketched out in the discussion to his paper. Geometrical curvature involves inner products and tangent vectors; the corresponding statistical quantity involves expectations of the score and its derivative. Efron established a correspondence between these quantities by expressing γ for a curved exponential family both geometrically using the natural parameter space and statistically, thereby obtaining two definitions for statistical curvature. The definitions are identical for curved exponential families but the immediate benefit of the statistical definition is that it can be used for the general statistical model where there is no natural parameter space. An equally important consequence is the foreshadowing of the geometry of exponential families described by Efron (1978) and Amari (1982). Amari and others (see the books Amari (1990); Amari and Nagaoka (2000); Kass and Vos (1997); Marriott and Salmon (2000); Murray and Rice (1993) and the references therein) have taken this geometric structure beyond exponential families and followed the natural progression of associating higher derivatives and cumulants of the score with geometric quantities resulting in statistical fiber bundles.

These extensions were aided by the insights to differential geometry provided by discussants Dawid and Reeds. In particular, Reeds outlined how results could be extended to multivariate curved exponential families. Dawid described the important role of a connection in geometry – the connection determines what curves are straight lines, also called geodesics – and identified three connections that would play a fundamental role in the forthcoming statistical geometry. Besides the exponential connection used by Efron which identified one-parameter exponential families as geodesics, Dawid defined an equally natural connection in which one-parameter mixture families are geodesics, the mixture connection. The third connection is the metric or Riemannian connection whose geodesics are curves of minimum length under the Fisher information metric. Reeds identified the statistical curvature as an imbedding curvature and how imbedding curvatures differ from Riemannian curvatures. Dawid described these connections on $\mathcal{M}$, the space of all distributions equivalent to a given carrier measure. He then showed that on $\mathcal{M}$ the Riemannian curvature vanished in the exponential and mixture connections and was constant for the Riemannian connection. Pistone and Sempi (1995) further developed the geometry on $\mathcal{M}$ using Orlitz spaces. On finite-dimensional manifolds, Efron (1978) and Amari (1982) would show that these connections are very important for curved exponential families since they provide two geometries that characterize when the Kullback–Leibler divergence is minimized which, for exponential families, is equivalent to maximizing the likelihood. The exponential and mixture connections are dual in a manner that provide a generalization of the Pythagorean theorem and projections. This allows facts that are derived from Euclidean geometry in the multivariate normal model to be extended to exponential families.

The two definitions of curvature led Efron to give two important invariance properties of the curvature described in Sect. 4. The geometric definition showed that curvature does not depend upon the parameter chosen for the model. Indeed, parameter invariance – meaning, the mutual consistency of quantities defined in terms of alternative parameterizations – is an important attribute of procedures defined geometrically. Fisher information and maximum likelihood estimation are parameter invariant, and curvatures and other geometric quantities called tensors are used to describe these quantities. Tensors also play a role in some non-parameter invariant quantities, such as asymptotic variance, in that they indicate which parts of the expansion cannot be changed by reparameterization. The second invariance property of statistical curvature is its invariance under any mapping to a sufficient statistic. Efron noted that this invariance would not hold for inner products other than the Fisher information, which helped establish the central role of Fisher information in statistical geometry.

A useful aspect of the geometry motivated by Efron's statistical curvature is the identification of the data (through the sufficient statistic) with the expectation parameter space of the exponential family. Since we are interested in making inferences about statistical models from observations, it is advantageous to have the data and model represented in the same space. On the other hand, it is important to recognize these are distinct entities; inferences obtained regarding the claim that the distributional mean is 4 when the sample mean is 2 need not coincide with inferences about the claim that the distributional mean is 2 when the sample mean is 4. Geometry with a (symmetric) distance used to compare points cannot distinguish a point that represents data from one that represents a statistical model. However, the dual geometries for the asymmetric Kullback–Leibler divergence do preserve this distinction.

In generalizing statistical curvature beyond exponential families Efron showed (5.9) that any smooth one-parameter family can be approximated by a curved exponential family imbedded in a k-dimensional space provided k is large enough. This result further enhanced the role of curved exponential families in understanding statistical inference and, in particular, limiting relative information loss. Efron noticed that the curvature was a two-dimensional property in that its value for the approximating exponential family was the same for all $k \geq 2$. Approximating exponential families differed not in terms of γ but in terms of other invariant geometric quantities which would be identified by Amari (1982) in his decomposition of higher-order information loss. Marriott and Vos (2004) use the dual geometries to pursue a global approach to information recovery.

Careful not to neglect the practical importance of curvature, Efron argued in Sect. 8 that values for γ^2 greater than $1/8$ should be considered large. He did this by considering the locally most powerful level α test of a simple null hypothesis ($\theta = \theta_0$) against a one-sided alternative ($\theta > \theta_0$) and asking when this locally most powerful test will perform well for θ_1, a "statistically reasonable" alternative to θ_0 (θ_1 is approximately two standard errors from θ_0). The test statistic for the locally most powerful test at θ_0 is the score U_{θ_0} and if this differs substantially from U_{θ_1}, one would question the use of the locally most powerful test. Geometrically, the score is the tangent vector to the model and if the curvature is large, the angle a_{θ_1} between U_{θ_0} and U_{θ_1} is large. To establish what constitutes a large angle for statistical matters, Efron further exploited the interplay between statistical and geometric quantities. He showed that $\sin^2 a_{\theta_1}$ is the fraction of unexplained variance after linear regression of U_{θ_1} on U_{θ_0} and the condition that $1/2$ the variation being unexplained for a statistically reasonable alternative leads to curvatures greater than $1/\sqrt{8}$ as large. To check these heuristic calculations Efron used a bivariate normal example and the power envelope determined by the most powerful test given by the Neyman–Pearson lemma.

Barndorff-Nielsen (1986) developed a parallel geometric theory defining the metric tensor in terms of the observed Fisher information rather than the expected Fisher information.

3.3. Inference

Several of the other discussants made comments that led to additional research. In his discussion, Cox raised a question regarding the role curvature might play in understanding conditional inference and its relationship to unconditional inference. The discussion by Pierce addressed conditioning on an ancillary statistic and the relationship between observed and expected Fisher information. He suggested that the curvature described the standard deviation of the observed Fisher information which provided information about the precision of the MLE. Furthermore, he showed that the variance of the MLE conditional on the observed information compared to the unconditional variance is expanded (or contracted) by the ratio of the expected-to-observed Fisher information. Pierce provided a geometrical argument that related the magnitude of the ratio of these information to the curvature of the model. His description for the approximate relationship among these quantities is reminiscent of the exact relationship Fisher (1956) gave in his circle model. The role of observed and expected Fisher information was more fully developed in Efron and Hinkley (1978).

In his comments, Lindley was critical of the curvature in that integrations are taken over the sample space and wrote that "estimation is solved by describing the likelihood function or posterior distribution." Ghosh, too, emphasized the role of likelihood, noting that statistical curvature is related to Sprott's work on approximate Normal likelihoods. Sprott (1973) had constructed two measures: one indicates how different an observed likelihood is from being Normal and another indicates the corresponding difference between expected likelihoods. Ghosh suggested that for large statistical curvatures the two measures can be expected to differ significantly.

These queries about the role of curvature from the point of view of likelihood-based inference were taken up in our own research. One result was a theorem relating order $O(n^{-(q-1)})$ information loss to order $O(n^{-q})$ local approximability of the likelihood function (Theorem 3.3.10 in Kass and Vos 1997). Another involved the extension of Sprott's notion of deviation from Normal likelihoods to multiparameter families using geometrical methods (Kass and Slate 1994), and this also produced a Bayesian interpretation of likelihood-based curvature measures (Sect. 6.4 of Kass and Vos 1997).

Discussants Le Cam and Pfanzagl expressed some reservations about the role of curvature in statistics. Le Cam gave examples where the maximum likelihood estimator performed poorly and suggested that there are other important measures of information loss that would not be covered by this theory. Pfanzagl questioned whether any single quantity can measure departures from "exponentiality." Keiding provided an example illustrating the role of statistical curvature in modeling birth processes. Many more examples were given in our book (Kass and Vos 1997).

3.4. Conclusion

The most striking finding from our assessment of Efron's paper is that it generated a great deal of enthusiasm and much subsequent research. In fact, it spawned a small subdomain of Statistics, complete with multiple conferences spread across many years. While geometrical methods have not had a large impact on statistical practice, the geometrical approach to inference that Efron initiated has provided valuable intuition and insight to many who have invested the time to study it. Readers should see this work as an important indicator of the state of statistical research in 1975, and should enjoy it as an informative excursion into an unusually intriguing realm of statistical theory.

References

Amari, S.-I. (1982). Differential geometry of curved exponential families – Curvatures and information loss. *Annals of Statistics* **10**, 357–387.

Amari, S.-I. (1990). *Differential–Geometrical Methods in Statistics*. Springer, New York.

Amari, S.-I., Barndorff-Nielsen, O. E., Kass, R. E., Lauritzen, S. L. and Rao, C. R. (1987). *IMS Lecture Notes – Monograph Series: Differential Geometry in Statistical Inference* **10**. Institute of Mathematical Statistics, Hayward, CA.

Amari, S.-I. and Nagaoka, H. (2000). *Translations of Mathematical Monographs: Methods of Information Geometry* **191**. American Mathematical Society, Oxford University Press, Oxford.

Barndorff-Nielsen, O. E. (1986). Likelihood and observed geometries. *Annals of Statistics* **14**, 856–873.

Efron, B. (1975). Defining the curvature of a statistical problem (with applications to second order efficiency). *Annals of Statistics* **3**, 1189–1242 with discussion and reply.

Efron, B. (1978). The geometry of exponential families. *Annals of Statistics* **6**, 362–376.

Efron, B. and Hinkley, D. V. (1978). Assessing the accuracy of the maximum likelihood estimator: Observed versus expected Fisher information. *Biometrika* **65**, 457–487.

Eguchi, S. and Yanagimoto, T. (2008). Asymptotical improvement of maximum likelihood estimators on Kullback–Leibler loss. To appear *Journal of Statistical Planning and Inference*.

Fisher, R. (1956). *Statistical Methods and Scientific Inference*. Oliver and Boyd, Edinburgh.

Ghosh, J. and Subramanyam, K. (1974). Second order efficiency of maximum likelihood estimators. *Sankhya* **36**, 325–358.

Kass, R. E. and Slate, E. H. (1994). Some diagnostics of maximum likelihood and posterior nonnormality. *Annals of Statistics* **22**, 668–695.

Kass, R. E. and Vos, P. W. (1997). *Geometrical Foundations of Asymptotic Inference*. Wiley-Interscience, New York.

Kumagai, E. and Inagaki, N. (1996). Comment on Efron's counterexample. *Mathematica Japonica* **44**, 449–454.

Marriott, P. K. and Salmon, M. (2000). *Applications of Differential Geometry to Econometrics*. Cambridge University Press, London.

Marriott, P. K. and Vos, P. W. (2004). On the global geometry of parametric models and information recovery. *Bernoulli* **10**, 639–649.

Murray, M. K. and Rice, J. W. (1993). *Differential Geometry and Statistics*, Chapman and Hall, London.

Pistone, G. and Sempi, C. (1995). An infinite dimensional geometric structure on the space of all probability measures equivalent to a given one. *Annals of Statistics* **23**, 1543–1561.

Rao, C. (1961). Asymptotic efficiency and limiting information. In *Proceedings of the Fourth Berkeley Symposium on Mathematical Statistics and Probability* **1**, 531–545. Edited by J. Neyman. University of California Press, Berkeley.

Rao, C. (1962). Efficient estimates and optimum inference procedures in large samples. *Journal of the Royal Statistical Society, Series B* **24**, 46–72.

Rao, C. (1963). Criteria of estimation in large samples. *Sankhya* **25**, 189–206.

Sprott, D. A. (1973). Normal likelihoods and their relation to large sample theory of estimation. *Biometrika* **60**, 457–465.

The Annals of Statistics
1975, Vol. 3, No. 6, 1189-1242

DEFINING THE CURVATURE OF A STATISTICAL PROBLEM (WITH APPLICATIONS TO SECOND ORDER EFFICIENCY)

BY BRADLEY EFRON

Stanford University

Statisticians know that one-parameter exponential families have very nice properties for estimation, testing, and other inference problems. Fundamentally this is because they can be considered to be "straight lines" through the space of all possible probability distributions on the sample space. We consider arbitrary one-parameter families $\mathscr{F}$ and try to quantify how nearly "exponential" they are. A quantity called "the statistical curvature of $\mathscr{F}$" is introduced. Statistical curvature is identically zero for exponential families, positive for nonexponential families. Our purpose is to show that families with small curvature enjoy the good properties of exponential families. Large curvature indicates a breakdown of these properties. Statistical curvature turns out to be closely related to Fisher and Rao's theory of second order efficiency.

1. Introduction. Suppose we have a statistical problem involving a one-parameter family of probability density functions $\mathscr{F} = \{f_\theta(x)\}$. Statisticians know that if $\mathscr{F}$ is an exponential family then standard linear methods will usually solve the problem in neat fashion. For example, the locally most powerful test of $\theta = \theta_0$ versus $\theta > \theta_0$ is uniformly most powerful in an exponential family. The maximum likelihood estimator for θ is a sufficient statistic in an exponential family, and achieves the Cramér-Rao lower bound if we have chosen the right function of θ to estimate.

In this paper we consider arbitrary one-parameter families $\mathscr{F}$ and try to quantify how nearly "exponential" they are. A quantity γ_θ called "*the statistical curvature of* $\mathscr{F}$ *at* θ" is introduced such that γ_θ is identically zero if $\mathscr{F}$ is exponential and greater than zero, for at least some θ values, otherwise.

Our purpose is to show that families with small curvature enjoy, nearly, the good statistical properties of exponential families. Large curvature indicates a breakdown of this favorable situation. For example, if γ_{θ_0} is large, the locally most powerful test of $\theta = \theta_0$ versus $\theta > \theta_0$ can be expected to have poor operating characteristics. Similarly the variance of the maximum likelihood estimator (MLE) exceeds the Cramér-Rao lower bound in approximate proportion to γ_θ^2. (See Sections 8 and 10.)

For nonexponential families the MLE is not, in general, a sufficient statistic. How much information does it lose, compared with all the data x? The answer

Received March 1974; revised May 1975.

AMS 1970 *subject classifications.* 62B10, 62F20.

Key words and phrases. Curvature, exponential families, Cramér-Rao lower bound, locally most powerful tests, Fisher information, second order efficiency, deficiency, maximum likelihood estimation.

can be expressed in terms of γ_θ^2. This theory goes back to Fisher (1925) and Rao (1961, 1962, 1963). They attempted to show that if $\mathscr{F}$ is a one-parameter subset of the k-category multinomial distributions, indexed say by the vector of probabilities $f_\theta(x) = P_\theta(X \in \text{category } x)$, $x = 1, 2, \cdots, k$, the following result holds: let $\mathbf{i}_\theta$ be the Fisher information in an independent sample of size n from f_θ, $\mathbf{i}_\theta^{\hat\theta}$ the Fisher information in the maximum likelihood estimator $\hat\theta(x_1, x_2, \cdots, x_n)$ based on that sample, and i_θ the Fisher information in a sample of size one (so $\mathbf{i}_\theta = ni_\theta$). Then

$$\lim_{n\to\infty} (\mathbf{i}_\theta - \mathbf{i}_\theta^{\hat\theta}) = i_\theta \left\{\frac{\mu_{02} - 2\mu_{21} + \mu_{40}}{i_\theta^2} - 1 - \frac{\mu_{11}^2 + \mu_{30}^2 - 2\mu_{11}\mu_{30}}{i_\theta^3}\right\} \tag{1.1}$$

where

$$\mu_{kj} \equiv E_\theta \left(\frac{\dot f_\theta(x)}{f_\theta(x)}\right)^k \left(\frac{\ddot f_\theta(x)}{f_\theta(x)}\right)^j, \tag{1.2}$$

the dot indicating differentiation with respect to θ. Moreover, for any other consistent, efficient estimator $T(x_1, x_2, \cdots, x_n)$ the asymptotic loss of information $\lim_{n\to\infty} (\mathbf{i}_\theta - \mathbf{i}_\theta^T)$ is equal or greater than the right side of (1.1). Rao has coined the term "*second order efficiency*" for this property of the MLE which gives it a preferred place in the class of "first order efficient" estimators T, those which satisfy the weaker condition $\lim_{n\to\infty} \mathbf{i}_\theta^T/\mathbf{i}_\theta = 1$.

It turns out that the unpleasant looking bracketed term in (1.1) equals γ_θ^2. This leads to a straightforward geometrical "proof" of (1.1). The quotes are necessary here since, as the counter-example of Section 9 shows, the result is actually not true for multinomial families. However, the difficulty arises only because of the discrete nature of the multinomial, and can be overcome by dealing with less lumpy distributions. More importantly, a similar result of Rao's for squared error estimation risk holds even for the multinomial, as discussed in Section 10.

Under our definition an exponential family has zero curvature everywhere so in some sense it is a "straight line through the space of possible probability distributions." (This is intuitively plausible since linear methods, that is, methods based on linear approximations to the log likelihood function, tend to work perfectly in exponential families. The fact that locally most powerful tests are uniformly most powerful is an example of this.) We will make this notion precise by considering families $\mathscr{F}$ which are subsets of multi-parameter exponential families. If the subset is a straight line in the natural parameter space of the bigger family then $\mathscr{F}$ is a one-parameter exponential family. If the subset is a curved line through the natural parameter space then $\mathscr{F}$ is not exponential, and it turns out that the statistical curvature exactly equals the ordinary geometric curvature of the line, the rate of change of direction with respect to arc-length. For the sake of exposition we actually start with this latter definition in Section 3 and show in Section 5 how it leads to a sensible definition of statistical curvature in the general case.

There are really two halves to this paper. Sections 3–7 introduce the notion of statistical curvature, Sections 8–10 apply curvature to hypothesis testing, partial sufficiency, and estimation. Section 2 consists of a brief review of the notion of the geometrical curvature of a line.

2. Curvature. If $Y = Y(X)$ defines a curved line $\mathscr{L}$ in the (X, Y) plane then

$$\gamma_X = \left[\frac{(Y'')^2}{[1 + (Y')^2]^3}\right]^{\frac{1}{2}} \tag{2.1}$$

is defined to be the curvature of $\mathscr{L}$ at X, where $Y' \equiv dY/dX$, $Y'' \equiv d^2Y/dX^2$ are assumed to exist continuously in a neighborhood of the value X where the curvature is being evaluated. In particular if $Y' = 0$ then $\gamma_X = |Y''|$. An exercise in differential calculus shows that γ_X is the rate of change of direction of $\mathscr{L}$ with respect to arc-length along the curve. The "*radius of curvature*", $\rho_X \equiv 1/\gamma_X$, is the radius of the circle tangent to $\mathscr{L}$ at (X, Y) whose Taylor expansion about (X, Y) agrees up to the quadratic term with that of $\mathscr{L}$. Struik (1950) is a good elementary reference for curvature and related concepts.

The concept of curvature extends to curved lines in Euclidean k-space, E^k, say $\mathscr{L} = \{\eta_\theta, \theta \in \Theta\}$, where Θ is an interval of the real line. For each θ, η_θ is a vector in E^k whose componentwise derivatives with respect to θ we denote $\dot{\eta}_\theta \equiv (\partial/\partial\theta)\eta_\theta$, $\ddot{\eta}_\theta \equiv (\partial^2/\partial\theta^2)\eta_\theta$. These derivatives are assumed to exist continuously in a neighborhood of a value of θ where we wish to define the curvature. Suppose also that a $k \times k$ symmetric nonnegative definite matrix $\mathbf{\Sigma}_\theta$ is defined continuously in θ. Let M_θ be the 2×2 matrix, with entries denoted $\nu_{20}(\theta)$, $\nu_{11}(\theta)$, $\nu_{02}(\theta)$ as shown, defined by

$$M_\theta \equiv \begin{pmatrix} \nu_{20}(\theta) & \nu_{11}(\theta) \\ \nu_{11}(\theta) & \nu_{02}(\theta) \end{pmatrix} \equiv \begin{pmatrix} \dot{\eta}_\theta'\mathbf{\Sigma}_\theta\dot{\eta}_\theta & \dot{\eta}_\theta'\mathbf{\Sigma}_\theta\ddot{\eta}_\theta \\ \ddot{\eta}_\theta'\mathbf{\Sigma}_\theta\dot{\eta}_\theta & \ddot{\eta}_\theta'\mathbf{\Sigma}_\theta\ddot{\eta}_\theta \end{pmatrix} \tag{2.2}$$

and let

$$\gamma_\theta \equiv (|M_\theta|/\nu_{20}^3(\theta))^{\frac{1}{2}} . \tag{2.3}$$

Then γ_θ is "*the curvature of $\mathscr{L}$ at θ with respect to the inner product $\mathbf{\Sigma}_\theta$*". (If we take $k = 2$, $\theta = X$, $\eta_\theta = (X, Y(X))$, and $\mathbf{\Sigma}_\theta \equiv \mathbf{I}$, then (2.3) reduces to (2.1).)

Again it can be shown that γ_θ is the rate of change of direction of η_θ with respect to arc-length along $\mathscr{L}$. The relevant quantities are illustrated in Figure 1, where the arc-length from a given point η_{θ_0} to η_θ is called "s_θ" and the angle

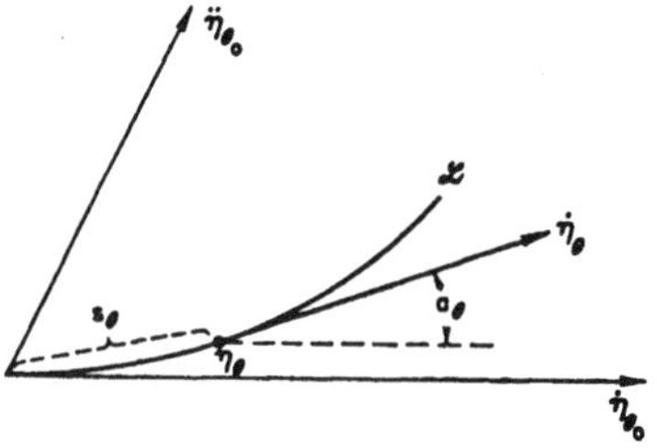

FIG. 1. The curvature of $\mathscr{L}$ at θ_0 is $da_\theta/ds_\theta|_{\theta=\theta_0}$.

between $\dot{\eta}_{\theta_0}$ and $\dot{\eta}_\theta$ called "a_θ". Then

$$\gamma_{\theta_0} = \frac{da_\theta}{ds_\theta}\bigg|_{\theta_0} \tag{2.4}$$

or equivalently $\gamma_{\theta_0} = d \sin a_\theta / ds_\theta|_{\theta_0}$. Both s_θ and a_θ are defined relative to the inner product Σ_θ,

$$\frac{ds_\theta}{d\theta} \equiv (\dot{\eta}_\theta' \Sigma_\theta \dot{\eta}_\theta)^{\frac{1}{2}} \tag{2.5}$$

$$\sin a_\theta \equiv \left[1 - \frac{(\dot{\eta}_{\theta_0}' \Sigma_{\theta_0} \dot{\eta}_\theta)^2}{(\dot{\eta}_{\theta_0}' \Sigma_{\theta_0} \dot{\eta}_{\theta_0})(\dot{\eta}_\theta' \Sigma_{\theta_0} \dot{\eta}_\theta)}\right]^{\frac{1}{2}}. \tag{2.6}$$

(Σ_{θ_0} can be replaced by Σ_θ anywhere in (2.6).) As Figure 1 indicates, for the purpose of evaluating γ_{θ_0} the k-dimensional curve $\mathscr{L}$ can be considered locally as a two-dimensional curve in the plane through η_{θ_0} spanned by $\dot{\eta}_{\theta_0}$ and $\ddot{\eta}_{\theta_0}$.

3. Curved exponential families. In this section we define statistical curvature for one parameter families $\mathscr{F}$ which are curved subsets of a larger k-parameter exponential family, "curved exponential families" for short. Denote the multiparameter family by

$$g_\eta(x) \equiv g(x) e^{\eta' x - \psi(\eta)} \tag{3.1}$$

a family of densities with respect to some given measure $m(\cdot)$, possibly discrete, on Euclidean k-space E^k. Here $\eta \in \mathscr{N}$, the subset of E^k for which $\int_{E^k} g(x) e^{\eta' x}\, dm(x) < \infty$. The convex set $\mathscr{N}$ is called *the natural parameter space* of the exponential family. If we define

$$\lambda(\eta) \equiv E_\eta x \tag{3.2}$$

the components of λ can be obtained by differentiation of ψ, $\lambda_i(\eta) = (\partial/\partial\eta_i)\psi(\eta)$. Moreover the covariance matrix $\Sigma(\eta)$ of x under g_η has ijth element equal to $\partial^2\psi(\eta)/\partial\eta_i\partial\eta_j$. We denote by Λ the set of all mean vectors λ,

$$\Lambda = \{\lambda(\eta) : \eta \in \mathscr{N}\} \tag{3.3}$$

The mapping (3.2) from $\mathscr{N}$ to Λ is one-to-one, and we will often write λ instead of $\lambda(\eta)$, recognizing that λ indexes the exponential family as well as η does. $\Sigma(\eta)$ has the same rank r for all η, and *we will assume rank* $r \geq 2$ *to avoid trivialities.*

Now suppose that

$$\mathscr{L} \equiv \{\eta_\theta : \theta \in \Theta\} \tag{3.4}$$

is a one-parameter subset in the interior of $\mathscr{N}$, where η_θ is a continuously twice differentiable function of $\theta \in \Theta$, an interval of the real line. Define the density f_θ to be

$$f_\theta(x) \equiv g_{\eta_\theta}(x) = g(x) e^{\eta_\theta' x - \psi_\theta}, \tag{3.5}$$

where $\psi_\theta = \psi(\eta_\theta)$. (Likewise $\lambda_\theta = \lambda(\eta_\theta)$, $\Sigma_\theta \equiv \Sigma(\eta_\theta)$.) It is easy to verify that

$$\dot{\lambda}_\theta = \Sigma_\theta \dot{\eta}_\theta, \qquad \dot{\psi}_\theta = \dot{\eta}_\theta' \lambda_\theta = E_\theta \dot{\eta}_\theta' x. \tag{3.6}$$

$\mathscr{F}$ will stand for the family of densities $\{f_\theta(x) : \theta \in \Theta\}$, our curved exponential family.

DEFINITION. γ_θ, the statistical curvature of $\mathscr{F}$ at θ, is the geometrical curvature of $\mathscr{L} = \{\eta_\theta : \theta \in \Theta\}$ at θ with respect to the covariance inner product Σ_θ, as defined in (2.2) and (2.3).

EXAMPLE 1. *Bivariate normal.* x is a bivariate normal random vector with covariance matrix I and mean vector $\eta_\theta = (\theta, (\gamma_0/2)\theta^2)'$, $\theta \in \Theta = (-\infty, \infty)$,

$$x \sim \mathscr{N}_2(\eta_\theta, \mathrm{I}) . \tag{3.7}$$

Then $\dot\eta_\theta = (1, \gamma_0\theta)'$, $\ddot\eta_\theta = (0, \gamma_0)'$, and

$$M_\theta = \begin{pmatrix} 1 + \gamma_0^2\theta^2 & \gamma_0^2\theta \\ \gamma_0^2\theta & \gamma_0^2 \end{pmatrix} \tag{3.8}$$

so

$$\gamma_\theta^2 = \frac{\gamma_0^2}{(1 + \gamma_0^2\theta^2)^3} . \tag{3.9}$$

In particular $\gamma_0^2 = \gamma_0^2$, justifying the notation. This artificial but very simple curved exponential family will be used for illustrative purposes in Section 8.

EXAMPLE 2. *Poisson regression.* $x_1, x_2, \cdots, x_k$ are independent Poisson random variables, x_i having mean $a + \theta b_i$, $b_1, b_2, \cdots, b_k$ and $a > 0$ being known parameters. Θ is the interval of θ values such that $a + \theta b_i > 0$ for $i = 1, 2, \cdots, k$. Since $x = (x_1, \cdots, x_k)'$ has a k parameter exponential family of distributions if the k means are unconstrained, we apply definition (2.2) to get the elements of M_θ,

$$\nu_{20}(\theta) = \sum_{i=1}^k \frac{b_i^2}{a + \theta b_i} , \qquad \nu_{11}(\theta) = -\sum_{i=1}^k \frac{b_i^3}{(a + \theta b_i)^2} , \tag{3.10}$$
$$\nu_{02}(\theta) = \sum_{i=1}^k \frac{b_i^4}{(a + \theta b_i)^3} .$$

The formula (2.3) for γ_θ^2 simplifies at $\theta = 0$ to

$$\gamma_0^2 = \frac{1}{a}\left[\frac{\sum_{i=1}^k b_i^4}{(\sum_{i=1}^k b_i^2)^2} - \frac{(\sum_{i=1}^k b_i^3)^2}{(\sum_{i=1}^k b_i^2)^3}\right] \tag{3.11}$$

That the entries of M_θ are summations follows from the independence of $x_1, x_2, \cdots, x_k$, as mentioned in Section 6. A very similar formula holds for the analogous binomial regression model.

The *Neyman–Davies model*, $x_1, x_2, \cdots, x_k$ independent scaled χ_1^2 random variables, $x_i \overset{\text{ind}}{\sim} (1 + \theta\delta_i)\chi_1^2$, $\delta_1, \delta_2, \cdots, \delta_k$ known constants, has the same structure. (Davies (1969) uses this model, which originates in an application due to Neyman, to investigate the power of the locally most powerful test of $\theta = 0$ versus $\theta > 0$. We compare our results with his in Section 8.) By direct calculation or by the remark at the end of Section 6 we get that M_0 has elements

$$\nu_{20}(0) = \tfrac{1}{2} \sum_{i=1}^k \delta_i^2, \qquad \nu_{11}(0) = -\sum_{i=1}^k \delta_i^3, \qquad \nu_{02}(0) = 2 \sum_{i=1}^k \delta_i^4 \tag{3.12}$$

and so

$$\gamma_0^2 = 8\left[\frac{\sum_{i=1}^k \delta_i^4}{(\sum_{i=1}^k \delta_i^2)^2} - \frac{(\sum_{i=1}^k \delta_i^3)^2}{(\sum_{i=1}^k \delta_i^2)^3}\right]. \tag{3.13}$$

EXAMPLE 3. *Autoregressive process.* $y_0, y_1, \cdots, y_T$ are observations of the autoregressive process $y_0 = u_0$, $y_{t+1} = \theta y_t + (1-\theta^2)^{\frac{1}{2}} u_{t+1}$, $t = 1, 2, \cdots, T$. Here $u_t \overset{\text{ind}}{\sim} \mathscr{N}(0, 1)$, $t = 0, 1, \cdots, T$ and $\Theta = (-1, 1)$. Writing out the likelihood function of $(y_0, \cdots, y_T)$ shows that this is a curved exponential family with $k = 3$, the η vector being $\eta_\theta' = (-(1+\theta^2)/a, \theta, -\frac{1}{2})/(1-\theta^2)$, with corresponding sufficient statistics $x' = (\sum_1^{T-1} y_t^2, \sum_1^T y_t y_{t-1}, y_0^2 + y_T^2)$. For $\theta = 0$ the calculations are easy, yielding

$$M_0 = \begin{pmatrix} T & 0 \\ 0 & 8T-6 \end{pmatrix}, \qquad \gamma_0^2 = \frac{8T-6}{T^2}. \tag{3.14}$$

Much messier expressions are found for other values of θ. γ_θ^2 is of the form $c_\theta/T + O(1/T^2)$ as $T \to \infty$, with $c_0 = 8$, $c_{.25} = 6.25$, $c_{.5} = 3.07$, $c_{.75} = .96$. (For any T, $\gamma_{-\theta} = \gamma_\theta$, $i_{-\theta} = i_\theta$ since the mapping $(y_0, y_1, y_2, \cdots) \to (y_0, -y_1, y_2, \cdots)$ takes θ into $-\theta$ while preserving the curvature and Fisher information.) This family is least like a one-parameter exponential family at $\theta = 0$.

If $\mathscr{L}$ is a straight line through $\mathscr{N}$, $\eta_\theta = a + b\tau(\theta)$ where a and b are known vectors and $\tau(\theta)$ some real-valued twice differentiable function of θ, then $\gamma_\theta = 0$ for all θ since the curvature of a straight line is zero. In this case $f_\theta(x) = (g(x)e^{a'x}) \exp[\tau(\theta)b'x - \psi_\theta]$ is a one-parameter exponential family with natural parameter $\tau(\theta)$ and sufficient statistic $b'x$. Under our definition all one-parameter exponential families $\mathscr{F}$, and only such families, have statistical curvature everywhere equal to zero. This desirable property would still hold if we defined the curvature with respect to an inner product other than Σ_θ, say Σ_θ^{-1} or $\mathbf{I}$. The following discussion and Section 4 add support to the choice Σ_θ.

Let $l_\theta(x)$ denote the logarithm of $f_\theta(x)$,

$$l_\theta(x) \equiv \log f_\theta(x) \tag{3.15}$$

and denote the first and second partial derivations with respect to θ by

$$\dot{l}_\theta(x) \equiv \frac{\partial}{\partial\theta} l_\theta(x), \qquad \ddot{l}_\theta(x) \equiv \frac{\partial^2}{\partial\theta^2} l_\theta(x). \tag{3.16}$$

The moment relationships

$$E_\theta \dot{l}_\theta = 0, \qquad E_\theta \dot{l}_\theta^2 = -E_\theta \ddot{l}_\theta \equiv i_\theta, \tag{3.17}$$

where i_θ is Fisher's information, hold because the exponential family structure (3.1)—(3.5) allows us to differentiate under integral signs with impunity. (We will suppress the random element "x" in much of the subsequent notation.)

Notice that $l_\theta(x) = \eta_\theta' x - \psi_\theta + \log g(x)$ so that

$$\dot{l}_\theta(x) = \dot{\eta}_\theta'(x - \lambda_\theta), \qquad \ddot{l}_\theta(x) = \ddot{\eta}_\theta'(x - \lambda_\theta) - \dot{\eta}_\theta' \mathbf{\Sigma}_\theta \dot{\eta}_\theta, \tag{3.18}$$

where we have made use of (3.6) in taking the derivatives. Remembering that $\mathbf{\Sigma}_\theta$ is the covariance matrix of x, we see that (3.17) holds with

$$i_\theta = \dot{\eta}_\theta' \mathbf{\Sigma}_\theta \dot{\eta}_\theta. \tag{3.19}$$

As a matter of fact the covariance matrix of $(\dot{l}_\theta, \ddot{l}_\theta)$ is

$$E_\theta \begin{pmatrix} \dot{l}_\theta \\ \ddot{l}_\theta + i_\theta \end{pmatrix} (\dot{l}_\theta, \ddot{l}_\theta + i_\theta) = \begin{pmatrix} \dot{\eta}_\theta' \mathbf{\Sigma}_\theta \dot{\eta}_\theta & \dot{\eta}_\theta' \mathbf{\Sigma}_\theta \ddot{\eta}_\theta \\ \ddot{\eta}_\theta' \mathbf{\Sigma}_\theta \dot{\eta}_\theta & \ddot{\eta}_\theta' \mathbf{\Sigma}_\theta \ddot{\eta}_\theta \end{pmatrix} \tag{3.20}$$

which is just the matrix M_θ defined at (2.2). Therefore

$$\nu_{20}(\theta) = i_\theta = E_\theta \dot{l}_\theta^2, \qquad \nu_{11}(\theta) = E_\theta \dot{l}_\theta \ddot{l}_\theta = \operatorname{Cov}_\theta(\dot{l}_\theta, \ddot{l}_\theta), \tag{3.21}$$
$$\nu_{02}(\theta) = E_\theta \ddot{l}_\theta^2 - i_\theta^2 = \operatorname{Var}_\theta \ddot{l}_\theta.$$

These definitions make no explicit reference to the geometrical structure of the curved exponential family. We will use them in Section 5 to provide the curvature definition for an arbitrary one-parameter family.

4. Invariance properties of the curvature. The two definitions of M_θ, the geometrical one following (3.6) and the statistical one (3.21) give two useful invariance properties of the curvature γ_θ.

i) Statistical curvature is an intrinsic property of the family $\mathscr{F}$ and does not depend on the particular parameterization used to index $\mathscr{F}$. If we let $\tilde{\theta} \equiv g(\theta)$, where g is any strictly monotone twice differentiable function, and $\tilde{f}_{\tilde{\theta}}(x) \equiv f_{g^{-1}(\tilde{\theta})}(x)$, then $\tilde{\gamma}_{\tilde{\theta}} = \gamma_{g^{-1}(\tilde{\theta})}$ for every $\tilde{\theta} \in \tilde{\Theta} \equiv g(\Theta)$. This follows from the same property of the geometrical curvature (2.3). [Note: this is not true for the Fisher information: $\tilde{\imath}_{\tilde{\theta}} = i_{g^{-1}(\tilde{\theta})}(d\theta/d\tilde{\theta})^2$.]

ii) If $t = T(x)$, is sufficient for θ then $l_\theta^T(t) \equiv \partial/\partial\theta \log f_\theta^T(t) = l_\theta(x)$, where f_θ^T indicates the density of T, implying by (3.18) that $M_\theta^T = M_\theta$ and $\gamma_\theta^T = \gamma_\theta$. The statistical curvature is invariant under any mapping to a sufficient statistic, including of course all one-to-one mappings of the sample space. This property would not hold if we had chosen an inner product other than $\mathbf{\Sigma}_\theta$ in the definition of statistical curvature.

We can use property (ii) to transform an arbitrary curved exponential family into a form particularly convenient for theoretical calculations. Let θ_0 be some value of θ at which we wish to investigate the local behavior of $\mathscr{F}$. Write $\mathbf{\Sigma}_{\theta_0} \equiv \mathbf{\Lambda}'\mathbf{D}\mathbf{\Lambda}$, $\mathbf{D}$ an $r \times r$ diagonal matrix with positive diagonal elements and $\mathbf{\Lambda}$ an $r \times k$ matrix with orthonormal rows, $\mathbf{\Lambda}\mathbf{\Lambda}' = \mathbf{I}_r$ (rank $\mathbf{\Sigma}_\theta = r$, $\mathbf{I}_r$ the $r \times r$ identity matrix). Let $\tilde{x} \equiv \mathbf{\Gamma}\mathbf{D}^{-\frac{1}{2}}\mathbf{\Lambda}(x - \lambda_{\theta_0})$ where $\mathbf{\Gamma}$ is an as yet unspecified $r \times r$ orthogonal matrix. $\tilde{x}$ is an r-dimensional sufficient statistic for the family (3.1). For $\theta \in \Theta$ it has a curved exponential family of densities where we can take $\tilde{\eta}_\theta = \mathbf{\Gamma}\mathbf{D}^{\frac{1}{2}}\mathbf{\Lambda}(\eta_\theta - \eta_{\theta_0})$. (These statements are easily shown in the full rank case $r = k$ and are not difficult for $r < k$.)

Notice that $\tilde{\eta}_{\theta_0} = \mathbf{0}$, $\tilde{\lambda}_{\theta_0} = \mathbf{0}$, and $\tilde{\mathbf{\Sigma}}_{\theta_0} = I_r$. Proper choice of the rotation matrix Γ makes $\dot{\tilde{\eta}}_{\theta_0}$ proportional to $\mathbf{e}_1 = (1, 0, \cdots, 0)'$ and $\ddot{\tilde{\eta}}_{\theta_0}$ a linear combination of $\mathbf{e}_1$ and $\mathbf{e}_2 = (0, 1, 0, \cdots, 0)'$. By (3.6), $\dot{\tilde{\lambda}}_{\theta_0}$ is then also proportional to $\mathbf{e}_1$.

DEFINITION. The family $\mathscr{F}$ is in *standard form* at $\theta = \theta_0$ if $k = r$, the dimension of $\mathscr{F}$,

$$\eta_{\theta_0} = \lambda_{\theta_0} = 0\,, \qquad \mathbf{\Sigma}_{\theta_0} = \mathbf{I}_r \tag{4.1}$$

and

$$\dot{\eta}_{\theta_0} = \dot{\lambda}_{\theta_0} = i_{\theta_0}^{\frac{1}{2}}\mathbf{e}_1\,, \qquad \ddot{\eta}_{\theta_0} = \frac{\nu_{11}(\theta_0)}{i_{\theta_0}^{\frac{1}{2}}}\,\mathbf{e}_1 + i_{\theta_0}\gamma_{\theta_0}\mathbf{e}_2\,. \tag{4.2}$$

(The constants in (4.2) are necessary to satisfy (2.2).) We will use standard form to simplify proofs in Sections 9 and 10. If $\mathscr{F}$ is not in standard form at θ_0 the above transformation makes it so, and by property (ii) M_θ and hence all information and curvature properties remain unchanged. We could use property (i) to further standardize the situation so that $i_{\theta_0} = 1$, $\nu_{11}(\theta_0) = 0$, but that does not simplify any of the theoretical calculations which follow. Property (i) is useful for calculating curvatures, as will be shown in Section 7.

5. General definition of statistical curvature. Leaving exponential families, let

$$\mathscr{F} \equiv \{f_\theta(x),\, \theta \in \Theta\} \tag{5.1}$$

be an arbitrary family of density functions indexed by the single parameter $\theta \in \Theta$, a possibly infinite interval of the real line. The sample space $\mathscr{X}$ and carrier measure for the densities can be anything at all so we have not excluded the possibility that $\mathscr{F}$ consists of discrete distributions. Let

$$l_\theta(x) \equiv \log f_\theta(x)\,, \qquad \dot{l}_\theta(x) \equiv \frac{\partial}{\partial\theta}\, l_\theta(x)\,, \qquad \ddot{l}_\theta(x) \equiv \frac{\partial^2}{\partial\theta^2}\, l_\theta(x) \tag{5.2}$$

as in (3.15), (3.16). We assume the derivatives exist continuously and can be uniformly dominated by integrable functions in a neighborhood of the given θ, so that $E_\theta \dot{l}_\theta = 0$, $E_\theta \dot{l}_\theta^2 = -E_\theta \ddot{l}_\theta \equiv i_\theta$ as in (3.17). Finally, as in (3.20)—(3.21) we let M_θ be the covariance matrix of $(\dot{l}_\theta, \ddot{l}_\theta)$,

$$M_\theta \equiv \begin{pmatrix} \nu_{20}(\theta) & \nu_{11}(\theta) \\ \nu_{11}(\theta) & \nu_{02}(\theta) \end{pmatrix} \equiv \begin{pmatrix} E_\theta \dot{l}_\theta^2 & E_\theta \dot{l}_\theta \ddot{l}_\theta \\ E_\theta \ddot{l}_\theta \dot{l}_\theta & E_\theta \ddot{l}_\theta^2 - i_\theta^2 \end{pmatrix} \tag{5.3}$$

and define the *statistical curvature of* $\mathscr{F}$ *at* θ to be

$$\gamma_\theta \equiv (|M_\theta|/i_\theta^3)^{\frac{1}{2}} = \left[\frac{\nu_{02}(\theta)}{i_\theta^2} - \frac{\nu_{11}^2(\theta)}{i_\theta^3}\right]^{\frac{1}{2}}. \tag{5.4}$$

In making this definition we assume $0 < i_\theta < \infty$ and $\nu_{02}(\theta) < \infty$. Properties (i) and (ii) of Section 4 are verified to hold for γ_θ as defined in (5.4). Substituting $\dot{l}_\theta = \dot{f}_\theta/f_\theta$, $\ddot{l}_\theta = \ddot{f}_\theta/f_\theta - (\dot{f}_\theta/f_\theta)^2$ into (5.3), (5.4) shows that γ_θ^2 equals the bracketed term in (1.1), the crucial quantity in the Fisher–Rao theory.

What does γ_θ measure in this general situation? It is a measure of how quickly Fisher's score statistic is changing (more precisely, "turning") as θ changes. An argument along those lines is given next, further support coming in the calculations of Section 8.

Comparing (5.3) with (2.2), we can connect the two definitions by thinking of $\mathscr{L} \equiv \{\dot{l}_\theta, \theta \in \Theta\}$ as a curve through the space of random variables on $\mathscr{X}$. The inner product $\langle u, v\rangle_\theta \equiv u'\mathbf{\Sigma}_\theta v$ of (2.2) is taken to be the covariance inner product in (5.3). (Section 3 makes the analogy precise in the exponential family case.) All of the quantities in Figure 1 can now be given a statistical interpretation.

The element of arc length along $\mathscr{L}$, by analogy with (2.5), is $ds_\theta/d\theta = (E_\theta \dot{l}_\theta^2)^{\frac{1}{2}} = i_\theta^{\frac{1}{2}}$. Define

$$U_\theta(x) \equiv \frac{\dot{l}_\theta(x)}{i_\theta} + \theta\,. \tag{5.5}$$

U_{θ_0} is the version of Fisher's score statistic $\dot{l}_{\theta_0}$ that is the best locally unbiased estimator for θ near θ_0: $\mathrm{Var}_{\theta_0} U_{\theta_0} = 1/i_{\theta_0}$, the Cramér–Rao lower bound, and $E_{\theta_0} U_{\theta_0} = \theta_0$, $dE_\theta U_{\theta_0}/d\theta|_{\theta=\theta_0} = 1$. Therefore

$$\left.\frac{(d/d\theta)E_\theta U_{\theta_0}}{(\mathrm{Var}_{\theta_0} U_{\theta_0})^{\frac{1}{2}}}\right|_{\theta=\theta_0} = \left.\frac{ds_\theta}{d\theta}\right|_{\theta=\theta_0}. \tag{5.6}$$

(The quantity on the left of (5.6) is called the "efficacy" of the statistic U_{θ_0}.) We see that

$$(\theta - \theta_0)\cdot \left.\frac{ds_\theta}{d\theta}\right|_{\theta=\theta_0} = \frac{E_\theta U_{\theta_0} - E_{\theta_0} U_{\theta_0}}{(\mathrm{Var}_{\theta_0} U_{\theta_0})^{\frac{1}{2}}} + O(\theta - \theta_0)^2\,. \tag{5.7}$$

Therefore s_θ of Figure 1 can be interpreted locally as the number of (θ_0) standard deviations from $E_{\theta_0} U_{\theta_0}$ to $E_\theta U_{\theta_0}$.

By analogy with (2.6)

$$\sin a_\theta = \left[1 - \frac{[\mathrm{Cov}_{\theta_0}(\dot{l}_{\theta_0}, \dot{l}_\theta)]^2}{\mathrm{Var}_{\theta_0} \dot{l}_{\theta_0}\, \mathrm{Var}_{\theta_0} \dot{l}_\theta}\right]^{\frac{1}{2}} = [1 - \mathrm{corr}^2_{\theta_0}(\dot{l}_{\theta_0}, \dot{l}_\theta)]^{\frac{1}{2}}\,, \tag{5.8}$$

so $\sin^2 a_\theta$ is interpreted as the unexplained fraction of the variance in $U_\theta(x)$ after linear regression on $U_{\theta_0}(x)$, under density f_{θ_0}.

From (2.4) we get the following interpretation of the statistical curvature: *γ_{θ_0} is the derivative at $\theta = \theta_0$ of the unexplained fraction of the standard deviation of U_θ given U_{θ_0}, the derivative being taken with respect to the efficacy distance $(E_\theta U_{\theta_0} - E_{\theta_0} U_{\theta_0})/(\mathrm{Var}_{\theta_0} U_{\theta_0})^{\frac{1}{2}}$ along $\mathscr{L}$.* If this quantity is large then the locally best estimator (also the locally best test statistic) is changing quickly as θ changes and $\mathscr{F}$ is highly curved in a statistical sense. At the opposite extreme are one-parameter exponential families for which $a_\theta \equiv 0$, so U_θ is statistically equivalent to U_{θ_0} for all θ and θ_0. We pursue this interpretation of γ_θ in Section 8 to decide what constitutes a seriously large value of the curvature.

In a certain sense any smooth one-parameter family $\mathscr{F}$ can be embedded in a suitably large exponential family. Suppose at some point θ_0 in Θ, l_θ is k times differentiable. Consider the k-parameter exponential family

$$(5.9)\qquad g_\eta(x) \equiv \exp[l_{\theta_0}(x) + \eta_1 \dot{l}_{\theta_0}(x) + \eta_2 \ddot{l}_{\theta_0}(x) + \cdots + \eta_k l_{\theta_0}^{(k)}(x) - \psi(\eta)],$$

$l_{\theta_0}^{(k)}(x) \equiv (\partial^k/\partial\theta^k) l_\theta(x)|_{\theta=\theta_0}$, $\psi(\eta)$ being chosen to make (5.9) integrate to one over $\mathscr{X}$ with respect to the carrying measure for $\mathscr{F}$. Choosing

$$\eta_\theta = \left((\theta - \theta_0), \frac{(\theta - \theta_0)^2}{2}, \cdots, \frac{(\theta - \theta_0)^k}{k!}\right)$$

gives a one-parameter family of densities $\tilde{f}_\theta \equiv g_{\eta_\theta}$ approximating f_θ near $\theta = \theta_0$. If the Taylor expansion for l_θ converges at θ_0 this approximation becomes increasingly accurate as $k \to \infty$. For any value of $k \geq 2$ definitions (5.3) and (3.21) show that $\tilde{M}_{\theta_0} = M_{\theta_0}$, so $\tilde{\imath}_{\theta_0} = i_{\theta_0}$ and $\tilde{\gamma}_{\theta_0} = \gamma_{\theta_0}$. It is reasonable to expect results proved in the context of curved exponential families to hold for sufficiently smooth nonexponential families, though no justifying theorem has been proved to this effect. This is in the same spirit as approximating an arbitrary family by a multinomial with a large number of categories, as in Fisher (1925) and Barnett (1966), but seems to make the approximation in a smoother way.

6. Repeated sampling. Suppose we sample $x_1, x_2, \cdots, x_n$ independently and identically distributed with density f_θ. We will use boldface letters to indicate quantities connected with the repeated sample, $\mathbf{x} \equiv (x_1, x_2, \cdots, x_n)'$, $\mathbf{l}_\theta(\mathbf{x}) \equiv \sum_{i=1}^n l_\theta(x_i)$, $\mathbf{U}_\theta(\mathbf{x}) = \dot{\mathbf{l}}_\theta(\mathbf{x})/\mathbf{i}_\theta + \theta$, etc, In particular

$$(6.1)\qquad \mathbf{M}_\theta = nM_\theta$$

since $\mathbf{M}_\theta$ is the covariance matrix of $(\dot{\mathbf{l}}_\theta(\mathbf{x}), \ddot{\mathbf{l}}_\theta(\mathbf{x})) = \sum_{i=1}^n (\dot{l}_\theta(x_i), \ddot{l}_\theta(x_i))$. Besides the familiar relationship $\mathbf{i}_\theta = ni_\theta$ this gives

$$(6.2)\qquad \boldsymbol{\gamma}_\theta = \frac{\gamma_\theta}{n^{\frac{1}{2}}}.$$

The curvature goes to zero at rate $1/n^{\frac{1}{2}}$ under repeated sampling. This makes sense since we know that linear methods work better in large samples.

In curved exponential families, (3.18)—(3.19) combine with $\mathbf{l}_\theta(\mathbf{x}) = \sum_{i=1}^n l_\theta(\mathbf{x}_i)$ to give

$$(6.3)\qquad \dot{\mathbf{l}}_\theta(\mathbf{x}) = n\dot{\eta}_\theta'(\bar{x} - \lambda_\theta), \qquad \ddot{\mathbf{l}}_\theta(\mathbf{x}) = n\{\ddot{\eta}_\theta'(\bar{x} - \lambda_\theta) - ni_\theta\},$$

$\bar{x} \equiv \sum_{i=1}^n x_i/n$ being the sufficient statistic for the complete family (3.1).

If the x_i are independent but not necessarily identically distributed we still have $\mathbf{l}_\theta(\mathbf{x}) = \sum_{i=1}^n l_\theta^{(i)}(x_i)$, the superscript indicating the distribution for x_i, and so $\mathbf{M}_\theta = \sum_{i=1}^n M_\theta^{(i)}$. This explains the simple form of $\mathbf{M}_\theta$ in Example 2 of Section 3.

7. Some examples. Before discussing the statistical properties of γ_θ we will expand our catalog of examples to include several nonexponential families. Those results illustrate some simple principles that make γ_θ easy to calculate in familiar statistical situations. In the first three examples we assume that the densities given are with respect to Lebesgue measure on the real line, i.e., that we have just one observation of a continuous variable. For an independent, identically distributed (i.i.d) sample of size n the curvature is obtained from formula (6.2). This last remark applies also to Example 7, and to the examples of Section 3.

EXAMPLE 4. *Translation families.* Let $g(x)$ be a probability density function and $f_\theta(x) \equiv g(x-\theta)$. Also let $h(x) \equiv \log g(x)$. Then $\dot{l}_\theta(x) = -h^{(1)}(x-\theta)$, $\ddot{l}_\theta(x) = h^{(2)}(x-\theta)$, where $h^{(i)}(x) = d^i h(x)/dx^i$, so $E_\theta \dot{l}_\theta^{\,i}\ddot{l}_\theta^{\,j} = \int_{-\infty}^{\infty} [-h^{(1)}(x)]^i \times [h^{(2)}(x)]^j g(x)\,dx$. *Obviously M_θ and γ_θ do not depend on θ in a translation family.*

For the *t translation family*, f degrees of freedom,

$$g(x) = \frac{\Gamma\left(\frac{f+1}{2}\right)}{f^{\frac{1}{2}}\Gamma(\frac{1}{2})\Gamma\left(\frac{f}{2}\right)}\left(1+\frac{x^2}{f}\right)^{-(f+1)/2} \tag{7.1}$$

we calculate

$$\nu_{20}(\theta) = i(\theta) = \frac{f+1}{f+3}, \tag{7.2}$$

$$\nu_{02}(\theta) = \frac{f+1}{f+3}\left[\frac{(f+2)(f^2+8f+19)}{f(f+5)(f+7)} - \frac{f+1}{f+3}\right]$$

and $\nu_{11}(\theta) = 0$ (by symmetry), giving

$$\gamma_\theta^2 = \frac{6[3f^2+18f+19]}{f(f+1)(f+5)(f+7)}, \tag{7.3}$$

a monotone decreasing function of f. Some values are as follows:

(7.4)

f	1	2	5	10	20	$\to\infty$
γ_θ^2	2.5	1.063	.306	.107	.0334	$\sim 18/f^2$

The case $f = 1$ is the *Cauchy translation family*, and the value $\frac{5}{2}$ for γ_θ^2 agrees with a closely related calculation in Fisher (1925).

For the *Gamma translation family*

$$f_\theta(x) = \frac{(x-\theta)^{a-1}e^{-(x-\theta)}}{\Gamma(a)}, \qquad x \geqq \theta \tag{7.5}$$

$a > 4$ a fixed constant, we calculate

$$M_\theta = \frac{1}{a-2}\begin{pmatrix} 1 & \frac{2}{(a-3)} \\ \frac{2}{(a-3)} & \frac{4a-10}{(a-2)(a-3)(a-4)} \end{pmatrix}, \tag{7.6}$$

$$\gamma_\theta^2 = \frac{2}{(a-3)^2}\,\frac{a-1}{a-4}.$$

(For $a \leqq 4$, ν_{02} is infinite.)

EXAMPLE 5. *Scale families.* $x \sim \theta \cdot z$ where z has a known density, $\theta \in \Theta = (0, \infty)$. If z is a positive random variable then $\log x = \log \theta + \log z$ is a translation family. By Section 4 the curvature will be the same for this family as for the original one, and by Example 4 it will not depend on θ: *For scale families* γ_θ *does not depend on* θ. (The argument above applied separately to the positive and negative axes gives the result in general. It can also be derived directly from (5.3).)

A particular example is the *normal with known coefficient of variation*, $x \sim \mathcal{N}(\theta, c\theta^2)$, c known. Here $x \sim \theta z$, $z \sim \mathcal{N}(1, c)$. We calculate $i_\theta = 2(c + \frac{1}{2})/(c\theta^2)$ and

$$\gamma_\theta^2 = \frac{c^2}{4(c + \frac{1}{2})^3}\,. \tag{7.7}$$

(Notice that $x \sim \mathcal{N}(\theta, c\theta)^2$ is a curved exponential family, $k = 2$.) The curvature is near 0 for all values of c, taking its maximum at $c = 1$:

(7.8)

c	$\frac{1}{4}$	$\frac{1}{2}$	1	2	4	$\to \infty$
γ_θ^2	.0370	.0625	.0740	.0640	.0439	$\sim 1/4c$

EXAMPLE 6. *Weibull shape parameter.* $f_\theta(x) = \theta x^{\theta-1} e^{-x^\theta}$ for $x \geqq 0$, $\theta \in \Theta = (0, \infty)$. That is $x \sim z^{1/\theta}$ where $P\{z < z_0\} = 1 - e^{-z_0}$ for $z_0 \geqq 0$. The transformation $\log x = 1/\theta \log z$ makes this a scale family in $1/\theta$, so once again γ_θ^2 does not depend on Θ. Taking $\theta = 1$ for convenience gives $\dot{l}_1(x) = (1 - x)\log x + 1$, $\ddot{l}_1(x) = -(x \log^2 x + 1)$, $E_1 \dot{l}_1^i \ddot{l}_1^j = \int_0^\infty [\dot{l}_1(x)]^i [\ddot{l}_1(x)]^j e^{-x}\,dx$. Numerical integration gives

$$\gamma_\theta^2 = .704\,. \tag{7.9}$$

EXAMPLE 7. *Mixture problems.* $f_\theta(x) = (1 - \theta)g(x) + \theta h(x)$, g and h known densities on an arbitrary space $\mathscr{X}$. The parameter space Θ contains $[0, 1]$. We see that

$$\dot{l}_\theta = \frac{h - g}{g + \theta(h - g)}\,, \qquad \ddot{l}_\theta = -\dot{l}_\theta^2 \tag{7.10}$$

and for $\theta = 0$

$$\dot{l}_0 = r - 1\,, \qquad \ddot{l}_0 = -(r - 1)^2 \tag{7.11}$$

where $r(x) \equiv h(x)/g(x)$. Defining $\alpha_j \equiv E_0(r - 1)^j$ gives

$$M_0 = \begin{pmatrix} \alpha_2 & -\alpha_3 \\ -\alpha_3 & \alpha_4 - \alpha_2^2 \end{pmatrix}, \qquad \gamma_0^2 = \frac{\alpha_4 - \alpha_2^2}{\alpha_2^2} - \frac{\alpha_3^2}{\alpha_2^3}\,. \tag{7.12}$$

If g and h are normal densities, say $g(x) = \varphi(x) = e^{-x^2/2}/(2\pi)^{\frac{1}{2}}$, $h(x) = \varphi(x - \mu)$, we have $r(x) = \exp(\mu x - \mu^2/2)$ and

$$M_0 = \begin{pmatrix} \xi - 1 & -[\xi^3 - 3\xi + 2] \\ -[\xi^3 - 3\xi + 2] & \xi^6 - 4\xi^3 - \xi^2 + 8\xi - 4 \end{pmatrix} \tag{7.13}$$

when $\xi \equiv e^{\mu^2}$. Therefore $i_0 = \xi - 1$ and

$$\gamma_0^2 = \xi^3(\xi + 1) . \tag{7.14}$$

The curvature approaches 2 for μ near 0 but becomes enormous as μ increases,

(7.15)

μ	.5	.832	1	1.048	1.180	$\to \infty$
γ_0^2	4.84	24	74.68	108	320	$\sim e^{4\mu^2}$.
i_0	.284	1	1.718	2	3	$\sim e^{\mu^2}$

8. Hypothesis testing. So far we have not tried to say what constitutes a "large" curvature—a value of γ_θ (or, in repeated sampling situations of γ_θ, the curvature based on all the data) of sufficient magnitude to undermine techniques based on linear approximations to the log likelihood function. We can obtain a rough idea of this value by considering the problem of testing $H_0: \theta = \theta_0$ versus $A_0: \theta > \theta_0$.

Define

$$\theta_1 \equiv \theta_0 + \frac{2}{i_{\theta_0}^{\frac{1}{2}}} \tag{8.1}$$

so that $i_{\theta_0}^{\frac{1}{2}}(\theta_1 - \theta_0) = 2$. From the discussion (5.5)—(5.7) this means that, approximately,

$$\frac{E_{\theta_1} U_{\theta_0} - E_{\theta_0} U_{\theta_0}}{(\operatorname{Var}_{\theta_0} U_{\theta_0})^{\frac{1}{2}}} = 2 \tag{8.2}$$

(where in (5.7) we have used $ds_\theta/d\theta|_{\theta=\theta_0} = i_{\theta_0}^{\frac{1}{2}}$). The locally most powerful level α test of H_0 versus A_0, LMP$_\alpha$, for short, rejects for large values of U_{θ_0}. From (8.2) we would expect LMP$_\alpha$ to have reasonable power at θ_1 for the customary values of α. That is θ_1 should be a "statistically reasonable" alternative to θ_0.

The discussion following (5.8) shows that the unexplained fraction of the variance of U_{θ_1} after linear regression on U_{θ_0}, calculated under f_{θ_0}, is approximately $4\gamma_{\theta_0}^2$. If this quantity is large, say $4\gamma_{\theta_0}^2 \geqq \frac{1}{2}$, then U_{θ_1} differs considerably from U_{θ_0}, and the test of H_0 based on U_{θ_1} will substantially differ from that based on U_{θ_0}. Under these circumstances it is reasonable to question the use of LMP$_\alpha$. *Based on those very rough calculations a value of $\gamma_{\theta_0}^2 \geqq \frac{1}{8}$ is "large".*

In the repeated sampling situation of Section 6 a sample of size $n > n_0$,

$$n_0 = 8\gamma_{\theta_0}^2 , \tag{8.3}$$

makes $\gamma_{\theta_0}^2 = \gamma_{\theta_0}^2/n < \frac{1}{8}$, and therefore reduces the curvature below the worrisome point. For the Cauchy translation family, Example 4, $n_0 = 20$. For the Weibull shape parameter, Example 6, $n_0 = 5.6$. For the normal with known coefficient of variation, Example 5, $n_0 < 1$ for all c. At the opposite extreme we have the normal mixture problem, Example 7, with $\mu = 1$, for which $n_0 = 597.4$. We expect linear methods to work well in Example 5 for any sample size, and poorly in the last example, even for large samples.

Moving from the vague to the specific, consider Example 1, Section 3, a bivariate normal vector $x = (x_1, x_2)'$ with mean $(\theta, \gamma_0\theta^2/2)$ and covariance matrix I. Assume we wish to test $H_0: \theta = 0$ versus $A_0: \theta > 0$ on the basis of observing x. The LMP_α, which rejects for large values of x_1, has power function (probability of rejection) $1 - \beta_0(\theta) = \Phi(\theta - z_\alpha)$, where z_α and Φ are the upper α point and cdf of a standard normal variate.

The Neyman–Pearson lemma says that the most powerful level α test of $\theta = 0$ versus some specific positive alternative $\theta = \theta_1$, $\mathrm{MP}_\alpha(\theta_1)$ for short, rejects for large values of $\eta'_{\theta_1}x$. It has power function

$$1 - \beta_{\theta_1}(\theta) = \Phi(\theta(1 + \gamma_0^2\theta^2/4)^{\frac{1}{2}} \cos(A_{\theta_1} - A_\theta) - z_\alpha), \tag{8.4}$$

A_θ being the angle from the x_1 axis to η_θ, illustrated in Figure 2. As θ_1 approaches 0, $\beta_{\theta_1}(\theta)$ approaches $\beta_0(\theta)$ for all θ, justifying the notation $1 - \beta_0(\theta)$ for the power function of LMP_α.

For a given value of $\theta > 0$ the power is maximized by taking $\theta_1 = \theta$, giving "power envelope"

$$1 - \beta^*(\theta) = \Phi(\theta(1 + \gamma_0^2\theta^2/4)^{\frac{1}{2}} - z_\alpha). \tag{8.5}$$

Figure 3 compares the power envelope function, for four values of γ_0, with the power function of LMP_α, $\alpha = .05$ (which does not depend on γ_0). As predicted the difference between $1 - \beta^*(\theta)$ and $1 - \beta_0(\theta)$ increases with the curvature γ_0. In this case we can actually see that γ_{θ_0} measures how fast the locally optimum test statistic $U_{\theta_0}(x)$ becomes nonoptimal as the alternative θ increases from 0. Also according to prediction the LMP_α has reasonable power properties for $\gamma_0^2 = \frac{1}{16}$ and poor properties for $\gamma_0^2 \geq \frac{1}{4}$.

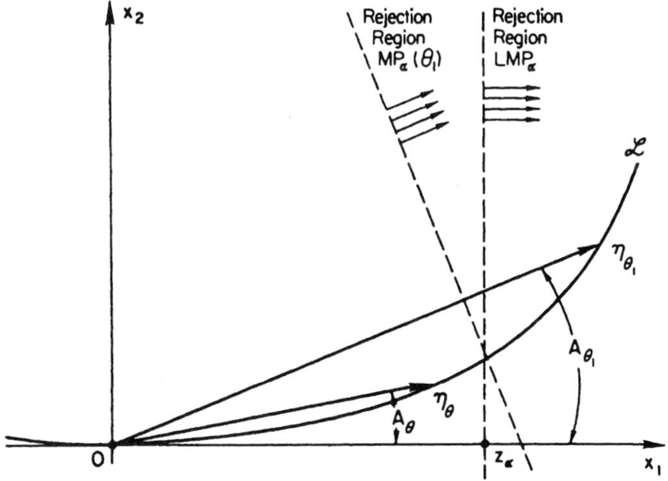

FIG. 2. Bivariate Normal, Example 1, testing $\theta = 0$ versus $\theta > 0$. The rejection region for the locally most powerful level α test, LMP_α, is compared with that for the most powerful level α test of θ versus θ_1, $\mathrm{MP}_\alpha(\theta_1)$.

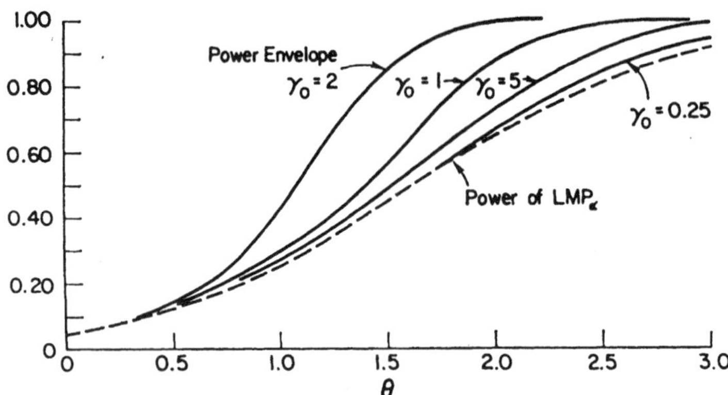

FIG. 3. Power of LMP$_\alpha$, $\alpha = .05$, compared with power envelope function, Example 1.

TABLE 1
Power comparison, Example 1
a) *Power envelope* b) *Power* MP$_{.05}(2)$ c) *Power* LMP$_{.05}$

γ_0	θ							
	0	.5	1.0	1.5	2.0	2.5	3.0	3.5
.25	.05[a]	.126	.262	.453	.662	.835	.941	.985
.25	.05[b]	.125	.260	.452	.662	.834	.938	.983
.25	.05[c]	.126	.260	.442	.639	.804	.912	.968
.5	.05	.127	.269	.483	.723	.904	.982	.999
.5	.05	.121	.261	.479	.723	.901	.980	.998
.5	.05	.126	.260	.442	.639	.804	.912	.968
1.0	.05	.129	.300	.591	.882	.991	1.000	1.000
1.0	.05	.115	.280	.583	.882	.990	1.000	1.000
1.0	.05	.126	.260	.442	.639	.804	.912	.968
2.0	.05	.139	.409	.855	.998	1.000	1.000	1.000
2.0	.05	.115	.381	.850	.998	1.000	1.000	1.000
2.0	.05	.126	.260	.442	.639	.804	.912	.968

Of course no level α test can achieve the power envelope for more than one value of $\theta > 0$. MP$_\alpha(\theta_1)$ achieves it for $\theta = \theta_1$ while LMP$_\alpha$ optimizes for θ near 0 in the sense that $d\beta_0(\theta)/d\theta|_{\theta=\theta_0} = d\beta^*(\theta)/d\theta|_{\theta=\theta_0}$. By following prescription (8.1) in choosing θ_1 we get a test which matches the power envelope at what should be a statistically interesting value of θ, one where the power is reasonably but not unreasonably high. In our example this means choosing $\theta_1 = 2$, since $i_0 = 1$. Table 1 shows that $1 - \beta_2(\theta)$ stays remarkably close to $1 - \beta^*(\theta)$, and that MP$_{.05}(2)$ has better power characteristics than LMP$_{.05}$, especially for large values of γ_0.

Davies performs similar evaluations for the Neyman–Davies model of Example 2. The curvatures for the upper and lower cases graphed on page 532 of Davies (1969) are $\gamma_0^2 = .488$ and $\gamma_0^2 = .244$ respectively, while the two on page 533 are $\gamma_0^2 = .00629$ and $\gamma_0^2 = .0364$. Ignoring the "Wald's test" curve, one sees that the magnitude of γ_0^2 is indeed a good predictor of the relative performance of LMP_α compared to $\mathrm{MP}_\alpha(\theta_1)$. His results are quite similar to those for our Example 1. (Davies chooses θ_1 so that $1 - \beta^*(\theta_1) = .8$. This is a more precise way of accomplishing what (8.1) is intended to do, but is computationally difficult in most situations.)

Section 10 shows that the Cramér-Rao lower bound for the variance of an unbiased estimator errs roughly by a factor of $1 + \gamma_\theta^2$, rejustifying the definition of $\gamma_\theta^2 \geqq \frac{1}{8}$ as a "large" curvature.

9. The Fisher-Rao theorem. We again assume an i.i.d. sample $x_1, x_2, \cdots, x_n$, as in Section 6. Result (1.1), originally stated by Fisher in his fundamental paper on estimation theory (1925) can be restated as

$$\lim_{n\to\infty} (\mathbf{i}_\theta - \mathbf{i}_\theta^{\hat\theta}) = i_\theta \gamma_\theta^2 \tag{9.1}$$

since γ_θ^2 equals the bracketed term in (1.1). (9.1) is derived from (1.1) and (5.4) by means of the relationships $\nu_{20}(\theta) = \mu_{20} = i_\theta$, $\nu_{11}(\theta) = \mu_{11} - \mu_{30}$, and $\nu_{02}(\theta) = \mu_{02} - 2\mu_{21} + \mu_{40} - \mu_{20}^2$, these following from (1.2), (5.3) and

$$\dot{l}_\theta = \dot{f}_\theta / f_\theta\,, \qquad \ddot{l}_\theta = \ddot{f}_\theta / f_\theta - (\dot{f}_\theta / f_\theta)^2 \tag{9.2}$$

To use Fisher's evocative language, asymptotically the MLE $\hat\theta(x_1, x_2, \cdots, x_n)$ extracts all but $i_\theta \gamma_\theta^2$ of the information in the sample $\mathbf{x} = (x_1, \cdots, x_n)'$. Since a single observation contains an amount i_θ of information this is equivalent to a reduction in effective sample size from n to $n - \gamma_\theta^2$, for example from n to $n - \frac{5}{2}$ in the Cauchy translation parameter problem. This is the price one pays for a one-dimensional summary of the data and, also according to Fisher, any summary statistic other than the MLE would pay a greater price. (Rao's substantial contributions to this argument are discussed toward the end of the section.)

The geometrical argument which follows shows clearly why the curvature γ_θ plays the role that it does in (9.1). It also leads quickly to a counterexample to (9.1) and shows that by working within multinomial families, Fisher and Rao chose perhaps the *least* tractable curved exponential families. We will work with a general curved exponential family in the *standard form* (4.1)—(4.2). For notational convenience we let θ_0, a particular value of θ where we wish to evaluate $\lim_{n\to\infty} (\mathbf{i}_\theta - \mathbf{i}_\theta^{\hat\theta})$, equal 0. Then we have $\eta_0 = \lambda_0 = \mathbf{0}$, $\mathbf{\Sigma}_0 = \mathbf{I}_r$, $\dot\eta_0 = \dot\lambda_0 = i_0^{\frac{1}{2}} \mathbf{e}_1$, and $\ddot\eta_0 = (\nu_{11}(0)/i_0^{\frac{1}{2}})\mathbf{e}_1 + i_0 \gamma_0 \mathbf{e}_2$.

Fisher's argument depends on two useful results which we borrow:

1) If $T(\mathbf{x})$ is any statistic, with density say $\mathbf{f}_\theta^T(t)$ and score function (log derivative) $\dot{\mathbf{l}}_\theta^T(t) \equiv \partial/\partial\theta \log f_\theta^T(t)$, then $\dot{\mathbf{l}}_\theta^T(t) = E_\theta\{\dot{\mathbf{l}}_\theta(\mathbf{x}) \mid T = t\}$ (where we recall from Section 6 that $\dot{\mathbf{l}}_\theta(\mathbf{x})$ is the score based on all the data). This implies that the loss of information in going from $\mathbf{x}$ to $T(\mathbf{x})$ is

$$\mathbf{i}_\theta - \mathbf{i}_\theta^T = E_\theta \operatorname{Var}_\theta \{\dot{\mathbf{l}}_\theta(\mathbf{x}) \mid T\} \tag{9.3}$$

since $\mathbf{i}_\theta - \mathbf{i}_\theta^T = \operatorname{Var}_\theta \dot{\mathbf{l}}_\theta - \operatorname{Var}_\theta \dot{\mathbf{l}}_\theta^T$.

2) Let L_θ be the set of values of $\bar{x} \equiv \sum_{i=1}^n x_i/n$ for which $\dot{\mathbf{l}}_\theta(\mathbf{x}) = 0$; L_θ consists

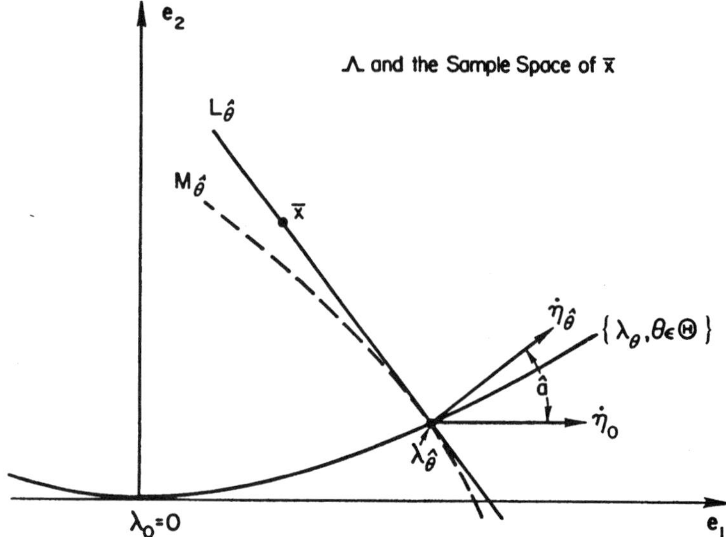

FIG. 4. A curved exponential family of dimension $r = 2$. $L_{\hat\theta}$ is the set of $\bar x$ for which $\hat\theta$ is a solution to the maximum likelihood equations. $M_{\hat\theta}$ is the level curve for another consistent efficient estimator.

of those values of the sufficient statistic $\bar x$ for which θ is a solution to the likelihood equations $\dot{\mathbf{l}}_\theta(\mathbf{x}) = 0$. Then, since $\dot{\mathbf{l}}_\theta = n\dot\eta_\theta'(\bar x - \lambda_\theta)$,

$$L_\theta = \{\bar x : \dot\eta_\theta'(\bar x - \lambda_\theta) = 0\} \tag{9.4}$$

the $r - 1$-dimensional hyperplane through λ_0, orthogonal to $\dot\eta_\theta$.

Figure 4 illustrates the situation for the case $r = 2$. (Notice that the sample space, the space of possible $\bar x$ values, has been superimposed on Λ, the space of possible mean vectors λ.) Actually this two-dimensional picture is appropriate for any dimension since curvature is locally a two-dimensional property, as pointed out at the end of Section 2. A heuristic proof of (9.1) based on this picture now follows in five easy steps:

(i) $\dot{\mathbf{l}}_0(\mathbf{x}) = n(i_0)^{\frac12}\bar x_1$ (by (6.3)).

(ii) $n^{\frac12}\bar x \to \mathscr{N}_r(\mathbf{0}, \mathbf{I})$ as $n \to \infty$ if $\theta = 0$ (since $\lambda_0 = \mathbf{0}$, $\mathbf{\Sigma}_0 = \mathbf{I}$, and central limit conditions are satisfied inside an exponential family).

(iii) Let $\hat\theta$ be the MLE and $\hat a$ the angle between $\dot\eta_{\hat\theta}$ and $\dot\eta_0 = i_0^{\frac12}\mathbf{e}_1$. Then $\hat a = i_0^{\frac12}\gamma_0\hat\theta + O(\hat\theta^2)$. (Since $\eta_{\hat\theta} = \hat\theta\dot\eta_0 + O(\hat\theta^2) = i_0^{\frac12}\hat\theta\mathbf{e}_1 + O(\hat\theta^2)$, the element of arclength in Figure 1 is $s_{\hat\theta} = ||\eta_{\hat\theta}|| + O(\hat\theta^2) = i_0^{\frac12}\hat\theta + O(\hat\theta^2)$. By (2.4) we have $a_{\hat\theta} \equiv \hat a = i_0^{\frac12}\gamma_0\hat\theta + O(\hat\theta^2)$.)

(iv) $\mathrm{Var}_0\{\dot{\mathbf{l}}_0(x) \mid \hat\theta\} = n^2 i_0 \tan^2 \hat a \cdot \mathrm{Var}_0\{\bar x_2 \mid \hat\theta\}$. (In the case $r = 2$ this follows immediately from (i) and the geometry of the situation. For $r > 2$, $\bar x_2$ is replaced by $\nu'\bar x/||\nu||$ where $\nu = \dot\eta_{\hat\theta} - ||\dot\eta_{\hat\theta}|| \cos \hat a \cdot \dot\eta_0$, the part of $\dot\eta_{\hat\theta}$ orthogonal to $\dot\eta_0$.)

(v) $\mathrm{Var}_0\{\bar x_2 \mid \hat\theta\} = 1/n + o(1/n)$. (This is plausible because of (ii) and the fact that near $\theta = \hat\theta$ the partition of the sample space generated by the "lines" L_θ looks like the partition generated by lines parallel to $L_{\hat\theta}$.)

Steps (iii) and (iv) together give $\text{Var}_0\{\dot{\text{l}}_0 \mid \hat{\theta}\} = n^2 i_0^2 \gamma_0^2 \hat{\theta}^2(1 + O(\hat{\theta}))\,\text{Var}_0\{\bar{x}_2 \mid \hat{\theta}\}$, which, combined with (v), gives

$$\text{Var}_0\{\dot{\text{l}}_0(x) \mid \hat{\theta}\} = n i_0^2 \gamma_0^2 \hat{\theta}^2(1 + O(\hat{\theta}))(1 + o_n(1)) \tag{9.5}$$

$o_n(1) \to 0$ as $n \to \infty$, $O(\hat{\theta}) \to 0$ as $\hat{\theta} \to 0$. The heuristic proof of (9.1) is completed by (9.3), giving

$$\lim_{n\to\infty} \text{i}_0 - \text{i}_0^{\hat{\theta}} = \lim_{n\to\infty} E_0 \,\text{Var}_0\{\dot{\text{l}}_0 \mid \hat{\theta}\} = i_0 \gamma_0^2\,. \tag{9.6}$$

Here we have used

$$\lim_{n\to\infty} nE_0|\hat{\theta}|^3 = 0\,, \qquad \lim_{n\to\infty} nE_0\hat{\theta}^2 = i_0^{-1}\,, \tag{9.7}$$

which one might hope for in view of $n^{\frac{1}{2}}\hat{\theta} \to \mathscr{N}(0, i_0^{-1})$.

All of the weak links in this chain of reasoning can be made solid except for (v). Its fatal flaw is shown by a counterexample to (9.1) based on the trinomial distribution

$$P\{\text{observed object is in category } j\} = \lambda_j\,, \qquad j = 1, 2, 3 \tag{9.8}$$
$$(\text{so } \lambda_j \geqq 0, \lambda_1 + \lambda_2 + \lambda_3 = 1)\,.$$

The trinomial can be considered as an exponential family of form (3.1) with $k = r = 2$; $\lambda = (\lambda_1, \lambda_2)'$, $\eta = (\eta_1, \eta_2)'$, $\eta_j = \log[\lambda_j/(1 - \lambda_1 - \lambda_2)]$, $j = 1, 2$, and $\psi(\eta) = \log(1 + e^{\eta_1} + e^{\eta_2})$. The x vector takes on three possible values: (1, 0), (0, 1), (0, 0), corresponding to the observed object being in the first, second, or third categories respectively. The carrier measure $m(\cdot)$ puts mass one at each of these three x values.

The counterexample is a one parameter family $\mathscr{F}$ with L_θ passing through the fixed point $c = (-2^{\frac{1}{2}}, -1)$ as illustrated in Figure 5, the parameter θ being

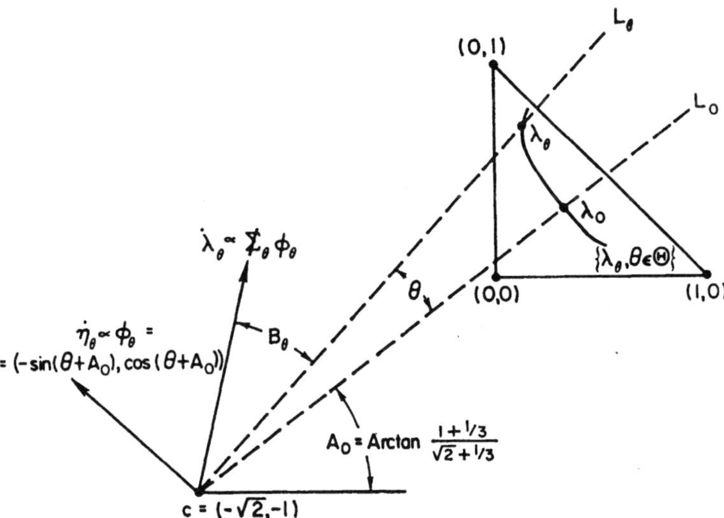

FIG. 5. Counterexample to (9.1) based on trinomial. Each line L_θ contains at most one possible sample point $\bar{x}$.

the angle between L_0 and L_θ. Such a family does exist, as the following construction shows: let $\lambda_0 \equiv (\frac{1}{3}, \frac{1}{3})$ and

$$\lambda_{\theta_0} \equiv \lambda_0 + \textstyle\int_0^{\theta_0} \mu_\theta(\mathbf{\Sigma}_\theta \phi_\theta)\, d\theta \tag{9.9}$$

where $\mu_\theta \equiv \|\lambda_\theta - c\|/(\|\mathbf{\Sigma}_\theta \phi_\theta\| \sin B_\theta)$, $\mathbf{\Sigma}_\theta$ is the covariance matrix of x under f_θ, the vector ϕ_θ and the angle B_θ being defined as in Figure 5. Definition (9.9) gives $\lambda_\theta \in L_\theta$ and also that, by (3.6), $\dot{\eta}_\theta \propto \phi_\theta$, the normal vector to L_θ, as necessitated by (9.4).

$\mathscr{F}$ is a curved exponential family having the following property: if $\bar{x}_{(1)}$ and $\bar{x}_{(2)}$ are two values of $\bar{x} \equiv \sum_1^n x_i/n$ giving the same MLE $\hat{\theta}$, then both $\bar{x}_{(1)}$ and $\bar{x}_{(2)}$ lie on $L_{\hat{\theta}}$. But $\bar{x}_{(i)} = (n_{1(i)}/n, n_{2(i)}/n)$, $i = 1, 2$ the $n_{j(i)}$ being nonnegative integers. This implies either $\bar{x}_{(1)} = \bar{x}_{(2)}$ or

$$\frac{n_{2(2)} - n_{2(1)}}{n_{1(2)} - n_{1(1)}} = \frac{n_{2(1)} + 1 \cdot n}{n_{1(1)} + 2^{\frac{1}{2}} \cdot n}. \tag{9.10}$$

Since (9.10) would make $2^{\frac{1}{2}}$ a rational number, $\bar{x}_{(1)}$ must equal $\bar{x}_{(2)}$. In short there is at most one possible $\bar{x}$ value corresponding to any $\hat{\theta}$, and so the MLE is a sufficient statistic in $\mathscr{F}$, implying $\mathbf{i} - \mathbf{i}^{\hat{\theta}} = 0$ for all n. But γ_θ^2 must be positive for all θ values since $\dot{\eta}_\theta$ is always changing direction. This completes the counterexample.

Remark 1. Let $\varphi(t) \equiv E_0 e^{it'x}$ be the characteristic function of f_0. If $|\varphi(t)|^p$ is integrable for some $p \geqq 1$ then $n^{\frac{1}{2}}\bar{x}$ has a density function converging uniformly to $(2\pi)^{-k/2} \exp(-\|x\|^2/2)$. See Efron and Truax (1968), Gnedenko and Kolmogorov (1954). Under those conditions (9.1) can be verified. The technical details, which depend on an exponential bound to the density of $\bar{x}$, are indicated in the Appendix.

Remark 2. Instead of working with the MLE $\hat{\theta}$ itself we can consider the coarser statistic which only records which interval $\hat{\theta}$ lies in, among intervals of the form $(i\varepsilon_n, (i+1)\varepsilon_n)$, $i = 0, \pm 1, \pm 2, \cdots$. The line $L_{\hat{\theta}}$ in Figure 4 is now replaced by a pair of lines $L_{i\varepsilon_n}$, $L_{(i+1)\varepsilon_n}$, and step (v) can be weakened to say only that the conditional distribution of $\bar{x}_2$, given that $\bar{x}$ is between the two lines, has variance $1/n + o(1/n)$. However in order for statement (iv) to still have meaning we need to take $\varepsilon_n = o(1/n)$ (so that the conditional variance of $\dot{\mathbf{l}}_0$ will still be due mainly to the slope of the lines $L_{i\varepsilon_n}$, $L_{(i+1)\varepsilon_n}$, and not to the distance between them). It turns out (Efron and Truax (1968)) to be possible to choose ε_n in this way and to get the proper convergence of the conditional variance if f_0 is nonlattice, $|\varphi(t)| < 1$ for all $t \neq 0$. (This excludes the multinomial.) In this case it is possible to show that $\limsup_{n\to\infty} (\mathbf{i}_\theta - \mathbf{i}_\theta^{\hat{\theta}}) \leqq i_\theta \gamma_\theta^2$.

Remark 3. If $\tilde{\theta}(\bar{x})$ is any other consistent efficient estimator of θ, and M_θ is the set of $\bar{x}$ values having $\tilde{\theta}(\bar{x}) = \theta$, then as in Figure 4, $M_{\hat{\theta}}$ passes through $\lambda_{\hat{\theta}}$ and is tangent to $L_{\hat{\theta}}$ at that point. See Section 10. The increment of $[\lim_{n\to\infty} (\mathbf{i}_\theta - \mathbf{i}_\theta^T) - i_\theta \gamma_\theta^2]$ above zero is due to the quadratic term in the expansion of $M_{\hat{\theta}}$ near $\lambda_{\hat{\theta}}$. The details are almost identical to those of Section 10 and will not be given here. (See (10.25).)

REMARK 4. It is possible for two of the surfaces (9.4), say L_0 and $L_{\hat{\theta}}$, to intersect. If $\bar{x} \in L_0 \cap L_{\hat{\theta}}$ then both 0 and $\hat{\theta}$ are solutions to the likelihood equation. As $\hat{\theta}$ decreases to zero in Figure 4, $L_0 \cap L_{\hat{\theta}}$ converges to a point (in general an $r-2$ dimensional flat) on $L_0 = \{c\mathbf{e}_2\}$ a distance $\rho_0 \equiv 1/\gamma_0$ above $\mathbf{0}$. Values of $\bar{x}$ on L_0 which lie above this point are local *maxima* of the likelihood function, while those lying below are local *minima*.

REMARK 5. Rao (1961, 1962, 1963) uses a different definition of the information which avoids the difficulty illustrated by the counterexample. (9.3) can be written as $\mathbf{i}_\theta - \mathbf{i}_\theta{}^T = \inf E_\theta\{\dot{\mathbf{l}}_\theta(x) - h(T(x))\}^2$, the infimum being over all choices of the function $h(\cdot)$. Rao redefines $\mathbf{i}_\theta{}^T$ by restricting the function h to be quadratic. Rao states that he believes the two definitions to be equivalent, but the counterexample can be used to show that they are not.

REMARK 6. Is (9.1) a useful fact, assuming it is true? Fisher seemed to think of Fisher information as a perfect measure of the amount of information available to the statistician. For ordinary "first order efficiency" calculations in large samples this is true enough, in the following sense: let $T(x)$ be a statistic having Fisher information $\mathbf{i}_\theta{}^T$. Then in a neighborhood of any given value θ_0 of θ we can construct, under suitable regularity conditions, a function $\tilde{T}(T)$, that is approximately $\mathcal{N}(\theta, 1/\mathbf{i}_\theta{}^T)$, as compared with $\mathcal{N}(\theta, 1/\mathbf{i}_\theta{}^{\hat{\theta}})$ for the MLE. If $\mathbf{i}_\theta{}^T/\mathbf{i}_\theta{}^{\hat{\theta}} = .8$ for example, then any statistic $h(\tilde{T})$ will have almost the same distribution as $h(\hat{\theta})$ with $\hat{\theta}$ based on a sample 80% as large.

This argument breaks down for information discrepancies as small as those contemplated in (9.1), since the central limit theorem is in general not capable of supporting such fine distinctions. To give substance to Fisher and Rao's theorem we must demonstrate that in specific statistical problems the Fisher information determines relative performance at the level of accuracy suggested by (9.1). Rao (1963) showed that this indeed was the case for the problem of estimating θ with squared error loss. We review his results from the point of view of this paper in Section 10.

10. Estimation with squared error loss. Suppose we wish to estimate the parameter θ in a curved exponential family on the basis of an i.i.d. sample $x_1, x_2, \cdots, x_n$, using a squared error loss function to evaluate possible estimators. We will only consider estimators that are smooth functions of the sufficient statistic $\bar{x}$ and are consistent and efficient in the usual sense (see (10.5)—(10.7) below). The following result will be discussed: let $\tilde{\theta}(\bar{x})$ be such an estimator, the form of $\tilde{\theta}$ not depending on n, and let $\phi(\theta) \equiv E_\theta \mathbf{U}_{\theta_0}(\bar{x})$ where as before $\mathbf{U}_{\theta_0}(\bar{x}) \equiv \dot{\mathbf{l}}_{\theta_0}/\mathbf{i}_{\theta_0} + \theta_0$ is the best locally unbiased estimator of θ near θ_0. Also let $b_\theta \equiv E_\theta\tilde{\theta}(\bar{x}) - \theta$ be the bias of $\tilde{\theta}$, a quantity which will turn out to be of order

$O(1/n)$ in the theory below. Then

$$\text{(10.1)} \qquad \text{Var}_{\theta_0}\tilde{\theta} = \frac{1}{ni_{\theta_0}} + \frac{1}{n^2 i_{\theta_0}}\left\{\gamma_{\theta_0}^2 + 4\frac{\Gamma_{\theta_0}^2}{i_{\theta_0}} + \Delta_{\theta_0}^{\tilde{\theta}}\right\} + 2\frac{\dot{b}_{\theta_0}}{ni_{\theta_0}} + o\left(\frac{1}{n^2}\right)$$

where $\Delta_{\theta_0}^{\tilde{\theta}} \geqq 0$ and for the MLE $\hat{\theta}$, $\Delta_{\theta_0}^{\hat{\theta}} \equiv 0$. The quantity Γ_{θ_0} is the ordinary curvature at $\theta = \theta_0$ of the two-dimensional curve $(\theta, \phi(\theta))$ as defined at (2.1).

Before verifying (10.1) several remarks are pertinent.

1) The term $1/ni_{\theta_0}$ is the Cramér–Rao lower bound for the variance of an unbiased estimator. The bracketed quantity in (10.1) expresses the coefficient of the $1/n^2 i_{\theta_0}$ term as the sum of three nonnegative quantities: $\gamma_{\theta_0}^2$, the statistical curvature, which is invariant under transformations of θ; $4\Gamma_{\theta_0}^2/i_{\theta_0}$, the "naming curvature", which depends on how $\mathscr{F}$ is parametrized (however, notice that $4\Gamma_{\theta_0}^2/i_{\theta_0}$ is invariant under *linear* reparametrizations $\theta \to \alpha + \beta\theta$); and $\Delta_{\theta_0}^{\tilde{\theta}}$, which can be made zero by using the MLE. Taken literally (10.1) says that the MLE is superior to other efficient estimations with the same bias structure.

2) The estimators $\tilde{\theta}$ will generally be biased by an amount of order $1/n$. This affects mean square error to order $1/n^2$. A simple adjustment, noted below at Remark 11, produces estimators biased only to order $1/n^2$; (10.1), with the bias term $2\dot{b}_{\theta_0}/ni_{\theta_0}$ removed, is valid for such estimators. Among such bias corrected estimators, (10.1) says that the MLE has asymptotically smallest variance.

3) The Fisher information is essentially invariant under reparametrizations of $\mathscr{F}$, in the sense that if $\mu = \mu(\theta)$ is a differentiable monotonic function then $\mathbf{i}_\mu{}^T = \mathbf{i}_\theta{}^T (d\theta/d\mu)^2$ for every statistic $T(x)$. The squared error estimation problem is *not* invariant under reparametrization and this accounts for the presence of the $4\Gamma_{\theta_0}^2$ term in (10.1). For a given θ_0 the "best" parametrization is in terms of $\phi(\theta)$, the expectation of the best locally unbiased estimator of θ. (Notice that ϕ will be the same, except for scale and translation constants, no matter what "θ" we begin with.) It will turn out that if the MLE $\hat{\theta}$ is unbiased for θ then $\phi \equiv \theta$ for all choices of θ_0, so we are automatically using the best parametrization.

4) (10.1) is not a special case of the Bhattacharyya lower bounds. The second Bhattacharyya bound, applying to estimators biased by amount $O(1/n^2)$ or less, is of the form

$$\text{(10.2)} \qquad \text{Var}_{\theta_0}\tilde{\theta} \geqq \frac{1}{ni_{\theta_0}} + \frac{1}{n^2 i_{\theta_0}}\left\{\frac{4\Gamma_{\theta_0}^2}{i_{\theta_0}}\right\} + O\left(\frac{1}{n^3}\right),$$

and the higher Bhattacharyya bounds are identical until order $O(1/n^3)$, so these bounds relate only to the naming part of the estimation problem. It is possible for an estimator to achieve equality in (10.2), but then it cannot be efficient in a neighborhood of θ_0, so (10.1) is not contradicted.

5) Even if $\mathscr{F}$ is not a curved exponential family we can use (10.1) to get an improved approximation to $\text{Var}_{\theta_0}\tilde{\theta}$, compared with the Cramér–Rao lower bound $1/ni_{\theta_0}$. The Cauchy translation family discussed at (7.4) has $i_{\theta_0} = \frac{1}{2}$, $\gamma_{\theta_0}^2 = \frac{5}{2}$. The MLE $\hat{\theta}$ is unbiased in this case, so $\Gamma_{\theta_0}^2 = 0$ and (10.1) is of the form $\text{Var}_{\theta_0}\hat{\theta} = 1/ni_0 + \gamma_{\theta_0}^2/n^2 i_{\theta_0} + O(1/n^3)$. Numerical comparisons of this formula with the

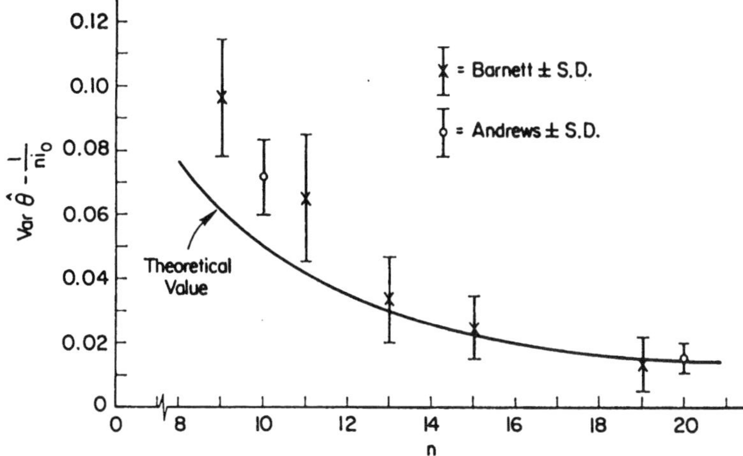

FIG. 6. Variance of MLE minus Cramér-Rao lower bound, for estimating the Cauchy translation parameter. Theoretical value from (10.1) compared with Monte Carlo results.

Monte Carlo studies of Barnett (1966) and also of Andrews et al. (1972) are shown in Figure 6. The theoretical values are obviously too small for $n \leqq 11$, but seem to be more accurate than the Monte Carlo results for $n \geqq 13$. For $n = 40$ Andrews et al. estimate $\text{Var}_{\theta_0} \hat{\theta} - 1/ni_{\theta_0} = .0025 \pm .0017$ while (10.1) gives .0031.

6) For estimating a translation parameter Pitman's estimator is known to have smaller variance than the MLE. However, (10.1) suggests that this effect must be of magnitude at most $O(1/n^3)$.

7) Nothing in (10.1), except the application to general curved exponential families, is new. Rao (1963) states the result for curved multinomial families, and notes that for the MLE it was previously derived by Haldane and Smith (1956). The identification of the bracketed terms with curvatures is new, as well as the line of proof which leads to a rigorous verification.

8) The similarity of (9.1) and (10.1) can be viewed as a vindication of the belief that Fisher information is an accurate measure of the information contained in a given statistic. This conclusion is premature; the squared error estimation problem is very closely related to the information calculation, a fact which would be more obvious if we had presented a geometric argument below, as in Section 9, instead of using analytic methods. It is more reasonable to say that the curvature γ_θ is the leading term defining the nonlinearity of a family $\mathscr{F}$, and must play a central role in all calculations like (9.1) and (10.1). On the other hand in the absence of evidence to the contrary it seems difficult to dispute Fisher and Rao's assertion that the MLE provides the most informative one-dimensional summary statistic even when there is no one-dimensional sufficient statistic.

Our derivation of (10.1) will be done with the curved exponential family $\mathscr{F}$

in standard form, and assuming $\theta_0 = 0$. Neither of these assumptions affects the generality of the result. (The transformation to standard form maps any estimator into an estimator having the same variance, and leaves the quantities i_{θ_0}, γ_{θ_0}, and Γ_{θ_0} unchanged.) We assume that the estimator $\tilde{\theta}(\bar{x})$ has continuous third partial derivatives with respect to the components of $\bar{x}$, so that around $\bar{x} = 0$ it has the Taylor's series expansion

$$\tilde{\theta}(\bar{x}) = a_0 + \mathbf{a}'\bar{x} + (\bar{x}'\mathbf{A}\bar{x})/2 + O(\bar{x}^3)\,, \tag{10.3}$$

where a_0 is a scalar, $\mathbf{a}$ is a $r \times 1$ vector, and $\mathbf{A}$ an $r \times r$ matrix, r being the dimension of the full exponential family containing $\mathscr{F}$.

Here $O(\bar{x}^3)$ indicates a term that near the origin is bounded in absolute value by some polynomial in the components of $\bar{x}$ containing only terms of order 3.

Differentiating (10.3) with respect to the components of $\bar{x}$ gives the gradient vector

$$\nabla\tilde{\theta}(\bar{x}) = \mathbf{a} + \mathbf{A}\bar{x} + O(\bar{x}^2)\,. \tag{10.4}$$

In order for $\tilde{\theta}$ to be consistent and efficient, (10.3) must have the special form shown in the lemma:

LEMMA. *A consistent, efficient estimator $\tilde{\theta}(\bar{x})$, having continuous third partial derivatives near $\bar{x} = \mathbf{0}$, has the Taylor series expansion*

$$\tilde{\theta}(\bar{x}) = \frac{\bar{x}_1}{i_0^{\frac{1}{2}}} - \frac{\mu_{11}}{i_0^2}\frac{\bar{x}_1^2}{2} + \frac{\gamma_0}{i_0^{\frac{1}{2}}}\bar{x}_1\bar{x}_2 + \frac{\bar{x}_{(1)}'\mathbf{A}_{(1)}\bar{x}_{(1)}}{2} + O(\bar{x}^3) \tag{10.5}$$

assuming $\mathscr{F}$ is in standard form at $\theta = 0$. Here $\bar{x}_i$ indicates the ith *component of $\bar{x}$, $\bar{x}_{(1)} \equiv (\bar{x}_2, \bar{x}_3, \cdots, \bar{x}_r)$, and $\mathbf{A}_{(1)}$ is the matrix $\mathbf{A}$ with its first row and column removed. For the* MLE *$\hat{\theta}(\bar{x})$, $\mathbf{A}_{(1)} = \mathbf{0}$. As in* (1.1), *$\mu_{11} = E_0\dot{f}_0\ddot{f}_0/f_0^2$.*

The proof of the lemma is based on two simple facts: in order for a continuous estimator $\tilde{\theta}(\bar{x})$ to be consistent it must have "Fisher consistency",

$$\tilde{\theta}(\lambda_\theta) = \theta\,, \tag{10.6}$$

since $\bar{x} \to_p \lambda_\theta$ under repeated independent sampling from f_θ. Moreover, letting

$$\nabla_\theta \equiv \nabla\tilde{\theta}(\bar{x})|_{\bar{x}=\lambda_\theta}\,,$$

$$\lim_{n\to\infty}\frac{1}{\mathbf{i}_\theta \operatorname{Var}_\theta\tilde{\theta}} = \frac{(\dot{\eta}_\theta'\mathbf{\Sigma}_\theta\nabla_\theta)^2}{(\dot{\eta}_\theta'\mathbf{\Sigma}_\theta\dot{\eta}_\theta)(\nabla_\theta'\mathbf{\Sigma}_\theta\nabla_\theta)} \tag{10.7}$$

so $\tilde{\theta}$ will be first order efficient at θ, ($\lim_{n\to\infty}\mathbf{i}_\theta \operatorname{Var}_\theta\tilde{\theta} = 1$), if and only if

$$\nabla_\theta \equiv \nabla\tilde{\theta}(\bar{x})|_{\bar{x}=\lambda_\theta} = c_\theta\dot{\eta}_\theta \tag{10.8}$$

for some scalar c_θ. Taken together (10.6) and (10.8) say that the level surface $M_\theta \equiv \{\bar{x}: \tilde{\theta}(\bar{x}) = \theta\}$ of an efficient consistent estimator $\tilde{\theta}$ must cross $\{\lambda_\theta, \theta \in \Theta\}$ at λ_θ, and at that point must be parallel to the level surface (9.4) of the MLE, as shown in Figure 4. (10.7) merely says that the linear term in the expansion of $\tilde{\theta}(\bar{x})$ about λ_θ, $\tilde{\theta} = \theta + \nabla_\theta'(\bar{x} - \lambda_\theta) + O((\bar{x} - \lambda_\theta)^2)$, must be proportional to the score statistic $\dot{\mathbf{l}}_\theta = n\dot{\eta}_\theta'(\bar{x} - \lambda_\theta)$ in order to get first order efficiency. A proof follows from a greatly simplified version of the argument below, but the result is well known and will not be derived here.

The proof of (10.5) is obtained by seeing what form of (10.3) is necessary in order that (10.6) and (10.8) hold for λ_θ near 0. We will need the Taylor series expansions

$$\dot{\eta}_\theta = i_0^{\frac{1}{2}}\mathbf{e}_1 + \left[\frac{\nu_{11}}{i_0^{\frac{1}{2}}}\mathbf{e}_1 + i_0\gamma_0\mathbf{e}_2\right]\theta + o(\theta), \tag{10.9}$$

$$\lambda_\theta = i_0^{\frac{1}{2}}\mathbf{e}_1\theta + O(\theta^2)$$

and a more accurate expansion for the first component of λ_θ,

$$\mathbf{e}_1'\lambda_\theta = i_0^{\frac{1}{2}}\theta + \frac{\mu_{11}}{i_0^{\frac{1}{2}}}\frac{\theta^2}{2} + o(\theta^2). \tag{10.10}$$

(10.9) follows from the standard form relationships (4.1)—(4.2). To prove (10.10) notice that $\mathbf{e}_1'\lambda_\theta = E_\theta x_1 = (1/i_0^{\frac{1}{2}})E_\theta \dot{l}_0(x)$ (see (3.18)). Formally

$$E_\theta \dot{l}_0 = \int_{\mathscr{X}} \frac{\dot{f}_0}{f_0}(x)\left[f_0(x) + \theta\dot{f}_0(x) + \frac{\theta^2}{2}\ddot{f}_0(x) + o(\theta^2)\right]dm(x) \tag{10.11}$$

$$= i_0\theta + \frac{\mu_{11}\theta^2}{2} + o(\theta^2),$$

a result which is easy to verify rigorously in an exponential family.

(10.4), (10.9), and (10.8) combine to give (writing $c_\theta = c_0 + \dot{c}_0\theta + o(\theta)$)

$$\mathbf{a} + \mathbf{A}(i_0^{\frac{1}{2}}\mathbf{e}_1\theta) + O(\theta^2) \tag{10.12}$$

$$= c_0 i_0^{\frac{1}{2}}\mathbf{e}_1 + \left[\dot{c}_0 i_0^{\frac{1}{2}}\mathbf{e}_1 + \frac{c_0\nu_{11}(0)}{i_0^{\frac{1}{2}}}\mathbf{e}_1 + c_0 i_0\gamma_0\mathbf{e}_2\right]\theta + o(\theta)$$

implying

$$\mathbf{a} = c_0 i_0^{\frac{1}{2}}\mathbf{e}_1 \tag{10.13}$$

and

$$i_0^{\frac{1}{2}}\mathbf{A}\mathbf{e}_1 = \left(\dot{c}_0 i_0^{\frac{1}{2}} + \frac{c_0\nu_{11}(0)}{i_0^{\frac{1}{2}}}\right)\mathbf{e}_1 + c_0 i_0\gamma_0\mathbf{e}_2. \tag{10.14}$$

Notice that (10.14) shows that

$$A_{31} = A_{41} = \cdots = A_{r1} = 0. \tag{10.15}$$

(10.9), (10.10), (10.13), (10.6), and (10.3) combine to give

$$\theta = a_0 + c_0 i_0^{\frac{1}{2}}\left[i_0^{\frac{1}{2}}\theta + \frac{\mu_{11}}{i_0^{\frac{1}{2}}}\frac{\theta^2}{2}\right] + \frac{i_0 A_{11}}{2}\theta^2 + o(\theta^2), \tag{10.16}$$

implying

$$a_0 = 0, \tag{10.17}$$

$c_0 = 1/i_0$, and $c_0\mu_{11} + i_0 A_{11} = 0$. Therefore

$$\mathbf{a} = \frac{1}{i_0^{\frac{1}{2}}}\mathbf{e}_1, \qquad A_{11} = -\frac{\mu_{11}}{i_0^2}, \qquad A_{21} = \frac{\gamma_0}{i_0^{\frac{1}{2}}}, \tag{10.18}$$

the first of these following from (10.13), the last from (10.14). Taken together, (10.15), (10.17) and (10.18) are equivalent to (10.5). Finally, for the MLE, $\hat{\theta}((0, \bar{x}_{(1)})') = 0$, implying $\mathbf{A}_{(1)} = \mathbf{0}$. This completes the proof of (10.5).

Two more simple results give (10.1) from (10.5). First of all, the Cramér–Rao lower bound for the variance of a possibly biased estimator $T(\bar{x})$ can be rewritten as an equality in the following useful form:

$$E_0 T^2 = \frac{1}{ni_0} + E_0\left(T - \frac{\bar{x}_1}{i_0^{\frac{1}{2}}}\right)^2 + 2\frac{\dot{b}_0}{ni_0}. \tag{10.19}$$

((10.19) follows from $\mathrm{Cov}_0(T, \dot{\mathbf{l}}_0) = 1 + \dot{b}_0$.) Notice that $\dot{\mathbf{l}}_0/\mathbf{i}_0 = \bar{x}_1/i_0^{\frac{1}{2}}$ by (6.3) so this statistic is just the best locally unbiased estimator of θ, $\mathbf{U}_0$, introduced at (5.5). For an unbiased estimator, (10.19) says that $\mathrm{Var}_0 T$ exceeds the Cramér–Rao lower bound by the expected squared error of T in predicting $\mathbf{U}_0$. In a curved exponential family the regularity conditions necessary for (10.19) are satisfied if $E_\theta T^2 < \infty$ for θ in a neighborhood of 0. The second fact needed is that if z is standard multivariate normal, $z \sim \mathscr{N}_r(\mathbf{0}, \mathbf{I})$, and $\mathbf{A}$ is an $r \times r$ symmetric matrix, then $E(z'\mathbf{A}z)/2 = \mathrm{tr}\,\mathbf{A}/2$ and

$$\mathrm{Var}\,\frac{z'\mathbf{A}z}{2} = \tfrac{1}{2}\,\mathrm{tr}\,\mathbf{A}^2. \tag{10.20}$$

As $n \to \infty$, $z_n \equiv n^{\frac{1}{2}}\bar{x} \to \mathscr{N}_r(\mathbf{0}, \mathbf{I})$, and because f_0 is inside an exponential family the moments of z_n converge to the moments of $z \sim \mathscr{N}_r(\mathbf{0}, \mathbf{I})$. Ignoring the $O(\bar{x}^3)$ term, an omission justified (under an additional restriction on $\tilde{\theta}$) in Remark 12 below, (10.3) and (10.5) give

$$E_0\tilde{\theta} = E_0\frac{\bar{x}'\mathbf{A}\bar{x}}{2} = \frac{1}{n}\frac{\mathrm{tr}\,\mathbf{A}}{2} = \frac{1}{n}\left(-\frac{\mu_{11}}{2i_0^2} + \frac{\mathrm{tr}\,\mathbf{A}_{(1)}}{2}\right). \tag{10.21}$$

Moreover (10.5) combines with (10.19) and (10.20) to give

$$E_0\tilde{\theta}^2 = \frac{1}{ni_0} + \frac{1}{n^2}\left(\frac{\gamma_0^2}{i_0} + \frac{\mu_{11}^2}{2i_0^4} + \frac{\mathrm{tr}\,\mathbf{A}_{(1)}^2}{2} + \frac{\mathrm{tr}^2\,\mathbf{A}}{4}\right) + \frac{2\dot{b}_0}{ni_0} + o\left(\frac{1}{n^2}\right). \tag{10.22}$$

Therefore,

$$\mathrm{Var}_0\,\tilde{\theta} = \frac{1}{ni_0} + \frac{1}{n^2}\left(\frac{\gamma_0^2}{i_0} + \frac{\mu_{11}^2}{2i_0^4} + \frac{\mathrm{tr}\,\mathbf{A}_{(1)}^2}{2}\right) + \frac{2\dot{b}_0}{ni_0} + o\left(\frac{1}{n^2}\right). \tag{10.23}$$

Finally, (10.11) gives $\phi(\theta) = \theta + (\mu_{11}/2i_0)\theta^2 + o(\theta^2)$, where $\phi(\theta) = E_\theta \dot{\mathbf{l}}_0/\mathbf{i}_0 = E_\theta \dot{l}_0(x)/i_0$, and then (2.1) gives the curvature squared of $(\theta, \phi(\theta))$ equal to $\mu_{11}^2/8i_0^2$ at $\theta = 0$. This completes the proof of (10.1). We see that the term $\Delta_{\theta_0}^{\tilde{\theta}}$ is

$$\Delta_0^{\tilde{\theta}} = i_0\,\mathrm{tr}\,\mathbf{A}_{(1)}^2/2 \tag{10.24}$$

and so equals 0 for the MLE.

Several more remarks can now be made about (10.1).

9) The bias of the MLE up to $O(1/n)$ is, by (10.21), equal to $-\mu_{11}/(2i_0 n)$. If $\hat{\theta}$ is unbiased to $O(1/n)$, as it is for example in any translation parameter estimation problem involving a symmetric density, then we must have $\mu_{11} = 0$. By

(10.23) we then have $\mathrm{Var}_0 \hat{\theta} = 1/ni_0 + \gamma_0^2/n^2 i_0 + o(1/n^2)$. The naming curvature term disappears from (10.1) in this case, so θ must be equivalent to the best name, ϕ, at every point in $\mathscr{F}$.

10) The expression (10.24) for the excess variance of $\tilde{\theta}$ over the MLE also occurs in the theory of Section 9,

$$\lim_{n\to\infty} \mathrm{i}_0 - \mathrm{i}_0^{\tilde{\theta}} = i_0 \gamma_0^2 + \Delta_0^{\tilde{\theta}}\,, \tag{10.25}$$

see Rao (1963).

11) Let $\mathrm{A}(\theta_0)$ be the matrix A in the Taylor expansion (10.3) when we have put $\mathscr{F}$ into standard form at $\theta = \theta_0$, and define $B_\theta^{\tilde{\theta}} \equiv \mathrm{tr}\, \mathrm{A}(\theta)/2$. Then up to $O(1/n)$, $B_{\theta_0}^{\tilde{\theta}}/n$ is the bias of $\tilde{\theta}$ when $\theta = \theta_0$. It is easy to show, by calculations similar to those in Remark 12 below, that $\tilde{\theta}_n \equiv \tilde{\theta} - B_{\tilde{\theta}(\bar{x})}^{\tilde{\theta}}/n$ has bias of order $O(1/n^2)$ and variance as given in (10.23) but with the term $2b_0/ni_0$ removed. See Rao (1963). For the MLE $\hat{\theta}$, $B_\theta^{\hat{\theta}} = -(\mu_{11}(\theta)/2i_\theta^2)$. The estimator $\hat{\theta} - B_{\hat{\theta}(\bar{x})}^{\hat{\theta}}/n + B_{\hat{\theta}(\bar{x})}^{\tilde{\theta}}/n$ has variance as given in (10.1) but with the term $\Delta_{\theta_0}^{\tilde{\theta}}$ removed. The point is that by modifying the MLE we can obtain an estimator with the same bias structure and smaller variance than any other consistent, efficient estimator $\tilde{\theta}$.

12) We have ignored the $O(\bar{x}^3)$ term in (10.3) in the derivation of (10.23) and (10.1). To justify this requires the following result: let $\mathscr{C}_n$ be the cube $\{z : |z_i| \leq n^\alpha, i = 1, 2, \cdots, r\}$, $0 < \alpha < \frac{1}{6}$, and $I_n(z)$ the indicator function of $\mathscr{C}_n$. Define $z_n \equiv n^{\frac{1}{2}}\bar{x}$ (so $z_n \to_{\mathscr{L}} \mathscr{N}_r(\mathbf{0}, \mathbf{I})$) and let $p(z_n)$ be a polynomial of degree l in the coordinates of z_n. Then

$$E_0 p(z_n)[1 - I_n(z_n)] = O(n^{l\alpha} \exp\{-\tfrac{1}{2}n^{2\alpha}\}) \tag{10.26}$$

as discussed in the Appendix.

Now write (10.5) as $\tilde{\theta} - \bar{x}_1/i_0^{\frac{1}{2}} = Q + R$ where Q is the quadratic term $\bar{x}'\mathrm{A}\bar{x}$, A having the special form indicated in the lemma, and R is the remainder term $O(\bar{x}^3)$. Also define $S(\bar{x}) \equiv Q(\bar{x})I_n(n^{\frac{1}{2}}\bar{x})$, $T(\bar{x}) \equiv Q(\bar{x})[1 - I_n(n^{\frac{1}{2}}\bar{x})]$, and $V = T + R$ (so $Q = S + T$, $\tilde{\theta} - \bar{x}_1/i_0^{\frac{1}{2}} = S + V$). Notice that

$$|V| = |O(\bar{x}^3)| < Kn^{-3(\frac{1}{2}-\alpha)} \quad \text{for} \quad n^{\frac{1}{2}}\bar{x} \in \mathscr{C}_n \tag{10.27}$$

for some positive constant K. (We use below the same symbol K to represent any bounding constant.) To the assumptions of the lemma we now add *that* $|\tilde{\theta} - \bar{x}_1/i_0^{\frac{1}{2}}|$ *is uniformly bounded*, giving

$$|V| < K\,, \qquad n^{\frac{1}{2}}\bar{x} \notin \mathscr{C}_n\,. \tag{10.28}$$

(With only slightly greater effort below, the boundedness condition can be relaxed to $|\tilde{\theta}| \leq K(n^{\frac{1}{2}}||\bar{x}||)^k$ for $n^{\frac{1}{2}}\bar{x} \notin \mathscr{C}_n$ for some positive constants K, k.) By (10.26) and (10.27),

$$E_0|V|^l = O(n^{-3l(\frac{1}{2}-\alpha)}) \tag{10.29}$$

for any $l \geq 0$, while

$$E_0|T|^l = O(n^{2\alpha l}e^{-n^{2\alpha}/2})\,. \tag{10.30}$$

Formulas (10.21) and (10.23) were derived assuming $\tilde{\theta} - \bar{x}_1/i_0^{\frac{1}{2}} = Q$. But

$$|E_0Q - E_0S| \leqq E_0|T| = O(n^{\alpha}e^{-n^{2\alpha}/2})$$

and

$$|E_0(\tilde{\theta} - \bar{x}_1/i_0^{\frac{1}{2}}) - E_0S| \leqq E_0|V| = O(n^{-3(\frac{1}{2}-\alpha)}) .$$

Since $\alpha < \frac{1}{6}$ this shows that $E_0\tilde{\theta} = E_0Q + o(1/n)$, so (10.21) is valid. Likewise

$$\begin{aligned}|E_0(\tilde{\theta} - \bar{x}_1/i_0^{\frac{1}{2}})^2 - E_0Q^2| &= |E_0(S + V)^2 - E_0(S + T)^2| \\ &= |E_0[2SV + V^2 - T^2]|\end{aligned}$$

(since $ST \equiv 0$), which is $\leqq 2E_0|SV| + E_0|V|^2 + E_0|T|^2$. The last two terms are $o(n^{-2})$ by (10.29) and (10.30). Notice that $SV = O(\bar{x}^5)$ and $SV = 0$ for $n^{\frac{1}{2}}\bar{x} \notin \mathscr{C}_n$, so

$$|SV| < Kn^{-5(\frac{1}{2}-\alpha)} . \tag{10.31}$$

Taking $\alpha < \frac{1}{10}$ makes $E_0|SV| = o(n^{-2})$, completing the proof that (10.28) is valid. We remark that a more careful proof, assuming $\tilde{\theta}$ four times continuously differentiable, allows one to replace $o(1/n^2)$ bs $O(1/n^3)$ in (10.1).

Acknowledgment. Much of this work was done while I was visiting Imperial College, London, Department of Mathematics. I appreciate the assistance of Margaret Ansell in carrying out the more difficult numerical computations. The Associate Editor provided extensive help, especially with the Appendix.

APPENDIX

Complete proofs of the statements made in Sections 9 and 10 require large deviation results of the type discussed in Chernoff (1952) and the references therein. Suppose $x_1, x_2, \cdots, x_n, \cdots$ are independent, identically distributed real valued random variables such that $Ex_i = 0$, Var $x_i = 1$, and $\psi(s) \equiv Ee^{sx}$ exists for $|s| < s_0$, s_0 some positive constant. Then $\psi(s) = 1 + s^2/2 + O(s^3)$ for s near 0, so

$$\log \psi(s) = s^2/2 + O(s^3) . \tag{A1}$$

Define $I_{[y,\infty)}(z) = 1$ or 0 for $z \geqq y$ or $z < y$, respectively. Because $e^{ns(\bar{x}_n - y)} \geqq I_{[y,\infty)}(\bar{x}_n)$ for all values of $\bar{x}_n \equiv \sum_{i=1}^n x_i/n$ we have, for $|s| < s_0$,

$$P\{\bar{x}_n \geqq y\} \leqq Ee^{ns(\bar{x}_n - y)} = [\psi(s)e^{-sy}]^n . \tag{A2}$$

LEMMA. *For c_n a sequence of numbers going to infinity, $c_n \doteq o(n^{\frac{1}{6}})$, and l a nonnegative integer,*

$$E\{(n^{\frac{1}{2}}\bar{x}_n)^l I_{[c_n,\infty)}(n^{\frac{1}{2}}\bar{x}_n)\} \leqq c_n{}^l e^{-c_n{}^2/2 + o_n(1)} . \tag{A3}$$

PROOF. Let $\bar{F}_n(y) \equiv P\{\bar{x}_n \geqq y\}$, so $\bar{F}_n(y) \leqq [\psi(s)e^{-sy}]^n$ for $|s| < s_0$ by (A2). We have

$$E\{(n^{\frac{1}{2}}\bar{x}_n)^l I_{[c_n,\infty)}(n^{\frac{1}{2}}\bar{x}_n)\} = -n^{l/2} \int_{c_n/n^{\frac{1}{2}}}^{\infty} x^l \, d\bar{F}_n(x)$$

and integration by parts gives

$$-\int_{c_n/n^{\frac12}}^{\infty} x^l\, d\bar F_n(x) = \left(\frac{c_n}{n^{\frac12}}\right)^l \bar F_n\left(\frac{c_n}{n^{\frac12}}\right) + l\int_{c_n/n^{\frac12}}^{\infty} x^{l-1}\bar F_n(x)\, dx$$
$$\leq \left(\frac{c_n}{n^{\frac12}}\right)^l \left[\psi\left(\frac{c_n}{n^{\frac12}}\right) e^{-sc_n/n^{\frac12}}\right]^n + l\int_{c_n/n^{\frac12}}^{\infty} x^{l-1}[\psi(s)e^{-sx}]^n\, dx\,.$$

Taking $s = c_n/n^{\frac12}$ gives

$$E\{n^{\frac12}\bar x_n)^l I_{[c_n,\infty)}(n^{\frac12}\bar x_n)\} \leq \psi^n\left(\frac{c_n}{n^{\frac12}}\right) e^{-c_n^2}\left[c_n^{\ l} + \frac{l}{c_n} E\left(c_n + \frac{G}{c_n}\right)^{l-1}\right], \tag{A4}$$

where G has density e^{-g} for $g \geq 0$, 0 otherwise. Finally

$$\psi^n\left(\frac{c_n}{n^{\frac12}}\right) = e^{n\log\psi(c_n/n^{\frac12})} = e^{c_n^2/2 + O(c_n^3/n^{\frac12})} \tag{A5}$$

by (A1). Combining (A4) and (A5) gives (A3) with

$$o_n(1) = O([c_n/n^{\frac12}]^3) + \log\{1 + lc_n^{-2}E(1 + G/c_n^2)^{l-1}\}, \tag{A6}$$

where we now use $c_n = o(n^{\frac16})$, $c_n \to \infty$.

Now let $x_1, x_2, \cdots, x_n, \cdots$ be independent identically distributed random vectors, dimension k, $Ex_i = \mathbf{0}$, Cov $x_i = \mathbf{I}$, such that $\psi(t) \equiv Ee^{t'x_i}$ exists for $||t|| < t_0$, some positive constant. For any unit vector v define $x_i^{\,v} \equiv v'x_i$. Then (A3) holds with $\bar x_n$ replaced by $\bar x_n^{\,v}$. The term $o_n(1)$ is defined as in (A6), with the big O term being the one in the expression $\log\psi(t) = ||t||^2/2 + O(t^3)$. (Notice that $o_n(1)$ does not depend on v.) (10.26) now follows easily.

LEMMA. *If $|E_0 e^{it'x}|^p$ is integrable as a function of t for some $p \geq 1$ then $g_n(z)$, the density of $z \equiv n^{\frac12}\bar x_n$, exists and satisfies*

$$g_n(z) < \frac{2^{\frac32}}{(2\pi)^{k/2}}\, e^{-(||z||/4)\min\{c_n, ||z||\} + o_n(1)}, \tag{A7}$$

$c_n = o(n^{\frac16})$, $c_n \to \infty$.

PROOF. Consider the univariate case, with n even. Define

$$\begin{aligned} h(z) &\equiv \int_{-\infty}^{\infty} g_{n/2}(w)g_{n/2}(z-w)\, dw \\ &= \int_{-\infty}^{z/2} g_{n/2}(w)g_{n/2}(z-w)\, dw + \int_{z/2}^{\infty} g_{n/2}(w)g_{n/2}(z-w)\, dw\,. \end{aligned} \tag{A8}$$

Here $g_{n/2}(z)$, the density of $(n/2)^{\frac12}\bar x_{n/2}$, is known to exist and to converge uniformly to $(2\pi)^{-\frac12}\exp(-z^2/2)$, see page 244 of Gnedenko and Kolmogorov (1954). Then $M_n = \sup_z |g_n(z)| = (2\pi)^{-\frac12} + o_n(1)$, so for $0 \leq z \leq c_n$

$$\begin{aligned} h(z) &\leq M_{n/2}\{\int_{-\infty}^{z/2} g_{n/2}(z-w)\, dw + \int_{z/2}^{\infty} g_{n/2}(w)\, dw\} \\ &\leq 2M_{n/2}e^{-(z/8)\min\{2c_n, z\} + o_1(1)} \end{aligned}$$

where we have used the bound $P\{n^{\frac12}\bar x_n \geq z\} \leq \exp[-z/2\min\{c_n, z\} + o_n(1)]$ obtained by setting $y = z/n^{\frac12}$ and $s = \min\{z_n/n^{\frac12}, c_n/n^{\frac12}\}$ in (A2). But $g_n(z) = 2^{\frac12}h(2^{\frac12}z)$, giving (A7). The same proof with trivial modifications works for n odd. For the multivariate case the integrals in (A8) are over the regions $R_1 = \{w : z'w < \|z\|^2/2\}$ and $R_2 = \{w : z'w > \|z\|^2/2\}$.

Remark 1 of Section 9 follows because (A7) makes step (v) of the heuristic proof valid. All the other approximations involved in the proof are handled by power series expansions and the bounding arguments of Remark 12, Section 10.

REFERENCES

[1] Andrews, F., Bickel, P., Hampel, P., Huber, P., Rogers, W., and Tukey, J. (1972). *Robust Estimates of Location*. Princeton Univ. Press.

[2] Barnett, V. D. (1966). Evaluation of the maximum likelihood estimator when the likelihood equation has multiple roots. *Biometrika* **53** 151-165.

[3] Chernoff, H. (1952). A measure of asymptotic efficiency for tests of a hypothesis based on the sum of observations. *Ann. Math. Statist.* **23** 493-507.

[4] Davies, R. B. (1969). Beta optimal tests and an application to the summary evaluation of experiments. *J. Roy. Statist. Soc. Ser. B* **31** 524-538.

[5] Davies, R. B. (1971). Rank tests for Lehmann's alternative. *J. Amer. Statist. Assoc.* **66** 879-883.

[6] Efron, B. and Truax, D. (1968). Large deviations theory in exponential families. *Ann. Math. Statist.* **39** 1402-1424.

[7] Fisher, R. A. (1925). Theory of statistical estimation. *Proc. Cambridge Philos. Soc.* **122** 700-725.

[8] Gnedenko, B. V., and Kolmogorov, A. N. (1954). (Translated by K. Chung.) *Limit Distributions for Sums of Independent Random Variables*. Addison-Wesley, Cambridge, Mass.

[9] Haldane, J. B. S. and Smith, S. M. (1956). The sampling distribution of a maximum likelihood estimate. *Biometrika* **43** 96-103.

[10] Rao, C. R. (1961). Asymptotic efficiency and limiting information. (J. Neyman, Ed.). *Proc. Fourth Berkeley Symp. Math. Statist. Prob.* **1** 531-545. Univ. of California Press.

[11] Rao, C. R. (1962). Efficient estimates and optimum inference procedures in large samples. *J. Roy. Statist. Soc. Ser. B.* **24** 46-72.

[12] Rao, C. R. (1963). Criteria of estimation in large samples. *Sankhyā* **25** 189-206.

[13] Struik, D. J. (1950). *Differential Geometry*. Addison-Wesley, Reading, Mass.

Department of Statistics
Stanford University
Stanford, California 94305

DISCUSSION ON PROFESSOR EFRON'S PAPER

Professor Efron's paper was presented at the 1974 Annual Meeting of the Institute of Mathematical Statistics at Edmonton, Alberta. Professors D. R. Cox, A. P. Dawid, J. K. Ghosh, N. Keiding, L. M. Le Cam, D. V. Lindley, J. Pfanzagl, D. A. Pierce, C. R. Rao and J. Reeds were invited discussants. The Editor greatly appreciates the willing assistance of Professor Efron as well as the discussants in arranging this discussion paper. Professor Rao's remarks arrived after the author's reply to the discussion was received and are not referred to for that reason.

C. R. RAO
Indian Statistical Institute, New Delhi

I am delighted to see the paper by Bradley Efron and also the paper by J. K. Ghosh and K. Subrahmaniam (*Sankhyā A*, 1975 **36** 325–358) on the subject of second order efficiency. Having worked for some time on second order efficiency of estimators, I was aware of the importance of measures of how closely a given model can be approximated by an exponential family $\{f_\theta = C(\theta) \exp[K(\theta)T(X)]\}$. Measures of this sort are of course closely related to what Professor Efron calls the curvature of a statistical problem. What is quite new about Professor Efron's measure is its invariance under smooth $1-1$ transformations and the elegant geometric interpretation which makes the term so apt and illuminating and provides new tools and insights into the subject.

My endeavour in this area was motivated by two results in the literature on estimation which seemed to contradict Fisher's claims about MLE's. (maximum likelihood estimators). One is the concept of super efficiency, according to which MLE is not efficient in the sense defined by Fisher. Another is the concept of BANE (best asymptotically normal estimator), according to which ML is only one out of a very wide class of estimation procedures.

The first task was to redefine the concept of efficiency of an estimator since its asymptotic variance is a poor indicator of its performance in statistical inference. To do this it is necessary to see how well an optimum inference procedure based on a given estimator T_n alone compares with that based on all the observations. Following Fisher's ideas, I thought it is relevant, at least in large samples, to consider the score function $\dot{l}(\theta)$ (see Efron's paper for notations) as basic to all inference problems. Then the problem reduces to examining how closely $\dot{l}(\theta)$ and T_n are related. Under the additional condition that T_n is consistent for θ, T_n was defined to be *first order efficient* if

$$\text{(1)} \qquad \operatorname{plim}_{n\to\infty} |n^{-\frac{1}{2}}\dot{l}(\theta) - \alpha - \beta n^{\frac{1}{2}}(T_n - \theta)| \to 0\,.$$

There are a large number of estimators which are first order efficient. To distinguish among them, it is natural to examine the rapidity of convergence in (1), which led to the consideration of the random variable (rv)

$$\text{(2)} \qquad |\dot{l}(\theta) - n^{\frac{1}{2}}\alpha - n\beta(T_n - \theta)|$$

which is $n^{\frac{1}{2}}$ times the rv in (1). The asymptotic variance of (2) was defined as the *second order efficiency*. Instead of (2) we may as well consider the rv

$$\text{(3)} \qquad |\dot{l}(\theta) - n^{\frac{1}{2}}\alpha - n\beta(T_n - \theta) - \lambda n(T_n - \theta)^2|$$

and define its minimum asymptotic variance for a proper choice of λ as the second order efficiency. Fisher suggested the use of

$$\text{(4)} \qquad \lim_{n\to\infty} n(i - i_{T_n})$$

to distinguish between alternate estimators, but the computation of (4) is extremely difficult.

The definition arising out of (3) was criticised as not being directly related to an inference problem, although it attempts to examine how close T_n is to $\dot{l}(\theta)$. This led to another definition of second order efficiency based on the expansion (under some conditions) of the variance of T_n after correcting for bias

$$V(T_n) = \frac{1}{in} + \frac{\psi(\theta)}{n^2} + o\left(\frac{1}{n^2}\right). \tag{5}$$

The quantity $\psi(\theta)$ was considered as a measure of second order efficiency. A major component of $\psi(\theta)$ was the measure based on (3).

With this background, the work of Efron is valuable in many ways.

(i) The results due to Fisher and me were confined to multinomial distributions. Efron, and also Ghosh and Subrahmaniam extend the results to a wider class of distributions.

(ii) Efron relates second order efficiency to what he calls curvature of a statistical problem, which appears to be natural and throws further light on problems of inference (providing, for instance, an intimate connection between curvature and properties of test criteria).

(iii) Efron provides a decomposition of $\psi(\theta)$ in (5), which is extremely interesting.

(iv) Efron suggests the use of a most powerful test at a suitably chosen alternative in preference to a locally most powerful test, which seems to be an attractive idea worth pursuing.

No doubt Efron's work has led to considerable clarification of second order efficiency and its relevance in problems of inference. However, there are many problems which require deeper investigation.

(i) Efron shows by an example that measures of second order efficiency based on (3) and (4) can be different. In fact, as he observes, it may be shown (from definition) that the measure based on (4) is smaller than that on (3). But the question remains: under what conditions are the two measures the same, and is the MLE efficient under the measure (4)?

(ii) I have considered Fisher's score function $\dot{l}(\theta)$ as a basic in problems of inference. Perhaps, following Barnard and Sprott, one should consider $1(\theta)$ itself. How should efficiency of T_n be defined in such a case?

(iii) How can the result based on quadratic loss function as in (5) be extended to more general loss functions?

DON A. PIERCE

Oregon State University

I think that I am not alone in having had great difficulty with the reasoning of Fisher's 1925 paper. Professor Efron's elegant contribution to clarifying these ideas is very helpful.

The part of Fisher's paper which has intrigued and puzzled me most is the final section in which he suggests the use of $\ddot{\mathbf{l}}_{\hat\theta}(\mathbf{x})$, in Efron's notation, as an ancillary statistic. I would like to indicate here how the geometry of this paper helps clarify this, although there are many details yet unclear to me.

It is characteristic of "curvature" that $-\ddot{\mathbf{l}}_{\hat\theta}(\mathbf{x}) \neq \mathbf{i}_{\hat\theta}$. In fact, one can always parameterize so that $\mathrm{Cov}_{\theta_0}(\dot{l}_{\theta_0}, \ddot{l}_{\theta_0}) = 0$, and then $\gamma_{\theta_0}^2 = \mathrm{Var}_{\theta_0}(\ddot{l}_{\theta_0})/i_{\theta_0}^2$. Fisher seems to suggest using $-\ddot{\mathbf{l}}_{\hat\theta}(\mathbf{x})$, rather than $\mathbf{i}_{\hat\theta}$, as a post-data measure of precision of $\hat\theta$. This is also suggested by standard asymptotic Bayesian arguments, but the sampling theory justification has never been clear to me. Such use of $\ddot{\mathbf{l}}_{\hat\theta}$ would be significant relative to the order of n^{-2} of approximation to Var $(\hat\theta)$ considered in this paper, for $-\ddot{\mathbf{l}}_{\hat\theta} = \mathbf{i}_{\hat\theta} + O_p(n^{\frac{1}{2}})$ and thus $-1/\ddot{\mathbf{l}}_{\hat\theta} = 1/\mathbf{i}_{\hat\theta} + O_p(n^{-\frac{3}{2}})$.

The geometrical structure exposed in this paper is indeed very helpful in understanding the role of $\ddot{\mathbf{l}}_{\hat\theta}$ as an ancillary statistic. For a curved exponential family of dimension k think of the projection from the sample point $\mathbf{x} \in E^k$ to the MLE $\lambda_{\hat\theta}$, where $\lambda_\theta = E(\mathbf{x})$ as an orthogonal projection (relative to $\sum_{\hat\theta}^{-1}$) first to $\hat\lambda$ in the local osculating plane of the curve λ_θ and then a projection from $\hat\lambda$ to $\lambda_{\hat\theta}$. The argument below suggests that $(-\ddot{\mathbf{l}}_{\hat\theta}(\mathbf{x}) - \mathbf{i}_{\hat\theta})/\mathbf{i}_{\hat\theta}$ is a useful measure of the signed distance from $\hat\lambda$ to the curve λ_θ, positive when $\hat\lambda$ is on the outside of the curve. This is useful ancillary information because the projection from $\hat\lambda$ to $\lambda_{\hat\theta}$ is a contraction (resp. expansion) mapping when $\hat\lambda$ is on the outside (resp. inside) of the curve λ_θ. The extent of this contraction is a function of the distance from $\hat\lambda$ to the curve λ_θ, as measured by the above statistic. Thus the conditional precision of $\lambda_{\hat\theta}$ given $\ddot{\mathbf{l}}_{\hat\theta}(\mathbf{x})$ is either greater or less than the unconditional precision. Furthermore, it appears plausible that the component of $\hat\lambda$ orthogonal to the curve λ_θ at $\lambda_{\hat\theta}$ is itself uninformative regarding the value of λ_θ.

More precisely, consider the situation of Figure 4 with the additional assumption that θ is a choice of parameter such that $\mathrm{Cov}_0(\dot{l}_0, \ddot{l}_0) = 0$. The point (x_1, x_2) corresponds to the λ of the above discussion. It follows directly from (6.3) and the relations given in the second paragraph after (9.2) that

$$\bar{x}_1 = \dot{\mathbf{l}}_0/n(i_0)^{\frac{1}{2}}\,, \qquad \bar{x}_2 = -[-\ddot{\mathbf{l}}_0 - ni_0]/ni_0\gamma_0\,.$$

Near the origin the curve λ_θ is approximately a segment of a circle with center at e_2/γ_θ, and the arc distance of $\lambda_{\hat\theta}$ from the origin is to first order $i_0^{\frac{1}{2}}\hat\theta$. Proportionality of arc lengths to radii gives

$$\begin{aligned} i_0^{\frac{1}{2}}\hat\theta/\bar{x}_1 &\doteq (1/\gamma_0)/(1/\gamma_0 - \bar{x}_2) \\ &= (1 - \gamma_0\bar{x}_2)^{-1}\,, \end{aligned}$$

so

$$\begin{aligned} \hat\theta &\doteq (\bar{x}_1/i_0^{\frac{1}{2}})(1 - \gamma_0\bar{x}_2)^{-1} \\ &= (\bar{x}_1/i_0^{\frac{1}{2}})[1 + (-\ddot{\mathbf{l}}_0 - ni_0)/ni_0]^{-1} \qquad (1) \\ &= (\bar{x}_1/i_0^{\frac{1}{2}})[ni_0/(-\ddot{\mathbf{l}}_0)]\,. \end{aligned}$$

Equation (1) can be seen to agree with the rigorously established (10.5) of the paper, where $\mu_{11} = 0$ since $\nu_{11} = 0$.

Thus we have

$$\text{(2)} \qquad \operatorname{Var}(\hat\theta \mid \ddot{\mathrm{l}}_0) \doteq (1/ni_0)[ni_0/(-\ddot{\mathrm{l}}_0)]^2 = [ni_0/(-\ddot{\mathrm{l}}_0)][1/(-\ddot{\mathrm{l}}_0)] .$$

Since $-\ddot{\mathrm{l}}_0 = ni_0 + O_p(n^{\frac{1}{2}})$ this expression can be *either greater or less* than $1/ni_0$ by an amount $O_p(n^{-\frac{3}{2}})$.

I do not know the effect of conditioning on $\ddot{\mathrm{l}}_{\hat\theta}$ rather than $\ddot{\mathrm{l}}_0$ nor can I see yet whether $1/(-\ddot{\mathrm{l}}_{\hat\theta})$ as suggested by Fisher is a good approximation to $\operatorname{Var}(\hat\theta \mid \ddot{\mathrm{l}}_{\hat\theta})$. Note that the expression in (2) differs by $O_p(n^{-\frac{3}{2}})$ from $1/(-\ddot{\mathrm{l}}_0)$. I also do not know the effect of relaxing the assumption that one has parameterized so that $\operatorname{Cov}_0(\dot l, \ddot l_0) = 0$.

It appears, then, that the curvature γ_θ is essentially the standard deviation of an approximately ancillary statistic. This interpretation might have a number of advantages over that furnished by relations such as (1.1) and (10.1). Loosely put, the degree of curvature relates to the amount of information in the sample which is not captured by the MLE; information in a sense regarding not θ but rather the precision of $\hat\theta$. Moreover, this information can be largely recovered through appropriate use of $\ddot{\mathrm{l}}_{\hat\theta}$.

D. R. Cox

Imperial College, London

Dr. Efron's impressive paper throws much light on a longstanding problem. I will confine my comments to one aspect that he has not treated. For an approach to statistical inference in which evidence in unique sets of data is interpreted via frequencies in hypothetical repetitions, appropriate conditioning is important, at least theoretically, in making the hypothetical repetitions relevant to the data under study. Thus for the translation family, Example 4, Fisher (1934) provided a simple definitive solution to inference about θ by conditioning on the ancillary statistic, the set of differences among order statistics. This leads to the use of normalized likelihood as giving confidence limits. Curvature here measures the variation among the different kinds of likelihood functions that can arise. It would be useful to make this more specific and to draw any implications about the comparison of conditional and unconditional inference.

More importantly, what are the implications of conditional inference for some of the other problems, for instance Example 1? Here, if $x = (x_1, x_2)$, $x_2 - \frac{1}{2}\gamma_0(x_1^2 - 1)$ is approximately ancillary in some sense, at least for small $\gamma_0\theta$. Existence of an approximate ancillary must be connected with the approximate constancy of γ_0 as a function of θ; it would be good to have the connexions explored.

REFERENCES

Fisher, R. A. (1934). Two new properties of mathematical likelihood. *Proc. Roy. Soc. Ser. A.* **144** 285-307.

D. V. LINDLEY

The University of Iowa

My first comment is to repeat the point made in discussing C. R. Rao's (1962) paper, namely that it is doubtful whether any general measure of second-order efficiency is possible. The reason for suggesting this is that an admissible estimate is typically, to order n^{-1}, equivalent to the maximum likelihood estimate, for a wide class of loss functions: but to order n^{-2} its asymptotic form depends on some features of the loss structure. Consequently the second-order "correction" to the maximum likelihood estimate typically depends on the loss structure, as does its efficiency. The point is discussed more fully in Lindley (1961).

Efron's thought-provoking paper does not introduce curvature solely for second-order efficiency properties; nevertheless the definition of curvature he proposes suffers from a defect in some statistical problems. The defect arises from the fact that it involves an integration over sample space and thereby violates the likelihood principle. Put it this way: suppose we have some data x and its associated likelihood function, $l_\theta(x)$, then, according to Efron, we have to consider what other data we might have had, but did not, before any inference can be made. These data are needed before the integrations, symbolized by E_θ in the paper, can be performed. That such data are needed is puzzling and any reasonable axiomatization of inference seems to deny their relevance. The author tacitly assumes that the other data are samples of the same size, but many practical problems do not naturally fit into this framework. Even the notation helps to reinforce this view. Likelihood is a function of θ for fixed x and yet Efron lowers the status of the variable to that of a subscript and the constant appears in the place customarily reserved for the argument. The notation $l(\theta \mid x)$ is surely to be preferred.

An example of the misuse of the integration is provided by the discussion of the t-translation family [Example 4 of Section 7: see also the remark after (8.3)]. If samples are taken from a t-distribution with low degrees of freedom, then it will be found that a substantial majority of them look very like samples from a normal distribution—the comparison being made through the t- and normal likelihoods. It is only rarely (how rarely depends on f and n) that a sample arises which is clearly nonnormal and its log-likelihood is markedly not quadratic. But because of the integration, or averaging, over all samples, these "peculiar" samples get put in with the "normal" ones and nonstandard estimates proposed. Looked at without prejudice, I think you will find this is a surprising thing to do. The argument can be extended to query whether it is reasonable to look for a point estimate in the "peculiar" cases: for example, when the likelihood is bimodal. I would go further and suggest that point estimation is not a good model for *any* inference procedure, though it does occasionally occur in a decision context. Estimation is solved by describing the likelihood function or the posterior distribution.

These criticisms have less force *before* the data, x, are to hand. If it is a question of experimental design, or choice of a survey sample, then naturally one has to consider what data *might* be obtained, and integration becomes natural and necessary. Hence curvature could have a place in these fields and it would be interesting to see whether, in some sense, linear designs were better than "curved" ones. However, the argument of my first paragraph would show that if a terminal (as distinct from design) decision problem is contemplated after the experimentation, then the choice of design would again involve a loss function, so that no general measure seems possible. Some experiments are not associated with terminal decisions and are genuinely inferential in character. In these one is collecting information about parameters and Shannon's measure is essentially the only one to use. I have tried to see whether some second-order expansion of it might lead to anything analogous to Efron's curvature, but without success.

REFERENCES

LINDLEY, D. V. (1961). The use of prior probability distributions in statistical inference and decisions. *Proc. Fourth Berkeley Symp. Math. Statist. Prob.* **1** 453–468.

LUCIEN LE CAM

University of California, Berkeley

Professor Bradley Efron is to be congratulated for a clear and informative discussion of the differential properties of families of measures. The paper is certainly a step in the right direction. However, as I shall try to explain below, much remains to be done.

The paper tends to give the impression that the curvature measures the loss of information sustained by using a one dimensional summary of the data. This is perhaps so if "information" is measured by Fisher's number. However, one can define other measures of loss of information more directly in terms of performance in testing or other decision problems. See for instance E. N. Torgersen (1970). These definitions are usable for arbitrary families, whether or not they are smoothly differentiable.

It can probably be shown that these other measures of loss of information are related to Fisher's numbers in certain special situations, but not in general. One could roughly say that Torgersen's formula for testing deficiencies relies on finite differences instead of relying on the first and second derivatives used to compute curvatures. Efron's curvature has the merit of being easily computable, but one should not take it for granted that computations with differences, which may be difficult, should not be attempted.

The part of the paper which relates to the presumed excellency of maximum likelihood estimates should be taken with a great deal of caution. It is easy to modify Bahadur's example (1958) to construct one parameter families of densities which are infinitely differentiable, satisfy all kinds of reasonable conditions locally but are such that, when the number of observations tends to infinity, the maximum likelihood estimate always converges to infinity, no matter what the true value of θ is.

It is also easy to find exponential families where, for reasonable numbers of observations, maximum likelihood estimates are difficult to compute and definitely worse (in the sense of expected square deviations) than some readily available alternatives. An example occurs in bioassay using the logit method (see Berkson (1951)). Another example with an interesting discussion is given by T. S. Ferguson (1958).

Finally, it seems that the entire asymptotic argument relies essentially on a replacement of the actual logarithm of likelihood ratio by a suitable approximation which is quadratic in θ.

If this is indeed the case, the technique of using a preliminary estimate, fitting a quadratic around the estimated value and then maximizing the quadratic should give the same asymptotic results. Preliminary considerations suggest that this technique may well work better than straight maximum likelihood estimation in the finite sample situation.

REFERENCES

[1] BAHADUR, R. R. (1958). Examples of inconsistency of maximum likelihood estimates. *Sankhyā* **20** 207-210.

[2] BERKSON, J. (1951). Relative precision of minimum chi-square and maximum likelihood estimates of regression coefficients. *Proc. Second Berkeley Symp. Math. Statist. Prob.* 471-479. Univ. of California Press.

[3] FERGUSON, T. S. (1958). A method of generating best asymptotically normal estimates with application to the estimation of bacterial densities. *Ann. Math. Statist.* **29** 1046-1062.

[4] TORGERSEN, E. N. (1970). Comparison of experiments when the parameter space is finite. *Z. Wahrscheinlichkeitstheorie und Verw. Gebiete* **16** 219-249.

J. K. GHOSH

Indian Statistical Institute, Calcutta

Thanks to my work on second order efficiency, I was aware of the significance of the quantity which Professor Efron calls the curvature of a statistical problem. What enhances the importance of it is the elegant geometric interpretation of it, which affords new techniques and deeper insight into the problem.

It is natural to expect that this quantity also plays an important role in asymptotic problems of testing hypotheses. By considering a number of examples of curved exponential families, Professor Efron has shown that this is indeed the case and unless curvature is small such commonly used methods as maximising the local power perform rather poorly for moderate sample sizes. Pfanzagl (1974) has arrived at the same conclusion. (Pfanzagl's $D =$ (curvature)^2/4.)

Probably even more interesting than this is the suggestion by both Pfanzagl and Efron to use a suitable most powerful test instead of a locally most powerful test when the curvature is appreciable. Following Davies, Professor Efron suggests the use of a most powerful test against an alternative θ_1 such that its power at θ_1 is about .8 and recommends the thumb rule of taking $\theta_1 = \theta_0 + 2/I_{\theta_0}^{\frac{1}{2}}$. These suggestions must be tried out in lots of problems involving nonexponential families to see if one does get reasonable tests this way even for moderate samples. (Pfanzagl (1974) provides some criteria for comparing two tests.) I report below some calculations for a curved nonexponential family, namely, the Cauchy with unknown location parameter. To make matters worse, I take sample size $N = 1$.

Suppose then that I have a random variable X with density $f_\theta(x) = 1/\pi \cdot 1/(1 + (x - \theta^2))$ and want to test $H_0(\theta = 0)$ vs. $H_1(\theta > 0)$. Let ϕ_0 be the most powerful test of Davies and ϕ_1 the test: reject H_0 iff $X > C$. The second test has its greatest power against $\theta = 2C$ and seems to me a reasonable one. For $\alpha = .05$, ϕ_0 is most powerful against $\theta = 5$ (approximately) and ϕ_1 is most powerful against $\theta = 13$ (approximately). The following table compares ϕ_0 and ϕ_1.

	$\theta = 5$	$\theta = 13$
ϕ_0	.8	.06
ϕ_1	.2	.95

If $\alpha = .2$, ϕ_0 and ϕ_1 are nearly the same and are most powerful against $\theta = 2(2)^{\frac{1}{2}}$ which is the alternative obtained by Efron's thumb rule. I refrain from drawing any conclusion.

It is not difficult to come up with analogues of curvature when one has more parameters than one. Extension of the results due to Rao and Fisher to multiparameter families is provided in Ghosh and Subramanyam (1974). But it is now necessary to study testing problems of composite hypotheses along the lines of investigation carried out by Efron and Pfanzagl for simple hypotheses.

How relevant is curvature for a Bayesian? Ghosh and Subramanyam (1974) have shown how one can construct a Bayesian proof of the second order efficiency of the MLE. What is lacking and would be useful to have is a study of relevance of curvature in Bayesian analysis. The difficulty here is that one cannot think of any simple and convincing reason why a Bayesian would prefer the linear exponential families to nonexponential ones. All is grist that comes to the mill of the lucky man who not only has a prior but knows what it is.

It is a little disappointing, though not really surprising in retrospect, that curvature has nothing to do with the geometrical curvature of the likelihood curves. Curvature is, however, useful in the problems that Sprott (1973) discusses. For it is easy to show that his two approaches of minimizing $F_E(\phi)$ or $F(\phi)$ (in his notations) coincide iff one has a linear exponential family. (This statement is true provided the MLE satisfies the likelihood equation with probability one for all θ.) For example (2.3) of Sprott (1973), the curvature is fairly large for x near .5 and so Sprott's transformation which minimizes $F_E(\phi)$ may not be efficient in normalising the likelihood for x near .5. Incidentally, I suspect that for small curvature one can reparametrize in such a way that the approach of a posterior to normality, guaranteed by the Bernstein–von Mises theorem, would be faster with the new parameter than with the original. (This may be an answer to the question of relevance of curvature for a Bayesian.)

It may be worth pointing out here that the results of Pfanzagl (1973) and those of Fisher and Rao (i.e. results like (10.1) of Efron) are not really comparable. In fact for all the efficient estimators considered by Efron or Ghosh and Subramanyam (1974), inequality (6.4) of Pfanzagl (1973, page 1005) reduces to an equality. This result, which is not very hard to show, will appear in Ghosh and Srinivasan (1975).

Finally, a question suggested by the beautiful counter example of Professor Efron. Is there any example such that among the Fisher consistent efficient estimators the MLE does not minimize the loss in Fisher's information for all values of θ? It seems reasonable to expect that such examples do exist.

REFERENCES

[1] Ghosh, J. K. and Subramanyam, K. (1974). Second order efficiency of maximum likelihood estimators. *Sankhya Ser. A*. (To appear).

[2] Ghosh, J. K. and Srinivasan, C. (1975). Asymptotic sufficiency and second order efficiency. Unpublished.

[3] Pfanzagl, J. (1973). Asymptotic expansions related to minimum contrast estimators. *Ann. Statist.* **1** 993–1026.

[4] Pfanzagl, J. (1974). Nonexistence of tests with deficiency zero. University of Cologne, preprint in Statistics #8.

[5] Sprott, D. A. (1973). Normal likelihoods and their relation to large sample theory of estimation. *Biometrika* **60** 457–465.

J. Pfanzagl

University of Cologne

In hypothesis testing, one-parameter exponential families are distinguished by the fact that for one-sided alternatives uniformly most powerful tests exist for arbitrary sample sizes. For other families, the test has to be chosen with particular alternatives in mind. It is intuitively clear that the dependence of the test on these particular alternatives will be weak if the family is close to an exponential one. Is it possible to measure "nonexponentiality" (for this and other purposes) by a single quantity? Mr. Efron's suggestion to use the "curvature" γ_θ for this purpose is based on a geometric analogy. Therefore, its usefulness for statistical theory is not obvious in advance. It is the purpose of this note to draw attention to some results of asymptotic theory where the function γ_θ has been in use already for some time. Whether curvature admits an easy statistical interpretation in nonasymptotic theory seems doubtful.

"Nonexponentiality" implies in particular that a LMP (locally most powerful) test will not be MP (most powerful) against the statistically reasonable alternatives. The author uses a particular example to support his claim (see end of Section 8) that $\gamma_{\theta_0}^2$ is a good predictor for the relative performance of the LMP test compared to the test which is MP against a specific alternative. In this connection he suggests that the difference in power can be neglected if $\gamma_{\theta_0}^2 \leqq \frac{1}{8}$. For the case of a sample of n i.i.d. variables this entails that the difference in power can be neglected if the sample size exceeds $8\gamma_{\theta_0}^2$ (see 8.3).

Since this rule is rather arbitrary, the reader should be aware of other results which make the role of γ_{θ_0} more clear. These results concern the case of n i.i.d. variables, the distribution of which is nonatomic and sufficiently regular (as a function of θ). To define for a given level α-test "deficiency at rejection level β" we determine first the alternative closest to the hypothesis which can be rejected with probability β by some level α-test. (The test for which this is achieved is called β-optimal.) In order to reach rejection probability β for this alternative with the given test, the sample size has to be increased. The additional number of observations needed for this purpose is the "deficiency at rejection level β."

For the LMP α-test the deficiency at rejection level β is asymptotically equal to

$$\frac{1}{4}\gamma_{\theta_0}^2(N_\beta - N_\alpha)^2 + o(n^0)\,, \tag{1}$$

where N_δ is the δ-quantile of the standard normal distribution. (See Chibisov (1973, Corollary 2) and Pfanzagl (1973, Section 8, formula 24) or Pfanzagl (1975, Proposition 1, formula 6.2).)

This result enables one to check whether the rule suggested by (8.3) is reasonable. For $\alpha = .01$ and $\beta = .99$ the deficiency is $5.4\gamma_{\theta_0}^2 + o(n^0)$. Mr. Efron suggests in (8.3) not to worry about curvature if $n \geqq 8\gamma_{\theta_0}^2$. To follow this suggestion and to use a LMP test instead of a β-optimal test could mean to waste more than half of the sample.

The following is another asymptotic result (for nonatomic families) illustrating the statistical relevance of curvature. If a sequence of tests is β_0-optimal, then its deficiency at rejection level β is at least

$$\frac{1}{4}\gamma_{\theta_0}^2(N_\beta - N_{\beta_0})^2 + o(n^0) \tag{2}$$

(see Pfanzagl 1975, Corollary 2, formula 6.5). Hence a sequence of tests having asymptotic deficiency zero for more than one alternative cannot exist unless the curvature is zero.

In another attempt to demonstrate the statistical relevance of "curvature," Mr. Efron refers to a result of Fisher (see (9.1)). Mr. Efron is careful enough not to follow Fisher's abuse of language using a suggestive word for a mathematical construct (such as "information" or "likelihood") without paying any attention to the question whether the interpretation thus suggested is meaningful from the operational point of view.

A statement like "Since a single observation contains an amount i_θ of information this [namely the use of a MLE instead of the whole sample] is equivalent to a reduction in effective sample size from n to $n - \gamma_\theta^2 \cdots$" (see beginning of Section 9) is misleading, at least, since for nonatomic families the level α-test based on the MLE has asymptotic deficiency zero at the rejection level $(1 - \alpha)$, and not asymptotic deficiency γ_θ^2, as the statement quoted above might suggest.

(See Chibisov 1973, Corollary 3 or Pfanzagl 1973, formula 23 for $t = -2N_\alpha L_{(0)(0)}^{-\frac{1}{2}}$ or Pfanzagl 1975, end of Section 6.) Probably the statement quoted above is meant as the interpretation Fisher himself would give to (9.1). Since this interpretation is unjustified, how can (9.1) convince the reader that "curvature" is statistically significant?

REFERENCES

[1] Chibisov, D. M. (1973). Asymptotic expansions for some asymptotically optimal tests. *Proc. Prague Symp. Asymptotic Statist.* **2** 37–68.

[2] Pfanzagl, J. (1973). Asymptotically optimum estimation and test procedures. *Proc. Prague Symp. Asymptotic Statist.* **1** 201–272.

[3] Pfanzagl, J. (1975). On asymptotically complete classes. *Statistical Inference.* **2** 1–43 M. Puri, ed. Academic Press.

Niels Keiding

University of Copenhagen

1. An important feature of Efron's paper is the study of the loss of information resulting from summarizing the data in n replications $X_1, \cdots, X_n$ of a multivariate random variable into a one-dimensional statistic $T(\mathbf{X}) = T(X_1, \cdots, X_n)$. In most of the paper it is assumed that the X_i's are observable and that their distribution belongs to an exponential family of which the statistical model forms a "curved subset", in the sense of the mean value parametrization. The basic result in this connection is formula (9.3), stating that the information loss from n replications is

$$\mathbf{i}_\theta - \mathbf{i}_\theta{}^T = E_\theta \operatorname{Var}_\theta \{\dot{\mathbf{l}}_\theta(\mathbf{X}) \mid T\},$$

where for $T = \hat{\theta}$, the right hand side is $i_\theta \gamma_\theta{}^2$, independent of n. (Notice that it is an implicit consequence of this that $\hat{\theta}$ cannot itself have the form $\Sigma t(X_i)$).

A somewhat related problem is that of incomplete observation of an exponential family, where the statistician is "forced" to work with nonsufficient reduction of data. It is here assumed that the statistical problem is specified in terms of an exponential family where only a function $Y = Y(X)$ of each component may be observed. If Y is a linear function of the canonical statistic X, there seems to be a canonical way of decomposing the parameter vector into an efficiently estimable part and a nonidentifiable part, using the concepts of "mixed parametrization" and "cut" introduced and further studied by Barndorff-Nielsen (1973, 1974) and Barndorff-Nielsen and Blaesild (1975), and in the case of continuously distributed random variables this seems to hold as soon as the level curves of Y are hyperplanes. Asymptotic results for arbitrary "curved" functions Y were given by Sundberg (1974) who points out that the same formula as above applies for the information loss, which here in general will be of order n.

It is clear that the two situations might be combined: a "curved" model with incomplete observation. An example of this was discussed by Fisher (1958, Section 57.1).

2. The relation (10.1) for the asymptotic variance of any consistent and efficient estimator $\hat{\theta}$ contains the term $\Delta_{\theta_0}^{\bar{\theta}}$, being always nonnegative and zero for the MLE. This quantity was computed by Rao (1963) for several estimation methods in the multinomial distribution, as noted by Efron. It would be interesting if some geometrical interpretation, or at least a bit more transparent expression than (10.24) could be given for this quantity, which must be related to the intuitive discussion by Fisher (1958, Section 57) of "the contribution to χ^2 of errors of estimation".

3. Curved exponential families occur frequently in population process and life testing models leading to occurrence/exposure estimates of birth or death intensities. One familiar example is that of estimating the mean μ^{-1} of an exponential distribution from a sample of n, censored at a fixed point t. If D is the number of variables less than t, and S the sum of these $+ (n - D)t$, then the likelihood function is $\mu^D e^{-\mu S}$, yielding $\hat{\mu} = D/S$.

We shall here comment a little upon the similar example of estimating the birth intensity λ in a pure (linear) birth process (X_u) from continuous observation of the process in $[0, t]$. See Keiding (1974) for details of the problem.

Assuming $X_0 = x_0$, degenerate, the likelihood is

$$\lambda^{X_t - x_0} e^{-\lambda S_t}$$

with $S_t = \int_0^t X_u \, du$. Setting $B_t = X_t - x_0$, the maximum likelihood estimator is $\hat{\lambda} = B_t/S_t$. It is readily seen that the Fisher information

$$i_\lambda = x_0(e^{\lambda t} - 1)/\lambda^2$$

and the statistical curvature γ_λ is given by

$$\gamma_\lambda^2 = \frac{1}{x_0}\left[\frac{1}{1 - e^{-\lambda t}} - \frac{(\lambda t)^2 e^{2\lambda t}}{(e^{\lambda t} - 1)^3}\right].$$

In the spirit of the paper, we quote some values of γ_λ^2 $(x_0 = 1)$ in Table 1.

Two asymptotic schemes are inviting: large initial population size $(x_0 \to \infty)$ for fixed t and large observation period $(t \to \infty)$ for fixed x_0. Being a branching process, a birth process with $X_0 = x_0$ may be interpreted as a sum of x_0 birth processes with $X_0 = 1$ and the same λ. Therefore the first scheme is still within the realm of independent identical replications, and may be treated with the methods of Efron's paper. This was done by Beyer, Keiding and Simonsen (1975) for this case as well as for the life-testing situation outlined above.

The second scheme, however, is a "real" stochastic process situation, and we encounter here the trouble that the minimal sufficient statistic is not consistent,

TABLE 1
Statistical curvature for the birth process with $x_0 = 1$

λt	0	0.1	0.5	1	2	5	∞
γ_λ^2	0	0.009	0.052	0.125	0.319	0.835	1

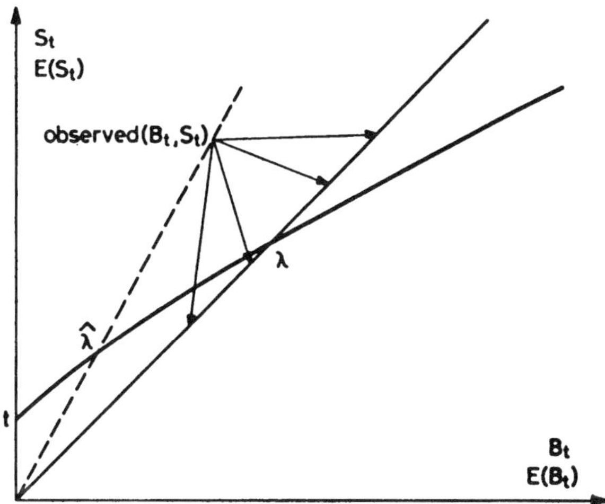

FIG. 1. The canonical sample space of the birth process estimation problem. The curve is the statistical model corresponding to $0 < \lambda < \infty$ (mean value parametrisation). The full-drawn line is the set of points for which $B_t = \lambda S_t$ where λ is the "true" value, and the broken line is the set where $B_t = \hat{\lambda} S_t$.

in fact, as $t \to \infty$

$$e^{-\lambda t}(B_t, S_t) \to (1, \lambda^{-1})W$$

almost surely, where the random variable W is gamma distributed with form parameter x_0 and expectation x_0. Nevertheless $\hat{\lambda} \to \lambda$ a.s., as illustrated in Figure 1. Here λ^{-1} is the slope of the full-drawn line, $\hat{\lambda}^{-1}$ is the slope of the broken line (connecting the observed (B_t, S_t) and the origin.) Normalising with $e^{-\lambda t}$, the minimal sufficient statistic will converge towards some $(1, \lambda^{-1})W$ (shown by arrows), but the empirical line will always converge towards the correct line.

In the standard situation the asymptotic normality of $\hat{\theta}$ is based upon the asymptotic normality of the minimal sufficient statistic combined with pure differential geometry, as noted by Efron in Section 9. It is therefore no surprise that asymptotic normality breaks down here. Notice also that $\gamma_\lambda \to 1$ (not 0) as $t \to \infty$. However, for given "nuisance statistic" W, the minimal sufficient statistic *is* asymptotically normal with asymptotic variance proportional to W^{-1}, and hence also $\hat{\lambda}$ is asymptotically normal. (Marginally, the distribution of $e^{\lambda t/2}(\hat{\lambda} - \lambda)$ converges towards a Student distribution with $2x_0$ d.f., which may be interpreted as the mixture of the normal distributions over the gamma distributed inverse variances.)

It is thus tempting to investigate the problem obtained by conditioning on $W = w$, replacing the "nuisance statistic" W by a nuisance parameter w, see Keiding (1974). The resulting "conditional" maximum likelihood estimator λ^* has the same first-order efficiency properties as $\hat{\lambda}$. A comparison of second-order efficiencies is not yet completed.

4. A more general aspect of the last example is: can curved exponential families be "avoided"? In the birth process situation a stopping rule like "sample until $X_t = n$" will make the minimal sufficient statistic one-dimensional, in fact equal to S_τ, $\tau = \inf\{t \mid X_t = n\}$. Also it should be mentioned that conditioning on statistics which are in some sense ancillary (see Barndorff-Nielsen (1973) for a survey of ancillarity) may completely change the curvature properties of the problem.

REFERENCES

BARNDORFF-NIELSEN, O. (1973). *Exponential Families and Conditioning.* Univ. of Copenhagen.

BARNDORFF-NIELSEN, O. (1974). Factorization of likelihood functions for exponential families. *J. Roy. Statist. Soc. Ser. B.* (Submitted).

BARNDORFF-NIELSEN, O. and BLAESILD, P. (1975). *S*-ancillarity in exponential families. To appear in *Sankhyā A* **37**.

BEYER, J. E., KEIDING, N. and SIMONSEN, W. (1975). The exact behaviour of the maximum likelihood estimator in the pure birth process and the pure death process. *Scand. J. Statist.* **2**. To appear.

FISHER, R. A. (1958). *Statistical Methods for Research Workers.* 13th ed. Oliver & Boyd, Edinburgh.

KEIDING, N. (1974). Estimation in the birth process. *Biometrika* **61** 71–80 and 647.

RAO, C. R. (1963). Criteria of estimation in large samples. *Sankhyā A* **25** 189–206.

SUNDBERG, R. (1974). Maximum likelihood theory for incomplete data from an exponential family. *Scand. J. Statist.* **1** 49–58.

A. P. DAWID

University College London

With his introduction of the concept of statistical curvature, Professor Efron has provided, not merely a valuable theoretical tool, but a new way of looking at statistical problems which at once unifies what has gone before and opens up new territory.

The general study of curvature belongs to Differential Geometry, a subject which has proved an invaluable tool in Physics, both Newtonian and Einsteinian. It may have much to offer Statistics. A good introduction is Laugwitz (1965) while Hicks (1965) emphasises a coordinate-free approach more suitable for Statistics.

In general differentiable spaces, we cannot talk about curvature until we have chosen, somewhat arbitrarily, a *linear connexion*: this defines what we mean by "displacement of a vector *parallel to itself* along a curve." For example, consider an observer who lives and measures on a plane inverted in its unit circle. To him, a circle through the origin looks like a straight line, and he would consider its tangents as parallel; to us they are not. The need for the parallelism concept may be seen from Efron's Figure 1: a_θ is the angle between (i) $\dot{\eta}_\theta$ and (ii) $\dot{\eta}_{\theta_0}$ *displaced parallel to itself* along $\mathscr{L}$ to η_θ. This depends on our connexion.

Let us try to frame Statistics within Differential Geometry as follows (ignoring obvious technical difficulties): Let $\mathscr{P}$ be the family of all distributions over $\mathscr{X}$

equivalent to a carrier measure μ. A *curve* $\mathscr{C}$ in $\mathscr{P}$ is a 1-parameter family in $\mathscr{P}$, say $\mathscr{C} = \{P_\theta\}$ with densities $\{f_\theta\}$, having suitable regularity properties.

If $\mathscr{M}$ is the vector space of signed measures m on $\mathscr{X}$, with $m \ll \mu$ and $m(\mathscr{X}) = 0$, we may define the *tangent* to $\mathscr{C}$ at $P = P_\theta$ as $m_\theta^{\mathscr{C}} \in \mathscr{M}$, with $m_\theta^{\mathscr{C}} =$ "$\lim_{\delta \to 0}$" $(P_{\theta+\delta} - P_\theta)/\delta$. (Equivalently, $dm_\theta^{\mathscr{C}}/d\mu = \dot{f}_\theta$). Conversely $m \in \mathscr{M}$ is tangent to some curve at P.

Let $\mathscr{V}_P$ be the vector space of random variables $T(x)$ having $E_P[T(X)] = 0$. For given P, there is a natural isomorphism between $\mathscr{M}$ and $\mathscr{V}_P$: $dm = T(x)\, dP$. Then $m_\theta^{\mathscr{C}}$ maps into $\dot{l}_\theta(x)$, which may again be identified with the *tangent* to $\mathscr{C}$ at P_θ.

Now let $P_{\theta_0}, P_{\theta_1} \in \mathscr{C}$, with tangent spaces $\mathscr{V}_0$, $\mathscr{V}_1$, and let $T_0 \in \mathscr{V}_0$, $T_1 \in \mathscr{V}_1$. To be able to talk about the angle between T_0 and T_1 we must put them into the same space. We may do this by a *parallel displacement* of T_0 along $\mathscr{C}$ to θ_1, where it becomes $T_0' \in \mathscr{V}_1$.

The parallel displacement used implicitly by Efron—what I propose to call the "Efron connexion"—has

$$T_0' = T_0 - E_{\theta_1}(T_0) . \tag{1}$$

This happens to be independent of the curve $\mathscr{C}$, which is not always so. Noting $(d/d\theta)E_\theta[T] = E_\theta[T\dot{l}_\theta]$ for fixed T, we can generate (1) by the infinitesimal displacement rule (having $\theta_1 = \theta_0 + d\theta$):

$$T_0' = T_0 - E_{\theta_0}(T_0 \dot{l}_{\theta_0}) \cdot d\theta . \tag{2}$$

For curvature, we look at the angle between $\dot{l}_{\theta_0}' = \dot{l}_{\theta_0} - i_{\theta_0} \cdot d\theta$ and $\dot{l}_{\theta_1} = \dot{l}_{\theta_0} + \ddot{l}_{\theta_0} d\theta$. We may measure this by any convenient inner product, but in our statistical set-up there appears to be only one natural inner product in $\mathscr{V}_P$, namely $\langle T, U\rangle = E_P(TU)$. (For any parametric family $\{P_\phi\}$, this yields the information inner product, with matrix $(E_\phi[(\partial l/\partial\phi_i)(\partial l/\partial\phi_j)]$.) Hence we may call this the *information metric*). This leads to Efron's measurement of angle and of curvature.

The "straight lines" have a characterisation independent of the metric: $\dot{l}_{\theta_0}$ must displace to become a scalar multiple of $\dot{l}_{\theta_1}$. By reparametrisation, the multiple may be taken as unity. This leads to the differential equation

$$\ddot{l}_\theta + i_\theta = 0 \tag{3}$$

characterising exponential families.

The Efron connexion is not, however, the only available one (although it probably is the only one that fits in neatly with repeated sampling, as in Efron's Section 6). An alternative obvious definition of parallel displacement considers $\mathscr{M}$ as the tangent space and uses the identity transformation (again, independent of path). This is equivalent to transforming $\mathscr{V}_0$ into $\mathscr{V}_1$ with

$$T_0' = T_0 \cdot \left(\frac{dP_{\theta_0}}{dP_{\theta_1}}\right), \tag{4}$$

yielding the infinitesimal displacement

$$T_0' = T_0 - T_0 l_{\theta_0} \cdot d\theta \,. \tag{5}$$

To measure curvature with this connexion, using the information metric, M_θ in Efron's (2.3) must be replaced by the covariance matrix of $\dot{l}_\theta$ and $(\ddot{l}_\theta + \dot{l}_\theta^2)$. The "straight lines" now have $\ddot{l}_\theta + \dot{l}_\theta^2 = 0$, which yields mixture families: $P_\theta = (1-\theta)P_0 + \theta P_1$. Thus the above connexion may be termed the "mixture connexion".

Now the information metric makes $\mathscr{P}$ into a *Riemannian space*, and from this point of view there is a serious deficiency in both connexions above: they are *not compatible* with the metric. That is, the length of T_0 at P_{θ_0}(viz $[E_{\theta_0}(T_0^2)]^{\frac{1}{2}}$) is not the same as that of its parallel translate T_0' at P_{θ_1}. It may be checked that the infinitesimal displacement

$$T_0' = T_0 - \tfrac{1}{2}[T_0 l_{\theta_0} + E_{\theta_0}(T_0 l_{\theta_0})] \cdot d\theta \tag{6}$$

yields a connexion—the "information connexion"—that *is* compatible with the information metric. Curvature for this connexion (which is the *geodesic curvature* associated with the information metric) uses the covariance matrix of $\dot{l}_\theta$ and $\ddot{l}_\theta + \frac{1}{2}\dot{l}_\theta^2$.

We can calculate the *torsion* and *curvature tensors* (Hicks, page 59) for the above connexions. We find that all have zero torsion (equivalently: are *symmetric*, or *affine*). There is a unique affine connexion compatible with a given metric, hence (6) supplies it for the information metric.

We find zero curvature for the Efron and mixture connexions, while the curvature tensor R associated with the information connexion has

$$R(T, U)V = \tfrac{1}{4}[T \cdot E(UV) - U \cdot E(TV)] \,. \tag{7}$$

The Riemann–Christoffel curvature tensor K of type 0, 4 (Hicks, page 72) is then given by:

$$K(T, U, V, W) = \tfrac{1}{4}[E(TV)E(UW) - E(TW)E(UV)] \,. \tag{8}$$

From this we find that the space $\mathscr{P}$, with the information metric, has constant, positive, Riemannian curvature $\frac{1}{4}$.

The *geodesics* (shortest paths) for the information metric are the "straight lines" of the information connexion, satisfying

$$\ddot{l}_\theta + \tfrac{1}{2}\dot{l}_\theta^2 + \tfrac{1}{2}i_\theta = 0 \,. \tag{9}$$

Solutions of (9) are *closed* curves, parametrized by an angle θ, having an angle-valued sufficient statistic t, with density of the form

$$f(t \mid \theta) = 1 + \cos(t - \theta) \tag{10}$$

with respect to a probability measure ν over the unit circle for which $\int_0^{2\pi} e^{it}\, d\nu(t) = 0$. Such curves have $i_\theta \equiv 1$, and total length 2π. Thus $\mathscr{P}$ looks rather like the surface of a sphere of radius 2, opposite points being identified.

The nonvanishing of (7) means that the information parallel displacement depends on path, which makes it less immediately intelligible than the Efron and mixture displacements. Can we give any interesting statistical interpretation to the information connexion, and its associated families (10)?

REFERENCES

[1] HICKS, N. J. (1965). *Notes on Differential Geometry*. Van Nostrand, Princeton.
[2] LAUGWITZ, D. (1965). *Differential and Riemannian Geometry*. Academic Press, New York.

JIM REEDS
Harvard University

1. Ideas of geometrical curvature are not completely new to statistics. Efron's paper is the logical successor to papers applying the differential geometric point of view to statistical estimation. Rao (1945) and Bhattacharyya (1943) viewed the multiparameter Fisher information as defining a local (Riemannian) metric (Eisenhart (1926 and 1960), Spivak (1970)) on the parameter space; the integrated arc length of a geodesic connecting two parameter values then defines a global metric or distance function on parameter space. Holland (1973), Huzurbazar (1950 and 1956) and Mitchell (1962) exploited transformation properties of the Fisher information viewed as a Riemannian metric. Holland, for instance, sought covariance stabilizing transformations (like the square root transformation of univariate Poissons). Such a transformation makes the Fisher information matrix, expressed in transformed coordinates, a constant matrix. "When can it be found?" is the question "When is a given Riemannian manifold locally isometric to a Euclidean space?" Riemann gave the answer: "When the Riemannian curvature (or, in two dimensions, the Gaussian curvature) vanishes identically." This always happens only in dimension one. In all higher dimensions non-Euclidean manifolds—and noncovariance stabilizable parameter spaces—occur.

Recent unpublished work of Tadashi Yoshizawa (1971) makes explicit use of the inherent Riemannian structure in parameter estimation problems. He shows how one can isometrically embed the parameter space into a Euclidean space of sufficiently high dimension, and then read off the (first order) asymptotic properties of the estimation problem by inspecting the parameter space as a curved submanifold of a Euclidean space.

Thus curvature of one sort is not new to statistics. But Efron's curvature is of a different sort—not the Riemannian or "intrinsic" curvature but instead the curvature of embedding, associated with the particular way a parameter space is placed inside a higher-dimensional "natural parameter" space. Riemann curvature—measured by the curvature tensor—is determined solely by the "first fundamental form" or metric tensor, the physicists' metric ground form, the statisticians' Fisher information matrix. Efron's curvature, curvature of embedding, is measured by the "second fundamental form" and depends on more than

Fisher information. The distinction is illustrated by a cylinder embedded in Euclidean 3-space. This surface has curvature of embedding but no Riemann curvature, for any piece of it can be unrolled without distorting lengths. A sphere in 3-space has both sorts of curvature; a parabola in the plane has only curvature of embedding. No submanifolds of Euclidean space have Riemann curvature without curvature of embedding.

Efron takes the *natural* parameter space as Euclidean, with *constant* metric given by the Fisher information evaluated at the true value of the parameter, θ_0. The actual parameter space is a submanifold of natural parameter space; its curvature of embedding is calculated with respect to this constant Euclidean structure on the natural parameter space. Efron's discussion in the second paragraph of Section 2 is unclear; one might falsely assume that the natural parameter space was endowed with the (nonconstant) metric provided by the Fisher information as a function of θ.

The point of Efron's paper is that the curvature of embedding, calculated in this way, has an effect on statistical procedures, an effect amenable to quantitative study.

2. The main result of Section 10 may be generalized to a multivariate curved exponential family. Both this result and Efron's suffer from a defect which might be overcome in future work. The defect is that both make statements about the coefficients of the asymptotic expansions of the variance, *not about the variance itself*. Thus, the conclusions are of the form

$$\text{“Var}\,(T_n) = \frac{a}{n} + \frac{b}{n^2} + o(n^{-2}) \quad (\text{or } O(n^{-3}))\,,$$

$$\text{and} \qquad a \geqq \alpha\,, \quad \text{and if} \quad a = \alpha\,, \quad b \geqq \beta\,,\text{”}$$

where α and β are certain theoretical lower bounds. This should be contrasted with a stronger type of conclusion:

$$\text{“Var}\,(T_n) \geqq \frac{\alpha}{n} + \frac{\beta}{n^2}\,,\text{”}$$

where α and β have the same meaning as above. (If T_n is such that Var (T_n) has an asymptotic expansion at all, the second conclusion implies the first.) Both the Cramér–Rao and the Bhattacharyya inequalities provide conclusions of the second type. In a sense, we can trace the difference to the different methods used to prove the various inequalities. The classical proof of the Cramér–Rao bound proceeds by constructing a certain variance-covariance matrix, and using its positive semidefiniteness to get the desired results. This is to be contrasted with the method used in the present theorems: Taylor expansions of the functional form of the estimate, coupled with systematic discarding of negligible terms.

It is conceivable that a proof of the theorem of Section 10 could be constructed by the classical method, by considering the joint covariance of the estimate, the first derivative of the log likelihood, the square of the first derivative of the log likelihood, and the product of the first and second derivatives of the log likelihood. This is conjectured on the grounds of the simple form the covariance matrix takes, when only terms of order up through $1/n^2$ are considered.

We may define a curved q-parameter exponential family by means of a smooth map $\eta: \Theta \to H$, where Θ is some open subset of $\mathbb{R}^q$, and H is the natural parameter space of a k-variate exponential family. To simplify the discussion that follows, we will assume that η is an embedding in the sense of differential geometry: η is a C^∞ injection, with differential of full rank at each point, and that "smooth"—whenever it appears in this discussion—means C^∞. Note that according to this set-up, Θ is not a submanifold of H; but $\eta(\Theta)$ is. An estimate is a function $T: \mathscr{X} \to \Theta$, mapping the space of the sufficient statistic to the parameter space.

If we restrict ourselves to estimates T that depend only on the sufficient statistic $\bar{x}_n = n^{-1}(x_1 + \cdots + x_n)$ (and not on n), and which satisfy certain regularity conditions, we may prove:

THEOREM. *Let T depend only on $\bar{x}_n$, the sufficient statistic for a curved q-parameter exponential family. Suppose T is smooth in some neighborhood of $E(\bar{x}_n)$, and suppose T grows (as a function of $\bar{x}_n$) no faster than exponentially.*

If T is a consistent and first order efficient estimate of θ, the variance of T possesses an asymptotic expansion

$$\operatorname{Var}(T(\bar{x}_n)) = \text{CRLB} + \frac{A}{n^2} + \frac{B}{n^2} + \frac{C}{n^2} + O(n^{-3}) .$$

(Here CRLB denotes the Cramér-Rao Lower Bound,

- A denotes the "naming" or "Bhattacharyya" curvature, which can be made zero by an appropriate reparameterization of parameter space. It is independent of T.
- B is the "Efron excess", or statistical curvature term and is independent of T.
- C depends only on the function T, and vanishes for the particular choice $T =$ the maximum likelihood estimate.

All these quantities are q by q positive semidefinite symmetric matrices.)

The proof of this multivariate theorem parallels Efron's univariate arguments. It shares the use of affine transformations to bring the problem into "standard form," calculations with Taylor expansions to exhibit the consequences of consistency and first order efficiency, and finally, replacement of T by a Taylor approximation, and the calculation of expectations and variances of the Taylor approximant.

The key quantity of interest in the conclusion of this theorem is the term B, the "Efron" or "statistical curvature" excess. It is the multivariate generalization of γ^2/i, and (like γ^2/i) may be defined in several ways.

(1) Let $\eta(\theta)$, in the vicinity of θ_0, have an expansion

$$\eta^i(\theta) = a^i + \sum_j b_j^i(\theta^j - \theta_0^j) + \tfrac{1}{2}\sum_{jk} c^i_{jk}(\theta^j - \theta_0^j)(\theta^k - \theta_0^k) + \cdots$$

where θ has coordinates $(\theta^1, \theta^2, \cdots, \theta^q)$. Let (g^{ij}) denote the inverse of the Fisher information matrix for θ, and let (G_{rs}) denote the Fisher information matrix for the natural parameter η. Let

$$D_{ih} = \sum_{rs} b_i^r G_{rs} b_h^s ,$$

$$E_{i,mn} = \sum_{rs} b_i^r G_{rs} c^s_{mn}$$

and

$$F_{jk,mn} = \sum_{rs} c^r_{jk} G_{rs} c^s_{mn} .$$

Let the inverse of $D = (D_{ih})$ be $D^{-1} = (D^{ij})$. Let

$$\tilde{F}_{jk,mn} = F_{jk,mn} - \sum_{ih} E_{i,jk} E_{h,mn} D^{ih} .$$

Then

$$B^{ij} = \sum_{kl} \sum_{mn} g^{im} g^{jn} g^{kl} \tilde{F}_{nk,nl} .$$

If, at θ_0, the Fisher matrices of both θ and η are equal to identity matrices, this simplifies to

$$B^{ij} = \sum_{k,r} c^r_{ik} c^r_{kj} ,$$

where the summation extends over $r \geqq q + 1$.

(2) Let l be the log likelihood function. If

$$\dot{l}_i = \frac{\partial}{\partial \theta_i} l$$

and

$$\ddot{l}_{ij} = \frac{\partial^2}{\partial \theta_i \, \partial \theta_j} l ,$$

we may form the linear regression of $\ddot{l}$ on $\dot{l}$ as follows:

$$\hat{\ddot{l}}_{jk} = \sum_i \beta^i_{jk} \dot{l}_i ,$$

and we may calculate the regression-residual variance:

$$\mathrm{Cov}\,(\ddot{l}_{ij} - \hat{\ddot{l}}_{ij}, \ddot{l}_{mn} - \hat{\ddot{l}}_{mn}) = \alpha_{ij,mn} .$$

Then

$$B^{ij} = \sum_{mn} \sum_{kl} g^{im} g^{jn} \alpha_{mk,ln} g^{kl} .$$

(3) Let $\Omega_{r|ij}$ be the components of the second fundamental form of the imbedding $\eta: \Theta \to H$ (see Eisenhart (1926 and 1960)) where H has the Euclidean structure induced by the Fisher information evaluated at $\eta(\theta_0)$. Then

$$B^{ij} = \sum_{mn} \sum_{kl} \sum_r g^{im} \Omega_{r|mk} g^{kl} \Omega_{r|ln} g^{nj} .$$

Similar formulas hold for the naming curvature term A. In the special case where both the Fisher information matrices are equal to identity matrices (at θ_0) and where

$$b_j^i = \left.\frac{\partial \eta^i}{\partial \theta^j}\right|_{\theta_0} = \delta_{ij} ,$$

the ijth term of the naming curvature is given by

$$A^{ij} = \sum_{a,b} \left(c^i_{ab} + \frac{\partial G_{ab}}{\partial \theta_i}\bigg|_{\theta_0}\right)\left(c^j_{ab} + \frac{\partial G_{ab}}{\partial \theta_j}\bigg|_{\theta_0}\right),$$

where the summation extends over $1 \leqq a, b \leqq q$.

Notice that in the univariate case the naming curvature term A can always be made to vanish identically by a suitable reparameterization, but in the multivariate case this cannot in general be done. It *can* always be made to vanish at isolated points, but there need not in general exist reparameterizations which make the naming curvature vanish globally. This is related to the general nonexistence of multivariate covariance stabilizing transformations. In the univariate case, the naming curvature vanishes identically exactly when we parameterize the curve by arc length: that is, it vanishes when the variance is stabilized. In the multivariate setting, however, we cannot in general covariance stabilize, and we cannot in general make the naming curvature identically zero. Perhaps the easiest example is provided by the trivariate normal distribution, with unit covariance matrix, with the mean vector constrained to have unit length (and, to avoid global topological problems, with first coordinate positive). Thus, in the multivariate case the naming curvature term takes on added significance, and must be viewed as serious an object of study as the statistical curvature term itself.

REFERENCES

[1] BHATTACHARYYA, A. (1943). On a measure of divergence between two statistical populations. *Bull. Calcutta Math. Soc.* **35** 99–109.

[2] EISENHART, L. (1926 and 1960). *Riemannian Geometry*. Princeton Univ. Press.

[3] HOLLAND, P. (1973). Covariance stabilizing transformations. *Ann. Statist.* **1** 84–92.

[4] HURZURBAZAR, V. (1950). Probability distributions and orthogonal parameters. *Proc. Cambridge Philos. Soc.* **46** 281.

[5] HURZURBAZAR, V. (1956). Sufficient statistics and orthogonal parameters. *Sankhyā* **17** 217–220.

[6] MITCHELL, A. (1962). Sufficient statistics and orthogonal parameters. *Proc. Cambridge Philos. Soc.* **58** 326–337.

[7] RAO, C. R. (1945). Information and accuracy attainable in the estimation of statistical parameters. *Bull. Calcutta Math. Soc.* **37** 81–91.

[8] SPIVAK, M. (1970). *Differential Geometry*. Publish or Perish, Boston.

[9] YOSHIZAWA, T. (1971). Memorandum TYH-2, *A Geometrical Interpretation of Location and Scale Parameters*. Statist. Dept., Harvard Univ. Cambridge.

REPLY TO DISCUSSION

The discussants are (almost) uniformly constructive and informative in their comments. They point out many important facts, and even whole areas, that the paper misses. Only two of them consider me basically deranged in my thought processes. In what follows I have tried to answer a few specific points, without exploring much further the bigger questions raised.

Professors Cox and Pierce suggest that the distance from $(\bar{x}_1, \bar{x}_2)$, to $\lambda_{\hat{\theta}}$ is a useful approximate ancillary statistic. (See Figure 4. It is simplest to assume that the family $\mathscr{F}$ is in standard form at $\theta = 0$, and that we are considering θ values near zero.) I particularly like Pierce's suggestion that the ancillary information has to do with the *precision* of $\hat{\theta}$ and not its location. To make things really easy, consider repeated sampling in Example 1, and suppose that we happen to get $\hat{\theta} = 0$, that is $\bar{x}_1 = 0$. (See Figure 2.) The likelihood function for θ is proportional to $\exp\{-(n/2)[1 - \gamma_0\bar{x}_2 - \gamma_0^2\theta^2/4]\theta^2\}$ which for θ in the interval $\hat{\theta} \pm c/n^{\frac{1}{2}}$ behaves like $\exp\{-(n/2)[1 - \gamma_0\bar{x}_2]\theta^2\}$. That is, the likelihood function for θ is approximately $\mathscr{N}(\hat{\theta}, [1 - \gamma_0\bar{x}_2]/(ni_{\hat{\theta}}))$. The distance from $(\bar{x}_1, \bar{x}_2)$ to $\hat{\theta}$, $\bar{x}_2$ in this case, modifies the unconditional variance $(ni_{\hat{\theta}})^{-1}$ by the factor $[1 - \gamma_0\bar{x}_2]$. It is probably possible to extend this likelihood analysis to a genuine conditional variance statement, as Pierce suggests.

Bayesians and other nonfrequentist statisticians do not like averages taken over the sample space $\mathscr{X}$ with θ fixed. Professor Lindley raises this objection to the curvature γ_θ^2, as it has been raised to the Fisher information i_θ itself. Those who believe in direct interpretation of likelihood functions prefer $-\ddot{l}_{\hat{\theta}}(x)$, the actual curvature of the log likelihood function at its maximum, to the average value i_θ. (Incidentally, I use θ as a subscript rather than an argument to save writing parentheses!) I find some force in these kinds of considerations but, perhaps because of my training, can never be convinced without the support of some relevant averaging property, be it frequentist, conditional frequentist, Bayesian, or otherwise. (See my discussion following Blyth (1970).)

If a Cauchy translation sample of size 10 yields a very normal looking likelihood function, say $\mathscr{N}(0, .3)$, should we behave as if the MLE has variance .3? Professor Lindley answers "yes" on Bayesian grounds, in the absence of prior information. Professor Pierce's remarks indicate that the curvature may have something helpful to say to frequentists about such problems.

Returning to less slippery ground, here is a calculation of asymptotic Bayes risk that makes use of the curvature. In a curved exponential with an i.i.d. sample of size n, let θ have prior distribution $\mathscr{N}(\theta_0, c_n/n)$, where c_n is going sufficiently slowly to infinity. Then the Bayes risk is asymptotically

$$\frac{1}{n_{i_{\theta_0}}} + \frac{1}{n^2 i_{\theta_0}}\left\{\gamma_{\theta_0}^2 + \frac{4\Gamma_{\theta_0}^2}{i_{\theta_0}}\right\} + o\left(\frac{1}{n^2}\right)$$

which equals to order $1/n^2$ the squared error risk of the biased corrected MLE at $\theta = \theta_0$. (This result follows, with some effort, from (10.19).)

Professor Le Cam's warning about over-reliance on local methods is well taken. As a matter of fact, my paper is most concerned with curvature as a check on the appropriateness of first order local properties such as Fisher's information and the Cramér-Rao lower bound. In the situation of Figure 6, curvature can be used quantitatively to improve the first order approximation. I hope, but of course am not certain, that other situations will be similarly obliging.

Le Cam's criticism of the MLE as a point estimator should not be confused with Fisher's preference for it is an information gatherer. A function of the MLE may be better than the MLE itself for any specific estimation problem. This is the case in the Berkson example quoted. Berkson finds a "better" estimator than the MLE, which eventually is improved by Rao-Blackwellizing it on the sufficient statistics. This gives a function of the MLE! (It has to because the situation involves a genuine uncurved exponential family.) Figure 4 becomes more convincing the more you study it. Locally the straight level line $L_{\hat{\theta}}$ seems intuitively preferable to any curved competitor $M_{\hat{\theta}}$. (See Dr. Keiding's remarks and my reply.)

Quadratic approximations to the log likelihood function have been used successfully by many authors, notably Professor Le Cam himself. They are the basis of Rao's work in second order efficiency. They can be used to produce estimators other than the MLE which are second order efficient. Whether there is a corresponding theory of third order efficiency, and whether the MLE is still the champion, is an interesting open question.

After a long fallow period there seems to be a revival of interest in second order efficiency and related topics. I am eager to see Professor Ghosh's work with Subrahmaniam and Srinivasan. (Also, I must apologize for not having been aware of Pfanzagl and Chibisov's papers, which demonstrate rigorously the relevance of what I have called curvature to hypothesis testing problems, even outside an exponential family framework.) As Ghosh suggests and as I mentioned in discussing Pierce's comments, there is some connection between γ_θ^2 and the geometrical curvature of the likelihood function, but not one I understand clearly yet. Professor Ghosh's last question can be partially answered in the affirmative: in the counter-example of Figure 5, change c to $(-2^{\frac{1}{2}}, \frac{1}{3})$. Then the MLE of any $\bar{x}$ vector with $\bar{x}_1 = \frac{1}{3}$ is zero, but if $\bar{x}_1 \neq \frac{1}{3}$ each $\bar{x}$ corresponds to a unique $\hat{\theta}$. For n any multiple of 3, $\hat{\theta}$ will lose information because of the grouping of those $\bar{x}$ vectors with $\bar{x}_1 = \frac{1}{3}$. It is easy to curve the level lines of another consistent efficient estimator $\tilde{\theta}$, à la Figure 4, so that the vectors with $\bar{x}_1 = \frac{1}{3}$ are separated, and $\tilde{\theta}(\bar{x})$ is different for all different $\bar{x}$ vectors, so no information is lost. This works for any fixed n divisible by 3, but I am less certain about finding a $\tilde{\theta}$ that works for all values of n.

There is less difference between Professor Pfanzagl and me than the tone of his comments indicates. His results (1) and (2) follow from (8.4). I should have said earlier that a rescaled version of this equation holds as an approximation when testing $\theta = 0$ versus $\theta > 0$ under i.i.d. sampling in any curved exponential family,

$$1 - \beta_{\tilde{\theta}_1}(\tilde{\theta}) \approx \Phi(\tilde{\theta}(1 + \tilde{\gamma}_0^2\tilde{\theta}^2/4)^{\frac{1}{2}} \cos(A_{\tilde{\theta}_1} - A_{\tilde{\theta}}) - z_\alpha),$$

where $\tilde{\theta} \equiv (ni_0)^{\frac{1}{2}}\theta$, $\tilde{\gamma}_0 \equiv \gamma_0/n^{\frac{1}{2}}$, and $A_{\tilde{\theta}} \equiv \tan^{-1}(\tilde{\gamma}_0\tilde{\theta}/2)$. In order for this approximation to be sufficiently accurate to yield Pfanzagl's asymptotic results, the family must be nonatomic. However, the type of power comparisons presented in Table 3 are less sensitive as well as more familiar. For $\alpha = .01$, power $= .99$,

$\gamma_0^2/n = \frac{1}{8}$, the case Pfanzagl discusses, the locally most powerful test has approximate power .94 compared with the envelope value .99. I consider this borderline acceptable, and will stick to my suggestion of $\gamma_0^2/n > \frac{1}{8}$ as a rough indicator of nonnegligible curvature effects.

Fisher defined γ_θ^2 as the loss of information in using $\hat{\theta}$ instead of the whole sample. Rao's results on estimation with squared error loss partially vindicate this definition. Pfanzagl's own work shows that γ_θ^2 plays a key role in the loss of effective sample size in hypothesis testing problems. Then why does he seem to say that γ_θ^2 has no statistical significance? The fact that the level α test based on the MLE is asymptotically equivalent to the β optimal test with power $1 - \alpha$ has nothing to do with the existence of curvature effects. There still is no uniformly most powerful test. The global deviations of any attainable power curve from the power envelope are still ruled by the magnitude of γ_θ^2.

I was happy to see that Dr. Keiding had found a definite use for curved exponential families in his work on birth processes. Time series problems offer many other examples, of which my Example 3 is close to the simplest. (With Dr. Reeds' multiparameter theory available we are now in a position to analyze the second order asymptotics of higher autoregressive schemes.) The geometric interpretation of the penalty $\Delta_{\theta_0}^{\tilde{\theta}}$ for not using the MLE is simple in the case $r = 2$. Comparing (10.24) with (10.5) shows that it equals one-half of the squared curvature of the level curve $M_0 = \{\bar{x} : \tilde{\theta}(\bar{x}) = \theta_0\}$. See Figure 4.

Dr. Dawid raises a deep question: why have I chosen to represent families of probability distributions by their log densities rather than, say, the density functions themselves? This latter representation would make mixture families rather than exponential families straight lines, as he points out. What I have called the matrix M_θ then has elements μ_{hj} as at (1.2) rather than ν_{hj} as at (3.21). Dawid makes the interesting observation that still another definition is needed to make straight lines into geodesics in the information metric. (Rao 1945a and 1945b, has proposed using this type of geodesic distance to measure the separation of probability distributions. Atkinson and Mitchell have calculated Rao distances for many familiar distribution families.) I can't answer Dr. Dawid's deep question except to say that my definition was motivated by what seemed to be the most pressing statistical considerations. He makes a good case for other definitions also yielding useful results for the statistician.

My paper considers only one parameter families. Dr. Reeds gives a convincing extension to the multiparameter case. Having been frustrated myself by the intricacies of the higher order differential geometry, I am impressed! Hopefully, his "B", the analogue of γ_θ^2, will also play the correct corresponding role vis-à-vis Fisher information and hypothesis testing.

Two technical comments: (i) a version of the usual super-efficiency examples prevents Reeds' formula (2) from holding generally. In my Example 1, Figure 2, let $\tilde{\theta}(\bar{x}) = \bar{x}_1$ except in a band of width $\pm x_1^2$ on either side of $\mathscr{L}$. Within this band modify $\tilde{\theta}$ so that it is consistent and first order efficient. Then (10.19) can be used to show that $\tilde{\theta}$ satisfies (10.1) with the $1/n^2 i_{\theta_0}\{\ \}$ term set equal to zero at $\theta_0 = 0$. (ii) It is not true in general, even in the one parameter case, that the "arc-length parameter" has naming curvature equal to zero. Let $\sigma(\theta)$ be this parameter measured from $\theta = \theta_0 = 0$, where we assume for convenience that $i_0 = 1$. By definition $\sigma(\theta) = \int_0^\theta i_{\theta'}^{\frac{1}{2}} d\theta'$ so that $d\sigma(\theta)/d\theta = i_\theta^{\frac{1}{2}}$, $d^2\sigma(\theta)/d\theta^2 = (di_\theta/d\theta)/2(i_\theta)^{\frac{1}{2}}$. It is easy to show by an expansion similar to (10.10) that in terms of the quantities μ_{kj} defined at (1.2),

$$di_\theta/d\theta = 2\mu_{11} - \mu_{30}\,.$$

This gives the Taylor expansion about zero

$$\sigma(\theta) = \theta + \left(\mu_{11} - \frac{\mu_{30}}{2}\right)\frac{\theta^2}{2} + o(\theta^2)\,,$$

μ_{11} and μ_{00} being evaluated at $\theta = 0$.

The parameter $\phi(\theta)$ which figures in the definition of Γ_{θ_0} in (10.1) has Taylor expansion

$$\phi(\theta) = \theta + \mu_{11}\frac{\theta^2}{2} + o(\theta^2)$$

as given in (10.11). Therefore the naming curvature $\Gamma_{\theta_0}^2$ will not be zero for the arc-length parameter unless $\mu_{30} = 0$. (That is, Fisher's score function has third moment zero.)

It is not clear to me whether or not one can always choose a reparameterization for $\mathscr{F}$ which has naming curvature identically zero, even in the one-parameter case. We probably wouldn't want to estimate such a parameter anyway unless it had something more to recommend it than $\Gamma_\theta^2 = 0$. I didn't mean to imply that naming curvature is less important than statistical curvature, onyl that it depends on the name.

Finally, I would like to thank the Editor for arranging this discussion which involved a large amount of extra work on his part. I hope the Annals of Statistics will continue the entertaining and enlightening policy of providing occasional discussion papers.

REFERENCES

[1] ATKINSON, C., and MITCHELL, A. F. *Rao's Distance Measure.* Unpublished.

[2] BLYTH, C. R. (1970). On the inference and decision models of statistics. *Ann. Math. Statist.* 3 1034–1058.

[3] RAO, C. R. (1945 a). Information and the accuracy attainable in the estimation of statistical parameters. *Bull. Calcutta Math. Soc.* 37 81–91.

[4] RAO, C. R. (1945 b). On the distance between two populations. *Sankhyā* 9 246–248.

4

Data Analysis Using Stein's Estimator and its Generalizations

Introduction by John E. Rolph
University of Southern California

This paper is one of a series on empirical Bayes estimation by Brad Efron and Carl Morris that appeared in the early and mid-1970s. The starting point for their work was the fact that the James–Stein estimator improves on the maximum likelihood estimator (MLE) of several independent normal means in terms of total squared error loss. Efron and Morris developed a variety of generalizations of and improvements on the original James–Stein estimator.

Efron and Morris were among the first to recognize that Stein-type estimators can be derived as empirical Bayes estimators. And at the time, their empirical Bayes approach greatly clarified the understanding of these estimators. It was also helpful in motivating the improved shrinkage estimators that they and subsequent researchers devised.

The introduction to this paper describes its purpose as addressing two then-common objections to using Stein-type estimators in applied problems: (1) MLE estimators usually perform well and (2) gains from more complicated estimators cannot be worth the extra trouble. Efron's and Morris's other articles in this series derive properties of and propose improvements on the original James–Stein estimator. In contrast, as its title and introduction promise, this paper addresses a series of practical issues that must be addressed when actually applying Stein-type estimators to real problems. To achieve their stated goal, they develop and apply Stein-type estimators to predict batting averages, to estimate toxoplasmosis rates, and to improve the results of a computer simulation.

This paper is essentially a synthetic paper that pulls together and applies the results of the other papers in the series to real problems. One of its charms is that one can learn about Efron's and Morris's numerous improvements and generalizations of the James–Stein estimator without having read their earlier papers.

The improvements in Stein-type estimators that Efron and Morris use in analyzing these three data sets include the positive part version, shrinking to the grand mean, estimating unknown variances, handling unequal variances of the components, and limiting the component translation. They also do numerous computations of various risks and efficiencies to give the reader an understanding of how the performances of their Stein-type estimators compare to each other and to the MLE.

The result in this paper that I found most compelling is the trade-off achieved by limiting the translation of the estimators of individual component means. Stein's estimators have uniformly lower total (squared error) risk than the MLE, but they allow substantially increased risks in estimating the individual component means. Increased individual component risk might well give a prospective user of Stein-type estimators pause. However, to estimate batting averages, Efron and Morris introduce a limited translation Stein-type estimator that reduces the maximum component risk by 2/3 while retaining 80% of the savings of Stein's rule over the MLE. This paper, even today, offers valuable insights on how to tailor shrinkage estimators to applied settings.

I was at RAND during the 1970s. Brad Efron was a long-time RAND consultant who visited regularly, particularly during the summers. I had the good fortune to have an office next to Carl Morris at RAND and thus get in on the ground floor of their collaboration. To my great benefit, I was able to use some of their results in a RAND project of mine. As an aid to developing better New York City Fire Department dispatching policies, my statistical problem was to estimate the probability that an incoming alarm reported by pulling a street alarm box signaled a serious fire rather than a minor fire or a false alarm. This was a binomial situation like their baseball batting average prediction and had unequal variances because of differing sample sizes like their toxoplasmosis rate estimation. Carter and Rolph (1974) give details.

Soon after, Fay and Herriot (1979) also developed and applied empirical Bayes estimators for estimating component means that were geographical locations. Indeed borrowing strength across geography to improve small area estimates is now a classic application of Bayes and empirical Bayes methods. And Efron's and Morris's early work got the ball rolling.

Since this paper appeared in 1975, shrinkage estimation has developed in a variety of directions. Nonetheless, I found that reading this paper today still offers useful insights into the bias-variance trade-offs inherent in addressing estimation problems.

References

Carter, G. M. and Rolph, J. E. (1974). Empirical Bayes methods applied to estimating fire alarm probabilities. *Journal of the American Statistical Association* **69**, 880–885.

Fay, R. E. III and Herriot, R. A. (1979). Estimates of income for small places: An application of James–Stein procedures to census data. *Journal of the American Statistical Association* **74**, 269–277.

Data Analysis Using Stein's Estimator and Its Generalizations

BRADLEY EFRON and CARL MORRIS

Reprinted from the Journal of the American Statistical Association
June 1975, Volume 70, Number 350
Pages 311–319

Data Analysis Using Stein's Estimator and Its Generalizations

BRADLEY EFRON and CARL MORRIS*

In 1961, James and Stein exhibited an estimator of the mean of a multivariate normal distribution having uniformly lower mean squared error than the sample mean. This estimator is reviewed briefly in an empirical Bayes context. Stein's rule and its generalizations are then applied to predict baseball averages, to estimate toxomosis prevalence rates, and to estimate the exact size of Pearson's chi-square test with results from a computer simulation. In each of these examples, the mean square error of these rules is less than half that of the sample mean.

1. INTRODUCTION

Charles Stein [15] showed that it is possible to make a uniform improvement on the maximum likelihood estimator (MLE) in terms of total squared error risk when estimating several parameters from independent normal observations. Later James and Stein [13] presented a particularly simple estimator for which the improvement was quite substantial near the origin, if there are more than two parameters. This achievement leads immediately to a uniform, nontrivial improvement over the least squares (Gauss-Markov) estimators for the parameters in the usual formulation of the linear model. One might expect a rush of applications of this powerful new statistical weapon, but such has not been the case. Resistance has formed along several lines:

1. Mistrust of the statistical interpretation of the mathematical formulation leading to Stein's result, in particular the sum of squared errors loss function;
2. Difficulties in adapting the James-Stein estimator to the many special cases that invariably arise in practice;
3. Long familiarity with the generally good performance of the MLE in applied problems;
4. A feeling that any gains possible from a "complicated" procedure like Stein's could not be worth the extra trouble. (J.W. Tukey at the 1972 American Statistical Association meetings in Montreal stated that savings would not be more than ten percent in practical situations.)

We have written a series of articles [5, 6, 7, 8, 9, 10, 11] that cover Points 1 and 2. Our purpose here, and in a lengthier version of this report [12], is to illustrate the methods suggested in these articles on three applied problems and in that way deals with Points 3 and 4. Only one of the three problems, the toxoplasmosis data, is "real" in the sense of being generated outside the statistical world. The other two problems are contrived to illustrate in a realistic way the genuine difficulties and rewards of procedures like Stein's. They have the added advantage of having the true parameter values available for comparison of methods. The examples chosen are the first and only ones considered for this report, and the favorable results typify our previous experience.

To review the James-Stein estimator in the simplest setting, suppose that for given θ_i

$$X_i | \theta_i \overset{\text{ind}}{\sim} N(\theta_i, 1), \quad i = 1, \cdots, k \geq 3 , \qquad (1.1)$$

meaning the $\{X_i\}$ are independent and normally distributed with mean $E_{\theta_i} X_i \equiv \theta_i$ and variance $\text{Var}_{\theta_i}(X_i) = 1$. The example (1.1) typically occurs as a reduction to this canonical form from more complicated situations, as when X_i is a sample mean with known variance that is taken to be unity through an appropriate scale transformation. The unknown vector of means $\boldsymbol{\theta} \equiv (\theta_1, \cdots, \theta_k)$ is to be estimated with loss being the sum of squared component errors

$$L(\boldsymbol{\theta}, \hat{\boldsymbol{\theta}}) \equiv \sum_{i=1}^{k} (\hat{\theta}_i - \theta_i)^2 , \qquad (1.2)$$

where $\hat{\boldsymbol{\theta}} \equiv (\hat{\theta}_1, \cdots, \hat{\theta}_k)$ is the estimate of $\boldsymbol{\theta}$. The MLE, which is also the sample mean, $\boldsymbol{\delta}^0(\mathbf{X}) \equiv \mathbf{X} \equiv (X_1, \cdots, X_k)$ has constant risk k,

$$R(\boldsymbol{\theta}, \boldsymbol{\delta}^0) \equiv E_{\boldsymbol{\theta}} \sum_{i=1}^{k} (X_i - \theta_i)^2 = k , \qquad (1.3)$$

$E_{\boldsymbol{\theta}}$ indicating expectation over the distribution (1.1). James and Stein [13] introduced the estimator $\boldsymbol{\delta}^1(\mathbf{X}) = (\delta_1^1(\mathbf{X}), \cdots, \delta_k^1(\mathbf{X}))$ for $k \geq 3$,

$$\delta_i^1(\mathbf{X}) \equiv \mu_i + (1 - (k-2)/S)(X_i - \mu_i) , \quad i = 1, \cdots, k \qquad (1.4)$$

with $\boldsymbol{\mu} \equiv (\mu_1, \cdots, \mu_k)'$ any initial guess at $\boldsymbol{\theta}$ and $S \equiv \sum (X_j - \mu_j)^2$. This estimator has risk

$$R(\boldsymbol{\theta}, \boldsymbol{\delta}^1) \equiv E_{\boldsymbol{\theta}} \sum_{i=0}^{k} (\delta_i^1(\mathbf{X}) - \theta_i)^2 \qquad (1.5)$$

$$\leq k - \frac{(k-2)^2}{k - 2 + \sum (\theta_i - \mu_i)^2} < k , \qquad (1.6)$$

being less than k for all $\boldsymbol{\theta}$, and if $\theta_i = \mu_i$ for all i the risk is two, comparing very favorably to k for the MLE.

* Bradley Efron is professor, Department of Statistics, Stanford University, Stanford, Calif. 94305. Carl Morris is statistician, Department of Economics, The RAND Corporation, Santa Monica, Calif. 90406.

Reprinted from: © Journal of the American Statistical Association
June 1975, Volume 70, Number 350
Applications Section
Pages 311-319

The estimator (1.4) arises quite naturally in an empirical Bayes context. If the $\{\theta_i\}$ themselves are a sample from a prior distribution,

$$\theta_i \overset{\text{ind}}{\sim} N(\mu_i, \tau^2), \quad i = 1, \cdots, k, \tag{1.7}$$

then the Bayes estimate of θ_i is the *a posteriori* mean of θ_i given the data

$$\delta_i^*(X_i) = E\theta_i | X_i = \mu_i + (1 - (1 + \tau^2)^{-1})(X_i - \mu_i). \tag{1.8}$$

In the empirical Bayes situation, τ^2 is unknown, but it can be estimated because marginally the $\{X_i\}$ are independently normal with means $\{\mu_i\}$ and

$$S = \sum (X_j - \mu_j)^2 \sim (1 + \tau^2)\chi_k^2, \tag{1.9}$$

where χ_k^2 is the chi-square distribution with k degrees of freedom. Since $k \geq 3$, the unbiased estimate

$$E(k - 2)/S = 1/(1 + \tau^2) \tag{1.10}$$

is available, and substitution of $(k - 2)/S$ for the unknown $1/(1 + \tau^2)$ in the Bayes estimate δ_i^* of (1.8) results in the James-Stein rule (1.4). The risk of δ_i^1 averaged over both $\mathbf{X}$ and $\boldsymbol{\theta}$ is, from [6] or [8],

$$E_\tau E_\theta(\delta_i^1(\mathbf{X}) - \theta_i)^2 = 1 - (k - 2)/k(1 + \tau^2), \tag{1.11}$$

E_τ denoting expectation over the distribution (1.7). The risk (1.11) is to be compared to the corresponding risks of 1 for the MLE and $1 - 1/(1 + \tau^2)$ for the Bayes estimator. Thus, if k is moderate or large δ_i^1 is nearly as good as the Bayes estimator, but it avoids the possible gross errors of the Bayes estimator if τ^2 is misspecified.

It is clearly preferable to use $\min\{1, (k - 2)/S\}$ as an estimate of $1/(1 + \tau^2)$ instead of (1.10). This results in the simple improvement

$$\delta_i^{1+}(\mathbf{X}) = \mu_i + (1 - (k - 2)/S)^+(X_i - \mu_i) \tag{1.12}$$

with $a^+ \equiv \max(0, a)$. That $R(\boldsymbol{\theta}, \boldsymbol{\delta}^{1+}) < R(\boldsymbol{\theta}, \boldsymbol{\delta}^1)$ for all $\boldsymbol{\theta}$ is proved in [2, 8, 10, 17]. The risks $R(\boldsymbol{\theta}, \boldsymbol{\delta}^1)$ and $R(\theta, \delta^{1+})$ are tabled in [11].

2. USING STEIN'S ESTIMATOR TO PREDICT BATTING AVERAGES

The batting averages of 18 major league players through their first 45 official at bats of the 1970 season appear in Table 1. The problem is to predict each player's batting average over the remainder of the season using only the data of Column (1) of Table 1. This sample was chosen because we wanted between 30 and 50 at bats to assure a satisfactory approximation of the binomial by the normal distribution while leaving the bulk of at bats to be estimated. We also wanted to include an unusually good hitter (Clemente) to test the method with at least one extreme parameter, a situation expected to be less favorable to Stein's estimator. Batting averages are published weekly in the *New York Times*, and by April 26, 1970 Clemente had batted 45 times. Stein's estimator requires equal variances,[1] or in this situation, equal at bats, so the remaining 17 players are all whom either the April 26 or May 3 *New York Times* reported with 45 at bats.

Let Y_i be the batting average of Player i, $i = 1, \cdots, 18$ $(k = 18)$ after $n = 45$ at bats. Assuming base hits occur according to a binomial distribution with independence between players, $nY_i \overset{\text{ind}}{\sim} \text{Bin}(n, p_i)$ $i = 1, 2, \cdots, 18$ with p_i the true season batting average, so $EY_i = p_i$. Because the variance of Y_i depends on the mean, the arc-sin transformation for stabilizing the variance of a binomial distribution is used: $X_i \equiv f_{45}(Y_i)$, $i = 1, \cdots, 18$ with

$$f_n(y) \equiv (n)^{\frac{1}{2}} \arcsin(2y - 1). \tag{2.1}$$

Then X_i has nearly unit variance[2] independent of p_i. The mean[3] θ_i of X_i is given approximately by $\theta_i = f_n(p_i)$. Values of X_i, θ_i appear in Table 1. From the central limit theorem for the binomial distribution and continuity of f_n we have approximately

$$X_i | \theta_i \overset{\text{ind}}{\sim} N(\theta_i, 1), \quad i = 1, 2, \cdots, k, \tag{2.2}$$

the situation described in Section 1.

We use Stein's estimator (1.4), but we estimate the common unknown value $\mu = \sum \mu_i/k$ by $\bar{X} = \sum X_i/k$, shrinking all X_i toward $\bar{X}$, an idea suggested by Lindley [6, p. 285–7]. The resulting estimate of the ith component θ_i of $\boldsymbol{\theta}$ is therefore

$$\tilde{\delta}_i^1(\mathbf{X}) = \bar{X} + (1 - (k - 3)/V)(X_i - \bar{X}) \tag{2.3}$$

with $V \equiv \sum (X_i - \bar{X})^2$ and with $k - 3 = (k - 1) - 2$ as the appropriate constant since one parameter is estimated. In the empirical Bayes case, the appropriateness of (2.3) follows from estimating the Bayes rule (1.8) by using the unbiased estimates $\bar{X}$ for μ and $(k - 3)/V$ for $1/(1 + \tau)^2$ from the marginal distribution of $\mathbf{X}$, analogous to Section 1 (see also [6, Sec. 7]). We may use the Bayesian model for these data because (1.7) seems at least roughly appropriate, although (2.3) also can be justified by the non-Bayesian from the suspicion that $\sum (\theta_i - \bar{\theta})^2$ is small, since the risk of (2.3), analogous to (1.6), is bounded by

$$R(\boldsymbol{\theta}, \tilde{\boldsymbol{\delta}}^1) \leq k - \frac{(k - 3)^2}{k - 3 + \sum (\theta_i - \bar{\theta})^2}, \quad \bar{\theta} \equiv \sum \theta_i/k. \tag{2.4}$$

For our data, the estimate of $1/(1 + \tau^2)$ is $(k - 3)/V = .791$ or $\hat{\tau} = 0.514$, representing considerable *a priori* information. The value of $\bar{X}$ is -3.275 so

$$\tilde{\delta}_i^1(\mathbf{X}) = \hat{\theta}_i = .791\bar{X} + .209X_i = .209X_i - 2.59. \tag{2.5}$$

[1] The unequal variances case is discussed in Section 3.

[2] An exact computer computation showed that the standard deviation of X_i is within .036 of unity for $n = 45$ for all p_i between 0.15 and 0.85.

[3] For most of this discussion we will regard the values of p_i of Column 2, Table 1 and θ_i as the quantities to be estimated, although we actually have a prediction problem because these quantities are estimates of the mean of Y_i. Accounting for this fact would cause Stein's method to compare even more favorably to the sample mean because the random error in p_i increases the losses for all estimators equally. This increases the errors of good estimators by a higher percentage than poorer ones.

1. 1970 Batting Averages for 18 Major League Players and Transformed Values X_i, θ_i

i	Player	Y_i = batting average for first 45 at bats	p_i = batting average for remainder of season	At bats for remainder of season	X_i	θ_i
		(1)	(2)	(3)	(4)	(5)
1	Clemente (Pitts, NL)	.400	.346	367	−1.35	−2.10
2	F. Robinson (Balt, AL)	.378	.298	426	−1.66	−2.79
3	F. Howard (Wash, AL)	.356	.276	521	−1.97	−3.11
4	Johnstone (Cal, AL)	.333	.222	275	−2.28	−3.96
5	Berry (Chi, AL)	.311	.273	418	−2.60	−3.17
6	Spencer (Cal, AL)	.311	.270	466	−2.60	−3.20
7	Kessinger (Chi, NL)	.289	.263	586	−2.92	−3.32
8	L. Alvarado (Bos, AL)	.267	.210	138	−3.26	−4.15
9	Santo (Chi, NL)	.244	.269	510	−3.60	−3.23
10	Swoboda (NY, NL)	.244	.230	200	−3.60	−3.83
11	Unser (Wash, AL)	.222	.264	277	−3.95	−3.30
12	Williams (Chi, AL)	.222	.256	270	−3.95	−3.43
13	Scott (Bos, AL)	.222	.303	435	−3.95	−2.71
14	Petrocelli (Bos, AL)	.222	.264	538	−3.95	−3.30
15	E. Rodriguez (KC, AL)	.222	.226	186	−3.95	−3.89
16	Campaneris (Oak, AL)	.200	.285	558	−4.32	−2.98
17	Munson (NY, AL)	.178	.316	408	−4.70	−2.53
18	Alvis (Mil, NL)	.156	.200	70	−5.10	−4.32

The results are striking. The sample mean **X** has total squared prediction error $\sum (X_i - \theta_i)^2$ of 17.56, but $\bar{\delta}^1(\mathbf{X}) \equiv (\bar{\delta}_1{}^1(\mathbf{X}), \cdots, \bar{\delta}_k{}^1(\mathbf{X}))$ has total squared prediction error of only 5.01. The efficiency of Stein's rule relative to the MLE for these data is defined as $\sum (X_i - \theta_i)^2 / \sum (\bar{\delta}_i{}^1(\mathbf{X}) - \theta_i)^2$, the ratio of squared error losses. The efficiency of Stein's rule is 3.50 (=17.56/5.01) in this example. Moreover, $\bar{\delta}_i{}^1$ is closer than X_i to θ_i for 15 batters, being worse only for Batters 1, 10, 15. The estimates (2.5) are retransformed in Table 2 to provide estimates $\hat{p}_i{}^1 = f_n{}^{-1}(\hat{\theta}_i)$ of p_i.

Stein's estimators achieve uniformly lower aggregate risk than the MLE but permit considerably increased risk to individual components of the vector $\boldsymbol{\theta}$. As a function of $\boldsymbol{\theta}$, the risk for estimating θ_1 by $\bar{\delta}_1{}^1$, for example, can be as large as $k/4$ times as great as the risk of the MLE X_1. This phenomenon is discussed at length in [5, 6], where "limited translation estimators" $\bar{\delta}^s(\mathbf{X})$ $0 \le s \le 1$ are introduced to reduce this effect. The MLE corresponds to $s = 0$, Stein's estimator to $s = 1$. The estimate $\bar{\delta}_i{}^s(\mathbf{X})$ of θ_i is defined to be as close as possible to $\bar{\delta}_i{}^1(\mathbf{X})$ subject to the condition that it not differ from X_i by more than $[(k-1)(k-3)/kV]^{\frac{1}{2}} D_{k-1}(s)$ standard deviations of X_i, $D_{k-1}(s)$ being a constant taken from [6, Table 1]. If $s = 0.8$, then $D_{17}(s) = 0.786$, so $\bar{\delta}_i{}^{0.8}(\mathbf{X})$ may differ from X_i by no more than

$$0.786\ (17 \times 0.791/18)^{\frac{1}{2}} = .68\ .$$

This modification reduces the maximum component risk of 4.60 for $\bar{\delta}_i{}^1$ to 1.52 for $\bar{\delta}_i{}^{0.8}$ while retaining 80 percent of the savings of Stein's rule over the MLE. The retransformed values $\hat{p}_i{}^{0.8}$ of the limited translation estimates $f_n{}^{-1}(\bar{\delta}_i{}^{0.8}(\mathbf{X}))$ are given in the last column of Table 2, the estimates for the top three and bottom two batters being affected. Values for $s = 0.9$ are also given in Table 2.

2. Batting Averages and Their Estimates

i	Batting average for season remainder p_i	Maximum likelihood estimate Y_i	Retransform of Stein's estimator $\hat{p}_i{}^1$	Retransform of $\bar{\delta}^{0.9}$ $\hat{p}_i{}^{0.9}$	Retransform of $\bar{\delta}^{0.8}$ $\hat{p}_i{}^{0.8}$
1	.346	.400	.290	.334	.351
2	.298	.378	.286	.313	.329
3	.276	.356	.281	.292	.308
4	.222	.333	.277	.277	.287
5	.273	.311	.273	.273	.273
6	.270	.311	.273	.273	.273
7	.263	.289	.268	.268	.268
8	.210	.267	.264	.264	.264
9	.269	.244	.259	.259	.259
10	.230	.244	.259	.259	.259
11	.264	.222	.254	.254	.254
12	.256	.222	.254	.254	.254
13	.303	.222	.254	.254	.254
14	.264	.222	.254	.254	.254
15	.226	.222	.254	.254	.254
16	.285	.200	.249	.249	.242
17	.316	.178	.244	.233	.218
18	.200	.156	.239	.208	.194

Clemente ($i = 1$) was known to be an exceptionally good hitter from his performance in other years. Limiting translation results in a much better estimate for him, as we anticipated, since $\bar{\delta}_1{}^1(\mathbf{X})$ differs from X_1 by an excessive 1.56 standard deviations of X_1. The limited translation estimators are closer than the MLE for 16 of the 18 batters, and the case $s = 0.9$ has better efficiency (3.91) for these data relative to the MLE than Stein's rule (3.50), but the rule with $s = 0.8$ has lower efficiency (3.01). The maximum component error occurs for Munson ($i = 17$) with all four estimators. The Bayesian effect is so strong that this maximum error $|\hat{\theta}_{17} - \theta_{17}|$ decreased from 2.17 for $s = 0$, to 1.49 for $s = 0.8$, to 1.25 for $s = 0.9$ to 1.08 for $s = 1$. Limiting translation therefore increases the worst error in this example, just opposite to the maximum risks.

3. A GENERALIZATION OF STEIN'S ESTIMATOR TO UNEQUAL VARIANCES FOR ESTIMATING THE PREVALENCE OF TOXOPLASMOSIS

One of the authors participated in a study of toxoplasmosis in El Salvador [14]. Sera obtained from a total sample of 5,171 individuals of varying ages from 36 El Salvador cities were analyzed by a Sabin-Feldman dye test. From the data given in [14, Table 1], toxoplasmosis prevalence rates X_i for City i, $i = 1, \cdots, 36$ were calculated. The prevalence rate X_i has the form (observed minus expected)/expected, with "observed" being the number of positives for City i and "expected" the number of positives for the same city based on an indirect standardization of prevalence rates to the age distribution of City i. The variances $D_i = \text{Var}(X_i)$ are known from binomial considerations and differ because of unequal sample sizes.

These data X_i together with the standard deviations $D_i^{\frac{1}{2}}$ are given in Columns 2 and 3 of Table 3. The prevalence rates satisfy a linear constraint $\sum d_i X_i = 0$ with known coefficients $d_i > 0$. The means $\theta_i = EX_i$, which also satisfy $\sum d_i\theta_i = 0$, are to be estimated from the $\{X_i\}$. Since the $\{X_i\}$ were constructed as sums of independent random variables, they are approximately normal; and except for the one linear constraint on the $k = 36$ values of X_i, they are independent. For simplicity, we will ignore the slight improvement in the independence approximation that would result from applying our methods to an appropriate 35-dimensional subspace and assume that the $\{X_i\}$ have the distribution of the following paragraph.

3. Estimates and Empirical Bayes Estimates of Toxoplasmosis Prevalence Rates

i	X_i	$\sqrt{D_i}$	$\delta_i(\mathbf{X})$	$\hat{A}_i$	k_i	$\hat{B}_i$
1	.293	.304	.035	.0120	1334.1	.882
2	.214	.039	.192	.0108	21.9	.102
3	.185	.047	.159	.0109	24.4	.143
4	.152	.115	.075	.0115	80.2	.509
5	.139	.081	.092	.0112	43.0	.336
6	.128	.061	.100	.0110	30.4	.221
7	.113	.061	.088	.0110	30.4	.221
8	.098	.087	.062	.0113	48.0	.370
9	.093	.049	.079	.0109	25.1	.154
10	.079	.041	.070	.0109	22.5	.112
11	.063	.071	.045	.0111	36.0	.279
12	.052	.048	.044	.0109	24.8	.148
13	.035	.056	.028	.0110	28.0	.192
14	.027	.040	.024	.0108	22.2	.107
15	.024	.049	.020	.0109	25.1	.154
16	.024	.039	.022	.0108	21.9	.102
17	.014	.043	.012	.0109	23.1	.122
18	.004	.085	.003	.0112	46.2	.359
19	−.016	.128	−.007	.0116	101.5	.564
20	−.028	.091	−.017	.0113	51.6	.392
21	−.034	.073	−.024	.0111	37.3	.291
22	−.040	.049	−.034	.0109	25.1	.154
23	−.055	.058	−.044	.0110	28.9	.204
24	−.083	.070	−.060	.0111	35.4	.273
25	−.098	.068	−.072	.0111	34.2	.262
26	−.100	.049	−.085	.0109	25.1	.154
27	−.112	.059	−.089	.0110	29.4	.210
28	−.138	.063	−.106	.0110	31.4	.233
29	−.156	.077	−.107	.0112	40.0	.314
30	−.169	.073	−.120	.0111	37.3	.291
31	−.241	.106	−.128	.0114	68.0	.468
32	−.294	.179	−.083	.0118	242.4	.719
33	−.296	.064	−.225	.0111	31.9	.238
34	−.324	.152	−.114	.0117	154.8	.647
35	−.397	.158	−.133	.0117	171.5	.665
36	−.665	.216	−.140	.0119	426.8	.789

To obtain an appropriate empirical Bayes estimation rule for these data we assume that

$$X_i \mid \theta_i \overset{\text{ind}}{\sim} N(\theta_i, D_i), \quad i = 1, \cdots, k \tag{3.1}$$

and

$$\theta_i \overset{\text{ind}}{\sim} N(0, A), \quad i = 1, \cdots, k , \tag{3.2}$$

A being an unknown constant. These assumptions are the same as (1.1), (1.7), which lead to the James-Stein estimator if $D_i = D_j$ for all i, j. Notice that the choice of *a priori* mean zero for the θ_i is particularly appropriate here because the constant $\sum d_i\theta_i = 0$ forces the parameters to be centered near the origin.

We require $k \geq 3$ in the following derivations. Define

$$B_i \equiv D_i/(A + D_i) \ . \tag{3.3}$$

Then (3.1) and (3.2) are equivalent to

$$\theta_i \mid X_i \overset{\text{ind}}{\sim} N((1 - B_i)X_i, D_i(1 - B_i)), \quad i = 1, \cdots, k \ . \tag{3.4}$$

For squared error loss[4] the Bayes estimator is the *a posteriori* mean

$$\delta_i^*(X_i) = E\theta_i \mid X_i = (1 - B_i)X_i \ , \tag{3.5}$$

with Bayes risk $\text{Var}(\theta_i \mid X_i) = (1 - B_i)D_i$ being less than the risk D_i of $\hat{\theta}_i = X_i$.

Here, A is unknown, but the MLE $\hat{A}$ of A on the basis of the data $S_j \equiv X_j^2 \sim (A + D_j)\chi_1^2$, $j = 1, 2, \cdots, k$ is the solution to

$$\hat{A} = \sum_{j=1}^{k} (S_j - D_j)I_j(\hat{A}) / \sum_{j=1}^{k} I_j(\hat{A}) \tag{3.6}$$

with

$$I_j(A) \equiv 1/\text{Var}(S_j) = 1/[2(A + D_j)^2] \tag{3.7}$$

being the Fisher information for A in S_j. We could use $\hat{A}$ from (3.6) to define the empirical Bayes estimator of θ_i as $(1 - D_i/(\hat{A} + D_i))X_i$. However, this rule does not reduce to Stein's when all D_j are equal, and we instead use a minor variant of this estimator derived in [8] which does reduce to Stein's. The variant rule estimates a different value $\hat{A}_i$ for each city (see Table 3). The difference between the rules is minor in this case, but it might be important if k were smaller.

Our estimates $\delta_i(\mathbf{X})$ of the θ_i are given in the fourth column of Table 3 and are compared with the unbiased

[4] Or for any other increasing function of $|\theta_i - \hat{\theta}_i|$.

estimate X_i in Figure A. Figure A illustrates the "pull in" effect of $\delta_i(\mathbf{X})$, which is most pronounced for Cities 1, 32, 34, 35, and 36. Under the empirical Bayes model, the major explanation for the large $|X_i|$ for these cities is large D_i rather than large $|\theta_i|$. This figure also shows that the rankings of the cities on the basis of $\delta_i(\mathbf{X})$ differs from that based on the X_i, an interesting feature that does not arise when the X_i have equal variances.

A. Estimates of Toxoplasmosis Prevalence Rates

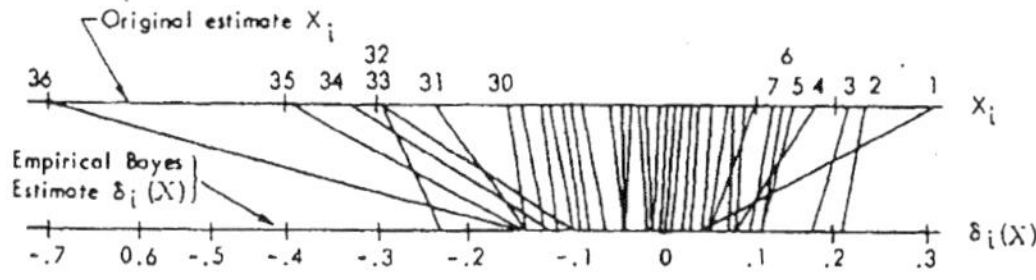

The values $\hat{A}_i$, $\hat{k}_i$, and $\hat{B}_i(S)$ defined in [8] are given in the last three columns of Table 3. The value $\hat{A}$ of (3.6) is $\hat{A} = 0.0122$ with standard deviation $\sigma(\hat{A})$ estimated as 0.0041 (if $A = 0.0122$) by the Cramér-Rao lower bound on $\sigma(\hat{A})$. The preferred estimates $\hat{A}_i$ are all close to but slightly smaller than $\hat{A}$, and their estimated standard deviations vary from 0.00358 for the cities with the smallest D_i to 0.00404 for the city with the largest D_i.

The likelihood function of the data plotted as a function of A (on a log scale) is given in Figures B and C as LIKELIHOOD. The curves are normalized to have unit area as a function of $\alpha = \log A$. The maximum value of this function of α is at $\hat{\alpha} = \log(\hat{A}) = \log(.0122) = -4.40 \equiv \mu_\alpha$. The curves are almost perfectly normal with mean $\hat{\alpha} = -4.40$ and standard deviation $\sigma_\alpha \equiv .371$. The likely values of A therefore correspond to a α differing from μ_α by no more than three standard deviations, $|\alpha - \mu_\alpha| \leq 3\sigma_\alpha$, or equivalently, $.0040 \leq A \leq .0372$.

In the region of likely values of A, Figure B also graphs two risks: BAYES RISK and EB RISK (for empirical Bayes risk), each conditional on the data $\mathbf{X}$. EB RISK[5] is the conditional risk of the empirical Bayes rule defined (with $D_0 \equiv (1/k) \sum_{i=1}^{k} D_i$) as

$$E_A \frac{1}{kD_0} \sum_{i=0}^{k} (\delta_i(\mathbf{X}) - \theta_i)^2 | \mathbf{X} , \tag{3.8}$$

and BAYES RISK is

$$E_A \frac{1}{kD_0} \sum_{i=1}^{k} \left(\frac{A}{A + D_i} X_i - \theta_i \right)^2 \Bigg| \mathbf{X} . \tag{3.9}$$

Since A is not known, BAYES RISK yields only a lower envelope for empirical Bayes estimators, agreeing with EB RISK at $A = .0122$. Table 4 gives values to supplement Figure B. Not graphed because it is too large to fit in Figure B is MLE RISK, the conditional risk of the MLE, defined as

$$E_A \frac{1}{kD_0} \sum_{i=1}^{k} (X_i - \theta_i)^2 | \mathbf{X} . \tag{3.10}$$

MLE RISK exceeds EB RISK by factors varying from 7 to 2 in the region of likely values of A, as shown in Table 4. EB RISK tends to increase and MLE RISK to decrease as A increases, these values crossing at $A = .0650$, about $4\frac{1}{2}$ standard deviations above the mean of the distribution of $\hat{A}$.

4. Conditional Risks for Different Values of A

Risk	A				
	.0040	.0122	.0372	.0650	∞
EB RISK	.35	.39	.76	1.08	2.50
MLE RISK	2.51	1.87	1.27	1.08	1.00
P(EB CLOSER)	1.00	1.00	.82	.50	.04

The remaining curve in Figure B graphs the probability that the empirical Bayes estimator is closer to $\boldsymbol{\theta}$ than the MLE $\mathbf{X}$, conditional on the data $\mathbf{X}$. It is defined as

$$P_A[\sum (\delta_i(\mathbf{X}) - \theta_i)^2 < \sum (X_i - \theta_i)^2 | \mathbf{X}] . \tag{3.11}$$

This curve, denoted P(EB CLOSER), decreases as A increases but is always very close to unity in the region of likely values of A. It reaches one-half at about $4\frac{1}{2}$ standard deviations from the mean of the likelihood function and then decreases as $A \to \infty$ to its asymptotic value .04 (see Table 4).

The data suggest that almost certainly A is in the interval $.004 \leq A \leq .037$, and for all such values of A, Figure B and Table 4 indicate that the numbers $\delta_i(\mathbf{X})$ are much better estimators of the θ_i than are the X_i. Non-Bayesian versions of these statements may be based on a confidence interval for $\sum \theta_i^2/k$.

Figure A illustrates that the MLE and the empirical Bayes estimators order the $\{\theta_i\}$ differently. Define the

B. Likelihood Function of A and Aggregate Operating Characteristics of Estimates as a Function of A, Conditional on Observed Toxoplasmosis Data

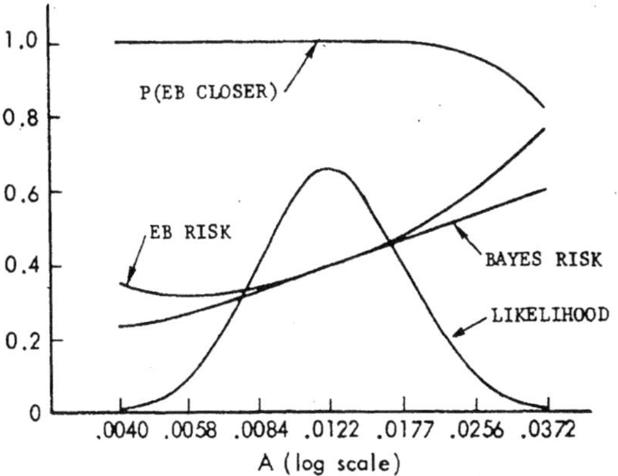

[5] In (3.8) the $\delta_i(\mathbf{X})$ are fixed numbers—those given in Table 3. The expectation is over the *a posteriori* distribution (3.4) of the θ_i.

correlation of an estimator $\hat{\theta}$ of θ by

$$r(\hat{\theta}, \theta) = \sum \hat{\theta}_i \theta_i / (\sum \hat{\theta}_i^2 \sum \theta_i^2)^{\frac{1}{2}} \quad (3.12)$$

as a measure of how well $\hat{\theta}$ orders θ. We denote $P(r^{EB} > r^{MLE})$ as the probability that the empirical Bayes estimate δ orders θ better than $\mathbf{X}$, i.e., as

$$P_A\{r(\delta, \theta) > r(\mathbf{X}, \theta) | \mathbf{X}\} \ . \quad (3.13)$$

The graph of (3.13) given in Figure C shows that $P(r^{EB} > r^{MLE}) > .5$ for $A \leq .0372$. The value at $A = \infty$ drops to .046.

C. Likelihood Function of A and Individual and Ordering Characteristics of Estimates as a Function of A, Conditional on Observed Toxoplasmosis Data

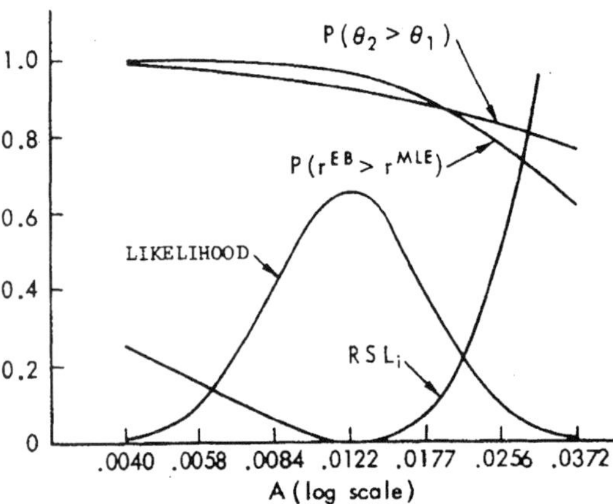

Although $X_1 > X_2$, the empirical Bayes estimator for City 2 is larger, $\delta_2(\mathbf{X}) > \delta_1(\mathbf{X})$. This is because $D_1 \gg D_2$, indicating that X_1 is large under the empirical Bayes model because of randomness while X_2 is large because θ_2 is large. The other curve in Figure C is

$$P_A(\theta_2 > \theta_1 | \mathbf{X}) \quad (3.14)$$

and shows that $\theta_2 > \theta_1$ is quite probable for likely values of A. This probability declines as $A \to \infty$, being .50 at $A = .24$ (eight standard deviations above the mean) and .40 at $A = \infty$.

4. USING STEIN'S ESTIMATOR TO IMPROVE THE RESULTS OF A COMPUTER SIMULATION

A Monte Carlo experiment is given here in which several forms of Stein's method all double the experimental precision of the classical estimator. The example is realistic in that the normality and variance assumptions are approximations to the true situation.

We chose to investigate Pearson's chi-square statistic for its independent interest and selected the particular parameters ($m \leq 24$) from our prior belief that empirical Bayes methods would be effective for these situations. Although our beliefs were substantiated, the outcomes in this instance did not always favor our pet methods.

The simulation was conducted to estimate the exact size of Pearson's chi-square test. Let Y_1 and Y_2 be independent binomial random variables, $Y_1 \sim \text{bin}(m, p')$, $Y_2 \sim \text{bin}(m, p'')$ so $EY_1 = mp'$, $EY_2 = mp''$. Pearson advocated the statistic and critical region

$$T = \frac{2m(Y_1 - Y_2)^2}{(Y_1 + Y_2)(2m - Y_1 - Y_2)} > 3.84 \quad (4.1)$$

to test the composite null hypothesis $H_0: p' = p''$ against all alternatives for the nominal size $\alpha = 0.05$. The value 3.84 is the 95th percentile of the chi-square distribution with one degree of freedom, which approximates that of T when m is large.

The true size of the test under H_0 is defined as

$$\alpha(p, m) \equiv P(T > 3.84 | p, m) \ , \quad (4.2)$$

which depends on both m and the unknown value $p \equiv p' = p''$. The simulation was conducted for $p = 0.5$ and the $k = 17$ values of m with $m_j = 7 + j$, $j = 1, \cdots, k$. The k values of $\alpha_j \equiv \alpha(0.5, m_j)$ were to be estimated. For each j we simulated (4.1) $n = 500$ times on a computer and recorded Z_j as the proportion of times H_0 was rejected. The data appear in Table 5. Since $nZ_j \sim \text{bin}(n, \alpha_j)$ independently, Z_j is the unbiased and maximum likelihood estimator usually chosen[6] to estimate α_j.

5. Maximum Likelihood Estimates and True Values for p = 0.5

	MLE		True values
j	m_j	Z_j	α_j
1	8	.082	.07681
2	9	.042	.05011
3	10	.046	.04219
4	11	.040	.05279
5	12	.054	.06403
6	13	.084	.07556
7	14	.036	.04102
8	15	.036	.04559
9	16	.040	.05151
10	17	.050	.05766
11	18	.078	.06527
12	19	.030	.05306
13	20	.036	.04253
14	21	.060	.04588
15	22	.052	.04896
16	23	.046	.05417
17	24	.054	.05950

Under H_0 the standard deviation of Z_j is approximately $\sigma = \{(.05)(.95)/500\}^{\frac{1}{2}} = .009747$. The variables $X_j \equiv (Z_j - .05)/\sigma$ have expectations

$$\theta_j \equiv EX_j = (\alpha_j - .05)/\sigma$$

[6] We ignore an extensive bibliography of other methods for improving computer simulations. Empirical Bayes methods can be applied simultaneously with other methods, and if better estimates of α_j than Z_j were available then the empirical Bayes methods could instead be applied to them. But for simplicity we take Z_j itself as the quantity to be improved.

and approximately the distribution

$$X_j|\theta_j \overset{\text{ind}}{\sim} N(\theta_j, 1), \quad j = 1, 2, \cdots, 17 = k \ , \quad (4.3)$$

described in earlier sections.

The average value $\bar{Z} = .051$ of the 17 points supports the choice of the "natural origin" $\bar{\alpha} = .05$. Stein's rule (1.4) applied to the transformed data (4.3) and then retransformed according to $\hat{\alpha}_j = .05 + \sigma\hat{\theta}_j$ yields

$$\hat{\alpha}_j = (1 - \hat{B})Z_j + .05\hat{B} \ , \quad \hat{B} = .325 \ , \quad (4.4)$$

where $\hat{B} \equiv (k - 2)/S$ and

$$S \equiv \sum_{j=1}^{17} (Z_j - .05)^2/\sigma^2 = 46.15 \ .$$

All 17 true values α_j were obtained exactly through a separate computer program and appear in Figure D and Table 5, so the loss function, taken to be the normalized sum of squared errors $\sum (\hat{\alpha}_j - \alpha_j)^2/\sigma^2$, can be evaluated.[7] The MLE has loss 18.9, Stein's estimate (4.4) has loss 10.2, and the constant estimator, which always estimates α_j as .05, has loss 23.4. Stein's rule therefore dominates both extremes between which it compromises.

Figure D displays the maximum likelihood estimates, Stein estimates, and true values. The true values show a surprising periodicity, which would frustrate attempts at improving the MLE by smoothing.

D. MLE, Stein Estimates, and True Values for p = 0.5

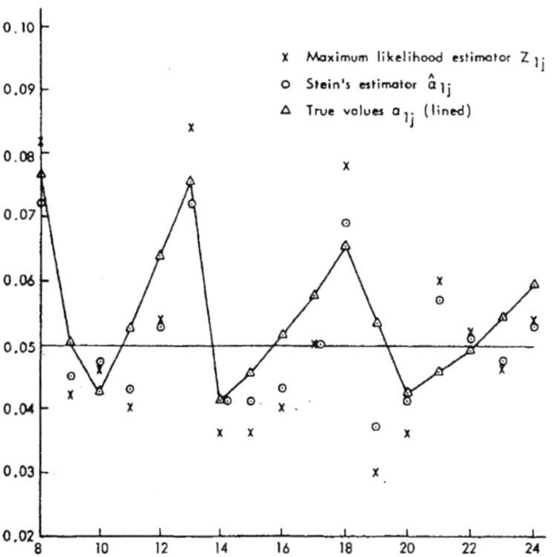

On theoretical grounds we know that the approximation $\alpha(p, m) = .05$ improves as m increases, which suggests dividing the data into two groups, say $8 \leq m \leq 16$ and $17 \leq m \leq 24$. In the Bayesian framework [9] this disaggregation reflects the concern that A_1, the expectation of $A_1^* \equiv \sum_{j=1}^{9} (\alpha_j - .05)^2/9\sigma^2$ may be much larger than A_2, the expectation of $A_2^* = \sum_{j=10}^{17} (\alpha_j - .05)^2/8\sigma^2$, or equivalently that the pull-in factor $B_1 = 1/(1 + A_1)$ for Group 1 really should be smaller than $B_2 = 1/(1 + A_2)$ for Group 2.

[7] Exact rejection probabilities for other values of p are given in [12].

The combined estimator (4.4), having $\hat{B}_1 = \hat{B}_2$, is repeated in the second row of Table 6 with loss components for each group. The simplest way to utilize separate estimates of B_1 and B_2 is to apply two separate Stein rules, as shown in the third row of the table.

6. Values of $\hat{B}$ and Losses for Data Separated into Two Groups, Various Estimation Rules

Rule	$8 \leq m \leq 16$ $\hat{B}_1$	Group 1 loss	$17 \leq m \leq 24$ $\hat{B}_2$	Group 2 loss	Total loss
Maximum Likelihood Estimator	.000	7.3	.000	11.6	18.9
Stein's rule, combined data	.325	4.2	.325	6.0	10.2
Separate Stein rules	.232	4.5	.376	5.4	9.9
Separate Stein rules, bigger constant	.276	4.3	.460	4.6	8.9
All estimates at .05	1.000	18.3	1.000	5.1	23.4

In [8, Sec. 5] we suggest using the bolder estimate

$$\hat{B}_i = (k_i - .66)/S_i \ , \quad S_1 \equiv \sum_{j=1}^{9} (Z_j - .05)^2/\sigma^2 \ ,$$

$$S_2 \equiv S - S_1 \ , \quad k_1 = 9 \ , \quad k_2 = 8 \ .$$

The constant $k_i - .66$ is preferred because it accounts for the fact that the positive part (1.12) will be used, whereas the usual choice $k_i - 2$ does not. The fourth row of Table 6 shows the effectiveness of this choice.

The estimate of .05, which is nearly the mean of the 17 values, is included in the last row of the table to show that the Stein rules substantially improve the two extremes between which they compromise.

The actual values are

$$A_1^* = \sum_{j=1}^{9} (\alpha_j - .05)^2/9\sigma^2 = 2.036$$

for Group 1 and

$$A_2^* = \sum_{j=10}^{17} (\alpha_j - .05)^2/8\sigma^2 = .635 \ ,$$

so $B_1^* = 1/(1 + A_1^*) = .329$ and $B_2^* = 1/(1 + A_2^*) = .612$. The true values of B_1^* and B_2^* are somewhat different, as estimates for separate Stein rules suggest. Rules with $\hat{B}_1$ and $\hat{B}_2$ near these true values will ordinarily perform better for data simulated from these parameters $p = 0.5, m = 8, \cdots, 24$.

5. CONCLUSIONS

In the baseball, toxoplasmosis, and computer simulation examples, Stein's estimator and its generalizations increased efficiencies relative to the MLE by about 350 percent, 200 percent, and 100 percent. These examples were chosen because we expected empirical Bayes methods to work well for them and because their efficiencies could be determined. But we are aware of other successful applications to real data[8] and have suppressed no negative results. Although blind application of these methods would gain little in most instances, the statistician who uses them sensibly and selectively can expect major improvements.

Even when they do not significantly increase efficiency, there is little penalty for using the rules discussed here because they cannot give larger total mean squared error than the MLE and because the limited translation modification protects individual components. As several authors have noted, these rules are also robust to the assumption of the normal distribution, because their operating characteristics depend primarily on the means and variances of the sampling distributions and of the unknown parameters. Nor is the sum of squared error criterion especially important. This robustness is borne out by the experience in this article since the sampling distributions were actually binomial rather than normal. The rules not only worked well in the aggregate here, but for most components the empirical Bayes estimators ranged from slightly to substantially better than the MLE, with no substantial errors in the other direction.

Tukey's comment, that empirical Bayes benefits are unappreciable (Section 1), actually was directed at a method of D.V. Lindley. Lindley's rules, though more formally Bayesian, are similar to ours in that they are designed to pick up the same intercomponent information in possibly related estimation problems. We have not done justice here to the many other contributors to multiparameter estimation, but refer the reader to the lengthy bibliography in [12]. We have instead concentrated on Stein's rule and its generalizations to illustrate the power of the empirical Bayes theory, because the main gains are derived by recognizing the applicability of the theory, with lesser benefit attributable to the particular method used. Nevertheless, we hope other authors will compare their methods with ours on these or other data.

The rules of this article are neither Bayes nor admissible, so they can be uniformly beaten (but not by much; see [8, Sec. 6]). There are several published, admissible, minimax rules which also would do well on the baseball data, although probably not much better than the rule used there, for none yet given is known to dominate Stein's rule with the positive part modification. For applications, we recommend the combination of simplicity, generalizability, efficiency, and robustness found in the estimators presented here.

The most favorable situation for these estimators occurs when the statistician wants to estimate the parameters of a linear model that are known to lie in a high dimensional parameter space H_1, but he suspects that they may lie close to a specified lower dimensional parameter space $H_0 \subset H_1$.[9] Then estimates unbiased for every parameter vector in H_1 may have large variance, while estimates restricted to H_0 have smaller variance but possibly large bias. The statistician need not choose between these extremes but can instead view them as endpoints on a continuum and use the data to determine the compromise (usually a smooth function of the likelihood ratio statistic for testing H_0 versus H_1) between bias and variance through an appropriate empirical Bayes rule, perhaps Stein's or one of the generalizations presented here.

We believe many applications embody these features and that most data analysts will have good experiences with the sensible use of the rules discussed here. In view of their potential, we believe empirical Bayes methods are among the most under utilized in applied data analysis.

[*Received October 1973. Revised February 1975.*]

REFERENCES

[1] Anscombe, F., "The Transformation of Poisson, Binomial and Negative-Binomial Data," *Biometrika*, 35 (December 1948), 246–54.

[2] Baranchik, A.J., "Multiple Regression and Estimation of the Mean of a Multivariate Normal Distribution," Technical Report No. 51, Stanford University, Department of Statistics, 1964.

[3] Carter, G.M. and Rolph, J.E., "Empirical Bayes Methods Applied to Estimating Fire Alarm Probabilities," *Journal of the American Statistical Association*, 69, No. 348 (December 1974), 880–5.

[4] Efron, B., "Biased Versus Unbiased Estimation," *Advances in Mathematics*, New York: Academic Press (to appear 1975).

[5] ——— and Morris, C., "Limiting the Risk of Bayes and Empirical Bayes Estimators—Part I: The Bayes Case," *Journal of the American Statistical Association*, 66, No. 336 (December 1971), 807–15.

[6] ——— and Morris, C., "Limiting the Risk of Bayes and Empirical Bayes Estimators—Part II: The Empirical Bayes Case," *Journal of the American Statistical Association*, 67, No. 337 (March 1972), 130–9.

[7] ——— and Morris, C., "Empirical Bayes on Vector Observations—An Extension of Stein's Method," *Biometrika*, 59, No. 2 (August 1972), 335–47.

[8] ——— and Morris, C., "Stein's Estimation Rule and Its Competitors—An Empirical Bayes Approach," *Journal of the American Statistical Association*, 68, No. 341 (March 1973), 117–30.

[9] ——— and Morris, C., "Combining Possibly Related Estimation Problems," *Journal of the Royal Statistical Society*, Ser. B, 35, No. 3 (November 1973; with discussion), 379–421.

[10] ——— and Morris, C., "Families of Minimax Estimators of the Mean of a Multivariate Normal Distribution," P-5170, The RAND Corporation, March 1974, submitted to *Annals of Mathematical Statistics* (1974).

[11] ——— and Morris, C., "Estimating Several Parameters Simultaneously," to be published in *Statistica Neerlandica*.

[12] ——— and Morris, C., "Data Analysis Using Stein's Estimator and Its Generalizations," R-1394-OEO, The RAND Corporation, March 1974.

[13] James, W. and Stein, C., "Estimation with Quadratic Loss,"

[8] See, e.g., [3] for estimating fire alarm probabilities and [4] for estimating reaction times and sunspot data.

[9] One excellent example [17] takes H_0 as the main effects in a two-way analysis of variance and $H_1 - H_0$ as the interactions.

Proceedings of the Fourth Berkeley Symposium on Mathematical Statistics and Probability, Vol. 1, Berkeley: University of California Press, 1961, 361–79.

[14] Remington, J.S., *et al.*, "Studies on Toxoplamosis in El Salvador: Prevalence and Incidence of Toxoplasmosis as Measured by the Sabin-Feldman Dye Test," *Transactions of the Royal Society of Tropical Medicine and Hygiene*, 64, No. 2 (1970), 252–67.

[15] Stein, C., "Inadmissibility of the Usual Estimator for the Mean of a Multivariate Normal Distribution," *Proceedings of the Third Berkeley Symposium on Mathematical Statistics and Probability*, Vol. 1, Berkeley: University of California Press, 1955, 197–206.

[16] ———, "Confidence Sets for the Mean of a Multivariate Normal Distribution," *Journal of the Royal Statistical Society*, Ser. B, 24, No. 2 (1962), 265–96.

[17] ———, "An Approach to the Recovery of Inter-Block Information in Balanced Incomplete Block Designs," in F.N. David, ed., *Festschrift for J. Neyman*, New York: John Wiley & Sons, Inc., 1966, 351–66.

5

Estimating the Number of Unseen Species: How Many Words did Shakespeare Know?

Introduction by Peter McCullagh
University of Chicago

This paper is the first of two written by Brad Efron and Ron Thisted studying the frequency distribution of words in the Shakespearean canon. The key idea due to Fisher et al. (1943) in the context of sampling of species is simple and elegant. When applied to Shakespeare, the idea appears to be preposterous: an author has a personal vocabulary of word species represented by a distribution G, and text is generated by sampling from this distribution. Most results do not require successive words to be sampled independently, which leaves room for individual style and context, but stationarity is needed for prediction and inference.

The expected number of words that occur $x \geq 1$ times in a large sample of n words is

$$\eta_x = E(m_x) = n \int e^{-\lambda} \lambda^x \, dG(\lambda)/x!,$$

and one target for inference is η_0, the number of unseen word species. Fisher employed the Gamma model for the distribution of λ, as do most subsequent workers such as Good and Toulmin (1956), Holgate (1969), and Mosteller and Wallace (1984). The gamma model implies that $\eta_x = \eta_1 \gamma^{x-1} \Gamma(x + \alpha)/(x! \, \Gamma(1 + \alpha))$ proportional to the negative binomial frequencies for some constant γ less than 1 but typically close to 1. Finiteness of η_0 requires $\alpha > 0$, but the expression for η_x with $x > 1$ makes sense for $\alpha > -1$, and the estimate for Shakespeare is negative. Presumably Shakespeare's vocabulary increased over time, so an infinite estimate is not unreasonable.

As Efron and Thisted point out, there is no compelling reason to suppose that Fisher's gamma model will fit Shakespeare's word frequencies. Stylistic and literary compositional arguments are not supportive of random sampling. In fact, the fit is extremely good. Despite the overwhelming textual evidence to the contrary, the word frequency data are consistent with the Bard choosing his words at random.

Fisher's negative binomial model is a precursor of the celebrated Ewens distribution (Ewens 1972) on partitions. One way to generate a random partition of the integer n is to generate a random permutation of $\{1, \ldots, n\}$ with weight function $\theta^{\#\pi}$ where $\#\pi$ is the number of cycles in the permutation π. The associated exponential family of distributions on permutations is $\Gamma(\theta)\theta^{\#\pi} / \Gamma(n + \theta)$. If π contains m_1 uni-cycles of length 1, m_2 bi-cycles, m_3 tri-cycles and so on, then $\#\pi = m_.$ and the induced integer partition is $1^{m_1} 2^{m_2} \cdots n^{m_n}$. The marginal distribution on integer partitions is the Ewens distribution

$$p_n(1^{m_1} 2^{m_2} \ldots n^{m_n}) = \frac{\Gamma(\theta)\, \theta^{m_.}\, n!}{\Gamma(n + \theta) \prod_{j=1}^{n} j^{m_j}\, m_j!},$$

where $m_. = \sum m_j$ is the number of parts or word species in a sample of $n = \sum_{x=1}^{n} x m_x$ words. In other words, the multiplicities $m_1, \ldots$ are independent Poisson random variables with mean $E(m_x) = \theta/x$ subject to the condition that $\sum x m_x = n$. This point is on the boundary of the Fisher model ($\alpha = 0, \gamma = 1$). It is a curious fact that in ecological, genetic and literary applications, the boundary points of the model appear to be the most natural. For further discussion in this direction, see Keener et al. (1987).

The modern literature on random partitions emphasizes exchangeable partition processes generated by sampling from a symmetric Dirichlet distribution or Dirichlet process. A partition process has a sequential representation in the form of a Chinese restaurant process (Pitman 2006, Sect. 3.2), specifying the conditional distribution for each new word or specimen. Fisher's gamma model coincides with the two-parameter Chinese restaurant process, and this process allows α to be negative.

As chance would have it, the opportunity arose in 1985 to apply the Efron–Thisted model to test the authorship of a poem discovered by Gary Taylor in the archives of the Bodleian Library. The new poem of 429 words includes nine words not previously seen in Shakespeare's writings. The Efron–Thisted model predicts 6.97 new words, a rare statistical bullseye predating the observation. Additional tests based on rare words indicated that the new poem is reasonably compatible with previous Shakespearean usage.

5.0.1. Acknowledgment

Support for this research was provided by NSF Grant DMS-0305009.

References

Ewens, W. J. (1972). The sampling theory of selectively neutral alleles. *Theoretical Population Biology* **3**, 87–112.

Fisher, R. A., Corbet, A. S. and Williams, C. B. (1943). The relation between the number of species and the number of individuals in a random sample of an animal population. *The Journal of Animal Ecology* **12**, 42–58.

Good, I. J. and Toulmin, G. H. (1956). The number of new species, and the increase in population coverage, when a sample is increased. *Biometrika* **43**, 45–63.

Holgate, P. (1969). Species frequency distributions. *Biometrika* **56**, 651–660.

Keener, R., Rothman, E. and Starr, N. (1987). Distributions on partitions. *Annals of Statistics* **15**, 1466–1481.

Mosteller, F. and Wallace, D. L. (1984). *Applied Bayesian and Classical Inference: The Case of the Federalist Papers*. Springer, New York.

Pitman, J. (2006). *Combinatorial Stochastic Processes: Ecole d'Eté de Probabilités de Saint-Flour XXXII-2002*. Edited by J. Picard. Springer, New York.

Biometrika (1976), **63**, 3, *pp*. 435–47
With 3 text-figures
Printed in Great Britain

Estimating the number of unseen species: How many words did Shakespeare know?

BY BRADLEY EFRON AND RONALD THISTED
Department of Statistics, Stanford University, California

SUMMARY

Shakespeare wrote 31 534 different words, of which 14 376 appear only once, 4343 twice, etc. The question considered is how many words he knew but did not use. A parametric empirical Bayes model due to Fisher and a nonparametric model due to Good & Toulmin are examined. The latter theory is augmented using linear programming methods. We conclude that the models are equivalent to supposing that Shakespeare knew at least 35 000 more words.

Some key words: Empirical Bayes; Euler transformation; Linear programming; Negative binomial; Vocabulary.

1. INTRODUCTION

Estimating the number of unseen species is a familiar problem in ecological studies. In this paper the unseen species are words Shakespeare knew but did not use. Shakespeare's known works comprise 884 647 total words, of which 14 376 are types appearing just one time, 4343 are types appearing twice, etc. These counts are based on Spevack's (1968) concordance and on the summary appearing in an unpublished report by J. Gani & I. Saunders. Table 1 summarizes Shakespeare's word type counts, where n_x is the number of word types appearing exactly x times ($x = 1, \ldots, 100$). Including the 846 word types which appear more than 100 times, a total of

$$\sum_{x=1}^{\infty} n_x = 31\,534$$

different word types appear. Note that 'type' or 'word type' will be used to indicate a distinct item in Shakespeare's vocabulary. 'Total words' will indicate a total word count including repetitions. The definition of type is any distinguishable arrangement of letters. Thus, 'girl' is a different type from 'girls' and 'throneroom' is a different type from both 'throne' and 'room'.

How many word types did Shakespeare actually know? To put the question more operationally, suppose another large quantity of work by Shakespeare were discovered, say 884 647t total words. How many new word types in addition to the original 31 534 would we expect to find? For the case $t = 1$, corresponding to a volume of new Shakespeare equal to the old, there is a surprisingly explicit answer. We will show that a parametric model due to Fisher, Corbet & Williams (1943) and a nonparametric model due to Good & Toulmin (1956) both estimate about 11 460 expected new word types, with an expected error of less than 150.

The case $t = \infty$ corresponds to the question as originally posed: how many word types did Shakespeare know? The mathematical model at the beginning of §2 makes explicit the sense of the question. No upper bound is possible, but we will demonstrate a lower bound of approximately 35000 more word types in addition to the 31534 already observed. Our bound involves the theory of empirical Bayes estimation (Robbins, 1956; Good, 1953). It also involves linear programming in both a computational and a theoretical sense. This approach is similar to that taken by Harris (1959). More details are given in an unpublished report of the same title, available from the authors on request.

Table 1. *Shakespeare's word type frequencies*

x	1	2	3	4	5	6	7	8	9	10	Row total
0+	14376	4343	2292	1463	1043	837	638	519	430	364	26305
10+	305	259	242	223	187	181	179	130	127	128	1961
20+	104	105	99	112	93	74	83	76	72	63	881
30+	73	47	56	59	53	45	34	49	45	52	513
40+	49	41	30	35	37	21	41	30	28	19	331
50+	25	19	28	27	31	19	19	22	23	14	227
60+	30	19	21	18	15	10	15	14	11	16	169
70+	13	12	10	16	18	11	8	15	12	7	122
80+	13	12	11	8	10	11	7	12	9	8	101
90+	4	7	6	7	10	10	15	7	7	5	78

Entry x is n_x, the number of word types used exactly x times. There are 846 word types which appear more than 100 times, for a total of 31534 word types.

2. THE BASIC MODEL

We use the species trapping terminology of Fisher's paper. Suppose that there exist S species and that after trapping for one unit of time we have captured x_s members of species s. Of course we only observe those values x_s which are greater than zero. The basic distributional assumption is that members of each species s enter the trap according to a Poisson process, the process for species s having expectation λ_s per unit time, so that x_s has a Poisson distribution of mean λ_s $(s = 1, \ldots, S)$. Most of the calculations in this paper do not require the S individual Poisson processes to be independent of one another. Whenever independence is required it will be specifically mentioned, and referred to as the 'independence assumption'.

Figure 1 gives a schematic representation of the situation. It is convenient to imagine the observation, or trapping, period as running from time -1 to time 0. We wish to extrapolate from the counts in $[-1, 0]$ to a time t in the future. Let $x_s(t)$ be the number of times species s appears in the whole period $[-1, t]$. The Poisson process assumption implies (i) that $x_s(t)$ has a Poisson distribution of mean $\lambda_s(1+t)$ and (ii) that, given $x_s(t)$, x conditionally is binomial $\{x_s(t), 1/(1+t)\}$. For the situation described in the introduction, t equals the total word count of newly discovered Shakespearean literature divided by 884647.

Assumption (i) is dispensable, but assumption (ii) is crucial. It says essentially that the time period $[-1, 0]$ is typical of the whole period $[-1, t]$. If the hypothetical newly discovered works of § 1 were to consist entirely of business letters, we would not expect our results to be valid.

Let $G(\lambda)$ be the empirical cumulative distribution function of the numbers $\lambda_1, \ldots, \lambda_S$. Also, if n_x is the number of species observed exactly x times in $[-1, 0]$, let

$$\eta_x = E(n_x) = S \int_0^\infty (e^{-\lambda} \lambda^x / x!) \, dG(\lambda), \tag{2·1}$$

and let $\Delta(t)$ be the expected number of species observed in $(0, t]$ but not in $[-1, 0]$, so that

$$\Delta(t) = S\int_0^\infty e^{-\lambda}(1-e^{-\lambda t})\,dG(\lambda). \tag{2·2}$$

We wish to estimate $\Delta(t)$, the expected number of new species to be found in the next t time units. By substituting the expansion

$$1-e^{-\lambda t} = \lambda t - \frac{\lambda^2 t^2}{2!} + \frac{\lambda^3 t^3}{3!} - \ldots$$

into (2·2), and comparing the result with (2·1), we obtain the formal equality

$$\Delta(t) = \eta_1 t - \eta_2 t^2 + \eta_3 t^3 - \ldots\ . \tag{2·3}$$

This intriguing result, which appears as formula (24) of Good & Toulmin (1956), is empirical Bayes in the sense Robbins (1956, 1968) originally attached to this term. It is related to an earlier result of Goodman (1949).

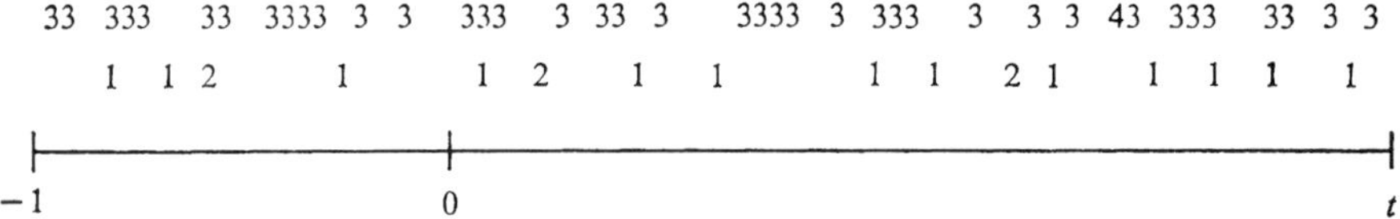

Fig. 1. The Poisson process model; $x_1 = 3$, $x_2 = 1$, $x_3 = 13$, $x_4 = 0$.

The right-hand side of (2·3) need not converge, but, if we assume that it does, expression (2·3) suggests the unbiased estimator for $\Delta(t)$

$$\hat{\Delta}(t) = n_1 t - n_2 t^2 + n_3 t^3 - \ldots\ . \tag{2·4}$$

For the Shakespeare data with $t = 1$ this estimate is

$$\hat{\Delta}(1) = 11\,430. \tag{2·5}$$

Under the independence assumption a reasonable approximation, erring on the conservative side, is to take the n_x themselves to be independent Poisson variates, with means η_x, in which case

$$\operatorname{var}\{\hat{\Delta}(1)\} = \sum_{x=1}^{\infty} \eta_x \doteq \sum_{x=1}^{\infty} n_x = 31\,534.$$

This gives $\hat{\Delta}(1)$ a standard deviation of 178.

The estimator $\Delta(t)$ is a function of the data only through the statistics $n_1, n_2, \ldots$; the quantity n_0 is unobservable, being in fact almost the same as $\Delta(\infty)$. We are disregarding the labels connected with the observations x_s. All the estimates considered in this paper are of this form, but other authors have attempted more refined models; see McNeil (1973) and an unpublished paper by J. Gani and I. Saunders.

Unfortunately (2·4) is useless for values of t larger than one. The geometrically increasing magnitude of t^x produces wild oscillations as the number of terms increases. Good & Toulmin suggest the use of Euler's transformation to force convergence of the series. This idea is discussed in detail in §4. First though, we will examine Fisher's parametric empirical Bayes model in §3.

3. FISHER'S NEGATIVE BINOMIAL MODEL

Fisher *et al.* (1943) added the following assumptions to those at the beginning of §2.

Fisher's assumption 1. The cumulative distribution function $G(\lambda)$ is approximated by a gamma distribution with density function, for $c_{\alpha\beta} = \{\beta^{\alpha}\Gamma(\alpha)\}^{-1}$,

$$g_{\alpha\beta}(\lambda) = c_{\alpha\beta}\lambda^{\alpha-1}e^{-\lambda/\beta}. \tag{3·1}$$

Fisher's assumption 2. The parameters $\lambda_1, \ldots, \lambda_S$ are independent and identically distributed with density $g_{\alpha\beta}(\lambda)$.

Fisher's assumption 1 by itself gives most of the useful conclusions from this model. We shall note explicitly whenever Fisher's assumption 2 is invoked. Assumptions 1 and 2 together constitute a parametric empirical Bayes model in the sense of Efron & Morris (1973).

From (2·1) we obtain, for $\gamma = \beta/(1+\beta)$,

$$\eta_x = \eta_1 \frac{\Gamma(x+\alpha)}{x!\,\Gamma(1+\alpha)}\gamma^{x-1}. \tag{3·2}$$

Expression (3·2) is proportional to the negative binomial distribution with parameters α and γ, written to take advantage of the fact that for the unseen species problem the case $x = 0$ need not be considered. This allows the parameter α to take values less than zero, any value greater than -1 giving finite values to $\eta_1, \eta_2, \ldots$. The density $g_{\alpha\beta}(\lambda)$ is improper at the origin for $\alpha < 0$, and the expression (3·1) for $c_{\alpha\beta}$ is meaningless. Fisher particularly liked the choice $\alpha = 0$, which gives (3·2) the form known as the logarithmic distribution; see also Engen (1974) and Holgate (1969) for extended discussion.

We can write (2·2) in the form

$$\Delta(t) = \eta_1 \frac{\int_0^{\infty} e^{-\lambda}(1-e^{-\lambda t})\,dG(\lambda)}{\int_0^{\infty} \lambda e^{-\lambda} dG(\lambda)} \tag{3·3}$$

to avoid ambiguities in the case where G is improper. By substituting (3·1) for $dG(\lambda)$ we obtain, in the obvious notation, $\Delta_{\alpha\gamma}(t) = -\eta_1\{(1+\gamma t)^{-\alpha}-1\}/(\gamma\alpha)$ unless $\alpha = 0$, in which case $\Delta_{0\gamma}(t) = (\eta_1/\gamma)\log(1+\gamma t)$.

If $\alpha > 0$, $\Delta_{\alpha\gamma}(t)$ approaches its limiting value η_1/α as t goes to infinity. The improper cases $\alpha \leqslant 0$ have $\Delta_{\alpha\gamma}(t)$ increasing without bound as t increases. The infinite spike of $g_{\alpha\beta}(\lambda)$ near $\lambda = 0$ produces an unbounded number of new species as longer and longer time periods are examined.

There is no reason to suppose that Fisher's parametric model will fit the Shakespeare data. It has only mathematical convenience and a limited amount of previous empirical successes to recommend it. In fact, the fit is extremely good. Substituting the values $\hat{\eta}_1 = 14376$, $\hat{\alpha} = -0{\cdot}3954$, $\hat{\gamma} = 0{\cdot}9905$, which are explained below, into (3·2) gives estimates $\hat{\eta}_x$ remarkably close to the observed n_x. To assess the accuracy of a fit such as Table 2 exhibits we need a theory of errors, and for that we need both the independence assumption mentioned at the beginning of §2 and also Fisher's assumption 2. Consider only the first x_0 values of n_x; $n_1, \ldots, n_{x_0}$. Denote their sum by N_0. Given N_0, the vector $(n_1, \ldots, n_{x_0})$ will have a multinomial distribution with N_0 trials and with vector of probabilities proportional to (3·2).

Table 3 shows the maximum likelihood fits, obtained by iterative search for various choices of x_0. The last column is Wilks's likelihood ratio statistic (Wilks, 1962, Chapter 13) for testing the adequacy of the two-parameter model based on (3·2). The sample sizes are enormous, the smallest being 23517, so that under the null hypothesis this statistic should be distributed as a χ^2 variate with $x_0 - 3$ degrees of freedom. We see that the fit is very good, even too good for $x_0 \leq 15$. With sample sizes of this magnitude, deviations of just a few percent from (3·2) would cause rejection.

All further calculations involving Fisher's model use the fitted parameter values for $x_0 = 40$,

$$\hat{\alpha} = -0{\cdot}3954, \quad \hat{\gamma} = 0{\cdot}9905. \tag{3·4}$$

We will use $\hat{\eta}_1 = n_1 = 14376$ rather than the fitted value $\hat{\eta}_1 = 14399$, which makes $\hat{\eta}_1 + \ldots + \hat{\eta}_{40}$ equal the observed sum 29660. However, in most of the calculations η_1 enters as a multiplicative constant, so that multiplication by $14399/14376 = 1{\cdot}0016$ converts the result. Note that the notation $\hat{\eta}_x$ will continue to mean any reasonable estimate of η_x. It will be mentioned when these are taken to be the maximum likelihood values from (3·2) and (3·4).

Table 2. *Maximum likelihood estimates for η_x from Fisher's negative binomial model, and observed frequencies*

	$x = 1$	$x = 2$	$x = 3$	$x = 4$	$x = 5$	$x = 6$	$x = 7$	$x = 8$	$x = 9$	$x = 10$
$\hat{\eta}_x$	14376	4305	2281	1471	1050	798	633	518	433	369
n_x	14376	4343	2292	1463	1043	837	638	519	430	364

Table 3. *Maximum likelihood fits of the negative binomial model to the first x_0 values of n_x; $\hat{\beta} = \hat{\gamma}/(1-\hat{\gamma})$*

x_0	$\sum_{x=1}^{x_0} n_x$	$\hat{\alpha}$	$\hat{\gamma}$	$\hat{\beta}$	$\chi^2_{x_0-3}$
5	23517	−0·3834	0·9795	47·82	0·027
10	26305	−0·3906	0·9884	85·44	2·024
15	27521	−0·3889	0·9861	70·78	3·815
20	28266	−0·3901	0·9875	78·77	8·832
30	29147	−0·3944	0·9899	97·92	16·874
40	29660	−0·3954	0·9905	104·26	30·437

Unfortunately $\hat{\alpha} = -0{\cdot}3954$ puts us in the case where $\Delta_{\alpha\gamma}(t)$ goes to infinity as t gets large. The data agree very well with a model we know must ultimately fail! However, we can still use (3·3) to estimate $\Delta(t)$ for finite values of t. For $t = 1$ we get

$$\hat{\Delta}(1) = \Delta_{-0{\cdot}3954,\,0{\cdot}9905}(1) = 11483.$$

This agrees with (2·5) to within 0·5 %.

For $t = 10$, $\Delta_{-0{\cdot}3954,\,0{\cdot}9905}(10) = 57704$, which is almost twice as large as Shakespeare's observed vocabulary. How accurate is this estimate? The hypothetical standard error from the negative binomial maximum likelihood estimation model, which we have not computed, is uninformative, since we know that that model must fail for large t. Sections 4 to 7 are devoted to finding nonparametric estimates of $\Delta(t)$ for large t, and assessing their accuracy.

4. Euler's transformation

Euler's transformation (Bromwich, 1955, p. 62) is a method of forcing oscillating series like (2·3) to converge rapidly. The substitution $t = u/(2-u)$ gives the formal relationship

$$\sum_{x=1}^{\infty} (-1)^{x+1} \eta_x t^x = \sum_{y=1}^{\infty} \xi_y u^y,$$

where

$$\xi_y = \sum_{x=1}^{y} \binom{y-1}{x-1} \frac{(-1)^{x+1}}{2^y} \eta_x = \frac{1}{2^y} \delta^y(\eta_1). \tag{4·1}$$

Here the backward difference operator is defined by

$$\delta^0(\eta_1) = \eta_1, \quad \delta^1(\eta_1) = \eta_1 - \eta_2, \quad \delta^2(\eta_1) = \eta_1 - 2\eta_2 + \eta_3, \quad \ldots .$$

Let

$$\Delta^{x_0}(t) = \sum_{x=1}^{x_0} (-1)^{x+1} \eta_x t^x, \quad \Delta^{x_0}(u) = \sum_{y=1}^{x_0} \xi_y u^y, \tag{4·2}$$

$$\Delta(t) = \lim_{x_0 \to \infty} \Delta^{x_0}(t), \quad \Delta(u) = \lim_{x_0 \to \infty} \Delta^{x_0}(u).$$

By definition $\Delta(t) = \Delta(u)$ if both limits exist. For η_x positive, as here, the partial sums $\Delta^{x_0}(u)$ will usually converge more quickly to the common limit than the sums $\Delta^{x_0}(t)$. For $\Delta_{\alpha\gamma}(t)$ as given in (3·3), $\Delta^{x_0}(t)$ does not even converge, while the series $\Sigma_y \xi_y u^y$ converges in the nicest possible way, having in fact all nonnegative terms if $\alpha \leqslant 1$.

LEMMA. *For* $-1 < \alpha \leqslant 1, \Delta_{\alpha\gamma}(u) = \Sigma_y \xi_y u^y$ *has* $\xi_y \geqslant 0$ *for all* y.

The proof will be omitted.

Good & Toulmin (1956) suggest estimating ξ_y by substituting $\hat{\eta}_x$ for η_x in (4·1), and then using the Euler transformed series to estimate $\Delta(t)$,

$$\hat{\Delta}^{x_0}(u) = \sum_{y=1}^{x_0} \hat{\xi}_y u^y, \quad u = \frac{2t}{1+t}. \tag{4·3}$$

We have computed the first 20 values of $\hat{\xi}_y$ from (4·1) using $\hat{\eta}_x = n_x$ and also by using the maximum likelihood values (3·4). The latter are all positive, in accordance with the Lemma. The former are positive for $y = 1, \ldots, 9$, and negative for $y = 10, \ldots, 20$. However, all the negative values are within one-half a standard deviation of zero.

This suggests not taking x_0 greater than 9 if we intend to compute $\hat{\Delta}^{x_0}(u)$ from $\hat{\eta}_x = n_x$. The estimates $\hat{\xi}_y$ for $y > 9$ are within noise distance of zero, and we have, admittedly weak, theoretical reasons for believing the ξ_y to be positive. The calculations of § 5 will show $x_0 = 9$ to be a reasonable choice. The corresponding estimate of $\Delta(1)$ is $\hat{\Delta}^9(1) = 11441 \pm 147$, the standard error 147 being computed from (5·2). Calculation of $\hat{\Delta}^9(1)$ from the maximum likelihood estimates of $\hat{\eta}_x$, (3·4), gives $\hat{\Delta}^9(1) = 11460$ as the estimate. The question of assigning a standard deviation to the second of these estimates is a difficult one, but it is reasonable to say that the estimate is at least as accurate as the first one, and perhaps considerably more so.

In the present notation, (2·5) can be written as $\hat{\Delta}^\infty(1) = 11430 \pm 178$. Comparison of this with $\hat{\Delta}^9(1)$ above shows that we have reduced the standard deviation considerably by reducing x_0 from ∞ to 9. The price we pay, as Good & Toulmin noted, is in terms of bias. Thus $\hat{\Delta}^{x_0}(t)$ is not an unbiased estimate of $\Delta(t)$ for $x_0 < \infty$ because of the truncated terms in the series. The calculations of § 5 will show that $\hat{\Delta}^9(1)$ can have a bias as large as $+8$ and as small as -62. This is with no assumptions on the form of $G(\lambda)$. Under the negative binomial model, the Lemma shows that $\hat{\Delta}^9(1)$ must have a negative bias, since all the terms we are ignoring are positive.

Taking both variance and bias into account, $\hat{\Delta}^9(1)$ is not noticeably superior to $\hat{\Delta}^\infty(1)$, except in computational effort. The choice of x_0 becomes far more crucial for values of $t > 1$, as § 5 will show.

5. General linear estimators

There is another expression of the Euler transformation which makes obvious its effect on oscillating series. Substitution of (4·1) into the right-hand side of (4·2) shows, after some rearrangement, that $\Delta^{x_0}(u)$ is just the average of the oscillating series $\Delta^x(t)$ over values of x distributed binomially $\{x_0, 1/(1+t)\}$. This averaging process is what smooths out the oscillations.

The estimator (4·3), with $\hat{\eta}_x = n_x$, is now seen to be of the form

$$\hat{\Delta} = \sum_{x=1}^{\infty} h_x n_x, \tag{5·1}$$

where, if Z denotes a binomially distributed random variable with index x_0 and parameter $1/(1+t)$,

$$h_x = \begin{cases} (-1)^{x+1} t^x \operatorname{pr}(Z \geqslant x) & (x = 1, \ldots, x_0), \\ 0 & (x > x_0). \end{cases}$$

Notice that the naive estimator

$$\hat{\Delta}^{x_0}(t) = \sum_{x=1}^{9} (-1)^{x+1} n_x t^x$$

has $h_x = (-1)^{x+1} 10^x$ in this case, so that $h_9 = 10^9$, compared with $h_9 = 0·424$ in Table 4! The Euler transformation drastically reduces h_x for large x.

Table 4. *Euler coefficients in the general linear estimator* (5·1) *for* $x_0 = 9$ *and* $t = 10$

x	1	2	3	4	5	6	7	8	9
h_x	5·759	−19·421	41·539	−59·152	57·155	−37·190	15·653	−3·859	0·424

We call estimators of the form (5·1) general linear estimators. We will calculate the variance of such an estimator from the independent Poisson assumption as

$$\operatorname{var}(\hat{\Delta}) = \sum_{x=1}^{\infty} h_x^2 \eta_x, \tag{5·2}$$

and note that this value may be somewhat large, as the calculations in an unpublished report by the authors demonstrate.

For each estimator (5·1) define the function

$$H(\lambda) = \sum_{x=1}^{\infty} h_x \lambda^x / x! \quad (0 < \lambda < \infty). \tag{5·3}$$

By (2·1) we have

$$E(\hat{\Delta}) = \sum_{x=1}^{\infty} h_x \eta_x = S \sum_{x=1}^{\infty} \int_0^{\infty} (h_x e^{-\lambda} \lambda^x / x!) \, dG(\lambda) = S \int_0^{\infty} e^{-\lambda} H(\lambda) \, dG(\lambda),$$

assuming, as will always be the case for the h_x used below, that summation and integration can be interchanged. The bias of $\hat{\Delta}$ for estimating $\Delta(t)$ is, by (2·2),

$$E\{\hat{\Delta} - \Delta(t)\} = S \int_0^{\infty} e^{-\lambda} \{H(\lambda) - (1 - e^{-\lambda t})\} \, dG(\lambda). \tag{5·4}$$

It is convenient to rewrite (5·4) in a form which depends on $\eta_+ = \eta_1 + \eta_2 + \ldots$, rather than S, since we always have an easy estimate for η_+ available, namely $n_+ = \Sigma n_x$. Define

$$P = \int_0^\infty (1-e^{-\lambda})\,dG(\lambda), \quad d\tilde{G}(\lambda) = P^{-1}(1-e^{-\lambda})\,dG(\lambda).$$

Notice that $\eta_+ = SP$, by summation of η_x in (2·1). That is, P is just the expected proportion of the λ_s having $x_s > 0$. Also $\tilde{G}$ can be thought of as the empirical cumulative distribution function of those λ_s having $x_s > 0$, although strictly speaking this interpretation is only justified in the limiting case $S \to \infty$.

By multiplying and dividing (5·4) by $(1-e^{-\lambda})/P$ we obtain

$$E\{\hat{\Delta} - \Delta(t)\} = \eta_+ \int_0^\infty \frac{e^{-\lambda}}{1-e^{-\lambda}} \{H(\lambda) - (1-e^{-\lambda t})\}\, d\tilde{G}(\lambda). \tag{5·5}$$

The integrand

$$B_t(\lambda) = \frac{e^{-\lambda}}{1-e^{-\lambda}} \{H(\lambda) - (1-e^{-\lambda t})\} \tag{5·6}$$

determines the bias of $\hat{\Delta}$ for any $G(\lambda)$ or $\tilde{G}(\lambda)$.

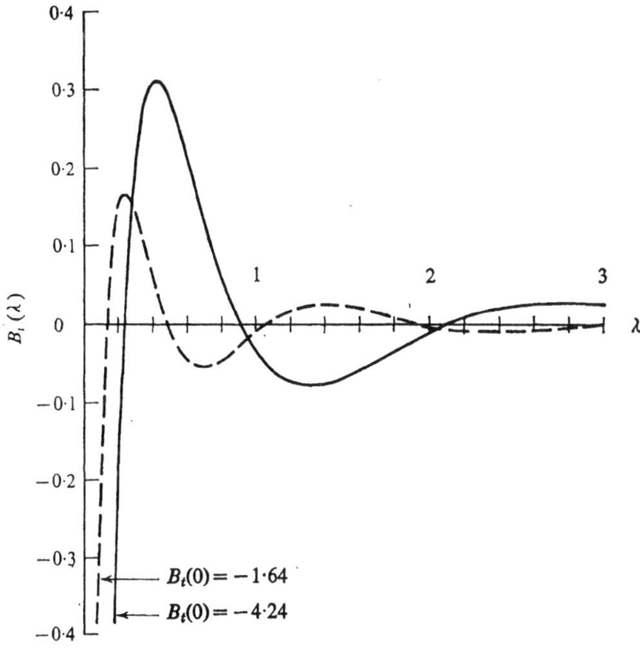

Fig. 2. The bias function $B_t(\lambda)$, equation (5·6), for $\hat{\Delta}^{x_0}(t)$, at $t = 10$; solid line, $x_0 = 9$, and dashed line, $x_0 = 19$.

For example, with $t = 1$, $x_0 = 9$ we compute $B_t(\lambda)$ to oscillate about zero, having its smallest value at $\lambda = 0$ and largest value at $\lambda = 0·6$, $B_t(0) = -0·00196$, $B_t(0·6) = 0·00024$. From (5·5) we see that the greatest negative bias occurs if $\tilde{G}$ puts all its mass at $\lambda = 0$, in which case the bias equals $-0·00196\eta_+ \doteq -0·00196 \times 31\,534 = -62$. The greatest positive bias occurs if $\tilde{G}$ puts all mass at $\lambda = 0·6$, in which case it equals $0·00024 \times 31\,534 = 8$. Of course, the data in Table 1 tell us that $\tilde{G}$ follows neither extreme for the Shakespeare word counts. In §§6 and 7 we employ such information to get better bounds in a systematic way.

Figure 2 shows $B_t(\lambda)$ at $t = 10$, for $x_0 = 9$ and for $x_0 = 19$. The bias situation is now much more serious. For $x_0 = 9$ the possible bias ranges from $-4·24\eta_+$ to $0·31\eta_+$. For $x_0 = 19$ the range is from $-1·64\eta_+$ to $0·15\eta_+$. This does not mean that $x_0 = 19$ is better than $x_0 = 9$. The respective estimators from equation (4·3) and their standard deviations from (5·2) are $\hat{\Delta}^9(10) = 45\,188 \pm 3994$ and $\hat{\Delta}^{19}(10) = 53\,867 \pm 702\,566$. Its huge variance makes $\hat{\Delta}^{19}(10)$

useless. The choice of x_0 must take into account both bias and variance. For this case $x_0 = 9$ seemed to be as good or better than any other choice, though admittedly the criterion of goodness is vague.

We need not restrict attention to linear estimators of the form (4·3). An attempt to choose a best linear estimator $\hat{\Delta}(10) = h_1 n_1 + \ldots + h_{x_0} n_{x_0}$ is described in unpublished work by the authors. This search yielded no estimator noticeably superior to $\hat{\Delta}^9(10)$.

6. Lower bounds for $\Delta(t)$

As t gets large it becomes more and more difficult to estimate a reasonable upper bound for $\Delta(t)$. Suppose that Shakespeare actually had 10^6 word types with $\lambda_s = 10^{-6}$. These types would have almost no effect on our data set. The expected number of them occurring in our sample is only 1. However, for $t = 10^6$ an expected fraction $1 - e^{-1} = 0{\cdot}632$ of them would be observed. This type of counterexample can be pushed arbitrarily far. Unfortunately, the trouble begins for t values much smaller than 10^6. We see this in Fig. 2, where the possible negative bias is already very large for $t = 10$.

Table 5. *Lower bound estimates for $\Delta(t)$ based on linear transformations of $\hat{\Delta}^{x_0}(u)$, $x_0 = 9$*

t	a	b	Lower bound estimate (6·2)	St. dev. (5·2)	Estimate −st. dev.
1	0·999	0·0001	11454	147	11307
3	0·979	0·002	25143	986	24157
5	0·939	0·007	31974	1966	30008
8	0·879	0·010	36554	2965	33588
10	0·850	0·012	38015	3397	34618
12	0·827	0·014	38927	3713	35214
15	0·801	0·016	39784	4048	35736
20	0·772	0·017	40580	4408	36172
30	0·742	0·019	41331	4793	36538
60	0·710	0·022	42061	5212	36848
120	0·694	0·023	42411	5433	36977

The situation is better for lower bounds. Equation (5·3) shows that $\hat{\Delta} = \Sigma h_x n_x$ satisfies $E(\hat{\Delta}) \leqslant \Delta(t)$ if, for all $\lambda \geqslant 0$,

$$H(\lambda) \leqslant 1 - e^{-\lambda t}. \tag{6·1}$$

In other words, the linear estimator $\hat{\Delta}$ will be a lower bound for $\Delta(t)$ in expectation, no matter what G happens to be, if $H(\lambda)$ is everywhere less than $1 - e^{-\lambda t}$.

As we saw in §5, the estimators based on Euler's transformation do not satisfy (6·1). However, given $\hat{\Delta} = \Sigma h_x n_x$ we can always make a linear transformation $h_x^0 = a h_x - b$ $(x = 1, 2, \ldots)$ which gives, through (5·3), $H^0(\lambda) = aH(\lambda) - b(e^{\lambda} - 1)$. The corresponding new estimator is

$$\hat{\Delta}^0 = \sum_{x=1}^{\infty} h_x^0 n_x = a\hat{\Delta} - b n_+, \tag{6·2}$$

where $n_+ = \Sigma n_x$ as before.

Table 5 shows the lower bounds obtained in this way from the Euler estimators (5·1), with $x_0 = 9$, for various choices of t. The constants a and b were chosen so that H^0 satisfied (6·1). Subject to this constraint, a and b were selected to maximize (6·2) with $\hat{\eta}_x$ in place of n_x, $\hat{\eta}_x$ the maximum likelihood estimates obtained from (3·2) and (3·4). The resulting value of $\hat{\Delta}^0$ is tabulated as the 'lower bound estimate'. The standard deviation from (5·2) appears in the next column, followed by the estimate minus one standard deviation.

Table 5 shows that this reasonably conservative lower bound for $\Delta(t)$ fails to get much larger as t grows from 10 to 120. As we shall see in §7, it is impossible to get a substantially larger lower bound for t approaching infinity, even using more general linear estimators. This seems to say that the Shakespeare data, unaided by parametric assumptions like Fisher's assumption 1, runs out of predictive power for t greater than 10.

A potential flaw in Table 5 is that the estimates and standard deviations are derived ignoring the fact that a and b depend on the data, since they are chosen so that (6·2) is maximized for the data set at hand. This point is considered more carefully in §7, and is shown not to make much difference.

7. Linear programming bounds

The method employed in §6 to find $\hat{\Delta}$ satisfying $E(\hat{\Delta}) \leqslant \Delta(t)$ can be approached more generally as a linear programming problem.

Program 1. Choose $h_1, \ldots, h_{x_0}, h_{x_0+1}$ to maximize

$$\hat{\Delta} = \sum_{x=1}^{x_0} h_x \hat{\eta}_x + h_{x_0+1} \sum_{x=x_0+1}^{\infty} \hat{\eta}_x \tag{7·1}$$

subject to the constraints, for $\lambda > 0$,

$$H(\lambda) = \sum_{x=1}^{x_0} h_x \lambda^x/x! + h_{x_0+1} \sum_{x=x_0+1}^{\infty} \lambda^x/x! \leqslant 1 - e^{-\lambda t}. \tag{7·2}$$

Condition (7·2) guarantees, by (6·1), that $E(\hat{\Delta}) \leqslant \Delta(t)$ for any G. Subject to this constraint, (7·1) requires maximization of the estimated value at a likely value of the true parameters $\eta_1, \eta_2, \ldots$. In this section we take $\hat{\eta}_x$ to be the maximum likelihood values from (3·2) and (3·4) for $x = 1, \ldots, x_0$ and set

$$\sum_{x=x_0+1}^{\infty} \hat{\eta}_x = \sum_{x=x_0+1}^{\infty} n_x.$$

Other reasonable choices of $\hat{\eta}_x$ give almost identical answers.

Program 1 was solved on the IBM 360/67 computer at Stanford using the IBM program MPS/360. The infinite number of constraints in (7·2) was replaced by the discrete set

$$H(\lambda_l) \leqslant 1 - e^{-\lambda_l t}, \quad \lambda_l = 2^{\frac{1}{16}l - 10} \quad (l = 0, \ldots, 272) \tag{7·3}$$

($\lambda_0 = 2^{-10}, \lambda_{272} = 128$). As before, $x_0 = 9$ was used for most of the calculations. These choices were based on a small amount of numerical experimentation.

For the case $t = \infty$ the resulting optimum coefficients h_x were substituted into (7·1) to obtain the lower bound estimate $\hat{\Delta}(\infty) = 59568$ for Shakespeare's total unobserved vocabulary. Unfortunately, the standard error for $\hat{\Delta}(\infty)$, calculated from (5·2) on the assumption that the h_x are fixed constants, is the enormous value, 204784. This might seem to render $\hat{\Delta}(\infty)$ useless, but we shall see that this is not actually so.

The linear programming problem dual to program 1 (Hillier & Lieberman, 1974, p. 90) or, rather, the dual to the discretized version (7·1) and (7·3), is as follows.

Program 2. Choose $S > 0$ and a discrete distribution function $G(\lambda)$ with support on the set $\{\lambda_1, \lambda_2, \lambda_l, \ldots, \lambda_L\}$ to minimize

$$\Delta(t) = S \int_0^{\infty} e^{-\lambda}(1 - e^{-\lambda t})\, dG(\lambda) \tag{7·4}$$

subject to the constraints

$$S\int_0^\infty (e^{-\lambda}\lambda^x/x!)\,dG(\lambda) = \hat{\eta}_x \quad (x = 1, \ldots, x_0), \quad S\int_0^\infty e^{-\lambda}\sum_{x=x_0+1}^{\infty}(\lambda^x/x!)\,dG(\lambda) = \sum_{x=x_0+1}^{\infty}\hat{\eta}_x. \tag{7·5}$$

The dual Program 2 finds the 'least favourable situation' in that it selects S and G to minimize $\Delta(t)$ subject to the constraint that the expected word counts $\eta_1, \ldots, \eta_{x_0}$ and the sum of their successors equal certain specified values. Program 2 is nearly identical to the problem considered by Harris (1959). With the $\hat{\eta}_x$ chosen as before we see that, by the duality

Table 6. *Points of support of minimizing distribution G in Program 2 at* $t = \infty$

	l = 120–121	l = 167–168	l = 191–192	l = 208–209	l = 226–227
λ	0·1806	1·384	3·914	8·175	17·830
dG	0·7444	0·1220	0·0507	0·0294	0·0535

Table 7. *Lower and upper bounds on* $\Delta(t)$ *calculated by solving the linear programming problem* (7·4) *and* (7·6); $x_0 = 9$, $c = 1$

t	Lower bound on $\Delta(t)$	Upper bound on $\Delta(t)$
1	11205	11732
3	23828	29411
5	29898	45865
10	34640	86600
20	35530	167454
∞	35554	∞

theorem, Program 2 has the same solution as Program 1. For $t = \infty$ solving Program 2 gives $\hat{\Delta}(\infty) = 59568$, as before. The minimizing distribution G has its support at 10 of the λ_l values, occurring in five adjacent pairs, given in Table 6. Of course we do not believe that $\eta_x = \hat{\eta}_x$ exactly, but we can loosen the constraints to take into account our uncertainty, say by taking

$$\hat{\eta}_x - c\sqrt{\hat{\eta}_x} \leqslant S\int_0^\infty (e^{-\lambda}\lambda^x/x!)\,dG(\lambda) \leqslant \hat{\eta}_x + c\sqrt{\hat{\eta}_x} \quad (x = 1, \ldots, x_0), \tag{7·6}$$

and similarly for the last constraint in (7·5). Here c measures approximately how many standard deviations we allow the fitted values of η_x to vary from $\hat{\eta}_x$.

Solution of (7·4) and (7·6) for $t = \infty$, $x_0 = 9$ and $c = 1$ gives a minimum value of $\hat{\Delta} = 35554$, which is quite consistent with the last column of Table 4.

We now have a believable lower bound on $\Delta(\infty)$. The choice $c = 1$ may seem optimistic, but we have reason to believe the true η_x to be nearer $\hat{\eta}_x$ than (2·6) indicates. The issue of concern to us in §6, namely, that of choosing the estimator from the data and then ignoring that selection process in setting confidence intervals, has disappeared. The linear programming method yields a lower bound directly as a function of the unknown parameters η_x. Confidence bounds on η_x of the type (7·6) then yield a bound on $\hat{\Delta}$ in the usual way. For those preferring a still more conservative bound, $c = 2$ gives $\hat{\Delta}(\infty) = 30845$ with $x_0 = 9$.

Table 7 gives lower and upper bounds on $\Delta(t)$ obtained from (7·4) and (7·6) with $x_0 = 9$, $c = 1$. For the upper bound the 'minimize' in (7·5) is simply changed to 'maximize'. The agreement of the lower bounds with the last column of Table 5 is remarkable. This is important since Table 5 is much easier to calculate than Table 7.

15-2

8. CONCLUSIONS

Figure 3 displays the different estimates of $\Delta(t)$. Our experience with the Shakespeare data can be summarized as follows.

(i) Estimate $\hat{\Delta}(\infty) = 35\,000$ is a reasonably conservative lower bound for the amount of vocabulary Shakespeare knew but did not use.

(ii) An estimate of $\Delta(t)$ can be made very accurately for $t \leqslant 1$, but the uncertainties magnify quickly as t grows larger. Without a parametric model the data give very little additional information for t larger than 10.

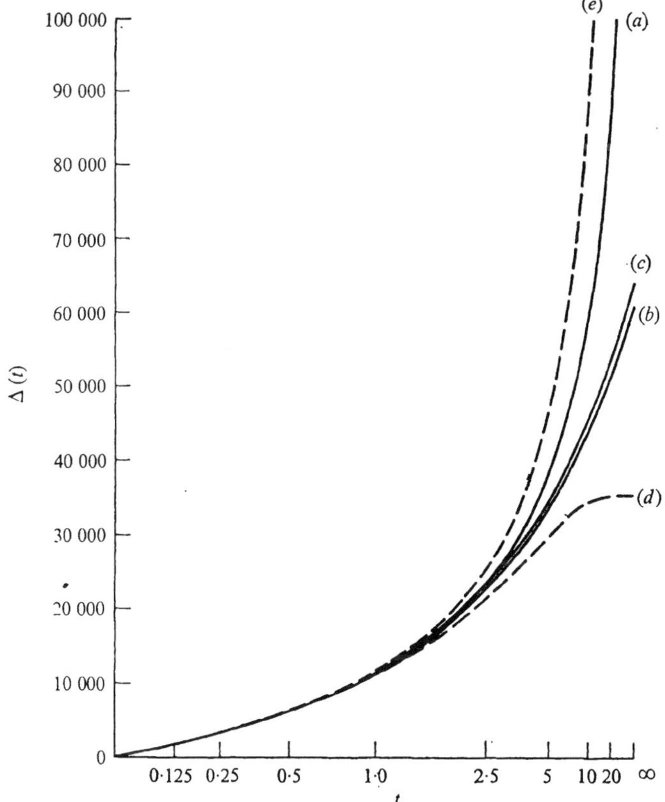

Fig. 3. Different estimates of $\Delta(t)$ for the Shakespeare data: (a) Fisher's negative binomial model with parameters (3·4); (b) Euler transformation (4·4), $x_0 = 9$, $\hat{\xi}_v$ from $\hat{\eta}_x = n_x$; (c) as (b), but with $\hat{\xi}_v$ from maximum likelihood values (3·2) and (3·4); (d) lower bound estimates from linear program (7·4) and (7·6), $c = 1$; (e) upper bound, as (d).

(iii) Fisher's negative binomial model fits the data extraordinarily well. However the linear programming approach produces other empirical Bayes solutions which also fit the observed data, and give smaller estimates of $\Delta(t)$ for $t > 1$.

(iv) All the methods give very similar answers for $t \leqslant 1$.

(v) Euler's transformation performs well compared to more elaborate techniques.

This paper was inspired by a lecture by J. Gani; P. Diaconis contributed many useful ideas and references.

References

Bromwich, T. (1955). *An Introduction to the Theory of Infinite Series*, 2nd edition. London: Macmillan.

Efron, B. & Morris, C. (1973). Stein's estimation rule and its competitors – an empirical Bayes approach. *J. Am. Statist. Assoc.* **68**, 117–30.

Engen, S. (1974). On species frequency models. *Biometrika* **61**, 263–70.

Fisher, R. A., Corbet, A. S. & Williams, C. B. (1943). The relation between the number of species and the number of individuals in a random sample of an animal population. *J. Anim. Ecol.* **12**, 42–58.

Good, I. J. (1953). The population frequencies of species and the estimation of population parameters. *Biometrika* **40**, 237–64.

Good, I. J. & Toulmin, G. H. (1956). The number of new species, and the increase in population coverage, when a sample is increased. *Biometrika* **43**, 45–63.

Goodman, L. A. (1949). On the estimation of the number of classes in a population. *Ann. Math. Statist.* **20**, 572–9.

Harris, B. (1959). Determining bounds on integrals with applications to cataloging problems. *Ann. Math. Statist.* **30**, 521–48.

Hillier, F. & Lieberman, G. (1974). *Introduction to Operations Research*, 2nd edition. San Francisco: Holden-Day.

Holgate, P. (1969). Species frequency distributions. *Biometrika* **56**, 651–60.

McNeil, D. (1973). Estimating an author's vocabulary. *J. Am. Statist. Assoc.* **68**, 92–6.

Robbins, H. (1956). An empirical Bayes approach to statistics. *Proc. 3rd Berkeley Symp.* **1**, 137–63.

Robbins, H. (1968). Estimating the total probability of the unobserved outcomes of an experiment. *Ann. Math. Statist.* **39**, 256–7.

Spevack, M. (1968). *A Complete and Systematic Concordance to the Works of Shakespeare*, Vols. 1–6. Hildesheim: George Olms.

Wilks, S. S. (1962). *Mathematical Statistics*. New York: Wiley.

[*Received June* 1975. *Revised December* 1975]

6

The Efficiency of Cox's Likelihood Function for Censored Data

Introduction by John D. Kalbfleisch
University of Michigan, Ann Arbor

This elegant paper appeared at a crucial time in the development of methods for the analysis of censored data subsequent to the 1972 landmark paper of Cox and the proposed Cox or relative risk regression model. In this model, the hazard function of the time to failure is $\lambda(t; z(t)) = \lambda_0(t)r(z(t), \theta)$ where $z(t)$ is a (possibly time-varying) covariate vector, t measures time from some well-defined time origin, $r(z(t), \theta)$ is the relative risk function of known from, and θ is a finite-dimensional (regression) parameter. Cox proposed leaving the baseline hazard $\lambda_0(t)$ unspecified so that the model had both parametric and nonparametric parts and proposed simple methods of analysis. One aspect of this was a series of conditional arguments that led to a simple derived likelihood, the Cox likelihood, for θ.

In the years following Cox's paper, there was considerable interest in the flexible and general methods proposed, and by 1977 these methods were becoming standard in many applications, especially in clinical trials. Correspondingly, there was a strong interest in the theoretical foundations and properties of these methods. In his follow-up paper on partial likelihood, Cox (1975) had formulated the basic arguments involved in constructing the likelihood for θ. He had shown that the partial likelihood (or Cox likelihood) shared many properties with the ordinary likelihood, and provided heuristic arguments that, under suitable regularity conditions, the approach would have the usual asymptotic properties associated with likelihood inference. An open and important question concerned the efficiency of these methods when compared with parametric analyses based on special cases such as exponential or Weibull regression. Some efficiency results were already known for the related log rank test (e.g., Peto and Peto 1972; Crowley and Thomas 1975) and Kalbfleisch (1974) had carried out some calculations regarding asymptotic and finite sample efficiency compared to a parametric exponential model. Oakes (1977) appeared in *Biometrika* at almost the same time as Efron's paper and, by an approach employing probability limits, independently derived some of the same results.

In this paper, Efron considered a sequence of parametric models for $\lambda_0(t)$ and showed that the Cox likelihood for the parameter θ was asymptotically nearly efficient when compared to a highly flexible parametric model. Technical details involved the development of a novel conditional approach to evaluate the expected or Fisher information in the presence of censoring, both in the fully parametric and Cox likelihoods; this approach and many of the comments and arguments anticipated the martingale approach developed later by Andersen and Gill (1982). Efron further argued that, as the span of the parametric model for $\lambda_0(t)$ increased, the asymptotic variance of the estimate of θ in the parametric model would approach that of the estimator from the Cox likelihood. Thus, he argued that the Cox likelihood should be viewed as efficient in a nonparametric sense. More importantly from a practical point of view, his arguments showed that the typical loss of asymptotic efficiency of the Cox likelihood compared to many parametric models would be fairly small, at least if the relative risks involved are not too large. The paper also includes a number of very perceptive comments in Sect. 6, one of which gives a widely used approximation to the likelihood for ties proposed by Kalbfleisch and Prentice (1973).

The influence of this paper extends beyond its important contribution to the analysis of censored data and our understanding of the Cox model. Efron's arguments were formalized and generalized in a fundamental paper on asymptotic efficiency in semiparametric models by Begun et al. (1983); see also Bickel et al. (1993, Chap. 5). This work by Efron stands as an early and elegant example of establishing a nonparametric information bound for the parametric part of an important regular semiparametric model.

References

Andersen, P. K. and Gill, R. D. (1982). Cox's regression model for counting processes: A large sample study. *Annals of Statistics* **10**, 1100–1120.

Begun, J. M., Hall, W. J., Huang, W.-M. and Wellner, J. A. (1983). Information and asymptotic efficiency in parametric–nonparametric models. *Annals of Statistics* **11**, 432–452.

Bickel, P. J., Klassen, C. A., Ritov, Y. and Wellner, J. A. (1993). *Efficient and Adaptive Estimation for Semiparametric Models.* Johns Hopkins University Press, Baltimore, MD.

Cox, D. R. (1975). Partial likelihood. *Biometrika* **62**, 269–279.

Crowley, J. and Thomas, D. R. (1975). Large sample theory for the log rank test. Technical Report 415, Department of Statistics, University of Wisconsin.

Kalbfleisch, J. (1974). Some efficiency calculations for survival distributions. *Biometrika* **61**, 31–38.

Kalbfleisch, J. and Prentice, R. (1973). Marginal likelihoods based on Cox's regression and life model. *Biometrika* **60**, 267–279.

Oakes, D. (1977). The asymptotic information in censored survival data. *Biometrika* **64**, 441–448.

Peto, R. and Peto, J. (1972). Asymptotically efficient rank invariant test procedures. *Journal of the Royal Statistical Society, Series A* **135**, 185–206 with discussion.

The Efficiency of Cox's Likelihood Function for Censored Data

BRADLEY EFRON*

D.R. Cox has suggested a simple method for the regression analysis of censored data. We carry out an information calculation which shows that Cox's method has full asymptotic efficiency under conditions which are likely to be satisfied in many realistic situations. The connection of Cox's method with the Kaplan-Meier estimator of a survival curve is made explicit.

KEY WORDS: Censored data; Cox likelihood; Survival curves.

1. INTRODUCTION

A recent California study investigated the survival times of residents at a senior citizens' facility. New arrivals joined the facility at various ages past 65, sometimes moved out of the facility, and of course not all had died by the end of the study. Complicated data-censoring patterns such as this are common in studies involving human beings. In a heavily censored situation standard regression techniques are inappropriate for analyzing the effects of covariates (such as race, sex, and blood pressure in the example above) on survival time.

D.R. Cox (1972) has suggested a regression analysis for survival data which cleverly finesses censoring difficulties. Cox's model assumes that the ith subject has hazard rate

$$h_i(t) = \theta_i(t, \beta)h(t, \gamma) , \qquad (1.1)$$

where the unobserved vector β, which parameterizes the regression of survival time on the observed covariates, is the main object of interest. Cox uses the parameterization $\theta_i(t, \beta) = \exp(\beta \cdot \mathbf{z}_i(t))$, where $\mathbf{z}_i(t)$ is the possibly time-varying vector of observed covariates, but this particular form does not play a crucial role in the analysis. The unknown nuisance function $h(t, \gamma)$ modifies all the individual hazard rates equally, depending for its form on another unobserved vector γ of parameters.

In order to visualize (1.1) more concretely, it helps to imagine the time axis divided into infinitesimal intervals of length ϵ. We have a collection of (time-varying) coins indexed by $i = 1, 2, \ldots, n$ corresponding to all the subjects ever observed in the study. During time interval $(t, t + \epsilon)$ a subset $\mathcal{R}(t)$ of these coins, called the "risk set at time t," are each flipped once, with the probability of heads (death, in the senior citizens' study) equal to $h_i(t)\epsilon$ for the ith coin, independently of all other coins. This process proceeds sequentially in time. Once a head is achieved that coin is removed from subsequent flippings. Coins may be removed from the risk set for reasons other than death, and new coins may come on risk, i.e., join $\mathcal{R}(t)$ as t increases.

Cox's analysis proceeds as follows: let $t_1 < t_2 < \ldots < t_J$ be the observed failure times, assuming no ties, say for items $i_1, i_2, \ldots, i_J$, respectively, and let $\mathcal{R}(t_j)$ be the risk set of items on test just before the jth failure. Given $\mathcal{R}(t_j)$ and the fact that one item failed at time t_j, the conditional probability that item i_j failed is

$$\theta_{i_j}(t_j, \beta) / \sum_{i \in \mathcal{R}(t_j)} \theta_i(t_j, \beta) .$$

Simply multiplying these factors together gives the "partial likelihood function"

$$\prod_{j=1}^{J} \{\theta_{i_j}(t_j, \beta) / \sum_{i \in \mathcal{R}(t_j)} \theta_i(t_j, \beta)\} . \qquad (1.2)$$

The coin-tossing model in the preceding paragraph clarifies the derivation of (1.2).

Cox treats (1.2) as an ordinary likelihood function for the purposes of inference on β. Maximum likelihood estimates, hypothesis tests, and asymptotic confidence intervals are then derived in the usual way. Cox's analysis relates to earlier work by many authors, in particular Mantel and Haenzel (1959) and Peto and Peto (1972). A "major outstanding problem," which is the main topic of this paper, is the efficiency of inferences about β based on (1.2) (Cox 1972).

There are three very attractive features of Cox's approach: (1) The nuisance function $h(t, \gamma)$ is completely removed from the inference process on β; (2) Covariate information on the different items is easily incorporated into (1.1), for example in the form $\theta_i(t, \beta) = \exp(\beta \cdot \mathbf{z}_i(t))$ suggested by Cox; and (3) Data censoring patterns often encountered in life tests, such as those in the senior citizens study, do not affect (1.2).

Qualms about (1.2) were expressed in the discussion following Cox's paper. It is not really a likelihood function since it ignores a factor in the likelihood, essentially that relating to the "nonfailure intervals," $t_1, t_2 - t_1, t_3 - t_2, \ldots, t_J - t_{J-1}$, nor is it a conditional or marginal likelihood, except in very special cases. (See Kalbfleisch

* Bradley Efron is Professor, Department of Statistics and Biostatistics, Stanford University, Stanford, CA 94305. Research was supported in part by National Science Foundation Grant MPS74-21416 and Public Health Service Grant 5 R01 GM1215-02.

Reprinted from: © Journal of the American Statistical Association
September 1977, Volume 72, Number 359
Theory and Methods Section
Pages 557–565

and Prentice 1973 and also Remark E, Section 6.) Cox's (1975) theory of *partial likelihood* shows among other things that (1.2) produces inferences similar to ordinary likelihood procedures. We use his results in Section 3.

In this article, the meaning of (1.2) is set in context by considering the complete likelihood function of all the observed data. The heuristic argument of Section 3 shows that if the class of nuisance functions $h(t, \gamma)$ is moderately large, then inferences about β based on (1.2) are asymptotically equivalent to those based on all the data. In a rough sense this solves Cox's "outstanding problem."

In practice, $h(t, \gamma)$ may be an important quantity in its own right rather than a nuisance. The connection between (1.2) and inferences about $h(t, \gamma)$ is considered briefly in Section 5, particularly as it concerns the Kaplan-Meier estimator. This analysis is closely related to that in Breslow (1974). There is also considerable overlap with Breslow and Crowley (1974), and the work of Aalen (1975) which concerns the efficiency of (1.2) for testing purposes.

We begin in Section 2 with the case of many identical items on test, to which the Kaplan-Meier estimator refers. The main result is in Section 3 with the proof deferred until Section 7. Section 4 illustrates the general theory in the special case of the two-sample problem. Section 6 consists of several brief remarks on Cox's likelihood and the Kaplan-Meier estimator.

2. IDENTICAL ITEMS ON TEST

Suppose several identical items are on test, each obeying the same hazard function $h(t)$. A typical item has lifetime T, a continuous positive random variable with

$$\text{Prob}\,\{T > t_2 | T > t_1\} = \exp\left\{-\int_{t_1}^{t_2} h(t)dt\right\} . \quad (2.1)$$

We wish to infer h from the observed failure times $t_1 < t_2 < \ldots < t_J$; h is assumed to belong to some parametric family, which for the moment won't be indicated in the notation; and the nonparametric case is the limit when the family is allowed to include all hazard functions. Let

$$n(t) \equiv \text{number of items on test just before time } t\ . \quad (2.2)$$

In what follows, $n(t)$ is assumed to be a step function continuous from the left, changing value (due to losses, failures, and introduction of new items) only finitely often in any finite interval. The likelihood of the observed data, considered as a function of the unknown hazard rate h, is

$$f_h(\text{data}) = \exp\left\{-\int_0^\infty n(t)h(t)dt\right\} \prod_{j=1}^{J} n(t_j)h(t_j) . \quad (2.3)$$

This is derived from standard Poisson process arguments by noting that the probability of no event between t_{j-1} and t_j is

$$\exp\left\{-\int_{t_{j-1}}^{t_j} n(t)h(t)dt\right\} ,$$

while the probability of the single event, "one out of $n(t_j)$ items fails at time t_j," is proportional to $n(t_j)h(t_j)$. A more careful derivation is obtained by dividing the time axis into infinitesimal discrete units as in the introduction, see also Aalen (1975). Formula (2.3) assumes that $h(t)$ is continuous at the failure times t_j.

It can be shown that in the nonparametric case, the unrestricted maximizer of (2.3), say $h^*(t)$, satisfies

$$\exp\left\{-\int_{t_j-}^{t_j+} h^*(t)dt\right\} = 1 - \frac{1}{n(t_j)}$$
$$j = 1, 2, \ldots, J . \quad (2.4)$$

This leads to the familiar "Kaplan-Meier estimate" of the survival function (1958).

$$\text{Prob}^*\,\{T > t\} = \prod_{t_j \le t} \left[1 - \frac{1}{n(t_j)}\right] . \quad (2.5)$$

There are some minor technical difficulties in deriving (2.4) from (2.3) because $h^*(t)$ does not refer to a continuous distribution for T. The discretization argument mentioned above avoids this difficulty.

3. COX'S PARTIAL LIKELIHOOD FUNCTION

We return to the situation where the different items on test have different hazard rates,

$$h_i(t) = \theta_i(t)h(t) \quad i = 1, 2, \ldots, n . \quad (3.1)$$

Here n is the number of items ever on test during the course of the experiment. The parameterization of the unknown functions θ_i and h introduced below is slightly different from (1.1); for the moment it will not be indicated in the notation.

The likelihood function of the observed data is now

$$f_{\theta,h}(\text{data}) = \exp\left\{-\int_0^\infty \left(\sum_{i\in\mathcal{R}(t)} \theta_i(t)\right)h(t)dt\right\} \cdot \prod_{j=1}^{J} \theta_{i_j}(t_j)h(t_j) , \quad (3.2)$$

where as before t_j is the jth ordered failure time, i_j the index of the failed item, and $\mathcal{R}(t)$ the risk set of items on test just before time t. This is derived in the same way as (2.3). Aalen (1975) gives a rigorous derivation. Equation (3.2) assumes that $h_{i_j}(t)$ is continuous at t_j and that the risk sets are continuous from the left and change only finitely often in any finite interval.

We will rewrite (3.2) to emphasize its relation to the partial likelihood (1.2) and the likelihood (2.3) for the identical items situation. Define

$$H(t) \equiv \left(\sum_{i=1}^{n} \theta_i(t)/n\right)h(t) , \quad (3.3)$$

the average hazard rate if all n items were on test at time t, and also

$$N(t) \equiv n\left\{\sum_{i\in\mathcal{R}(t)} \theta_i(t)/\sum_{i=1}^{n} \theta_i(t)\right\} . \quad (3.4)$$

If all the items are identical, i.e., if $\theta_i(t)$ doesn't depend on i, then $N(t) = n(t)$, the number at risk at time t. In general $N(t)/n$ is the proportion of the total possible hazard on test at time t. To put it another way, $N(t)$ identical items each with hazard rate $H(t)$ would have the same total hazard as the items actually in $\mathcal{R}(t)$.

The likelihood function (3.2) can now be written as

$$f_{\theta,h}(\text{data}) = \{\prod_{j=1}^{J} [\theta_{i_j}(t_j)/\sum_{\mathcal{R}(t)} \theta_i(t_j)]\} \cdot \left\{\left[\exp - \int_0^\infty N(t)H(t)dt\right] \prod_{j=1}^{J} N(t_j)H(t_j)\right\} . \quad (3.5)$$

The first factor is the Cox likelihood, while the second factor is similar to (2.3).

The parameterization we will use assumes that the relative value of $\theta_i(t)$ and $\theta_{i'}(t)$, for any two indices i and i', is

$$\frac{\theta_i(t)}{\theta_{i'}(t)} = \frac{\exp\{\boldsymbol{\beta}\mathbf{z}_i(t)\}}{\exp\{\boldsymbol{\beta}\mathbf{z}_{i'}(t)\}} , \quad (3.6)$$

where $\boldsymbol{\beta}$ is a $1 \times B$ unknown parameter vector, and $\mathbf{z}_i(t)$ is a $B \times 1$ possibly time-varying vector of observed covariates. This parameterization makes (1.2) equal to

$$\prod_{j=1}^{J} [\exp\{\boldsymbol{\beta}\mathbf{z}_{i_j}(t_j)\}/\sum_{i\in\mathcal{R}(t_j)} \exp\{\boldsymbol{\beta}\mathbf{z}_i(t_j)\}] , \quad (3.7)$$

as in Cox (1972); (3.4) becomes

$$N(t) \equiv N(t, \boldsymbol{\beta}) = n \sum_{\mathcal{R}(t)} \exp\{\boldsymbol{\beta}\mathbf{z}_i(t)\}/\sum_{1}^{n} \exp\{\boldsymbol{\beta}\mathbf{z}_i(t)\} . \quad (3.8)$$

Notice that (3.6) is weaker than the assumption $\theta_i(t) = \exp\{\boldsymbol{\beta}\mathbf{z}_i(t)\}$ mentioned in Section 1. We will work directly with (3.7) and (3.8), obviating the need to explicitly parameterize the functions $\theta_i(t)$.

The function $H(t)$ is assumed to be of the form

$$H(t, \boldsymbol{\gamma}) = \exp\{\boldsymbol{\gamma}\mathbf{w}(t)\} , \quad (3.9)$$

where $\boldsymbol{\gamma}$ is a $1 \times C$ unknown parameter vector functionally independent of $\boldsymbol{\beta}$, and $\mathbf{w}(t)$ is another time-varying $C \times 1$ vector of observed covariates. Substituting (3.7)–(3.9) into (3.5) gives the likelihood expression,

$$f_{\beta,\gamma}(\text{data}) = \left\{\prod_{j=1}^{J} \frac{\exp\{\boldsymbol{\beta}\mathbf{z}_{i_j}(t_j)\}}{\sum_{\mathcal{R}(t_j)} \exp\{\boldsymbol{\beta}\mathbf{z}_i(t_j)\}}\right\} \cdot \left\{\exp\left\{-\int_0^\infty N(t,\boldsymbol{\beta})H(t,\boldsymbol{\gamma})dt\right\} \prod_{j=1}^{J} N(t_j,\boldsymbol{\beta})H(t_j,\boldsymbol{\gamma})\right\} . \quad (3.10)$$

(See Remark H, Section 6.)

Cox (1975) shows that the first factor can be treated as an ordinary likelihood function for the purpose of large-sample inference. In particular, the "maximum likelihood estimator" of $\boldsymbol{\beta}$ obtained by maximizing (3.7) will asymptotically have mean $\boldsymbol{\beta}$ and a covariance matrix which is the inverse of the "Fisher information matrix," the covariance matrix of the partial derivatives of the log of (3.7) with respect to the components of $\boldsymbol{\beta}$. The quotation marks used here serve as a reminder that (3.7) is not really a likelihood function. (For example it it *not* in general the likelihood of the reduced data set $(\mathcal{R}(t_1), i_1), (\mathcal{R}(t_2), i_2), \ldots, (\mathcal{R}(t_J), i_J)$.)

In what follows we will calculate the actual Fisher information matrix for $\boldsymbol{\beta}$ from (3.10) and give a heuristic demonstration that asymptotically it equals the information matrix based just on (3.7) assuming that the class of hazards $H(t, \boldsymbol{\gamma})$ is moderately large. This equality shows that the maximum likelihood estimate of $\boldsymbol{\beta}$ based on (3.7) must be asymptotically equivalent to that based on all the data. Similar statements hold true for asymptotic testing and confidence procedures (see Aalen 1975).

For convenience we consider only the case where β and, therefore, $z_i(t)$ is a scaler rather than a vector. The vector case is discussed briefly in Remark A, Section 6. Define

$$E_\beta\{z \mid \mathcal{R}(t)\} \equiv \sum_{i\in\mathcal{R}(t)} z_i(t)\exp\{\beta z_i(t)\}/\sum_{i\in\mathcal{R}(t)} \exp\{\beta z_i(t)\} ,$$

$$E_\beta z \equiv \sum_{i=1}^{n} z_i(t)\exp\{\beta z_i(t)\}/\sum_{i=1}^{n} \exp\{\beta z_i(t)\} , \quad (3.11)$$

and

$$\text{var}_\beta\{z \mid \mathcal{R}(t)\} = \sum_{i\in\mathcal{R}(t)} [z_i(t) - E_\beta\{z \mid \mathcal{R}(t)\}]^2 \cdot \exp\{\beta z_i(t)\}/\sum_{i\in\mathcal{R}(t)} \exp\{\beta z_i(t)\} . \quad (3.12)$$

$E_\beta\{z \mid \mathcal{R}(t)\}$ and $\text{var}_\beta\{z \mid \mathcal{R}(t)\}$ are the conditional mean and variance of $z_i(t)$ with respect to a probability distribution proportional to $\exp\{\beta z_i(t)\}$ on $i \in \mathcal{R}(t)$. They are functions of β and the random variable $\mathcal{R}(t)$. The following lemma computes the Fisher information in (3.10) for estimating β, i.e., one over the Cramér-Rao lower bound for unbiased estimation.

Lemma: The Fisher information for estimating β in (3.10) is

$$\inf_{\mathbf{g}} \int_0^\infty \mathcal{E}(\{\text{var}_\beta\{z \mid \mathcal{R}(t)\} + [(E_\beta\{z \mid \mathcal{R}(t)\} - E_\beta z) - \mathbf{g}\mathbf{w}(t)]^2\}N(t,\beta)H(t,\boldsymbol{\gamma}))dt , \quad (3.13)$$

where the infimum is over all choices of the C dimensional vector $\mathbf{g}$, and $\mathcal{E}$ indicates expectation over the randomness in the risk sets $\mathcal{R}(t)$. The same expression without the term in square brackets is the Fisher information for β based just on Cox's partial likelihood (3.7). (The proof is given in Section 7.)

Recall that if A and B are any two random variables, B is nonnegative, and a is any constant, then

$$E(A - a)^2B = [\text{var}_B A + (a - E_BA)^2]EB ,$$

where

$$E_BA \equiv EAB/EB \text{ and } \text{var}_B A \equiv E(A - E_BA)^2B/EB . \quad (3.14)$$

Let $\eta(t, \beta)$ be the expectation, over the randomness in $\mathcal{R}(t)$, of $N(t, \beta)$,

$$\eta(t, \beta) \equiv \mathcal{E}N(t, \beta)$$
$$= n(\mathcal{E} \sum_{\mathcal{R}(t)} \exp\{\beta z_i(t)\}/\sum_{i=1}^{n} \exp\{\beta z_i(t)\}) \; ; \quad (3.15)$$

define

$$B \equiv N(t, \beta) \; , \quad A \equiv E_\beta\{z \mid \mathcal{R}(t)\} - E_\beta z \; ,$$

and $a \equiv \mathbf{g}\mathbf{w}(t)$. Using (3.14), the integrand of (3.13) can be expressed as

$$\{\mathcal{E}[N(t, \beta)/\eta(t, \beta)] \operatorname{var}_\beta \{z \mid \mathcal{R}(t)\} + \operatorname{var}_N E_\beta\{z \mid \mathcal{R}(t)\}$$
$$+ [e_\beta(t) - \mathbf{g}\mathbf{w}(t)]^2\}\eta(t, \beta)H(t, \gamma) \; , \quad (3.16)$$

where

$$e_\beta(t) \equiv \mathcal{E}_N(E_\beta\{z \mid \mathcal{R}(t)\} - E_\beta z) \; , \quad (3.17)$$

and $\operatorname{var}_N E_\beta\{z \mid \mathcal{R}(t)\}$ indicates a weighted variance, as in (3.14) with the random quantity being $\mathcal{R}(t)$.

A simple calculation shows that if $P_i(t)$ is the probability that item i is in $\mathcal{R}(t)$, then

$$e_\beta(t) = \sum_1^n P_i(t)z_i(t) \exp\{\beta z_i(t)\}/\sum_1^n P_i(t) \exp\{\beta z_i(t)\}$$
$$- \sum_1^n z_i(t) \exp\{\beta z_i(t)\}/\sum_1^n \exp\{\beta z_i(t)\} \; . \quad (3.18)$$

Notice that $P_i(t)$ is also a function of β and γ and possibly other extraneous random factors.

The principle implied by the lemma and (3.16), admittedly in a rough manner, is the following: *If, as the number items tested goes to infinity, the function $e_\beta(t)$ can be approximated arbitrarily well by a linear combination of the functions $w_1(t)$, $w_2(t)$, ..., $w_C(t)$, then the Cox likelihood is asymptotically fully efficient for the estimation of β.* In other words, the Fisher information for β based on the Cox likelihood has asymptotic ratio unity with that based on all the data. Section 4 illustrates this principle in a particularly simple special case.

Suppose for a moment that $e_\beta(t) = \mathbf{g}\mathbf{w}(t)$ for all t for some choice of $\mathbf{g}$. This eliminates the last term in square brackets from (3.16). The additional information for estimating β *not* in the Cox likelihood corresponds to the term $\operatorname{var}_N E_\beta\{z \mid \mathcal{R}(t)\}$. Intuitively this comes from local variations in $N(t, \beta)$ due to random fluctuations in the risk sets, which influence the observed times between failures. These random fluctuations can not be explained away by any possible choice of $H(t, \gamma)$ since this is necessarily a fixed (nonrandom) function of time. However, the magnitude of this term tends to be $0(1/\eta)$ compared to the term $(N/\eta) \operatorname{var}_\beta \{z \mid \mathcal{R}(t)\}$ from the partial likelihood, essentially because $E_\beta\{z \mid \mathcal{R}(t)\}$ is the average of about η random quantities. (See Remark I, Section 6.)

For asymptotic efficiency we don't need $e_\beta(t)$ to actually be in the linear space generated by $w_1, w_2, \ldots, w_C$,

$$\mathcal{L}(\mathbf{w}) \equiv \{\sum_{c=1}^{C} g_c w_c(t)\} \; , \quad (3.19)$$

but only that it be increasingly well approximated by some function in $\mathcal{L}(\mathbf{w})$ as the number of tested items grows large. In other words, we need to be able to ignore the term $[e_B(t) - \mathbf{g}\mathbf{w}(t)]^2$ in (3.16).

In order for the partial likelihood to estimate β with reasonable efficiency in finite samples it is necessary for $e_\beta(t)$ to be in or at least near $\mathcal{L}(\mathbf{w})$. Is this a realistic assumption? In many situations the answer is yes. For example, if the z_i are not functions of time, and if there is no censoring, then (3.18) shows that $e_\beta(t)$ is monotonic. For $\beta > 0$, $e_\beta(t)$ will decrease monotonically in time as those items with large values of z_i are selectively removed by earlier failure. Censoring can distort $e_\beta(t)$ but not seriously unless a large proportion of the items have the same fixed censoring time. (See Section 4.) In the absence of firm prior knowledge it may be reasonable to assume that $H(t, \gamma) = \exp\{\gamma\mathbf{w}(t)\}$ can be any smooth monotonic function, which in this case guarantees the asymptotic efficiency of the partial likelihood.

Of course there are situations in which the partial likelihood by itself produces seriously inefficient inferences. For example $\mathcal{L}(\mathbf{w})$ might be known to be the class of linear functions $w_1 + w_2 t$ while $e_\beta(t)$ is some considerably more complicated function. In theory at least, the statistician can always calculate the actual maximum likelihood estimator (MLE) of β from (3.10) in such cases. Kalbfleisch (1974) gives an efficiency calculation in one such case, which reinforces faith in using (3.7) by itself, as do the calculations of Section 4.

4. THE TWO-SAMPLE PROBLEM

The general calculations of Section 3 are more understandable in special cases, the most special of which we consider now: the two-sample problem with exponentially distributed lifetimes. Let

$$e^\beta \equiv \alpha \quad (4.1)$$

be the ratio of expectations for the two samples, and let β be the parameter to be estimated. The two samples are of sizes, say, n_0 and n_1, respectively, $n_0 + n_1 = n$, with sample membership being indicated by the dummy variable

$$z_i = 0 \text{ if item } i \text{ is in sample } 0, \; i = 1, 2, \ldots, n \; ;$$
$$= 1 \text{ if item } i \text{ is in sample } 1, \; i = 1, 2, \ldots, n \; . \quad (4.2)$$

Also let

$$q \equiv n_0/n \; , \quad p \equiv n_1/n \; , \quad \text{and} \quad D_\alpha \equiv q + p\alpha \; . \quad (4.3)$$

To parameterize this situation as in Section 3 the hazard rates (3.1) are written as

$$h_i(t) = \alpha^{z_i} e^\gamma / D_\alpha \; , \quad i = 1, 2, \ldots, n \; . \quad (4.4)$$

This makes $H(t, \gamma)$ defined at (3.3) equal e^γ, and, as will be apparent, there is no loss of generality in assuming $\gamma = 0$, $H(t, \gamma) \equiv 1$. We see that the probability of item

i's lifetime exceeding t equals

$$P_i(t) = \exp\{-t\alpha^{z_i}/D_\alpha\}, \quad (4.5)$$

assuming that there is no censoring.

In the absence of censoring, (3.13) and (3.16) give a simple expression for the asymptotic variance of β^*, which is the MLE based on Cox's partial likelihood function (3.7). We will consider the effects of censoring later. Let

$$n_\ell(t) \equiv \text{number of sample } \ell \text{ members in } \mathcal{R}(t),\ \ell = 0, 1,$$

and

$$n(t) \equiv n_0(t) + n_1(t), \quad q_t \equiv n_0(t)/n(t), \quad p_t \equiv n_1(t)/n(t). \quad (4.6)$$

Then

$$N(t, \beta) = n(t)(q_t + p_t\alpha)/D_\alpha, \quad (4.7)$$
$$\text{var}_\beta\{z \mid \mathcal{R}(t)\} = q_t p_t \alpha/(q_t + p_t\alpha)^2,$$

by substitution in (3.8), (3.12). As n gets large, the random quantities $n(t)/n$, $q(t)$, and $p(t)$ approach constants easily determined from (4.5) giving

$$\lim_{n\to\infty} \frac{1}{n}\mathcal{E}[\text{var}_\beta\{z \mid \mathcal{R}(t)\}]N(t,\beta)H(t,\gamma) = \frac{pq\alpha\exp\{-t\alpha/D_\alpha\}}{D_\alpha[q + p\alpha\exp\{-t(\alpha-1)/D_\alpha\}]}. \quad (4.8)$$

The lemma gives the expression

$$\lim_{n\to\infty}\frac{1}{n\,\text{var}\,\beta^*} = \int_0^1 \frac{pq\,du}{q + p\alpha u^{(\alpha-1)/\alpha}} \quad (4.9)$$

for the limiting variance of β^*, where the substitution $u = \exp\{-t\alpha/D_\alpha\}$ has been made in (3.13), (3.16). The limiting variance of β^{**}, the MLE based on all the data, is

$$\lim_{n\to\infty} n\,\text{var}\,\beta^{**} = 1/pq \quad (4.10)$$

under (4.4), as shown by standard Fisher information arguments. The asymptotic relative efficiency (ARE) of the partial likelihood estimate compared to full maximum likelihood is

$$\text{ARE} \equiv \lim_{n\to\infty}\frac{\text{var}\,\beta^{**}}{\text{var}\,\beta^*} = \int_0^1 \frac{du}{q + p\alpha u^{(\alpha-1)/\alpha}}. \quad (4.11)$$

The first line of the table tabulates (4.11) for various choices of α with $p = q = \frac{1}{2}$.

The Asymptotic Relative Efficiency of the Partial Likelihood Estimate of β Compared to the Full Maximum Likelihood Estimate, p = q = ½.

$\alpha \equiv e^\beta$	1	2	4	8	16
ARE from (4.11)	1.000	.901	.705	.502	.334
ARE under model (4.15)–(4.17)					
No censoring	1.000	.982	.959	.914	.819
Censoring pattern 1	.991	.978	.950	.912	.819
Censoring pattern 2	.994	.987	.967	.915	.816

NOTE: The ARE increases under the larger model (4.15)–(4.17) for $H(t,\gamma)$. Censoring has little effect on these calculations. Pattern 1 has half of sample 1 censored at the median of the distribution for sample 0. Pattern 2 has one quarter of sample 1 censored at each quartile of the distribution for sample 0.

Any inefficiency of β^* compared to β^{**} comes from the last term in (3.16), $[e_\beta(t) - \mathbf{g}\mathbf{w}(t)]^2$. By assuming $H(t,\gamma)$ constant we have restricted $\mathcal{L}(\mathbf{w})$, (3.19), to involve only constant functions. The inefficiency, $1 -$ ARE, equals

$$(pq)^{-1}\min_g \int_0^\infty [e_\beta(t) - g]^2 \frac{\eta(t,\beta)}{n}\,dt \quad (4.12)$$

(the factor $(pq)^{-1}$ coming from (4.10)), where

$$\frac{\eta(t,\beta)}{n} = \frac{q\exp\{-t/D_\alpha\} + p\alpha\exp\{-t\alpha/D_\alpha\}}{D_\alpha} \quad (4.13)$$

by (3.15) and (4.5).

It is easy to evaluate $e_\beta(t)$ from (3.18),

$$e_\beta(t) = \frac{p\alpha}{D_\alpha}\left[\frac{D_\alpha}{q\exp\{t(\alpha-1)/D_\alpha\} + p\alpha} - 1\right]. \quad (4.14)$$

The figure graphs $e_\beta(t)$ for $\alpha = e^\beta = 1, 2, 4, 8$, and 16 in the case $p = q = .5$ showing a smooth monotonic decrease to the asymptote $-p\alpha/D_\alpha$ (for $\alpha > 1$). Notice that in the absence of censoring, $e_1(t) \in \mathcal{L}(\mathbf{w})$ which explains the full asymptotic efficiency in this case.

The Function $e_\beta(t)$, (4.14) for $\alpha = e^\beta = 1,2,4,8,16$ ($p = q = .5$)[a]

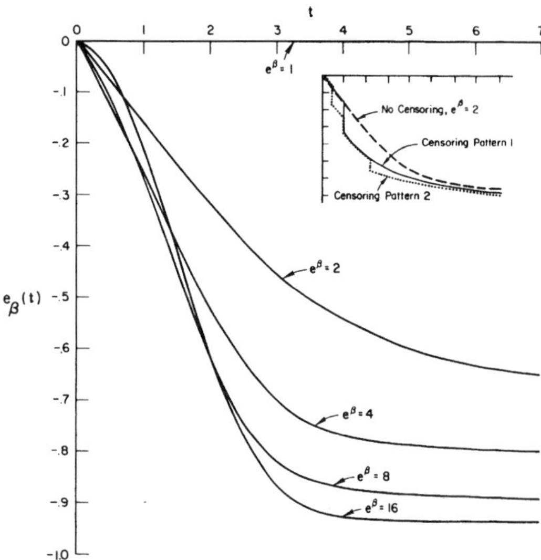

[a] The insert shows $e_\beta(t)$, $e^\beta = 2$, for the two censoring patterns mentioned in the text.

Suppose now that we are not willing to assume an exponential model for the lifetimes, but are willing to assume that the relative hazard rate between the two groups is constant, $h_1(t)/h_0(t) = e^\beta$. This type of "Lehmann alternative" is more in the spirit of Cox's article. As a first step, (4.4) can be expanded to

$$h_i(t) = \frac{\alpha^{z_i}\exp\{\gamma_0 + \gamma_1 w_1(t)\}}{D_\alpha} \quad i = 1, 2, \ldots, n, \quad (4.15)$$

making

$$H(t, \gamma) = \exp\{\gamma_0 + \gamma_1 w_1(t)\}\ ,$$
$$\mathcal{L}(\mathbf{w}) = \{g_0 + g_1 w_1(t)\}\ . \tag{4.16}$$

Here $w_1(t)$ is some specified function we are willing to use to expand the class of possible hazard rates. For example,

$$w_1(t) = e^{-t} \tag{4.17}$$

allows $h_i(t)$ to vary monotonically by a finite factor as t goes from zero to infinity.

The partial likelihood (3.7) is unaffected by changes in $H(t, \gamma)$, so (4.9) remains valid when sampling under $\gamma = (0, 0)$. The full MLE β^{**} now has greater limiting variance, so the ARE of β^* to β^{**} is larger, as shown in the fourth line of the table. The asymptotic inefficiency of β^* relative to β^{**} depends on the magnitude of

$$\min_{g_0, g_1} \int_0^\infty [e_\beta(t) - g_0 - g_1 w_1(t)]^2 \frac{\eta(t, \beta)}{n} dt\ . \tag{4.18}$$

The choice $w_1(t) = e^{-t}$ gives high ARE in this case because it closely matches the shape of $e_\beta(t)$, at least for $e^\beta \leq 8$.

It is easiest to interpret the table in terms of the asymptotic variance of the MLE based on all the data, relative to that of the MLE based just on the partial likelihood (3.7). The numbers also have a testing interpretation as Pitman efficiencies. For example, under Censoring pattern 1, the locally most powerful test of $\alpha = 1$ vs $\alpha > 1$ based on (3.7) has Pitman efficiency .991 compared to that based on all the data. (The test based on (3.7) is a generalization of the Savage rank test, described in Cox (1972) and also in Thomas (1971).) A more general interpretation is that using (3.7) rather than the full likelihood asymptotically wastes nine out of 1,000 observations in this particular situation, for any inferential purpose at all.

Of course there is no real reason behind the choice (4.17). In most practical problems there isn't any obvious choice, beyond perhaps a qualitative preference for monotonic reasonably smooth hazard rates. The functions $e_\beta(t)$ in the figure fit this description. If we take $w_1(t) = e\beta(t)$, the ARE of β^* to β^{**} is one. It is the author's opinion that Cox's method will usually give high efficiency under any reasonably realistic assumptions on the class of possible hazard rates.

Censoring seems to have little effect on the efficiency calculations. The insert to the figure shows $e_\beta(t)$, $e^\beta = 2$, for two censoring patterns: (1) sample 0 uncensored, 50 percent of sample 1 censored at the median of the distribution for sample 0; (2) sample 0 uncensored, 25 percent of sample 1 censored at each quartile of the distribution for sample 0. The discontinuities in $e_\beta(t)$ come from the $P_i(t)$ in (3.18) going suddenly to zero as the fixed censoring times are encountered. Nevertheless, the ARE stays almost constant as the last two lines of the table show.

All of these calculations are asymptotic in nature. In finite samples there is a further loss of efficiency for β^* compared to β^{**} coming from the term $\text{var}_N\, E_\beta\{z \mid \mathcal{R}(t)\}$ in (3.16). The calculations in Kalbfleisch (1974), in particular his Table 1 and equation (15), suggest an additional efficiency loss of about 10 percent for $n = 10$, 6 percent for $n = 20$, and 5 percent for $n = 40$.

5. ESTIMATING THE HAZARD RATES

Suppose we are willing to rely on the first factor in (3.10), the Cox likelihood, for the estimation of $\boldsymbol{\beta}$. We can treat the estimate obtained in this way, say $\boldsymbol{\beta}^*$, as if it were the true value of $\boldsymbol{\beta}$ and then maximize the second factor in (3.10) to estimate $\boldsymbol{\gamma}$.

Let $\boldsymbol{\gamma}^*$ be the "maximum likelihood" estimator of $\boldsymbol{\gamma}$ obtained in this way, the quotes indicating that $\boldsymbol{\gamma}^*$ is really only the conditional maximizer given the value $\boldsymbol{\beta}^*$ obtained from the Cox likelihood. From (3.1), (3.3), and (3.6), we get

$$h_i(t) = n[\theta_i(t) / \sum_{i'=1}^n \theta_{i'}(t)]H(t)$$
$$= n[\exp\{\boldsymbol{\beta}\mathbf{z}_i(t)\} / \sum_{i'=1}^n \exp\{\boldsymbol{\beta}\mathbf{z}_{i'}(t)\}]H(t, \boldsymbol{\gamma})\ ; \tag{5.1}$$

therefore, the corresponding estimate of the hazard rate for item i is

$$h_i^*(t) = n[\exp\{\boldsymbol{\beta}^*\mathbf{z}_i(t)\} / \sum_{i'=1}^n \exp\{\boldsymbol{\beta}^*\mathbf{z}_{i'}(t)\}]H(t, \boldsymbol{\gamma}^*)\ . \tag{5.2}$$

In the Kaplan-Meier nonparametric situation, $H(t, \boldsymbol{\gamma}^*)$ approaches $H^*(t)$, a sum of delta functions at $t_1, t_2, \ldots, t_J$ satisfying

$$\exp\left\{-\int_{t_j^-}^{t_j^+} H^*(t)dt\right\} = 1 - \frac{1}{N(t_j, \boldsymbol{\beta}^*)}\cdot \tag{5.3}$$

Assuming that the functions $\mathbf{z}_i(t)$, $i \in \mathcal{R}(t_j)$ are continuous at t_j, this gives

$$\exp\left\{-\int_{t_j^-}^{t_j^+} h_i^*(t)dt\right\} = \left[1 - \frac{1}{N(t_j, \boldsymbol{\beta}^*)}\right]^{\phi_{ij}^*}\ , \tag{5.4}$$

where

$$\phi_{ij}^* \equiv n \exp\{\boldsymbol{\beta}^*\mathbf{z}_i(t_j)\} / \sum_1^n \exp\{\boldsymbol{\beta}^* z_{i'}(t_j)\}\ .$$

The estimate of the ith cdf is

$$F_i^*(t) = \prod_{t_j \leq t} \left[1 - \frac{1}{N(t_j, \boldsymbol{\beta}^*)}\right]^{\phi_{ij}^*}$$
$$\approx \exp - [\sum_{t_j \leq t} (\exp\{\boldsymbol{\beta}^*\mathbf{z}_i(t_j)\} /$$
$$\sum_{i \in \mathcal{R}(t)} \exp\{\boldsymbol{\beta}^*\mathbf{z}_{i'}(t_j)\})]\ ; \tag{5.5}$$

this last form is essentially the same as that derived in Breslow (1974) and also in Kalbfleisch and Prentice (1973) for the case which is not time-dependent. (See Remark c, Section 6 of this article.)

6. SOME REMARKS

A. The information calculations of Section 3 carry over directly to the case where β is a vector. The expression for the information matrix for estimating β is the multivariate analog of (3.13),

$$\inf_{\mathbf{G}} \int_0^\infty \mathcal{E}(\{\text{cov}_\beta\{\mathbf{z} \mid \mathcal{R}(t)\} + [(E_\beta\{\mathbf{z} \mid \mathcal{R}(t)\} - E_\beta \mathbf{z}) - \mathbf{G}\mathbf{w}(t)][(E_\beta\{\mathbf{z} \mid \mathcal{R}(t)\} - E_\beta \mathbf{z}) - \mathbf{G}\mathbf{w}(t)]'\} \cdot N(t, \beta)H(t, \gamma))dt , \quad (6.1)$$

with the infimum being taken over all $B \times C$ matrices $\mathbf{G}$.

B. There is no particular advantage to the exponential forms $\exp\{\beta \mathbf{z}_i(t)\}$, $\exp\{\gamma \mathbf{w}(t)\}$ used in Section 3. Any other simple positive function serves just as well and may be more natural in some situations. Suppose, e.g., that the event $T < 1$ is hypothesized to follow a linear logistic law in terms of β and the (non-time-varying) covariate $\mathbf{z}_i$,

$$\text{Prob}\{T_i < 1\} = \exp\{\beta \mathbf{z}_i\}/1 + \exp\{\beta \mathbf{z}_i\} . \quad (6.2)$$

This implies

$$\theta_1(\beta) \propto \log[1 + \exp\{\beta \mathbf{z}_i\}] \quad (6.3)$$

rather than $\theta_i(\beta) \propto \exp\{\beta \mathbf{z}_i\}$.

C. If m is a large positive number then

$$\log(1 - 1/m) = -1/[m - c(m)] , \quad (6.4)$$

where $c(m) = \frac{1}{2} - 1/12m + \ldots$. Expression (2.5) for the Kaplan-Meier estimator can be written as

$$\text{Prob}^*\{T > t\} = \exp\{-\sum_{t_j \le t} 1/[n(t_j) - c(n(t_j))]\} . \quad (6.5)$$

Ignoring the correction term $c(n(t_j))$ leads to the last expression in (5.5).

D. The Kaplan-Meier estimator corresponds to the limit of continuous hazard functions putting mass $1/[n(t_j) - c(n(t_j))]$ at t_j, *not* mass $1/n(t_j)$

$$(\text{since } \exp - \{\text{mass at } t_j\} = 1 - 1/n(t_j)) .$$

E. The likelihood expressions (3.2), (3.5), and (3.10) assume that the risk sets $\mathcal{R}(t)$ are themselves uninformative for β and γ. It is allowable for $\mathcal{R}(t)$ to depend on all data observed before time t, plus random elements whose distributions don't depend on β or γ. Subject to these restrictions, a malevolent censorer trying to confuse the statistician cannot affect the likelihood function or any Bayesian/likelihood based inferences, though he can affect expectations connected with the likelihood such as the Fisher information.

Kalbfleisch and Prentice (1973) tacitly make a stronger assumption about the censoring mechanism; it in no way depends on the real time axis except through the ordering of the observed events. Otherwise, their marginal likelihood interpretation of Cox's likelihood can easily be contradicted. Take $n = 3$, and suppose that z_1, z_2, z_3 do not depend on time, so that $\theta_1, \theta_2, \theta_3$ are time independent. Suppose also that no observations are censored if $\min\{T_1, T_2, T_3\} \le 1.5$, but if the first observation is T_1 and it exceeds 1.5, then further observation on T_2 is immediately censored. An easy calculation gives the probability of observing the partial ordering "T_1 less than $\min\{T_2, T_3\}$" to be

$$\text{Prob}\{(1, 2, 3,) \cup (1, 3, 2)\} = \exp\left\{-(\theta_1 + \theta_2 + \theta_3)\int_0^{1.5} h(t)dt\right\} \cdot \theta_1/(\theta_1 + \theta_2 + \theta_3) , \quad (6.6)$$

which does not equal the Cox likelihood $\theta_1/(\theta_1 + \theta_2 + \theta_3)$.

F. Another hidden assumption in (3.2) is that once an item leaves the experiment due to censoring it does not return on test at a later time. Suppose an item did drop out at time $t = a$ and returned at $t = b$; then either it will be known to have failed during that interval, multiplying the likelihood function by the ungainly factor $1 - \exp\{-\int_a^b \theta_i(t)h(t)dt\}$, or it will be seen not to have failed during that interval, in which case it really was observed. This point does not arise in the Kaplan-Meier situation of Section 2 unless we add labels to the identical test items in order to make them identifiable.

The two types of allowable changes in the risk sets, aside from failure, are illustrated in the senior citizen study. These are caused by items entering the study late, without any information on those failing before entry (*left truncation*), and items leaving the study before failure (*right censoring*).

G. Real censored data problems are often *discrete*; items are reported to fail during intervals, not by exact times. (In the senior citizen study, e.g., deaths and changes in the risk sets were reported by day but not by minute and second.) Let us add the assumption that the ratio of hazards (3.6) is constant during any one such reporting interval, and that no changes in $\mathcal{R}(t)$ occur within such an interval except those due to failure. Then given the information that the m items $i_{j1}, i_{j2}, \ldots, i_{jm}$ failed during the jth reporting interval, we know that the Cox likelihood for the (unobservable) continuous data takes on one of $m!$ possible values, corresponding to the $m!$ possible orderings of $i_{j1}, i_{j2}, \ldots, i_{jm}$, each with equal probability. It is notationally messy to average these $m!$ quantities, but an obvious approximation for the jth factor in the Cox likelihood is

$$\frac{\theta_{i_{j1}}(t_j)\theta_{i_{j2}}(t_j)\ldots\theta_{i_{jm}}(t_j)}{\prod_{\ell=0}^{m-1}[\sum_{i\in\mathcal{R}(t_j)} \theta_i(t_j) - \frac{\ell}{m}\sum_{h=1}^{m}\theta_{i_{jh}}(t_j)]} . \quad (6.7)$$

This is a slightly more accurate approximation than those suggested in the discussion following Cox's 1972 article, but as Peto suggests there, it probably doesn't make much difference.

H. The parameterization, (3.6)-(3.9), which leads to the likelihood expression (3.10) assumes that the relative hazard rates for the different items in the experiment do not functionally determine the total hazard rate. More precisely, the information calculations at, say, $\boldsymbol{\beta}^{(0)}$, $\boldsymbol{\gamma}^{(0)}$ require that the possible $\boldsymbol{\gamma}$ vectors corresponding to $\boldsymbol{\beta} = \boldsymbol{\beta}^{(0)}$ include an open set around $\boldsymbol{\gamma}^{(0)}$.

An alternative parameterization which seems appealing is to let $\bar{h}(t, \gamma) \equiv (\sum_{\mathcal{R}(t)} \theta_i(t, \boldsymbol{\beta})/n(t))h(t)$ be the average hazard rate of those items on test at time t, where $n(t)$ is the number of items in $\mathcal{R}(t)$, and to assume $\theta_i(t, \boldsymbol{\beta}) = \exp\{\boldsymbol{\beta}\mathbf{z}_i(t)\}$, $h(t, \boldsymbol{\gamma}) = \exp\{\boldsymbol{\gamma}\mathbf{w}(t)\}$. This makes the second factor in (5.10) equal to

$$\exp\left\{-\int_0^\infty n(t)\bar{h}(t, \gamma)dt\right\} \prod_{j=1}^{J} n(t_j)\bar{h}(t_j, \gamma) \,, \quad (6.8)$$

which is much simpler since it does not involve $\boldsymbol{\beta}$ at all. However, this parameterization is untenable. The function $\bar{h}$ must depend on $\boldsymbol{\beta}$ in some way, since if $\boldsymbol{\beta}$ is not zero, $\bar{h}$ changes value discontinuously whenever the risk set changes. This is impossible for any function of the form $\exp\{\boldsymbol{\gamma}\mathbf{w}(t)\}$, except in very restricted situations.

I. Suppose that all the items act independently of each other in terms of failures and censoring. Then standard expansion methods would show that the quantity $\text{var}_N\, E_\beta\{z \mid \mathcal{R}(t)\}$, which figures in (3.16), approximately equals

$$(1/\eta)\left(\sum_{i=1}^{n} P_i Q_i \phi_i^2 (z_i - R)^2/\eta\right) \,, \quad (6.9)$$

where t, β, and γ have been dropped from the notation, $Q_i \equiv 1 - P_i$, and

$$\phi_i \equiv n \exp\{\beta z_i\} / \sum_{i'=1}^{n} \exp\{\beta z_{i'}\} \,,$$

$$R \equiv \sum_{i=1}^{n} P_i z_i \exp\{\beta z_i\} / \sum_{i=1}^{n} P_i \exp\{\beta z_i\} \,. \quad (6.10)$$

Assuming the $z_i(t)$ are bounded, (6.9) is $O(1/\eta)$ as η goes to infinity.

7. PROOF OF THE LEMMA

To prove the lemma of Section 3, we calculate the score functions for β and $\gamma_1, \gamma_2, \ldots, \gamma_C$ from (3.8)–(3.10),

$$S_\beta \equiv \frac{\partial \log f_{\beta,\gamma}}{\partial \beta} = \sum_{j=1}^{J} [z_{ij}(t_j) - E_\beta\{z \mid \mathcal{R}(t)\}] + \int_0^\infty [E_\beta\{z \mid \mathcal{R}(t)\} - E_\beta z] \cdot [\mathbf{J}(t) - N(t, \beta)H(t, \gamma)]dt \,, \quad (7.1)$$

and

$$S_{\gamma_c} \equiv \frac{\partial \log f_{\beta,\gamma}}{\partial \gamma_c} = \int_0^\infty w_c(t)[\mathbf{J}(t) - N(t, \beta)H(t, \gamma)]dt \,,$$

where $\mathbf{J}(t) = \sum_{j=1}^{J} \delta(t - t_j)$, which is the sum of delta functions at $t_1, t_2, \ldots, t_\delta$. For an arbitrary choice of $\mathbf{g} = (g_{,1}\, g_2, \ldots, g_C)$ we can write

$$S_\beta - \sum_{c=1}^{C} g_c S_{\gamma_c} = \int_0^\infty [U(t) + (D_\beta(t) - \mathbf{g}\mathbf{w}(t))]dK(t) \,, \quad (7.2)$$

where

$$\begin{aligned} U(t) &= z_{ij}(t_j) - E_\beta\{z \mid \mathcal{R}(t_j)\} \,, \quad \text{if } t = t_j \\ &= 0 \,, \quad \text{if } t \notin \{t_1, t_2, \ldots, t_j\} \,, \end{aligned}$$

$$D_\beta(t) = E_\beta\{z \mid \mathcal{R}(t)\} - E_\beta z \,, \quad (7.3)$$

and

$$dK(t) = [\mathbf{J}(t) - N(t, \beta)H(t, \gamma)]dt \,.$$

Given the observed value of $\mathcal{R}(t)$, $D_\beta(t)$ is a fixed number while $U(t)$ is a random variable with mean zero and variance

$$\begin{aligned} \text{var}\{U(t) \mid \mathcal{R}(t)\} &= \text{var}_\beta\{z \mid \mathcal{R}(t)\} \,, \quad \text{if } t = t_j \\ &= 0 \,, \quad \text{if } t \notin \{t_1, t_2, \ldots, t_j\} \,, \end{aligned} \quad (7.4)$$

with $\text{var}_\beta\{z \mid \mathcal{R}(t)\}$ as defined in (3.12). Also, still assuming $\mathcal{R}(t)$ given,

$$\begin{aligned} dK(t) &= 1 - N(t, \beta)H(t, \gamma)dt \\ &\qquad \text{with Prob } N(t, \beta)H(t, \gamma)dt \\ &= -N(t, \beta)H(t, \gamma)dt \\ &\qquad \text{with Prob } 1 - N(t, \beta)H(t, \gamma)dt \,. \end{aligned} \quad (7.5)$$

Notice that the two cases for $dK(t)$ correspond to the two cases for $U(t)$ given in (7.3). Expressions (7.4)–(7.5) are easier to understand in the coin-tossing formulation of Section 1.

Putting (7.3)–(7.5) together gives

$$\begin{aligned} E\{([U(t) &+ (D_\beta(t) - \mathbf{g}\mathbf{w}(t))]dK(t))^2 \mid \mathcal{R}(t)\} \\ &= \{\text{var}_\beta\{z \mid \mathcal{R}(t)\} + [(E_\beta\{z \mid \mathcal{R}(t)\} \\ &\quad - E_\beta z) - \mathbf{g}\mathbf{w}(t)]^2\} N(t, \beta)H(t, \gamma)dt \,, \end{aligned} \quad (7.6)$$

and, for $t' < t$,

$$\begin{aligned} E\{([U(t') &+ (D_\beta(t') - \mathbf{g}\mathbf{w}(t'))]dK(t')) \\ &\cdot ([U(t) + (D_\beta(t) - \mathbf{g}\mathbf{w}(t))]dK(t)) \mid \mathcal{R}(s) \,, \\ &\qquad 0 \le s \le t\} = 0 \,. \end{aligned} \quad (7.7)$$

(In deriving (7.7) we have used $E\{dK(t) \mid \mathcal{R}(t)\} = E\{U(t)dK(t) \mid \mathcal{R}(t)\} = 0$.) Therefore, writing the integral (7.2) as a Reimann sum and conditioning successively on $\mathcal{R}(t)$ as t increases from 0 to ∞ gives the expected value of $(S_\beta - \sum_{c=1}^{C} g_c S_{\gamma_c})^2$ to be the integral in (3.13). But the reciprocal of the Cramér-Rao lower bound for β, by definition the Fisher information for β, is the infimum of the expected value of $(S_\beta - \sum_{c=1}^{C} g_c S_{\gamma_c})^2$ over all choices of $\mathbf{g}$. This proves the first part of the lemma. The second part follows by a similar argument which is made easier by the fact that (3.7) does not involve $\boldsymbol{\gamma}$.

[*Received May 1976. Revised November 1976.*]

REFERENCES

Aalen, Odd O. (1975), "Statistical Inference for a Family of Counting Processes," Ph.D. dissertation, Department of Statistics, University of California, Berkeley.

Breslow, N. (1974), "Covariance Analysis of Censored Data," *Biometrics*, 30, 89–99.

Breslow, N., and Crowley, J. (1974), "A Large Sample Study of the Life Table and Product Limit Estimates Under Random Censorship," *Annals of Statistics*, 2, 437–53.

Cox, D.R. (1972), "Regression Models and Life-Tables (with Discussion)," *Journal of the Royal Statistical Society*, Ser. B, 34, 187–220.

——— (1975), "Partial Likelihood," *Biometrika*, 62, 269–79.

Kalbfleisch, J., and Prentice, R. (1973), "Marginal Likelihoods Based on Cox's Regression and Life Model," *Biometrika*, 60, 267–79.

———, (1974), "Some Efficiency Calculations for Survival Distributions," *Biometrika*, 61, 31–8.

Kaplan, E.L., and Meier, P. (1958), "Nonparametric Estimation from Incomplete Observations," *Journal of the American Statistical Association*, 53, 457–81.

Mantel, N., and Haenzel, W. (1959), "Statistical Aspects of the Analysis of Data From Retrospective Studies of Disease," *Journal of the National Cancer Institute*, 22, 719–48.

Peto, R., and Peto, J. (1972), "Asymptotically Efficient Rank Invariant Procedures," *Journal of the Royal Statistical Society*, Ser. A, 135, 185–206.

Thomas, D.R. (1971), "On the Asymptotic Normality of a Generalized Savage Statistic for Comparing Two Arbitrarily Censored Samples," Technical Report, Department of Statistics, Oregon State University.

7

Stein's Paradox in Statistics

Introduction by Jim Berger
Duke University

Discussion of the Stein paradox – that combining three or more unrelated estimation problems can yield superior estimates from a frequentist perspective – was pervasive in statistics at the time of this article, in part due to the stimulating series of articles by Efron and Morris that appeared between 1971 and 1975 on the relationship between empirical Bayes analysis and Stein estimation. The time was thus ripe to bring this issue to the larger scientific community, and this article did so in spectacular fashion.

The article does an excellent job of highlighting the two key issues in understanding the Stein paradox:

1. The limits of the meaning of the paradox
2. The practical implementation of shrinkage estimation

The article very clearly points out the primary limit to the paradox that the goal must be average performance in estimation across the problems; there is no guarantee of improvement for an individual problem and, indeed, considerable potential for harm for any problem that is *unlike* the other problems. Thus one could imagine a company utilizing Stein estimation for three problems from different company divisions, since presumably it is overall company performance that is the goal; but it would be difficult to imagine three scientists in separate fields agreeing that their goal was overall performance in estimation across their problems.

For practical implementation of shrinkage, the authors turn to empirical Bayes analysis, and present two compelling examples from their JASA article, *Data Analysis Using Stein's Estimator and Its Generalizations*. The authors' summary of the benefits arising from shrinkage in these examples was:

> The rationale of the method is to reduce the overall risk by assuming that the true means are more similar to one another than the observed data. That assumption can degrade the estimation of a genuinely atypical mean.

In support of this last point, they give the delightful example of introducing estimation of the proportion of imported cars in Chicago with the estimation of the eighteen baseball players' batting averages.

It is of interest, in looking back, that the practical implementation recommended by the authors was empirical Bayes estimation, not Stein estimation. Indeed, in their second example involving toxoplasmosis, the estimator they recommend is not an estimator that uniformly dominates the least squares estimator and, hence, is not an example of a Stein estimator. This presaged the reality of practice today: empirical Bayes – and the very similar hierarchical Bayes or multilevel analyses – dominate statistical practice involving simultaneous estimation. That this is so, however, owes a great deal historically to the comfort that many non-Bayesians felt in moving towards such analyses because of the Stein effect, and because of the articles of Efron and Morris that showed how related empirical Bayes ideas could be used to great practical advantage.

Stein's Paradox in Statistics

The best guess about the future is usually obtained by computing the average of past events. Stein's paradox defines circumstances in which there are estimators better than the arithmetic average

by Bradley Efron and Carl Morris

Sometimes a mathematical result is strikingly contrary to generally held belief even though an obviously valid proof is given. Charles Stein of Stanford University discovered such a paradox in statistics in 1955. His result undermined a century and a half of work on estimation theory, going back to Karl Friedrich Gauss and Adrien Marie Legendre. After a long period of resistance to Stein's ideas, punctuated by frequent and sometimes angry debate, the sense of paradox has diminished and Stein's ideas are being incorporated into applied and theoretical statistics.

Stein's paradox concerns the use of observed averages to estimate unobservable quantities. Averaging is the second most basic process in statistics, the first being the simple act of counting. A baseball player who gets seven hits in 20 official times at bat is said to have a batting average of .350. In computing this statistic we are forming an estimate of the player's true batting ability in terms of his observed average rate of success. Asked how well the player will do in his next 100 times at bat, we would probably predict 35 more hits. In traditional statistical theory it can be proved that no other estimation rule is uniformly better than the observed average.

The paradoxical element in Stein's result is that it sometimes contradicts this elementary law of statistical theory. If we have three or more baseball players, and if we are interested in predicting future batting averages for each of them, then there is a procedure that is better than simply extrapolating from the three separate averages. Here "better" has a strong meaning. The statistician who employs Stein's method can expect to predict the future averages more accurately no matter what the true batting abilities of the players may be.

Baseball is a sport with a large and carefully compiled body of statistics, which supplies convenient material for illustrating the workings of Stein's method. As our primary data we shall consider the batting averages of 18 major-league players as they were recorded after their first 45 times at bat in the 1970 season. These were all the players who happened to have batted exactly 45 times the day the data were tabulated. A batting average is defined, of course, simply as the number of hits divided by the number of times at bat; it is always a number between 0 and 1. We shall denote each such average by the letter y.

The first step in applying Stein's method is to determine the average of the averages. Obviously this grand average, which we give the symbol $\bar{y}$, must also lie between 0 and 1. The essential process in Stein's method is the "shrinking" of all the individual averages toward this grand average. If a player's hitting record is better than the grand average, then it must be reduced; if he is not hitting as well as the grand average, then his hitting record must be increased. The resulting shrunken value for each player we designate z. This value is the James-Stein estimator of that player's batting ability, named for Stein and W. James, who together proposed a particularly simple version of the method in 1961. Stein's paradox is simply that the z values, the James-Stein estimators, give better estimates of true batting ability than the individual batting averages.

The James-Stein estimator for each player is found through the following equation: $z = \bar{y} + c(y - \bar{y})$. The quantity $(y - \bar{y})$ is the amount by which the player's batting average differs from the grand average. The equation thus states that the James-Stein estimator z differs from the grand average by this same quantity $(y - \bar{y})$ multiplied by a constant, c. The constant c is the "shrinking factor." If it were equal to 1, then the equation would state that the James-Stein estimator for a given player is identical with that player's batting average; in other words, y equals z. Stein's theorem states that the shrinking factor is always less than 1. Its actual value is determined by the collection of all the observed averages.

In the case of the baseball data, the grand average $\bar{y}$ is .265 and the shrinking factor c is .212. Substituting these values in the equation, we find that for each player z equals $.265 + .212(\bar{y} - .265)$. Because c is about .2, each average will shrink about 80 percent of the distance to the grand average, and the total spread of the averages will be reduced about 80 percent.

As an example consider the late Roberto Clemente, who was the leading batter in the major leagues when our statistics were compiled. For Clemente y is equal to .400, and z can be determined by evaluating the expression $z = .265 + .212(.400 - .265)$. The result is .294. In other words, Stein's theorem states that Clemente's true batting ability is best estimated not by .400 but lies closer to .294. Thurman Munson, in a batting slump early in the 1970 season, had an average of only .178. Substituting this value in the equation, we find that his estimated batting ability is substantially increased: the James-Stein estimator for Munson is .247.

Which set of values, y or z, is the better indicator of batting ability for the 18 players in our example? In order to answer that question in a precise way one would have to know the "true batting ability" of each player. This true average we shall designate with θ (the Greek letter theta). Actually it is an unknowable quantity, an abstraction representing the probability that a player will get a hit on any given time at bat. Although θ is unobservable, we have a good approximation to it: the subsequent performance of the batters. It is sufficient to consider just the remainder of the 1970 season, which includes about nine times as much data as the preliminary averages were based on. The expected statistical error in such a sample is small enough for us to neglect it and proceed as if the seasonal average were the "true batting ability" θ of a player. That is one reason for choosing batting averages for this example. In most problems the true value of θ cannot be determined.

One method of evaluating the two es-

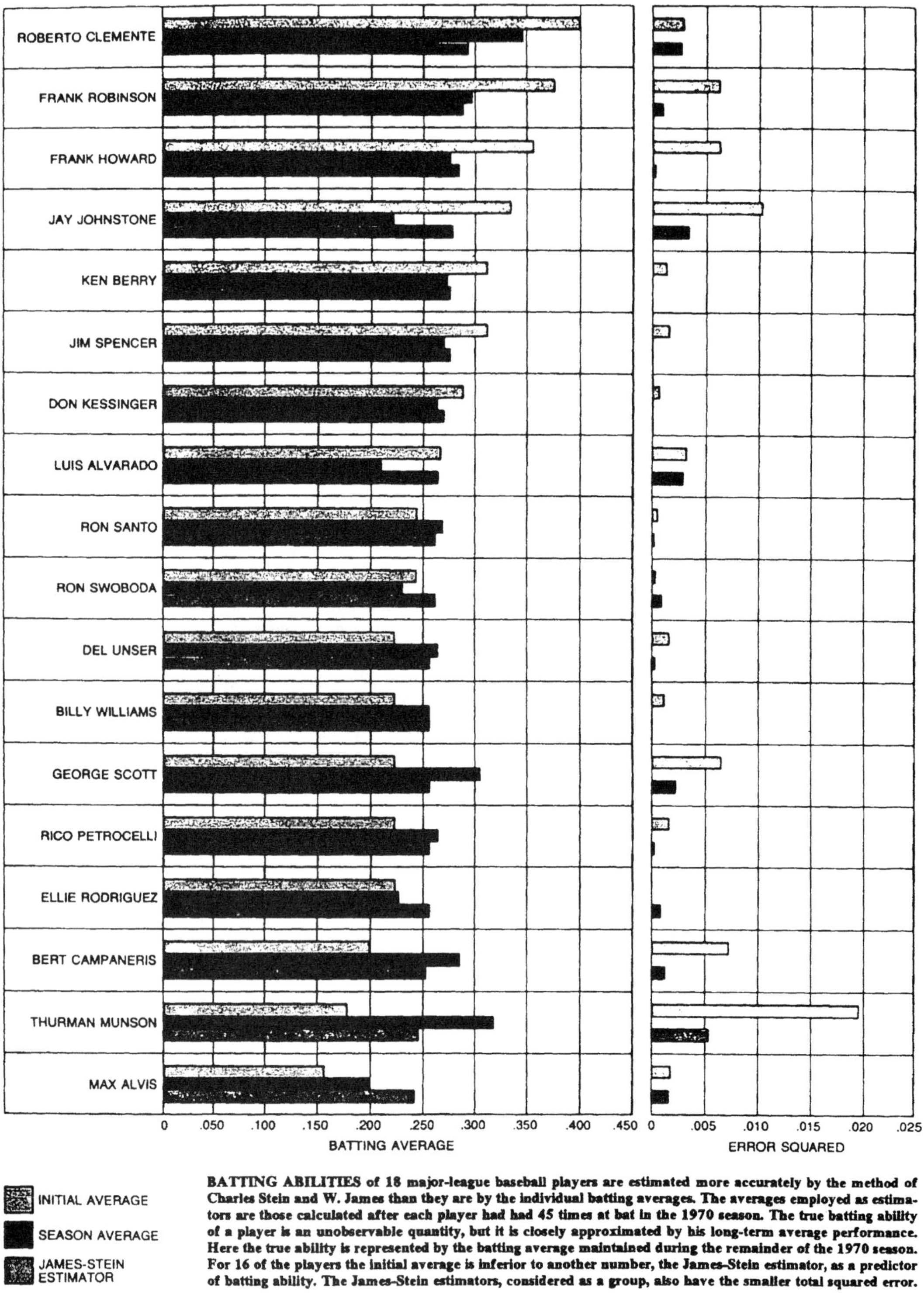

ROBERTO CLEMENTE
FRANK ROBINSON
FRANK HOWARD
JAY JOHNSTONE
KEN BERRY
JIM SPENCER
DON KESSINGER
LUIS ALVARADO
RON SANTO
RON SWOBODA
DEL UNSER
BILLY WILLIAMS
GEORGE SCOTT
RICO PETROCELLI
ELLIE RODRIGUEZ
BERT CAMPANERIS
THURMAN MUNSON
MAX ALVIS
0 .050 .100 .150 .200 .250 .300 .350 .400 .450
BATTING AVERAGE
0 .005 .010 .015 .020 .025
ERROR SQUARED
INITIAL AVERAGE
SEASON AVERAGE
JAMES-STEIN ESTIMATOR

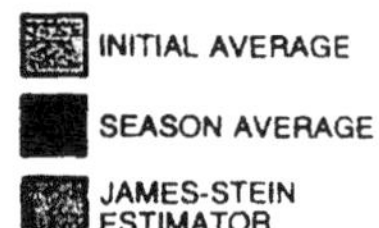

BATTING ABILITIES of 18 major-league baseball players are estimated more accurately by the method of Charles Stein and W. James than they are by the individual batting averages. The averages employed as estimators are those calculated after each player had had 45 times at bat in the 1970 season. The true batting ability of a player is an unobservable quantity, but it is closely approximated by his long-term average performance. Here the true ability is represented by the batting average maintained during the remainder of the 1970 season. For 16 of the players the initial average is inferior to another number, the James-Stein estimator, as a predictor of batting ability. The James-Stein estimators, considered as a group, also have the smaller total squared error.

120

timates is by simply counting their successes and failures. For 16 of the 18 players the James-Stein estimator z is closer than the observed average y to the "true," or seasonal, average θ. A more quantitative way of comparing the two techniques is through the total squared error of estimation. This is measured by first determining the actual error of each prediction, given by $(\theta - y)$ and $(\theta - z)$, for each player. Each of these quantities is then squared and the squared values are added up. The observed averages y have a total squared error of .077, whereas the squared error of the James-Stein estimators is only .022. By this comparison, then, Stein's method is 3.5 times as accurate. It can be shown that for the data given 3.5 is close to the expected ratio of the total squared errors of the two methods. We have not just been lucky.

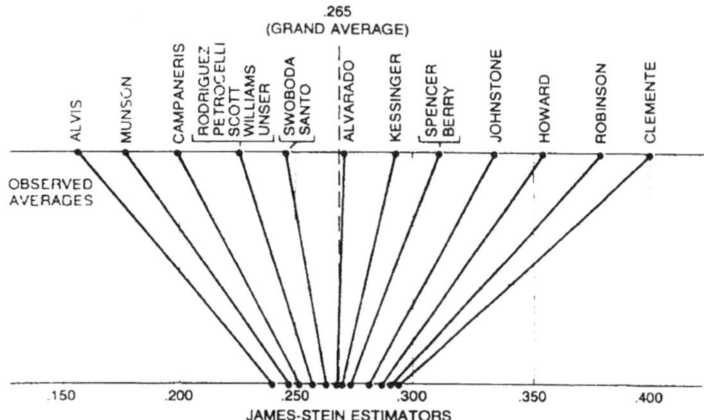

JAMES-STEIN ESTIMATORS for the 18 baseball players were calculated by "shrinking" the individual batting averages toward the overall "average of the averages." In this case the grand average is .265 and each of the averages is shrunk about 80 percent of the distance to this value. Thus the theorem on which Stein's method is based asserts that the true batting abilities are more tightly clustered than the preliminary batting averages would seem to suggest they are.

Suppose a statistician makes a random sampling of automobiles in Chicago and finds that of the first 45 recorded nine are foreign-made and the remaining 36 are domestic. We want to estimate the true proportion of imported cars in Chicago, a quantity represented by another unobservable θ. The observed average of $9/45 = .200$ is one estimate. Another can be obtained by simply lumping this problem together with that of the 18 baseball players. Substituting the value .200 in the equation used in that problem gives a James-Stein estimator of .251 for the imported-car ratio. (Actually the addition of a 19th value changes the grand average $\bar{y}$ and also slightly alters the shrinking factor c. The changes are small, however; the amended value of z is .249.)

In this case intuition argues strongly that the observed average and not the James-Stein estimator must be the better predictor. Indeed, the entire procedure seems silly: what could batting averages have to do with imported cars? It is here that the paradoxical nature of Stein's theorem is most uncomfortably apparent. The theorem applies as well to the 19 problems as it did to the original 18. There is nothing in the statement of the theorem that requires the component problems to have some sensible relation to one another.

The same disconcerting indifference to common sense can be demonstrated in another way. What does Clemente's .400 observed average have to do with Max Alvis, who was poorest in batting among the 18 players? If Alvis had had an early-season hitting streak, batting say .444 instead of his actual .156, the James-Stein estimator for Clemente's average would have been increased from .294 to .325. Why should Alvis' success or lack of it have any influence on our estimate of Clemente's ability? (They were not even in the same league.)

It is questions of this kind that have been raised by critics of Stein's method. In order to reply to them it will be necessary to describe the method rather more carefully.

Taking an average is an easy and familiar process that seems to need no justification. Actually it is not obvious why the average is so often useful in estimating the true center of gravity of a random process. The explanation lies in the distribution that the values of the random variable tend to assume.

The distribution most common in scientific work is the "normal" distribution, described by a bell-shaped curve; it was first investigated in depth by Gauss and is sometimes called the Gaussian distribution. It is constructed by assuming that the random variable can take on any value along some axis; the probability that it falls within any given interval is then made equal to the area under the same interval of the bell-shaped curve. The curve is completely specified by two parameters: the mean, θ, which lies at the peak of the curve, and the standard deviation, which measures how closely the values are distributed around the mean. It is customary to assign the standard deviation the symbol σ (sigma). The larger the standard deviation is, the more widely dispersed the data are.

In probability theory a known mean and standard deviation are employed to predict future behavior. A problem in statistics proceeds in the opposite direction: from observed data the statistician must infer the mean θ and the standard deviation σ.

Suppose, for example, the measurement of some random variable x yields the five successive values 10.0, 9.4, 10.3, 8.6 and 9.7. Suppose further the values are known to be part of a normal distribution with a standard deviation of 1. What is the value of the true mean θ? In principle the mean could have any value, but some values are more likely than others. A mean of 6.5, for example, would require that all five values be under the extreme tail of the curve and that none be found near the center. Gauss showed that among all possible choices for the mean, the average $\bar{x}$ of the observed data (which in this case has a value of 9.6) maximizes the probability of obtaining the data actually seen. In this sense the average is the most likely estimate of the mean; in fact, Gauss constructed the normal distribution just so that it would have this property.

There is a further justification, also pointed out by Gauss, for choosing the average as the best estimator of the unobservable mean θ. Gauss noted that the average of the data is an "unbiased" estimator of the mean, in the sense that it favors no selected value of θ. To be more precise, the average is unbiased because the expected value of $\bar{x}$ equals the true θ no matter what θ may be. There are infinitely many unbiased estimators of θ, none of which estimates θ perfectly. Gauss showed that the expected squared error of estimation for the average $\bar{x}$ is lower than that for any other linear, unbiased function of the observations. In the 1940's it was demonstrated that no other unbiased function of the data, whether it is linear or nonlinear, can estimate θ more accurately than the average, in terms of expected squared error. An essential contribution to that proof had been made in the 1920's by

R. A. Fisher, who showed that all the information about θ that can possibly be found in the data is contained in the average $\bar{x}$.

In the 1930's a mathematically more rigorous approach to statistical inference was undertaken by Jerzy Neyman, Egon S. Pearson and Abraham Wald; the ideas they developed are part of what is now known as statistical decision theory. They discarded the requirement of unbiased estimation and examined all functions of the data that could serve as estimators of the unknown mean θ. These estimators were compared through a risk function, defined as the expected value of the squared error for every possible value of θ.

Consider three competing estimators: the average of the data, $\bar{x}$; half that average, $\bar{x}/2$, and the median of the data, or middle value. For both the average and the median the risk function is constant; that is merely another way of saying that their expected squared error in predicting the mean θ is the same no matter what the value of θ really is. Of the two constant risk functions, the one for the average $\bar{x}$ is uniformly smaller by a factor of about two-thirds; clearly the average is the preferred estimator. In the language of decision theory the median is said to be "inadmissible" as an estimator of θ, since there is another estimator that has a smaller risk (expected squared error) no matter what θ is. (It should be mentioned, however, that when the data have a distribution other than the normal one, it is possible for the order of preference to be reversed.)

For the estimator $\bar{x}/2$, which is biased toward the value $\theta = 0$, the risk function is not constant; this estimator is accurate if θ happens be close to zero, but the expected squared error increases rapidly as the true mean departs from zero. The risk function describes a parabola, with the minimum point at $\theta = 0$; if the mean does happen to be zero, then the risk function for $\bar{x}/2$ is four times smaller than that for the average itself. At large values of the mean, however, the average $\bar{x}$ regains its superiority. With other estimators we can poke down the risk function below that of the average at any point we wish to, but it always pops up again somewhere else.

There remains the possibility that some other estimator has a risk that is uniformly lower than that of the average. In 1950 Colin R. Blyth, Erich L. Lehmann and Joseph L. Hodges, Jr., proved that no such estimator exists. In other words, the average $\bar{x}$ is admissible, at least when it is applied to one set of observations for the purpose of estimating one unknown mean.

Stein's theorem is concerned with the estimation of several unknown means. No relation between the means need be assumed; they can be batting abilities or proportions of imported cars. On the other hand, the means are assumed to be independent of one another. In evaluating estimators for these means it is once again convenient to employ a risk function defined as the sum of the expected values of the squared errors of estimation for all the individual means.

The obvious first choice of an estimator for each of several means is the average of the data related to that mean. The entire historical development of statistical theory from Gauss through decision theory argues that the average is an admissible estimator as long as there is just one mean, θ, to be estimated. Stein showed in 1955 that the average is also admissible for estimating two means. Stein's paradox is simply his proof that when the number of means exceeds two, estimating each of them by its own average is an inadmissible procedure. No matter what the values of the true means, there are estimation rules with smaller total risk.

In 1955 Stein was able to prove this proposition only in those cases where the number of means, a quantity we shall designate k, was very large. Stein's 1961 paper written in collaboration with James extended the result to all values of k greater than 2; moreover, it did so in a constructive manner. Stein and James not only showed that estimators must exist that are everywhere superior to the

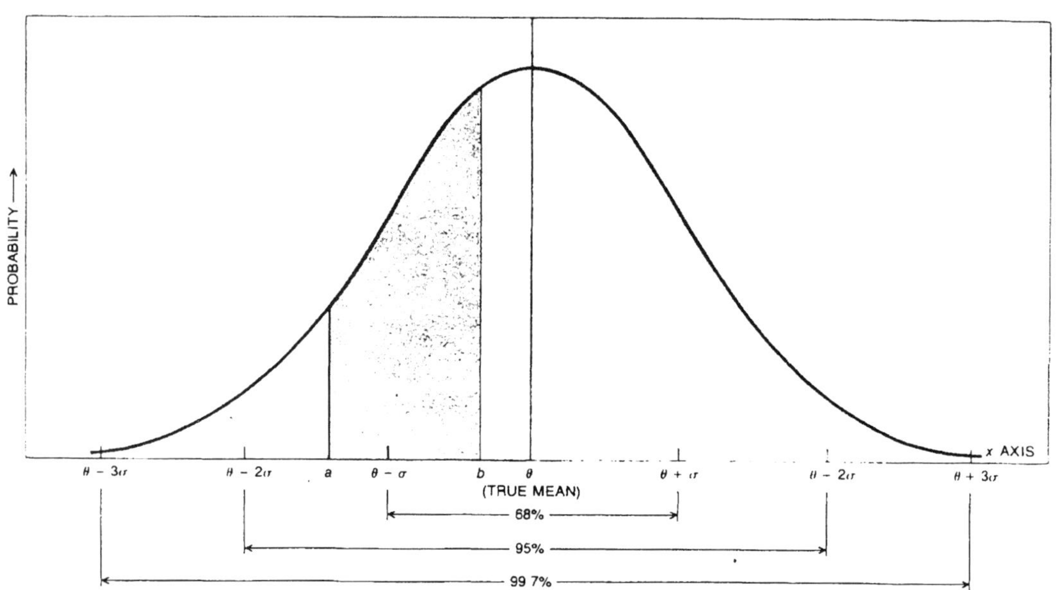

NORMAL DISTRIBUTION of a random variable around the mean value of that variable provides the fundamental justification for estimation by averaging. The distribution is defined by two parameters, the mean, θ, which locates the central peak of the distribution, and the standard deviation, σ, which measures how widely scattered the data points are. It is assumed in defining the distribution that the variable x can take on any value on the x axis. The most likely value of x is, by definition, the mean θ. The probability that x lies within any given interval on the axis, such as that between the points a and b, is equal to the area under the bell-shaped curve between those points.

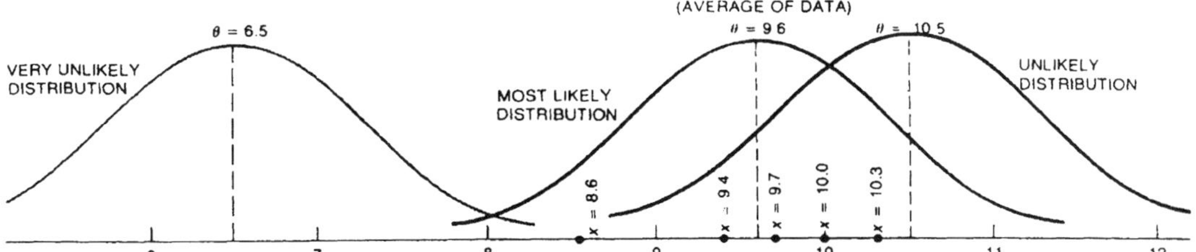

PROBLEM IN STATISTICS is to deduce from a set of data the true mean and standard deviation of the distribution. Even when it is known that the distribution is a normal one and that the standard deviation is 1, the mean could in principle have any value. Some values, however, are more likely than others. For example, the five data points (x) given here could be described by a normal distribution with a mean of 6.5 only if all five points were more than two standard deviations above the mean. It can be shown that the data are most likely to be generated by a distribution with a mean equal to the observed average of the data, denoted $\bar{x}$. In this case the average is equal to 9.6.

averages; they were also able to provide an example of such an estimator.

The James-Stein estimator has already been defined in our investigation of batting averages. It is given by the equation $z = \bar{y} + c(y - \bar{y})$, where y is the average of a single set of data, $\bar{y}$ is the grand average of averages and c is a "shrinking factor." There are several other expressions for the James-Stein estimator, but they differ mainly in detail. All of them have in common the shrinking factor c; it is the definitive characteristic of the James-Stein estimator.

In the baseball problem c was treated as if it were a constant. Actually it is determined by the observed averages and therefore is not a constant. The shrinking factor is given by the equation

$$c = 1 - \frac{(k-3)\sigma^2}{\Sigma(y - \bar{y})^2}.$$

Here k is again the number of unknown means, σ^2 is the square of the standard deviation and $\Sigma(y - \bar{y})^2$ is the sum of the squared deviations of the individual averages y from the grand average $\bar{y}$.

Let us briefly explore the meaning of this rather forbidding equation. With k and σ^2 fixed, we find that the shrinking factor c becomes smaller (and the predicted means are more severely affected by it) as the expression $\Sigma(y - \bar{y})^2$ gets smaller. On the other hand, c increases, approaching unity, and the shrinking is less drastic as the expression $\Sigma(y - \bar{y})^2$ increases.

What do these equations mean in terms of the behavior of the estimator? In effect the James-Stein procedure makes a preliminary guess that all the unobservable means are near the grand average $\bar{y}$. If the data support that guess in the sense that the observed averages are themselves not too far from $\bar{y}$, then the estimates are all shrunk further toward the grand average. If the guess is contradicted, then not much shrinking is done. These adjustments to the shrinking factor are accomplished through the effect the distribution of averages around the grand average $\bar{y}$ has on the equation that determines c. The number of means being estimated also influences the shrinking factor, through the term $(k - 3)$ appearing in this same equation. If there are many means, the equation allows the shrinking to be more drastic, since it is then less likely that variations observed represent mere random fluctuations.

With c calculated in this manner, the risk function for the James-Stein estimator is less than that for the sample averages no matter what the true values of the means θ happen to be. The reduction of risk can be substantial, particularly when the number of means is larger than five or six. The risk function is not constant for all values of the true mean θ, as it is for the observed averages. The risk of the James-Stein estimator is smallest when all the true means are the same. As the true means depart from one another the risk of the estimator increases, approaching that of the observed averages but never quite equaling it. The James-Stein estimator does substantially better than the averages only if the true means lie near each other, so that the initial guess involved in the technique is confirmed. What is surprising is that the estimator does at least marginally better no matter what the true means are.

The expression for the James-Stein estimator that we have employed refers all observed averages to the grand average $\bar{y}$. This procedure is not the only one possible; other expressions for the estimator dispense with $\bar{y}$ entirely. What cannot be avoided is the introduction of some more or less arbitrary initial guess or point of origin for the estimator. The observed averages, it will be noted, do not depend on a choice of origin. Before Stein discovered his method it was felt that such "invariant" estimators must be preferable to those whose predictions change with each choice of an origin. The theory of invariance, to which Stein had been a principal contributor, was badly shaken by the James-Stein counterexample. From the standpoint of mathematics this is the most unsettling aspect of Stein's theorem. Indeed, the paradox was not discovered earlier largely because of a strong prejudice that the estimation problem, being stated without reference to any particular origin, should be solved in a similar way.

Applications of Stein's method tend to involve large sets of data with many unknown parameters. Some of the difficulties of such problems, as well as the practical potential of the method itself, can be illustrated by an example: an analysis of the distribution of the disease toxoplasmosis in the Central American country of El Salvador.

Toxoplasmosis is a disease of the blood that is endemic in much of Central America and in other regions of the Tropics. In El Salvador roughly 5,000 people drawn in varying numbers from 36 cities were tested for toxoplasmosis. The observed rate of incidence for each city can conveniently be expressed by comparison with the national rate (that is, with the grand average $\bar{y}$). A measured rate of .050, for example, denotes a city with an incidence of the disease 5 percent higher than the national average. The measured rates have an approximately normal distribution. The standard deviations of these distributions are known, but they differ from city to city, depending inversely on how large a sample population was tested in that city. It is the task of the statistician to estimate the true mean θ of the distribution for each city from the measured incidence y.

In this case the appropriate form of the James-Stein estimator is $z = cy$. The simplification, which was introduced by us, is made possible by the chosen manner of expressing the observations y. They are defined in such a way that the grand average $\bar{y}$ is zero, and terms containing $\bar{y}$ therefore drop out of the equation. On the other hand, the estimation

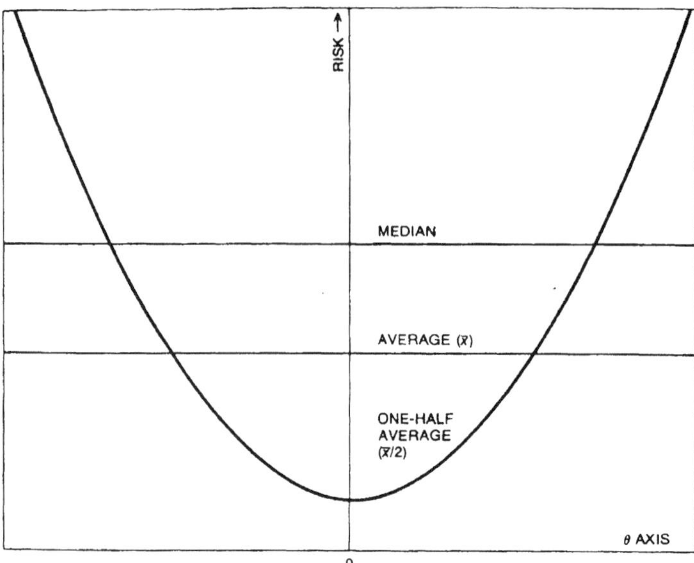

VARIOUS ESTIMATORS of a single true mean, θ, can be evaluated by way of a risk function. The risk is defined as the expected value of the squared error of estimation, considered as a function of the mean θ. The average of the data, $\bar{x}$, is an estimator with a constant risk function: no matter what the true mean is, the expected value of the squared error is the same. The median, or middle value, of the data also has constant risk, but it is everywhere greater (by a factor of 1.57) than the risk of the average. Half the average ($\bar{x}/2$) is an estimator whose risk depends on the actual value of the mean; the risk is smallest when the mean is near zero and increases rapidly when the mean departs from zero. For the estimation of a single mean there is no estimator with a risk function that is everywhere less than the risk function of the average $\bar{x}$.

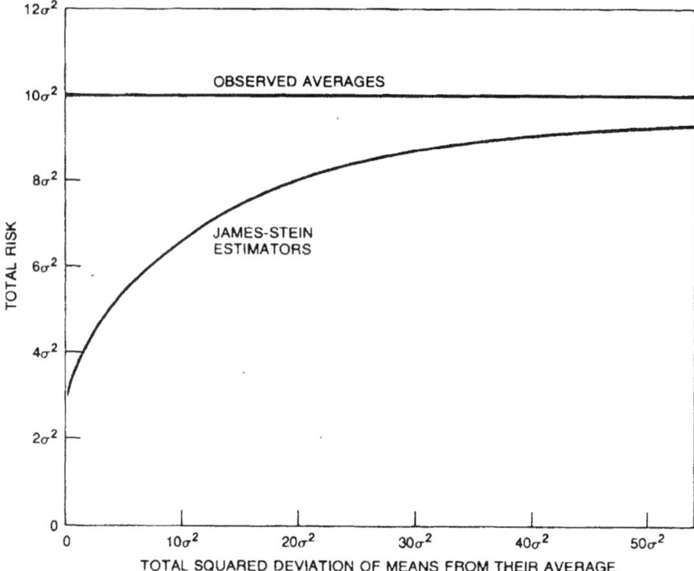

TOTAL RISK FUNCTION for the James-Stein estimators is everywhere less than that for the individual observed averages, as long as the number of means being estimated is greater than two. In this case there are 10 unknown means. The risk is smallest when all the means are clustered at a single point. As the means depart from one another the risk of the James-Stein estimators increases, approaching that of the observed averages but never quite reaching it.

procedure is now complicated by the fact that the shrinking factor c is different for each city, varying inversely as the standard deviation of y for that city. This dependence of the shrinking factor on the standard deviation has a simple intuitive rationale. A large standard deviation implies a high degree of randomness or uncertainty in a measurement. If the measured incidence is unusually large, it can therefore be attributed more reasonably to random fluctuations within the normal distribution than to a genuinely large value of the true mean θ. It is thus proper to reduce this value drastically, that is, to apply a small shrinking factor.

The same argument can be made even more forcefully by returning for a moment to baseball. Frank O'Connor pitched for Philadelphia in the 1893 season. He batted twice in his major-league career, hitting successfully both times. His observed batting average is hence 1.000. The James-Stein rule for the 18 players considered above estimates O'Connor's true batting ability to be $.265 + .212(1.000 - .265) = .421$ (ignoring the effect of the new data on the grand average and on the shrinking factor). This is a silly estimate, although not as silly as 1.000. A perfect average after two times at bat is not at all inconsistent with a true value in the range from .242 to .294 that is estimated for the other players. The shrinking constant c applied to O'Connor's average should be severer in order to compensate for the smaller amount of data available for him.

For the El Salvador observations, most of the shrinking factors are quite gentle, between .6 and .9, but a few are in the range from .1 to .3. Which set of numbers should we prefer, the James-Stein estimators or the measured rates of incidence? That depends largely on what we want to use the numbers for.

If the Minister of Health for El Salvador intends to build local hospitals for people suffering from toxoplasmosis, the James-Stein estimators probably offer the more reliable guidance. The reason is that the expected value of the total squared error is smaller for the James-Stein estimators; in fact, it is smaller by a factor of about three. The important point in this calculation is that the expected error is added up for all the cities. Any particular hospital might be the wrong size or in the wrong place, but the sum of all such mismatches would be smaller for the James-Stein estimators than for the observed rates.

The James-Stein estimators are also likely to be preferable for determining the ordering of the true means. In this regard it is notable that the city with the highest apparent incidence (according to the measured rates y) is ranked 12th according to the James-Stein estimators.

The estimate is drastically reduced because the sample was very small in that city. This information might be useful if there were funds for only one hospital.

Suppose an epidemiologist wants to investigate the correlation of the true incidence in each city with attributes such as rainfall, temperature, elevation or population? Once again the James-Stein estimators are preferred; a rough calculation shows that they would give a closer approximation in about 70 percent of the cases.

There is one purpose for which the measured incidence may well be superior to the James-Stein estimator: when a single city is considered in isolation. As we have seen, the James-Stein method gives better estimates for a majority of cities, and it reduces the total error of estimation for the sum of all cities. It

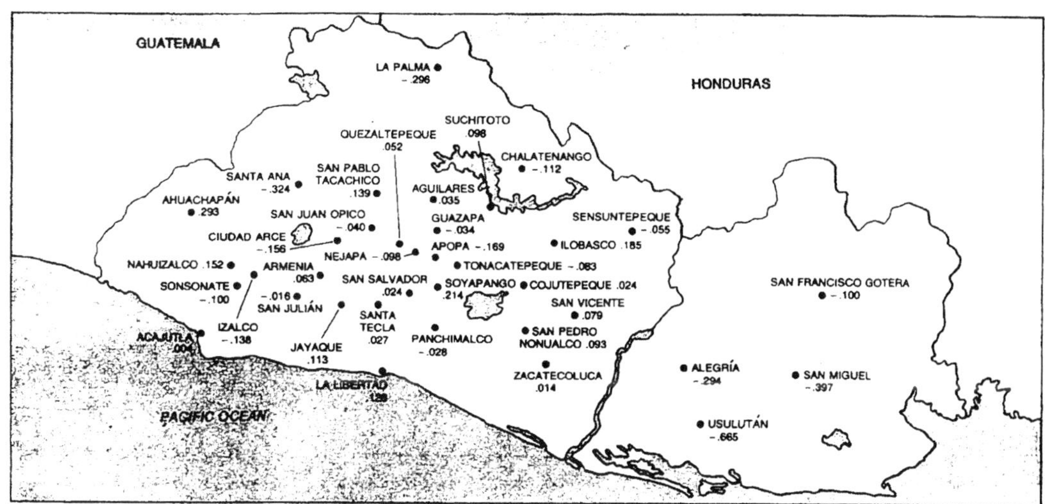

INCIDENCE OF TOXOPLASMOSIS, a disease of the blood, was surveyed in 36 cities in the Central American country El Salvador. The measured incidence in each city can be regarded as an estimator of the true incidence, which is unobservable. The measured incidence has a normal distribution whose standard deviation is determined by the number of people surveyed in that city. The measured rates are expressed in terms of deviation from the national incidence (the average of the rates observed in all the cities). Thus zero denotes exactly the national rate, and a city with a measured incidence of −.040 would have an observed rate 4 percent lower than the country as a whole.

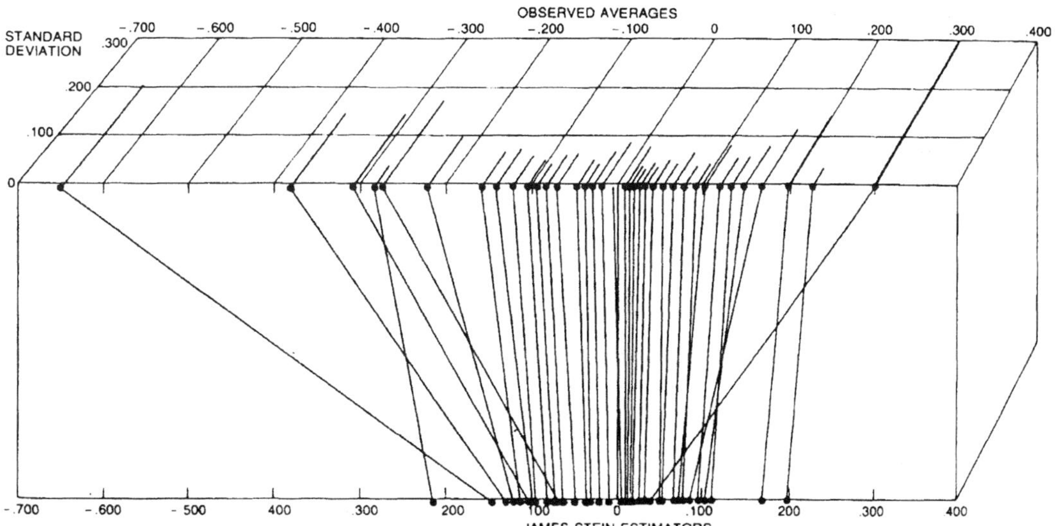

SHRINKING of the observed toxoplasmosis rates to yield a set of James-Stein estimators substantially alters the apparent distribution of the disease. The shrinking factor is not the same for all the cities but instead depends on the standard deviation of the rate measured in that city. A large standard deviation implies that a measurement is based on a small sample and is subject to large random fluctuations; that measurement is therefore compressed more than the others are. In the El Salvador data the most extreme observations tend to be correlated with the largest standard deviations, again suggesting the unreliability of those measurements. Compared with the observed rates, the James-Stein estimators can be proved to have a smaller total error of estimation. They also provide a more accurate ranking of the cities.

cannot be demonstrated, however, that Stein's method is superior for any particular city; in fact, the James-Stein prediction can be substantially worse.

Estimating the true mean for an isolated city by Stein's method creates serious errors when that mean has an atypical value. The rationale of the method is to reduce the overall risk by assuming that the true means are more similar to one another than the observed data. That assumption can degrade the estimation of a genuinely atypical mean. Now we see why imported cars should not be included in the same calculations with the 18 baseball players. There is a substantial probability that the automobiles will be atypical.

Suppose we ignore this hazard and lump together all 19 problems; we can then calculate the total expected squared error as a function of the true percentage of imported cars. It turns out that the risk for both the baseball players and the automobiles is reduced only if the percentage of imported cars happens to lie in the same range as the estimated batting averages; otherwise the risk of error for both kinds of problem is increased.

The question of whether or not a particular mean is "typical" is a subtle one whose implications are not yet fully understood. Returning to the problem of toxoplasmosis in El Salvador, let us single out for attention the city of Alegría, which has the fifth-smallest measured incidence of the disease: $-.294$. It is one of four cities included in the survey that are east of the Rio Lempa; all four have distinctly negative values of measured incidence y. It is plausible to suppose that this is no coincidence and that the rate of toxoplasmosis east of the Lempa is genuinely lower. A James-Stein estimator that consolidates information from the entire country therefore may be less than optimal in these cities. We have developed techniques for taking advantage of extra information of this kind, but the theory underlying those techniques remains rudimentary.

An astute follower of baseball might be aware that just as each player's batting ability can be represented by a Gaussian curve, so too the true batting abilities of all major-league players have an approximately normal distribution. This distribution has a mean of .270 and a standard deviation of .015. With this valuable extra information, which statisticians call a "prior distribution," it is possible to construct a superior estimate of each player's true batting ability. This new estimator, which we shall give the label Z, is defined by the equation $Z = m + C(y - m)$. Here y is again the observed batting average of the player, but $\bar{y}$, the grand average, has been replaced by m, the mean of the prior distribution, which is known to have the value .270. In addition there is a different shrinking factor, C, which depends in a simple way on the standard deviation of the prior distribution (equal to .015).

This procedure is not a refinement of Stein's method; on the contrary, it predates Stein's method by 200 years. It is the mathematical expression of a theorem published (posthumously) in 1763 by the Reverend Thomas Bayes.

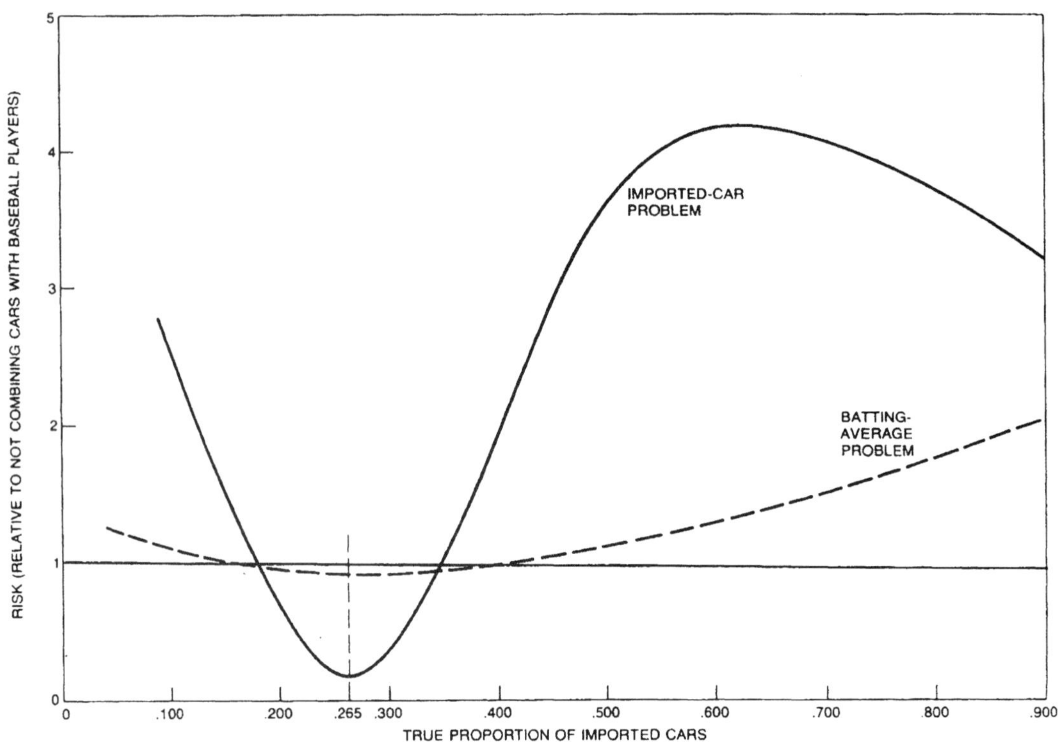

UNRELATED PROBLEMS can be lumped together for analysis by Stein's method, but only at the risk of increasing error. To the 18 batting averages computed earlier, for example, one might add a 19th number representing the proportion of imported cars observed in Chicago. New James-Stein estimators could then be calculated for both the baseball players and the automobiles, based on the grand average of all 19 numbers. Nothing in the statement of Stein's theorem prohibits such a procedure, but the evident illogic of it has justifiably been criticized. In fact, including the unrelated data can reduce the risk function only if the proportion of imported cars happens to be near the mean batting average of .265; otherwise the expected error of estimation for both the cars and the baseball players is increased.

He was able to show that this estimator minimizes the expected squared error associated with the randomness in both the observed averages (y) and in the true means (θ).

The formula for the James-Stein estimator is strikingly similar to that of Bayes's equation. Indeed, as the number of means being averaged grows very large, the two equations become identical. The two shrinking factors c and C converge on the same value, and the grand average $\bar{y}$ becomes equal to the mean m precisely when all players are included in the calculation. The James-Stein procedure, however, has one important advantage over Bayes's method. The James-Stein estimator can be employed without knowledge of the prior distribution; indeed, one need not even suppose the means being estimated are normally distributed. On the other hand, ignorance has a price, which must be paid in reduced accuracy of estimation. We have shown that the James-Stein method increases the risk function by an amount proportional to $3/k$, where k is again the number of means being estimated. The additional risk is therefore negligible when k is greater than 15 or 20, and it is tolerable for k as small as 9.

In this historical context the James-Stein estimator can be regarded as an "empirical Bayes rule," a term coined by Herbert E. Robbins of Columbia University. In work begun in about 1951 Robbins demonstrated that it is possible to achieve the same minimum risk associated with Bayes's rule without knowledge of the prior distribution, as long as the number of means being estimated is very large. Robbins' theory was immediately recognized as a fundamental breakthrough; Stein's result, which is closely related, has been much slower in gaining acceptance.

The James-Stein estimator is not the only one that is known to be better than the sample averages. Indeed, the James-Stein estimator is itself inadmissible! Its failure lies in the fact that the shrinking factor c can assume negative values, and it then pulls the means away from the grand average rather than toward it. When that happens, simply replacing c with zero produces a better estimator. This estimator in turn is also inadmissible, but no uniformly better estimator has yet been found.

The search for new estimators continues. Recent efforts have been concentrated on achieving results like those obtained with Stein's method for problems involving distributions other than the normal distribution. Several lines of work, including Stein's and Robbins' and more formal Bayesian methods seem to be converging on a powerful general theory of parameter estimation.

8

Assessing the Accuracy of the Maximum Likelihood Estimator: Observed Versus Expected Fisher Information

Introduction by Thomas J. DiCiccio
Cornell University

This stimulating and influential paper appeared at the forefront of the wave of interest in conditional inference and asymptotics for parametric models that burgeoned in the 1980s. The setting for the paper is the most basic one of a scalar parameter model, and the asymptotic technique primarily used is Laplace approximation. Thus, this paper is akin to another landmark contribution of the same vintage, Barndorff-Nielsen and Cox (1979), which used saddlepoint approximations in the context of conditional inference for exponential families.

The paper begins by recalling Fisher's (1934) demonstration that the information loss for maximum likelihood estimation in location-scale families can be recovered completely by basing inference on the conditional distribution of the maximum likelihood estimator $\widehat{\theta}$ given an exact ancillary statistic a for which $(\widehat{\theta}, a)$ is sufficient. A main motivation is to make sense of Fisher's enigmatic suggestion that the information loss for $\widehat{\theta}$ in more general models can be progressively alleviated by successively taking into account ancillary statistics that reflect the values of the second- and higher-order derivatives of the log likelihood function at its maximum. The inferential framework is that of presenting the maximum likelihood estimate with a suitable standard error. In light of Fisher's remarks, an approximation to the conditional variance given an ancillary a is relevant, and such an approximation should involve log likelihood derivatives.

The primary purpose of the paper is to demonstrate the appropriateness of using the reciprocal of the observed Fisher information, now typically denoted by $\widehat{j}^{-1}$, as an approximation to $\text{var}(\hat{\theta}|a)$. In doing so, the paper touches upon several important issues and themes that have been central to much subsequent research.

After revisiting Cox's (1958) classic example of two measuring instruments and providing numerical evidence supporting the use of $\widehat{j}^{-1}$ as a conditional variance approximation in the Cauchy location family, the paper presents detailed theoretical results for translation families, where an exact ancillary is available. Laplace's method is used to derive expansions for the conditional mean and variance of $\widehat{\theta}$: the leading term in the asymptotic formula for the variance is $\widehat{j}^{-1}$, and the next term, an order of magnitude smaller in the sample size n, involves the third- and fourth-order log likelihood derivatives. The accuracy of the conditional variance formula is illustrated in two examples from Fisher (1973): the normal circle and the gamma hyperbola, where the first term $\widehat{j}^{-1}$ is shown to be an adequate approximation except for very extreme values of the ancillary.

Another approximation to the conditional variance is the reciprocal of the expected Fisher information, $\widehat{i}^{-1}$, and a key result is presented for general models that facilitates comparison of the two approximations. For scalar parameter models, not necessarily of the translation family form, the distribution of $n^{1/2}(\widehat{j}/\widehat{i} - 1)$ is shown to be asymptotically normal with mean 0 and standard deviation equal to Efron's (1975) statistical curvature γ. For the translation family case, combining this result with the expansion for the conditional variance shows that the ratio $\{\text{var}(\hat{\theta}|a) - \widehat{j}^{-1}\}/\{\text{var}(\hat{\theta}|a) - \widehat{i}^{-1}\}$ is of order $n^{-1/2}$ and hence, in this context, $\widehat{j}^{-1}$ is established to be the more accurate approximation.

In addition to investigating the accuracy of $\widehat{j}^{-1}$ as an estimator of the conditional variance, the paper concerns the accuracy of conditional inferences, particularly the coverage accuracy of approximate confidence intervals. Initially, the confidence intervals considered are those constructed from a standard normal approximation to the conditional distribution of $\widehat{j}^{1/2}(\widehat{\theta} - \theta)$ or equivalently, through a χ_1^2 approximation for $\widehat{j}(\widehat{\theta} - \theta)^2$; however, intervals derived from the chi-squared approximation for the likelihood ratio statistic $2\{l(\widehat{\theta}) - l(\theta)\}$ are also discussed. The translation family setting is again used for preliminary theoretical and numerical investigation. The numerical results are for the Cauchy location problem, and they demonstrate strikingly that intervals based on $\widehat{i}(\widehat{\theta} - \theta)^2$ have inferior conditional properties to those based on either $\widehat{j}(\widehat{\theta} - \theta)^2$ or $2\{l(\widehat{\theta}) - l(\theta)\}$; furthermore, the intervals from $2\{l(\widehat{\theta}) - l(\theta)\}$ are found to be more accurate than those from $\widehat{j}(\widehat{\theta} - \theta)^2$. The latter finding is explained by the theoretical results, in which Laplace's method is used to develop expansions to order n^{-1} for the conditional distribution functions of $\widehat{j}(\widehat{\theta} - \theta)^2$ and $2\{l(\widehat{\theta}) - l(\theta)\}$ in terms of chi-squared distributions. It turns out that the order n^{-1} term in the expansion for $2\{l(\widehat{\theta}) - l(\theta)\}$ involves χ_3^2 while that for $\widehat{j}(\widehat{\theta} - \theta)^2$ involves χ_5^2 and χ_7^2. Hinkley (1978) showed similar results for location-scale families.

The account of Laplace's method given in Sect. 8 of the paper, by which the theoretical results for translation families are derived, is very illuminating, although perhaps not meeting the standards for rigor of the *Annals of Statistics* at the time. Fisher (1934) showed how the conditional density of $\widehat{\theta}$ can be expressed in terms of the likelihood function, and the connections between conditional and Bayesian inference for translation models are mentioned in the paper. In particular, the close relationship between the normal shape of the likelihood and the accuracy of the normal approximation to the conditional distribution of $\widehat{j}^{1/2}(\widehat{\theta} - \theta)$ is emphasized. A more thorough treatment of Laplace approximations in Bayesian contexts was given by Tierney and Kadane (1986); see also Tierney et al. (1989a,b) and Barndorff-Nielsen and Cox (1989).

The expansion for the conditional distribution of $2\{l(\widehat{\theta}) - l(\theta)\}$, (4.4) of the paper, implies the efficacy of conditional Bartlett correction in the translation family case. Moreover, it provides an example of Bayesian Bartlett correction. Conditional Bartlett correction has been studied by Barndorff-Nielsen and Cox (1984); see also Jensen (1993). Bayesian Bartlett correction has been studied by Bickel and Ghosh (1990); see also DiCiccio and Stern (1993, 1994).

It is noted in the paper that the expansions for $\widehat{j}(\widehat{\theta} - \theta)^2$ and $2\{l(\widehat{\theta}) - l(\theta)\}$ imply that their conditional distributions are χ_1^2 to error of order n^{-1}, and this observation is an early example of what has come to be known as a stability property. A statistic is now said to be second-order stable if an asymptotic expansion of the conditional distribution of that statistic given an arbitrary ancillary depends on the ancillary only in the terms of order n^{-1} or smaller. The concept of stability has been developed by McCullagh (1987); see also Barndorff-Nielsen and Cox (1994).

One obstacle to extending the results for translation families to more general scalar parameter models is that typically no exact ancillary statistic a exists such that $(\widehat{\theta}, a)$ is sufficient. The approach taken in the paper is to regard $(\widehat{\theta}, \widehat{j})$ as an approximate sufficient statistic and then determine a function of this statistic that is approximately ancillary. The function identified for this purpose is $q = (1 - \widehat{j}/\widehat{i})\widehat{\gamma}^{-1}$, where $\widehat{i} = i(\widehat{\theta})$ and $\widehat{\gamma} = \gamma(\widehat{\theta})$; q is asymptotically ancillary in the sense that $n^{1/2}q$ is asymptotically standard normal to error of order $n^{-1/2}$. A potential course of inquiry might have been to investigate how well $\widehat{j}^{-1}$ approximates the conditional variance $\text{var}(\hat{\theta}|q)$ and how closely the standard normal distribution approximates the conditional distribution of $\widehat{j}^{1/2}(\widehat{\theta} - \theta)$. However, a different approach is taken in the paper: to bring the problem closer to translation family form, and thereby improve the accuracy of the approximations, especially in small samples where the effect of conditioning is most pronounced, an ancillary dependent variance-stabilizing transformation of the form $\phi_q = n^{-1/2} \int^{\theta} [i(t)\{1 - q\gamma(t)\}]^{1/2} dt$ is introduced. Note that the usual variance-stabilizing transformation based on expected Fisher information is obtained from the case $q = 0$, i.e., $\phi_0 = n^{-1/2} \int^{\theta} \{i(t)\}^{1/2} dt$.

The observed information in the transformed problem ϕ_q for given q is $\widehat{j}_q = n^{-1}$, the same as the expected information for ϕ_0. The key issues for investigation then become how near the conditional variance $\text{var}(\widehat{\phi}_q|q)$ is to n^{-1} and how near the conditional distribution of $n^{1/2}(\widehat{\phi}_q - \phi_q)$ is to standard normal. In particular, the paper speculates that both $n(\widehat{\phi}_q - \phi_q)^2$ and $2\{l(\widehat{\theta}) - l(\theta)\}$ are conditionally distributed as χ_1^2 to error of order n^{-1}. The extent to which observed information is superior to expected information for approximating the conditional variance can be assessed by comparing $\text{var}(\widehat{\phi}_q|q)$ and $\text{var}(\widehat{\phi}_0|q)$. The paper also speculates that the ratio $\{\text{var}(\widehat{\phi}_q|q) - n^{-1}\}/\{\text{var}(\widehat{\phi}_0|q) - n^{-1}\}$ is of order $n^{-1/2}$. Although the paper offers no theoretical justification for these conjectures, substantial numerical evidence is provided. The missing theory was supplied by Peers (1978).

The two non-translation examples considered in the paper involve the spiral model, which is a generalization of Fisher's (1973) normal circle, and a bivariate normal model with unknown correlation. In the spiral model, the approximate ancillary q has a natural geometrical interpretation, and the numerical results show that the marginal distribution of $n^{1/2}q$ is nearly standard normal and that the conditional means and variances of $n^{1/2}(\widehat{\phi}_q - \phi_q)$ are nearly 0 and 1, respectively. Analogous numerical results are given for the correlation example; moreover, in this example, the conditional variances $\text{var}(\widehat{\phi}_q|q)$ and $\text{var}(\widehat{\phi}_0|q)$ are compared, as are the conditional coverage levels of confidence intervals derived from the χ_1^2 approximation for $n(\widehat{\phi}_q - \phi_q)^2$, $2\{l(\widehat{\theta}) - l(\theta)\}$, and $n(\widehat{\phi}_0 - \phi_0)^2$. The effects of conditioning on q are dramatic. As in the translation family case, the intervals from $2\{l(\widehat{\theta}) - l(\theta)\}$ have superior conditional coverage accuracy to those from $n(\widehat{\phi}_q - \phi_q)^2$. It is reported that $n^{1/2}(\widehat{\phi}_q - \phi_q)$ tends to be conditionally more normally distributed than is $\widehat{j}^{1/2}(\widehat{\theta} - \theta)$, and graphical evidence shows that the variance-stabilizing transformation ϕ_q has the effect of making the likelihood function more normally shaped.

With the exception of the Cauchy location problem, all of the examples in the paper are curved exponential families. The geometrical interpretation of q given for the spiral model is extended to general curved exponential models, and a prescription is described for constructing a vector asymptotic ancillary that takes higher-order derivatives of the log likelihood function into account.

Much of the work that followed this paper has focused on constructing approximate ancillaries and on approximating the conditional distributions of estimators and test statistics; it has largely been a synthesis of the approach taken in the paper and the types of approximations developed by Barndorff-Nielsen and Cox (1979) and Durbin (1980).

Cox (1980) introduced the notion of a local ancillary: a statistic a, possibly depending on θ_0, is rth-order locally ancillary if an expansion of its density at $\theta = \theta_0 + n^{-1/2}\delta$ is the same to relative error of order $n^{-r/2}$ as that at $\theta = \theta_0$. Cox (1980) studied second-order local ancillaries for scalar parameter models, and Ryall (1981) extended Cox's (1980) construction of a second-order local ancillary to the vector parameter case and constructed a third-order local ancillary for scalar parameter models. For general vector parameter models, Skovgaard (1985) developed a second-order local ancillary analogous to the Efron–Hinkley ancillary, and he showed that the conditional distribution of the likelihood ratio statistic is chi-squared to error of order n^{-1} for testing problems with nuisance parameters. Amari (1982) considered higher-order local ancillaries for multiparameter curved exponential families. In the context of curved exponential families, Barndorff-Nielsen (1980) introduced various approximate ancillaries, including the affine ancillary, which coincides with the Efron–Hinkley ancillary in many examples, and the directed likelihood ancillary, which is based on likelihood ratio statistics for testing the curved exponential model within a larger ambient full exponential model. By comparing the

Efron–Hinkley ancillary with the directed likelihood ancillary in a problem similar to Fisher's gamma hyperbola, Pedersen (1981) showed that the choice of approximate ancillary can have considerable effect on inferences, at least in small samples. Asymptotic ancillaries were considered by Grambsch (1983) in a sequential sampling scheme and by Sweeting (1992) for stochastic processes.

Barndorff-Nielsen (1980) introduced the p^*-formula for approximating the conditional distribution of the maximum likelihood estimator, and Hinkley (1980) provided a similar formula for scalar parameter curved exponential families; see also McCullagh (1987), Davison (1988), Fraser and Reid (1988), and Skovgaard (1990) for related developments.

The emphasis in the paper on normally shaped likelihoods in relation to the normality of $\widehat{j}^{1/2}(\widehat{\theta} - \theta)$ is similar to the approach advocated by Sprott (1973), who considered marginal inference. In his discussion of the paper, Sprott (1978) suggested using transformations designed explicitly to normalize the likelihood instead of using variance-stabilizing transformations. Just as using the chi-squared approximation for the likelihood ratio statistic is highlighted in the paper as a means to avoid the cumbersome calculations required by variance-stabilizing transformations when constructing confidence intervals, Sprott (1973) has recommended using a normal approximation for the signed root of the likelihood ratio statistic to circumvent the need to specify explicitly likelihood-normalizing transformations when constructing one-sided confidence limits. Corrected versions of the signed root likelihood ratio statistic have been studied extensively in connection with conditional inference. For general scalar parameter problems, McCullagh (1984) showed that, given any second-order locally ancillary statistic, the conditional distribution of the mean corrected signed root statistic is standard normal to error of order n^{-1}, and he extended his results to vector parameter models; see also Severini (1990, 2000). Barndorff-Nielsen (1986) derived a corrected signed root statistic r^* that is conditionally standard normal to error of order $n^{-3/2}$, although calculation of r^* requires that the conditioning statistic be explicitly specified. See Skovgaard (1987), Jensen (1992, 1995), Pierce and Peters (1992), Fraser et al. (1999), and Davison et al. (2006) for related developments. Much recent work has focused on versions of r^* that do not require specification of the ancillary; typically, these are conditionally standard normal to error of order n^{-1}.

The primary message of this paper, that variance estimates for maximum likelihood estimators should be constructed from observed information, is now widely accepted; see, for example, McLachlan and Peel (2000), Agresti (2002), and Lawless (2002). In their justification for this practice, Efron and Hinkley addressed several topics that have been the focus of subsequent work on inference for parametric models: asymptotic ancillarity; approximate sufficiency; stable inference; transformation theory; approximate conditional inference based on the likelihood ratio statistic; geometry of curved exponential families; Laplace approximation; normally shaped likelihoods; and conditional and Bayesian Bartlett correction. For these many contributions, this paper has rightly become a key reference, and it has clearly had a remarkable legacy.

References

Agresti, A. (2002). *Categorical Data Analysis*, 2nd edition. Wiley, New York.

Amari, S. (1982). Geometrical theory of asymptotic ancillarity and conditional inference. *Biometrika* **69**, 1–18.

Barndorff-Nielsen, O. E. (1980). Conditionality resolutions. *Biometrika* **67**, 293–310.

Barndorff-Nielsen, O. E. (1986). Inference on full and partial parameters based on the standardized signed log likelihood ratio. *Biometrika* **73**, 307–322.

Barndorff-Nielsen, O. E. and Cox, D. R. (1979). Edgeworth and saddlepoint approximations with statistical applications. *Journal of the Royal Statistical Society, Series B* **41**, 279–312 with discussion.

Barndorff-Nielsen, O. E. and Cox, D. R. (1984). Bartlett adjustments to the likelihood ratio statistic and the distribution of the maximum likelihood estimator. *Journal of the Royal Statistical Society, Series B* **46**, 483–495.

Barndorff-Nielsen, O. E. and Cox, D. R. (1989). *Asymptotic Techniques for Use in Statistics*. Chapman and Hall, London.

Barndorff-Nielsen, O. E. and Cox, D. R. (1994). *Inference and Asymptotics*. Chapman and Hall, London.

Bickel, P. J. and Ghosh, J. K. (1990). A decomposition for the likelihood ratio statistic and the Bartlett correction: A Bayesian argument. *Annals of Statistics* **3**, 1070–1090.

Cox, D. R. (1958). Some problems connected with statistical inference. *Annals of Mathematical Statistics* **29**, 357–372.

Cox, D. R. (1980). Local ancillarity. *Biometrika* **67**, 279–286.

Davison, A. C. (1988). Approximate conditional inference in generalized linear models. *Journal of the Royal Statistical Society, Series B* **50**, 445–461.

Davison, A. C., Fraser, D. A. S. and Reid, N. (2006). Improved likelihood inference for discrete data. *Journal of the Royal Statistical Society, Series B* **68**, 495–508.

DiCiccio, T. J. and Stern, S. E. (1993). On Bartlett adjustments for approximate Bayesian inference. *Biometrika* **80**, 731–740.

DiCiccio, T. J. and Stern, S. E. (1994). Frequentist and Bayesian Bartlett correction of test statistics based on adjusted profile likelihoods. *Journal of the Royal Statistical Society, Series B* **56**, 397–408.

Durbin, J. (1980). Approximations for densities of sufficient estimators. *Biometrika* **67**, 311–333.

Efron, B. (1975). Defining the curvature of a statistical problem (with applications to second order efficiency). *Annals of Statistics* **3**, 1189–1242 with discussion and Reply.

Fisher, R. A. (1934). Two new properties of mathematical likelihood. In *Proceedings of the Royal Society of London, Series A* **144**, 285–307.

Fisher, R. A. (1973). *Statistical Methods and Scientific Inference*, 3rd edition. Hafner Press, New York.

Fraser, D. A. S. and Reid, N. (1988). On conditional inference for a real parameter: A differential approach on the sample space. *Biometrika* **75**, 251–264.

Fraser, D. A. S., Reid, N. and Wu, J. (1999). A simple general formula for tail probabilities for frequentist and Bayesian inference. *Biometrika* **86**, 249–264.

Grambsch, P. (1983). Sequential sampling based on the observed Fisher information to guarantee the accuracy of the maximum likelihood estimator. *Annals of Statistics* **11**, 68–77.

Hinkley, D. V. (1978). Likelihood inference about location and scale parameters. *Biometrika* **65**, 253–261.

Hinkley, D. V. (1980). Likelihood as approximate pivotal distribution. *Biometrika* **67**, 287–292.

Jensen, J. L. (1992). The modified signed likelihood ratio and saddlepoint approximations. *Biometrika* **79**, 693–703.

Jensen, J. L. (1993). A historical sketch and some new results on the improved log likelihood ratio statistic. *Scandinavian Journal of Statistics* **20**, 1–15.

Jensen, J. L. (1995). *Saddlepoint Approximations*. Oxford University Press, Oxford.

Lawless, J. F. (2002). *Statistical Models and Methods for Lifetime Data*, 2nd edition. Wiley, New York.

McCullagh, P. (1984). Local sufficiency. *Biometrika* **71**, 233–244.

McCullagh, P. (1987). *Tensor Methods in Statistics*. Chapman and Hall, London.

McLachlan, G. and Peel, D. (2000). *Finite Mixture Models*. Wiley, New York.

Pedersen, B. V. (1981). A comparison of the Efron–Hinkley ancillary and the likelihood ratio ancillary in a particular example. *Annals of Statistics* **9**, 1328–1333.

Peers, H. W. (1978). Second-order sufficiency and statistical invariants. *Biometrika* **65**, 489–496.

Pierce, D. A. and Peters, D. (1992). Practical use of higher order asymptotics for multiparameter exponential families. *Journal of the Royal Statistical Society, Series B* **81**, 977–986 with discussion.

Ryall, T. A. (1981). Extensions of the concept of local ancillarity. *Biometrika* **68**, 677–683.

Severini, T. A. (1990). Conditional properties of likelihood-based significance tests. *Biometrika* **77**, 343–352.

Severini, T. A. (2000). *Likelihood Methods in Statistics*. Oxford University Press, Oxford.

Skovgaard, I. M. (1985). A second-order investigation of asymptotic ancillarity. *Annals of Statistics* **13**, 534–551.

Skovgaard, I. M. (1987). Saddlepoint expansions for conditional distributions. *Journal of Applied Probability* **24**, 875–887.

Skovgaard, I. M. (1990). On the density of minimum contrast estimators. *Annals of Statistics* **18**, 779–789.

Sprott, D. A. (1973). Normal likelihoods and their relation to large sample theory of estimation. *Biometrika* **60**, 457–465.

Sprott, D. A. (1978). Comments on a paper by B. Efron and D. V. Hinkley. *Biometrika* **65**, 485–486.

Sweeting, T. J. (1992). Asymptotic ancillarity and conditional inference for stochastic processes. *Annals of Statistics* **20**, 580–589.

Tierney, L. and Kadane, J. B. (1986). Accurate approximations for posterior moments and marginal densities. *Journal of the American Statistical Association* **81**, 82–86.

Tierney, L., Kass, R. E. and Kadane, J. B. (1989a). Fully exponential Laplace approximations to expectations and variances of nonpositive functions. *Journal of the American Statistical Association* **84**, 710–715.

Tierney, L., Kass, R. E. and Kadane, J. B. (1989b). Approximate marginal densities of nonlinear functions. *Biometrika* **76**, 425–433. Amendment (1991). *Biometrika* **78**, 233–234.

Biometrika (1978), 65, 3, pp. 457–87
With 11 *text-figures*
Printed in Great Britain

Assessing the accuracy of the maximum likelihood estimator: Observed versus expected Fisher information

BY BRADLEY EFRON

Department of Statistics, Stanford University, California

AND DAVID V. HINKLEY

School of Statistics, University of Minnesota, Minneapolis

SUMMARY

This paper concerns normal approximations to the distribution of the maximum likelihood estimator in one-parameter families. The traditional variance approximation is $1/\mathscr{I}_{\hat\theta}$, where $\hat\theta$ is the maximum likelihood estimator and $\mathscr{I}_\theta$ is the expected total Fisher information. Many writers, including R. A. Fisher, have argued in favour of the variance estimate $1/I(x)$, where $I(x)$ is the observed information, i.e. minus the second derivative of the log likelihood function at $\hat\theta$ given data x. We give a frequentist justification for preferring $1/I(x)$ to $1/\mathscr{I}_{\hat\theta}$. The former is shown to approximate the conditional variance of $\hat\theta$ given an appropriate ancillary statistic which to a first approximation is $I(x)$. The theory may be seen to flow naturally from Fisher's pioneering papers on likelihood estimation. A large number of examples are used to supplement a small amount of theory. Our evidence indicates preference for the likelihood ratio method of obtaining confidence limits.

Some key words: Ancillary; Asymptotics; Cauchy distribution; Conditional inference; Confidence limits; Curved exponential family; Fisher information; Likelihood ratio; Location parameter; Statistical curvature.

1. INTRODUCTION

In 1934, Sir Ronald Fisher's work on likelihood reached its peak. He had earlier advocated the maximum likelihood estimator as a statistic with least large sample information loss, and had computed the approximate loss. Now, in 1934, Fisher showed that in certain special cases, namely the location and scale models, all of the information in the sample is recoverable by using an appropriately conditioned sampling distribution for the maximum likelihood estimator. This marks the beginning of exact conditional inference based on exact ancillary statistics, although the notion of ancillary statistics had appeared in Fisher's 1925 paper on statistical estimation.

Beyond the explicit details of exact conditional distributions for special cases, the 1934 paper contains on p. 300 the following intriguing claim about the general case

> When these [log likelihood] functions are differentiable successive portions of the [information] loss may be recovered by using as ancillary statistics, in addition to the maximum likelihood estimate, the second and higher differential coefficients at the maximum.

To this may be coupled an earlier statement (Fisher, 1925, p. 724)

> The function of the ancillary statistic is analogous to providing a true, in place of an approximate, weight for the value of the estimate.

There are no direct calculations by Fisher to clarify the above remarks, other than calculations of information loss. But one may infer that approximate conditional inference based on the maximum likelihood estimate is claimed to be possible using observed properties of the likelihood function. To be specific, if we take for granted that inference is accomplished by attaching a standard error to the maximum likelihood estimate, then Fisher's remarks suggest that we use a conditional variance approximation based on the observed second derivative of the log likelihood function, as opposed to the usual unconditional variance approximation, the reciprocal of the Fisher information. Our main topics in this paper are (i) the appropriateness and easy calculation of such a conditional variance approximation and (ii) the ramifications of this for statistical inference in the single parameter case.

We begin with a simple illustrative example borrowed from Cox (1958). An experiment is conducted to measure a constant θ. Independent unbiased measurements y of θ can be made with either of two instruments, both of which measure with normal error: instrument k produces independent errors with a $N(0,\sigma_k^2)$ distribution ($k = 0, 1$), where σ_0^2 and σ_1^2 are known and unequal. When a measurement y is obtained, a record is also kept of the instrument used, so that after a series of n measurements the experimental results are of the form $(a_1, y_1), \ldots, (a_n, y_n)$, where $a_j = k$ if y_j is obtained using instrument k. The choice between instruments for the jth measurement is made at random by the toss of a fair coin,

$$\operatorname{pr}(a_j = 0) = \operatorname{pr}(a_j = 1) = \tfrac{1}{2}.$$

Throughout this paper, x will denote the entire set of experimental results available to the statistician, in this case $(a_1, y_1), \ldots, (a_n, y_n)$.

The log likelihood function $l_\theta(x)$, l_θ for short, is the log of the density function, thought of as a function of θ. In this example

$$l_\theta(x) = \text{const} - \sum_{j=1}^{n} \log \sigma_{a_j} - \frac{1}{2}\sum_{j=1}^{n} (y_j - \theta)^2/\sigma_{a_j}^2, \qquad (1{\cdot}1)$$

from which we obtain the maximum likelihood estimator as the weighted mean

$$\hat\theta = (\Sigma\, y_j/\sigma_{a_j}^2)(\Sigma\, 1/\sigma_{a_j}^2)^{-1}.$$

If we denote first and second derivatives of $l_\theta(x)$ with respect to θ by $\dot l_\theta(x)$ and $\ddot l_\theta(x)$, $\dot l_\theta$ and $\ddot l_\theta$ for short, then the total Fisher information for this experiment is

$$\mathscr{I}_\theta = \operatorname{var}\{\dot l_\theta(x)\} = E\{-\ddot l_\theta(x)\} = \tfrac{1}{2}n(1/\sigma_0^2 + 1/\sigma_1^2).$$

Standard theory shows that $\hat\theta$ is asymptotically normally distributed with mean θ and variance

$$\operatorname{var}(\hat\theta) \simeq 1/\mathscr{I}_\theta. \qquad (1{\cdot}2)$$

In this particular example $\mathscr{I}_\theta$ does not depend on θ, so that the variance approximation (1·2) is known. If this were not so we would use one of the two approximations (Cox & Hinkley, 1974, p. 302)

$$1/\mathscr{I}_{\hat\theta}, \quad 1/I(x), \qquad (1{\cdot}3)$$

where

$$I(x) = -\ddot l_{\hat\theta} = \left[-\frac{\partial^2 l_\theta(x)}{\partial\theta^2}\right]_{\theta=\hat\theta(x)}$$

The quantity $I(x)$ is aptly called the observed Fisher information by some writers, as distinguished from $\mathscr{I}_{\hat\theta}$, the expected Fisher information. This last name is useful even though $E(\mathscr{I}_{\hat\theta}) \neq \mathscr{I}_\theta$ in general. In the example above

$$I(x) = a/\sigma_1^2 + (n-a)/\sigma_0^2,$$

where $a = \Sigma\, a_j$, the number of times instrument 1 was used.

Approximation (1·2), one over the expected Fisher information, would presumably never be applied in practice, because after the experiment is carried out it is known that instrument 1 was used a times and that instrument 0 was used $n-a$ times. With the ancillary statistic a fixed at its observed value, $\hat{\theta}$ is normally distributed with mean θ and variance

$$\operatorname{var}(\hat{\theta} \mid a) = \{a/\sigma_1^2 + (n-a)/\sigma_0^2\}^{-1} \tag{1·4}$$

not (1·2). But now notice that, whereas (1·2) involves an average property of the likelihood, the conditional variance (1·4) is a corresponding property of the observed likelihood: (1·4) is equal to the reciprocal of the observed Fisher information $I(x)$.

It is clear here that the conditional variance $\operatorname{var}(\hat{\theta} \mid a)$ is more meaningful than $\operatorname{var}(\hat{\theta})$ in assessing the precision of the calculated value $\hat{\theta}$ as an estimator of θ, and that the two variances may be quite different in extreme situations.

This example is misleadingly neat in that $\operatorname{var}(\hat{\theta} \mid a)$ exactly equals $1/I(x)$. Nevertheless, a version of this relationship applies, as an approximation, to general one parameter estimation problems. A central topic of this paper is the accuracy of the approximation

$$\operatorname{var}(\hat{\theta} \mid a) \simeq 1/I(x), \tag{1·5}$$

where a is an ancillary or approximately ancillary statistic which affects the precision of $\hat{\theta}$ as an estimator of θ. To a first approximation, a will be equivalent to $I(x)$ itself. It is exactly so in Cox's example. The approximation (1·5) was suggested, never too explicitly, by Fisher in his fundamental papers on ancillarity and estimation. In complicated situations, such as that considered by Cox (1958), it is a good deal easier to compute $I(x)$ than $\mathscr{I}_\theta$. There are also philosophical advantages to (1·5). It is 'closer to the data' than $1/\mathscr{I}_\theta$, and tends to agree more closely with Bayesian and fiducial analyses. In Cox's example of the two measuring instruments, for instance, an improper uniform prior for θ on $(-\infty, \infty)$ gives $\operatorname{var}(\theta \mid x) = 1/I(x)$, in agreement with (1·5).

To demonstrate that (1·5) has validity in more realistic contexts, consider the estimation of the centre θ of a standard Cauchy translation family. For random samples of size n the Fisher information is $\mathscr{I}_\theta = \frac{1}{2}n$. When $n = 20$, then $\hat{\theta}$ has approximate variance 0·1, in accordance with (1·2); the exact variance is about 0·115 according to Efron (1975, p. 1210). In a Monte Carlo experiment 14,048 Cauchy samples of size 20, with $\theta = 0$, were obtained, and Fig. 1 plots the resulting estimated conditional variances of $\hat{\theta}$ given $I(x)$ versus $1/I(x)$. Samples were grouped according to interval values of $1/I(x)$. For example, 224 of the 14,048 samples had $1/I(x)$ in the range 0·170–0·180, averaging 0·175, and the 224 values of $\hat{\theta}^2$ had mean 0·201 and standard error 0·023. This gives the estimate

$$\operatorname{var}\{\hat{\theta} \mid 1/I(x) = 0{\cdot}175\} = 0{\cdot}201 \pm 0{\cdot}023$$

plotted in Fig. 1, since we know $E\{\hat{\theta} \mid I(x)\} = 0$ by symmetry.

Figure 1 strongly suggests the relationship

$$\operatorname{var}\{\hat{\theta} \mid I(x)\} \simeq 1/I(x). \tag{1·6}$$

This is a weakened version of (1·5). In translation families $I(x)$ is ancillary, but it is only a function of the maximal ancillary a, the configuration statistic, i.e. the $n-1$ spacings between the ordered values $x_{(1)} < \ldots < x_{(n)}$. The stronger statement (1·5) is verified for translation families in §§2 and 8. We prefer (1·5) to (1·6) because $\operatorname{var}(\hat{\theta} \mid a)$ is more relevant than $\operatorname{var}\{\hat{\theta} \mid I(x)\}$ as a measure of precision for $\hat{\theta}$; in principle (Fisher, 1934) inference about θ is conditional on a.

The implications of (1·5) are considerable. If $I(x) = 15$ in the Cauchy example, a not very remarkable event since $\text{pr}\{I(x) > 15\} \simeq 0{\cdot}05$, then the approximate 95% confidence interval for θ is

$$\hat{\theta} \pm 1{\cdot}96/\surd 15, \tag{1·7}$$

rather than

$$\hat{\theta} \pm 1{\cdot}96/\surd 10 \tag{1·8}$$

as suggested by (1·2). The latter interval is too wide, having conditional coverage probability of 98% rather than the normal 95%. Given the equally unremarkable event $I(x) = 6$, interval (1·8) is too narrow, having conditional coverage probability of only 87%. These numerical comparisons presuppose accuracy of normal approximations, which is justified in § 4.

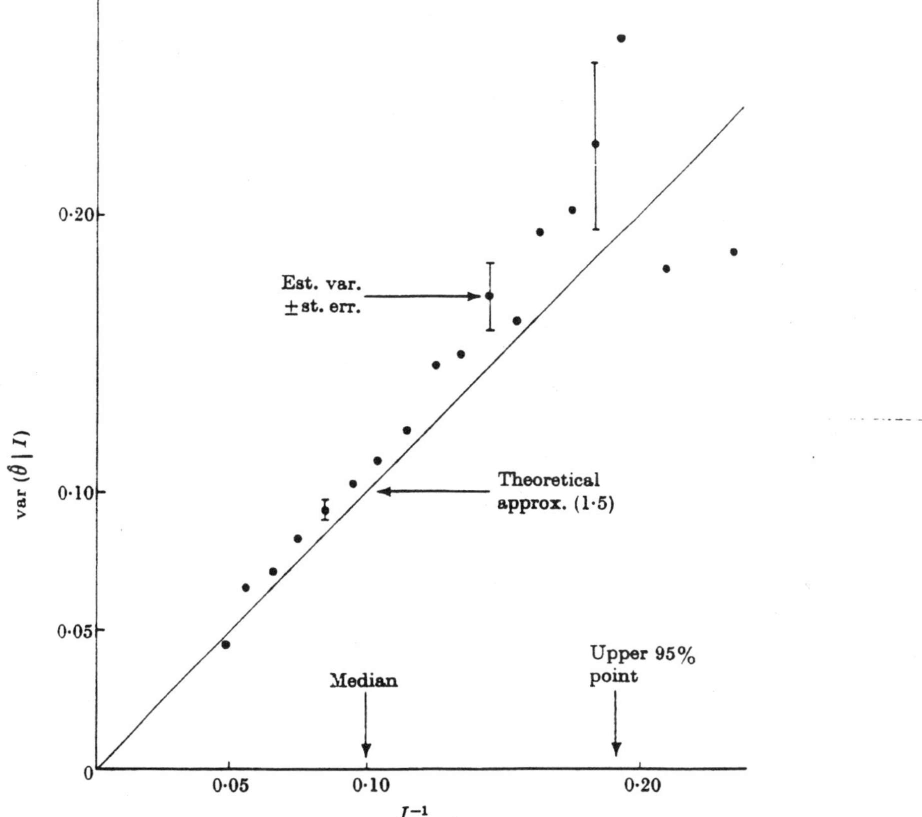

Fig. 1. Cauchy location θ. Monte Carlo estimates of conditional variance of maximum likelihood estimate $\hat{\theta}$ given the observed information $I(x)$. Sample size $n = 20$; 14,048 samples.

The purpose of this paper is to justify (1·5) for a wide variety of one-parameter problems. The justification consists of detailed numerical results for several special examples involving moderate sample sizes, in addition to the general asymptotic theory. The results are presented in the following order: § 2 gives an outline of the theory for translation families; § 3 contains two detailed examples of this theory; § 4 deals with confidence interval interpretations for the results of § 2; § 5 outlines the more complicated theory appropriate for nontranslation problems; § 6 follows with an example; §§ 7 and 8 present details of the asymptotic theory; § 9 contains brief concluding remarks, together with some further references and historical notes.

2. Translation families

2·1. *Conditional variance approximations*

The theory of ancillarity and conditional inference for translation families was developed by Fisher (1934). Here we will use Fisher's theory to justify (1·5), and its higher order corrections, in translation families. Section 4 contains the analogous results for approximate normal confidence limits based on (1·5). In § 5 the general one-parameter problem is reduced to approximate translation form by a transformation argument. The treatment in this section is presented in outline form, more careful calculations being reserved for § 8.

Suppose then that $x_1, \ldots, x_n$ are independent and identically distributed with density $f_\theta(x_1) = f_0(x_1 - \theta)$. The data vector x can be reduced to the sufficient statistic $(\hat\theta, a)$, where $\hat\theta(x)$ is the maximum likelihood estimator and $a(x)$ is the ancillary configuration statistic, representable as the spacings between successive order statistics, $x_{(2)} - x_{(1)}, \ldots, x_{(n)} - x_{(n-1)}$.

Because a is ancillary, its density $g(a)$ does not depend on θ. The conditional density, with respect to Lebesgue measure, of $\hat\theta$ given a is of the translation form

$$f_\theta(\hat\theta \mid a) = h_a(\hat\theta - \theta). \tag{2·1}$$

The Jacobian of the transformation from the ordered x_i values to $(\hat\theta, a)$ is a constant not depending on x, which implies that the density of x can be written

$$f_\theta(x) = cg(a)\, h_a(\hat\theta - \theta) \tag{2·2}$$

for some constant c.

The likelihood function $\text{lik}_x(\theta)$ is $f_\theta(x)$ thought of as a function of θ, with x fixed at its observed value. Fisher's (1934) main result relates $f_\theta(\hat\theta \mid a) = h_a(\hat\theta - \theta)$ to $\text{lik}_x(\theta)$: for any value of $t \equiv \hat\theta(x) - \theta$,

$$\frac{h_a(t)}{h_a(0)} = \frac{\text{lik}_x\{\hat\theta(x) - t\}}{\text{lik}_x\{\hat\theta(x)\}}. \tag{2·3}$$

This result, which is derived immediately from (2·2), looks simple but is in fact a powerful computational tool. Given the data vector x, it is computationally easy to plot the shape of the likelihood function $\text{lik}_x(\theta)$. Reflection of this curve about its maximum point $\hat\theta(x)$ then gives the conditional sampling density $f_\theta(\hat\theta \mid a)$, which might otherwise be thought difficult to compute. The word 'shape' is necessary here since (2·3) determines $h_a(t)$ only relative to its maximum $h_a(0)$. Integration is necessary to determine the correct multiple, that which integrates to one. Fisher's tour de force was completed by noting that fully informative frequentist inferences about θ should certainly be made conditional on the ancillary a, so that the likelihood theory leads easily and naturally to the appropriate frequentist theory.

To see how (2·3) applies to the phenomenon pictured in Fig. 1 suppose, for the moment, that $\text{lik}_x(\theta)$ happens to be perfectly normal shaped; that is,

$$\frac{\text{lik}_x(\theta)}{\text{lik}_x\{\hat\theta(x)\}} = \exp[-\tfrac{1}{2}c_2\{\theta - \hat\theta(x)\}^2] \tag{2·4}$$

for some positive constant c_2. Then (2·4) and (2·3) quickly give

$$f_\theta(\hat\theta \mid a) = h_a(\hat\theta - \theta) = (2\pi/c_2)^{-\frac{1}{2}} \exp\{-\tfrac{1}{2}c_2(\hat\theta - \theta)^2\}.$$

In other words, a normal-shaped likelihood function implies that, conditional on a, $\hat\theta$ is normally distributed with mean θ and variance

$$\text{var}(\hat\theta \mid a) = 1/c_2. \tag{2·5}$$

In the notation of § 1, where $l_\theta(x)$ is the log likelihood function $\log \text{lik}_x(\theta)$, and dots indicate differentiation with respect to θ, (2·4) gives $c_2 = -\ddot{l}_\theta = I(x)$, so that (2·5) is an exact form of (1·5),

$$\text{var}(\hat{\theta} \mid a) = 1/I(x). \tag{2·6}$$

What if the likelihood function is not perfectly normal shaped? As n gets large the likelihood will approach normality, assuming some mild regularity conditions on the form of $f_\theta(x)$, and we can use this fact to obtain asymptotic expansions for the conditional mean and variance of $\hat{\theta}$ given a. These expansions involve the higher derivatives of the log likelihood function, say for $j = 3, 4, \ldots$

$$l_\theta^{(.j)} = \left[\frac{\partial^j l_\theta(x)}{\partial \theta^j}\right]_{\theta=\hat{\theta}(x)},$$

all of which are zero in the normal case (2·4). We also use the notation $l_\theta^{(.2)} = \ddot{l}_\theta$ where convenient. Notice that (2·2) implies

$$l_\theta^{(.j)} = \left[\frac{\partial^j \log h_a(\hat{\theta}-\theta)}{\partial \theta^j}\right]_{\theta=\hat{\theta}} = (-1)^j \left[\frac{\partial^j \log h_a(t)}{\partial t^j}\right]_{t=0},$$

which says that the $l_\theta^{(.j)}$ are functions of x only through a and are themselves ancillary statistics. This same statement applies to the observed Fisher information $I(x) = -\ddot{l}_\theta$.

LEMMA 1. *In translation families satisfying the regularity conditions stated in* § 8,

$$\text{var}(\hat{\theta} \mid a) = I^{-1}\{1 + (J^2/I^3 - \tfrac{1}{2}K/I^2) + o_p(n^{-1})\}, \quad E(\hat{\theta} \mid a) = \theta - \tfrac{1}{2}J/I^2 + o_p(n^{-1}), \tag{2·7}$$

$$E\{(\hat{\theta}-\theta)^2 \mid a\} = \frac{1}{I}\left\{1 + \left(\frac{5}{4}\frac{J^2}{I^3} - \frac{1}{2}\frac{K}{I^2}\right) + o_p(n^{-1})\right\}, \tag{2·8}$$

where

$$I = I(x) = -\ddot{l}_\theta(x), \quad J = l_\theta^{(.3)}(x), \quad K = -l_\theta^{(.4)}(x). \tag{2·9}$$

The proof of Lemma 1, which is an elaboration of the argument leading from a normal-shaped likelihood to (2·6), is given in § 8, along with appropriate regularity conditions.

The terms in round brackets in (2·7)–(2·8) are of order $O_p(n^{-1})$ or smaller. In particular $\text{var}(\hat{\theta} \mid a)$ in (2·7) can be written

$$\text{var}(\hat{\theta} \mid a) = I^{-1}\{1 + O_p(n^{-1})\}, \tag{2·10}$$

which verifies (1·5). Lemma 2, in § 2·2, provides the final justification for (1·5) being an improvement over $1/\mathscr{I}_\theta$. The approximate normality of the likelihood function, which is used to prove Lemma 1, also ensures that the conditional distribution of $\hat{\theta}$ is approximately normal, given Fisher's result (2·3). Results directly related to conditional confidence intervals for θ are described in § 4.

In special cases the higher order terms in the left-hand side of (2·7) can be evaluated, giving expressions for $\text{var}(\hat{\theta} \mid a)$ more accurate than (1·5). This is demonstrated by the two examples of § 3 and the brief discussion of the Cauchy translation problem in § 8.

Even though the maximal ancillary a consists of $I(x)$ plus the higher order derivatives $l_\theta^{(.j)}$ ($j = 3, 4, \ldots$), the conditional variance $\text{var}(\hat{\theta} \mid a)$ is asymptotically equivalent, to within terms of order n^{-2}, to a function of just $I(x)$, namely $1/I(x)$. To put it another way, $I(x) = -\ddot{l}_\theta$ recaptures most of the information lost by considering only $\hat{\theta}(x)$ instead of the full sample x. Some informal calculations to this effect were carried out by Fisher (1925). Roughly speaking, the pair $(\hat{\theta}, \ddot{l}_\theta)$ is the sufficient statistic for the two-parameter exponential family which best approximates the family $f_\theta(x)$ near the true value of θ; see § 7 below.

2·2. *Statistical curvature and comparison of variance approximations*

How different, numerically, is the conditional variance approximation $1/I(x)$ from the unconditional approximation $1/\mathscr{I}_\theta$? An asymptotic answer can be given in terms of the statistical curvature γ_θ, as defined by Efron (1975, §§ 3, 5). Suppose that each x_i has density function $f_\theta(x_1)$, not necessarily of translation form, and define for $j,k = 1,2$ the moments

$$\nu_{jk}(\theta) = E\left(\left\{\frac{\partial \log f(x_1)}{\partial\theta}\right\}^j \left[\frac{\partial^2 \log f_\theta(x_1)}{\partial\theta^2} + E\left\{\frac{\partial \log f_\theta(x_1)}{\partial\theta}\right\}^2\right]^k\right),$$

assuming these exist. Then the statistical curvature of $f_\theta(x_1)$ is

$$\gamma_\theta = (\nu_{02}\nu_{20} - \nu_{11}^2)^{1/2}/\nu_{20}^{3/2}. \tag{2·11}$$

This curvature is a measure of the deviation of $f_\theta(x_1)$ from exponential family form and is invariant under monotone reparameterization. One interpretation is that $\gamma_\theta^2 \mathscr{I}_\theta^2$ is the residual variation in $\ddot{l}_\theta$ after linear regression on $\dot{l}_\theta$. For one-parameter exponential families $\gamma_\theta = 0$, and in such families $I(x) = \mathscr{I}_\theta$, so that the two variance approximations being considered are equal. The statistical curvature of $l_\theta(x_1, \ldots, x_n)$ is $\gamma_\theta/\sqrt{n}$.

The relationship between γ_θ and the variance approximations is given by the following result.

LEMMA 2. *If $x_1, \ldots, x_n$ are independent and identically distributed with density function $f_\theta(x_1)$ satisfying the regularity conditions stated in* § 8, *then as $n \to \infty$*

$$\sqrt{n}\{I(x)/\mathscr{I}_\theta - 1\} \sim N(0, \gamma_\theta^2). \tag{2·12}$$

A proof is given in § 8. Fisher (1925) indirectly suggests (2·12).

In a translation family $\mathscr{I}_\theta = \mathscr{I}$ and $\gamma_\theta = \gamma$ are constants, so that (2·12) can then be written as $I(x)/\mathscr{I}_\theta = 1 + O_p(n^{-\frac{1}{2}})$. Combined with (2·10), this easily leads to

$$\frac{\operatorname{var}(\hat\theta \mid a) - 1/I(x)}{\operatorname{var}(\hat\theta \mid a) - 1/\mathscr{I}} = O_p(n^{-\frac{1}{2}}), \tag{2·13}$$

which shows that (1·5) is a valid and useful asymptotic approximation. Granting that $\operatorname{var}(\hat\theta \mid a)$ is a more meaningful measure of variance than $\operatorname{var}(\hat\theta)$, we see that $1/I(x)$ is a better variance approximation than $1/\mathscr{I}$ by a half order of magnitude, in the usual exaggerated sense of asymptotic comparisons. The numerical results for the Cauchy translation problem in § 1 and the two examples in § 3 show that the improvement can be substantial even for moderate sample sizes.

Suppose that we are interested in estimating a monotone function of θ, say $\sigma = \sigma(\theta)$, rather than θ itself. It is easy to verify that the observed Fisher information for σ, say $I^{(\sigma)}(x)$, is related to that for θ by

$$I^{(\sigma)}(x) = I(x)\,(d\hat\theta/d\hat\sigma)^2. \tag{2·14}$$

The expected Fisher information transforms in the same way. Since the maximum likelihood estimator maps the same way as does the parameter, $\hat\sigma = \sigma(\hat\theta)$, the notation $d\hat\theta/d\hat\sigma$ is unambiguous, and equals $[d\theta/d\sigma]_{\sigma=\hat\sigma}$. A standard expansion argument proceeding from Lemmas 1 and 2 shows that (1·5) is valid for $\hat\sigma$, in the sense of (2·13), that is

$$\frac{\operatorname{var}(\hat\sigma \mid a) - 1/I^{(\sigma)}(x)}{\operatorname{var}(\hat\sigma \mid a) - 1/\mathscr{I}_\theta^{(\sigma)}} = O_p(n^{-\frac{1}{2}}). \tag{2·15}$$

If we wish to compare confidence intervals for $\hat{\theta}$, conditional versus unconditional as at (1·7) and (1·8), the ratio of the lengths of conditional and unconditional intervals is

$$\{I(x)/\mathscr{I}_\theta\}^{\frac{1}{2}}. \tag{2·16}$$

Lemma 2 implies that $\{I(x)/\mathscr{I}_\theta\}^{\frac{1}{2}}$ has, asymptotically, a normal distribution with mean 1 and standard deviation $\gamma_\theta/(2\sqrt{n})$. For the Cauchy translation family $\gamma_\theta^2 = 2{\cdot}5$, so that with $n = 20$ the standard deviation of $\{I(x)/\mathscr{I}_\theta\}^{\frac{1}{2}}$ is approximately 0·18. We expect large variability in $\{I(x)/\mathscr{I}_\theta\}^{\frac{1}{2}}$ in this situation, which is indeed the case. Increasing n to 80 in the Cauchy problem reduces the standard deviation of $\{I(x)/\mathscr{I}_\theta\}^{\frac{1}{2}}$ to 0·09, so that conditioning effects become considerably less important.

3. EXAMPLES OF TRANSLATION FAMILIES

We illustrate the theory of the preceding section using two particularly simple examples of symmetric translation families due to Fisher.

Example 3·1: *Fisher's normal circle.* The first example is the circle model (Fisher, 1974, p. 138), where the data consist of a two-dimensional normally distributed vector x, covariance matrix the identity, whose mean vector is known to lie on a circle of known radius ρ_0 centred at the origin. That is,

$$E(x^{\mathrm{T}}) = \rho_0(\cos\theta, \sin\theta). \tag{3·1}$$

Having observed x, we wish to estimate the unknown θ. Note that given n independent observations $x_1, \ldots, x_n$ on this model the sufficient statistic $x = \Sigma x_i/\sqrt{n}$ satisfies (3·1) with $\rho_0\sqrt{n}$ in place of ρ_0.

If the data vector x has polar coordinates $(\hat{\theta}, r\rho_0)$ then $\hat{\theta}$ is the maximum likelihood estimate of θ, and $r = \|x\|/\rho_0$ is ancillary. The density is of the form (2·2), with a replaced by r, so that we can apply the theory of § 2, even though (3·1) does not look like a standard translation problem. The density $g(r)$ is noncentral chi, while the conditional density

$$f_\theta(\hat{\theta}\,|\,r) = h_r(\hat{\theta}-\theta)$$

is the 'circular normal',

$$c^{-1}\exp\{\rho_0^2 r\cos(\hat{\theta}-\theta)\}. \tag{3·2}$$

Here we are assuming that $\hat{\theta}$ given θ ranges from $\theta-\pi$ to $\theta+\pi$, for the sake of symmetric definition. The constant c equals $2\pi I_0(\rho_0^2 r)$, in the standard Bessel function notation.

Now we can apply Lemma 1 of § 2. From (3·2) we calculate

$$l_\theta^{(.j)} = (-1)^{\frac{1}{2}j}\rho_0^2 r \quad (j = 2, 4, \ldots), \tag{3·3}$$

and $l_\theta^{(.j)} = 0$ for j odd. The Fisher information $\mathscr{I}_\theta$ is constant,

$$\mathscr{I}_\theta = \mathscr{I} = \rho_0^2. \tag{3·4}$$

Using $I(x) = -\ddot{l}_\theta = r\mathscr{I}$, $l_\theta^{(.3)} = 0$, $l_\theta^{(.4)} = r\mathscr{I}$, from (3·3) and (3·4), (2·7) can be written

$$\operatorname{var}(\hat{\theta}\,|\,r) \simeq \frac{1}{r\mathscr{I}}\{1 + 1/(2r\mathscr{I})\}. \tag{3·5}$$

The exact conditional variance of $\hat{\theta}$ given the ancillary statistic r is calculated from (3·2) to be

$$\operatorname{var}(\hat{\theta}\,|\,r) = \left\{\int_{-\pi}^{\pi} t^2\exp(r\mathscr{I}\cos t)\,dt\right\}\Big/\left\{\int_{-\pi}^{\pi}\exp(r\mathscr{I}\cos t)\,dt\right\}. \tag{3·6}$$

Figure 2 compares (3·6) with (3·5). For values of $\rho_0^2 = \mathscr{I} \geqslant 8$ it can be shown that at least 95% of the realizations of $r\mathscr{I} = I(x)$ will be greater than 4. We see that approximation (1·5), $\text{var}(\hat{\theta}\,|\,r) \simeq 1/(r\mathscr{I})$, is quite acceptable in the range $1/(r\mathscr{I}) \leqslant 0{\cdot}25$, and that the improved approximation (3·5) derived from Lemma 1 is very accurate.

The exact variance (3·6) can be expressed in terms of Bessel functions, whose expansions lead to (3·5).

Example 3·2: *Fisher's gamma hyperbola.* Fisher's hyperbola model, introduced in connexion with his famous 'problem of the Nile' (Fisher, 1974, p. 169), involves two independent scaled gamma variables whose means are restricted to lie on an hyperbola. Thus we observe $x = (x_1, x_2)$ such that $x_1 = e^{\theta} G_{m1}$ and $x_2 = e^{-\theta} G_{m2}$, where G_{mi} indicates a variable with density $x^{m-1} e^{-x}/\Gamma(m)$ on $(0, \infty)$. The Fisher information for θ is $\mathscr{I} = 2m$. The maximum likelihood estimate of θ is $\hat{\theta} = \frac{1}{2}\log(x_1/x_2)$. This is illustrated in Fig. 3.

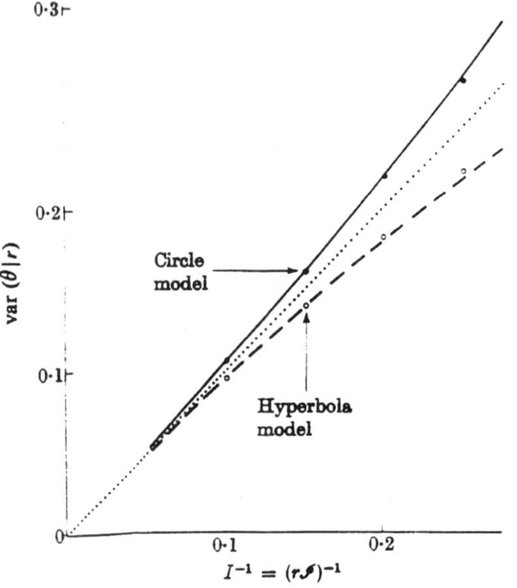

Fig. 2 (left). Exact conditional variances, circles, of $\hat{\theta}$ given r compared with approximations (2·7), curves, for the circle and hyperbola models. Dotted line, approximation (1·5).

Fig. 3 (right). Fisher's gamma hyperbola model. $x = (x_1, x_2)$, a pair of independent gamma variables with index m, whose means lie on solid curve. Broken curve, one orbit hyperbola for ancillary statistic r.

The ancillary statistic in the hyperbola model is $r = \surd(x_1 x_2)/m$, the level curves of which are hyperbolae 'parallel' to the curve of possible mean vectors, as shown in Fig. 3. It has density

$$g(r) = \frac{2m^{2m}}{\{\Gamma(m)\}^2} r^{2m-1} \int_{-\infty}^{\infty} \exp\{-2mr\cosh(t)\}\,dt,$$

the conditional density of $\hat{\theta}$ given r being

$$f_\theta(\hat{\theta}\,|\,r) = \exp\{-2mr\cosh(\hat{\theta}-\theta)\} \Big/ \int_{-\infty}^{\infty} \exp\{-2mr\cosh(t)\}\,dt. \tag{3·7}$$

In other words, this is another nonobvious example of form (2·2).

The log likelihood derivatives, from (3·7), are

$$l_{\theta}^{(\cdot j)} = -2mr = -r\mathscr{I} \quad (j = 2, 4, \ldots), \tag{3·8}$$

$l_{\theta}^{(\cdot j)} = 0$ for j odd, so that $I(x) = r\mathscr{I}$ as in the circle model, and (2·7) gives

$$\operatorname{var}(\hat{\theta} \mid r) \simeq \frac{1}{r\mathscr{I}}\{1 - 1/(2r\mathscr{I})\}. \tag{3·9}$$

This differs only in the sign of the second term from the corresponding formula (3·5) for the circle model. The actual conditional variance $\operatorname{var}(\hat{\theta} \mid r)$, obtained by integrating (3·7), can also be expressed as a function of $r\mathscr{I}$. The comparison of (3·9) with the actual conditional variance, Fig. 2, is almost exactly the same as for the circle model, except that here the deviations from the line $\operatorname{var}(\hat{\theta} \mid r) = 1/I(x) = 1/(r\mathscr{I})$ go in the opposite direction.

4. Conditional confidence intervals for the location parameter

Our results so far have been presented mainly in terms of variances, it being understood that these are of most interest in conjunction with a normal approximation for $\hat{\theta} - \theta$. The expansion theory of § 2 can be expressed directly in terms of conditional confidence intervals, an idea we now pursue explicitly.

As before, consider first the situation where $\text{lik}_x(\theta)$ happens to be perfectly normal shaped, so that (2·4) holds with $c_2 = I(x)$. There are two consequences of this relating to standard confidence interval methods. First, $\hat{\theta}$ has an exact normal distribution conditional on a, so that

$$u(x) = I(x)(\hat{\theta} - \theta)^2 \tag{4·1}$$

is exactly a χ_1^2 variable conditional on a. If the upper p point of χ_1^2 is denoted $\chi_1^2(p)$, then level p conditional limits on θ are $\hat{\theta} \pm \{\chi_1^2(p)/I(x)\}^{\frac{1}{2}}$. The other standard method of setting confidence limits is based on

$$v(x) = 2\{l_{\hat{\theta}(x)}(x) - l_{\theta}(x)\}, \tag{4·2}$$

which also has an exact χ_1^2 distribution conditional on a.

Although in general the likelihood function is not exactly normal shaped, it is approximately so for large n, and the same expansion methods used to confirm (1·5) also show that $u(x)$ and $v(x)$ defined above are asymptotically χ_1^2 conditional on a. More formally, we have the following result, proved in § 8.

LEMMA 3. *For translation families satisfying the regularity conditions in* § 8, *the statistics* $u(x)$ *and* $v(x)$ *defined by* (4·1) *and* (4·2) *satisfy*

$$\operatorname{pr}\{u(x) \geqslant u_0 \mid a\} = (1 - d_1 - d_2)\operatorname{pr}(\chi_1^2 \geqslant u_0) + d_1 \operatorname{pr}(\chi_5^2 \geqslant u_0) + d_2 \operatorname{pr}(\chi_7^2 \geqslant u_0) + o_p(n^{-1}), \tag{4·3}$$

$$\operatorname{pr}\{v(x) \geqslant u_0 \mid a\} = (1 - d_1 - d_2)\operatorname{pr}(\chi_1^2 \geqslant u_0) + (d_2 + d_1)\operatorname{pr}(\chi_3^2 \geqslant u_0) + o_p(n^{-1}), \tag{4·4}$$

where $d_1 = -K/(8I^2)$ *and* $d_2 = (5J^2)/(24I^3)$, *with* I, J *and* K *as defined in* (2·9).

Because d_1 and d_2 are both $O_p(n^{-1})$, (4·3) and (4·4) imply

$$I(x)(\hat{\theta} - \theta)^2 \mid a = \chi_1^2 + O_p(n^{-1}), \quad 2(l_{\hat{\theta}} - l_{\theta}) \mid a = \chi_1^2 + O_p(n^{-1}). \tag{4·5}$$

Note that the latter result is a conditional version of Wilks's famous theorem, and establishes that a standard method has the correct conditional properties.

The results (4·5) are superior to the unconditional result

$$\mathscr{I}_\theta(\hat{\theta}-\theta)^2 = \chi_1^2 + O_p(n^{-1}) \tag{4·6}$$

in a sense similar to (2·13). As we pointed out in § 2·2, the degree of superiority is determined by the curvature.

To investigate the practical validity of (4·5), we return to the Cauchy translation problem discussed in § 1. We have generated 20,000 samples of size $n = 20$ and computed the empirical frequencies with which $\mathscr{I}_\theta(\hat{\theta}-\theta)^2$, $I(x)(\hat{\theta}-\theta)^2$ and $2(l_{\hat{\theta}}-l_\theta)$ exceeded $\chi_1^2(p)$ for $p = 0{\cdot}05$ and $0{\cdot}01$, broken down by interval values of $I(x)$. Figure 4 graphs the results which show convincing evidence in support of (4·5) and dramatic conditional effects on the unconditional statistic (4·6). Note that the likelihood ratio method agrees better numerically with the chi-squared approximation than does the method based on $\hat{\theta}$. The expansions (4·3) and (4·4) indicate that this may be true in general, since $\text{pr}(\chi_q^2 \geqslant u_0)$ is an increasing function of q.

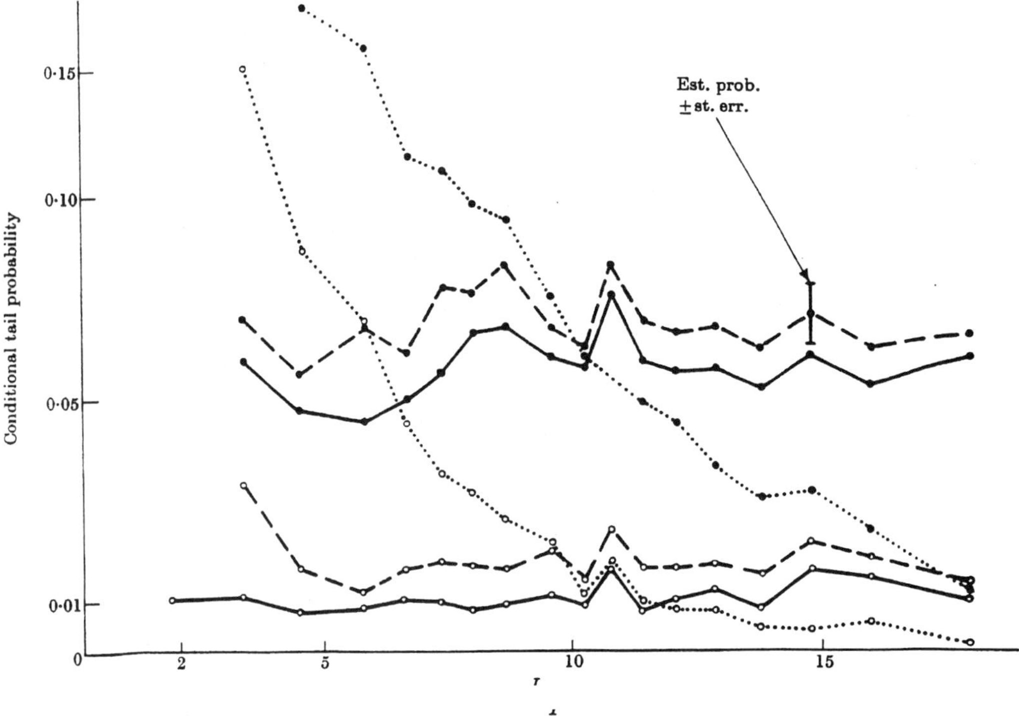

Fig. 4. Monte Carlo estimates of pr (statistic $\geqslant c \mid I$) with $c = 3{\cdot}84$, shown by closed circles, and $c = 5{\cdot}99$, open circles, for three likelihood statistics in the Cauchy location model, $n = 20$. Statistics: $2(l_{\hat{\theta}}-l_\theta)$ shown by solid curve; $I(\hat{\theta}-\theta)^2$, dashed curve; $\mathscr{I}(\hat{\theta}-\theta)^2$, dotted curve. Estimates from 20,000 samples.

5. Nontranslation families

This section discusses an example of a nontranslation problem in which a version of (1·5) can be seen to hold. We will use this example to introduce definitions appropriate for general nontranslation problems. The example is totally artificial, being in fact a simple variant of

Fisher's circle model, but furnishes a useful starting point because of its simplicity. A non-translation problem of a more realistic nature is discussed in § 6, again showing (1·5) at work.

We have not been able to provide a theoretical justification for these results in general, and pathological counterexamples are easy to construct, but nevertheless the examples suggest that (1·5), suitably interpreted, has wide validity.

Figure 5 illustrates a model in which the data vector is bivariate normal, covariance matrix the identity, and with mean vector constrained to lie on a spiral, instead of Fisher's circle; that is

$$E(x) = \beta_\theta = \begin{bmatrix} \cos\theta - \rho_\theta \sin\theta \\ \sin\theta + \rho_\theta \cos\theta \end{bmatrix}, \quad \rho_\theta = \rho_0 - \theta, \quad \theta \leqslant \rho_0. \tag{5·1}$$

The spiral is generated by the end of a thread unwinding from a circular spool of unit radius. By definition the thread has length ρ_0 at $\theta = 0$, which implies that it has length $\rho_\theta = \rho_0 - \theta$ at θ. At $\theta = +\rho_0$, we have $\rho_\theta = 0$, which accounts for the restriction in (5·1). We wish to assign some measure of accuracy to the maximum likelihood estimate $\hat{\theta}$ on the basis of the observed x. The sample size here is $n = 1$, but under repeated sampling essentially the same model holds with x replaced by $\bar{x} = \Sigma x_i/n$.

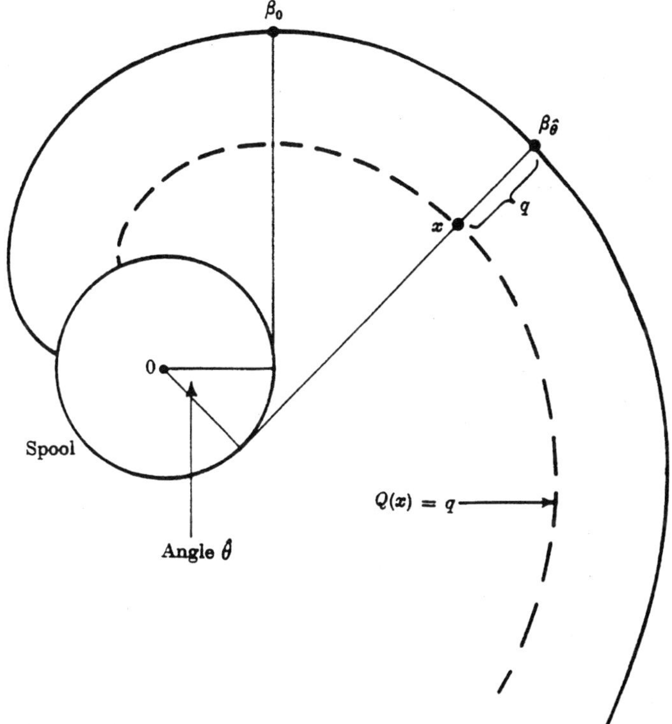

Fig. 5. Spiral model. $x = (x_1, x_2)$, bivariate normal with identity covariance matrix and mean β_θ on a logarithmic spiral shown by solid curve. Maximum likelihood estimate $\hat{\theta}$ is angular coordinate of straight thread on which x lies. Ancillary statistic has constant value q on parallel spiral through x, dashed curve.

It is easy to calculate that the Fisher information and curvature for the spiral model are $\mathscr{I}_\theta = \rho_\theta^2$ and $\gamma_\theta = 1/\rho_\theta$, and that having observed x, the maximum likelihood estimate $\hat{\theta}(x)$ is the angular coordinate of the thread upon which x lies. The vector $\beta_{\hat{\theta}}$ is the closest point to x on the spiral of possible mean vectors. When ρ_θ is large the curvature is small, and we expect small conditioning effects, the reverse being true when ρ_θ is small.

The geometry of Fig. 5 and familiarity with the bivariate normal distribution suggest that $Q(x)$, the signed distance of x from β_θ, should be approximately ancillary, with a limiting $N(0,1)$ distribution as ρ_θ gets large. We take the sign of $Q(x)$ positive if x is closer to the spool than β_θ, and negative if it is farther from the spool. Figure 5 shows that the level curve $Q(x) = q$ is a parallel spiral having thread everywhere q units shorter than that for the mean vector.

We intend to use $Q(x)$ as an approximate ancillary, conditioning upon the observed value of Q as we did upon a in the translation case. Both D. A. Pierce and D. R. Cox suggest this use of Q in the discussion following Efron (1975). Table 1 displays the marginal density of $Q(x)$ for four values of θ and eight values of $Q = q$. The four θ values are chosen such that $\rho_\theta = \rho_0 - \theta = \surd 8, \surd 16, \surd 32, \surd 64$; it is irrelevant which combinations of ρ_0 and θ are used to get these values of ρ_θ. The density would be constant across rows if $Q(x)$ were a genuine ancillary. We see that it is nearly constant, tending toward the $N(0,1)$ density as $\rho_\theta \to \infty$.

The marginal densities in Table 1 were obtained by numerical integration of the bivariate normal density along the spiral $Q(x) = q$. To avoid certain problems of definition, for each ρ_θ the integration was restricted to points on this spiral with angular coordinate in the interval $\theta \pm \pi$. Notice that if $\rho_\theta - q < \pi$, then the spiral runs into the central spool before the lower limit $\theta - \pi$ is reached. This end effect seriously distorts a few of the more extreme calculations, as indicated in the tables.

Table 1. *Exact marginal density of asymptotic ancillary statistic* $Q(x)$ *at* q *for* $\rho_\theta = \surd 8, \surd 16, \surd 32, \surd 64$

$Q = q$	$\rho_\theta = \surd 8$	$\rho_\theta = \surd 16$	$\rho_\theta = \surd 32$	$\rho_\theta = \surd 64$	$N(0,1)$ density
−2	0·07	0·07	0·06	0·06	0·05
−1·5	0·16	0·15	0·15	0·14	0·13
−1	0·29	0·27	0·26	0·26	0·24
−0·5	0·39	0·38	0·37	0·36	0·35
0	0·41	0·40	0·40	0·40	0·40
1	0·19	0·21	0·22	0·23	0·24
1·5	0·08*	0·10	0·11	0·12	0·13
2	0·02*	0·04	0·04	0·05	0·05

Last column gives the $N(0,1)$ density, corresponding to the limiting case $\rho_\theta \to \infty$. *: substantial distortion by end effects.

The observed Fisher information $I(x)$ is

$$I(x) = \{1 - \gamma_\theta Q(x)\} \mathscr{I}_\theta = \rho_\theta(\rho_\theta - q), \tag{5·2}$$

where $\hat{\theta} = \hat{\theta}(x)$ and $q = Q(x)$. We will also use the notation $I(q, \hat{\theta}) = I(x)$ to emphasize the partition of x into the approximate ancillary $Q(x) = q$ and the estimate $\hat{\theta}$. Rather than directly verifying that $\operatorname{var}(\hat{\theta} \mid q) \sim 1/I(q, \theta)$, which is in fact true, we will first make a 'variance stabilizing' transformation of parameter, to put the problem in an approximate translation form, where we can expect our approximation theory to work better. Fraser (1964) makes a similar effort using a different technique.

For a fixed value of q consider the transformation $\theta \to \phi_q$ defined by

$$\frac{d\phi_q}{d\theta} = \surd |I(q, \theta)| = \surd\{\rho_\theta(\rho_\theta - q)\}. \tag{5·3}$$

Equation (2·14) shows that

$$I^{(\phi_q)}(x) = I(x)/\{\rho_\theta(\rho_\theta - q)\} \tag{5·4}$$

for $Q(x) = q$. In terms of the new parameter ϕ_q, the observed Fisher information is one for every x on the level curve $Q(x) = q$. Since we intend to make conditional statements given $Q(x) = q$, it causes no trouble to use a different transformation for each value of q. Relation (1·5) then becomes simply $\operatorname{var}(\hat{\phi}_q|q) \sim 1$. Notice that if this is true, then transforming back to $\hat{\theta}$ gives

$$\operatorname{var}(\hat{\theta}|q) \sim \operatorname{var}(\hat{\phi}_q|q)\left(\frac{d\hat{\theta}}{d\hat{\phi}_q}\right)^2 \sim \left(\frac{d\hat{\theta}}{d\hat{\phi}_q}\right)^2 = \frac{1}{I(x)}, \tag{5·5}$$

which is (1·5). There is one more level of approximation in (5·5) than in $\operatorname{var}(\hat{\phi}_q|q) \sim 1$ which, to reiterate, is one reason for making the transformation (5·3).

Table 2 shows that the quantity $\hat{\phi}_q - \phi_q$ does indeed have nearly the right mean and variance, 0 and 1 respectively, for the cases considered. The worst case is $q = 0$, $\rho_\theta = \sqrt{8}$, for which the variance is 1·10. The case $q = 2$ with $\rho_\theta = \sqrt{8}$ looks terrible, but that is due to the end effect previously mentioned.

Table 2. *Mean and variance of* $\hat{\phi}_q - \phi_q$ *for* $\rho_\theta = \sqrt{8}, \sqrt{16}, \sqrt{32}, \sqrt{64}$.

q	$\rho_\theta = \sqrt{8}$		$\rho_\theta = \sqrt{16}$		$\rho_\theta = \sqrt{32}$		$\rho_\theta = \sqrt{64}$	
−2	−0·02	1·04	−0·01	1·02	−0·00	1·01	−0·00	1·01
−1	−0·02	1·06	−0·01	1·03	−0·00	1·01	−0·00	1·01
0	−0·02	1·10	−0·01	1·04	−0·00	1·02	−0·00	1·01
1	0·05*	0·99*	−0·01	1·05	−0·00	1·02	−0·00	1·01
2	0·49*	0·58*	−0·00	1·06	−0·00	1·03	−0·00	1·01

*: substantial distortion by end effects.

Other moments of $\hat{\phi}_q - \phi_q$ were calculated, all of which indicated good agreement with a standard normal distribution. For example, $E(|\hat{\phi}_q - \phi_q|)$ was within 4%, the worst case again being $q = 0$, $\rho_\theta = \sqrt{8}$. Another advantage of variance stabilizing transformations is that they tend to improve normality. In our two examples, $\hat{\phi}_q - \phi_q$ was more nearly normal than $\hat{\theta} - \theta$. This suggests forming conditional confidence intervals for θ by computing $\hat{\phi}_q \pm z_{\frac{1}{2}p}$, where $z_{\frac{1}{2}p}$ is the upper $\frac{1}{2}p$ point for $N(0, 1)$, and transforming back to the θ scale. This method agrees with $\hat{\theta} \pm z_{\frac{1}{2}p}/\{I(x)\}^{\frac{1}{2}}$ to first order, but can give quite different results for small sample sizes.

To summarize the results for the spiral model, $Q(x)$ contains very little direct information about θ, but its observed value considerably influences the variance of $\hat{\theta}$. For example, (5·2) shows that if $\rho_\theta = \sqrt{16}$, then $I(q, \hat{\theta})$ varies from 24 to 8 as q varies from −2 to 2, causing a threefold change in the variance approximation $1/I(x)$ for $\hat{\theta}$. In other words, $Q(x)$ acts as an effective ancillary statistic.

We now extend the definitions of $Q(x)$ and ϕ_q to an arbitrary one parameter family, say $\mathscr{F} = \{f_\theta(x), \theta \in \Theta\}$, Θ an interval of R^1, satisfying the regularity conditions of § 8.

Define

$$Q(x) = \frac{1 - I(x)/\mathscr{I}_\theta}{\gamma_\theta}, \tag{5·6}$$

which agrees with (5·2). Lemma 2 shows that $\sqrt{n}Q(x) \to N(0, 1)$ as $n \to \infty$, assuming that γ_θ is a continuous function of θ, that is $Q(x)$ is asymptotically ancillary. We have already mentioned that $(\hat{\theta}, \ddot{l}_\theta)$ acts like the sufficient statistic for the two parameter exponential family which best approximates $\mathscr{F}$ near any given point θ in Θ. The statistic $Q(x)$ is the function of $(\hat{\theta}, \ddot{l}_\theta)$ linear in $\ddot{l}_\theta$, for $\hat{\theta}$ fixed, which is asymptotically ancillary. The definition of $Q(x)$ is also motivated by the obvious geometrical considerations of Fig. 5, generalized in § 7.

From (5·6) we can write $I(x) = I(q, \hat{\theta}) = (1-\gamma_\theta q)\mathscr{I}_\theta$ as before, since $I(x)$ is a function of $\hat{\theta}$ and the observed value $q = Q(x)$.

The general definition we will use for ϕ_q is

$$\frac{d\phi_q}{d\theta} = \left(\frac{|I(q,\theta)|}{n}\right)^{\frac{1}{2}}, \tag{5·7}$$

where n is the sample size of a random sample $x_1, \ldots, x_n$ from some member of $\mathscr{F}$. In making this definition, q is considered fixed and θ variable. The mapping $\theta \to \phi_q$ is monotonic over intervals of Θ where $I(q, \theta)$ does not change sign. The possible difficulties of definition at points where $I(q, \theta) = 0$ do not cause trouble in our examples. A discussion of the special nature of such points is given in § 5 of Efron (1978).

In a translation family definition (5·7) automatically produces a linear function of the translation parameter, $\phi_q = c_q + d_q \theta$, if the original θ is any smooth monotonic function of the translation parameter.

By (2·14) and (5·7), the observed Fisher information for ϕ_q is

$$I^{(\phi_q)}(x) = nI(x)/I(q, \hat{\theta}) \tag{5·8}$$

and is n for $Q(x) = q$. That is, $I^{(\phi_q)}(x)$ is constant on the level surface $Q(x) = q$. The choice of the constant equal to n keeps ϕ_q and θ the same order of magnitude. In the example of § 6, as in the spiral example, we verify that $Q(x)$ is close to ancillary, and that $\text{var}(\hat{\phi}_q|q) \sim 1/n$ in accordance with (1·5).

The transformation $\theta \to \phi_q$ defined by (5·7) is mainly of theoretical and conceptual convenience. Practical evidence certainly suggests that the likelihood is often more normal on the ϕ_q scale. But the derivation of ϕ_q is often difficult and usually requires approximation; see § 6. Moreover, if, as we believe, the results of § 4 generalize, then

$$2(l_{\hat{\theta}} - l_\theta)|q = 2(l_{\hat{\phi}_q} - l_{\phi_q})|q = \chi_1^2 + O_p(n^{-1}), \tag{5·9}$$

so that confidence limits for θ can be derived directly from $l_\theta(x)$. We emphasize that (5·9) has not been proved for the general case. Confidence limits for θ can also be determined by taking the quadratic approximation in $(\hat{\theta}-\theta)$ to

$$n^{\frac{1}{2}}(\hat{\phi}_q - \phi_q) = \int_\theta^{\hat{\theta}} \sqrt{|I(q,t)|}\, dt.$$

The numerical results of § 6 suggest that the direct likelihood method based on (5·9) is preferable.

A corresponding treatment of locally most powerful tests of H_0: $\theta = \theta_0$ indicates that the appropriate standardized form of the score statistic is $\dot{l}_{\theta_0}/\{I(x)\}^{\frac{1}{2}}$, which is approximately $N(0, 1)$ conditional on $Q = q$. In this form the score statistic is no more convenient than its asymptotic equivalents, since $\hat{\theta}$ must be computed; for an example, see Hinkley (1977).

What happens when we have r independent sets of samples from the same model? How would we compare the estimates? How would we pool the estimates? Answers to these questions are essentially given by Fisher (1925). Suppose that we have only $(\hat{\theta}_j, I_j)$ for $j = 1, \ldots, r$. Then the appropriate pooled statistic is $(\hat{\theta}_., I_.)$, say, where $\hat{\theta}_. = \Sigma I_j \hat{\theta}_j / \Sigma I_j$ and $I_. = \Sigma I_j$; $\hat{\theta}_.$ is second-order efficient according to Fisher's informal argument. The effective part of $(Q_1, \ldots, Q_r)$, to first order, is presumably $Q_. = (1 + I_./\mathscr{I}_.)\gamma_{\hat{\theta}}^{-1}$, where $\mathscr{I}_.$ is the grand of $Q(x)$ is also motivated by the obvious geometrical considerations of Fig. 5, generalized in § 7.

From (5·6) we can write $I(x) = I(q, \hat{\theta}) = (1-\gamma_\theta q)\mathscr{I}_\theta$ as before, since $I(x)$ is a function of $\hat{\theta}$ and the observed value $q = Q(x)$.

The general definition we will use for ϕ_q is

$$\frac{d\phi_q}{d\theta} = \left(\frac{|I(q,\theta)|}{n}\right)^{\frac{1}{2}}, \tag{5·7}$$

where n is the sample size of a random sample $x_1, \ldots, x_n$ from some member of $\mathscr{F}$. In making this definition, q is considered fixed and θ variable. The mapping $\theta \to \phi_q$ is monotonic over intervals of Θ where $I(q, \theta)$ does not change sign. The possible difficulties of definition at points where $I(q, \theta) = 0$ do not cause trouble in our examples. A discussion of the special nature of such points is given in § 5 of Efron (1978).

In a translation family definition (5·7) automatically produces a linear function of the translation parameter, $\phi_q = c_q + d_q\theta$, if the original θ is any smooth monotonic function of the translation parameter.

By (2·14) and (5·7), the observed Fisher information for ϕ_q is

$$I^{(\phi_q)}(x) = nI(x)/I(q, \hat{\theta}) \tag{5·8}$$

and is n for $Q(x) = q$. That is, $I^{(\phi_q)}(x)$ is constant on the level surface $Q(x) = q$. The choice of the constant equal to n keeps ϕ_q and θ the same order of magnitude. In the example of § 6, as in the spiral example, we verify that $Q(x)$ is close to ancillary, and that $\operatorname{var}(\hat{\phi}_q | q) \sim 1/n$ in accordance with (1·5).

The transformation $\theta \to \phi_q$ defined by (5·7) is mainly of theoretical and conceptual convenience. Practical evidence certainly suggests that the likelihood is often more normal on the ϕ_q scale. But the derivation of ϕ_q is often difficult and usually requires approximation; see § 6. Moreover, if, as we believe, the results of § 4 generalize, then

$$2(l_{\hat{\theta}} - l_\theta)\,|\,q = 2(l_{\hat{\phi}_q} - l_{\phi_q})\,|\,q = \chi_1^2 + O_p(n^{-1}), \tag{5·9}$$

so that confidence limits for θ can be derived directly from $l_\theta(x)$. We emphasize that (5·9) has not been proved for the general case. Confidence limits for θ can also be determined by taking the quadratic approximation in $(\hat{\theta} - \theta)$ to

$$n^{\frac{1}{2}}(\hat{\phi}_q - \phi_q) = \int_\theta^{\hat{\theta}} \sqrt{|I(q,t)|}\,dt.$$

The numerical results of § 6 suggest that the direct likelihood method based on (5·9) is preferable.

A corresponding treatment of locally most powerful tests of $H_0\colon \theta = \theta_0$ indicates that the appropriate standardized form of the score statistic is $\dot{l}_{\theta_0}/\{I(x)\}^{\frac{1}{2}}$, which is approximately $N(0, 1)$ conditional on $Q = q$. In this form the score statistic is no more convenient than its asymptotic equivalents, since $\hat{\theta}$ must be computed; for an example, see Hinkley (1977).

What happens when we have r independent sets of samples from the same model? How would we compare the estimates? How would we pool the estimates? Answers to these questions are essentially given by Fisher (1925). Suppose that we have only $(\hat{\theta}_j, I_j)$ for $j = 1, \ldots, r$. Then the appropriate pooled statistic is $(\hat{\theta}_., I_.)$, say, where $\hat{\theta}_. = \Sigma I_j\hat{\theta}_j/\Sigma I_j$ and $I_. = \Sigma I_j$; $\hat{\theta}_.$ is second-order efficient according to Fisher's informal argument. The effective part of $(Q_1, \ldots, Q_r)$, to first order, is presumably $Q_. = (1 + I_./\mathscr{I}_.)\gamma_\theta^{-1}$, where $\mathscr{I}_.$ is the grand total Fisher information. A reasonable conjecture is that $(\hat{\theta}_., Q_.)$ is equivalent, to second order, to the statistics $(\hat{\theta}, Q)$ computed from the pooled likelihood function. Comparisons of the $\hat{\theta}_j$ would be made conditional on $(Q_1, \ldots, Q_r)$, and would be asymptotically equivalent to normal-theory comparisons with sample weights I_j. For example, in testing the equality of parameter values, the likelihood ratio statistic is

$$W = 2\sum_{j=1}^{r}(l_{j,\theta_j} - l_{j,\theta}) \sim 2\sum_{j=1}^{r}(l_{j,\theta_j} - l_{j,\theta.}) \sim \sum_{j=1}^{r} I_j(\hat{\theta}_j - \hat{\theta}_.)^2$$

under the hypothesis of equality. But the right-hand side is

$$\Sigma I_j(\hat{\theta}_j - \theta)^2 - I_.(\hat{\theta}_. - \theta)^2,$$

which by extension of earlier arguments is approximately χ^2_{r-1} conditional on

$$(Q_1, \ldots, Q_r) = (q_1, \ldots, q_r).$$

6. Example of a nontranslation family

We illustrate the theory of §5 for a simple nontranslation example, using Monte Carlo methods to estimate the conditional properties of the maximum likelihood estimate.

Let $x_i = (x_{1i}, x_{2i})$ for $i = 1, \ldots, n$ be independent bivariate normal pairs with zero mean, unit variances and correlation θ. The two-dimensional sufficient statistic is

$$s_1 = \sum_{i=1}^{n} x_{1i} x_{2i}, \quad s_2 = \sum_{i=1}^{n} (x_{1i}^2 + x_{2i}^2),$$

and the first derivative of the log likelihood function is

$$l_\theta = \frac{n\theta(1-\theta^2) - \theta s_2 + (1+\theta^2)s_1}{(1-\theta^2)^2}. \tag{6·1}$$

Calculations for the Fisher information and curvature (2·11) are straightforward, yielding

$$\mathscr{I}_\theta = n(1+\theta^2)/(1-\theta^2)^2, \quad \gamma_\theta^2 = 4(1-\theta^2)^2/(1+\theta^2)^3.$$

Some numerical values of both $\mathscr{I}_\theta$ and γ_θ^2 are given in Table 3. A qualitative interpretation of the curvature values is that our two-dimensional exponential family model is highly nonlinear for small $|\theta|$, but nearly linear as $\theta \to \pm 1$. The effect of replacing $\mathscr{I}_\theta$ by $I(x)$ is potentially large for small $|\theta|$.

Table 3. *Information, curvature, and parameter ϕ_θ for special bivariate model*

$\|\theta\|$	0	0·1	0·2	0·4	0·6	0·8	0·9	0·95	1
$\mathscr{I}_\theta/n$	1	1·03	1·13	1·64	3·32	12·65	50·14	200	∞
γ_θ^2	4·00	3·81	3·28	1·81	0·65	0·12	0·024	0·0055	0
ϕ_0	0	0·100	0·204	0·435	0·737	1·235	1·727	2·217	∞

The variance stabilizing transformation $\theta \to \phi_q$ defined by (5·7) is equivalent to

$$\phi_q = n^{-\frac{1}{2}} \int^\theta \mathscr{I}_t^{\frac{1}{2}}(1 - q\gamma_t)^{\frac{1}{2}}\, dt.$$

As in many examples it is difficult to evaluate this transformation exactly, but a good approximation can be obtained by substituting $(1-q\gamma_t)^{\frac{1}{2}} \simeq 1 - \frac{1}{2}q\gamma_t$; recall that q is $O(n^{-\frac{1}{2}})$.

In the present case this substitution leads simply to

$$\phi_q \simeq \phi_0 - q \tan^{-1}\theta, \tag{6·2}$$

where

$$\phi_0 = 2^{\frac{1}{2}} \tanh^{-1}(\xi_\theta 2^{\frac{1}{2}}) - \tanh^{-1}(\xi_\theta), \quad \xi_\theta = \theta(1+\theta^2)^{-\frac{1}{2}}.$$

The normalizing effect of the transformation $\theta \to \phi_q$ is illustrated by plots of likelihoods and their normal approximations in Fig. 6 for a small data set with $n = 20$, $s_1 = 12$, $s_2 = 35$. In each case the likelihood is graphed relative to its maximum. The observed informations, respectively $I(x)$ and n, are used as variance inverses in the normal approximations, which are centred on the maximum likelihood estimates.

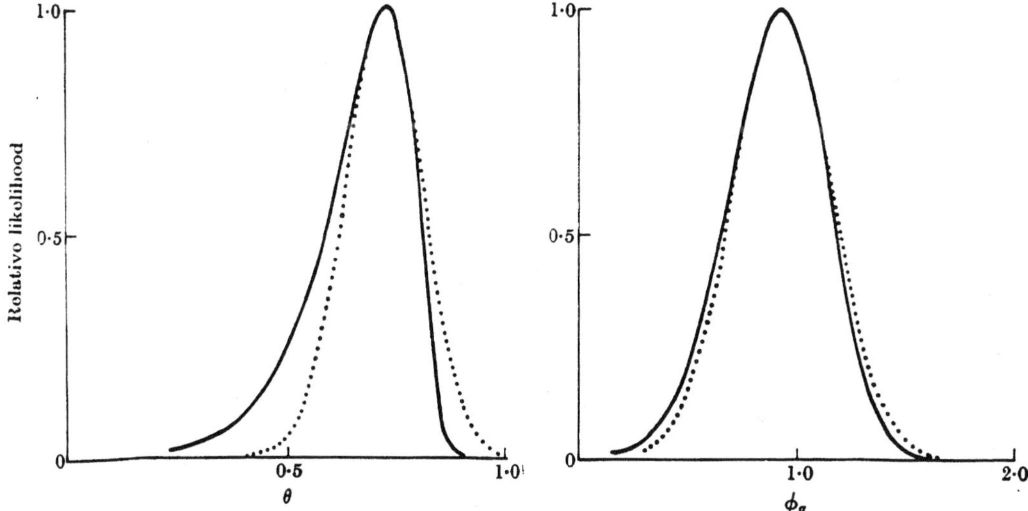

Fig. 6. Relative likelihood functions shown by solid curves, and normal approximations, dotted curves, for correlation θ and transformed parameter ϕ_q for bivariate normal sample with known $N(0, 1)$ marginal distributions. Data: $n = 20$, $s_1 = 12$, $s_2 = 35$, $\hat{\theta} = 0{\cdot}7185$, $I(x) = 122{\cdot}77$, $q = 0{\cdot}101$, $\hat{\phi}_q = 0{\cdot}928$.

Our interest is in whether $Q(x)$ defined by (5·2) is approximately ancillary, and whether $\text{var}(\hat{\phi}_q | q) \sim 1/n$ is accurate. Notice that ϕ_0 is the transformation which makes $\mathscr{I}_\theta = n$, so that the superiority of $I(x)$ over $\mathscr{I}_\theta$ as a measure of precision conditional on $Q(x) = q$ may be judged by comparing the conditional variances of $\hat{\phi}_q$ and $\hat{\phi}_0$. The preceding theory would indicate that conditional on q

$$\frac{\text{var}(\hat{\phi}_q | q) - 1/n}{\text{var}(\hat{\phi}_0 | q) - 1/n} = O_p(n^{-\frac{1}{2}}),$$

but we have not proved this.

The likelihood equation $l_\theta = 0$ has three solutions, two of which may be complex; the frequency of multiple real zeros increases with curvature and with q. We computed $\hat{\theta}$ as a solution to the likelihood equation by iterating from an efficient estimate of θ, not the sample correlation. We simulated samples for n between 15 and 40, with θ ranging between 0 and 0·9. The numbers of samples were 10,000, 50,000 and 10,000 for $n = 15$, 25 and 40 respectively. In each case results were recorded for twenty interval values of q in the 99% range $-2 \leqslant q \leqslant +2$, there being approximately the same number of samples for each q interval.

From the simulation results it was quickly apparent that the range of values of θ for which the approximate theory of § 5 is accurate depends markedly on sample size. For that

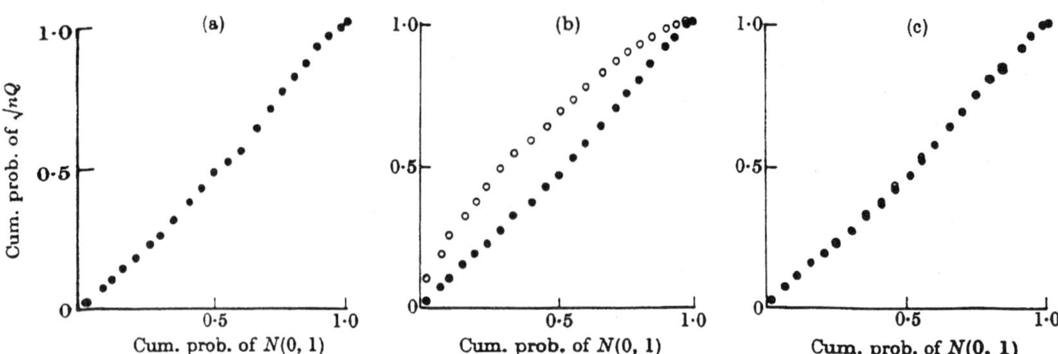

Fig. 7. Empirical cumulative probabilities of approximate ancillary Q in correlation model against $N(0, 1)$ cumulative probabilities. (a) $n = 15$, $\theta = 0, 0{\cdot}3$; (b) $n = 25$, $\theta = 0, 0{\cdot}3, 0{\cdot}5$, shown by closed circles, and $\theta = 0{\cdot}6$, open circles; (c) $n = 40$, $\theta = 0{\cdot}7$, closed circles, and $\theta = 0{\cdot}9$, open circles. Numbers of samples exceed 10,000.

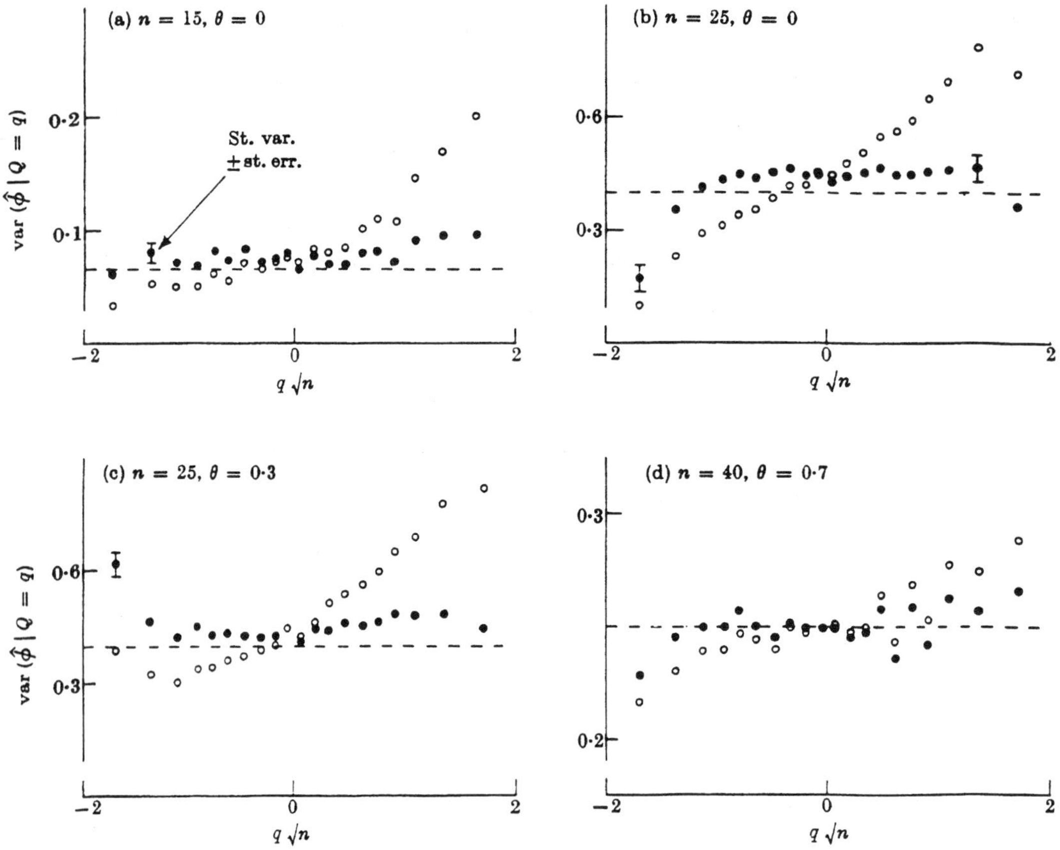

Fig. 8. Monte Carlo estimates of conditional variances of $\hat{\phi}_q$, shown by closed circles, and of $\hat{\phi}_0$, open circles, given $Q = q$ in correlation model. Dashed line, theoretical approximation, $1/n$. Numbers of samples exceed 10,000.

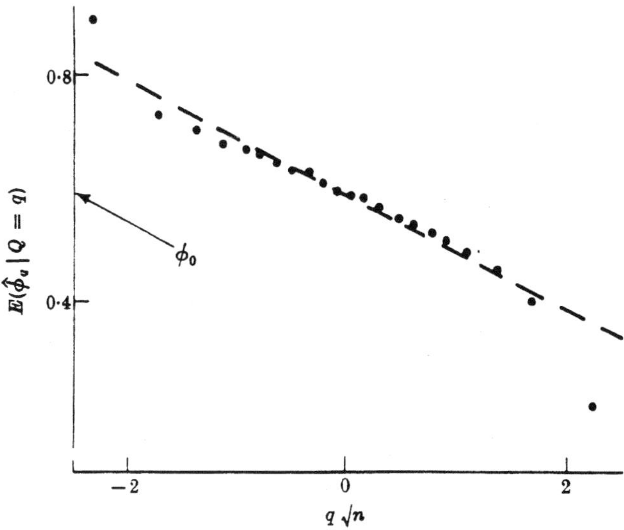

Fig. 9. Monte Carlo estimates of conditional mean of $\hat{\phi}_q$, circles, given $Q = q$, and theoretical approximation, dashed line, in the correlation model with $n = 25$, $\theta = 0{\cdot}5$, Estimates from 50,000 samples.

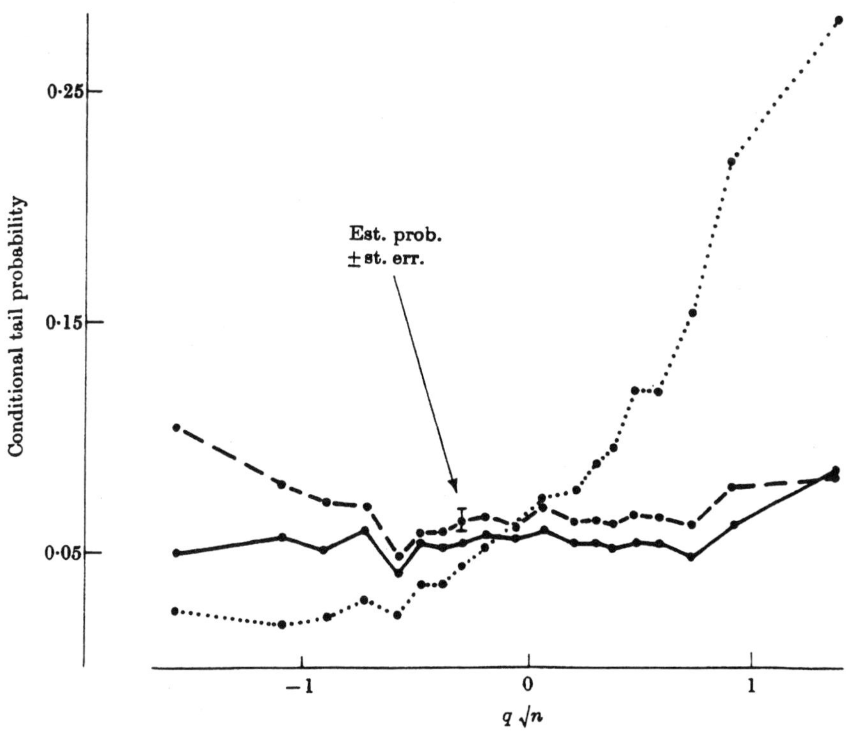

Fig. 10. Monte Carlo estimates of pr (statistic $\geqslant 3{\cdot}84 \mid Q = q$) for three likelihood statistics in correlation model with $n = 25$, $\theta = 0{\cdot}3$. Statistics: $2(l_{\hat{\theta}} - l_\theta)$, shown by solid curve; $n(\hat{\phi}_q - \phi_q)^2$, dashed curve; $n(\hat{\phi}_0 - \phi_0)^2$, dotted curve. Estimates from 50,000 samples.

reason we give a comprehensive set of illustrations here. First, Fig. 7 contains normal plots of the empirical distributions of $Q(x)$, a separate graph for each sample size. Several θ cases are indistinguishable, but clearly as $|\theta| \to 1$ the approximate ancillarity of $Q(x)$ breaks down.

Figure 8 contains plots of empirical conditional variances of both $\hat{\phi}_q$ and $\hat{\phi}_0$ for six representative cases. Standard errors for the estimated variances are indicated. These graphs confirm the theory to a remarkable degree. Particularly striking are the deviations from n^{-1} of the conditional variances of $\hat{\phi}_0$.

The approximation (6·2) is remarkably accurate for the conditional mean of $\hat{\phi}_q$, which implies that the conditional mean of $\hat{\phi}_0$ deviates from ϕ_0. Figure 9 illustrates a typical case.

The final numerical results are concerned with approximate methods for obtaining confidence limits for θ. According to our theory, both $n(\hat{\phi}_q - \phi_q)^2$ and $2(l_{\hat{\theta}} - l_\theta)$ are approximate χ_1^2 variables conditional on q. In contrast $n(\hat{\phi}_0 - \phi_0)^2$, which is an approximate χ_1^2 variable unconditionally, does not have this property conditionally. Figure 10 contains empirical conditional tail probabilities for all three of these statistics corresponding to the value 3·84, nominal 0·05 probability, for the case $n = 25$, $\theta = 0·3$. Our speculative theory is nicely confirmed by these and similar results. As in the Cauchy case, § 4, the likelihood ratio method gives the best agreement with the chi-squared approximation.

Note that even for $n = 40$ conditioning on q is likely to have an appreciable effect, because the coefficient of variation of $I(x)$ is as high as 0·3, its value at $\theta = 0$. Thus at $n = 40$ the unconditional variance approximation $1/\mathscr{I}_\theta$ can easily depart by a factor of two from the conditional variance approximation.

7. Curved exponential families

The definition of the asymptotic ancillary statistic $Q(x)$ at (5·6) is motivated by the geometry of curved exponential families. This section gives a brief description of the geometry involved. More details are given by Efron (1975, 1978).

We begin with a k-dimensional exponential family $\mathscr{G}$, with density functions of the form

$$g_\alpha(x) = \exp\{\alpha^{\mathrm{T}} x - \psi(\alpha)\} \quad (\alpha \in A, x \in \mathscr{X}), \tag{7·1}$$

$\psi(\alpha)$ being a normalizing constant. The natural parameter space A and the sample space $\mathscr{X}$ are both subsets of R^k, A being convex. Corresponding to each α is the mean vector and covariance matrix of x,

$$\beta = E_\alpha(x), \quad \Omega_\alpha = \mathrm{cov}_\alpha(x). \tag{7·2}$$

The mapping from α to β is one to one, so $\mathscr{G}$ can just as well be indexed by β as by α. The space $B = \{\beta(\alpha) : \alpha \in A\}$ is not necessarily convex.

A curved exponential family $\mathscr{F}$ is a one parameter subset of $\mathscr{G}$, with typical density function say

$$f_\theta(x) = \exp(\alpha_\theta^{\mathrm{T}} x - \psi_\theta), \quad \psi_\theta = \psi(\alpha_\theta). \tag{7·3}$$

Here θ is a real parameter contained in Θ, an interval of R^1, and the mapping $\theta \to \alpha_\theta$ is assumed to be continuously twice differentiable; $\mathscr{F}$ is fully described by the curve $\mathscr{F}_A = \{\alpha_\theta : \theta \in \Theta\}$ through A, or equivalently by the curve $\mathscr{F}_B = \{\beta_\theta = \beta(\alpha_\theta) : \theta \in \Theta\}$ through B. All of our examples, except for those in § 1, involve curved exponential families.

If $\mathscr{G}$ is two-dimensional, as in Figs 3 and 5, then for a given $\hat{\theta}$, the set $Q(x) = q$ is a single vector. It is shown in § 5 of Efron (1975) that this vector v has squared Mahalanobis distance q^2 from $\mathscr{F}_B$ in the inner product $\Omega_{\hat{\theta}}^{-1}$: $(v - \beta_{\hat{\theta}})^{\mathrm{T}} \Omega_{\hat{\theta}}^{-1} (v - \beta_{\hat{\theta}}) = q^2$. This generalizes the geometric description of $Q(x)$ given in Fig. 5, where Ω_θ is the identity. A similar interpretation holds for higher dimensional families $\mathscr{G}$. The Cauchy translation problem may be thought of as a limiting case of a curved exponential family, as remarked at the end of § 5 of Efron (1975).

We use the notation $\Omega_\theta = \Omega_{\alpha_\theta}$, and also $\dot\alpha_\theta = d\alpha_\theta/d\theta$, $\ddot\alpha_\theta = d^2\alpha_\theta/d\theta^2$.

Suppose $x_1, \ldots, x_n$ is a random sample from some member f_θ of $\mathscr{F}$. The average vector $\bar{x} = \Sigma x_i/n$ is a sufficient statistic for θ, and it is easy to verify that the derivatives of the log likelihood function are

$$\dot{l}_\theta(\bar{x}) = n\dot\alpha_\theta^{\mathrm{T}}(\bar{x}-\beta_\theta), \quad \ddot{l}_\theta(\bar{x}) = n\ddot\alpha_\theta^{\mathrm{T}}(\bar{x}-\beta_\theta) - \mathscr{I}_\theta, \tag{7·4}$$

where $\mathscr{I}_\theta = n\dot\alpha_\theta^{\mathrm{T}}\,\Omega_\theta\,\dot\alpha_\theta$ is the Fisher information.

Figure 11 illustrates two useful facts about solutions to the maximum likelihood equation $\dot{l}_\theta(\bar{x}) = 0$, both of which follow from (7·4).

(i) Given $\hat\theta$, the set of $\bar{x}$ vectors for which $\dot{l}_{\hat\theta}(\bar{x}) = 0$, that is for which $\hat\theta$ is a solution to the likelihood equations, is the $k-1$-dimensional hyperplane through $\beta_{\hat\theta}$ orthogonal to $\dot\alpha_{\hat\theta}$, say

$$\mathscr{L}_{\hat\theta} = \{\bar{x}\colon \dot\alpha_{\hat\theta}^{\mathrm{T}}(\bar{x}-\beta_{\hat\theta}) = 0\}. \tag{7·5}$$

(ii) From (5·6) and (7·4),

$$Q(x) = n\ddot\alpha_{\hat\theta}^{\mathrm{T}}(\bar{x}-\beta_{\hat\theta})/(\gamma_{\hat\theta}\mathscr{I}_{\hat\theta}). \tag{7·6}$$

Therefore for a given $\hat\theta$, the set of $\bar{x}$ vectors for which $Q(x) = q$ is the $k-2$-dimensional hyperplane contained in $\mathscr{L}_{\hat\theta}$ and orthogonal to $\delta_{\hat\theta}$, the projection of $\ddot\alpha_{\hat\theta}$ into $\mathscr{L}_{\hat\theta}$.

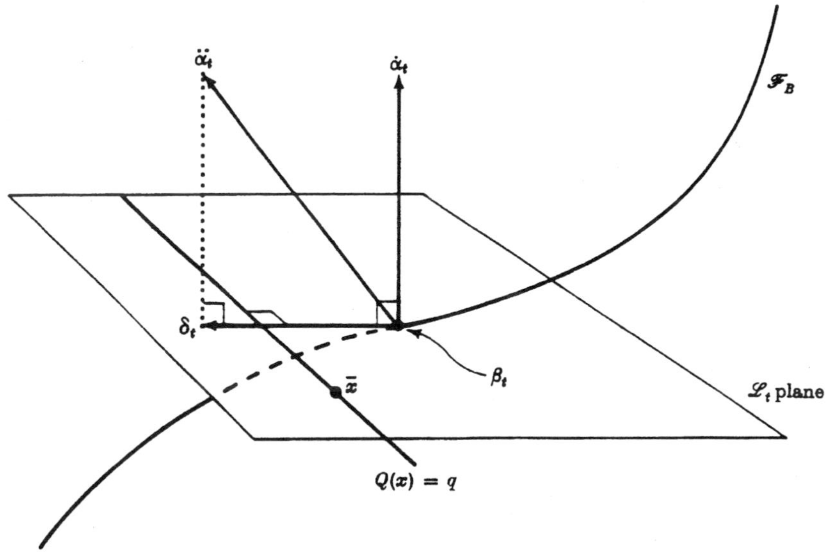

Fig. 11. Geometry of maximum likelihood estimation in a three-dimensional curved exponential family. Curve $\mathscr{F}_B$, values of $\beta_\theta = E_\theta(\bar{x})$; plane $\mathscr{L}_t$ orthogonal to $\dot\alpha_t$, vectors x for which $\hat\theta = t$; vector δ_t, projection of $\ddot\alpha_t$ in $\mathscr{L}_t$, is q axis when $\hat\theta = t$.

So far we have discussed two 'coordinates' of $\bar{x}$ of particular interest, namely $\hat\theta$ and q. The remaining $k-2$ coordinates necessary to specify $\bar{x}$ completely are higher order ancillaries, corresponding to the $l_{\hat\theta}^{(\cdot j)}$ for the translation problem. We can replace $\bar{x}$ by the sufficient statistic $(\hat\theta, a)$, where a represents the $k-1$ coordinates which locate $\bar{x}$ in $\mathscr{L}_{\hat\theta}$. The coordinate system for a rotates with $\mathscr{L}_{\hat\theta}$, so that the first coordinate of a always corresponds to $Q(x)$. The second coordinate of a is essentially the component of $\bar{x}$ along that part of $\alpha_{\hat\theta}^{(\cdot 3)}$ orthogonal to $\dot{\alpha}_\theta$ and $\ddot{\alpha}_\theta$, orthogonal being defined with respect to the inner product Ω_θ. This process of definition can be continued so that each successive component of a is less important in describing local behaviour of the likelihood function near $\hat{\theta}$, and so that $a\sqrt{n} \to N_{k-1}(0, V_a)$ as $n \to \infty$. In other words, a is asymptotically ancillary. In curved exponential families, Lemma 2 can be extended to give this stronger result.

8. Details of theoretical results

We describe here proofs of Lemmas 1 and 2 of § 2, and Lemma 3 of § 4, together with some incidental remarks about the Cauchy location example discussed in §§ 1 and 4.

Lemmas 1 and 3 relate to expansions for conditional expectations of the form $E\{k(t)\,|\,a\}$, where the conditional density of $t = \hat{\theta} - \theta$ is given by (2·1) and (2·3). These expansions are deterministic numerical approximations of the form

$$E\{k(t)\,|\,a\} = v(a) + r(a),$$

where for a given s and any $\varepsilon > 0$

$$\lim_{n\to\infty} \mathrm{pr}_\theta\{|n^s r(a)| < \varepsilon\} = 0.$$

We write $r(a) = o_p(n^{-s})$ to express this. Most of the theory relies on standard asymptotic results for regular likelihoods, a particularly useful reference being Walker (1969). Sufficient conditions for each result are given at the end of the proof.

For Lemma 1, consider the conditional mean squared error of $t = \hat{\theta} - \theta$. By (2·1) and (2·3) we may write

$$E(t^2\,|\,a) = \int_{-\infty}^{\infty} t^2 h_a(t)\,dt \Big/ \int_{-\infty}^{\infty} h_a(t)\,dt = \int_{-\infty}^{\infty} t^2 \exp(l_{\theta-t} - l_\theta)\,dt \Big/ \int_{-\infty}^{\infty} \exp(l_{\theta-t} - l_\theta)\,dt, \quad (8\cdot1)$$

where $l_\theta = l_{\theta(x)}(x)$. If both integrals here are finite, as we assume, then they may be approximated arbitrarily closely by the corresponding integrals truncated at $t = \pm b(a)$ for suitably large finite $b(a)$. We choose $b(a)$ so that the error incurred in (8·1) is $o_p(n^{-2})$.

The next simplification follows from the fact that for arbitrary $\delta > 0$ there is a $c_\delta > 0$ such that

$$\lim_{n\to\infty} \mathrm{pr}_\theta\{\sup_{|t|>\delta} n^{-1}(l_{\theta-t} - l_\theta) < -c_\delta\} = 1,$$

a result essentially given by Walker (1969, § 3). This result implies that the contributions to the integrals in (8·1) from $\delta < |t| \leqslant b(a)$ are $O_p(e^{-nc_\delta})$, certainly $o_p(n^{-k})$ for all k, for all $\delta > 0$. Our problem then reduces to computing for arbitrarily small δ the truncated integrals

$$N(\delta, a) = \int_{-\delta}^{\delta} t^2 \exp(l_{\theta-t} - l_\theta)\,dt, \quad D(\delta, a) = \int_{-\delta}^{\delta} \exp(l_{\theta-t} - l_\theta)\,dt. \quad (8\cdot2)$$

It will be convenient to write $c_j = (-1)^{j+1} l_{\hat{\theta}}^{(\cdot j)}$ for $j = 1, 2, \ldots$, where $c_2 = I(x)$ and $c_1 = 0$, assuming $\hat{\theta}$ to be a stationary point of l_θ. Then we have the Taylor expansion

$$l_{\theta-t} - l_\theta = -\tfrac{1}{2}c_2 t^2 - \tfrac{1}{6}c_3 t^3 - \tfrac{1}{24}c_4 t^4(1 + \varepsilon_n), \quad (8\cdot3)$$

where

$$\varepsilon_n = (l_{\theta_1}^{(\cdot 4)} - l_{\hat{\theta}}^{(\cdot 4)})/c_4, \quad \theta_1 \in (\hat{\theta} - t, \hat{\theta}). \quad (8\cdot4)$$

Under a continuity condition on $l_\theta^{(.4)}$, ε_n will be $o_p(1)$. We now use (8·3) to expand the integrals in (8·2) about the leading normal density term. To do this, let

$$z = t\sqrt{c_2}, \quad w = w(z) = \tfrac{1}{6}c_3 t^3 + \tfrac{1}{24}c_4 t^4(1+\varepsilon_n), \tag{8·5}$$

and note that

$$1 - w + \tfrac{1}{2}w^2 - \tfrac{1}{6}|w|^3 \leqslant e^{-w} \leqslant 1 - w + \tfrac{1}{2}w^2 + \tfrac{1}{6}|w|^3. \tag{8·6}$$

Substitution of (8·5) and use of (8·6) for the first integral in (8·2) gives

$$\int_{-\delta\sqrt{c_2}}^{\delta\sqrt{c_2}} z^2(1 - w + \tfrac{1}{2}w^2 - \tfrac{1}{6}|w|^3)\,\phi(z)\,dz \leqslant c_2^{3/2} N(\delta, a)/(2\pi)^{\frac{1}{2}}$$

$$\leqslant \int_{-\delta\sqrt{c_2}}^{\delta\sqrt{c_2}} z^2(1 - w + \tfrac{1}{2}w^2 + \tfrac{1}{6}|w|^3)\,\phi(z)\,dz, \tag{8·7}$$

where $\phi(z)$ is the $N(0,1)$ density. Next replace the integration limits $\pm\delta\sqrt{c_2}$ by $\pm\infty$, which incurs an error of $O_p(e^{-n})$ because $c_2^{-1} = O_p(n^{-1})$. Then the integrals in (8·7) simply involve the first twelve moments of the $N(0,1)$ distribution. Using the fact that $c_j/c_2^s = O_p(n^{1-s})$, calculation of the bounds in (8·7) immediately leads to

$$N(\delta, a) = \frac{(2\pi)^{\frac{1}{2}}}{c_2^{3/2}}\left\{1 - \frac{15}{24}\frac{c_4}{c_2^2} + \frac{105}{72}\frac{c_3^2}{c_2^3} + O_p(n^{-2}) + O_p(n^{-1}\varepsilon_n)\right\}. \tag{8·8}$$

The corresponding evaluation of the second integral in (8·2) gives

$$D(\delta, a) = \frac{(2\pi)^{\frac{1}{2}}}{c_2^{\frac{1}{2}}}\left\{1 - \frac{3}{24}\frac{c_4}{c_2^2} + \frac{5}{4}\frac{c_3^2}{c_2^3} + O_p(n^{-2}) + O_p(n^{-1}\varepsilon_n)\right\}. \tag{8·9}$$

Finally, noting that the magnitudes of earlier truncation errors are smaller than those in (8·8) and (8·9), and that $\varepsilon_n = o_p(1)$, we substitute for numerator and denominator in (8·1) and simplify to obtain (2·8).

The corresponding calculation for $E(t\,|\,a)$ is very similar and need not be given here. The conditional variance (2·7) is simply $E(t^2\,|\,a) - \{E(t\,|\,a)\}^2$.

The essential conditions for these results to hold are, first, that $E\{(\hat\theta - \theta)^2\} < \infty$, so that the integrals in (8·1) exist. Secondly, the first four derivatives of $\log f_\theta(x)$ with respect to θ exist in an open neighbourhood of the true θ and have finite expectations; the second derivative has strictly negative expectation, $\mathscr{I} > 0$. Also we require a condition on $l_\theta^{(.4)}(x)$ to ensure that ε_n in (8·4) is $o_p(1)$. Since in (8·4) we have $|\hat\theta_1 - \hat\theta| < \delta$, and $|\theta - \hat\theta| < \delta$ for suitably large n, it is sufficient to assume that for $|\theta_0 - \theta| < \delta$

$$|l_{\theta_0}^{(.4)}(x_1) - l_\theta^{(.4)}(x_1)| < M_\delta(x_1, \theta), \quad \lim_{\delta\to 0} E_\theta M_\delta(x_1, \theta) \to 0.$$

A similar condition was used by Walker (1969).

Under stronger regularity conditions the remainder terms in (2·7)–(2·8) can be shown to be $O_p(n^{-2})$ rather than $o_p(n^{-1})$.

Lemma 3 is proved in much the same way as Lemma 1, using Laplace transforms. Because of the very similar natures of (4·3) and (4·4), we discuss only the latter.

Consider, then, the log likelihood ratio statistic $v(x) = 2(l_{\hat\theta} - l_\theta) = 2(l_{\hat\theta} - l_{\hat\theta - t})$. The conditional moment generating function of $v(x)$ is, by (2·1) and (2·3),

$$E(e^{sv(x)}\,|\,a) = \int_{-\infty}^{\infty} \exp\{(1-2s)(l_{\hat\theta - t} - l_{\hat\theta})\}\,dt \Big/ \int_{-\infty}^{\infty} \exp(l_{\hat\theta - t} - l_{\hat\theta})\,dt. \tag{8·10}$$

We assume that $s < \frac{1}{2}$. Approximation to both integrals is accomplished just as in Lemma 1, after which (8·3) is used for $|t| < \delta$ with ε_n as in (8·4). A calculation parallel to that leading from (8·2) to (8·8) gives

$$\int_{-\delta}^{\delta} \exp\{(1-2s)(l_{\hat\theta-t}-l_{\hat\theta})\}\,dt = \left(\frac{2\pi}{c_2}\right)^{\frac{1}{2}} \frac{1}{(1-2s)^{\frac{1}{2}}}\left\{1-\left(\frac{1}{8}\frac{c_4}{c_2^2}-\frac{5}{24}\frac{c_3^2}{c_2^3}\right)\left(\frac{1}{1-2s}\right)+O_p(n^{-1}\varepsilon_n)\right\}, \tag{8·11}$$

with the c_j as in Lemma 1. Substitution of (8·11) in the numerator and denominator ($s = 0$) of (8·10) gives

$$E(e^{sv(x)}\,|\,a) = \frac{1}{(1-2s)^{\frac{1}{2}}}\left\{1+\frac{1}{8}\frac{c_4}{c_2^2}-\frac{5}{24}\frac{c_3^2}{c_2^3}-\left(\frac{1}{8}\frac{c_4}{c_2^2}-\frac{5}{24}\frac{c_3^2}{c_2^3}\right)\left(\frac{1}{1-2s}\right)+o_p(n^{-1})\right\} \tag{8·12}$$

for $s < \frac{1}{2}$; the $o_p(n^{-1})$ term is a bounded function of s for $s \leqslant \frac{1}{2}-\eta$, $\eta > 0$.

Formal inversion of (8·12) gives the result (4·4). The necessary regularity conditions are clearly the same as those given for Lemma 1, and in addition we assume that $E_\theta[\exp\{sv(x)\}] < \infty$ for $s < \frac{1}{2}$.

The $o_p(n^{-1})$ terms in (4·3) and (4·4) will be $O_p(n^{-2})$ under stronger regularity conditions.

The second-order expansions in Lemmas 1 and 3 help to explain the deviations from first-order approximations apparent in Figs 1 and 4 for the Cauchy case. To show this informally we argue as follows. By symmetry $E_\theta\{l_\theta^{(\cdot 3)}\,|\,I(x)\} = 0$, so that $(l_\theta^{(\cdot 3)})^2\,|\,I(x) = O_p(n)$. Therefore, taking expectations with respect to a conditional on $I(x)$ in Lemmas 1 and 3, we have that

$$\operatorname{var}(\hat\theta\,|\,I) \sim \frac{1}{I}\left\{1+\frac{1}{2}\frac{K(I)}{I^2}\right\}, \tag{8·13}$$

$$\begin{aligned} \operatorname{pr}\{u(x)\leqslant c\,|\,I(x)\} &\sim \left\{1+\frac{1}{8}\frac{K(I)}{I^2}\right\}\operatorname{pr}(\chi_1^2\leqslant c)+\frac{1}{8}\frac{K(I)}{I^2}\operatorname{pr}(\chi_5^2\leqslant c),\\ \operatorname{pr}\{v(x)\leqslant c\,|\,I(x)\} &\sim \left\{1+\frac{1}{8}\frac{K(I)}{I^2}\right\}\operatorname{pr}(\chi_1^2\leqslant c)+\frac{1}{8}\frac{K(I)}{I^2}\operatorname{pr}(\chi_3^2\leqslant c), \end{aligned} \tag{8·14}$$

where $K(I) = E\{l_\theta^{(\cdot 4)}\,|\,I(x)\}$.

Now suppose that $K(I) \sim b_1 n + b_2 I$, so that

$$E\{l_\theta^{(\cdot 4)}\} = E\{K(I)\} \sim b_1 n + b_2 E\{I(x)\} \sim b_1 n + b_2 \mathscr{I}.$$

For the Cauchy distribution $E\{l_\theta^{(\cdot 4)}\} = \mathscr{I}$, so that $b_1 \sim 0$ and $b_2 \sim 1$. The implied form of (8·13) is $\operatorname{var}(\hat\theta\,|\,I) \sim I^{-1}\{1+1/(2I)\}$, which is a very good approximation to the empirical variances in Fig. 1, and is for the majority of cases at $n = 10$. The implied forms of (8·14) are also very accurate. Note that (8·14) also explains the tendency for $v(x)$ to be closer to χ_1^2 than is $u(x)$, because χ_3^2 is stochastically smaller than χ_5^2.

For Lemma 2 sufficient regularity conditions are stated in the last paragraph, following the formal derivation. First notice that since $I(x)/\mathscr{I}_\theta$ is invariant under monotone reparameterizations, by (2·14), we can change to the parameter σ defined by $d\sigma/d\theta = \mathscr{I}_\theta/n$, for which $\mathscr{I}^{(\sigma)} = n$. We might as well assume this parameterization to begin with, so $\mathscr{I}_\theta = n$ for all θ. This implies $\nu_{20}(\theta) = 1$ since, by definition, $\nu_{20}(\theta)$ is the Fisher information in a single observation. For notational convenience, let $\theta = 0$ be the true value of the parameter. Then we wish to show that

$$\sqrt{n}\left(\frac{I(x)}{n}-1\right) = \frac{I(x)-n}{\sqrt{n}} \to N(0,\gamma_0^2) \tag{8·15}$$

under independent sampling from $f_0(x)$.

Let $S(x) = \{-\ddot{l}_0(x) - n - \nu_{11}(0)\,\dot{l}_0(x)\}/\sqrt{n}$. Because $\nu_{11}(0)$ is the regression coefficient of $-\ddot{l}_0$ on $\dot{l}_0$, and by definition (2·11) and the preceding definitions, it is easy to see that $S(x) \to N(0, \gamma_0^2)$. The proof is completed by showing that $S(x)$ is asymptotically equivalent to $\{I(x) - n\}/\sqrt{n}$.

Notice that by differentiating $\nu_{20}(\theta) = 1 = E_\theta\{-\ddot{l}_\theta(x)\}$ with respect to θ we get that

$$E_\theta\{-l_\theta^{(\cdot 3)}(x)\} = \nu_{11}(\theta). \tag{8·16}$$

The strong law of large numbers then implies that $-l_\theta^{(\cdot 3)}(x)/n = \nu_{11}(\theta)\,(1+\varepsilon_n)$, where $\varepsilon_n \to 0$ almost everywhere. In what follows, ε_n will stand for any sequence of random variables converging to zero almost everywhere. The standard proof for the asymptotic normality of $\hat\theta$, as in Rao (1973, § 5f), shows that $\hat\theta = \{\dot{l}_0(x)/n\}\,(1+\varepsilon_n)$. These results imply that the expansion $\ddot{l}_{\hat\theta}(x) = \ddot{l}_0(x) + \hat\theta l_0^{(\cdot 3)}(x) + \frac{1}{2}\hat\theta^2 l_{\theta_1}^{(\cdot 4)}(x)$, $\theta_1 \in (0, \hat\theta)$, can be rewritten as

$$\frac{-\ddot{l}_{\hat\theta}(x) - n}{\sqrt{n}} = S(x) - \varepsilon_n \nu_{11}(0)\frac{\dot{l}_0(x)}{\sqrt{n}} - \hat\theta^2\sqrt{n}\,\frac{l_{\theta_1}^{(\cdot 4)}(x)}{2n}. \tag{8·17}$$

Since $\dot{l}_0(x)/\sqrt{n} \to N(0,1)$, the term $\varepsilon_n \nu_{11}\dot{l}_0(x)/\sqrt{n} \to 0$. The last term in (8·17) is also negligible under a boundedness condition on $l_\theta^{(\cdot 4)}(x)$, completing the proof of (8·15).

The regularity conditions needed in this proof are (i) the usual conditions for the asymptotic normality of the maximum likelihood estimate, see Rao (1973, § 5f); (ii) equality (8·16), or any regularity conditions justifying the differentiation under the integral sign leading to (8·16); (iii) $|l_{\theta_1}^{(\cdot 4)}(x)| < M(x)$ for θ_1 in a neighbourhood of 0, where $E_0 M(x) < \infty$. Then $|l_{\theta_1}^{(\cdot 4)}(x)/n| < \Sigma M(x)/n \to E_0\{M(x)\}$ justifying the last step in the proof. We remark that these conditions are always satisfied in curved exponential families. Less stringent conditions are required for (2·14).

9. Concluding remarks

The thrust of this paper has been to offer what we believe to be convincing evidence that there is a meaningful approximate conditional inference for single parameters based on the maximum likelihood estimate $\hat\theta$ and the observed information $I(x)$. For the most part the evidence has been empirical, although in the location case the theory flows directly from Fisher's exact calculations. In the nonlocation case the discussions of §§ 5 and 7 demonstrate the existence of a dominant approximate ancillary, together with a convenient framework of curved exponential families for further work. We have not obtained a formal proof generalizing the results of §§ 2 and 4, although we do not doubt that this is possible.

A careful reading of Fisher's work on likelihood estimation suggests that the emphasis on $\mathscr{I}_\theta$ is due to the emphasis on *a priori* comparisons among estimators. The use of $I(x)$ in the interpretation of given data sets is recommended, and some readers of Fisher's work have inferred (1·5); see, for example, the biography by Yates & Mather (1963, p. 100). Relevant work in the context of nonlinear regression may be found in papers by Beale (1960) and Bliss & James (1966), both of which mention some of the geometric ideas underlying our own approach in §§ 5 and 7. A different approach to the problem is due to Fraser (1964). A useful general article on likelihood inference is by Sprott (1975), which summarizes much of the work by G. A. Barnard, J. D. Kalbfleisch, Sprott himself and others.

Not all of the examples that we considered have been included here. We also looked at the $N(\theta, c\theta^2)$ case discussed by Hinkley (1977); a two-parameter linear model version of Fisher's hyperbola; and the double-exponential location model, where the theory of § 2 must fail, but does so in an intriguing manner.

We have not attempted to discuss the case of several parameters, which raises certain problems of definition. However, the extension of our results to the case of the general regular location-scale model is straightforward, again because of duality between conditional distribution and likelihood (Hinkley, 1978).

References

BEALE, E. M. L. (1960). Confidence regions in non-linear estimation (with discussion). *J. R. Statist. Soc.* B **22**, 41–88.

BLISS, C. I. & JAMES, A. T. (1966). Fitting the rectangular hyperbola. *Biometrics* **22**, 573–602.

COX, D. R. (1958). Some problems connected with statistical inference. *Ann. Math. Statist.* **29**, 357–72.

COX, D. R. & HINKLEY, D. V. (1974). *Theoretical Statistics*. London: Chapman and Hall.

EFRON, B. (1975). Defining the curvature of a statistical problem (with applications to second order efficiency) (with discussion). *Ann. Statist.* **3**, 1189–242.

EFRON, B. (1978). The geometry of exponential families. *Ann. Statist.* **6**, 362–76.

FISHER, R. A. (1925). Theory of statistical estimation. *Proc. Camb. Phil. Soc.* **22**, 700–25.

FISHER, R. A. (1934). Two new properties of mathematical likelihood. *Proc. R. Soc.* A **144**, 285–307.

FISHER, R. A. (1974). *Statistical Methods and Scientific Inference*, 3rd edition. Edinburgh: Oliver & Boyd.

FRASER, D. A. S. (1964). Local conditional sufficiency. *J. R. Statist. Soc.* B **26**, 52–62.

HINKLEY, D. V. (1977). Conditional inference about a normal mean with known coefficient of variation. *Biometrika* **64**, 105–8.

HINKLEY, D. V. (1978). Likelihood inference about location and scale parameters. *Biometrika* **65**, 253–61.

RAO, C. R. (1973). *Linear Statistical Inference and its Applications*, 2nd edition. New York: Wiley.

SPROTT, D. A. (1975). Application of maximum likelihood methods to finite samples. *Sankhyā* **37** B, 259–70.

WALKER, A. M. (1969). On the asymptotic behaviour of posterior distributions. *J. R. Statist. Soc.* B **31**, 80–8.

YATES, F. & MATHER, K. (1963). Ronald Aylmer Fisher 1890–1962. *Biog. Mem. Fellows R. Soc., London* **9**, 91–120.

[*Received February* 1978. *Revised June* 1978]

Comments on paper by B. Efron and D. V. Hinkley

BY OLE BARNDORFF-NIELSEN

Department of Theoretical Statistics, Aarhus University

Ever since Fisher's (1925) first discussion of ancillarity it has been a reigning impression that the observed information $I(x)$ is, in general, an approximate ancillary statistic. Parts of Efron & Hinkley's (1978) paper seem prone to perpetuate this impression; see the remarks immediately after formulae (1·5) and (1·6). It is, however, false. The statistic $I(x)$ may be ancillary and may capture most or all of the relevant ancillary information available for inference on θ, such as is the case in the Cauchy example and for Fisher's circle and hyperbola models. Note that the possible ancillarity properties of $I(x)$ depend on the parameterization chosen. Thus in Fisher's hyperbola model the observed information relative to the parameter e^{θ}, which was the original parameter in Fisher's discussion, is not ancillary. At the other extreme, $I(x)$ may be a minimal sufficient statistic and this is often the case for linear exponential families: the $\sigma^2\chi^2$ distribution affords an example of this. In relation to this latter example, note that if $x_1, ..., x_n$ is a sample from the $\sigma^2\chi^2$ distribution with one degree of freedom, then $I(x_1, ..., x_n)$ is minimal sufficient for any sample size even though the distribution sampled is, in effect, a translation family. It can also happen that $I(x)$ is constant while there exists a simple and significant ancillary statistic. This can be illustrated for instance by means of the hyperbolic distribution (Barndorff-Nielsen, 1977, 1978).

Incidentally, the conditional distribution (3·7) of the maximum likelihood estimator in Fisher's hyperbola model is closely related to the hyperbolic distribution and, moreover, its properties are quite analogous to those of the von Mises–Fisher distribution; see Barndorff-Nielsen (1978) and Blæsild (1978a, b). In cases like this, the statistic $Q(x) = \{1 - I(x)/\mathscr{I}_\theta\}/\gamma_\theta$, which is proposed in the paper as an approximate ancillary statistic, acts no better than $I(x)$ in capturing the relevant ancillary information.

Furthermore, if t is any, say positive, statistic which is asymptotically normally distributed with mean value $\tau(\theta)$ and variance $\sigma^2(\theta)$, then it will often be the case that $\{1 - t/\tau(\hat{\theta})\}/\delta(\hat{\theta})$, where $\delta^2(\theta) = \sigma^2(\theta)/\tau^2(\theta)$ equals the variance of $t/\tau(\theta)$, is an approximate or asymptotic ancillary statistic. To illustrate this, suppose that in the Fisher hyperbola model we take t equal to either of the two observations, say $t = x_1$; then the exact ancillary $\surd(x_1 x_2)$ is recovered by the construction $\{1 - t/\tau(\hat{\theta})\}/\delta(\hat{\theta})$. Another example is provided by a sample

$$(x_1, y_1), \ldots, (x_n, y_n)$$

from the bivariate normal distribution with $E(x_i) = E(y_i) = 0$, $\operatorname{var}(x_i) = \operatorname{var}(y_i) = 1$ and $E(x_i y_i) = \theta$. After minimal sufficient reduction, $t = \Sigma x_i^2 + \Sigma y_i^2$ is exactly ancillary in mean, and $\{1 - t/(2n)\}/\delta(\hat{\theta})$ is approximately ancillary in distribution. Thus it seems important to investigate whether there exists a likelihood functional t, other than $I(x)$, such that $\{1 - t/\tau(\hat{\theta})\}/\delta(\hat{\theta})$ is generally preferable to $Q(x)$ above as an approximate ancillary statistic.

Fisher repeatedly stressed the function of ancillary statistics as indices of the precision of maximum likelihood estimates. In fact, it may well be that Fisher considered it to be one of the requirements for calling a statistic ancillary that it should allow of an interpretation as such an index. But with the now generally adopted definition of an ancillary statistic, as a function of the observations whose distribution does not depend on the parameter, there is no guarantee that in any given model an ancillary statistic can be meaningfully thought of as a precision index, and counterexamples are easily constructed.

For these and other general reasons it seems best to discuss the basic issues of the paper in a broader setting.

It is broadly useful to consider likelihood functions as specifiable by a position and a shape, the position being given by the maximum likelihood estimate $\hat{\theta}$. In order to find, for a given model, an appropriate approximately ancillary statistic which can serve as an index of precision of the maximum likelihood estimate $\hat{\theta}$ one may first classify the log likelihood functions according to the value of $\hat{\theta}$ and then compare the classes thus determined to see whether they contain approximately the same range of shapes. If so, a 'shape statistic' s, say, which indexes the shapes of this common range is a solution to the problem provided s is, in fact, ancillary to a suitable degree of approximation.

References

Barndorff-Nielsen, O. (1977). Exponentially decreasing distributions for the logarithm of particle size. *Proc. R. Soc., Lond.* A **353**, 401–19.

Barndorff-Nielsen, O. (1978). Hyperbolic distributions and distributions on hyperbolae. *Scand. J. Statist.* **5**. To appear.

Blæsild, P. (1978a). Conditioning with conic sections in the two-dimensional normal distribution. *Ann. Statist.* **6**, To appear.

Blæsild, P. (1978b). A generalization of the exponential distribution to convex sets in R^k. *Scand. J. Statist.* **5**. To appear.

Efron, B. & Hinkley, D. V. (1978). Assessing the accuracy of the maximum likelihood estimator: Observed versus expected Fisher information. *Biometrika* **65**, 457–82.

Fisher, R. A. (1925). Theory of statistical information. *Proc. Camb. Phil. Soc.* **122**, 700–25.

[*Received July* 1978]

Comments on paper by B. Efron and D. V. Hinkley

By A. T. JAMES

Department of Statistics, University of Adelaide

The authors (Efron & Hinkley, 1978) give a number of examples in which the observed information $I(x)$ is preferable to the expected information $\mathscr{I}_\theta$. To investigate whether this is always so, consider an example deliberately chosen so that no conditioning is possible, namely a single sufficient statistic.

We take a slight modification of an example of Cox (1958), but for a different purpose.

Suppose that a measurement is to be made. Ten instruments are available for use, with unbiased normally distributed error, of which nine have unit error variance and one has $\sigma^2 = 0{\cdot}1$. However, we assume that the more accurate instrument has no outward identification and hence may be picked up unrecognized with probability 1/10. When a single measurement is made, the error distribution is then the mixture $0{\cdot}9N(0,1)+0{\cdot}1N(0,0{\cdot}1)$.

According to formula (4) below, the observed information, $I(x) \simeq 3{\cdot}34$, is considerably greater than the reciprocal of the error variance, $\{\text{var}(x)\}^{-1} = (0{\cdot}91)^{-1} \simeq 1{\cdot}10$. On the other hand, the expected information $\mathscr{I}_\theta \simeq 1{\cdot}18$, obtained by numerical integration, is not too far from the reciprocal of the error variance.

An upper bound for the average information

$$\mathscr{I}_\theta = \int [\{(f'(x)\}^2/f(x)]\,dx, \tag{1}$$

where $f(x)$ is the probability density function of the mixture,

$$(1-p)N(0,1)+pN(0,\sigma^2) \quad (0 \leqslant p \leqslant 1,\ \sigma^2 \ll 1), \tag{2}$$

may be obtained by omitting the second term of the mixture from the denominator of the integrand and expanding the square in the numerator. We obtain

$$\mathscr{I}_\theta \leqslant 1+p+p^2(1-p)^{-1}\{\sigma^2(2-\sigma^2)\}^{-3/2}. \tag{3}$$

In the case above of $p = 0{\cdot}1$, $\sigma^2 = 0{\cdot}1$ the bound is approximately 1·23.

Table 1. *Values of: the observed information* $I(x)$, *first entry; the expected information* $\mathscr{I}_\theta$, *middle entry; and the reciprocal of the error variance*, $\{\text{var}(x)\}^{-1}$, *lowest entry*

$(\sigma^2)^{-1}$	$p^{-1} = 30$	$p^{-1} = 25$	$p^{-1} = 20$	$p^{-1} = 15$	$p^{-1} = 10$	$p^{-1} = 5$
5	1·29	1·34	1·42	1·55	1·80	2·43
	1·03	1·04	1·05	1·07	1·11	1·27
	1·03	1·03	1·04	1·06	1·09	1·19
10	1·88	2·05	2·28	2·66	3·34	4·97
	1·04	1·05	1·07	1·10	1·18	1·53
	1·03	1·04	1·05	1·06	1·10	1·22
30	5·61	6·39	7·49	9·16	11·97	17·76
	1·09	1·12	1·18	1·28	1·57	2·84
	1·03	1·04	1·05	1·07	1·11	1·24

The distribution is the mixture given in equation (2). $\mathscr{I}_\theta$ was computed by numerical integration by Dr W. N. Venables who kindly supplied this table; a more extensive version is available from the author.

Dr W. N. Venables has kindly calculated the expected information for various values of p and σ^2 by numerical integration. They may be compared with the observed information

$$I(x) = \frac{(1-p)+p/\sigma^3}{(1-p)+p/\sigma} \tag{4}$$

and the inverse

$$\{\text{var}(x)\}^{-1} = \{(1-p)+p\sigma^2\}^{-1} \tag{5}$$

of the variance in Table 1, which was kindly supplied by him.

The example and inspection of the table show that $\mathscr{I}_\theta$, which equals $\mathscr{I}_\theta$ in this case, is preferable to $I(x)$. Possibilities such as this may explain the equivocation of R. A. Fisher (1925, 1974) concerning $I(x)$ versus $\mathscr{I}_\theta$.

There may be similar cases involving a single conditionally sufficient statistic in which a conditional $\mathscr{I}_\theta$ is preferable to $I(x)$.

References

Cox, D. R. (1958). Some problems connected with statistical inference. *Ann. Math. Statist.* **29**, 357–72.

Efron, B. & Hinkley, D. (1978). Assessing the accuracy of the maximum likelihood estimator: Observed versus expected Fisher information. *Biometrika* **65**, 457–82.

Fisher, R. A. (1925). Theory of statistical estimation. *Proc. Camb. Phil. Soc.* **22**, 700–25.

Fisher, R. A. (1974). *Statistical Methods and Scientific Inference*, 3rd edition. Edinburgh: Oliver & Boyd.

[*Received June* 1978]

Comments on paper by B. Efron and D. V. Hinkley

By G. K. ROBINSON

Australian Road Research Board, Nunawading, Australia

The quoting of observed Fisher information by Efron & Hinkley (1978) rather than expected Fisher information clearly makes a procedure more nearly Bayesian. 'Why not use a proper Bayesian procedure?' is a question likely to be asked.

I think that the various arguments for the necessity of Bayesianity need to be tempered by the practical requirements of data analysis; the image associated with Bayesianity ought to be changed. Adopting what may be described as a pragmatic Bayesian viewpoint, the question of whether it is reasonable to quote the maximum likelihood estimate and the observed information as an inference is straightforward to answer. It is reasonable whenever a reasonable prior distribution gives an approximately normal posterior with mean and variance approximately the maximum likelihood estimate and the reciprocal of the observed information. In particular, if a uniform improper prior distribution seems reasonable then it is only necessary to check the nearness to normality of the posterior distribution. The procedure does not have the whole-hearted support of the arguments for the necessity of Bayesianity, but it does seem sensible. I prefer this way of looking at the procedure to Efron & Hinkley's presentation, because it is easier to see the weaknesses of the procedure and so be guided as to when not to use it.

Reference

Efron, B. & Hinkley, D. V. (1978). Assessing the accuracy of the maximum likelihood estimator: Observed versus expected Fisher information. *Biometrika* **65**, 457–82.

[*Received June* 1978]

Comments on paper by B. Efron and D. V. Hinkley

By D. A. SPROTT

Department of Statistics, University of Waterloo, Ontario

Space limitations confine me to raising my main query concerning this interesting paper (Efron & Hinkley, 1978), namely the emphasis on variance. It appears that the authors are attempting to justify the use of $I(x)$ over $\mathscr{I}_\theta$ by appealing to variance. In following **Fisher's** work so closely the authors appear to have missed Fisher's point about the irrelevance of variance in finite samples; see, for example, Fisher (1950, p. 11.699a; 1925, p. 714; 1956, pp. 135, 136; 1973, p. 158). The problem of the truth or falsity of (1·5) thus appears to be somewhat academic. For example, the distinction between (1·7) and (1·8) and the implications thereof, as well as the accuracy of the confidence intervals of §§4 and 5, are unaffected by the truth or falsity of (1·5). Also, for a single observation from the well-worn Cauchy example, $\mathscr{I}_\theta = \frac{1}{2}$, $I(x) = 2$, and $\text{var}(\hat{\theta}) = \infty$. Here $\mathscr{I}$ is preferable to I, although both are inadequate, while more importantly, $\text{var}(\hat{\theta})$ is clearly irrelevant. Indeed, Fisher's concept of intrinsic accuracy, discussed at length in the above references, was designed explicitly to replace variance in finite samples.

Given the population deemed relevant, i.e. conditional, surely the question of using $\mathscr{I}_\theta$, or I, or either, is assessed by examining the numerical accuracy of the inferences to which they give rise. By numerical accuracy is meant that an alleged confidence interval, for example of 99%, should have an actual confidence coefficient of approximately 99%. And variance considerations would appear to be of no use in attaining or assessing this. Consider a single observation from $\alpha^{-1}e^{-x/\alpha}$. Using the authors' location parameter $\theta = \log \alpha$ and $y = \log x$, $\mathscr{I}_\theta = I_\theta = 1$. Use of $(\hat{\theta} - \theta) \sim N(0, 1)$ assigns a confidence coefficient of 95·2% to an exact 99% confidence interval. In this regard the use of such a location parameter (5·7) is suboptimal. The parameter $\phi = \alpha^{1/3}$, for which $\mathscr{I}_\phi = I_\phi = 9/\hat{\phi}^2$, is preferable. It leads to $3(1 - \phi/\hat{\phi}) \sim N(0, 1)$ giving a confidence coefficient of 98·1% to the above 99% confidence interval. But where does the variance have any relevance to these procedures? Use of $\text{var}(\hat{\theta})$ or $\text{var}(\hat{\phi})$ in place of I or $\mathscr{I}$ results in considerable loss of accuracy and information as can easily be verified. Nor is it even a question of I versus $\mathscr{I}$; it is a question of $(\hat{\phi}, 9/\hat{\phi}^2)$ being a better representation of the data, yielding better numerical accuracy, than $(\hat{\theta}, 1)$.

In general, use of a parameter ϕ for which the family of 'probable' likelihoods is approximately normal, not merely the single observed one as might be inferred, implies that $u = (\hat{\phi} - \phi)\sqrt{I_\phi}$ yields reasonably accurate frequency inferences when used as an $N(0, 1)$ variate (Sprott, 1973, 1975). The more nearly normal are the likelihoods, the greater is this accuracy. This is why $\phi = \alpha^{-1/3}$ is preferable to $\theta = \log \alpha$ in the previous example. To this degree of approximation the estimation problem in such a case would then appear to have been solved. The approximate normality of u in finite samples is closely connected with the normality of the likelihoods. But it is not necessarily closely connected with $\text{var}(\hat{\phi})$. Thus, of what relevance is the truth or falsity of (1·5) and similar formulae?

The foregoing has concentrated on my difficulty with the paper. I do agree, however, that where a quadratic representation of log likelihood is adequate, $I(x)$ would be preferable to $\mathscr{I}$.

References

EFRON, B. & HINKLEY, D. (1978). Assessing the accuracy of the maximum likelihood estimation: Observed versus expected Fisher information. *Biometrika* **65**, 457–82.

FISHER, R. A. (1925). Theory of statistical estimation. *Proc. Camb. Phil. Soc.* **22**, 700–25.

FISHER, R. A. (1950). *Contributions to Mathematical Statistics*. New York: Wiley.

FISHER, R. A. (1956). The analysis of variance with various binomial transformations. *Biometrics* **10**, 130–9.

FISHER, R. A. (1973). *Statistical Methods and Scientific Inference*, 3rd edition. Edinburgh: Oliver & Boyd.

SPROTT, D. A. (1973). Normal likelihood and their relation to large sample theory of estimation. *Biometrika* **60**, 457–65.

SPROTT, D. A. (1975). Application of maximum likelihood methods to finite samples. *Sankhyā* **37**, 259–70.

[*Received May* 1978)

Reply to comments

BY B. EFRON AND D. V. HINKLEY

Professors Barndorff-Nielsen, James and Sprott all introduce examples with sample size $n = 1$ to show that I^{-1} can be a poor variance approximation, to caution against over-emphasis on variance, and to remark on the value of looking at the likelihood function. All are valid general points. Actually, by appropriate choices of h_a in our (2·2) one can have $\text{var}(\hat{\theta}\,|\,a)$ anywhere in $(0, \infty)$ while having $I = 1$. But such examples seem remote from what we think is the context of our paper, namely the derivation of confidence limits when normal approximations to the likelihood are available. In this context variance is undeniably relevant. In §§ 2 and 4 we show that the more nearly normal is the likelihood function, the better is the agreement between I^{-1} and the conditional variance of $\hat{\theta}$, and the more accurate are the resulting normal confidence limits. Incidentally, I approximates the conditional version of $\mathscr{I}_{\theta}$.

If likelihood functions such as that in James's example were the rule, rather than the exception, then the normal approximation theory would not be useful. However, this was not our experience. Repeated sampling, with n as low as 10, seems to induce normality of the likelihood rather quickly. Our examples show this, but further study would be welcome. Nonnormal cases can be handled using higher derivatives, as in Lemma 1, although other approximations may be useful; for example, gamma-type approximations are sometimes useful for scale parameter problems. In the general case we have pointed to the role of Q and the associated parameter transformation in obtaining normal approximations. On the subject of transformations, we refer to the transformation-invariant method of obtaining confidence limits directly from the likelihood ratio (4·2).

Barndorff-Nielsen questions the use of Q. The optimal ancillary A is presumably such that the approximately sufficient $(\hat{\theta}, I)$ is in one–one correspondence to $(\hat{\theta}, A)$. The choice $A = Q$ satisfies this, whereas $A = S_2$ in the correlation model does not and hence loses information. Admittedly a thorough mathematical treatment is lacking.

For repeated sampling of Barndorff-Nielsen's pretty hyperbola location model, the theory of § 2 can be extended to cover the nonhomogeneous conditional distributions that arise. We look forward to seeing numerical results for this and other examples.

To end on an optimistic note, the results of our paper do suggest greater similarity between Fisherian, frequentist and Bayesian theories than was previously granted. We hope that our paper and the discussion will encourage rereading of Fisher's original work.

We thank the discussants.

[*Received July* 1978]

9

Bootstrap Methods: Another Look at the Jackknife

Introduction by David Hinkley
University of California, Santa Barbara

This first paper on the bootstrap is Brad's most-cited work, currently with some 2,500 citations. A Google search throws up more than 300,000 entries for "Efron, bootstrap." I think it not an exaggeration to say that the paper started a revolution in statistical methodology. Much of the success of that revolution was due to the rapidity and wide range of consequent developments by Brad and those around him, including several of his Ph.D. students.

As Brad has recounted in Efron (2003), his thinking about this topic started in 1972–1973 when he and Rupert Miller left the tranquility of Stanford for sabbatical visits to Imperial College in London. Rupert was finishing his review paper (Miller 1974) and giving talks on jackknife methods, which aroused genuine interest in many of us there. Cross-validation methods were also being talked about – Mervyn Stone presented a paper (Stone 1974) on this topic at the Royal Statistical Society at the end of 1973, but this was focussed on choice[1] rather than assessment, which was Brad's interest. The jackknife and cross-validation methods had both been advocated in a book on data analysis by Mosteller and Tukey (1968), and were again in a second book (Mosteller and Tukey 1977) in the same year that Brad's paper was presented as the Rietz Lecture at the IMS meeting.

Also in the early to mid-1970s there was considerable interest in robust estimation, partly because of Peter Huber's theoretical work that built on John Tukey's data analytic inspiration. As Brad reminds us in the 1979 paper, one key product from the robustness area was Louis Jaeckel's influential 1972 manuscript which elaborated on the idea of an estimate T of a parameter θ as a statistical function $T = t(\hat{F})$ estimating $\theta = t(F)$, with $\hat{F}$ the empirical distribution function. This was used in Jaeckel's development of the "infinitesimal jackknife," including an extension of the Fisher delta method by which an approximate variance for T could be found if T were a differentiable function of a finite-dimensional average: the empirical distribution function $\hat{F}$ is just a more complicated average.

The original jackknife developed by Maurice Quenouille in the 1950s was a beautifully simple procedure for direct numerical approximation of the bias of a statistical estimate. Then in 1958, John Tukey intriguingly provided the jackknife formula for calculating the variance of a statistical estimate. In the simplest context of one sample, say that the estimate $T = t(\mathbf{X})$ is based on data vector $\mathbf{X} = (X_1, \ldots, X_n)$ where the X_i's are outcomes of independent random sampling. Write $\mathbf{X}_{(-i)}$ for $\mathbf{X}$ with

[1] I mention this partly because, by an odd coincidence, Stone's paper introduced cross-validation as a way of arriving at the Efron–Morris estimates, the subject of another paper in this volume: *Data Analysis Using Stein's Estimator and Its Generalizations*.

the ith case omitted, and then define $D_i = (n-1)\{t(\mathbf{X}) - t(\mathbf{X}_{(-i)})\}$. The bias and variance of T are, respectively, approximated by

$$B = -\frac{1}{n}\bar{D}, \quad V = \frac{1}{n^2}\sum_{i=1}^{n} D_i^2.$$

Somewhat mysteriously at the time, Tukey saw V as the natural consequence of a linear approximation for T and what the values $T + D_i$ then represented. This variance formula is model-free, except insofar as its applicability assumes independence of the X_i's.

The most ambitious use of the jackknife suggested by Tukey was to set confidence limits for the parameter θ estimated by T by assuming a Student-t approximation, with degrees of freedom that would "usually" be $n-1$, for the Studentized estimate $R = \frac{T-B}{\sqrt{V}}$, hence mimicking the conventional Student-t confidence limits for a Normal mean. There were four things wrong with this. First, a relatively minor point, the order of accuracy of the approximation is not improved by bias correction. Secondly, and for the same reason, the "improvement" of the Student-t approximation over the standard Normal approximation was not justified in general. Thirdly, even when T is approximately Normal (more precisely, $\sqrt{n}(T-\theta)$ has distribution tending to a $N(0, \tau^2)$ distribution as $n \to \infty$), there is no guarantee that the jackknife variance estimate V "works." Fourthly, even if the Student-t approximation did work, the advice for selecting degrees of freedom seemed somewhat mystical.

A prominently awkward case was that of the sample median, where the variance formula clearly did not work – and somewhat confusingly Tukey's ad hoc rule for calculating degrees of freedom gave value 1. I think that this case intrigued Brad, and it played a very prominent role in this first bootstrap paper. Brad points out that the bootstrap variance estimate *does* work in this case, although he leaves us with a little exercise in "standard asymptotic theory" to show this – which I hope the referees did! I can just picture Brad in the end office of the old Sequoia Hall, the large sheets of blotting paper filling up with his Biro calculations.

Beyond that, a natural question to ask was for what distribution the jackknife bias and variance formulae were approximations. As Brad acknowledged, the Jaeckel paper was central to his thinking about this. And so we get the simple bootstrap solution, that the distribution of $T - \theta = t(\hat{F}) - t(F)$ is being approximated by that of $T^* - \theta^*$, with θ^* the parameter of $\hat{F}$ and T^* the estimator calculated from a random sample $\mathbf{X}^*$ drawn from $\hat{F}$ – we just replace F by $\hat{F}$ everywhere in the carefully stated original problem.

One particularly valuable point is reinforced by the extension to regression in Sect. 7 of this paper, namely that the totally model-free nature of jackknife methods was a potential weakness, by not allowing appropriate assumptions to modify the methods. This can often result in loss of efficiency – whereas with bootstrap calculations, such assumptions *can* be used to modify the EDF as choice for $\hat{F}$. In the same vein, the use of a fitted parametric model for $\hat{F}$ is briefly mentioned right at the end of the paper. This immediately presents the bootstrap as a much more flexible tool, in addition to going beyond just the bias and variance calculations. This aspect of how to correctly choose $\hat{F}$ continues to intrigue researchers as they look at ever-more-complex applications, as evidenced by the recent work of Brad's colleague Art Owen (Owen 2007).

As was often the case with Brad's papers, a wealth of information was in the *Remarks* section. Here in Remarks D and E we get a brief discussion of confidence intervals, and reference to the important notion of (approximate) pivots. Not long after this paper was published, Brad introduced the ingenious yet simple percentile method (Efron 1981) for confidence intervals. The subsequent experience and valuable criticism in the literature, much of it motivated by challenging and important applications, led to Brad's introduction of a clever Normal approximation to incorporate deviation from exact pivotality and the landmark confidence interval paper of 1987.

This paper also makes a mysterious but useful detour to introduce the idea (Sect. 4) that considerable improvement on cross-validation may be possible in the problem of estimating error rate for statistical classification methods. Subsequent developments of this notion by Brad and others culminated in Efron and Tibshirani (1997).

The valuable numerical illustrations in this paper are persuasive, as far as the potential of the bootstrap idea. Not that the bootstrap made life quite as simple as the Stanford students suggest in the song "The Bootstrap Begins" (Finkelman et al. 2004), but their enthusiasm coupled with Brad's insights certainly led to a valuable series of developments in statistical methodology.

References

Efron, B. (1979). Bootstrap methods: Another look at the jackknife. *Annals of Statistics* **7**, 1–26.

Efron, B. (1981). Nonparametric standard errors and confidence intervals. *Canadian Journal of Statistics* **9**, 139–172 with discussion.

Efron, B. (1987). Better bootstrap confidence intervals. *Journal of the American Statistical Association* **82**, 171–200 with discussion and Rejoinder.

Efron, B. (2003). Second thoughts on the bootstrap. *Statistical Science* **18**, 135–140.

Efron, B. and Tibshirani, R. (1997). Improvements on cross-validation: The .632+ bootstrap method. *Journal of the American Statistical Association* **92**, 548–560.

Finkelman, M., Hooker, G. and Schwartzman, A. (2004). *The Stanford Statistics Songbook: A Musical Tribute*. Department of Statistics Technical Report, Stanford University. [Available at `www.bcsb.cornell.edu/~hooker/`]

Jaeckel, L. A. (1972). The infinitesimal jackknife. Bell Telephone Laboratories Memorandum. [Available at `www.stat.washington.edu/fritz/`]

Miller, R. G. (1974). The jackknife: A review. *Biometrika* **61**, 1–15.

Mosteller, F. and Tukey, J. W. (1968). Data analysis including statistics. In *Handbook of Social Psychology*, Vol.2: Research Methods. Edited by G. Lindzey and E. Aronson. Addison-Wesley, Reading, MA.

Mosteller, F. and Tukey, J. W. (1977). *Data Analysis and Regression*. Addison-Wesley, Reading, MA.

Owen, A. (2007). The pigeonhole bootstrap. Department of Statistics Technical Report, Stanford University.

Quenouille, M. (1956). Notes on bias in estimation. *Biometrika* **43**, 353–360.

Stone, M. (1974). Cross-validatory choice and assessment of statistical predictions (with Discussion). *Journal of the Royal Statistical Society, Series B* **36**, 111–147.

Tukey, J. W. (1958). Bias and confidence in not-quite large samples (abstract). *Annals of Mathematical Statistics* **29**, 614.

The Annals of Statistics
1979, Vol. 7, No. 1, 1–26

THE 1977 RIETZ LECTURE

BOOTSTRAP METHODS: ANOTHER LOOK AT THE JACKKNIFE

BY B. EFRON

Stanford University

We discuss the following problem: given a random sample $\mathbf{X} = (X_1, X_2, \ldots, X_n)$ from an unknown probability distribution F, estimate the sampling distribution of some prespecified random variable $R(\mathbf{X}, F)$, on the basis of the observed data $\mathbf{x}$. (Standard jackknife theory gives an approximate mean and variance in the case $R(\mathbf{X}, F) = \theta(\hat{F}) - \theta(F)$, θ some parameter of interest.) A general method, called the "bootstrap," is introduced, and shown to work satisfactorily on a variety of estimation problems. The jackknife is shown to be a linear approximation method for the bootstrap. The exposition proceeds by a series of examples: variance of the sample median, error rates in a linear discriminant analysis, ratio estimation, estimating regression parameters, etc.

1. Introduction. The Quenouille–Tukey jackknife is an intriguing nonparametric method for estimating the bias and variance of a statistic of interest, and also for testing the null hypothesis that the distribution of a statistic is centered at some prespecified point. Miller [14] gives an excellent review of the subject.

This article attempts to explain the jackknife in terms of a more primitive method, named the "bootstrap" for reasons which will become obvious. In principle, bootstrap methods are more widely applicable than the jackknife, and also more dependable. In Section 3, for example, the bootstrap is shown to (asymptotically) correctly estimate the variance of the sample median, a case where the jackknife is known to fail. Section 4 shows the bootstrap doing well at estimating the error rates in a linear discrimination problem, outperforming "cross-validation," another nonparametric estimation method.

We will show that the jackknife can be thought of as a linear expansion method (i.e., a "delta method") for approximating the bootstrap. This helps clarify the theoretical basis of the jackknife, and suggests improvements and variations likely to be successful in various special situations. Section 3, for example, discusses jackknifing (or bootstrapping) when one is willing to assume symmetry or smoothness of the underlying probability distribution. This point reappears more emphatically in Section 7, which discusses bootstrap and jackknife methods for regression models.

The paper proceeds by a series of examples, with little offered in the way of general theory. Most of the examples concern estimation problems, except for Remark F of Section 8, which discusses Tukey's original idea for t-testing using the

Received June 1977; revised December 1977.
AMS 1970 *subject classifications*. Primary 62G05, 62G15; Secondary 62H30, 62J05.
Key words and phrases. Jackknife, bootstrap, resampling, subsample values, nonparametric variance estimation, error rate estimation, discriminant analysis, nonlinear regression.

jackknife. The bootstrap results on this point are mixed (and won't be reported here), offering only slight encouragement for the usual jackknife t tests.

John Hartigan, in an important series of papers [5, 6, 7], has explored ideas closely related to what is called bootstrap "Method 2" in the next section, see Remark I of Section 8. Maritz and Jarrett [13] have independently used bootstrap "Method 1" for estimating the variance of the sample median, deriving equation (3.4) of this paper and applying it to the variance calculation. Bootstrap "Method 3," the connection to the jackknife via linear expansions, relates closely to Jaeckel's work on the infinitesimal jackknife [10]. If we work in a parametric framework, this approach to the bootstrap gives Fisher's information bound for the asymptotic variance of the maximum likelihood estimator, see Remark K of Section 8.

2. Bootstrap methods. We discuss first the one-sample situation in which a random sample of size n is observed from a completely unspecified probability distribution F,

$$X_i = x_i, \qquad X_i \sim_{\text{ind}} F \qquad i = 1, 2, \ldots, n. \tag{2.1}$$

In all of our examples F will be a distribution on either the real line or the plane, but that plays no role in the theory. We let $\mathbf{X} = (X_1, X_2, \ldots, X_n)$ and $\mathbf{x} = (x_1, x_2, \ldots, x_n)$ denote the random sample and its observed realization, respectively.

The problem we wish to solve is the following. Given a specified random variable $R(\mathbf{X}, F)$, possibly depending on both $\mathbf{X}$ and the unknown distribution F, estimate the sampling distribution of R on the basis of the observed data $\mathbf{x}$.

Traditional jackknife theory focuses on two particular choices of R. Let $\theta(F)$ be some parameter of interest such as the mean, correlation, or standard deviation of F, and $t(\mathbf{X})$ be an estimator of $\theta(F)$, such as the sample mean, sample correlation, or a multiple of the sample range. Then the sampling distribution of

$$R(\mathbf{X}, F) = t(\mathbf{X}) - \theta(F), \tag{2.2}$$

or more exactly its mean (the bias of t) and variance, is estimated using the standard jackknife theory, as described in Section 5. The bias and variance estimates say $\widehat{\text{Bias}}\,(t)$ and $\widehat{\text{Var}}\,(t)$, are cleverly constructed functions of $\mathbf{X}$ obtained by recomputing $t(\cdot)$ n times, each time removing one component of $\mathbf{X}$ from consideration. The second traditional choice of R is

$$R(\mathbf{X}, F) = \frac{t(\mathbf{X}) - \widehat{\text{Bias}}\,(t) - \theta(F)}{(\widehat{\text{Var}}\,(t))^{\frac{1}{2}}}. \tag{2.3}$$

Tukey's original suggestion was to treat (2.3) as having a standard Student's t distribution with $n - 1$ degrees of freedom. (See Remark F, Section 8.) Random variables (2.2), (2.3) play no special role in the bootstrap theory, and, as a matter of fact, some of our examples concern other choices of R.

The bootstrap method for the one-sample problem is extremely simple, at least in principle:

1. Construct the sample probability distribution $\hat{F}$, putting mass $1/n$ at each point $x_1, x_2, x_3, \ldots, x_n$.

2. With $\hat{F}$ fixed, draw a random sample of size n from $\hat{F}$, say

$$X_i^* = x_i^*, X_i^* \sim_{\text{ind}} \hat{F} \qquad i = 1, 2, \cdots, n. \tag{2.4}$$

Call this the *bootstrap sample*, $\mathbf{X}^* = (X_1^*, X_2^*, \cdots, X_n^*)$, $\mathbf{x}^* = (x_1^*, x_2^*, \cdots, x_n^*)$. Notice that we are not getting a permutation distribution since the values of $\mathbf{X}^*$ are selected *with* replacement from the set $\{x_1, x_2, \ldots, x_n\}$. As a point of comparison, the ordinary jackknife can be thought of as drawing samples of size $n - 1$ *without* replacement.

3. Approximate the sampling distribution of $R(\mathbf{X}, F)$ by the *bootstrap distribution* of

$$R^* = R(\mathbf{X}^*, \hat{F}), \tag{2.5}$$

i.e., the distribution of R^* induced by the random mechanism (2.4), with $\hat{F}$ held fixed at its observed value.

The point is that the distribution of R^*, which in theory can be calculated exactly once the data $\mathbf{x}$ is observed, equals the desired distribution of R if $F = \hat{F}$. Any nonparametric estimator of R's distribution, i.e., one that does a reasonably good estimation job without prior restrictions on the form of F, must give close to the right answer when $F = \hat{F}$, since $\hat{F}$ is a central point amongst the class of likely F's, having observed $\mathbf{X} = \mathbf{x}$. Making the answer exactly right for $F = \hat{F}$ is *Fisher consistency* applied to our particular estimation problem.

Just how well the distribution of R^* approximates that of R depends upon the form of R. For example, $R(\mathbf{X}, F) = t(\mathbf{X})$ might be expected to bootstrap less successfully than $R(\mathbf{X}, F) = [t(\mathbf{X}) - E_F t]/(\text{Var}_F t)^{\frac{1}{2}}$. This is an important question, related to the concept of pivotal quantities, Barnard [2], but is discussed only briefly here, in Section 8. Mostly we will be content to let the varying degrees of success of the examples speak for themselves.

As the simplest possible example of the bootstrap method, consider a probability distribution F putting all of its mass at zero or one, and let the parameter of interest be $\theta(F) = \text{Prob}_F\{X = 1\}$. The most obvious random variable of interest is

$$R(\mathbf{X}, F) = \bar{X} - \theta(F) \qquad \bar{X} = (\textstyle\sum_{i=1}^n X_i/n). \tag{2.6}$$

Having observed $\mathbf{X} = \mathbf{x}$, the bootstrap sample $\mathbf{X}^* = (X_1^*, X_2^*, \ldots, X_n^*)$ has each component independently equal to one with probability $\bar{x} = \theta(\hat{F})$, zero with probability $1 - \bar{x}$. Standard binomial results show that

$$R^* = R(\mathbf{X}^*, \hat{F}) = \bar{X}^* - \bar{x} \tag{2.7}$$

has mean and variance

$$E_*(\bar{X}^* - \bar{x}) = 0, \qquad \text{Var}_*(\bar{X}^* - \bar{x}) = \bar{x}(1 - \bar{x})/n. \tag{2.8}$$

(Notations such as "E_*," "Var_*," "Prob_*," etc. indicate probability calculations relating to the bootstrap distribution of $\mathbf{X}^*$, with $\mathbf{x}$ and $\hat{F}$ fixed.) The implication that $\bar{X}$ is unbiased for θ, with variance approximately equal to $\bar{x}(1 - \bar{x})/n$, is universally familiar.

As a second example, consider estimating $\theta(F) = \text{Var}_F X$, the variance of an arbitrary distribution on the real line, using the estimator $t(\mathbf{X}) = \sum_{i=1}^n (X_i - \bar{X})^2/(n-1)$. Perhaps we wish to know the sampling distribution of

$$R(\mathbf{X}, F) = t(\mathbf{X}) - \theta(F). \tag{2.9}$$

Let $\mu_k(F)$ indicate the kth central moment of F, $\mu_k(F) = E_F(X - E_F X)^k$, and $\hat{\mu}_k = \mu_k(\hat{F})$, the kth central moment of $\hat{F}$. Standard sampling theory results, as in Cramér [3], Section 27.4, show that

$$R^* = R(\mathbf{X}^*, \hat{F}) = t(\mathbf{X}^*) - \theta(\hat{F})$$

has

$$E_* R^* = 0, \qquad \text{Var}_* R^* = \frac{\hat{\mu}_4 - ((n-3)/(n-1))\hat{\mu}_2^2}{n}. \tag{2.10}$$

The approximation $\text{Var}_F t(\mathbf{X}) \approx \text{Var}_* R^*$ is (almost) the jackknife estimate for $\text{Var}_F\, t$.

The difficult part of the bootstrap procedure is the actual calculation of the bootstrap distribution. Three methods of calculation are possible:

Method 1. Direct theoretical calculation, as in the two examples above and the example of the next section.

Method 2. Monte Carlo approximation to the bootstrap distribution. Repeated realizations of $\mathbf{X}^*$ are generated by taking random samples of size n from $\hat{F}$, say $\mathbf{x}^{*1}, \mathbf{x}^{*2}, \ldots, \mathbf{x}^{*N}$, and the histogram of the corresponding values $R(\mathbf{x}^{*1}, \hat{F}), R(\mathbf{x}^{*2}, \hat{F}), \ldots, R(\mathbf{x}^{*N}, \hat{F})$ is taken as an approximation to the actual bootstrap distribution. This approach is illustrated in Sections 3, 4 and 8.

Method 3. Taylor series expansion methods can be used to obtain the approximate mean and variance of the bootstrap distribution of R^*. This turns out to be the same as using some form of the jackknife, as shown in Section 5.

In Section 4 we consider a two sample problem where the data consists of a random sample $\mathbf{X} = (X_1, X_2, \ldots, X_n)$ from F and an independent random sample $\mathbf{Y} = (Y_1, Y_2, \ldots, Y_n)$ from G, F and G arbitrary probability distributions on a given space. In order to estimate the sampling distribution of a random variable $R((\mathbf{X}, \mathbf{Y}), (F, G))$, having observed $\mathbf{X} = \mathbf{x}$, $\mathbf{Y} = \mathbf{y}$, the one-sample bootstrap method can be extended in the obvious way: $\hat{F}$ and $\hat{G}$, the sample probability distributions corresponding to F and G, are constructed; bootstrap samples $X_i^* \sim \hat{F}$, $i = 1, 2, \ldots, m$, $Y_j^* \sim \hat{G}$, $j = 1, 2, \ldots, n$, are independently drawn; and finally the bootstrap distribution of $R^* = R((\mathbf{X}^*, Y^*), (\hat{F}, \hat{G}))$ is calculated, for use as an approximation to the actual distribution of R. The calculation of the bootstrap distribution proceeds by any of the three methods listed above. (The third method makes clear the correct analogue of the jackknife procedure for nonsymmetric situations, such as the two sample problem; see the remarks of Section 6.)

So far we have only used nonparametric maximum likelihood estimators, $\hat{F}$ and $(\hat{F}, \hat{G})$, to begin the bootstrap procedure. This isn't crucial, and as the examples of Sections 3 and 7 show, it is sometimes more convenient to use other estimates of the underlying distributions.

3. Estimating the median. Suppose we are in the one-sample situation (2.1), with F a distribution on the real line, and we wish to estimate the median of F using the sample median. Let $\theta(F)$ indicate the median of F, and let $t(\mathbf{X})$ be the sample median,

$$t(\mathbf{X}) = X_{(m)}, \tag{3.1}$$

where $X_{(1)} \leqslant X_{(2)} \leqslant \cdots \leqslant X_{(n)}$ is the order statistic, and we have assumed an odd sample size $n = 2m - 1$ for convenience. Once again we take $R(\mathbf{X}, F) = t(\mathbf{X}) - \theta(F)$, and hope to say something about the sampling distribution of R on the basis of the observed random sample.

Having observed $\mathbf{X} = \mathbf{x}$, we construct the bootstrap sample $\mathbf{X}^* = \mathbf{x}^*$ as in (2.4). Let

$$N_i^* = \#\{X_i^* = x_i\}, \tag{3.2}$$

the number of times x_i is selected in the bootstrap sampling procedure. The vector $\mathbf{N}^* = (N_1^*, N_2^*, \cdots, N_n^*)$ has a multinomial distribution with expectation one in each of the n cells.

Denote the observed order statistic $x_{(1)} \leqslant x_{(2)} \leqslant x_{(3)} \leqslant \cdots \leqslant x_{(n)}$, and the corresponding N^* values $N_{(1)}^*, N_{(2)}^*, \cdots, N_{(n)}^*$. (Ties $x_i = x_{i'}$ can be broken by assigning the lower value of i, i' to the lower position in the order statistic.) The bootstrap value of R is

$$R^* = R(\mathbf{X}^*, \hat{F}) = X_{(m)}{}^* - x_{(m)}. \tag{3.3}$$

We notice that for any integer value l, $1 \leqslant l < n$,

$$\begin{aligned} \text{Prob}_*\{X^*_{(m)} > x_{(l)}\} &= \text{Prob}_*\{N^*_{(1)} + N_{(2)}{}^* + \cdots + N_{(l)}{}^* \leqslant m - 1\} \\ &= \text{Prob}\left\{\text{Binomial}\left(n, \frac{l}{n}\right) \leqslant m - 1\right\} \\ &= \textstyle\sum_{j=0}^{m-1} \binom{n}{j} \left(\frac{l}{n}\right)^j \left(\frac{n-l}{n}\right)^{n-j}. \end{aligned} \tag{3.4}$$

Therefore

$$\begin{aligned} \text{Prob}_*\{R^* = x_{(l)} - x_{(m)}\} &= \text{Prob}\left\{\text{Binomial}\left(n, \frac{l-1}{n}\right) \leqslant m - 1\right\} \\ &\quad - \text{Prob}\left\{\text{Binomial}\left(n, \frac{l}{n}\right) \leqslant m - 1\right\}, \end{aligned} \tag{3.5}$$

a result derived independently by Maritz and Jarrett [13].

The case $n = 13$ ($m = 7$) gives the following bootstrap distribution for R^*:

(3.6)

$l =$	2 or 12	3 or 11	4 or 10	5 or 9	6 or 8	7
(3.5) =	.0015	.0142	.0550	.1242	.1936	.2230

For any given random sample of size 13 we can compute

$$E_*(R^*)^2 = \Sigma_{l=1}^{13}\left[x_{(l)} - x_{(7)}\right]^2 \operatorname{Prob}_*\{R^* = x_{(l)} - x_{(7)}\}, \tag{3.7}$$

and use this number as an estimate of $E_F R^2 = E_F[t(\mathbf{X}) - \theta(F)]^2$, the expected squared error of estimation for the sample median. Standard asymptotic theory, applied to the case where F has a bounded continuous density $f(x)$, shows that as the sample size n goes to infinity, the quantity $nE_*(R^*)^2$ approaches $1/4f^2(\theta)$, where $f(\theta)$ is the density evaluated at the median $\theta(F)$. This is the correct asymptotic value, see Kendall and Stuart [11], page 237. The standard jackknife applied to the sample median gives a variance estimate which is not even asymptotically consistent (Miller [14], page 8, is incorrect on this point): $n\,\widehat{\operatorname{Var}}(R) \to (1/4f^2(\theta))[\chi_2^2/2]^2$. The random variable $[\chi_2^2/2]^2$ has mean 2 and variance 20.

Suppose we happened to know that the probability distribution F was symmetric. In that case we could replace $\hat{F}$ by the symmetric probability distribution obtained from $\hat{F}$ by reflection about the median,

$$\hat{F}_{\mathrm{SYM}}: \quad \text{probability mass } \frac{1}{2n-1} \text{ at } x_{(1)}, x_{(2)}, \ldots, x_{(n)} \quad \text{and} \quad 2x_{(m)} - x_{(1)}, \ldots, 2x_{(m)} - x_{(n)}. \tag{3.8}$$

This is not the nonparametric maximum likelihood estimator for F, but has similar asymptotic properties, see Hinkley [8]. Let $z_{(1)} \leqslant z_{(2)} \leqslant \cdots \leqslant z_{(2n-1)}$ be the ordered values appearing in the distribution of $\hat{F}_{\mathrm{SYM}}$. The bootstrap procedure starting from $\hat{F}_{\mathrm{SYM}}$ gives

$$\begin{aligned}\operatorname{Prob}_*\{R^* = z_{(l)} - x_{(m)}\} &= \operatorname{Prob}\left\{\operatorname{Binomial}\left(n, \frac{l-1}{2n-1}\right) \leqslant m-1\right\} \\ &\quad - \operatorname{Prob}\left\{\operatorname{Binomial}\left(n, \frac{l}{2n-1}\right) \leqslant m-1\right\},\end{aligned} \tag{3.9}$$

by the same argument leading to (3.5). For $n = 13$ the bootstrap probabilities (3.9) equal

(3.10)

$l =$	4 or 22	5 or 21	6 or 20	7 or 19	8 or 18
(3.9) =	.0016	.0051	.0125	.0245	.0414

$l =$	9 or 17	10 or 16	11 or 15	12 or 14	13
(3.9) =	.0614	.0820	.1002	.1125	.1170

The corresponding estimate of $E_F R^2$ would be $\Sigma_{l=1}^{25}[z_{(l)} - x_{(7)}]^2 \operatorname{Prob}_*\{R^* = z_{(l)} - x_{(7)}\}$.

Usually we would not be willing to assume F symmetric in a nonparametric estimation situation. However in dealing with continuous variables we might be willing to attribute a moderate amount of smoothness to F. This can be incorporated into the bootstrap procedure at step (2.4). Instead of choosing each X_i^* randomly from the set $\{x_1, x_2, \ldots, x_n\}$, we can take

$$X_i^* = \bar{x} + c[x_{I_i} - \bar{x} + \hat{\sigma} Z_i] \tag{3.11}$$

where the I_i are chosen independently and randomly from the set $\{1, 2, \ldots, n\}$, and the Z_i are a random sample from some fixed distribution having mean 0 and variance σ_Z^2, for example the uniform distribution on $[-\frac{1}{2}, \frac{1}{2}]$, which has $\sigma_Z^2 = 1/12$. The most obvious choice is a normal distribution for the Z_i, but this would be self-serving in the Monte Carlo experiment which follows, where the X_i themselves are normally distributed. The quantities $\bar{x}$, $\hat{\sigma}$, and c appearing in (3.11) are the sample mean, sample standard deviation ($= (\hat{\mu}_2)^{\frac{1}{2}}$), and $[1 + \sigma_Z^2]^{-\frac{1}{2}}$, respectively, so that X_i^* has mean $\bar{x}$ and variance $\hat{\sigma}^2$ under the bootstrap sampling procedure. In using (3.11) in place of (2.4), we are replacing $\hat{F}$ with a smoothed "window" estimator, having the same mean and variance as $\hat{F}$.

A small Monte Carlo experiment was run to compare the various bootstrap methods suggested above. Instead of comparing the squared error of the sample median, the quantity bootstrapped was

$$R(\mathbf{X}, F) = \frac{|t(\mathbf{X}) - \theta(F)|}{\sigma(F)}, \tag{3.12}$$

the absolute error of the sample median relative to the population standard deviation. (This quantity is more stable numerically, because the absolute value is less sensitive than the square and also because $R^* = |t(\mathbf{X}^*) - \theta(\hat{F})|/\hat{\sigma}$ is scale invariant, which eliminates the component of variation due to $\hat{\sigma}$ differing from $\sigma(F)$. The stability of (3.12) greatly increased the effectiveness of the Monte Carlo trial.)

The Monte Carlo experiment was run with $n = 13$, $X_i \sim_{\text{ind}} \mathfrak{N}(0, 1)$, $i = 1, 2, \ldots, n$. In this situation the true expectation of R is

$$E_F R = 0.95. \tag{3.13}$$

The first two columns of Table 1 show $E_F R^*$ for each trial, using the bootstrap probabilities (3.6), and then (3.10) for the symmetrized version. It is not possible to theoretically calculate $E_* R^*$ for the smoothed bootstrap (3.11), so these entries of Table 1 were obtained by a secondary Monte Carlo simulation, as described in "Method 2" of Section 2. A total of $N = 50$ replications $\mathbf{x}^{*j}$ were generated for each trial. This means that the values in the table are only unbiased estimates of the actual bootstrap expectations $E_* R^*$ (which could be obtained by letting $N \to \infty$); the standard error being about .15 for each entry. The effect of this approximation is seen in the column "$d = 0$," which would exactly equal column "(3.6)" if $N \to \infty$. (Within each trial, the same set of random numbers was used to generate the four different uniform distributions for Z_i, $d = 0, .25, .5, 1$.)

TABLE 1*

Trial #	Unsmoothed Bootstrap (3.6)	(3.10)	Smoothed Bootstrap (3.11): Z_i uniform dist. on $[-d/2, d/2]$, $d = 0$	$d = .25$	$d = .5$	$d = 1$	Z_i triangular dist., $\sigma_Z^2 = 1/12$
1	1.07	1.18	1.09	1.10	1.12	1.11	1.16
2	.96	.74	1.10	1.10	1.08	1.09	1.15
3	1.22	.74	1.36	1.35	1.33	1.43	1.52
4	1.38	1.51	1.44	1.41	1.38	1.28	1.30
5	1.00	.83	1.03	1.05	1.09	1.14	1.17
6	1.13	1.21	1.27	1.26	1.23	1.20	1.26
7	1.07	.98	1.01	.94	.83	.79	.92
8	1.51	1.40	1.40	1.45	1.47	1.51	1.50
9	.56	.64	.69	.71	.74	.80	.81
10	1.05	.86	1.14	1.17	1.20	1.13	1.22
Ave.	1.09	1.01	1.15	1.15	1.15	1.15	1.20
S.D.	.26	.30	.23	.23	.23	.23	.22

*Ten Monte Carlo trials of $X_i \sim_{\text{ind}} \mathcal{N}(0, 1)$, $i = 1, 2, \ldots, 13$ were used to compare different bootstrap methods for estimating the expected value of random variable (3.12). The true expectation is 0.95. The quantities tabled are $E_* R^*$, the bootstrap expectation for that trial. The values in the first two columns are for the bootstrap as described originally, and for the symmetrized version (3.8)–(3.10). The smoothed bootstrap expectations were approximated using a secondary Monte Carlo simulation for each trial, $N = 50$, as described in "Method 2," Section 2. Each of these entries estimates the actual value of $E_* R^*$ unbiasedly with a standard error of about .15. The column "$d = 0$" would exactly equal column "(3.6)" if $N \to \infty$.

The most notable feature of Table 1 is that the simplest form of the bootstrap, "(3.6)," seems to do just as well as the symmetrical or smoothed versions. A larger Monte Carlo investigation of the same situation as in Table 1, 200 trials, 100 bootstrap replications per trial, was just a little more favorable to the smoothed bootstrap methods:

	(3.6)	(3.10)	$d = 0$	$d = .25$	$d = .5$	$d = 1$	$d = 2$
AVE.:	1.01	1.00	1.00	1.01	1.00	.99	.93
S.D.:	.31	.33	.32 [.31]	.32 [.30]	.32 [.30]	.30 [.29]	.26 [.25].

(The figures in square brackets are estimated standard deviations if N were increased from 100 to ∞, obtained by a components of variance calculation.) Remembering that we are trying to estimate the true value $E_F R = .95$, these seem like good performances for a nonparametric method based on a sample size of just 13.

The symmetrized version of the bootstrap might be expected to do relatively better than the unsymmetrized version if R itself was of a less symmetric form than (3.12), e.g., $R(\mathbf{X}, F) = \exp\{X_{(m)} - \theta(F)\}$. Likewise, the smoothed versions of the bootstrap might be expected to do relatively better if R itself were less smooth, e.g., $R(\mathbf{X}, F) = \text{Prob}\{X_{(m)} > \theta(F) + \sigma(F)\}$. However no evidence to support these guesses is available at present.

4. Error rate estimation in discriminant analysis. This section discusses the estimation of error rates in a standard linear discriminant analysis problem. There is a tremendous literature on this problem, nicely summarized in Toussaint [17]. In the two examples considered below, bootstrap methods outperform the commonly used "leave-one-out," or *cross-validation*, approach (Lachenbruch and Mickey [12]).

The data in the discriminant problem consists of independent random samples from two unknown continuous probability distributions F and G on some k-dimensional space R^k,

$$
\begin{aligned}
&X_i = x_i, \qquad X_i \sim_{\text{ind}} F \qquad i = 1, 2, \ldots, m \\
&Y_j = y_j, \qquad Y_j \sim_{\text{ind}} G \qquad j = 1, 2, \ldots, n.
\end{aligned}
\tag{4.1}
$$

On the basis of the observed data $\mathbf{X} = \mathbf{x}$, $\mathbf{Y} = \mathbf{y}$ we use some method (linear discriminant analysis in the examples below) to partition R^k into two complementary regions A and B, the intent being to ascribe a future observation z to the F distribution if $z \in A$, or to the G distribution if $z \in B$.

The obvious estimate of the error rate, for the F distribution, associated with the partition (A, B) is

$$
\widehat{\text{error}}_F = \frac{\#\{x_i \in B\}}{m},
\tag{4.2}
$$

which will tend to underestimate the true error rate

$$
\text{error}_F = \text{Prob}_F\{X \in B\}.
\tag{4.3}
$$

(In probability calculation (4.3), B is considered fixed, at its observed value, even though it is originally determined by a random mechanism.) We will be interested in the distribution of the difference

$$
R((\mathbf{X}, \mathbf{Y}), (F, G)) = \text{error}_F - \widehat{\text{error}}_F,
\tag{4.4}
$$

and the corresponding quantity for the distribution G. We could directly consider the distribution of $\widehat{\text{error}}_F$, but concentrating on the difference (4.4) is much more efficient for comparing different estimation methods. This point is discussed briefly at the end of the section.

Given $\mathbf{x}$ and $\mathbf{y}$, we define the region B by

$$
B = \left\{ z : (\bar{y} - \bar{x})' S^{-1}\left(z - \frac{\bar{x} + \bar{y}}{2}\right) > \log \frac{m}{n} \right\},
\tag{4.5}
$$

where $\bar{x} = \Sigma x_i/m$, $\bar{y} = \Sigma y_j/n$, and $S = [\Sigma(x_i - \bar{x})(x_i - \bar{x})' + \Sigma(y_j - \bar{y})(y_j - \bar{y})']/(m + n)$. This is the maximum likelihood estimate of the optimum division under multivariate normal theory, and differs just slightly (in the definition of S) from the estimated version of the Fisher linear discriminant function discussed in Chapter 6 of Anderson [1].

"Method 2," the brute force application of the bootstrap via simulation, is implemented as follows: given the data **x**, **y**, bootstrap random samples

$$(4.6)\qquad \begin{array}{lll} X_i^* = x_i^*, & X_i^* \sim_{\text{ind}} \hat{F} & i = 1, 2, \ldots, m \\ Y_j^* = y_j^*, & Y_j^* \sim_{\text{ind}} \hat{G} & j = 1, 2, \ldots, n \end{array}$$

are generated, $\hat{F}$ and $\hat{G}$ being the sample probability distributions corresponding to F and G. This yields a region B^* defined by (4.5) with $\bar{x}^*, \bar{y}^*, S^*$ replacing $\bar{x}, \bar{y}, S$. The bootstrap random variable in this case is

$$(4.7)\qquad R^* = R((\mathbf{X}^*, \mathbf{Y}^*), (\hat{F}, \hat{G})) = \frac{\#\{x_i \in B^*\}}{m} - \frac{\#\{x_i^* \in B^*\}}{m}.$$

In other words, (4.7) is the difference between the actual error rate, actual now being defined with respect to the "true" distribution $\hat{F}$, and the apparent error rate obtained by counting errors in the bootstrap sample.

Repeated independent generation of $(\mathbf{X}^*, \mathbf{Y}^*)$ yields a sequence of independent realizations of R^*, say $R^{*1}, R^{*2}, \ldots, R^{*N}$, which are then used to approximate the actual bootstrap distribution of R^*, this hopefully being a reasonable estimate of the unknown distribution of R. For example, the bootstrap expectation $E_* R^* = \sum_{j=1}^{N} R^{*j}/N$ can be used as an estimate of the true expectation $E_{F,G}R$.

To test out this theory, bivariate normal choices of F and G were investigated,

$$(4.8)\qquad F\colon X \sim \mathfrak{N}_2\left(\begin{pmatrix} -\frac{1}{2} \\ 0 \end{pmatrix}, \mathbf{I}\right) \quad G\colon Y \sim \mathfrak{N}_2\left(\begin{pmatrix} \frac{1}{2} \\ 0 \end{pmatrix}, \mathbf{I}\right).$$

Two sets of sample sizes, $m = n = 10$ and $m = n = 20$, were looked at, with the results shown in Table 2. (The entries of Table 2 were themselves estimated by averaging over repeated Monte Carlo trials, which should not be confused with the

TABLE 2*

	$m = n = 10$			$m = n = 20$			
Random Variable	Mean	(S.E.)	S.D.	Mean	(S.E.)	S.D.	Remarks
Error Rate Diff. (4.4) R	**.062**	(.003)	**.143**	**.028**	(.002)	**.103**	Based on 1000 trials
Bootstrap Expectation $E_* R^*$	**.057**	(.002)	**.026** [**.023**]	**.029**	(.001)	**.015** [**.011**]	Based on 100 trials; N = 100 Bootstrap Replications per trial. (Figure in brackets is S.D. if $N = \infty$.)
Bootstrap Standard Deviation $SD_*(R^*)$	.131	(.0013)	**.016**	**.097**	(.002)	**.010**	
Cross-Validation Diff. $\tilde{R}$	**.054**	(.009)	**.078**	**.032**	(.002)	**.043**	Based on 40 trials

* The error rate difference (4.4) for linear discriminant analysis, investigated for bivariate normal samples (4.8). Sample sizes are $m = n = 10$ and $m = n = 20$. The values for the bootstrap method were obtained by Method 2, $N = 100$ bootstrap replications per trial. The bootstrap method gives useful estimates of both the mean and standard deviation of R. The cross-validation method was nearly unbiased for the expectation of R, but had about three times as large a standard deviation. All of the quantities in this table were estimated by repeated Monte Carlo trials; standard errors are given for the means.

Monte Carlo replications used in the bootstrap process. "Replications" will always refer to the bootstrap process, "trials" to repetitions of the basic situation.) Because situation (4.8) is symmetric, only random variable (4.4), and not the corresponding error rate for G, need be considered.

Table 2 shows that with $m = n = 10$, the random variable (4.4) has mean and standard deviation approximately (.062, .143). The apparent error rate underestimates the true error rate by about 6%, on the average, but the standard deviation of the difference is 14% from trial to trial, so bias is less troublesome than variability in this situation. The bootstrap method gave an average of .057 for $E_* R^*$, which, allowing for sampling error, shows that the statistic $E_* R^*$ is nearly an unbiased estimator for $E_{F,G} R$. Unbiasedness is not enough, of course; we want $E_* R^*$ to have a small standard deviation, ideally zero, so that we can rely on it as an estimate. The actual value of its standard deviation, .026, is not wonderful, but does indicate that most of the trials yielded $E_* R^*$ in the range [.02, .09], which means that the statistician would have obtained a reasonably informative estimate of the true bias $E_{F,G} R = .062$.

As a point of comparison, consider the cross-validation estimate of R, say $\tilde{R}$, obtained by: deleting one x value at a time from the vector $\mathbf{x}$; recomputing B using (4.5), to get a new region $\tilde{B}$ (it is important *not* to change m to $m - 1$ in recomputing B—doing so results in badly biased estimation of R); seeing if the deleted x value is correctly classified by $\tilde{B}$; counting the proportion of x values misclassified in this way to get a cross-validated error rate $\widetilde{\text{error}}_F$; and finally, defining $\tilde{R} = \widetilde{\text{error}}_F - \widehat{\text{error}}_F$. The last row of Table 2 shows that $\tilde{R}$ has mean and standard deviation approximately (.054, .078). That is, *$\tilde{R}$ is three times as variable as $E_* R^*$ as an estimator of $E_{F,G} R$.*

The bootstrap standard deviation of R^*, $SD_*(R^*) = \{\sum_{j=1}^{N} [R^{*j} - E_* R^*]^2/(N - 1)\}^{\frac{1}{2}}$, can be used as an estimate of $SD_{F,G}(R)$, the actual standard deviation of R. Table 2 shows that $SD_*(R^*)$ had mean and standard deviation (.131, .016) across the 100 trials. Remembering that $SD_{F,G}(R) = .143$, the bootstrap estimate $SD_*(R^*)$ is seen to be a quite useful estimator of the actual standard deviation of R.

How much better would the bootstrap estimator $E_* R^*$ perform if the number of bootstrap replications N were increased from 100 to, say, 10,000? A components of variance analysis of all the data going into Table 2 showed that only moderate further improvement is possible. As $N \to \infty$, the trial-to-trial standard deviation of $E_* R^*$ would decrease from .026 to about .023 (from .015 to .011 in the case $m = n = 20$).

The reader may wonder which is the best estimator of the error rate error_F itself, rather than of the difference R. In terms of expected squared error, the order of preference is $\widehat{\text{error}}_F + E_* R^*$ (the bias-corrected value based on the bootstrap), $\widehat{\text{error}}_F$, and lastly $\widetilde{\text{error}}_F$, but the differences are quite small in the two situations of Table 2. The large variability of $\widehat{\text{error}}_F$, compared to its relatively small bias, makes bias correction an almost fruitless chore in these two situations. (Of course, this might not be so in more difficult discriminant problems.) *The bootstrap estimates of* $E_{F,G}R$ *and* $SD_{F,G}(R)$ *considered together make it clear that this is the case*, which is a good recommendation for the bootstrap approach.

5. Relationship with the jackknife. This section concerns "Method 3" of approximating the bootstrap distribution, Taylor series expansion (or the *delta method*), which turns out to be the same as the usual jackknife theory. To be precise, it is the same as Jaeckel's *infinitesimal jackknife* [10, 14], a useful mathematical device which differs only in detail from the standard jackknife. Many of the calculations below, and in Remarks G—K of Section 8, can be found in Jaeckel's excellent paper, which offers considerable insight into the workings of jackknife methods.

Returning to the one-sample situation, define $P_i^* = N_i^*/n$, where $N_i^* = \#\{X_i^* = x_i\}$ as at (3.2), and the corresponding vector

$$\mathbf{P}^* = (P_1^*, P_2^*, \cdots, P_n^*). \tag{5.1}$$

By the properties of the multinomial distribution, $\mathbf{P}^*$ has mean vector and covariance matrix

$$E_*\mathbf{P}^* = \mathbf{e}/n, \qquad \mathrm{Cov}_*\mathbf{P}^* = \mathbf{I}/n^2 - \mathbf{e}'\mathbf{e}/n^3 \tag{5.2}$$

under the bootstrap sampling procedure, where $\mathbf{I}$ is the identity matrix and $\mathbf{e} = (1, 1, 1, \ldots, 1)$.

Given the observed data vector $\mathbf{X} = \mathbf{x}$, and therefore $\hat{F}$, we can use the abbreviated notation

$$R(\mathbf{P}^*) = R(\mathbf{X}^*, \hat{F}) \tag{5.3}$$

for the bootstrap realization of R corresponding to $\mathbf{P}^*$. In making this definition we assume that the random variable of interest, $R(\mathbf{X}, F)$, is symmetrically defined in the sense that its value is invariant under any permutation of the components of X, so that it is sufficient to know $\mathbf{N}^* = n\mathbf{P}^*$ in order to evaluate $R(\mathbf{X}^*, \hat{F})$. This is always the case in standard applications of the jackknife.

We can approximate the bootstrap distribution of $R(\mathbf{X}^*, \hat{F})$ by expanding $R(\mathbf{P}^*)$ in a Taylor series about the value $\mathbf{P}^* = \mathbf{e}/n$, say

$$R(\mathbf{P}^*) \doteq R(\mathbf{e}/n) + (\mathbf{P}^* - \mathbf{e}/n)\mathbf{U} + \tfrac{1}{2}(\mathbf{P}^* - \mathbf{e}/n)\mathbf{V}(\mathbf{P}^* - \mathbf{e}/n)'. \tag{5.4}$$

Here

$$\mathbf{U} = \begin{bmatrix} \vdots \\ \dfrac{\partial R(\mathbf{P}^*)}{\partial P_i^*} \\ \vdots \end{bmatrix}_{\mathbf{P}^* = \mathbf{e}/n} \qquad \mathbf{V} = \begin{bmatrix} & \vdots & \\ \cdots & \dfrac{\partial^2 R(\mathbf{P}^*)}{\partial P_i^* \partial P_j^*} & \cdots \\ & \vdots & \end{bmatrix}_{\mathbf{P}^* = \mathbf{e}/n}. \tag{5.5}$$

Expansion (5.4), and definitions (5.5), assume that the definition of $R(\mathbf{P}^*)$ can be smoothly interpolated between the lattice point values originally contemplated for $\mathbf{P}^*$. How to do so will be obvious in most specific cases, but a general recipe is difficult to provide. See Remarks G and H of Section 8.

The restriction $\Sigma P_i^* = 1$ has been ignored in (5.4), (5.5). This computational convenience is justified by extending the definition of $R(\mathbf{P}^*)$ to all vectors $\mathbf{P}^*$ having nonnegative components, at least one positive, by the homogeneous extension

$$R(\mathbf{P}^*) = R\left(\frac{\mathbf{P}^*}{\Sigma_{i=1}^n P_i^*}\right). \tag{5.6}$$

It is easily shown that the homogeneity of definition (5.6) implies

$$\mathbf{eU} = 0, \qquad \mathbf{eV} = -n\mathbf{U}', \qquad \mathbf{eVe}' = 0. \tag{5.7}$$

From (5.2) and (5.4) we get the approximation to the bootstrap expectation

$$E_* R(\mathbf{P}^*) \doteq R(\mathbf{e}/n) + \frac{1}{2}\text{trace } \mathbf{V}\left[\mathbf{I}/n^2 - \mathbf{e}'\mathbf{e}/n^3\right] = R(\mathbf{e}/n) + \frac{1}{2n}\bar{V}, \tag{5.8}$$

where

$$\bar{V} = \Sigma_{i=1}^n V_{ii}/n. \tag{5.9}$$

Ignoring the last term in (5.4) gives a cruder approximation for the bootstrap variance,

$$\text{Var}_* R(\mathbf{P}^*) \doteq \mathbf{U}'\left[\mathbf{I}/n^2 - \mathbf{e}'\mathbf{e}/n^3\right]\mathbf{U} = \Sigma_{i=1}^n U_i^2/n^2. \tag{5.10}$$

(Both (5.8) and (5.10) involve the use of (5.7).)

Results (5.8) and (5.10) are essentially the jackknife expressions for bias and variance. The usual jackknife theory considers $R(\mathbf{X}, F) = \theta(\hat{F}) - \theta(F)$, the difference between the obvious nonparametric estimator of some parameter $\theta(F)$ and $\theta(F)$ itself. In this case $R(\mathbf{X}^*, F) = \theta(\hat{F}^*) - \theta(\hat{F})$, $\hat{F}^*$ being the empirical distribution of the bootstrap sample, so that $R(\mathbf{e}/n) = \theta(\hat{F}) - \theta(\hat{F}) = 0$. Then (5.8) becomes $E_*[\theta(\hat{F}^*) - \theta(\hat{F})] \doteq (1/2n)\bar{V}$, suggesting $E_F[\theta(\hat{F}) - \theta(F)] \approx (1/2n)\bar{V}$; likewise (5.10) becomes $\text{Var}_*[\theta(\hat{F}^*) - \theta(\hat{F})] \doteq \Sigma U_i^2/n^2$, suggesting $\text{Var}_F\theta(\hat{F}) \approx \Sigma U_i^2/n^2$.

The approximations

$$\text{Bias}_F\ \theta(\hat{F}) \approx \frac{1}{2n}\bar{V}, \qquad \text{Var}_F\ \theta(\hat{F}) \approx \Sigma_{i=1}^n U_i^2/n^2 \tag{5.11}$$

exactly agree with those given by Jaeckel's infinitesimal jackknife [10], which themselves differ only slightly from the ordinary jackknife expressions. Without going into details, which are given in Jaeckel [10] and Miller [14], the ordinary jackknife replaces the derivatives $U_i = \partial R(\mathbf{P}^*)/\partial P_i$ with finite differences

$$\tilde{U}_i = (n-1)(R_\cdot^* - R_{(i)}^*) \tag{5.12}$$

where $R^*_{(i)} = R(\mathbf{e}_{(i)}/(n-1))$, $\mathbf{e}_{(i)}$ being the vector with zero in the ith coordinate and ones elsewhere, and $R^*_{\cdot} = \Sigma_{i=1}^n R^*_{(i)}/n$. Expansion (5.4) combines with (5.7) to give

$$\tilde{U}_i \doteq \frac{n-2}{n-1} U_i - \frac{1}{2(n-1)}\left(V_{ii} - \bar{V}\right), \tag{5.13}$$

so that $\tilde{U}_i/U_i = 1 + O(1/n)$. The ordinary jackknife estimate of variance is $\Sigma_{i=1}^n \tilde{U}_i^2/n\cdot(n-1)$, differing from the variance expression in (5.11) by a factor $1 + O(1/n)$, the same statement being true for the bias. (In the familiar case $R = \theta(\hat{F}) - \theta(F)$, definition (5.12) becomes $\tilde{U}_i = (n-1)(\hat{\theta}_{\cdot} - \hat{\theta}_{(i)})$, where $\hat{\theta}_{(i)}$ is the estimate of θ with x_i removed from the sample, and $\hat{\theta}_{\cdot} = \Sigma_i \hat{\theta}_{(i)}/n$; the jackknife estimate of θ is $\tilde{\theta} = \hat{\theta} + (n-1)(\hat{\theta} - \hat{\theta}_{\cdot})$, and $\tilde{\theta}_i = \hat{\theta} + \tilde{U}_i$ is the ith *pseudo-value*, to use the standard terminology.)

As an example of Method 3, consider *ratio estimation*, where the X_i are bivariate observations, say $X_i = (Y_i, Z_i)$, and we wish to estimate $\theta(F) = E_F Y/E_F Z$. (Take $Y, Z > 0$ for convenience.) Let $t(\mathbf{X}) = \bar{Y}/\bar{Z}$, and $R(\mathbf{X}, F) = t(\mathbf{X})/\theta(F)$. It is easily verified that

$$U_i = \frac{y_i}{\bar{y}} - \frac{z_i}{\bar{z}}, \qquad V_{ij} = 2\frac{z_i}{\bar{z}}\frac{z_j}{\bar{z}} - \left(\frac{y_i}{\bar{y}}\frac{z_j}{\bar{z}} + \frac{y_j}{\bar{y}}\frac{z_i}{\bar{z}}\right), \tag{5.14}$$

and that (5.8), (5.10) give

$$E_* R^* \doteq 1 - \frac{1}{n^2}\left\{\Sigma_i\left(\frac{y_i}{\bar{y}} - 1\right)\left(\frac{z_i}{\bar{z}} - 1\right) - \Sigma_i\left(\frac{z_i}{\bar{z}} - 1\right)^2\right\}, \tag{5.15}$$

$$\mathrm{Var}_* R^* \doteq \frac{1}{n^2}\Sigma_i\left[\frac{y_i}{\bar{y}} - \frac{z_i}{\bar{z}}\right]^2.$$

The biased corrected estimate for $\theta(F)$ is $t(\mathbf{X})/E_* R^*$, with approximate variance $(\hat{\theta}/n)^2\Sigma[y_i/\bar{y} - z_i/\bar{z}]^2$. If the statistician feels uneasy about expressions (5.15) for any particular data set, perhaps because of outlying values, Method 2 can be invoked to check the bootstrap distribution of $t(\mathbf{X}^*)$ directly.

The infinitesimal jackknife and the ordinary jackknife can both be applied starting from $\hat{F}_{\mathrm{SYM}}$, (3.8), rather than from $\hat{F}$. It is easiest to see how for the infinitesimal jackknife. Expansion (5.4) is still valid except that $\mathbf{U}$ is now a $(2n-1)\times 1$ vector, $\mathbf{V}$ is a $(2n-1)\times(2n-1)$ matrix, and $\mathbf{P}^*$ has bootstrap mean $\mathbf{e}/(2n-1)$, covariance matrix $(1/n)[\mathbf{I}/(2n-1) - \mathbf{e}'\mathbf{e}(2n-1)^2]$. The variance approximation corresponding to (5.10) is

$$\mathrm{Var}_{*\mathrm{SYM}} R(\mathbf{P}^*) = \frac{\Sigma_{l=1}^{2n-1} U_l^2}{n(2n-1)}. \tag{5.16}$$

6. Wilcoxon's statistic. We again consider the two-sample situation (4.1), this time with F and G being continuous probability distributions on the real line. The parameter of interest will be

$$\theta(F, G) = P_{F,G}(X < Y), \tag{6.1}$$

estimated by Wilcoxon's statistic

$$\hat{\theta} = \theta(\hat{F}, \hat{G}) = \frac{1}{mn} \Sigma_{i=1}^{m} \Sigma_{j=1}^{n} I(X_i, Y_j), \tag{6.2}$$

where

$$I(a, b) = 1 \quad a < b \qquad = 0 \quad a \geqslant b. \tag{6.3}$$

The bootstrap variance of $\hat{\theta}$ can be calculated directly by Method 1, and will turn out below to be the same as the standard variance approximation for Wilcoxon's statistic. The comparison with Method 3, the infinitesimal jackknife, illustrates how this theory works in a two-sample situation. More importantly, *it suggests the correct analogue of the ordinary jackknife for such situations.*

There has been considerable interest in extending the ordinary jackknife to "unbalanced" situations, i.e., those where it is not clear what the correct analogue of "leave one out" is, see Miller [15], Hinkley [9]. In the two-sample problem, for example, should we leave out one x_i at a time, then one y_j at a time, or should we leave out all mn pairs (x_i, y_j) one at a time? (The former turns out to be correct.) This problem gets more crucial in the next section, where we consider regression problems.

Let $R((\mathbf{X}, \mathbf{Y}), (F, G))$ be $\hat{\theta}$ itself, so that the bootstrap value of R corresponding to $(\mathbf{X}^*, \mathbf{Y}^*)$ is $R((\mathbf{X}^*, \mathbf{Y}^*), (\hat{F}, \hat{G})) = \hat{\theta}^*$,

$$\hat{\theta}^* = \frac{1}{mn} \Sigma_i \Sigma_j I(X_i^*, Y_j^*). \tag{6.4}$$

Letting $I_{ij}^* = I(X_i^*, Y_j^*)$, straightforward calculations familiar from standard nonparametric theory, give

$$E_* I_{ij}^* = \hat{\theta}, \qquad \mathrm{Var}_* I_{ij}^* = \hat{\theta}(1 - \hat{\theta}), \qquad E_* I_{ij}^* I_{i'j'}^* = \hat{\theta}^2 \quad i \neq i', j \neq j' \tag{6.5}$$

and

$$E_* I_{ij}^* I_{ij'}^* = \int_{-\infty}^{\infty} [1 - \hat{G}(z)]^2 d\hat{F}(z) \equiv \hat{\alpha}, \qquad j \neq j' \tag{6.6}$$

$$E_* I_{ij}^* I_{i'j}^* = \int_{-\infty}^{\infty} \hat{F}^2(z) d\hat{G}(z) \equiv \hat{\beta}, \qquad i \neq i'.$$

Using these results in (6.4) gives

$$\mathrm{Var}_* \hat{\theta}^* = \frac{(n-1)(\hat{\alpha} - \hat{\theta}^2) + (m-1)(\hat{\beta} - \hat{\theta})^2 + \hat{\theta}(1 - \hat{\theta})}{mn}, \tag{6.7}$$

which is the usual estimate for the variance of the Wilcoxon statistic, see Noether [16], page 32.

Method 3, the Taylor series or infinitesimal jackknife, proceeds as in Section 5, with obvious modifications for the two-sample situation. Let $\mathbf{N}_F^* = (N_{F1}^*, N_{F2}^*, \cdots, N_{Fm})$ be the numbers of times $x_1, x_2, \cdots, x_m$ occur in the bootstrap sample $\mathbf{X}^*$, likewise $\mathbf{N}_G^* = (N_{G1}^*, N_{G2}^*, \cdots, N_{Gn}^*)$ for $\mathbf{Y}^*$, and define $\mathbf{P}_F{}^* = \mathbf{N}_F^*/m$, $\mathbf{P}_G^* = \mathbf{N}_G^*/n$, these being independent random vectors with mean and covariance as in (5.2). The expansion corresponding to (5.4) is

$$(6.8)\qquad R(\mathbf{P}_F^*, \mathbf{P}_G^*) \doteq R(\mathbf{e}/m, \mathbf{e}/n) + (\mathbf{P}_F^* - \mathbf{e}/m)\mathbf{U}_F + (\mathbf{P}_G^* - \mathbf{e}/n)\mathbf{U}_G$$
$$+ \frac{1}{2}\big[(\mathbf{P}_F^* - \mathbf{e}/m)V_F(\mathbf{P}_F^* - \mathbf{e}/m)'$$
$$+ 2(\mathbf{P}_F^* - \mathbf{e}/m)V_{FG}(\mathbf{P}_G^* - \mathbf{e}/n)'$$
$$+ (\mathbf{P}_G^* - \mathbf{e}/n)V_G(\mathbf{P}_G^* - \mathbf{e}/n)'\big],$$

where

$$(6.9)\qquad U_{Fi} = \partial R/\partial P_{Fi}^*, \qquad V_{Fii'} = \partial^2 R/\partial P_{Fi}^*\partial P_{Fi'}^*, \qquad V_{FGij} = \partial^2 R/\partial P_{Fi}^*\partial P_{Gj}^*,$$

all the derivatives being evaluated at $(\mathbf{P}_F^*, \mathbf{P}_G^*) = (\mathbf{e}/m, \mathbf{e}/n)$, analogous definitions applying to $\mathbf{U}_G$ and $\mathbf{V}_G$.

The results corresponding to (5.8) and (5.10) are

$$(6.10)\qquad E_*R^* \doteq R(\mathbf{e}/m, \mathbf{e}/n) + \frac{1}{2}\left[\frac{\bar{V}_F}{m} + \frac{\bar{V}_G}{n}\right]$$

and

$$(6.11)\qquad \mathrm{Var}_* R^* \doteq \Sigma_{i=1}^m U_{Fi}^2/m^2 + \Sigma_{j=1}^n U_{Gj}^2/n^2,$$

$\bar{V}_F = \Sigma_i V_{Fii}/m$, $\bar{V}_G = \Sigma_j V_{Gjj}/n$. For $R = \theta(\hat{F}, \hat{G}) - \theta(F, G)$, the approximations corresponding to (5.11) are

(6.12)

$$\mathrm{Bias}_{F,G}\theta(\hat{F}, \hat{G}) \approx \frac{1}{2}\left[\frac{\bar{V}_F}{m} + \frac{\bar{V}_G}{n}\right], \qquad \mathrm{Var}_{F,G}\theta(F, G) \approx \frac{\Sigma_{i=1}^m U_{Fi}^2}{m^2} + \frac{\Sigma_{j=1}^n U_{Gj}^2}{n^2}.$$

For the case of the Wilcoxon statistic (6.11) (or (6.12)) gives

$$(6.13)\qquad \mathrm{Var}_*\hat{\theta}^* \doteq \frac{n[\hat{\alpha} - \hat{\theta}^2] + m[\hat{\beta} - \hat{\theta}^2]}{mn},$$

which should be compared with (6.7).

How can we use the ordinary jackknife to get results like (6.12)? A direct analogy of (5.12) can be carried through, but it is simpler to change definitions slightly, letting

$$(6.14)\qquad D_{(i,\ )} = R(\mathbf{e}/m, \mathbf{e}/n) - R(\mathbf{e}_{(i)}/(m-1), \mathbf{e}/n)$$
$$D_{(\ ,j)} = R(\mathbf{e}/m, \mathbf{e}/n) - R(\mathbf{e}/m, \mathbf{e}_{(j)}/(n-1)),$$

the difference from $R((\mathbf{x}, \mathbf{y}), (\hat{F}, \hat{G}))$ obtained by deleting x_i from $\mathbf{x}$ or y_j from $\mathbf{y}$. Expansion (6.8) gives

$$(6.15) \qquad D_{(i,\ )} \doteq \frac{m-2}{(m-1)^2} U_{Fi} - \frac{1}{2(m-1)^2} V_{Fii}$$

$$D_{(\ ,j)} \doteq \frac{(n-2)^2}{(n-1)^2} U_{Gj} - \frac{1}{2(n-1)^2} V_{Gjj}.$$

From (6.15) it is easy to obtain approximations for the bias and variance expressions in terms of the D's:

$$(6.16) \qquad -\left[\Sigma_{i=1}^m D_{(i,\ )} + \Sigma_{j=1}^n D_{(\ ,j)}\right] \doteq \frac{1}{2}\left[\left(\frac{m}{m-1}\right)^2 \bar{V}_F + \left(\frac{n}{n-1}\right)^2 \bar{V}_G\right],$$

which, as m and n grow large, approaches the second term in (6.10). (For $R = \hat{\theta} - \theta$, this gives the bias-corrected estimate $\tilde{\theta} = (m + n - 1)\hat{\theta} - \Sigma_i \hat{\theta}_{(i,\ )} - \Sigma_j \hat{\theta}_{(\ ,j)}$.) Likewise, just using the first line of (6.8) gives

$$(6.17) \ \Sigma_{i=1}^m D_{(i,\ )}^2 + \Sigma_{j=1}^n D_{(\ ,j)}^2 \doteq \frac{m^2(m-2)^2}{(m-1)^4} \frac{\Sigma_{i=1}^m U_{Fi}^2}{m^2} + \frac{n^2(n-2)^2}{(n-1)^2} \frac{\Sigma_{j=1}^n U_{Gj}^2}{n^2},$$

which approaches (6.11) as $m, n \to \infty$.

The advantage of the D's over expressions like (5.12) is that no group averages, such as $R_{\cdot}^*$, need be defined. Group averages are easy enough to define in the two-sample problem, but are less clear in more complicated situations such as regression. Expressions (6.16) and (6.17) are easy to extend to any situation (which doesn't necessarily mean they give good answers—see the remarks of the next section!).

7. Regression models. A reasonably general regression model is

$$(7.1) \qquad X_i = g_i(\beta) + \epsilon_i \qquad i = 1, 2, \ldots, n,$$

the $g(\cdot)$ being known functions of the unknown parameter vector β, and

$$(7.2) \qquad \epsilon_i \sim_{\text{ind}} F \qquad i = 1, 2, \ldots, n.$$

All that is assumed known about F is that it is centered at zero in some sense, perhaps $E_F\epsilon = 0$ or $\text{Median}_F\epsilon = 0$. Having observed $\mathbf{X} = \mathbf{x}$, we use some fitting technique to estimate β, perhaps least squares,

$$(7.3) \qquad \hat{\beta} : \min_\beta \Sigma_{i=1}^n [x_i - g_i(\beta)]^2,$$

and wish to say something about the sampling distribution of $\hat{\beta}$.

Method 2, the brute force application of the bootstrap, can be carried out by defining $\hat{F}$ as the sample probability distribution of the residuals $\hat{\epsilon}_i$,

$$(7.4) \qquad \hat{F} : \text{mass} \frac{1}{n} \quad \text{at} \quad \hat{\epsilon}_i = x_i - g_i(\hat{\beta}), \qquad i = 1, 2, \ldots, n.$$

(If one of the components of β is a translation parameter for the functions $g(\cdot)$, then $\hat{F}$ has mean zero. If not, and if the assumption $E_F\epsilon = 0$ is very firm, one might still modify $\hat{F}$ by translation to achieve zero mean.) The bootstrap sample, given $(\hat{\beta}, \hat{F})$, is

$$X_i^* = g_i(\hat{\beta}) + \epsilon_i^*, \qquad \epsilon_i^* \sim_{\text{ind}} \hat{F} \qquad i = 1, 2, \ldots, n. \tag{7.5}$$

Each realization of (2.5) yields a realization of $\hat{\beta}^*$ by the same minimization process that gave $\hat{\beta}$,

$$\hat{\beta}^* : \min_\beta \Sigma_{i=1}^n [x_i^* - g_i(\beta)]^2. \tag{7.6}$$

Repeated independent bootstrap replications give a random sample $\hat{\beta}^{*1}, \hat{\beta}^{*2}, \hat{\beta}^{*3}, \ldots, \hat{\beta}^{*N}$ which can be used to estimate the bootstrap distribution of $\hat{\beta}^*$.

A handy test case is the familiar linear model, $g_i(\beta) = c_i\beta$, c_i a known $1 \times p$ vector, with first coordinate $c_{i1} = 1$ for convenience. Let $\mathbf{C}$ be the $n \times p$ matrix whose ith row is c_i, and $\mathbf{G}$ the $p \times p$ matrix $\mathbf{C'C}$, assumed nonsingular. Then the least squares estimator $\hat{\beta} = \mathbf{G}^{-1}\mathbf{C'X}$ has mean β and covariance matrix $\sigma_F^2\mathbf{G}^{-1}$ by the usual theory.

The bootstrap values ϵ_i^* used in (7.5) are independent with mean zero and variance $\hat{\sigma}^2 = \Sigma_{i=1}^n [x_i - g(\hat{\beta})]^2/n$. This implies that $\hat{\beta}^* = \mathbf{G}^{-1}\mathbf{C'X^*}$ has bootstrap mean and variance

$$E_*\hat{\beta}^* = \hat{\beta}, \qquad \text{Cov}_*\hat{\beta}^* = \hat{\sigma}^2\mathbf{G}^{-1}. \tag{7.7}$$

The implication that $\hat{\beta}$ is unbiased for β, with covariance matrix approximately equal to $\hat{\sigma}^2\mathbf{G}^{-1}$, agrees with traditional theory, except perhaps in using the estimate $\hat{\sigma}^2$ for σ^2.

Miller [15] and Hinkley [9] have applied, respectively, the ordinary jackknife and infinitesimal jackknife to the linear regression problem. They formulate the situation as a one-sample problem, with (c_i, x_i) as the ith observed data point, essentially removing one row at a time from the model $\mathbf{X} = \mathbf{C}\beta + \epsilon$. The infinitesimal jackknife gives the approximation

$$\text{Cov}\,\hat{\beta} \approx \mathbf{G}^{-1}[\Sigma_{i=1}^n c_i'c_i\hat{\epsilon}_i^2]\mathbf{G}^{-1}, \tag{7.8}$$

(and the ordinary jackknife a quite similar expression) for the estimated covariance matrix. This doesn't look at all like (7.7)!

The trouble lies in the fact that the jackknife methods as used above ignore an important aspect of the regression model, namely that the errors ϵ_i are assumed to have the same distribution for every value of i. To make (7.8) agree with (7.7) it is only necessary to "symmetrize" the data set by adding hypothetical data points, corresponding to all the possible values of the residual $\hat{\epsilon}$, at each value of i, say

$$x_{ij} = c_i\hat{\beta} + \hat{\epsilon}_j \qquad j = 1, 2, \ldots, n \quad (i = 1, 2, \ldots, n). \tag{7.9}$$

Notice that the bootstrap implicitly does this at step (7.5). Applying the infinitesimal jackknife to data set (7.9), and remembering to take account of the artificially increased amount of data as at step (5.16), gives covariance estimate (7.7).

Returning to the nonlinear regression model (7.1), (7.2), where bootstrap-jackknife methods may really be necessary in order to get estimates of variability for $\hat{\beta}$, we now suspect that jackknife procedures like "leave out one row at a time" may be inefficient unless preceded by some form of data symmetrization such as (7.9). To put things the other way, as in Hinkley [9], such procedures tend to give consistent estimates of Cov $\hat{\beta}$ without assumption (7.2) that the residuals are identically distributed. The price of such complete generality is low efficiency. Usually assumption (7.2) can be roughly justified, perhaps after suitable transformations on X, in which case the bootstrap should give a better estimate of Cov $\hat{\beta}$.

8. Remarks.

REMARK A. Method 2, the straightforward calculation of the bootstrap distribution by repeated Monte Carlo sampling, is remarkably easy to implement on the computer. Given the original algorithm for computing R, only minor modifications are necessary to produce bootstrap replications $R^{*1}, R^{*2}, \ldots, R^{*N}$. The amount of computer time required is just about N times that for the original computations. For the discriminant analysis problem reported in Table 2, each trial of $N = 100$ replications, $m = n = 20$, took about 0.15 seconds and cost about 40 cents on Stanford's 370/168 computer. For a single real data set with $m = n = 20$, we might have taken $N = 1000$, at a cost of \$4.00.

REMARK B. Instead of estimating $\theta(F)$ with $t(\mathbf{X})$, we might make a transformation $\phi = g(\theta)$, $s = g(t)$, and estimate $\phi(F) = g(\theta(F))$ with $s(\mathbf{X}) = g(t(\mathbf{X}))$. That is, we might consider the random variable $S(\mathbf{X}, \mathbf{F}) = s(\mathbf{X}) - \phi(F)$ instead of $R(\mathbf{X}, \mathbf{F}) = t(\mathbf{X}) - \theta(F)$. The effect of such a transformation on the bootstrap is very simple: a bootstrap realization $R^* = R^*(\mathbf{X}^*, \hat{F}) = t(\mathbf{X}^*) - \theta(F)$ transforms into $S = S(\mathbf{X}^*, \hat{F}) = g(t(\mathbf{X}^*)) - g(\theta(\hat{F}))$, or more simply

$$S^* = g(R^* + \hat{\theta}) - g(\hat{\theta}), \tag{8.1}$$

so the bootstrap distribution of R^* transforms into that of S^* by (8.1).

Figure 1 illustrates a simple example. Miller [14], page 12, gives 9 pairs of numbers having sample Pearson correlation coefficient $\hat{\rho} = .945$. The top half of Figure 1 shows the histogram of $N = 1000$ bootstrap replications of $\hat{\rho}^* - \hat{\rho}$, the bottom half the corresponding histogram of $\tanh^{-1} \hat{\rho}^* - \tanh^{-1} \hat{\rho}$. The first distribution straggles off to the left, the second distribution to the right. The median is above zero, but only slightly so compared to the spread of the distributions, indicating that bias correction is not likely to be important in this example.

The purpose of making transformations is, presumably, to improve the inference process. In the example above we might be willing to believe, on the basis of normal theory, that $\tanh^{-1} \hat{\rho} - \tanh^{-1} \rho$ is more nearly *pivotal* than $\hat{\rho} - \rho$ (see

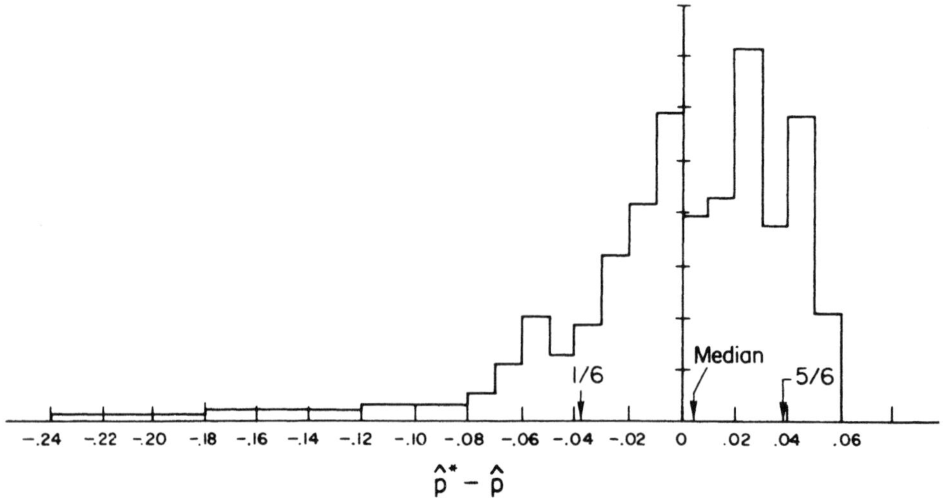

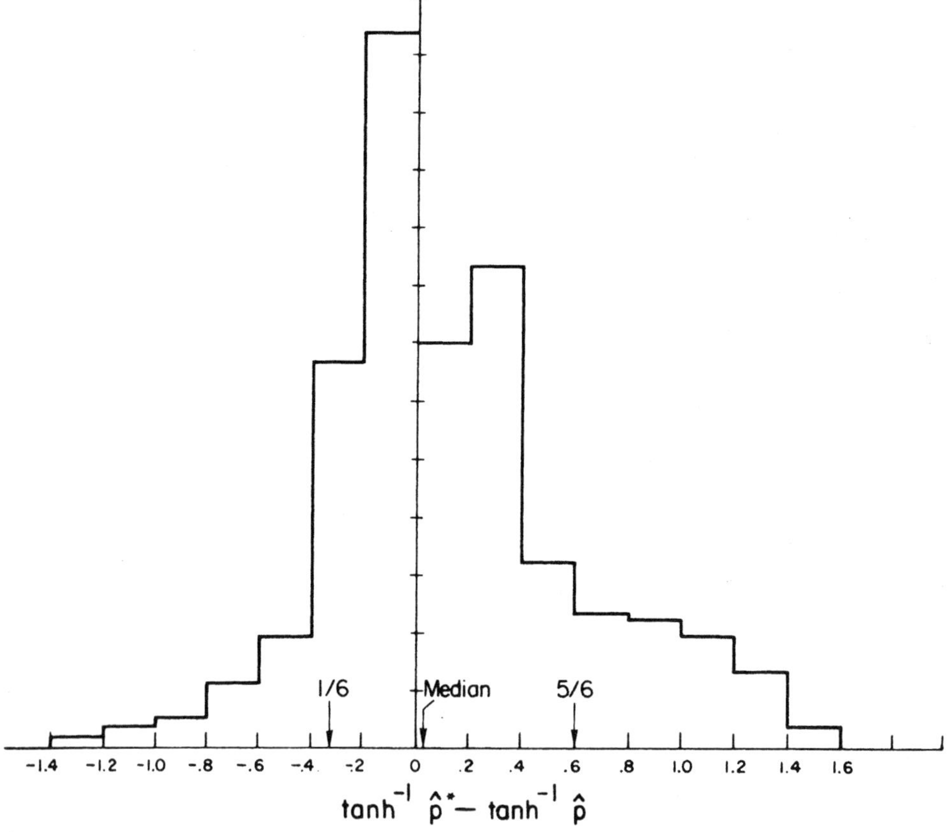

FIG. 1. The top histogram shows $N = 1000$ bootstrap replications of $\hat{\rho}^* - \hat{\rho}$ for the nine data pairs from Miller [10]: (1.15, 1.38), (1.70, 1.72), (1.42, 1.59), (1.38, 1.47), (2.80, 1.66), (4.70, 3.45), (4.80, 3.87), (1.41, 1.31), (3.90, 3.75). The bottom histogram shows the corresponding replications for $\tanh^{-1} \hat{\rho}^* - \tanh^{-1} \hat{\rho}$. The 1/6, 1/2, and 5/6 quantiles are shown for both distributions. All quantiles transform according to equation (8.1).

Remark E) and so more worthwhile investigating by the bootstrap procedure. This does not mean that the bootstrap gives more accurate results, only that the results are more useful. Notice that if $g(\cdot)$ is monotone, then any quantile of the bootstrap distribution of R^* maps into the corresponding quantile of S^* via (8.1), and vice-versa. In particular, if we use the median (rather than the mean) to estimate the center of the bootstrap distribution, then we get the same answer working directly with $\hat{\theta}^* - \hat{\theta}$ ($\hat{\rho}^* - \hat{\rho}$ in the example), or first transforming to $\hat{\phi}^* - \hat{\phi}$ ($\tanh^{-1}\hat{\rho}^* - \tanh^{-1}\hat{\rho}$), taking the median, and finally transforming back to the original scale.

REMARK C. The bias and variance expressions (5.11) suggested by the infinitesimal jackknife transform exactly as in more familiar applications of the "delta method." That is, if $\phi = g(\theta)$, $\hat{\phi} = g(\hat{\theta})$ as above, and $\widehat{\text{Bias}}_F\,\hat{\theta}$, $\widehat{\text{Var}}_F\,\hat{\vartheta}$ are as given in formula (5.11), then it is easy to show that

$$(8.2)\qquad \widehat{\text{Bias}}_F\,\hat{\phi} = g'(\hat{\theta})\widehat{\text{Bias}}_F\,\hat{\theta} + \frac{g''(\hat{\theta})}{2}\widehat{\text{Var}}_F\,\hat{\theta},$$

$$\widehat{\text{Var}}_F\,\hat{\phi} = \left[\,g'(\hat{\theta})\right]^2\widehat{\text{Var}}_F\,\hat{\theta}.$$

In the context of this paper, the infinitesimal jackknife *is* the delta method; starting from a known distribution, that of $\mathbf{P}^*$, approximations to the moments of an arbitrary function $R(\mathbf{P}^*)$ are derived by Taylor series expansion. See Gray et al. [4] for a closely related result.

REMARK D. A standard nonparametric confidence statement for the median $\theta(F)$, $n = 13$, is

$$(8.3)\qquad \text{Prob}_F\{x_{(4)} < \theta < x_{(10)}\} = \text{Prob}\{4 \leqslant \text{Bi}(13, \tfrac{1}{2}) \leqslant 9\} = .908.$$

If we make the continuity correction of halving the end point probabilities, (3.6) gives

$$(8.4)\qquad \text{Prob}_*\{x_{(4)} < \hat{\theta}^* < x_{(10)}\} = .914,$$

where $\hat{\theta}^* = X^*_{(m)}$, the bootstrap value of the sample median. The agreement of (8.4) with (8.3) looks striking, until we try to use (8.4) for inference about θ; (8.4) can be rewritten as $\text{Prob}_*\{x_{(4)} - x_{(7)} < \hat{\theta}^* - \hat{\theta} < x_{(10)} - x_{(7)}\}$ (remembering that $\hat{\theta} = x_{(7)}$), which suggests

$$(8.5)\qquad \text{Prob}_F\{x_{(4)} - x_{(7)} < \hat{\theta} - \theta < x_{(10)} - x_{(7)}\} \approx .914.$$

The resulting confidence interval statement for θ, again using $\hat{\theta} = x_{(7)}$, is

$$(8.6)\qquad \text{Prob}_F\{2x_{(7)} - x_{(10)} < \theta < 2x_{(7)} - x_{(4)}\} \approx .914,$$

which is the reflection of interval (8.3) about the median!

The trouble here has nothing in particular to do with the bootstrap, and does not arise from the possibly large approximation error in statement (8.5), but rather in the inferential step from (8.5) to (8.6), which tries to use $\hat{\theta} - \theta$ as a *pivotal quantity*.

The same difficulty can be exhibited in parametric families: suppose we know that F is a translation of a standard exponential distribution (density e^{-x}, $x > 0$). Then there exist two positive numbers a and b, $a < b$, such that $\text{Prob}_F\{-a < \hat{\theta} - \theta < b\} = .91$. The corresponding interval statement $\text{Prob}_F\{x_{(7)} - b < \theta < x_{(7)} + a\} = .91$ will tend to look more like (8.6) than (8.3).

REMARK E. The difficulty above is a reminder that the bootstrap, and the jackknife, provide approximate *frequency* statements, not approximate *likelihood* statements. Fundamental inference problems remain, no matter how well the bootstrap works. For example, even if the bootstrap expectation $E_*(\hat{\theta}^* - \theta)^2$ very accurately estimates $E_F(\hat{\theta} - \theta)^2$, the resulting interval estimate for θ given $\hat{\theta}$ will be useless if small changes in F (or more exactly, in $\theta(F)$) result in large changes in $E_F(\hat{\theta} - \theta)^2$.

For the correlation coefficient, as discussed in Remark B, Fisher showed that $\tanh^{-1} \hat{\rho} - \tanh^{-1} \rho$ is nearly pivotal when sampling from bivariate normal populations. That is, its distribution is nearly the same for all bivariate normal populations, at least in the range $-.9 < \rho < .9$. This property tends to ameliorate inference difficulties, and is the principal reason for transforming variables, as in Remark B. The theory of pivotal quantities is well developed in parametric families, see Barnard [2], but not in the nonparametric context of this paper.

REMARK F. The classic pivotal quantity is Student's t-statistic. Tukey has suggested using the analogous quantity (2.3) for hypothesis testing purposes, relying on the standard t tables for significance points. This amounts to treating (2.3) as a pivotal quantity for all choices of F, $\theta(F)$, and $t(\mathbf{X})$. The only theoretical justifications for this rather optimistic assumption apply to large samples, where the Student t effect rapidly becomes negligible, see Miller [14]. Given the current state of the theory, one is as well justified in comparing (2.3) to a $\mathfrak{N}(0, 1)$ distribution as to a Student's t distribution (except when $t(\mathbf{X}) = \bar{X}$).

An alternative approach is to bootstrap (2.3) by Method 2 to obtain a direct estimate of its distribution, instead of relying on the t distribution, and then compare the observed value of (2.3) to the bootstrap distribution.

REMARK G. The rationale for bootstrap methods becomes particularly clear when the sample space $\mathfrak{X}$ of the X_i is a finite set, say

$$\mathfrak{X} = \{1, 2, 3, \ldots, L\}. \tag{8.7}$$

The distribution F can now be represented by the vector of probabilities $\mathbf{f} = (f_1, f_2, \ldots, f_L)$, $f_l = \text{Prob}_F\{X_i = l\}$. For a given random sample $\mathbf{X}$ let $\hat{f}_l = \#\{X_i = l\}/n$ and $\hat{\mathbf{f}} = (\hat{f}_1, \hat{f}_2, \ldots, \hat{f}_L)$, so that if $R(\mathbf{X}, F)$ is symmetrically defined in the components of $\mathbf{X}$ we can write it as a function of $\hat{\mathbf{f}}$ and $\mathbf{f}$, say

$$R(\mathbf{X}, F) = Q(\hat{\mathbf{f}}, \mathbf{f}). \tag{8.8}$$

Likewise $R(\mathbf{X}^*, \hat{F}) = Q(\hat{\mathbf{f}}^*, \hat{\mathbf{f}})$, where $\hat{f}_l^* = \#\{X_i^* = l\}/n$ and $\hat{\mathbf{f}}^* = (\hat{f}_1^*, \hat{f}_2^*, \ldots, \hat{f}_L^*)$.

Bootstrap methods estimate the sampling distribution of $Q(\hat{\mathbf{f}}, \mathbf{f})$, given the true distribution $\mathbf{f}$, by the conditional distribution of $Q(\hat{\mathbf{f}}^*, \hat{\mathbf{f}})$ given the observed value of $\hat{\mathbf{f}}$. This is plausible because

$$\hat{\mathbf{f}}|\mathbf{f} \sim \mathfrak{M}_L(n, \mathbf{f})/n \quad \text{and} \quad \hat{\mathbf{f}}^*|\hat{\mathbf{f}} \sim \mathfrak{M}_L(n, \hat{\mathbf{f}})/n \tag{8.9}$$

where $\mathfrak{M}_L(n, \mathbf{f})$ is the L-category multinomial distribution with sample size n, probability vector $\mathbf{f}$. In large samples we expect $\hat{\mathbf{f}}$ to be close to $\mathbf{f}$, so that for reasonable functions $Q(\cdot, \cdot)$ (8.9) should imply the approximate validity of the bootstrap method.

The *asymptotic* validity of the bootstrap is easy to verify in this framework, assuming some regularity conditions on $Q(\cdot, \cdot)$. Suppose that $Q(\mathbf{f}, \mathbf{f}) = 0$ for all $\mathbf{f}$ (as it does in the usual jackknife situation where $R(\mathbf{X}, F) = \theta(\hat{F}) - \theta(F)$); that the vector $\mathbf{u}(\hat{\mathbf{f}}^*, \hat{\mathbf{f}})$ with lth component equal to $\partial Q(\hat{\mathbf{f}}^*, \hat{\mathbf{f}})/\partial \hat{f}_l^*$ exists continuously for $(\hat{\mathbf{f}}^*, \hat{\mathbf{f}})$ in an open neighborhood of $(\mathbf{f}, \mathbf{f})$; and that $\mathbf{u} = \mathbf{u}(\mathbf{f}, \mathbf{f})$ does not equal zero. By Taylor's theorem, and the fact that $\hat{\mathbf{f}}^*$ and $\hat{\mathbf{f}}$ converge to $\mathbf{f}$ with probability one,

$$Q(\hat{\mathbf{f}}, \mathbf{f}) = (\hat{\mathbf{f}} - \mathbf{f})(\mathbf{u} + \boldsymbol{\epsilon}_n) \quad \text{and} \quad Q(\hat{\mathbf{f}}^*, \mathbf{f}) = (\hat{\mathbf{f}}^* - \hat{\mathbf{f}})(\mathbf{u} + \hat{\boldsymbol{\epsilon}}_n), \tag{8.10}$$

both $\boldsymbol{\epsilon}_n$ and $\hat{\boldsymbol{\epsilon}}_n$ converging to zero with probability one. From (8.9) and the fact that $\hat{\mathbf{f}}$ converges to $\mathbf{f}$ with probability one, we have

$$n^{\frac{1}{2}}(\hat{\mathbf{f}} - \mathbf{f})|\mathbf{f} \to \mathfrak{N}_L(\mathbf{0}, \Sigma_f) \quad \text{and} \quad n^{\frac{1}{2}}(\hat{\mathbf{f}}^* - \hat{\mathbf{f}})|\hat{\mathbf{f}} \to \mathfrak{N}_L(\mathbf{0}, \Sigma_f), \tag{8.11}$$

where Σ_f is the matrix with element (l, m) equal to $f_l(\delta_{lm} - f_m)$. Combining (8.10) and (8.11) shows that the bootstrap distribution of $n^{\frac{1}{2}}Q(\hat{\mathbf{f}}^*, \hat{\mathbf{f}})$, given $\hat{\mathbf{f}}$, is asymptotically equivalent to the sampling distribution of $n^{\frac{1}{2}}Q(\hat{\mathbf{f}}, \mathbf{f})$, given the true probability distribution $\mathbf{f}$. Both have the limiting distribution $\mathfrak{N}(0, \mathbf{u}'\Sigma_f\mathbf{u})$.

The argument above assumes that the form of $Q(\cdot, \cdot)$ does not depend upon n. More careful considerations are necessary in cases like (2.3) where $Q(\cdot, \cdot)$ does depend on n, but in a minor way. Some nondifferentiable functions such as the sample median (3.3) can also be handled by a smoothing argument, though direct calculation of the limiting distribution is easier in that particular case.

REMARK H. Taylor expansion (5.4) looks suspicious because the dimension of the vectors involved increases with the sample size n. However in situation (8.7), (8.8), it is easy to verify that (5.4) is the same as the second order Taylor expansion of $Q(\hat{\mathbf{f}}^*, \hat{\mathbf{f}})$, for $\hat{\mathbf{f}}^*$ near $\hat{\mathbf{f}}$,

$$Q(\hat{\mathbf{f}}^*, \hat{\mathbf{f}}) \doteq Q(\hat{\mathbf{f}}, \hat{\mathbf{f}}) + (\hat{\mathbf{f}}^* - \hat{\mathbf{f}})\hat{\mathbf{u}} + \tfrac{1}{2}(\hat{\mathbf{f}}^* - \hat{\mathbf{f}})\hat{\mathbf{v}}(\hat{\mathbf{f}}^* - \hat{\mathbf{f}}). \tag{8.12}$$

Here $\hat{\mathbf{u}}$ has lth element $\partial Q(\hat{\mathbf{f}}^*, \hat{\mathbf{f}})/\partial \hat{\mathbf{f}}_l^*|_{\hat{\mathbf{f}}^* = \hat{\mathbf{f}}}$ and $\hat{\mathbf{v}}$ has l,mth element $\partial^2 Q(\hat{\mathbf{f}}^*, \hat{\mathbf{f}})/\partial \hat{f}_l^* \partial \hat{f}_m^*|_{\hat{\mathbf{f}}^* = \hat{\mathbf{f}}}$. The dimension of the vectors in (8.12) is L, and does not increase with sample size n. Expressions (5.8), (5.10) are the standard delta theory approximation for the mean and variance of $Q(\hat{\mathbf{f}}^*, \hat{\mathbf{f}})$, given $\hat{\mathbf{f}}$, obtained from (8.12) and the distributional properties of $\hat{\mathbf{f}}^*|\hat{\mathbf{f}} \sim \mathfrak{M}_L(n, \hat{\mathbf{f}})$.

REMARK I. Hartigan [5, 7] has suggested using *subsample* values to obtain confidence statements for an estimated parameter. His method consists of choosing a vector $\mathbf{x}^*$ whose components are a nonempty subset of the observed data vector $\mathbf{X} = \mathbf{x}$ (so each component x_i appears either zero or one time in $\mathbf{x}^*$). This process is repeated N times, where N is small compared to 2^n, giving vectors $\mathbf{x}^{*1}, \mathbf{x}^{*2}, \cdots, \mathbf{x}^{*N}$ and corresponding subsample values $t(\mathbf{x}^{*1}), t(\mathbf{x}^{*2}), \cdots, t(\mathbf{x}^{*N})$ for some symmetric estimator $t(\cdot)$ defined for samples of an arbitrary size. By a clever choice of the vectors $\mathbf{x}^{*j}$, and for certain special estimation problems, the $t(\mathbf{x}^{*j})$ can be used to make precise confidence statements about an unknown parameter. More importantly in the context of this paper, Hartigan shows that by choosing the $\mathbf{x}^{*j}$ randomly, without replacement, from the $2^n - 1$ possible nonempty subsamples of x, asymptotically valid confidence statements can be made under fairly general conditions. This is very similar to bootstrap Method 2, except that the $\mathbf{x}^{*j}$ are selected by subsampling rather than bootstrapping.

In the finite case (8.7), let $\mathbf{x}^*$ be a randomly selected subsample vector, and let $\hat{f}_l^* = \#\{x_i^* = l\}/(\text{number of components of } \mathbf{x}^*)$, so $\hat{\mathbf{f}}^* = (\hat{f}_1^*, \hat{f}_2^*, \cdots, \hat{f}_L^*)$, as before, is the vector of proportions in the artificially created sample. It is easy to show that $n^{\frac{1}{2}}(\hat{\mathbf{f}}^* - \hat{\mathbf{f}})|\hat{\mathbf{f}} \to \mathfrak{N}_L(\mathbf{0}, \Sigma_f)$, as at (8.11), which is all that is needed to get the same asymptotic properties obtained for the bootstrap. (Conversely, it can be shown that bootstrap samples have the same asymptotic "typicality" properties Hartigan discusses in [5, 7].) The bootstrap *may* give better small sample performance, because the similarity in (8.9), which is unique to the bootstrap, is a stronger property than the asymptotic equivalence (8.11), and also because the artificial samples used by the bootstrap are the same size as the original sample. However, no evidence one way or the other is available at the present time.

Hartigan's 1971 paper [6] introduces another method of resampling, useful for constructing prediction intervals, which only involves artificial samples of the same size as the real sample. Let $\{x_1^*, x_2^*, \cdots, x_n^*\}$ be a set of size n, each element of which is selected with replacement from $\{x_1, x_2, \cdots, x_n\}$. There are $\binom{2n-1}{n-1}$ distinct such sets, not counting differences in the order of selection. (For example $\{x_1, x_2\}$ yields the three sets $\{x_1, x_1\}$, $\{x_2, x_2\}$, $\{x_1, x_2\}$.) The random version of Hartigan's second method selects $\mathbf{x}^*$, or more exactly the set of components of $\mathbf{x}^*$, with equal probability from among these $\binom{2n-1}{n-1}$ possible choices. It can be shown that this results in $n^{\frac{1}{2}}(\hat{\mathbf{f}}^* - \hat{\mathbf{f}})|\hat{\mathbf{f}} \to \mathfrak{N}_L(\mathbf{0}, 2\Sigma_f)$, so that the asymptotic covariance matrix is twice what it is in (8.11). Looking at (8.10), one sees that for this resampling scheme, $2^{-\frac{1}{2}} Q(\hat{\mathbf{f}}^*, \hat{\mathbf{f}})$ has the same asymptotic distribution as $Q(\hat{\mathbf{f}}, \mathbf{f})$.

It is not difficult to construct other resampling schemes which give correct asymptotic properties. The important question, but one which has not been investigated, is which scheme is most efficient and reliable in small samples.

REMARK J. In situation (8.7), (8.8), the ordinary jackknife depends on evaluating $Q(\hat{\mathbf{f}}^*, \hat{\mathbf{f}})$ for vectors $\hat{\mathbf{f}}^*$ of the form $\hat{\mathbf{f}}^*_{(l)}$,

$$(\hat{\mathbf{f}}^*_{(l)} - \hat{\mathbf{f}}) = \frac{1}{n-1}(\hat{\mathbf{f}} - \mathbf{e}_l), \tag{8.13}$$

$\mathbf{e}_l = (0, 0, \ldots, 1, 0, \ldots, 0)$, 1 in the lth place. (The values of l needed are those occurring in the observed sample $(x_1, x_2, \ldots, x_n)$; a maximum of $\min(n, L)$ different l values are possible.) Notice that

$$\|\hat{f}^*_{(l)} - \hat{f}\| \leqslant \frac{2^{\frac{1}{2}}}{n-1}. \tag{8.14}$$

The "resampling" vectors $\hat{f}^*_{(l)}$ are distance $O(1/n)$ away from $\hat{\mathbf{f}}$, as compared to $O_p(n^{-\frac{1}{2}})$ for the bootstrap vectors $\hat{\mathbf{f}}^*$, as seen in (8.11). In the case of the median, (3.3), the jackknife fails because of its overdependence on the behavior of $Q(\hat{\mathbf{f}}^*, \hat{\mathbf{f}})$ for $\hat{\mathbf{f}}^*$ very near $\hat{\mathbf{f}}$. In this case the derivative of the function $Q(\cdot, \cdot)$ is too irregular for the jackknife's quadratic extrapolation formulas to work. The grouped jackknife, in which the $\hat{\mathbf{f}}^*$ vectors are created by removing observations from $\mathbf{x}$ in groups of size g at a time, see page 1 of Miller [14], overcomes this objection if g is sufficiently large. (The calculations above suggest $g = O(n^{\frac{1}{2}})$.) As a matter of fact, the grouped jackknife gives the correct asymptotic variance for the median. If g is really large, say $g = n/2$, and the removal groups are chosen randomly, then this resampling method is almost the same as Hartigan's subsampling plan, discussed in Remark I.

REMARK K. We have applied the bootstrap in a nonparametric way, but there is no reason why it cannot be used in parametric problems. The only change necessary is that at (2.4), $\hat{F}$ is chosen to be the parametric m.l.e. for F, rather than the nonparametric m.l.e. As an example, suppose that F is known to be normal, with unknown mean and variance, and that we are interested in the expectation of $R(\mathbf{X}, F) = I_{[a,b]}(\bar{X})$, i.e., the probability that $\bar{X}$ occurs in a prespecified interval $[a, b]$. Then the *nonparametric* bootstrap estimate is $E_* R^* = \hat{G}^{(n)}(b) - \hat{G}^{(n)}(a)$, where $\hat{G}^{(n)}$ is the cdf of $\Sigma_{i=1}^n X_i^*/n$, obtained by convoluting the sample distribution $\hat{F}$ n times and then rescaling by division by n. The *parametric* bootstrap estimate is $E_* R^* = \Phi\big((b - \bar{x})/(\hat{\sigma} - n^{\frac{1}{2}})\big) - \Phi\big((a - \bar{x})/(\hat{\sigma}/n^{\frac{1}{2}})\big)$, where $\hat{\sigma} = \hat{\mu}_2^{\frac{1}{2}}$ and $\Phi(\cdot)$ is the standard normal cdf. If F is really normal and if n is moderately large, $n \geqslant 20$ according to standard Edgeworth series calculations, then the two estimates will usually be in close agreement.

It can be shown that the parametric version of Method 3 of the bootstrap, applied to estimating the variance of the m.l.e. in a one parameter family, gives the usual approximation: one over the Fisher information. The calculation is almost the same as that appearing in Section 3 of Jaeckel [10].

Acknowledgments. I am grateful to Professors Rupert Miller and David Hinkley for numerous discussions, suggestions and references, and to Joseph Verducci for help with the numerical computations. The referees contributed several helpful ideas, especially concerning the connection with Hartigan's work, and the large sample theory. I also wish to thank the many friends who suggested names more colorful than *Bootstrap*, including *Swiss Army Knife, Meat Axe, Swan-Dive, Jack-Rabbit*, and my personal favorite, the *Shotgun*, which, to paraphrase Tukey, "can blow the head off any problem if the statistician can stand the resulting mess."

REFERENCES

[1] Anderson, T. W. (1958). *An Introduction to Multivariate Statistical Analysis*. Wiley, New York.

[2] Barnard, B. (1974). Conditionality, pivotals, and robust estimation. *Proceedings of the Conference on Foundational Questions in Statistical Inference*. Memoirs No. 1, Dept. of Theoretical Statist., Univ. of Aarhus, Denmark.

[3] Cramér, H. (1946). *Mathematical Methods in Statistics*. Princeton Univ. Press.

[4] Gray, H., Schucany, W. and Watkins, T. (1975). On the generalized jackknife and its relation to statistical differentials. *Biometrika* **62** 637–642.

[5] Hartigan, J. A. (1969). Using subsample values as typical values. *J. Amer. Statist. Assoc.* **64** 1303–1317.

[6] Hartigan, J. A. (1971). Error analysis by replaced samples. *J. Roy. Statist. Soc. Ser. B* **33** 98–110.

[7] Hartigan, J. A. (1975). Necessary and sufficient conditions for asymptotic joint normality of a statistic and its subsample values. *Ann. Statist.* **3** 573–580.

[8] Hinkley, D. (1976[a]). On estimating a symmetric distribution. *Biometrika* **63** 680.

[9] Hinkley, D. (1976[b]). On jackknifing in unbalanced situations. Technical Report No. 22, Division of Biostatistics, Stanford Univ.

[10] Jaeckel, L. (1972). The infinitesimal jackknife. Bell Laboratories Memorandum ♯MM 72-1215-11.

[11] Kendall, M. and Stuart, A. (1950). *The Advanced Theory of Statistics*. Hafner, New York.

[12] Lachenbruch, P. and Mickey, R. (1968). Estimation of error rates in discriminant analysis. *Technometrics* **10** 1–11.

[13] Maritz, J. S. and Jarrett, R. G. (1978). A note on estimating the variance of the sample median. *J. Amer. Statist. Assoc.* **73** 194–196.

[14] Miller, R. G. (1974[a]). The jackknife—a review. *Biometrika* **61** 1–15.

[15] Miller, R. G. (1974[b]). An unbalanced jackknife. *Ann. Statist.* **2** 880–891.

[16] Noether, G. (1967). *Elements of Nonparametric Statistics*. Wiley, New York.

[17] Toussaint, G. (1974). Bibliography on estimation of misclassification. *IEEE Trans. Information Theory* **20** 472–479.

Department of Statistics
Stanford University
Stanford, California 94305

10

The Jackknife Estimate of Variance

Introduction by Jun Shao
University of Wisconsin, Madison

and C. F. Jeff Wu
Georgia Institute of Technology

In his seminal 1979 paper on the "Bootstrap methods," Efron used "Another look at the jackknife" as a subtitle. As the bootstrap was still a novice, this subtitle allowed it to be connected with a known commodity called "the jackknife." One may also say that it was Efron's modest way of introducing his new and ground-breaking idea to the world. Judging from his writing in this period and his long-term association with Rupert Miller and David Hinkley (two active researchers on the jackknife), it is a reasonable speculation that Efron's work on the bootstrap came at least in part from his attempt to understand the somewhat mysterious (at least up to the early 1980s) method of jackknife. Efron has gone far beyond that. He has single-handedly invented a new methodology for assessing error in statistical analysis which is unrivalled except by Neyman's work on confidence intervals.

The Efron–Stein (ES) paper was written about the same time as his 1979 bootstrap paper. This timing further lends support to our contention that his bootstrap work was initially influenced by the jackknife. So how does the ES paper fit into the picture? A bit of the history on the jackknife will provide a good perspective. The original work by Quenouille (1949) was proposed for bias reduction motivated by a problem in time series. It was Tukey who recognized the potentially general use of this resampling scheme for estimation of variance. He also coined the catchy term "jackknife." (Efron was similarly talented in coining the term "bootstrap" for his more general method.) As Tukey's original work on the jackknife did not appear in any journal, one can only refer to his short abstract for an IMS talk in 1958 for information. When he proposed the jackknife variance estimator $\widehat{\mathrm{VAR}}$ in (1.2) of the ES paper, the main justification is that it matches the usual variance estimate (sample variance divided by sample size) for the sample mean (a linear statistic). For a smooth nonlinear statistic, it can be shown by Taylor expansion or other asymptotic arguments that $\widehat{\mathrm{VAR}}$ captures the leading term in the unknown variance of the statistic. Other than the linear statistics and in spite of the work in an active research area for more than 20 years, there have not been any *exact* results obtained for the jackknife variance estimators until the Efron–Stein work.

When ES proved in this paper that the jackknife variance estimator overestimates (slightly) the unknown variance, the results came as a surprise. At first glance, one could not believe it was possible to prove such results for very general statistics. The trick is to exploit the ANOVA decomposition for a general statistic S, which allows the expectation of $\widehat{\mathrm{VAR}}$ and the unknown variance $\mathrm{VAR}S$ to be expanded in finite sums of terms in descending orders. By comparing the leading terms on the two sides, the upward-biased results follow easily. These expansions also lead to exact formulas for the bias of the variance estimator. The proofs themselves are textbook examples of clarity and elegance. ES was generous in crediting past work by Hoeffding, Hajek, and Mallows for using similar techniques. The

ANOVA-type decomposition has since been successfully used in proving asymptotic results in areas ranging from function estimation to space-filling designs and estimation.

The result on statistical functionals in Sect. 3 of the ES paper has had a great influence on the theoretical (asymptotic) studies of the jackknife over the last 25 years. Prior to the ES paper, the asymptotic properties of the jackknife variance estimator were investigated for functions of sample means, U-statistics, and parametric maximum likelihood estimators. Although the result in the ES paper is limited to some simple functionals, it provides insight for the later development in application of the jackknife for a very general class of statistical functionals (Parr 1985; Shao and Wu 1989; Shao 1993a). Furthermore, it also leads into the study of why jackknifing the sample median does not work and why a more general jackknife, the delete-d jackknife, rectifies the problem (Shao and Wu 1989).

The discussion of sample size adjustment in the ES paper is another one that has had impact on subsequent research on the jackknife. A rather mysterious factor, $\frac{n-1}{n}$, where n is the sample size, is used in the jackknife formulae (1.4) and (1.5) of the ES paper. Tukey (1958) provided an explanation on the use of this factor by showing that $\widehat{\text{VAR}}$ reduces to the usual sample variance divided by n when the statistic of concern is the sample mean. For a general statistic S, results (2.11) and (2.14) of the ES paper gave a convincing argument on why this factor is needed. Since the leading term in $\text{VAR}S$ is σ_α^2/n for some σ_α^2 and the leading term in $E\widetilde{\text{VAR}}$ is $\sigma_\alpha^2/(n-1)$, where $\widetilde{\text{VAR}} = \frac{n}{n-1}\widehat{\text{VAR}}$ (the jackknife estimate without the factor $\frac{n-1}{n}$), multiplying by $\frac{n-1}{n}$ gives a natural adjustment. The result also reveals the fact that the jackknife estimate $\widetilde{\text{VAR}}$ is an estimate of the variance of S with a sample of size $n-1$, rather than n. Later work on the delete-d cross-validation (Shao 1993b) and subsampling (Politis et al. 1999) has been, to a certain extent, influenced by this result.

References

Efron, B. (1979). Bootstrap methods: Another look at the jackknife. *Annals of Statistics* **7**, 1–26.

Parr, W. C. (1985). Jackknifing differentiable statistical functionals. *Journal of the Royal Statistical Society, Series B* **47**, 56–66.

Politis, D. N., Romano, J. P., and Wolf, M. (1999). *Subsampling*. Springer, New York.

Quenouille, M. (1949). Approximation tests of correlation in time series. *Journal of the Royal Statistical Society, Series B* **11**, 18–84.

Shao, J. (1993a). Differentiability of statistical functionals and consistency of the jackknife. *Annals of Statistics* **21**, 61–75.

Shao, J. (1993b). Linear model selection by cross-validation. *Journal of the American Statistical Association* **88**, 486–494.

Shao, J. and Wu, C. F. J. (1989). A general theory for jackknife variance estimation. *Annals of Statistics* **17**, 1176–1197.

Tukey, J. (1958). Bias and confidence in not-quite large samples (Abstract). *Annals of Mathematical Statistics* **29**, 614.

The Annals of Statistics
1981, Vol. 9, No. 3, 586–596

THE JACKKNIFE ESTIMATE OF VARIANCE

BY B. EFRON AND C. STEIN

Stanford University

Tukey's jackknife estimate of variance for a statistic $S(X_1, X_2, \cdots, X_n)$ which is a symmetric function of i.i.d. random variables X_i, is investigated using an ANOVA-like decomposition of S. It is shown that the jackknife variance estimate tends always to be biased upwards, a theorem to this effect being proved for the natural jackknife estimate of Var $S(X_1, X_2, \cdots, X_{n-1})$ based on $X_1, X_2, \cdots, X_n$.

1. Introduction. The Quenouille-Tukey jackknife, as described in Miller (1974), gives useful nonparametric estimates of bias and variance. Suppose $S(X_1, X_2, \cdots, X_n)$ is a statistic of interest,where $X_1, X_2, \cdots, X_n$ are independent and identically distributed observable random variables, and $S(X_1, X_2, \cdots, X_n)$ is invariant under permutation of the arguments. The jackknife estimate of variance, $\widehat{\text{VAR}}\, S(X_1, X_2, \cdots, X_n)$, is defined in terms of the quantities

$$S_{(i)} \equiv S(X_1, X_2, \cdots, X_{i-1}, X_{i+1}, \cdots, X_n), \tag{1.1}$$

the value of S when X_i is deleted from the sample,

$$\widehat{\text{VAR}}\, S(X_1, X_2, \cdots, X_n) \equiv \frac{n-1}{n} \sum_{i=1}^{n} [S_{(i)} - S_{(\cdot)}]^2, \tag{1.2}$$

where

$$S_{(\cdot)} \equiv \sum_{i=1}^{n} S_{(i)}/n. \tag{1.3}$$

Formula (1.2) is often used as a variance estimate for the jackknife version of S, defined as $nS - (n-1)S_{(\cdot)}$, but here we will be thinking of it either as a variance estimate for S itself, or perhaps more appropriately for $S_{(\cdot)}$.

Notice that $\widehat{\text{VAR}}$ is defined entirely with respect to samples of size $n-1$, rather than n. It is useful to think of $\widehat{\text{VAR}}\, S(X_1, X_2, \cdots, X_n)$ as estimating the true variance Var $S(X_1, X_2, \cdots, X_n)$ in two distinct steps, (i) a direct estimate of Var $S(X_1, X_2, \cdots, X_{n-1})$ the variance for sample size $n-1$, and (ii) a modification to go from sample size $n-1$ to sample size n. The direct estimate is

$$\widetilde{\text{VAR}}\, S(X_1, X_2, \cdots, X_{n-1}) \equiv \sum_{i=1}^{n} [S_{(i)} - S_{(\cdot)}]^2, \tag{1.4}$$

and the sample size modification is

$$\widehat{\text{VAR}}\, S(X_1, X_2, \cdots, X_n) = \frac{n-1}{n} \widetilde{\text{VAR}}\, S(X_1, X_2, \cdots, X_{n-1}), \tag{1.5}$$

which together give (1.2). Notice that $\widetilde{\text{VAR}}\, S(X_1, X_2, \cdots, X_{n-1})$ is a function of all n variables $S(X_1, X_2, \cdots, X_n)$.

Our main result is that the jackknife estimate of variance (1.4) is conservative in expectation,

---

Received March, 1979; revised October, 1979.

AMS 1970 *subject classification.* 62G05.

Key words and phrases. Jackknife, variance estimation, ANOVA decomposition, bootstrap, U statistics.

$$E\{\widetilde{\text{VAR}}\, S(X_1, X_2, \cdots, X_{n-1})\} \geq \text{Var}\, S(X_1, \cdots, X_{n-1}) \tag{1.6}$$

for any symmetric function $S(X_1, X_2, \cdots, X_{n-1})$. As a matter of fact, neither symmetry nor identical distributions for the X_i are essential, as shown in Comment 4 of Section 2. We will also discusss (1.5), and show that under certain conditions, in particular if S is a U statistic, this step produces a further conservative bias in the jackknife variance estimate, though the results are not as satisfactory here.

The main tool in verifying (1.6) is an ANOVA-like decomposition of $S(X_1, X_2, \cdots, X_n)$, described in Section 2, which is a simple extension of the "Hajek projection", Hajek (1968). Colin Mallows, in lectures and an unpublished paper (1975), has developed closely related methods. All of these ideas connect with Hoeffding's (1948) well known work on U statistics. (See also Rubin and Vitale (1980) for a similar development.) A simple formula for the bias in (1.6) is derived from the ANOVA decomposition.

Much of jackknife theory concerns statistics S which are smooth functions of the empirical probability distribution. Section 3 relates this concept to the ANOVA decomposition of S, particularly focusing on *quadratic functionals*, which are useful in understanding the approximations involved in jackknife estimates, both for bias and for variance. This approach is quite similar to that of Hinkley (1978), as are the results of Section 5. The rationale behind (1.5) is examined in Section 4. Section 5 suggests a bias correction technique for the jackknife variance estimate. The "bootstrap", which is a generalization of the jackknife described in Efron (1979a), is examined briefly in Section 6.

2. ANOVA decomposition of $S(X_1, X_2, \cdots, X_n)$. A random variable $S(X_1, X_2, \cdots, X_n)$ which is a function of n independent random variables $X_1, X_2, \cdots, X_n$ can be decomposed into "main effects", "interactions", "higher order interactions", etc., in a manner directly analogous to the decomposition of a complete n-way ANOVA table. Here we do *not* have to assume that the X_i are identically distributed, nor that S is symmetrically defined with respect to its n arguments. The only assumption is that $ES^2 < \infty$. Taking advantage of this wide generality, an extended version of the main result (1.6) is given in Comment 4.

The quantities involved in the decomposition, and their corresponding ANOVA names are

$$\mu = ES, \tag{2.1}$$

grand mean;

$$A_i(x_i) = E\{S \mid X_i = x_i\} - \mu, \tag{2.2}$$

ith main effect;

$$B_{ii'}(x_i, x_{i'}) = E\{S \mid X_i = x_i, X_{i'} = x_{i'}\} - E\{S \mid X_i = x_i\} - E\{S \mid X_{i'} = x_{i'}\} + \mu, \tag{2.3}$$

ii'th second order interaction; etc.

DECOMPOSITION LEMMA. *The random variable $S(X_1, X_2, \cdots, X_n)$ can be expressed as*

$$S(X_1, X_2, \cdots, X_n) = \mu + \textstyle\sum_i A_i(X_i) + \sum_{i<i'} B_{ii'}(X_i, X_{i'}) \tag{2.4}$$
$$+ \textstyle\sum_{i<i'<i''} C_{i,i',i''}(X_i, X_{i'}, X_{i''}) + \cdots + H(X_1, X_2, \cdots, X_n),$$

where all $2^n - 1$ random variables on the right side of (2.5) *have mean zero and are mutually uncorrelated with each other.*

PROOF. Following through definitions (2.1)–(2.4), the coefficient of μ on the right side of (2.4) is

$$1-\binom{n}{1}+\binom{n}{2}-\binom{n}{3}\cdots=(1-1)^n=0. \tag{2.5}$$

Likewise the coefficient of $E\{S|X_i\}$ is $(1-1)^{n-1}=0$, the coefficient of $E\{S|X_i, X_{i'}\}$ is $(1-1)^{n-2}=0$, etc. The last term in (2.4) $H(X_1, \cdots, X_n)$, itself has first term $S(X_1, X_2, \cdots, X_n)$, which is the only term not cancelling out. This verifies expression (2.4).

Notice that

$$EA_i(X_i)=E\{E(S|X_i)-\mu\}=0. \tag{2.6}$$

A similar calculation shows that

$$\begin{aligned} E\{B_{ii'}(X_i, X_{i'})|X_i\} &= E\{C_{ii'i''}(X_i, X_{i'}, X_{i''})|X_i, X_{i'}\} \\ &= \cdots E\{H(X_1, X_2, \cdots, X_n)|X_1, X_2, \cdots, X_{n-1})=0, \end{aligned} \tag{2.7}$$

and likewise $E\{B_{ii'}(X_i, X_{i'})|X_{i'}\}=0$, etc. Together, (2.6) and (2.7) imply that all the random variables on the right side of (2.4) have mean zero and correlation zero, which completes the proof of the lemma. Note that $\mu+\sum_{i=1}^n A_i(X_i)$ is the Hajek projection of $S(X_1, X_2, \cdots, X_n)$, Hajek (1960). Expansion (2.4) is unique in the sense that once given properties (2.6)–(2.7), the terms μ, A_i, $B_{ii'}$, $C_{ii'i''}$, $\cdots$ must be given by expressions (2.1)–(2.3).

We now return to the situation where $X_1, X_2, \cdots, X_n$ are i.i.d., and $S(X_1, X_2, \cdots, X_n)$ is symmetrically defined in its arguments. In this case the functions $A_i(\cdot)$, $B_{ii'}(\cdot,\cdot)$, $\cdots$ do not depend upon the subscripts $i, i', i'', \cdots$, and we can indicate them as $A(\cdot)$, $B(\cdot,\cdot)$, $\cdots$. It will be helpful, for reasons stated in Section 3, to rescale and rename quantities as follows,

$$\begin{aligned} &\alpha_i \equiv \alpha(X_i) \equiv nA(X_i), \qquad \beta_{ii'} \equiv \beta(X_i, X_{i'}) \equiv n^2B(X_i, X_{i'}), \\ &\gamma_{ii'i''} \equiv \gamma(X_i, X_{i'}, X_{i''}) \equiv n^3\, C(X_i, X_{i'}, X_{i''}), \cdots . \end{aligned} \tag{2.8}$$

Expansion (2.4) now becomes

$$\begin{aligned} S(X_1, X_2, \cdots, X_n) &= \mu+\frac{1}{n}\textstyle\sum_i \alpha_i+\frac{1}{n^2}\sum_{i<i'}\beta_{ii'} \\ &\quad+\frac{1}{n^3}\textstyle\sum_{i<i'<i''}\gamma_{ii'i''}+\cdots+\frac{1}{n^n}\eta_{1,2,3,\cdots,n}. \end{aligned} \tag{2.9}$$

Example. If $S(X_1, X_2, \cdots, X_n)=\sum_{i=1}^n (X_i-\bar{X})^2/n$, where $\bar{X}=\sum_{i=1}^n X_i/n$, and if the X_i have mean ξ and variance σ^2, then $\mu=\dfrac{n-1}{n}\sigma^2$, $\alpha(X_i)=\dfrac{n-1}{n}[(X_i-\xi)^2-\sigma^2]$, $\beta(X_i, X_{i'})=-2(X_i-\xi)(X_{i'}-\xi)$, and all the higher order terms equal zero.

Expansion (2.9), which is similar to a form Colin Mallows has suggested in unpublished lectures, leads to an easy proof of (1.6). Define

$$\sigma_\alpha^2 \equiv \operatorname{Var}\alpha(X_i), \qquad \sigma_\beta^2 \equiv \operatorname{Var}\beta(X_i, X_{i'}), \qquad \sigma_\gamma^2 \equiv \operatorname{Var}\gamma(X_i, X_{i'}, X_{i''}), \cdots. \tag{2.10}$$

Then using the fact that the terms in (2.9) are uncorrelated, we get

$$\operatorname{Var} S(X_1, X_2, \cdots, X_n)=\frac{\sigma_\alpha^2}{n}+\binom{n-1}{1}\frac{\sigma_\beta^2}{2n^3}+\binom{n-1}{2}\frac{\sigma_\gamma^2}{3n^5}+\cdots+\frac{\sigma_\eta^2}{n^{2n}}. \tag{2.11}$$

For any two indices i and i', the difference of the deleted sample values $S_{(i)}$ and $S_{(i')}$ is, by (2.9), equal to

$$S_{(i)}-S_{(i')}=\frac{1}{n-1}[\alpha_{i'}-\alpha_i]+\frac{1}{(n-1)^2}\textstyle\sum_j^{(i,i')}[\beta_{i'j}-\beta_{ij}]$$

(2.12)

$$+\frac{1}{(n-1)^3}\sum_{j<j'}^{(i,i')}[\gamma_{i'jj''}-\gamma_{ijj'}]+\cdots,$$

where the notation $\sum^{(i,i')}$ indicates summation *avoiding* the values i and i'. This implies that

$$E[S_{(i)}-S_{(i')}]^2=2\left[\frac{\sigma_\alpha^2}{(n-1)^2}+\binom{n-2}{1}\frac{\sigma_\beta^2}{(n-1)^4}+\binom{n-2}{2}\frac{\sigma_\gamma^2}{(n-1)^6}+\cdots\right]. \tag{2.13}$$

Since, by elementary algebra, $\widetilde{\text{VAR}}\ S(X_1, X_2, \cdots, X_{n-1}) \equiv \sum_{i=1}^n [S_{(i)} - S_{(\cdot)}]^2 = (1/n)\sum_{i<i'}[S_{(i)} - S_{(i')}]^2$, equation (2.13) gives

$$E\{\widetilde{\text{VAR}}\ S(X_1, X_2, \cdots, X_{n-1})\}=\frac{\sigma_\alpha^2}{n-1}+\binom{n-2}{1}\frac{\sigma_\beta^2}{(n-1)^3}+\binom{n-2}{2}\frac{\sigma_\gamma^2}{(n-1)^5}+\cdots. \tag{2.14}$$

This, when compared with formula (2.11), for sample size $n-1$, verifies inequality (1.6), and gives a simple expression for the difference between the two sides.

THEOREM 1.

$$E\{\widetilde{\text{VAR}}\ S(X_1, X_2, \cdots, X_{n-1})\}-\text{Var}\ S(X_1, X_2, \cdots, X_{n-1})=\frac{1}{2}\binom{n-2}{1}\frac{\sigma_\beta^2}{(n-1)^3}+\frac{2}{3}\binom{n-2}{2}\frac{\sigma_\gamma^2}{(n-1)^5}+\cdots, \tag{2.15}$$

there being $n-2$ terms on the right side of (2.15).

Comment 1. The variances $\sigma_\beta^2, \sigma_\gamma^2, \cdots$ appearing in (2.15) refer to the expansion (2.9) for $S(X_1, X_2, \cdots, X_{n-1})$, *not* for $S(X_1, X_2, \cdots, X_n)$. Section 3 discusses this point in more detail.

Comment 2. For *linear functionals*, i.e., S statistics such that the higher order terms $\beta_{ii'}, \gamma_{ii'i''}, \cdots$ are all zero, the right-hand side of (2.15) equals zero, and so the jackknife variance estimate is unbiased. For a *quadratic functional*, one having all third and higher order terms equal to zero, the bias is $\frac{n-2}{2}\sigma_\beta^2/(n-1)^3$, which equals the contribution of the quadratic term, in (2.11), to Var $S(X_1, X_2, \cdots, X_{n-1})$. In other words, the jackknife variance estimate doubles the quadratic term in expectation. A correction is suggested in Section 5. In general, the quadratic term is doubled, the cubic term tripled. etc.

Comment 3. Let $\mu = ES(X_1, X_2, \cdots, X_{n-1})$, and consider the identity $\sum[S_{(i)} - S_{(\cdot)}]^2 = \sum[S_{(i)} - \mu]^2 - n[S_{(\cdot)} - \mu]^2$. Taking expectations gives

$$E\sum[S_{(i)}-S_{(\cdot)}]^2=n\ \text{Var}\ S_{(i)}-n\ \text{Var}\ S_{(\cdot)}. \tag{2.16}$$

Equation (5.21) of Hoeffding (1948), applied with $m = n-1$, gives

$$n\ \text{Var}\ S_{(\cdot)}\le(n-1)\ \text{Var}\ S_{(i)}. \tag{2.17}$$

Together, (2.16) and (2.17) imply (1.6).[1] Expansion (2.9) is closely related to Hoeffding's theory; the quantities $\delta_1, \delta_2, \delta_3, \cdots$ which play a crucial role in his proofs are multiples of

[1] We are indebted to Akimichi Takemura for pointing out this connection, and also to Mark Chesters for discussions relating to Comment 4.

$\sigma_\alpha^2, \sigma_\beta^2, \sigma_\gamma^2, \cdots$, respectively. Expansion (2.9) is somewhat more convenient to work with, and yields one line proofs of Hoeffding's important theorems 5.1 and 5.2.

Comment 4. We can use expansion (2.4) to prove a more general version of (1.6):

THEOREM 2. *For any statistic $S(X_1, X_2, \cdots, X_n)$ having finite second moment, where S is not necessarily symmetrically defined in its arguments and the X_i are independent but not necessarily identically distributed,*

$$E \sum_{i=1}^n [S_{(i)} - S_{(\cdot)}]^2 \geq \frac{1}{n} \sum_{i=1}^n \operatorname{Var} S_{(i)} \geq \frac{n}{n-1} \operatorname{Var} S_{(\cdot)}. \tag{2.18}$$

PROOF. First assume that $\mu_i \equiv ES_{(i)} = 0$, $i = 1, 2, \cdots, n$. Define Diff $\equiv (n-1) \sum \operatorname{Var} S_{(i)} - n^2 \operatorname{Var} S_{(\cdot)}$ and let I, II, III be the three terms, from left to right, in (2.18). Taking expectations in $\sum [S_{(i)} - S_{(\cdot)}]^2 - \frac{1}{n} \sum S_{(i)}^2 = \{(n-1) \sum S_{(i)}^2 - n^2 S_{(\cdot)}^2\}/n$ gives I-II = Diff$/n$, while, directly, II-III = Diff$/\{n(n-1)\}$. We now show that Diff ≥ 0.

Still assuming $\mu_i = 0$, expansion (2.4) for $S_{(i)}$ can be written

$$S_{(i)} = \sum_{\mathscr{C}} S_{i\mathscr{C}} \tag{2.19}$$

where $\mathscr{C}$ indexes the $2^n - 2$ nonempty proper subsets of $\{1, 2, \cdots, n\}$. For example, with $i = 1$ and $\mathscr{C} = \{2, 3\}$, $S_{1\mathscr{C}} = B_{23}(X_2, X_3)$ in the expansion (2.4) of $S_{(1)}$. The random variables $S_{i\mathscr{C}}$ satisfy (i) $ES_{i\mathscr{C}} = 0$, (ii) $S_{i\mathscr{C}} = 0$ if $i \in \mathscr{C}$, and by (2.8), (iii) $ES_{i\mathscr{C}} S_{i'\mathscr{C}'} = 0$ *if* $\mathscr{C} \neq \mathscr{C}'$.

Define $S_{+\mathscr{C}} \equiv \sum_i S_{i\mathscr{C}}$, and notice that $ES_{+\mathscr{C}} S_{+\mathscr{C}'} = 0$ for $\mathscr{C} \neq \mathscr{C}'$. Therefore $E\, n^2 S_{(\cdot)}^2 = E \sum_{\mathscr{C}} S_{+\mathscr{C}}^2$, and likewise $E(n-1) \sum_i S_{(i)}^2 = E \sum_{\mathscr{C}} [(n-1) \sum_i S_{i\mathscr{C}}^2]$, so

$$\text{Diff} = E \sum_{\mathscr{C}} [(n-1) \textstyle\sum_i S_{i\mathscr{C}}^2 - S_{+\mathscr{C}}^2]. \tag{2.20}$$

Letting $n_{\mathscr{C}}$ be the number of elements in $\mathscr{C}$, and $\bar{S}_{\mathscr{C}} \equiv S_{+\mathscr{C}}/(n - n_{\mathscr{C}})$,

$$\text{Diff} = E \sum_{\mathscr{C}} [(n_{\mathscr{C}} - 1) \textstyle\sum_i S_{i\mathscr{C}}^2 + (n - n_{\mathscr{C}}) \sum_{i \notin \mathscr{C}} (S_{i\mathscr{C}} - \bar{S}_{\mathscr{C}})^2], \tag{2.21}$$

which verifies Diff ≥ 0.

Finally, notice that if we drop the assumption that the $S_{(i)}$ have means $\mu_i = 0$, the second and third terms in (2.18) are unchanged, while the first term is increased by the amount $\sum (\mu_i - \mu_\cdot)^2$, $\mu_\cdot \equiv \sum \mu_i/n$. This completes the proof of Theorem 2.

Essentially the same proof yields (2.18) in the following more general context: Let $(\Omega, \mathscr{B}, P)$ be a probability space and $\mathscr{B}_1, \cdots, \mathscr{B}_n$ independent sub σ-algebras of $\mathscr{B}$. For $i \in \{1, \cdots, n\}$ let $\mathscr{B}^{(i)}$ be the smallest σ-algebra containing all $\mathscr{B}_j$ for $j \neq i$, and let $S_{(i)}$ be a $\mathscr{B}^{(i)}$-measurable real random variable with $E\, S_{(i)}^2 < \infty$.

3. Functions of the empirical probability distribution. Most of the theoretical work relating to the jackknife concerns statistics which are smooth functions $s(\hat{F})$ of the empirical probability distribution $\hat{F}$, putting mass $1/n$ at each value X_i. A typical example is the sample variance $S(X_1, X_2, \cdots, X_n) = \sum_{i=1}^n (X_i - \bar{X})^2/n$, while the "unbiased" version $\sum_{i=1}^n (X_i - \bar{X})^2/(n-1)$ is not of form $s(\hat{F})$.

By considering the case where the sample space $\mathscr{X}$ of the X_i is finite, say $\mathscr{X} = \{1, 2, \cdots, L\}$, we can describe the condition $S(X_1, \cdots, X_n) = s(\hat{F})$ in concrete terms. Define $f_l \equiv \operatorname{Prob}\{X_i = l\}$, $\sum_{l=1}^L f_l = 1$, and let $\mathbf{f} = (f_1, f_2, \cdots, f_L)$ be the vector of probabilities. Likewise, let the empirical probabilities be the observed proportions $\hat{f}_l = \#\{X_i = l\}/n$, and let $\hat{\mathbf{f}} = (\hat{f}_1, \hat{f}_2, \cdots, \hat{f}_L)$ be the empirical probability vector. The possible values of $\mathbf{f}$ compose the L-dimensional simplex

$$\mathscr{S} \equiv \{\mathbf{v} : v_l \geq 0, \textstyle\sum_{l=1}^L v_l = 1\}, \tag{3.1}$$

while $\mathbf{f}$ can occur only at certain lattice points of $\mathscr{S}$.

In this case, a "smooth function of the empirical probability distribution" is a statistic

$$S(X_1, X_2, \cdots, X_n) = s(\hat{\mathbf{f}}), \tag{3.2}$$

where $s(\cdot)$ is defined continuously on $\mathscr{S}$. The statistic $S \sum_{i=1}^{n} (X_i - \bar{X})^2/(n-1)$ cannot be of this form since doubling the number of observed X_i's at each value of l changes S without changing $\hat{\mathbf{f}}$. Huber (1977) gives an enlightening description of the continuity properties desirable in a good statistic. A statistic of the form $s(\hat{\mathbf{f}})$ is automatically defined for every sample size, not just for the n we happen to have. This is a handy property for jackknife calculations, where it is necessary to change the sample size, at least by one, to get the variance and bias estimates.

The simplest form of (3.2) is a linear functional,

$$s(\hat{\mathbf{f}}) = s(\mathbf{f}) + (\hat{\mathbf{f}} - \mathbf{f})\mathbf{u} \tag{3.3}$$

where $\mathbf{u} = (u_1, u_2, \cdots, u_L)'$ is a fixed, known vector. Since

$$(\hat{\mathbf{f}} - \mathbf{f})\mathbf{u} = \sum_{l=1}^{L} (\hat{f}_l - f_l) u_l = \frac{1}{n} \sum_{i=1}^{n} (u_{X_i} - Eu) \tag{3.4}$$

where $Eu \equiv \sum_{l=1}^{L} f_l u_l$, (3.3) can be written as

$$S(X_1, X_2, \cdots, X_n) = s(\hat{\mathbf{f}}) = \mu + \frac{1}{n} \sum_{i=1}^{n} \alpha_i, \tag{3.5}$$

using the definitions

$$\mu = s(\mathbf{f}) \qquad \text{and} \qquad \alpha_i = \alpha(X_i = l) = u_l - Eu. \tag{3.6}$$

The α_i have mean zero under $\mathbf{f}$, and so (3.5) is of the form (2.9).

The jacknife works perfectly for linear functionals (3.3), in the sense that it produces the obvious unbiased estimate of variance, $\widehat{\text{VAR}}\ S(X_1, X_2, \cdots, X_n) = \sum_{l=1}^{L} \hat{f}_l (u_l - \bar{u})^2/(n-1)$, where $\bar{u} = \sum_{l=1}^{L} \hat{f}_l u_l$. In order to examine the effects of nonlinearity on the jackknife, it is natural to consider quadratic functionals, say

$$s(\hat{\mathbf{f}}) = s(\mathbf{f}) + (\hat{\mathbf{f}} - \mathbf{f})\mathbf{u} + \frac{1}{2} (\hat{\mathbf{f}} - \mathbf{f})\mathbf{v}(\hat{\mathbf{f}} - \mathbf{f})', \tag{3.7}$$

$\mathbf{v}$ being a given symmetric $L \times L$ matrix. Hinkley (1978) considers a similar class of functionals.

Some straightforward algebraic manipulation gives expansion (2.9) for a quadratic functional (3.7). Let $\mathbf{1} \equiv (1, 1, \cdots, 1)$, and define

$$\begin{aligned} \mathbf{U} &\equiv \mathbf{u} - \mathbf{1}'\mathbf{f}\mathbf{u}, \qquad \mathbf{V} \equiv \mathbf{v} - \mathbf{1}'\mathbf{f}\mathbf{v} - \mathbf{v}\mathbf{f}'\mathbf{1} + \mathbf{1}'(\mathbf{f}\mathbf{v}\mathbf{f}')\mathbf{1}, \\ \Delta(l) &\equiv V_{ll}/2, \qquad E\Delta \equiv \textstyle\sum_{l=1}^{L} f_l \Delta(l). \end{aligned} \tag{3.8}$$

Then (3.7) can be written as

$$S(X_1, X_2, \cdots, X_n) = s(\hat{\mathbf{f}}) = \mu^{(n)} + \frac{1}{n} \sum_i \alpha^{(n)}(X_i) + \frac{1}{n^2} \sum_{i<i'} \beta(X_i, X_{i'}), \tag{3.9}$$

where

$$\mu^{(n)} = s(\mathbf{f}) + \frac{E\Delta}{n}, \qquad \alpha^{(n)}(l) = U_l + \frac{\Delta(l) - E\Delta}{n}, \qquad \beta(l, l') = V_{l,l'}. \tag{3.10}$$

Comment 5. Letting $\hat{\mathbf{f}} = (1 - \epsilon)\mathbf{f} + \epsilon \mathbf{e}_l$, $\mathbf{e}_l$ the lth coordinate vector, we get

$$\left. \frac{ds(\hat{\mathbf{f}})}{d\epsilon} \right|_{\epsilon=0} = U_l, \tag{3.11}$$

so that the coordinates of $\mathbf{U}$ are the *influence function* of $s(\cdot)$, see Huber (1977). Likewise the coordinates of $\mathbf{V}$ are the second order influence function. The normalization in (2.8), by powers of n, results in $\alpha(\cdot)$ approaching the first order influence function as $n \to \infty$, $\beta(\cdot, \cdot)$ approaching the second order influence function, etc. In other words, (2.9) approaches the standard von Mises expansion as $n \to \infty$, see Hinkley (1978).

Comment 6. The jackknife estimate of bias is

$$\widehat{\text{BIAS}}\ S(X_1, X_2, \cdots, X_n) = (n-1)(S_{(\cdot)} - S). \tag{3.12}$$

For the quadratic functional (3.7), equation (2.9) implies that

$$\widehat{\text{BIAS}}\ S(X_1, X_2, \cdots, X_n) = \frac{\Delta_\cdot}{n} - \frac{\beta_{\cdot\cdot}}{2n}, \tag{3.13}$$

where

$$\Delta_i \equiv \Delta(X_i), \qquad \Delta_\cdot \equiv \textstyle\sum_{i=1}^n \Delta_i/n, \qquad \beta_{\cdot\cdot} \equiv \textstyle\sum_{i<i'} \beta_{ii'}\Big/\binom{n}{2}. \tag{3.14}$$

Equation (3.13) says that the expectation of $\widehat{\text{BIAS}}\ S(X_1, X_2, \cdots, X_n)$, for a quadratic functional, equals $E\Delta/n = \mu^{(n)} - s(\mathbf{f}) = ES(X_1, X_2, \cdots, X_n) - s(\mathbf{f})$, so that $\widehat{\text{BIAS}}$ is itself unbiased for the bias of $S(X_1, X_2, \cdots, X_n)$ in estimating $s(\mathbf{f})$. The variance of $\widehat{\text{BIAS}}$ is

$$\text{Var}\ \widehat{\text{BIAS}} = \frac{\sigma_\Delta^2}{n^3} + \frac{\sigma_\beta^2}{2n^3(n-1)} \tag{3.15}$$

where

$$\sigma_\Delta^2 \equiv \text{Var}\ \Delta_i = \textstyle\sum_{l=1}^L f_l[\Delta(l) - E\Delta]^2. \tag{3.16}$$

Expression (3.15) follows from (3.13) because the Δ_i are mutually uncorrelated with each other and also with all the $\beta_{ii'}$.

Comment 7. Following through definitions (3.8)–(3.10), we see that (3.9) can be written as

$$\begin{aligned} &\mu^{(n)} = \mu^{(\infty)} + \frac{E\beta(X, X)}{2n}, \\ &\alpha^{(n)}(X) = \alpha^{(\infty)}(X) + \frac{\Delta(X) - E\Delta}{n} = \alpha^{(\infty)}(X) + \frac{\beta(X, X) - E\beta(X, X)}{2n} \end{aligned} \tag{3.17}$$

A quadratic functional S can be defined as any statistic having the form (3.10), with $\mu^{(n)}$ and $\alpha^{(n)}(\cdot)$ obeying (3.17). This definition avoids mentioning the discrete sample space $\mathcal{X}$, and so is preferred for general discussion.

4. Variance relationships between different sample sizes. The rationale behind the sample size modification (1.5) is that for many familiar situations S, the true variance satisfies to a useful degree the approximation

$$\text{Var}^{(n)} \doteq \frac{n-1}{n} \text{Var}^{(n-1)}, \tag{4.1}$$

where $\text{Var}^{(j)} \equiv \text{Var}\ S(X_1, X_2, \cdots, X_j)$. For linear functionals $S = \mu + \frac{1}{n} \sum_{i=1}^n \alpha_i$, (4.1) is an exact equality. Here we will discuss (4.1) for three classes of nonlinear functionals, (i) U statistics, (ii) quadratic functionals (3.9), (3.17), and (iii) "Jth order von Mises series,"[2] $S = \mu + \frac{1}{n}\sum_i \alpha_i + \frac{1}{n^2}\sum_{i<i'} \beta_{ii'} + \cdots$, where the highest term of the series has coefficient $1/n^J$, and the functions $\alpha(\cdot)$, $\beta(\cdot, \cdot)$, $\cdots$ do not change form as the sample size changes.

Theorem 1 can be rewritten, using definition (1.5), as

$$E\,\widehat{\mathrm{VAR}} = \mathrm{Var}^{(n)} + \left\{\frac{n-1}{n}\mathrm{Var}^{(n-1)} - \mathrm{Var}^{(n)}\right\} + O\left(\frac{1}{n^2}\right), \tag{4.2}$$

where $O(1/n^2) \geq 0$. For our three classes of nonlinear functionals, it turns out that $O(1/n^2)$ is actually of order $1/n^2$, or smaller, as n goes to infinity, and that the bracketed term in (4.2) is of order $O(1/n^3)$; see Efron and Stein (1979), Section 4. If S is a U statistic of fixed degree J, then for $n \geq J + 1$, this follows from Section 5 of Hoeffding (1948). As a matter of fact, Hoeffding's results show that the bracketed term is always nonnegative, so that for U statistics we have

$$E\,\widehat{\mathrm{VAR}} \geq \mathrm{Var}^{(n)}. \tag{4.3}$$

Inequality (4.3) also holds for von Mises series (a proof is given in Efron and Stein (1979)), though in this case the bracketed term in (4.2) can be negative. Note: a slight modification of our definition of a von Mises series, to $S = \mu + \frac{1}{n}\sum_i \alpha_i + \frac{1}{n(n-1)}\sum_{i<i'} \beta_{ii'} + \cdots$, makes it into a U statistic. The only reason for not beginning this way at definitions (2.9)–(2.10) is that it makes the connection with polynomial functionals, as at (3.9), (3.17), slightly more complicated.

For quadratic functionals (3.9), (3.17),

$$E\,\widehat{\mathrm{VAR}} = \mathrm{Var}^{(n)} + \frac{1}{n(n-1)}\left\{\frac{n^3 - n^2 - 3n + 1}{n^3 - n^2}\frac{\sigma_\beta^2}{2} + \frac{2\sigma_{\alpha\Delta}}{n} + \frac{\sigma_\Delta^2}{n^2(n-1)}\right\}, \tag{4.4}$$

where $\sigma_\Delta^2 \equiv \mathrm{Var}\,\Delta(X)$ as at (3.16), and $\sigma_{\alpha\Delta} \equiv E\alpha(X)\Delta(X)$, see Efron and Stein (1979). If $\sigma_{\alpha\Delta} \geq 0$ then (4.3) holds for all n; if $\sigma_{\alpha\Delta} < 0$ then (4.3) holds for sufficiently large n, a sufficient condition being $n - [3/(n-1)] > -4\sigma_{\alpha\Delta}/\sigma_\beta^2$.

It is not true, then, that the usual jackknife variance formula (1.2) is always nonnegatively biased for $\mathrm{Var}\,S(X_1, X_2, \cdots, X_n)$. However for smooth functionals the bias terms are of high order, $O(1/n^2)$, and are positive for sufficiently large n. (The results for quadratic functionals can be extended to higher polynomial forms.) Specific analytical and numerical results, for the case of ratio estimation, are given in Rao and Rao (1971). A more important question, which this paper does not address, is the variance of $\widehat{\mathrm{VAR}}$ itself; see Efron (1979).

5. Correcting the bias of the variance estimate. Knowing that the jackknife variance estimate is always biased upwards, it is natural to look for some correction to remove this bias. We will consider only quadratic functionals, (3.9), (3.17), and omit algebraic details, which are straightforward. Hinkley (1978) provides a similar development.

Define $S_{(i,i')}$ to be the value of S when both X_i and $X_{i'}$ are removed from the original sample, and let

$$Q_{ii'} \equiv nS - (n-1)(S_{(i)} + S_{(i')}) + (n-2)S_{(ii')}, \qquad i \neq i'. \tag{5.1}$$

Then

$$Q_{ii'} = \frac{1}{n-2}\left[(\beta_{ii'} - \beta_{i\cdot} - \beta_{i'\cdot} + \beta_{\cdot\cdot}) - \frac{(\Delta_i - \Delta_\cdot) + (\Delta_{i'} - \Delta_\cdot)}{n-1}\right], \tag{5.2}$$

[2] A name coined by Colin Mallows, in unpublished lectures.

the dot indicating averaging as at (3.14): $\beta_{i\cdot} \equiv \sum_{j\neq i} \beta_{ij}/(n-1)$, etc. The $Q_{ii'}$ can be used to estimate σ_β^2, and thereby eliminate the leading term in the bias of the jackknife variance estimate, (2.15):

LEMMA. *For quadratic functionals,*

$$\frac{1}{2}E[Q_{12}-Q_{34}]^2 = \frac{\sigma_\beta^2}{(n-1)^2}\left\{1-\frac{3}{(n-2)^2}\right\} + \frac{2\sigma_\Delta^2}{(n-1)^2}\frac{1}{(n-2)^2} \tag{5.3}$$

and

$$\frac{1}{2}E[Q_{12}-Q_{23}]^2 = \frac{\sigma_\beta^2}{(n-1)^2}\left\{1-\frac{n-3}{(n-2)^2}\right\} + \frac{\sigma_\Delta^2}{(n-1)^2}\frac{1}{(n-2)^2}. \tag{5.4}$$

The lemma says that

$$\frac{(n-1)^2}{2}E[Q_{ii'}-Q_{jj'}]^2 = \sigma_\beta^2 + O(1/n) \tag{5.5}$$

for any two distinct pairs $(i, i') \neq (j, j')$. Suppose we evaluate $Q_{ii'}$ for all $N = n(n-1)/2$ distinct pairs, and let $\bar{Q}$ be the average of the N values of $Q_{ii'}$. Then

$$\hat{\sigma}_\beta^2 \equiv \frac{(n-1)^2}{N-1}\sum [Q_{ii'}-\bar{Q}]^2 \tag{5.6}$$

is an estimator of σ_β^2 having bias $O(1/n^2)$. The bias corrected estimate of VAR $S(X_1, X_2, \cdots, X_n)$ is

$$\widehat{\mathrm{VAR}}^{(\mathrm{Corr})} S(X_1, X_2, \cdots, X_n) = \widehat{\mathrm{VAR}}\, S(X_1, X_2, \cdots, X_n) - \frac{1}{n(n+1)}\sum_{i<i'} [Q_{ii'}-\bar{Q}]^2, \tag{5.7}$$

where $\widehat{\mathrm{VAR}}$ is given by (1.2), and $\bar{Q} = \sum_{i<i'} Q_{ii'}/[n(n-1)/2]$.

Example. Efron (1979b), pages 462–463, considers the sample correlation coefficient of 15 pairs of numbers, each pair referring to two characteristics of the 1973 entering class at an American law school. The data are in Table 1. In this case the statistic of interest, $S(X_1, X_2, \cdots, X_n)$ is the sample correlation coefficient between the two characteristics. $S = .776$, and (3.12), (1.2), and (5.7) give

$$\begin{aligned} \widehat{\mathrm{BIAS}}\, S(X_1, X_2, \cdots, X_{15}) &= 0.0065 \\ \widehat{\mathrm{VAR}}\, S(X_1, X_2, \cdots, X_{15}) &= .0203 \\ \widehat{\mathrm{VAR}}^{(\mathrm{Corr})} S(X_1, X_2, \cdots, X_{15}) &= .0179. \end{aligned} \tag{5.8}$$

The referee has pointed out that the jackknife itself could be used to remove the bias in $\widehat{\mathrm{VAR}}$. Doing so gives an estimate similar, but not identical, to $\widehat{\mathrm{VAR}}^{(\mathrm{Corr})}$. More ambitious unbiasing methods are also available, see Gray and Schucany (1972).

6. The bootstrap. A more general approach to jackknife-like calculations is de-

TABLE 1.
The average LSAT *score and undergraduate* GPA *at 15 American law schools, entering classes of 1973.*

School #	1	2	3	4	5	6	7	8	9	10	11	12	13	14	15
LSAT	576	635	558	578	666	580	555	661	651	605	653	575	545	572	594
GPA	3.39	3.30	2.81	3.03	3.44	3.07	3.00	3.43	3.36	3.13	3.12	2.74	2.76	2.88	2.96

scribed in Efron (1979a), under the rubric "bootstrap". The jackknife and the bootstrap are both examples of *resampling schemes*, in which the statistician attempts to learn the sampling properties of a statistic $S(X_1, X_2, \cdots, X_n)$ by recomputing its value for artificial samples, obtained by distorting the actual sample $X_1, X_2, \cdots, X_n$. Hartigan (1969, 1971, 1975) and Mallows (1975) have described several other interesting resampling schemes.

Using convenient notation, a vector of weights $\mathbf{P}^* = (P_1^*, P_2^*, \cdots, P_n^*)$, with the $P_i^* \geq 0$, $\sum_{i=1}^n P_i^* = 1$ leads to a resampled value of S in which the ith observation has weight P_i^*, compared to its weight $1/n$ in the real sample. We denote the resampled value as $S^* = s(\mathbf{P}^*)$. Here we are assuming that $S(X_1, X_2, \cdots, X_n)$ is of functional form, see Efron (1979a); $\mathbf{P}^*$ is abbreviated notation for the empirical probability distribution $\hat{F}^*$ putting mass P_i^* at X_i. Notice that $s(\mathbf{1}/n) = S(X_1, X_2, \cdots, X_n)$, the observed value.

The bootstrap considers vectors $\mathbf{P}^*$ having the distribution

$$\mathbf{P}^* \sim \frac{\text{Mult}(n, \mathbf{1}/n)}{n}. \tag{6.1}$$

(This should be compared with the jackknife, which uses $\mathbf{P}^*$ equal to all permutations of $(0, 1/(n-1), 1/(n-1), \cdots, 1/(n-1))$.) Here $\text{Mult}(n, \mathbf{1}/n)$ indicates a multinomial distribution, n draws, probability $1/n$ for each of the n categories. The vector $\mathbf{P}^*$ has mean vector and covariance matrix

$$E_*\mathbf{P}^* = \mathbf{1}/n, \qquad \text{Cov}_*\mathbf{P}^* = \mathbf{I}/n^2 - \mathbf{1}'\mathbf{1}/n^3. \tag{6.2}$$

The asterisk is a reminder that these calculations have nothing to do with the inherent randomness in the data, but rather with probabilities imposed by the statistician.

The bootstrap estimates of bias and variance are

$$\widehat{\text{BIAS}}^{(B)}\, S(X_1, X_2, \cdots, X_n) = E_*s(\mathbf{P}^*) - s(\mathbf{1}/n) \tag{6.3}$$

and

$$\widehat{\text{VAR}}^{(B)}\, S(X_1, X_2, \cdots, X_n) = \text{VAR}_*s(\mathbf{P}^*),$$

$E_*s(\mathbf{P}^*)$ and $\text{Var}_*s(\mathbf{P}^*)$ being taken with respect to distribution (6.1). The rationale behind these estimates is simply this: if the true probability distribution of the X_i happens to equal the empirical distribution (the distribution which puts mass $1/n$ at each observed X_i) then (6.3) gives exactly the correct bias and variance. The jackknife can be thought of as a "delta method" approximation to the bootstrap, see Section 5 of Efron (1979a). The bootstrap idea can be used to give different, more robust, estimates of bias and variance, see Efron (1979b), but here we will restrict our attention to (6.3), and demonstrate results similar to those obtained for the jackknife. Proofs are contained in Efron and Stein (1979) and will not be given here.

Once again we consider quadratic functionals $S(X_1, X_2, \cdots, X_n) = \mu + 1/n \sum_i \alpha_i + \frac{1}{n^2} \sum_{i<i'} \beta_{ii'}$. Since the bootstrap only involves samples of size n, the same size as the genuine sample, there is no need to consider how μ and α_i depend on n.

THEOREM 3. *For a quadratic functional* $S(X_1, X_2, \cdots, X_n)$,

$$\widehat{\text{BIAS}}^{(B)}\, S(X_1, X_2, \cdots, X_n) = \frac{n-1}{n} \widehat{\text{BIAS}}\, S(X_1, X_2, \cdots, X_n), \tag{6.4}$$

where $\widehat{\text{BIAS}}\, S(X_1, X_2, \cdots, X_n)$ *is the jackknife bias estimate* (3.13), *discussed in Comment* 6, *Section* 3.

THEOREM 4. *For a quadratic functional,*

$$E\left\{\frac{n}{n-1}\widehat{\mathrm{VAR}}^{(B)} S(X_1, X_2, \cdots, X_n) - \mathrm{Var}\, S(X_1, X_2, \cdots, X_n)\right\} \tag{6.5}$$
$$= \frac{c_1(n)\sigma_\beta^2 + 4c_2(n)\,\sigma_{\alpha\Delta}}{n^2} + \frac{6c_3(n)\sigma_\Delta^2 + c_4(n)(E\Delta)^2}{n^3},$$

where, as $n \to \infty$,

$$c_i(n) \to 1, \qquad i = 1, 2, 3, 4. \tag{6.6}$$

Specific values for $c_1(n)$, $c_2(n)$, $c_3(n)$, $c_4(n)$ are given in Efron and Stein (1979).

Comment 8. For a linear functional S, (6.5) equals 0. The form of (6.5) facilitates comparison with the corresponding jackknife result (4.4). The right-hand side of (6.5) can be either positive or negative, depending mainly on the sign of $\sigma_{\alpha\Delta}$ and the latter's relative magnitude compared to σ_β^2, cf. the remarks following (4.4).

REFERENCES

EFRON, B. (1979a). Bootstrap methods: Another look at the jackknife. *Ann. Statist.* **6** 1-26.

EFRON, B (1979b). Computers and the theory of statistics: Thinking the unthinkable. *SIAM Rev.* **21** 460–480.

EFRON, B and STEIN, C. (1979). The jackknife estimate of variance. Tech. Report No. 120, Dept. Statist., Stanford Univ.

GRAY, H. L. and SCHUCANY, W. R. (1972). *The Generalized Jackknife Statistic*. Dekker, New York.

HAJÉK, J. (1968). Asymptotic normality of simple linear rank statistics under alternatives. *Ann. Math. Statist.* **39** 325–346.

HARTIGAN, J. A. (1969). Using subsample values as typical values. *J. Amer. Statist. Assoc.* **64** 1303–1317.

HARTIGAN, J. A. (1971). Error analysis by replaced samples. *J. Roy. Statist. Soc. Ser. B* **33** 98–110.

HARTIGAN, J. A. (1975). Necessary and sufficient conditions for asymptotic joint normality of a statistic and its subsample values. *Ann. Statist.* **3** 573–580.

HINKLEY, D. V. (1978). Improving the jackknife with special reference to correlation estimation. *Biometrika* **65** 13–21.

HOEFFDING, W. (1948). A class of statistics with asymptotically normal distributions. *Ann. Math. Statist.* **19** 293–325.

HUBER, P. J. (1977). *Robust Statistical Procedures, SIAM*, Philadelphia.

MALLOWS, C. L. (1975). On some topics in robustness. Unpublished report, Bell Laboratories.

MILLER, R. G. (1974). The jackknife—a review. *Biometrika* **61** 1-15.

RAO, S. R. S. and RAO, J. N. K. (1971). Small sample results for ratio estimators. *Biometrika* **58** 625–630.

RUBIN, H. and VITALE, R. A. (1980). Asymptotic distribution of symmetric statistics. *Ann. Statist.* **8** 165–170.

DEPARTMENT OF STATISTICS
STANFORD UNIVERSITY
STANFORD, CALIFORNIA 94305

1956: High school graduation photo

1968: The new Associate Professor of Statistics at Stanford University

1978: Brad's first term as Chair of the Statistics Department, Stanford

1985: At his induction into the National Academy of Sciences, with fellow awardee Bryce Crawford, Jr. (inducted 1956, Chemistry)

1988: The Associate Dean of Stanford's School of Humanities and Sciences

1996: In his office during the last days of the old Sequoia Hall

1997: Brad as a big name in Taiwan, on tour with the Fisher paper

1998: Chairman of the Faculty Senate at Stanford University

1998: Presenting remarks at the School of Humanities & Sciences 50th Anniversary Convocation

2000: Lecturing at the RSS Annual Meeting at the University of Reading in England, Brad is pictured with members of the Fisher family; daughter Margaret is on Brad's right, and son Harry with his wife and son are to Brad's left

2002: Brad in campaign-mode for the post of President of the American Statistical Association

2003: Accepting the inaugural C.R. and Bhargarvi Rao Prize at Penn State University, with Dr. and Mrs. Rao in attendence

2003: The Chairman at a talk for students of the interdisciplinary Mathematical and Computational Science program at Stanford

June, 2007: Fun in the sun at the department reception in Brad's honor after the National Medal of Science announcement; Brad is joined by Trevor Hastie and Rob Tibshirani

June, 2007: Brad with his long-time friend and colleague, Charles Stein, at the Stanford Statistics Department roof-top reception celebrating his National Medal of Science award

July 26, 2007: Brad and Donna on their way to the National Medal of Science Reception on Thursday afternoon

July 26, 2007: Brad and Miles attending the NMS Reception at the National Science Foundation

July 27, 2007: Brad and Donna at the Medals Ceremony in the East Room of the White House on Friday afternoon

July 27, 2007: Wearing the 2005 National Medal of Science presented by President Bush

July 27, 2007: Meeting up with former-Stanford Provost Condoleezza Rice after the Medals Ceremony

July 27, 2007: Brad and Miles going to the black-tie dinner for new Laureates at the Ritz-Carlton Hotel on Friday evening

11

The Jackknife, the Bootstrap and Other Resampling Plans

Introduction by Peter Hall
The University of Melbourne

Brad Efron's introduction and development of bootstrap methods came at just the right time. By the early 1980s the powerful computing hardware needed to experiment with the bootstrap was starting to arrive on our desks, rather than in a separate computer center several blocks away. As Efron noted on page 30 of his CBMS–NSF monograph, "The bootstrap is a theory well suited to an age of plentiful computation." For many of us, that age had just dawned.

As it happened, Efron's monograph played the role of a wisely prepared plan for a moderate-sized town that was to grow rapidly into a sprawling metropolis. However, the plan was so carefully conceived that it remains relevant today, even though its original purpose has been completely outgrown.

In the early 1980s, when I first began to take a close interest in the bootstrap, the monograph was essential reading for anyone who wanted to glimpse the future of nonparametric statistical methods. It enjoyed this status for a little over a decade, at which point subsequent research monographs, including that by Efron and Tibshirani (1993), began to consolidate the rapid advances that were being made in methodology and theory and via extensive practical experience. However, the directions outlined by Efron back in 1982 are still of interest. We know very much more now about the possibilities offered by the bootstrap, but in important ways we are still advancing down the path onto which Efron guided us a quarter-century ago.

As the monograph's title makes clear, Efron's vision for the bootstrap encompassed "other resampling plans," including the jackknife and cross-validation. He defined a resampling method as one that addressed a parameter estimator computed using one of many possible "reweighted versions of the empirical probability distribution" (p. 37). Even today this breadth is to be applauded. But 25 years ago it was remarkable. It encompassed not only older, established methods, but also many subsequent ones which, as we shall see, were to come into existence at least partly because they were logical developments of Efron's visionary perspective. When I first read his definition of a resampling method, I didn't genuinely understand what he had in mind; it took several years to sink in.

Efron's outlook reflected his own thought processes during the 1970s. His deep understanding of both the strengths and drawbacks of the jackknife, particularly in the contexts of bias and variance estimation, had underpinned his approach to the development of bootstrap methods some time before he set out his ideas in the monograph. In a *Biometrika* paper the previous year (Efron 1981), he had shown that the jackknife, the infinitesimal jackknife and the bootstrap all gave methods for variance estimation that are instances of a more general approach, which could be considered to be of greater interest than any one of those ideas. While the monograph does not duplicate the technical details of the *Biometrika* treatment, it repeats the main message, with added emphasis and elucidation: the bootstrap is a member of a much larger class of techniques, many of which have important uses; but the bootstrap has distinct advantages. For example, the bootstrap enjoys greater breadth of application, and often produces estimators with lower bias or less stochastic variability.

This drawing together of resampling methods was achieved in Chap. 7 of the monograph, in the context of discrimination and classification. As Efron noted, the relationship among jackknife, cross-validation and bootstrap estimators of classification error, reported in his numerical studies, is both complex and ambiguous. He was subsequently to devote considerable effort to elucidating these issues. See, for example, Efron (1983) and Efron and Tibshirani (1997).

From a theoretical viewpoint this problem turns out to be surprisingly complex, and therefore especially interesting. For example, although cross-validation gives an almost-unbiased estimator of error, the estimator is highly stochastically variable. The bootstrap reduces this variability, although sometimes at the expense of significantly increased bias; but nevertheless the bootstrap offers substantially more attractive opportunities than cross-validation for tuning the classifier so as to optimize performance. See, for instance, Ghosh and Hall (2007).

Efron's view, that the bootstrap is part of a much larger class of resampling methods, made the development of improved but related techniques more straightforward. In particular, as we shall argue later at greater length, the broad outlook that Efron adopted in his monograph pointed the way towards the subsequent development of "tilted" or "biased" bootstrap methods, including empirical likelihood. However, his perspective was to prove more popular in the US than in other parts of the world, where the bootstrap stood relatively alone as a new statistical idea.

I still have clear memories of a seminar I gave at a British university in 1985, where I discussed the bootstrap and its potential for quantifying sampling variability, especially in terms of confidence intervals. I was taken aback by the suspicion with which the bootstrap was viewed. The audience's scepticism was epitomized by a concern that ties were very likely to occur in a bootstrap resample, whereas they might not exist at all in the sample. Indeed, my listeners seemed to focus on the ways in which resamples differed from the sample. They argued that, in view of these qualitative and quantitative departures, bootstrap resampling did not adequately reflect sampling from the population, and therefore was of dubious value. However, they were relatively happy with a "smoothed bootstrap" algorithm; they liked the idea of adding a little white noise to each resampled value, so that it became different from all the others.

My listeners' worries were a consequence of a particular view of resampling. Despite its failings, in broad terms that view coincided (and still does) with my own: resampling from the sample is a way of "modeling" empirically the operation of sampling from the original sample. My audience was sceptical of the adequacy of the model, but I doubt they had similar concerns about the jackknife, which they wouldn't have seen as trying to emulate the sampling process. Of course, it is now widely accepted that ties, and other artifacts of "sampling from the sample," generally have little impact on the bootstrap's performance. (Exceptions arise, for example, in bootstrap inference about "local" quantities, such as quantiles or distributions of extreme order statistics.)

This view of the bootstrap, as an empirical tool for modeling the operation of sampling from a population, seems today to be more prevalent than the perspective that sees it as a member of a broad class of methods that encompasses the jackknife, cross-validation and tilting. However, if scientific work on the bootstrap had started from the viewpoint that it is a distinctly new method, without the close linkages to the pre-existing techniques that Efron noted, then the potential for major new technologies that are not based directly on the "resample from a sample" idea would have been significantly reduced. Moreover, acceptance of the bootstrap might have been slower, as the reaction of the audience for my seminar indicated.

It may be helpful at this point to briefly survey the monograph, providing more detail about the way in which Efron introduced and motivated the bootstrap. Chapters 2, 3 and 4 developed properties of the jackknife, drawing significantly on the early work of M.H. Quenouille and of J.W. Tukey but pointing to difficulties with the method. Chapter 2 gave a particularly insightful treatment of the jackknife as a tool for bias reduction, and suggested a modification of Quenouille's formula. Efron also noted there that, in the problem of estimating the bias of an estimator $\hat{\theta}$ when sampling from a population with distribution F, the jackknife estimator is "roughly speaking ... the true bias of $\hat{\theta}$ if F were actually equal to the observed $\widehat{F}$." Here $\widehat{F}$ is the empirical distribution function, and Efron was setting the scene for his introduction of bootstrap methods in Chap. 5, where the bootstrap bias estimator was shown to be *exactly* equal to the true bias of $\hat{\theta}$ when sampling from $\widehat{F}$.

A further pointer from the jackknife to the bootstrap emerged in Efron's treatment (in Chap. 3) of Tukey's notion of "pseudo [data] values." Tukey had noticed that the jackknife idea suggested new data values that were, in a sense, respective approximations to the leverage that the actual data had on the value of $\hat{\theta}$; and he had proposed that the empirical variance of these pseudo-values be taken as an estimator of the variance of $\hat{\theta}$. The pseudo-data play the role that the actual data would if $\hat{\theta}$ were a mean, but as Efron noted, "the analogy does not seem to go deep enough" (p. 14).

The bootstrap removed this impediment by taking the variance estimator to be the empirical variance of values of the test statistic computed from many resampled versions of $\hat{\theta}$. As Efron observed in Chap. 4, in its conventional form the jackknife tends to overestimate variance, and it fails to consistently estimate the variance of the sample median. The success of the bootstrap in the latter setting was demonstrated by Maritz and Jarrett (1978), a little before Efron's seminal paper (Efron 1979) appeared. However, in 1978 the technique was regarded as a method for solving a particular problem, rather than as a utilitarian technology.

Chapter 5 formally defined a variety of bootstrap methods although, as we have seen, the bootstrap idea had already enjoyed a substantial presence in earlier parts of the monograph. Efron developed the bootstrap in both parametric and nonparametric settings, paying particular attention to the problems of estimating bias and variance.

His treatment in Chap. 5 of the case of regression involved conditioning on the design variables, which has always seemed to me to be the appropriate approach to what is formally defined as regression. Therefore he resampled from the set of residuals, rather than from the set of pairs of explanatory and response variables. However, at the very end of Chap. 5 and in Chap. 7, in settings more like model selection than prediction, he suggested resampling the pairs. Reading today's research papers, I'm troubled that more than a few authors treat these two techniques as interchangeable, as though choosing between them depended on little more than the statistician's whim. In the context of prediction, where we expect to condition on at least the current explanatory variable, resampling only the residuals is surely the right approach.

Interestingly, Efron did not center the residuals before resampling, although doing so would usually be recommended today. However, especially for a theoretician such as myself, it is a little fussy even to mention this point; the effects of centering are of second order.

Chapter 6 developed the relationship between the bootstrap and L.A. Jaeckel's "infinitesimal jackknife" or equivalently, the delta method. Using simple but deep geometric arguments, Efron described the connections among bootstrap, jackknife and infinitesimal jackknife estimators of bias and variance. A geometric approach to mathematical problems has always characterized Efron's theoretical work in statistics. For example, I've seen him derive terms in Edgeworth expansions via geometric constructions, a feat which still astounds me.

As we have already noted, one of the endearing features of the monograph is that, in addition to offering a forward-looking view of developments, it provided perspective by connecting the bootstrap to earlier resampling ideas. If one takes the view that the bootstrap gives us an empirical model for sampling from a population, then it is unsurprising that some of its major, early progenitors were developed by statisticians with special interests in sampling, especially survey sampling. This was one source of the bootstrap antecedents that Efron discussed.

Some of these linkages were delineated in Chap. 5, where the case of finite sample spaces was developed. The notions of sampling with replacement, including half-sampling, were taken up in Chap. 8, where Efron also pointed to an ingenious way of balancing resamples. These approaches had their roots decades earlier, in the context of survey sampling (see, e.g., Jones 1956; Kish 1957; McCarthy 1969). Balanced resampling substantially reduced computational labor, and was related to Latin hypercube sampling, which, in the guise of the balanced bootstrap (Davison et al. 1986), was to enjoy well-deserved popularity in the 1980s and 1990s. The fact that such labor-saving devices tend to be largely ignored today reflects the abundant computing riches that Moore's Law has delivered to statistics since Efron's monograph was written.

Chapter 9 connected Efron's ideas to the pioneering work of J.A. Hartigan in the late 1960s and 1970s (see, e.g., Hartigan 1969, 1971). Unlike antecedents in the area of survey sampling, Hartigan's contributions resembled Efron's in being designed for the case of simple random sampling. Hartigan (1969) introduced the notion of "typical values," and used them to construct a type of percentile-method bootstrap confidence interval for a mean. As Efron noted, his own broad definition of resampling methods encompassed Hartigan's notion of random subsampling, as well as half-sampling and balanced sampling ideas.

In Chap. 10, Efron saved the best for last. But even there, as though creating a light but exquisite dessert to complement a marvelous meal, he revealed the future only modestly. This was at least partly because the work outlined in his last chapter, on nonparametric confidence intervals, was very much in its infancy back in 1982. However, it can fairly be said that the material in the 16-page Chap. 10 was to prove the most obviously influential out of all the contributions in this exceptional book.

That is not to say that the last chapter was the most important. Like a tower that reaches skywards, Efron's discussion of bootstrap confidence regions relied critically on a firm foundation, without which his proposals might not have found acceptance. That foundation was provided by earlier chapters. But the concepts of percentile and percentile-t confidence regions, and the notion of the tilted bootstrap, were each introduced in Chap. 10 and have all been extremely important.

Of course, the roots of these ideas can be found in Efron (1979). It is clear from that paper (see, e.g., Remarks D, E, and F), as well as from Chap. 10 of the monograph, that Efron appreciated from the beginning that working with $\hat{\theta}$ or $\hat{\theta} - \theta$, rather than their Studentized form, posed potential difficulties. In particular, he was never a stranger to the issues of statistical pivoting that were to underpin subsequent comparisons of percentile and percentile-t methods. His aversion to the latter was spelled out on page 88 of the monograph: "A drawback is that the [percentile-t] method seems specific to translation problems. An attempt to use it for the correlation coefficient . . . gave poor results."

This remark captured the union of two of the main drawbacks to percentile-t methods: they require a good (in the sense of "relatively low variability") estimator of estimator variance, and they are not transformation-respecting. The percentile method suffers from neither of these deficiencies. Indeed, it does not need a variance estimator at all, and in many problems this is a substantial asset. However, the percentile method generally does not enjoy good coverage accuracy.

To remedy this difficulty, Chap. 10 introduced a bias correction for percentile-method confidence intervals. This certainly improved coverage accuracy. However, as the statistical community gained a better understanding of the issues that underpinned performance (see, e.g., Schenker 1985; Efron and Tibshirani 1986; Efron 1987; Hall 1988), it was to become clear that still further advances could be achieved. In particular, Efron (1987) was to combine his bias correction with an acceleration constant, in effect a further correction for skewness, to produce a method which achieved the second-order correctness of percentile-t methods but was at the same time transformation-respecting.

Today, the method of choice for constructing bootstrap confidence intervals, in general problems, is arguably to take Efron's basic percentile method, introduced in his 1982 monograph, and coverage-correct it, thereby using the bootstrap to check up on itself and diagnose its own shortcomings. That technique does not require a variance estimator or an estimator of skewness, it is transformation-respecting, it is second-order accurate, and the length and true coverage of its confidence intervals are monotone functions of the nominal coverage level. It usually works well, but it needs an order of magnitude more computational grunt. When Efron wrote his 1982 monograph, this "double bootstrap" technique, as it is often called, was out of practical reach, in addition to relying on ideas that were still several years away (see, e.g., Efron 1983; Hall 1986; Beran 1987).

A short section at the end of Chap. 10 included mention of bootstrap tilting methods, where data are resampled with different weights determined by specific, desired outcomes. "Without going into details," Efron remarked, it was possible to choose the weights so as to make the new resampling distribution "the closest distribution to $\widehat{F}$ (in a certain plausible metric)." This proposal, that resampling weights be chosen to optimize a criterion of choice, subject to minimizing the distance from the case of uniform weights, was to prove extraordinarily influential. The weights indicated in Efron's monograph were relatively specific, but alternative, more general suggestions were to form the basis for very important new classes of methods, for example Owen's (1988, 1990) empirical likelihood proposal, methods for simulating under a null hypothesis in order to implement bootstrap-based tests, and techniques for constructing estimators so that they might enjoy specific qualitative or quantitative constraints (see, e.g., Hall and Presnell 1999).

In summary, Efron's monograph has had a truly remarkable impact on the future directions of statistics, well out of proportion to its length. Today, Efron's bootstrap methods are so omnipresent that they touch the lives of virtually everyone in a developed country. For example, the data gathered on us by our national statistical offices are bootstrapped in order to assess the reliability of predictions made about our behavior; finance companies use bootstrap methods to help quantify the risks of lending us money; and the bootstrap plays key roles in data analysis for improving our health and monitoring our climate. To a large extent these methods are the direct outgrowth of ideas outlined in just 90 pages in this short volume. It is very hard to find another 90 pages, anywhere in the statistics literature of the last half-century, that has been so influential.

References

Beran, R. (1987). Prepivoting to reduce level error of confidence sets. *Biometrika* **74**, 457–468.

Davison, A. C., Hinkley, D. V. and Schechtman, E. (1986). Efficient bootstrap simulation. *Biometrika* **73**, 555–566.

Efron, B. (1979). Bootstrap methods: Another look at the jackknife. *Annals of Statistics* **7**, 1–26.

Efron, B. (1981). Nonparametric estimates of standard error: The jackknife, the bootstrap, and other methods. *Biometrika* **68**, 589–599.

Efron, B. (1983). Estimating the error rate of a prediction rule: Improvement on cross-validation. *Journal of the American Statistical Association* **78**, 316–331.

Efron, B. (1987). Better bootstrap confidence intervals. *Journal of the American Statistical Association* **82**, 171–200 with discussion and Rejoinder.

Efron, B. and Tibshirani, R. (1986). Bootstrap methods for standard errors, confidence intervals, and other measures of statistical accuracy. *Statistical Science* **1**, 54–77 with comment and Rejoinder.

Efron, B. and Tibshirani, R. (1993). *An Introduction to the Bootstrap*. Chapman and Hall, New York.

Efron, B. and Tibshirani, R. (1997). Improvements on cross-validation: The .632+ bootstrap method. *Journal of the American Statistical Association* **92**, 548–560.

Ghosh, A. K. and Hall, P. (2007). On error-rate estimation in nonparametric classification. To appear *Statistica Sinica*: preprint no. SS-06-184.

Hall, P. (1986). On the bootstrap and confidence intervals. *Annals of Statistics* **14**, 1431–1452.

Hall, P. (1988). Theoretical comparison of bootstrap confidence intervals. *Annals of Statistics* **16**, 927–985 with discussion and Reply.

Hall, P. and Presnell, B. (1999). Biased bootstrap methods for reducing the effects of contamination. *Journal of the Royal Statistical Society, Series B* **61**, 661–680.

Hartigan, J. A. (1969). Using subsample values as typical values. *Journal of the American Statistical Association* **64**, 1303–1317.

Hartigan, J. A. (1971). Error analysis by replaced samples. *Journal of the Royal Statistical Society, Series B* **33**, 98–110.

Jones, H. L. (1956). Investigating the properties of a sample mean by employing random subsample means. *Journal of the American Statistical Association* **51**, 54–83.

Kish, L. (1957). Confidence intervals for clustered samples. *American Sociological Review* **22**, 154–165.
Maritz, J. S. and Jarrett, R. G. (1978). A note on estimating the variance of the sample median. *Journal of the American Statistical Association* **73**, 194–196.
McCarthy, P. L. (1969). Pseudo-replication: Half samples. *Review of the International Statistical Institute* **37**, 239–264.
Owen, A. B. (1988). Empirical likelihood ratio confidence intervals for a single functional. *Biometrika* **75**, 237–249.
Owen, A. B. (1990). Empirical likelihood ratio confidence regions. *Annals of Statistics* **18**, 90–120.
Schenker, N. (1985). Qualms about bootstrap confidence intervals. *Journal of the American Statistical Association* **80**, 360–361.

Contents

12

Estimating the Error Rate of a Prediction Rule: Improvement on Cross-Validation

Trevor Hastie
Stanford University

This pioneering paper is the first in a series of several by Brad Efron on the thorny problem of estimating prediction error. Why are we so interested in estimating prediction error? It turns out that error estimates are a crucial ingredient of state-of-the-art methods for predictive modeling.

Most modern supervised learning techniques are set up to deliver a sequence of models indexed by a tuning parameter ν. Examples include:

- The "lasso" (Tibshirani 1996) regularizes a linear model (e.g., linear, logistic, Cox) by restricting the L_1 norm of the coefficient vector; the sequence of solutions $\hat{\beta}(\nu)$ is indexed by $\nu = ||\hat{\beta}(\nu)||_1$.
- The "support vector machine" (Boser et al. 1992) is regularized by the L_2 norm of the coefficients, or equivalently by the width of the "soft" margin (Hastie et al. 2001), either of which index the sequence of fits.
- "Boosting" (Schapire and Freund 1997; Hastie et al. 2001) fits a nested sequence of models increasing in complexity, by a forward stagewise strategy of adding small corrections to the current model: $\hat{f}_m(x) = \hat{f}_{m-1}(x) + \delta_m(x)$. Here $\nu = m$, the number of steps.

Typically the models are fit by optimizing a numerically convenient (e.g., differentiable) criterion for each value of ν, using the training data. Given the sequence of fitted models $\hat{\eta}_\nu$, we would then choose ν and hence a model, by evaluating its prediction performance using a different, more relevant loss function Q, such as misclassification error. If we were data rich, we could simply set aside a large test set for this purpose. Typically this is not the case, and so we have to devise clever ways to use the same training data at this testing stage. $\overline{\text{err}}$ is one such estimate – the average of the Q-errors over the training data – but is clearly not too clever, since it will be overly optimistic.

For some classical models, such as subset selection in linear regression, closed form expressions are available for unbiased estimates of the *in-sample* prediction error (making predictions at the same sites t_i observed in the training sample). These rely on the inherent linearity (in the training y_i) of the modeling methods, which in turn let us calculate the degrees of freedom of the fit. This leads to error estimates such as C_p and AIC, which make simple corrections to $\overline{\text{err}}$ based on the df.

However, most of the modern methods are nonlinear, and this theory falls by the wayside. As an aside, the lasso method above is an exception; Efron and coauthors discovered a simple and exact expression for the df of the lasso (Efron et al. 2004).

It turns out that it is very hard in general to estimate prediction error well. Efron takes up the challenge in this and a subsequent series of papers (Efron 1986; Efron and Tibshirani 1997). His

Bootstrap approaches, along with cross-validation, remain the state of the art. Early on in the paper, Efron makes clear what we are estimating:

$$\mathrm{Err}(\mathbf{x}) = E_F Q[Y_0, \eta(T_0, \mathbf{x})]. \tag{12.1}$$

This is the expected cost in terms of the loss function Q when we use our learned rule $\eta(\cdot, \mathbf{x})$ to make a prediction at a randomly chosen site T_0 with outcome Y_0. My notation emphasizes that this is a conditional error, since it depends on the training data $\mathbf{x}$. The dependence of η on the tuning parameter ν above is suppressed.

Efron's approach is to focus on the optimism $\mathrm{op}(\mathbf{x}) = \mathrm{Err}(\mathbf{x}) - \overline{\mathrm{err}}$, and to develop estimates for its expected value, the "ideal constant" $\omega(F) = E_F\{\mathrm{op}(\mathbf{x})\}$. This then leads to estimates of the form

$$\widehat{\mathrm{Err}}(\mathbf{x}) = \overline{\mathrm{err}} + \hat{\omega}. \tag{12.2}$$

One might wonder, as I did, why leave-one-out cross-validation $\widehat{\mathrm{Err}}^{(CV)}(\mathbf{x})$ is not the natural candidate for estimating this conditional error. After all, since each one-left-out training set is almost the same as the original training set, $\widehat{\mathrm{Err}}^{(CV)}(\mathbf{x})$ should have low bias for $\mathrm{Err}(\mathbf{x})$. This is true, but it gets killed by the variance. Efron's simulations show this, and also demonstrate the curious *negative correlation* between $\mathrm{Err}(\mathbf{x})$ and $\widehat{\mathrm{Err}}^{(CV)}(\mathbf{x})$ (noted also by Hall and Johnstone 1992).

It is easy to see that (12.2) also estimates the expected error

$$E_F \mathrm{Err}(\mathbf{x}) \tag{12.3}$$

and given the definition of ω, one might argue that (12.3) is more the target than is (12.1). What emerges from this and Efron's subsequent papers is that most reasonable estimators, such as $\widehat{\mathrm{Err}}^{(.632)}$ and ten-fold cross-validation, do a much better job estimating this expected error (12.3) than the conditional error (12.1); see also Kohavi (1995). In general terms this is achieved by shaking the data up before averaging, and hence reducing the variance.

A simple bootstrap estimate of error $\hat{\epsilon}^{(0)}$ shakes the data up as well, but suffers from upward bias. The effective training samples are much smaller than the original (on average 63.2% of the original samples occur in a bootstrap sample), and so are estimating an error rate for procedure with higher variance. The .632 estimator is designed to correct this bias, by shrinking toward $\overline{\mathrm{err}}$, which is biased in the opposite direction:

$$\widehat{\mathrm{Err}}^{(.632)} = .368\ \overline{\mathrm{err}} + .632\ \hat{\epsilon}^{(0)}. \tag{12.4}$$

[Efron's explanation is deeper and more intricate; mine is lifted from our data-mining book (Hastie et al. 2001, Sect. 7.11, first edition), and any criticisms can be directed at one of the editors of this volume!]

It turns out that the estimate (12.4) fails for methods that fit the data heavily, such as one-nearest-neighbor classification ($\overline{\mathrm{err}} = 0$). This lead Efron and Tibshirani (1997) to develop a correction, the $.632^{+}$.

This paper is an Efron classic, filled with clever insights and delicate derivations, which I had pleasure in revisiting. He approaches the problem thoroughly and systematically, from the ground up – a wonderful example of how statistics research should be done. And again, after 25 years, it remains the state of the art.

Acknowledgments
This research was supported by grant DMS-0505676 from the National Science Foundation, and grant 2R01 CA 72028-07 from the National Institutes of Health.

References

Boser, B., Guyon, I. and Vapnik, V. (1992). A training algorithm for optimal margin classifiers. In *Proceedings of COLT II*, Philadelphia, PA.

Efron, B. (1986). How biased is the apparent error rate of a prediction rule? *Journal of the American Statistical Association* **81**, 461–470.

Efron, B., Hastie, T., Johnstone, I. and Tibshirani, R. (2004). Least angle regression. *Annals of Statistics* **32**, 407–499 with discussion and Rejoinder.

Efron, B. and Tibshirani, R. (1997). Improvements on cross-validation: The 632+ bootstrap method. *Journal of the American Statistical Association* **92**, 548–560.

Hall, P. and Johnstone, I. (1992). Empirical functionals and efficient smoothing parameter selection. *Journal of the Royal Statistical Society, Series B* **54**, 475–530.

Hastie, T., Tibshirani, R. and Friedman, J. (2001). *The Elements of Statistical Learning: Data Mining, Inference and Prediction*. Springer, Berlin Heidelberg New York.

Kohavi, R. (1995). A study of cross-validation and bootstrap for accuracy estimation and model selection. In *Proceedings of the International Joint Conference on Artificial Intelligence* **2**, 1137–1143. Montreal, Canada.

Schapire, R. and Freund, Y. (1997). Boosting the margin: A new explanation for the effectiveness of voting methods. In *Proceedings of the 14th International Conference on Machine Learning*, 322–330. Nashville, TN.

Tibshirani, R. (1996). Regression shrinkage and selection via the lasso. *Journal of the Royal Statistical Society, Series B* **58**, 267–288.

Estimating the Error Rate of a Prediction Rule: Improvement on Cross-Validation

BRADLEY EFRON

Reprinted from the Journal of the American Statistical Association
June 1983, Volume 78, Number 382

Estimating the Error Rate of a Prediction Rule: Improvement on Cross-Validation

BRADLEY EFRON*

We construct a prediction rule on the basis of some data, and then wish to estimate the error rate of this rule in classifying future observations. Cross-validation provides a nearly unbiased estimate, using only the original data. Cross-validation turns out to be related closely to the bootstrap estimate of the error rate. This article has two purposes: to understand better the theoretical basis of the prediction problem, and to investigate some related estimators, which seem to offer considerably improved estimation in small samples.

KEY WORDS: Bootstrap; Prediction problem; ANOVA decomposition; Logistic regression.

1. INTRODUCTION

In the prediction problem the statistician has available a set of cases $x_1, x_2, \ldots, x_n$ collectively called the *training set* $\mathbf{x}$. Each case consists of two parts $x_i = (t_i, y_i)$, where t_i is a vector of predictors and y_i is a response variable. For example, t_i might describe a medical patient's age, weight, sex, race, previous disease history, and so on, and y_i might indicate whether the patient survived a certain operation. On the basis of the training set, a prediction rule $\eta(t, \mathbf{x})$ is constructed. The intention is to use $\eta(t_0, \mathbf{x})$ to predict a future unobserved response y_0 on the basis of its predictor vector t_0.

We are mainly concerned with the situation where y_i is a dichotomy, such as "survived" or "didn't survive," and the prediction $\eta_i = \eta(t_i, \mathbf{x})$ is likewise dichotomous. Let $Q[y_i, \eta_i]$ indicate the correctness of the ith prediction,

$$Q[y_i, \eta_i] = 0 \quad \text{if } \eta_i = y_i$$
$$= 1 \quad \text{if } \eta_i \neq y_i. \tag{1.1}$$

The *true error rate* (Err) of the prediction rule $\eta(t, \mathbf{x})$ is its probability of incorrectly classifying a randomly selected future case $X_0 = (T_0, Y_0)$, in other words the expectation $E\, Q[Y_0, \eta(T_0, \mathbf{x})]$.

Our goal is to estimate Err on the basis of the training set $\mathbf{x}$. The most obvious estimate is the *apparent error rate* $\bar{err} = \sum_{i=1}^{n} Q[y_i, \eta(t_i, \mathbf{x})]/n$, which is the proportion of observed errors made by $\eta(t, \mathbf{x})$ on its own training set $\mathbf{x}$. Usually $\bar{err}$ tends to be smaller than Err, because the same data have been used both to construct and to evaluate $\eta(t, \mathbf{x})$. This is a familiar fact in ordinary linear regression, where $\bar{err}$ = (residual sum of squares from fitted model)/n underestimates the true residual variance, and so the denominator n is usually reduced.

Cross-validation circumvents this difficulty by removing each x_i from the data set used in its own prediction. Let $\mathbf{x}_{(i)}$ be the training set with x_i removed, and $\eta(t, \mathbf{x}_{(i)})$ be the corresponding prediction rule. The cross-validated error rate is

$$\hat{\text{Err}}^{(\text{CV})} = \frac{1}{n} \sum_{i=1}^{n} Q[y_i, \eta(t_i, \mathbf{x}_{(i)})]. \tag{1.2}$$

The well-known paper by Lachenbruch and Mickey (1968) is a good reference. Cross-validation is discussed in a wider context by Stone (1974) and Geisser (1975).

In the next section we introduce another estimate of Err, based on the *bootstrap*, Efron (1979): $\hat{\text{Err}}^{(\text{BOOT})}$ is essentially the nonparametric maximum likelihood estimate of Err, assuming only that the training cases x_i are a random sample from some unknown distribution F on the space of possible vectors $x = (t, y)$. We will see that in some ways $\hat{\text{Err}}^{(\text{BOOT})}$ outperforms $\hat{\text{Err}}^{(\text{CV})}$ as an estimator of Err, though the comparison is not totally one-sided. Other estimators introduced in later sections outperform both $\hat{\text{Err}}^{(\text{CV})}$ and $\hat{\text{Err}}^{(\text{BOOT})}$, in an admittedly small catalog of five sampling experiments.

This article has two main purposes: to understand the theoretical basis of the prediction problem, especially as it relates to cross-validation and the bootstrap; and to investigate some related estimators, which seem to offer considerably improved estimation of err when the training set is small. The discussion is actually in the opposite order. The related estimators, all of which are simple variants of the bootstrap, are introduced in Sections 3 through 6. Sections 7 and 8 concern a decomposition of the prediction problem, based on the ANOVA description of Efron and Stein (1981), clarifying the theoretical connections between the various methods. The article ends with some remarks and a summary of recommendations.

2. THE BOOTSTRAP AND CROSS-VALIDATION

This section describes the bootstrap estimate of the true error rate, and relates it to $\hat{\text{Err}}^{(\text{CV})}$, the cross-vali-

* Bradley Efron is Professor of Statistics and Biostatistics, Department of Statistics, Stanford University, Stanford, CA 94305. In addition to the referees and the editor, Stephen Stigler, the author is grateful to Professors Samprit Chatterjee, Jerome Friedman, Gail Gong, and Charles Stone for helpful discussions on the prediction problem. Financial support was provided by Public Health Service Grant 5 R01 GM21215, and National Science Foundation Grant MCS80-24649.

Reprinted from: © Journal of the American Statistical Association
June 1983, Volume 78, Number 382
Theory and Methods Section

dation estimate. It is taken from a longer discussion in Chapter 7 of Efron (1982). We begin with a more careful description of the prediction problem.

Each case $x_i = (t_i, y_i)$ in the training set is the realization of a random quantity $X_i = (T_i, Y_i)$, where T_i is a p-dimensional row vector of predictors and Y_i is a real-valued response variable. We assume that there is some unknown distribution F on the $p+1$-dimensional sample space $\mathscr{X} = \mathscr{R}^{p+1}$ such that the training set $\mathbf{x} = (x_1, x_2, \ldots, x_n)$ is a random sample from F,

$$X_1, X_2, \ldots, X_n \overset{\text{iid}}{\sim} F, \tag{2.1}$$

iid abbreviating "independent and identically distributed." For convenience let $X_0 = (T_0, Y_0)$ denote a future observation from F, independent of the training set.

We have at hand a specific recipe for constructing a prediction rule $\eta(t, \mathbf{x})$ on the basis of the training set. A familiar example is the ordinary least squares rule $\eta(t, \mathbf{x}) = t(\mathbf{t}'\mathbf{t})^{-1}\mathbf{t}'\mathbf{y}$, where $\mathbf{y} = (y_1, y_2, \ldots, y_n)'$ and $\mathbf{t}'$ is the $p \times n$ matrix $(t'_1, t'_2, \ldots, t'_n)$. Notice that this rule remains the same under any reordering of the cases $x_1, x_2, \ldots, x_n$ constituting $\mathbf{x}$. This is the usual case and will be assumed to hold in what follows. We use $\eta(t_0, \mathbf{x})$ to predict y_0 from t_0, and measure the prediction error according to some function $Q[y_0, \eta(t_0, \mathbf{x})]$. In the dichotomous case both y and η are either 0 or 1, and Q is described by (1.1). The true error rate $\text{Err}(\mathbf{x}, F)$ is the expected value

$$\text{Err} = E_F\, Q[Y_0, \eta(T_0, \mathbf{x})], \tag{2.2}$$

the expectation being taken over $X_0 = (T_0, Y_0) \sim F$, with $\mathbf{x}$ fixed at its observed value. We will sometimes write $Q(x_0, \mathbf{x})$ for $Q[y_0, \eta(t_0, \mathbf{x})]$.

The apparent error rate $\bar{err}(\mathbf{x})$ is the statistic

$$\bar{err} = \frac{1}{n} \sum_{i=1}^{n} Q[y_i, \eta(t_i, \mathbf{x})] \tag{2.3}$$

usually an underestimate of Err. Let $\text{op}(\mathbf{x}, F)$, "op" being short for optimism, indicate the random variable

$$\text{op} = \text{Err} - \bar{err} \tag{2.4}$$

with expectation ω,

$$\omega(F) = E_F\, \text{op}(\mathbf{X}, F) = E_F\{\text{Err}(\mathbf{X}, F) - \bar{err}(\mathbf{X})\}. \tag{2.5}$$

If ω were known, we could estimate Err with

$$\hat{\text{Err}}^{(\text{IC})} = \bar{err} + \omega, \tag{2.6}$$

IC standing for "ideal constant." In most cases ω is not known, and must itself be estimated from the training set $\mathbf{x}$. Cross-validation as described in Section 1 amounts to using the estimate

$$\hat{\omega}^{(\text{CV})} = \frac{1}{n} \sum_i Q[y_i, \eta(t_i, \mathbf{x}_{(i)})] - \bar{err}, \tag{2.7}$$

so that $\hat{\text{Err}}^{(\text{CV})} = \bar{err} + \hat{\omega}^{(\text{CV})}$ equals $\sum Q[y_i, \eta(t_i, \mathbf{x}_{(i)})]/n$.

The bootstrap estimate $\hat{\text{Err}}^{(\text{BOOT})}$ equals $\bar{err} + \hat{\omega}^{(\text{BOOT})}$, where $\hat{\omega}^{(\text{BOOT})}$ if the nonparametric maximum likelihood estimate $\omega(\hat{F})$, $\hat{F}$ being the empirical probability distribution putting mass $1/n$ on each observed case,

$$\hat{F}: \text{mass } \frac{1}{n} \text{ on } x_i, \quad i = 1, 2, \ldots, n. \tag{2.8}$$

We can describe $\omega(\hat{F})$ more explicitly, in a way that suggests how to actually evaluate it. Let $\mathbf{X}^*$ indicate a random sample of size n from $\hat{F}$,

$$X_i^*, X_2^*, \ldots, X_n^* \overset{\text{iid}}{\sim} \hat{F}, \tag{2.9}$$

and E_* indicate expectation with respect to the random mechanism (2.9), $\hat{F}$ fixed at its observed value. Then

$$\begin{aligned}\hat{\omega}^{(\text{BOOT})} &= \omega(\hat{F}) = E_*\, \text{op}(\mathbf{X}^*, \hat{F}) \\ &= E_* \sum_i \left(\frac{1}{n} - P_i^*\right) Q[y_i, \eta(t_i, \mathbf{X}^*)], \end{aligned}\tag{2.10}$$

with P_i^* indicating the proportion of the bootstrap sample on x_i,

$$P_i^* = \frac{\#\{X_j^* = x_i\}}{n}, \quad i = 1, 2, \ldots, n. \tag{2.11}$$

The last expression in (2.10) is obtained by following through definition (2.5) (see Efron 1982).

Usually $\hat{\omega}^{(\text{BOOT})}$ must be evaluated by Monte Carlo: independent bootstrap training sets $\mathbf{x}^{*1}, \mathbf{x}^{*2}, \ldots, \mathbf{x}^{*B}$ are generated according to (2.9), and for each $\mathbf{x}^{*b}$ the prediction rule $\eta(t, \mathbf{x}^{*b})$ is calculated. This gives a bootstrap replication of op according to (2.10), $\text{op}^{*b\cdot} = \sum_{i=1}^{n} (1/n - P_i^{*b})\, Q[y_i, \eta(t_i, \mathbf{x}^{*b})]$, and we approximate $\hat{\omega}^{(\text{BOOT})}$ by the average $\sum_{b=1}^{B} \text{op}^{*b}/B$. As $B \to \infty$ this approaches definition (2.10). For practical purposes B in the range 25–200 seems quite adequate. *A better Monte Carlo method is given in Section 8.*

As an example, suppose $p = 2$, $n = 14$, and that each T_i is bivariate normal with mean vector either $\pm(\frac{1}{2}, 0)$,

$$Y_i = \begin{matrix} 0 \\ 1 \end{matrix} \ \text{prob} \ \begin{matrix} \frac{1}{2} \\ \frac{1}{2} \end{matrix} \ \text{and } T_i \mid Y_i = y_i \sim N_2((y_i - \tfrac{1}{2}, 0), I). \tag{2.12}$$

The prediction rule is Fisher's estimated linear discriminant,

$$\eta(t_0, \mathbf{x}) = \begin{matrix} 0 \\ 1 \end{matrix} \ \text{if } \hat{\alpha} + t_0\hat{\beta}' \text{ is } \begin{matrix} <0 \\ \geq 0 \end{matrix}. \tag{2.13}$$

The coefficients $\hat{\alpha}$ and $\hat{\beta}$ are given in terms of $n_y = \#\{y_j = y\}$, $\bar{t}_y = \sum_{y_i = y} t_i/n_y$, and $S = [\sum_i t'_j t_j - n_1\bar{t}'_1\bar{t}_1 - n_2\bar{t}'_2\bar{t}_2]/n$: $\hat{\alpha} = [\bar{t}_1 S^{-1}\bar{t}'_1 - \bar{t}_2 S^{-1}\bar{t}'_2]/2$ and $\hat{\beta} = [\bar{t}_2 - \bar{t}_1]S^{-1}$.

In the sampling experiment subsequently called (2, 14), 100 independent trials of situation (2.12), (2.13) were generated. Results for the first 10 of these, and summary statistics for all 100, appear in Table 1. We see that op

Table 1. The First 10 Trials of Experiment (2,14), Described at (2.12), (2.13), and Summary Statistics for all 100 Trials; $\hat{\omega}^{(CV)}$ is Less Biased Than $\hat{\omega}^{(BOOT)}$, But Much More Variable. The Bootstrap Was Run With B = 200 Replications per Trial

Trial	Err	$\bar{err}$	op	$\hat{\omega}^{(CV)}$	$\hat{\omega}^{(JACK)}$	$\hat{\omega}^{(BOOT)}$
1	.458	.286	.172	.214	.214	.083
2	.312	.357	−.045	.000	.066	.098
3	.313	.357	−.044	.071	.066	.110
4	.351	.429	−.078	.071	.066	.107
5	.330	.357	−.027	.143	.148	.102
6	.318	.143	.175	.214	.194	.073
7	.310	.071	.239	.071	.066	.047
8	.380	.286	.094	.071	.056	.097
9	.360	.429	−.069	.071	.087	.127
10	.335	.143	.192	.000	.010	.048
100 Trials Exp	.360	.264	ω = .096	.091	.093	.080
100 Trials (Sd)	(.045)	(.123)	(.113)	(.073)	(.068)	(.028)

averaged .096, which is ω except for sampling error, and so $\bar{err}$ tends to seriously underestimate Err in this case, $E\,\text{Err}/E\,\bar{err} = 1.36$. For each trial, $\hat{\omega}^{(CV)}$ was calculated according to (2.7), and $\hat{\omega}^{(BOOT)}$ calculated according to the Monte Carlo algorithm, using $B = 200$ bootstrap replications per trial. Err was calculated theoretically from (2.12), (2.13). The bootstrap estimate of ω was biased slightly downwards, averaging .080, but was far less variable than $\hat{\omega}^{(CV)}$.

The jackknife, or more precisely the jackknife estimate of bias, relates cross-validation to the bootstrap. It uses a quadratic expansion for op$(\mathbf{X}^*, \hat{F})$ and properties of the multinomial distribution to show that $E_*\,\text{op}(\mathbf{X}^*, \hat{F})$ can be approximated by

$$\hat{\omega}^{(JACK)} = \frac{1}{n}\sum_i Q[y_i, \eta(t_i, \mathbf{x}_{(i)})] - \frac{1}{n}\sum_i \{\sum_j Q[y_i, \eta(t_i, \mathbf{x}_{(j)})]/n\}. \quad (2.14)$$

Comparing (2.14) with (2.7), it is not surprising that $\hat{\omega}^{(JACK)}$ is usually close in value to $\hat{\omega}^{(CV)}$, as seen in Table 1. Gong (1982) shows that they have asymptotic correlation 1.00, under smoothness conditions on Q. The correlation over the 100 trials of experiment (2, 14) was .93. Derivation of (2.14) appears in Section 7.3 of Efron (1982). For an interesting application of almost the same idea to density estimation see Wong (1983), discussed here in Remark B of Section 9.

To summarize, $\hat{\omega}^{(BOOT)}$ is the obvious nonparametric MLE for ω; $\hat{\omega}^{(JACK)}$ is a quadratic approximation to $\hat{\omega}^{(BOOT)}$; and $\hat{\omega}^{(CV)}$ is similar in form and value to $\hat{\omega}^{(JACK)}$. All of this indicates that there is only one basic idea operating here: the substitution of $\hat{F}$ for F in whatever we are trying to estimate, that being $\omega(F)$ in the problem at hand.

3. FIVE SAMPLING EXPERIMENTS

Table 2 reports on five sampling experiments comparing cross-validation, the bootstrap, and several other methods of estimating Err, the true error rate. The first four experiments are (2, 14), described in Section 2, and three simple variations, (2, 20), (5, 14), and (5, 20). Experiment (2, 20) is exactly the same as (2, 14) except that the sample size n is increased from 14 to 20. Each trial of experiment (5, 14) involves $n = 14$ cases in $p = 5$ dimensions. The distribution of (T_i, Y_i) is as given at (2.12), except that $T_i \mid Y_i = y_i \sim N_5((2y_i - 1, 0, 0, 0, 0), I)$ a five-dimensional normal distribution. The prediction rule $\eta(t_0, \mathbf{x})$ is Fisher's estimated linear discriminant function, as described in Section 2. Experiment (5, 20) is the same as (5, 14) except that n is increased from 14 to 20. Each of these four experiments comprised 100 trials.

Experiment GG is taken from the Ph.D thesis of Gong (1982). The sample size is $n = 20$, the prediction dimension $p = 4$. Predictor vectors T_i have a four-dimensional normal distribution with mean vector 0, all standard deviations equal 1.0, and all correlations equal zero except for corr$(T_{i2}, T_{i3}) = .8$. Given $T_i = t_i$, the dichotomous response variable y_i equals 1 with probability $1/[1 + \exp -(\alpha + t_i\beta')]$, $\alpha = 0$, $\beta = (1, 2.25, 0, 0)$. The prediction rule $\eta(t, \mathbf{x})$ is based on a forward stepwise logistic regression, using a sequence of hypothesis tests to determine which of the components of t_i to include in making the prediction, and will not be further described here. Experiment GG comprised 171 trials.

We are estimating Err = $\bar{err}$ + op with statistics[1] of the form $\widehat{\text{Err}} = \bar{err} + \hat{\omega}$. The mean squared error (MSE) is $E(\widehat{\text{Err}} - \text{Err})^2 = E(\hat{\omega} - \text{op})^2$, or

$$\text{MSE} = (E\hat{\omega} - \omega)^2 + \text{var}(\hat{\omega}) - 2\,\text{cov}(\hat{\omega}, \text{op}) + \text{var}(\text{op}). \quad (3.1)$$

Notice that MSE measures how well, on the average, $\widehat{\text{Err}}$ estimates Err$(\mathbf{x}, F)$ *for the given training set* $\mathbf{x}$. In this sense it is a measure of average conditional risk. An un-

[1] The notation $\hat{\omega}$ could be changed to $\hat{op}$, but isn't for reasons given in the last paragraph of Section 2.

conditional measure such as $E[\text{Err} - E(\text{Err})]^2$ seems less appropriate.

Table 2 gives the MSE for each method, and also the information needed to compute the individual components on the right side of (3.1). For example, in experiment (2, 14), $\omega = E(\text{op}) = .096$ (100 trials), while the bootstrap estimate $\hat{\omega}^{(\text{BOOT})}$ has expectation .080. We see that the bias term in (3.1) contributes $(.080 - .096)^2 = .000256$, a negligible amount compared with the total MSE of .0179.

The large negative correlation of $\hat{\omega}^{(\text{BOOT})}$ with op, $-.64$ for experiment (2, 14), substantially increases MSE for the ordinary bootstrap. Cross-validation has correlations near zero, but suffers from high values of $\text{var}(\hat{\omega})$. The estimators performing well in Table 2 do so by reducing the negative correlation nearly to zero, without increasing $\text{var}(\hat{\omega})$ much above $\text{var}(\hat{\omega}^{(\text{BOOT})})$. It doesn't seem possible to make the correlation positive; see Remark G of Section 9.

Figure 1 graphically compares the performances of four of the estimators in Table 1. The MSE's are plotted on a relative inefficiency scale,

$$\text{REL} = \frac{\text{MSE} - \text{MSE}^{(\text{IC})}}{\text{MSE}^{(\text{ZERO})} - \text{MSE}^{(\text{IC})}}, \tag{3.2}$$

where $\text{MSE}^{(\text{IC})}$ is the mean squared error for the ideal constant estimator (2.6), $\widehat{\text{Err}} = \bar{err} + \omega$, and $\text{MSE}^{(\text{ZERO})}$ is the mean square error for the zero estimator $\widehat{\text{Err}} = \bar{err}$, that is, the apparent error rate. Large numbers are bad here: REL > 100% as for cross-validation in Experiments (2, 20) and GG, indicate estimators worse than $\bar{err}$.

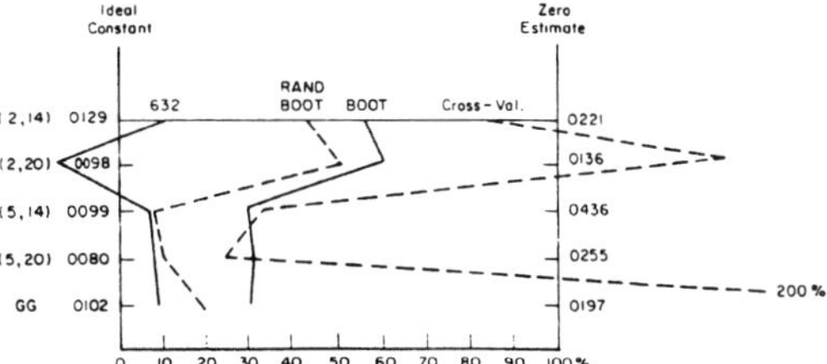

Figure 1. Relative inefficiencies (3.2) for four of the estimators in Table 1; 0% indicates performance equal to the ideal constant estimator, >100% indicates performance worse than the apparent error rate èrr.

The small sample sizes, $n = 14$ or 20, are an important factor in the large behavioral differences evident in Table 1 and Figure 1. The equivalent of Figure 1 for $n = 100$ would show all the estimators, doing much better (REL in the range 0–40%), but with the ordinary bootstrap and cross-validation still performing noticeably worst. Efron gives numerical results for all five experiments scaled up to have $n = 100$, Stanford Technical Report #78.

Table 2. Five Sampling Experiments, Comparing Several Different Methods for Estimating Err

	(2,14)	(2,20)	(5,14)	(5,20)	GG
	Exp(Sd Corr) MSE	Exp(Sd Corr) MSE	Exp(Sd Corr) MSE	Exp(Sd Corr) MSE	Exp(Sd Corr) MSE
100 Trials (171 for GG)					
True op	096(113)	059(099)	184(099)	130(090)	098(094)
Ideal Constant	096(0 0) 0129	059(0 0) 0099	184(0 0) 0099	130(0 0) 0080	098(0 0) 0088
Zero Correction	0(0 0) 0221	0(0 0) 0134	0(0 0) 0432	0(0 0) 0249	0(0 0) 0184
Cross-Validation	091(073 −15) 0206	067(070 +00) 0148	170(094 −15) 0216	139(070 +03) 0126	113(120 −21) 0280
Bootstrap (B = 200)	080(028 −64) 0179	061(020 −47) 0122	103(031 −58) 0210	086(025 −69) 0136	083(022 −57) 0118
Randomized Bootstrap	087(026 −55) 0169	062(020 −38) 0118	147(020 −31) 0129	109(017 −46) 0101	082(023 −28) 0108
Simple Randomized Bootstrap	097(023 −62) 0166	072(019 −51) 0123	157(021 −54) 0133	121(020 −67) 0109	100(020 −49) 0110
Double Bootstrap	097(038 −59) 0195	070(029 −40) 0132	184(054 −57) 0190	114(034 −61) 0132	106(033 −48) 0129
Randomized Double Bootstrap	097(036 −54) 0186	068(029 −43) 0133	186(038 −52) 0152	120(032 −62) 0128	NA NA
$632(\hat{\epsilon}^{(0)} - \bar{err})$	076(035 −09) 0138	059(032 +22) 0095	152(038 −04) 0126	112(035 +02) 0094	080(042 +14) 0097
$\hat{W}^{(0)} = \hat{\epsilon}^{(0)} - \hat{\mu}$	101(034 −56) 0184	071(024 −44) 0128	176(044 −54) 0167	124(030 −53) 0119	107(029 −43) 0120
1000 Trials: *op Err* *èrr* →	.093 .356 .262	.060 .340 .280	.178 .250 .072	.120 .219 .099	

NOTE: Entry MSE is the mean squared error $E(\widehat{Err} - Err)^2$; Exp = $E(\hat{\omega})$, Sd = Standard Dev $(\hat{\omega})$, Corr = correlation $(\hat{\omega}, \text{op})$. All bootstrap methods used B = 200 bootstrap replications per trial except in experiment GG, where B = 100.

The next three sections discuss the new estimators appearing in Table 2, some of which clearly outperform cross-validation and the bootstrap.

4. RANDOMIZED BOOTSTRAP

The randomized bootstrap is a particularly simple variant of the ordinary bootstrap appropriate when y is dichotomous. The two versions of the randomized bootstrap appearing in Table 2 performed well.

The empirical probability distribution $\hat{F}$, (2.8), concentrates all of its mass on the n points (t_i, y_i), $i = 1, 2, \ldots, n$. The idea of the randomized bootstrap is to assign some probability mass to the n complementary points $(t_i, \bar{y}_i)$, $\bar{y}_i \equiv 1 - y_i$. Given the training set $\mathbf{x}$, we have in mind some way of assigning probabilities to all $2n$ points (t_i, y_i), $(t_i, \bar{y}_i)$, say

$$\text{Assigned probability on } (t_i, y) = \frac{1}{n}\pi_i(y, \mathbf{x}), \quad (4.1)$$

where $\pi_i(y, \mathbf{x}) + \pi_i(\bar{y}, \mathbf{x}) = 1$. This last condition means that (t_i, y_i) and $(t_i, \bar{y}_i)$ are assigned total probability $1/n$, as with the ordinary bootstrap. For the *simple randomized bootstrap*, line 6 of Table 2,

$$\pi_i(y_i, \mathbf{x}) = .9, \quad \pi_i(\bar{y}_i, \mathbf{x}) = .1. \quad (4.2)$$

(To put it another way, this rule shrinks the maximum likelihood estimates $\pi(y_i, \mathbf{x}) = 1$, $\pi(\bar{y}_i, \mathbf{x}) = 0$ toward the central value .5.)

Let $\hat{F}^{(\text{RAND})}$ be the distribution on $2n$ points given by (4.1). Then the randomized bootstrap estimate of ω is

$$\hat{\omega}^{(\text{RAND})} = E_* \operatorname{op}(\mathbf{X}^*, \hat{F}^{(\text{RAND})}), \quad (4.3)$$

where now $\mathbf{X}^*$ is a random sample from $\hat{F}^{(\text{RAND})}$,

$$X_1^*, X_2^*, \ldots, X_n^* \overset{\text{iid}}{\sim} \hat{F}^{(\text{RAND})}, \quad (4.4)$$

and E_* indicates expectation with respect to this probability mechanism.

Defining $N_{i,y}^* = \#\{X_j^* = (t_i, y)\}$ and $N_i^* = N_{i0}^* + N_{i1}^*$, it is possible to express $\operatorname{op}(\mathbf{X}^*, \hat{F}^{(\text{RAND})})$ as a sum of n terms,

$$\operatorname{op}^* = \sum_{i=1}^{n} \frac{1}{n}\{[(2\pi_i(y_i, \mathbf{x}) - 1) - (2N_{i,y_i}^* - N_i^*)] \times Q[y_i, \eta(t_i, \mathbf{X}^*)] + [\pi_i(\bar{y}_i, \mathbf{x}) - N_{i,\bar{y}_i}^*]\}. \quad (4.5)$$

Notice that this reduces to the expression for op* in (2.9) if $\pi_i(y_i, \mathbf{x}) = 1$, $\pi_i(\bar{y}_i, \mathbf{x}) = 0$. Since $E_*[\pi_i(\bar{y}_i, \mathbf{x}) - N_{i,\bar{y}_i}^*] = 0$, (4.5) gives the quite simple expression

$$\hat{\omega}^{(\text{RAND})} = E_* \sum_{i=1}^{n} \frac{1}{n}\{(2\pi_i(y_i, \mathbf{x}) - 1) - (2N_{i,y_i}^* - N_i^*)\} \times Q[y_i, \eta(t_i, \mathbf{X}^*)]. \quad (4.6)$$

Both (4.5) and (4.6) remain valid if y_i is replaced by 1 and $\bar{y}_i$ is replaced by 0 everywhere.

Most often in dichotomous prediction problems, the prediction rule provides a probability assessment $\pi_i(y, \mathbf{x})$ as well as a specific prediction $\eta(t_i, \mathbf{x})$. For example Fisher's estimated linear discriminant is naturally associated with the probability assessment

$$\pi_i(1, \mathbf{x}) = 1/[1 + \exp - (\hat{\alpha} + t_i\hat{\beta}')], \quad (4.7)$$

(see Efron 1975, Sec. 1). Line 5 of Table 1, the randomized bootstrap, refers to the use of (4.7) in (4.1) through (4.4), except that the values $\pi_i(1, \mathbf{x})$ are restricted to lie in the range [.1, .9].

There is an obvious ad hoc component to the choice of the numbers .1, .9 for the two randomized bootstraps. In theory the statistician could make a subjective assessment of the uncertainty in each prediction $\eta(t_i, \mathbf{x})$, in order to assign $\pi_i(y_i, \mathbf{x})$ and $\pi_i(\bar{y}_i, \mathbf{x})$. In the sampling experiments the exact assignments seemed less important than keeping them away from 0 and 1. In particular the simple method (4.2) caused little bias (and as a matter of fact helped correct the bias in the ordinary bootstrap), and gave almost as much improvement as the more complicated method based on (4.7).

It is obvious that $\hat{F}$ can be a poor estimate of F, particularly if we know that F is smooth. Using $\hat{F}^{(\text{RAND})}$ in place of $\hat{F}$ is a form of smoothing. The smoothing is carried out entirely in the y direction. This is handy since in real applications t may be very complicated, having high dimensionality, censored components, missing values, qualitative and quantitative components, and so on.

5. THE DOUBLE BOOTSTRAP

The bootstrap estimate $\widehat{\text{Err}}^{(\text{BOOT})}$ was obtained in Section 2 by (a) writing Err as $\bar{err} + (\text{Err} - \bar{err})$ where $\bar{err} \equiv S(\mathbf{X})$ is an observable statistic and $(\text{Err} - \bar{err}) \equiv R(\mathbf{X}, F)$ is a random variable, and (b) estimating Err by $S + E_* R^*$, where $E_* R^*$ is the bootstrap expectation $E_* R(\mathbf{X}^*, \hat{F})$. There is no obvious theoretical reason for the choice $S = \bar{err}$. For any statistic S we could write $\text{Err} = S + (\text{Err} - S)$ and estimate Err by $S + E_*(\text{Err} - S)^*$. For example, choosing $S = 0$ gives the estimate $E_*\text{Err}^*$. It will turn out, in Section 8, that this is a poor estimate of Err.

This section concerns bootstrapping the bootstrap. We take $S = \bar{err} + \hat{\omega}^{(\text{BOOT})} = \widehat{\text{Err}}^{(\text{BOOT})}$, what we have called the ordinary bootstrap estimate of Err, write

$$\text{Err} = \widehat{\text{Err}}^{(\text{BOOT})} + (\text{Err} - \widehat{\text{Err}}^{(\text{BOOT})}) \equiv S + R, \quad (5.1)$$

and estimate Err by the "double bootstrap" estimate

$$\widehat{\text{Err}}^{(\text{DOUB})} = S + E_* R^*. \quad (5.2)$$

One motivation for doing so is the downward bias of the ordinary bootstrap evident in Table 2, in particular for experiment (5, 14). If $\widehat{\text{Err}}^{(\text{BOOT})}$ is an underestimator of Err, then we can correct it by bootstrapping, as in (5.1), in the same spirit as we originally corrected $\bar{err}$. (Sec. 7 discusses the downward bias of the bootstrap.)

We can rewrite (5.1), (5.2) as

$$\begin{aligned}\widehat{\text{Err}}^{(\text{DOUB})} &= (\bar{\text{e}}\text{rr} + \hat{\omega}^{(\text{BOOT})}) \\ &\quad + E_*(\text{Err} - \bar{err} - \hat{\omega}^{(\text{BOOT})})^* \\ &= (\bar{\text{e}}\text{rr} + \hat{\omega}^{(\text{BOOT})}) + E_*(\text{Err} - \bar{err})^* \\ &\quad - E_*(\hat{\omega}^{(\text{BOOT})})^* \\ &= \bar{err} + 2\hat{\omega}^{(\text{BOOT})} - E_*(\hat{\omega}^{(\text{BOOT})})^*, \end{aligned} \tag{5.3}$$

the last line following from $E_*(\text{Err} - \bar{err})^* = E_*(\text{op})^* = \hat{\omega}^{(\text{BOOT})}$. Another way to say (5.3) is that $\hat{\omega}^{(\text{DOUB})} = 2\hat{\omega}^{(\text{BOOT})} - E_*(\hat{\omega}^{(\text{BOOT})})^*$. Assuming that $\hat{\omega}^{(\text{BOOT})}$ has already been computed, we need to calculate $E_*(\hat{\omega}^{(\text{BOOT})})^*$ in order to find $\hat{\omega}^{(\text{DOUB})}$ and $\widehat{\text{Err}}^{(\text{DOUB})}$. This looks as if it involves two layers of bootstrapping, perhaps B^2 recomputations of the rule η, which would be $200^2 = 40{,}000$ recomputations in our case. It turns out that a total of $2B$ recomputations, 400 in our case, suffice, thanks to a Monte Carlo "swindle," so that the double bootstrap is computationally feasible, as well as properly named.

It is shown in the Appendix that

$$E_*(\hat{\omega}^{(\text{BOOT})})^* = E_{**}\left\{\sum_{i=1}^{n} e(P_i^{**})\, Q[y_i, \eta(t_i, \mathbf{X}^{**})]\right\}, \tag{5.4}$$

where $\mathbf{X}^{**}$ is a second-level bootstrap sample,

$$X_1^{**}, X_2^{**}, \ldots, X_n^{**} \overset{\text{iid}}{\sim} \hat{F}^*, \tag{5.5}$$

$\hat{F}^*$ indicating the empirical distribution function of a first-level bootstrap sample $X_1^*, \ldots, X_n^*$. The quantities $\mathbf{P}_i^{**}$ are the proportions of $\mathbf{X}^{**}$ on the various original cases x_i,

$$P_i^{**} = \#\{X_j^{**} = x_i\}/n, \tag{5.6}$$

and the function $e(P_i^{**})$ is given to a good approximation as follows:

nP_i^{**} =	0	1	2	3	4	5	6
$ne(P_i^{**})$ =	.37	.37	−.36	−1.02	1.65	−2.30	−2.97

(5.7)

The expectation E_* in (5.4) is with respect to the marginal distribution of $\mathbf{X}^{**}$, first taking a bootstrap sample as at (2.9), then constructing its empirical distribution function $\hat{F}^*$, and finally drawing $\mathbf{X}^{**}$ as at (5.5). The training set $\mathbf{x}$ is considered fixed during this entire process.

Expression (5.4) can be approximated by Monte Carlo just as was (2.9) for the ordinary bootstrap: (a) A sequence of independent double bootstrap vectors $\mathbf{x}^{**b}$, $b = 1, 2, \ldots, B$, is obtained, each one generated by the process described in the previous paragraph; (b) For each $\mathbf{x}^{**b}$ the rule $\eta(t, \mathbf{x}^{**b})$ is constructed and; (c) $(\hat{\omega}^{(\text{BOOT})})^{*b} = \sum_{i=1}^{n} e(P_i^{**b})\, Q[y_i, \eta(t_i, \mathbf{x}^{**b})]$ is evaluated. Then

$$E_*(\hat{\omega}^{(\text{BOOT})})^* \doteq \frac{1}{B}\sum_{b=1}^{B} (\hat{\omega}^{(\text{BOOT})})^{*b}, \tag{5.8}$$

with increasing accuracy as $B \to \infty$. For the entries in line 7 of Table 2, $B = 200$ in (5.8), taken in addition to the 200 replications used in calculating $\hat{\omega}^{(\text{BOOT})}$.

The double bootstrap nicely corrects the bias in the ordinary bootstrap, as can be seen by comparing $E(\hat{\omega}^{(\text{DOUB})})$ from line 7 of Table 2 with the actual values of ω. In terms of MSE its performance is about the same as the ordinary bootstrap.

Line 8 of Table 2 refers to a double bootstrap version of the simple randomized bootstrap defined at (4.2). This computation requires a formula like (5.4) referring to $(\hat{\omega}^{(\text{RAND})})^*$, from (4.6), rather than to $(\hat{\omega}^{(\text{BOOT})})^*$. No more will be said about the randomized double bootstrap here, except that it is not particularly difficult to compute and performs slightly better than the double bootstrap in Table 2.

6. THE .632 ESTIMATOR

The estimator of line 9, Table 2, $.632(\hat{\epsilon}^{(0)} - \bar{err})$, called "the .632 estimator" for short, was a clear winner in the sampling experiments. This section defines and motivates the .632 estimator. Unfortunately the motivation is weak, leaving open the possibility that the estimator's success here was a fluke. (It has continued to perform best in some additional, rather different, sampling experiments described in Gong 1982.)

Why does $\bar{err}$ tend to underestimate Err? Another answer, beside that it obviously does, can be given in terms of the distance of the point to be predicted from the training set: $\bar{err}$ is an error rate for points zero distance from the training set $\mathbf{x}$: Err is the expected error rate for a new point X_0 which may lie some distance away from $\mathbf{x}$. If the error rate of the prediction rule increases as the point being predicted moves away from $\mathbf{x}$, then $\bar{err}$ will underestimate Err.

To make this argument more concrete suppose that for each x and Δ the set $S(x, \Delta)$ is a neighborhood of x having probability content Δ under the true distribution F,

$$\text{Prob}_F\{X_0 \in S(x, \Delta)\} = \Delta. \tag{6.1}$$

The neighborhoods $S(x, \Delta)$ are assumed to grow smaller as Δ decreases, going to the single point $\{x\}$ as $\Delta \to 0$. Define

$$\delta(x_0, \mathbf{x}) = \inf_{\Delta}\left\{x_0 \in \bigcup_{i=1}^{n} S(x_i, \Delta)\right\}, \tag{6.2}$$

δ is large or small as x_0 is far from or near to the nearest point in the training set $\mathbf{x}$. Denote $\bar{Q}(\Delta)$ by the following:

$$\bar{Q}(\Delta) = E\{Q(X_0, \mathbf{X}) \mid \delta(X_0, \mathbf{X}) = \Delta\}, \tag{6.3}$$

where $Q(X_0, \mathbf{X}) = Q[Y_0, \eta(T_0, \mathbf{X})]$, so $\bar{Q}(\Delta)$ is the expected error rate given that X_0 is distance Δ from the nearest point in the training set.

The curve marked Actual in Figure 2 shows $\bar{Q}(\Delta)$ as a function of Δ for experiment (2, 20). The neighborhoods $S(x, \Delta)$, $x = (t, y)$, were taken as circles in the t space,

$$S((t, y), \Delta) = \{x_0 = (t_0, y_0): y_0 = y \text{ and } \| t_0 - t \| \le r_{x,\Delta}\}, \quad (6.4)$$

with $r_{x,\Delta}$ chosen to satisfy (6.1). The set $\cup_{i=1}^{n} S(x_i, \Delta)$ is a union of circles in the planes $y = 0$ and $y = 1$, each circle centered at x_i and having radius roughly inversely proportional to the density of model (2.12) at x_i.

As expected, $\bar{Q}(\Delta)$ is an increasing function of Δ. Notice that

$$\bar{Q}(0) = E\,\bar{err}, \quad (6.5)$$

$= .280$ for experiment (2, 20). Relation (6.5) is a consequence of the way we defined $\bar{Q}(\Delta)$ and $S(x_i, \Delta)$. As $\Delta \to 0$, the conditional distribution of a point X_0 in $\cup_{i=1}^{n} S(x_i, \Delta)$ approaches the empirical distribution $\hat{F}$, (2.8), so $\bar{Q}(\Delta) \to E\,\bar{err}$.

Let $D(\Delta)$ be the cumulative distribution function of $\delta(X_0, \mathbf{X})$, $D(\Delta) = \text{Prob}\{\delta(X_0, \mathbf{X}) \le \Delta\}$ for $0 \le \Delta \le 1$. Then the expected true error rate is $E(\text{Err}) = \int_0^1 \bar{Q}(\Delta)\,dD(\Delta)$ and, since $E\,\bar{err} = \bar{Q}(0)$,

$$\omega = \int_0^1 [\bar{Q}(\epsilon) - \bar{Q}(0)]dD(\Delta) \quad (6.6)$$

is the expected optimism. Looking at Figure 2, and at the insert, which shows $D(\Delta)$, one can see that most of $\omega = .060$ for experiment (2, 20) comes from values of Δ in the range (0, .1).

The curve marked Bootstrap in Figure 2 is

$$\bar{Q}^{(B)}(\Delta) = E\{Q(X_0^*, \mathbf{X}^*) \mid \delta(X_0^*, \mathbf{X}^*) = \Delta\}, \quad (6.7)$$

the expectation of $Q(X_0^*, \mathbf{X}^*) = Q[Y_0^*, \eta(T_0^*, \mathbf{X}^*)]$ given that the independent point $X_0^* \sim \hat{F}$ is distance Δ away from the bootstrap training set $X_1^*, X_2^*, \ldots, X_n^* \overset{\text{iid}}{\sim} \hat{F}$ used to construct $\eta(t, \mathbf{X}^*)$; E indicates marginal expectation over the choice of $X_1, X_2, \ldots, X_n \overset{\text{iid}}{\sim} F$ and then $X_0^*, X_1^*, X_2^*, \ldots, X_n^* \overset{\text{iid}}{\sim} \hat{F}$. *Notice the agreement between* $\bar{Q}(\Delta)$ *and* $\bar{Q}^{(B)}(\Delta)$. A point say $\Delta = .05$ away from a bootstrap training set has the same expected probability of misclassification as a point .05 away from an actual training set.

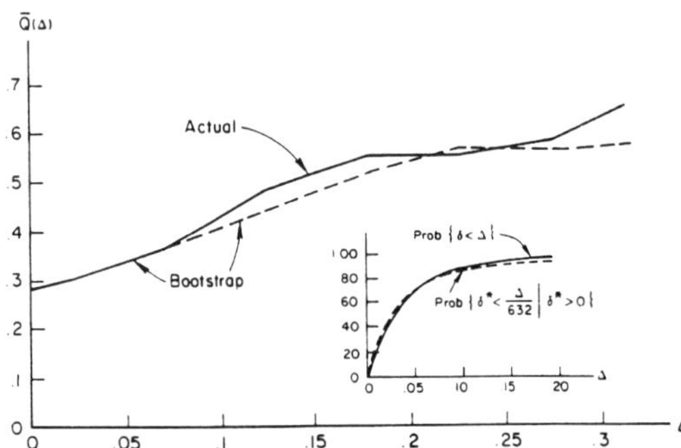

Figure 2. The actual and bootstrap expected error rates as a function of the distance from X_0 to the nearest point in the training set, experiment (2, 20). The insert shows the actual and bootstrap cumulative distribution function for the distance to the nearest point.

There is one big difference between the actual and the bootstrap situations: the distribution of the distance δ. The bootstrap distance $\delta(X_0^*, \mathbf{X}^*)$ has a high probability of equaling zero,

$$\text{Prob}\{\delta(X_0^*, \mathbf{X}^*) = 0\} = 1 - (1 - (1/n))^n \doteq 1 - e^{-1} = .632, \quad (6.8)$$

this being the probability that X_0^* falls on one of the support points of $\hat{F}$ occurring in the bootstrap sample $\mathbf{X}^*$ (i.e., that $X_0^* = x_i$ with $P_i^* > 0$). Given that $\delta(X_0^*, \mathbf{X}^*) > 0$ (i.e., $X_0^* = x_i$ with $P_i^* = 0$) we show in the Appendix that $\delta(X_0^*, \mathbf{X}^*)$ is roughly distributed as $\delta(X_0, \mathbf{X})/.632$,

$$\text{Prob}\left\{\delta(X_0^*, \mathbf{X}^*) > \frac{\Delta}{.632} \;\middle|\; \delta(X_0^*, \mathbf{X}^*) > 0\right\} \doteq \text{Prob}\{\delta(X_0, \mathbf{X}) > \Delta\}. \quad (6.9)$$

All the probabilities occurring in (6.8), (6.9) are marginal over the choice of $\mathbf{X}$ and $\mathbf{X}^*$, X_0^*.

For a given training set $\mathbf{x}$, let $\hat{\epsilon}^{(0)}$ be the bootstrap expected error rate at those points x_i not occurring in the bootstrap sample,

$$\hat{\epsilon}^{(0)} = E_*\{Q(X_0^*, \mathbf{X}^*) \mid X_0^* = x_i \text{ with } P_i^* = 0\}. \quad (6.10)$$

In any one trial of experiment (2, 20), $\hat{\epsilon}^{(0)}$ was computed by (a) examining all 4,000 entries (b, i), $b = 1, \ldots, 200$, $n = 1, \ldots, 20$; (b) looking at the approximately 36.8 percent of the entries having $P_i^{*b} = 0$; and (c) setting $\hat{\epsilon}^{(0)}$ equal to the observed error rate for these entries,

$$\hat{\epsilon}^{(0)} = \frac{\sum_{(b,i):P_i^{*b}=0} Q(x_i, \mathbf{x}^{*b})}{\#\{(b, i): P_i^{*b} = 0\}}. \quad (6.11)$$

Expression (6.11) approaches definition (6.10) as $B \to \infty$. Section 8 discusses $\hat{\epsilon}^{(0)}$ and also the conditional error rates for values of P_i^* besides 0.

The .632 estimator is $\hat{\omega}^{(.632)} = .632(\hat{\epsilon}^{(0)} - \bar{err})$. This gives the Err estimate

$$\widehat{\text{Err}}^{(.632)} = \bar{err} + \hat{\omega}^{(.632)} = .368\,\bar{err} + .632\,\hat{\epsilon}^{(0)}. \quad (6.12)$$

The weights .368 and .632 are suggested by (6.9). The points X_0^* contributing to $\hat{\epsilon}^{(0)}$ are about 1/.632 too far out along the δ axis in Figure 2, whereas the points contributing to $\bar{err}$ are at $\delta = 0$. The weighted average (6.12) therefore involves points with about the right expected value of δ. If $\bar{Q}(\Delta) \doteq \bar{Q}^{(B)}(\Delta)$, and both are roughly linear in Δ, this makes $\widehat{\text{Err}}^{(.632)}$ roughly unbiased for Err. In fact $\widehat{\text{Err}}^{(.632)}$ displays a moderate downward bias in Table 2.

The reason for its remarkably low MSE is its lack of negative correlation with op, term three of (3.1). This is discussed further in Section 9, Remark G.

It will turn out in Section 8 that $\hat{\epsilon}^{(0)}$ is almost the same as $\widehat{\text{Err}}^{(\text{HCV})}$, the estimated error rate based on a cross-validation that leaves out half of the sample at a time. This means that a good approximation $\widehat{\text{Err}}^{(.632)}$ can be written as $.368\ \overline{err} + .632\ \widehat{\text{Err}}^{(\text{HCV})}$.

7. ANOVA DECOMPOSITIONS

This section describes the ANOVA decomposition (Efron and Stein 1981) as it applies to $Q(X_0, \mathbf{X})$ and $Q(X_0^*, \mathbf{X}^*)$. It will give us a better theoretical basis for understanding the prediction problem, particularly the orders of magnitude involved, and the relationships between various estimates of err. The calculations are carried through formally and do not constitute valid asymptotic theorems, but nevertheless give quite accurate predictions in our numerical studies.

The ANOVA decomposition for $Q(x_0, \mathbf{x}) = Q[y_0, \eta(t_0, \mathbf{x})]$ is

$$Q(x_0, \mathbf{x}) = \mu_{x_0} + \frac{1}{n}\sum_i \alpha_{x_0}(x_i) + \frac{1}{n^2}\sum_{i<i'} \beta_{x_0}(x_i, x_{i'}) + \frac{1}{n^3}\sum_{i<i'<i''} \gamma_{x_0}(x_i, x_{i'}, x_{i''}) + \cdots \quad (7.1)$$

The quantities μ_{x_0}, $\alpha_{x_0}(x_i)$, $\beta_{x_0}(x_i, x_{i'})$, and so on correspond to the grand mean, main effects, second-order interactions, and so on in the standard ANOVA decomposition of an n-way table:

$$\mu_{x_0} = E\, Q(x_0, \mathbf{X}),$$
$$\alpha_{x_0}(x_i) = n[E\{Q(x_0, \mathbf{X}) \mid X_i = x_i\} - \mu_{x_0}],$$
$$\beta_{x_0}(x_i, x_{i'}) = n^2[E\{Q(x_0, \mathbf{X}) \mid X_i = x_i, X_{i'} = x_{i'}\} - \alpha_{x_0}(x_i) - \alpha_{x_0}(x_{i'}) + \mu_{x_0}] \quad (7.2)$$

and so on; x_0 is fixed in these expectations, with $\mathbf{X}$ random, subject to the indicated conditioning statements. The sums in (7.1) are over all integers $i, i', i'', \ldots$, in the range $1, 2, \ldots, n$, subject to ordering conditions as indicated. The right side of (7.1) terminates with a single nth order interaction term.

The factors of n in (7.2) give α, β, and so on, nondegenerate limiting distributions. As n grows large $\alpha_{x_0}(X_i)$ approaches the *influence function* for $Q(x_0, \mathbf{X})$, $\beta_{x_0}(X_i, X_{i'})$ approaches the second-order influence function, and so on. See Hampel (1974) for a good discussion of influence function ideas.

Expansion (7.1) is an orthogonal decomposition of $Q(x_0, \mathbf{x})$. The quantities $\alpha_{x_0}(X_i)$, $\beta_{x_0}(X_i, X_{i'})$, and so on have expectation zero and are mutually uncorrelated. In fact each of them has conditional expectation zero when conditioned upon all but one of its defining X_i,

$$E\, \alpha_{x_0}(X_i) = 0,\ E\, \beta_{x_0}(x_i, X_{i'}) = 0,$$
$$E\, \gamma_{x_0}(x_i, x_{i'}, X_{i''}) = 0 \quad (7.3)$$

and so on, $i < i' < i''$, only capitalized X's being random in these expectations.

Many quantities related to the prediction problem have simple expressions in terms of the ANOVA decomposition. As a first example we have

$$\omega = -E\, \alpha_{X_1}(X_1)/n. \quad (7.4)$$

This follows from

$$E\ \text{Err} = E\, Q(X_0, \mathbf{X}) = E\, \mu_{X_0},$$
$$E\, \overline{err} = E\frac{1}{n}\sum Q(X_i, \mathbf{X}) = E\, Q(X_1, \mathbf{X}) = E\, \mu_{X_1} + \frac{1}{n} E\, \alpha_{X_1}(X_1), \quad (7.5)$$

all other terms such as $E\, \beta_{X_1}(X_1, X_2)$ equaling zero by (7.3), so that $\omega = E(\text{Err} - \overline{err}) = -E\, \alpha_{X_1}(X_1)/n$.

As another example consider the following variant of cross-validation. Let $\mathbf{x}_{(i,j)}$ represent the modified training set with x_i removed and x_j included twice, and let $Q_{(i,j)} = Q(x_i, \mathbf{x}_{(i,j)})$. Then define

$$\hat{\omega}^{(\text{CV}+)} = \frac{1}{n(n-1)}\sum_{i\neq j} Q_{(i,j)} - \overline{err}. \quad (7.6)$$

This is a version of $\hat{\omega}^{(\text{CV})}$, (2.7), for which all the modified training sets $\mathbf{x}_{(i,j)}$ have sample size n. An easy calculation, very much like (7.4), shows that

$$E\, \hat{\omega}^{(\text{CV}+)} - \omega = (E\, \beta_{X_0}(X_1, X_1))/n^2. \quad (7.7)$$

(Because $\hat{\omega}^{(\text{CV})}$ involves samples of size $n - 1$, (7.1) cannot be applied to it; see Efron and Stein 1981.) Letting $\widehat{\text{Err}}^{(\text{CV}+)} = \overline{err} + \hat{\omega}^{(\text{CV}+)}$, (7.7) gives $E(\widehat{\text{Err}}^{(\text{CV}+)} - err) = E\, \beta_{X_0}(X_1, X_1)/n^2$, compared with $E(\text{Err} - \overline{err}) = -E\, \alpha_{X_1}(X_1)/n$ from (7.4). *Cross-validation reduces bias of the error estimate from $O(1/n)$ to $O(1/n^2)$.*

There is an analog of decomposition (7.1) that applies to the bootstrap quantity $Q(x_0^*, \mathbf{x}^*) = Q[y_0^*, \eta(t_0^*, \mathbf{x}^*)]$:

$$Q(x_0^*, \mathbf{x}^*) = \hat{\mu}_{x_0^*} + \frac{1}{n}\sum_i \hat{\alpha}_{x_0^*}(x_i^*) + \frac{1}{n^2}\sum_{i<i'} \hat{\beta}_{x_0^*}(x_i^*, x_{i'}^*) + \ldots \quad (7.8)$$

where

$$\hat{\mu}_{x_0^*} = E_*\, Q(x_0^*, \mathbf{X}^*),$$
$$\alpha_{x_0^*}(x_i^*) = E_*\{Q(x_0^*, \mathbf{X}^*) \mid X_i^* = x_i^*\} - \mu_{x_0^*} \quad (7.9)$$

and so on, as in (7.2). The bootstrap analogy of (7.3) is

$$E_*\, \hat{\alpha}_{x_0^*}(X_i^*) = 0, \quad E_*\, \hat{\beta}_{x_0^*}(x_1^*, X_2^*) = 0 \quad (7.10)$$

and so on. The random variables $X_0^*, X_1^*, \ldots, X_n^*$ take their values in the training set $\{x_1, x_2, \ldots, x_n\}$. We will use the shortened notation $\hat{\alpha}_{j_0}(j_1)$ for

$$\hat{\alpha}_{X_0^* = x_{j_0}}(X_1^* = x_{j_1}),$$

likewise $\hat{\beta}_{j_0}(j_1, j_2)$, and so on. Then (7.10) becomes

$$\frac{1}{n}\sum_{j_1=1}^{n} \hat{\alpha}_{j_0}(j_1) = 0, \quad \frac{1}{n}\sum_{j_2=1}^{n} \hat{\beta}_{j_0}(j_1, j_2) = 0, \quad (7.11)$$

and so on, the first relationship holding for all j_0, the second for all j_0, j_1, $1 \le j_0, j_1 \le n$.

We immediately get

$$\hat{\omega}^{(\mathrm{BOOT})} = -E_* \, \hat{\alpha}_{X_1^*}(X_1^*)/n = -\sum_{j=1}^{n} \hat{\alpha}_j(j)/n^2, \quad (7.12)$$

the proof being the same as for (7.4).

We can use (7.4), (7.12) to analyze the downward bias of $\hat{\omega}^{(\mathrm{BOOT})}$ as an estimator of ω noticed in Table 2 (details given in the Appendix):

$$E\,\hat{\omega}^{(\mathrm{BOOT})} - \omega = \frac{1}{n^2}[E\,\alpha_{X_1}(X_1) - E\,\beta_{X_1}(X_1, X_1) + E\,\beta_{X_0}(X_1, X_1) - \tfrac{1}{2}E\,\gamma_{X_0}(X_0, X_1, X_1)] + O\left(\frac{1}{n^3}\right). \quad (7.13)$$

Equation (7.13) shows that $\hat{\omega}^{(\mathrm{BOOT})}$, like $\hat{\omega}^{(\mathrm{CV}+)}$, estimates the $O(1/n)$ quantity ω with expected error $O(1/n^2)$. Comparing (7.13) and (7.7) shows that $\hat{\omega}^{(\mathrm{BOOT})}$ has three extra terms in the $O(1/n^2)$ expression, two of which turn out to be negative in our experiment.

Now we will evaluate the terms in (7.13). Let $\mathbf{X}_{(1,2)} = (X_2, X_2, X_3, \ldots, X_n)$. We already know that $E\,\alpha_{X_1}(X_1)/n^2 = -\omega/n$. Expressions (7.1), (7.3) give

$$E\,\beta_{X_0}(X_1, X_1)/n^2 = E\,Q(X_0, \mathbf{X}_{(1,2)}) - E\,Q(X_0, \mathbf{X}),$$

$$E\,\beta_{X_1}(X_1, X_1)/n^2 = E\,Q(X_2, \mathbf{X}_{(1,2)}) - E\,Q(X_1, \mathbf{X}) + E\,Q(X_0, \mathbf{X}),$$

$$E\,\gamma_{X_0}(X_0, X_1, X_1)/2n^2 = \binom{n}{2}[E\,Q(X_3, \mathbf{X}_{(1,2)}) - E\,Q(X_0, \mathbf{X}_{(1,2)}) - E\,Q(X_1, \mathbf{X}) + E\,Q(X_0, \mathbf{X})]. \quad (7.14)$$

Table 3 shows the components of $E\,\hat{\omega}^{(\mathrm{BOOT})} - \omega$ for experiments (5, 14), (5, 20), (2, 14) and, partially, (2, 20). These were obtained by Monte Carlo evaluation of the $E\,Q$ terms in (7.14). The sums (7.14) compare well with the actual biases $E\,\hat{\omega}^{(\mathrm{BOOT})} - \omega$, using 1,000 trials.

The first component $E\,\alpha_{X_1}(X_1)/n^2$ is negative, as we expect it to be since according to (7.4) it equals $-\omega/n$. (The modified estimator $n/(n-1)\,\hat{\omega}^{(\mathrm{BOOT})}$ has bias expression (7.13) except with the first component missing.) The third component $E\,\beta_{X_0}(X_1, X_1)/n^2$ is positive but small. This is also expected, from either (7.7) or the first line of (7.14).

The second component, $-E\,\beta_{X_1}(X_1, X_1)/n^2$, is negative and large. Its negativity amounts to a convexity relationship in the second line of (7.14): $E\,Q(X', \mathbf{X})$ is a convex decreasing function of the number of times, zero, once, or twice, that X' appears in the training set $\mathbf{X}$. We might say that Q is "deletion sensitive" in this case. All of our experiments were deletion sensitive, but artificial examples can be constructed going the other way. It is *not* a theorem that $E\,\hat{\omega}^{(\mathrm{BOOT})} - \omega < 0$, though that seems to be the usual case. In highly overfitted situations, where X' being in the training set even once makes $E\,Q(X', \mathbf{X})$ nearly zero, we expect $-E\,\beta_{X_1}(X_1, X_1)/n^2$ to be strongly negative because of (7.14). Experiment (5, 14) is a good example of this effect.

The fourth component, $-E\,\gamma_{X_0}(X_0, X_1, X_1)/2n^2$, is positive and large, though not as large as the second. The last line of (7.14) suggests that this component will always be positive in highly overfitted situations.

8. REPETITION ERROR RATES

The .632 estimator of Section 6 involves $\hat{\epsilon}^{(0)}$, the bootstrap error rate for cases having bootstrap weight zero. This section concerns the bootstrap error rates $\hat{\epsilon}^{(h)}$ for bootstrap weights $P_i^* = h/n$, $h = 0, 1, 2, \ldots$. The main result is a theorem relating $\hat{\epsilon}^{(h)}$ to the ANOVA expansion of Section 7. Among other things, this gives further information on how cross-validation relates to the bootstrap, and an improved Monte Carlo method for calculating the bootstrap. All proofs are deferred until the Appendix.

For a given training set $\mathbf{x}$, the hth repetition error rate $\hat{\epsilon}^{(h)}$ is defined to be the bootstrap error rate for values of

Table 3. Components of Bias for $\hat{\omega}^{(\mathrm{BOOT})}$, (7.13) (figures in parentheses are standard errors)

Exper.	$\frac{E\alpha_{X_1}(X_1)}{n^2}$	$\frac{-E\beta_{X_1}(X_1,X_1)}{n^2}$	$\frac{E\beta_{X_0}(X_1,X_1)}{n^2}$	$\frac{-E\gamma_{X_0}(X_0,X_1,X_1)}{2n^2}$	Sum (7.13)	$E\hat{\omega}^{(\mathrm{BOOT})} - \omega$ (1000 trials)
(5,14)	−.0127	−.1255	+.0083	+.0585	−.071	−.072
	(.0002)	(.0006)	(.0005)	(.0052)	(.006)	(.003)
(5,20)	−.0060	−.0639	+.0046	+.0355	−.030	−.033
	(.0001)	(.0009)	(.0002)	(.0092)	(.006)	(.003)
(2,14)	−.0066	−.0231	+.0047	+.0174	−.008	−.013
	(.0003)	(.0005)	(.0004)	(.0019)	(.002)	(.004)
(2,20)	−.0030	−.0106	+.0037	NA	NA	.000
	(.0002)	(.0009)	(.0004)			(.004)

the predicted point X_0^* equaling an x_i with bootstrap weight h/n,

$$\hat{\epsilon}^{(h)} = E_*\{Q(X_0^*, \mathbf{X}^*) \mid X_0^* = x_i \quad \text{with} \quad P_i^* = h/n\}. \tag{8.1}$$

Usually $\hat{\epsilon}^{(h)}$ must be calculated by Monte Carlo as described following (6.10),

$$\hat{\epsilon}^{(h)} = \frac{\sum_{(b,i):P_i^{*b}=h/n} Q(x_i, \mathbf{x}^{*b})}{\#\{(b,i):P_i^{*b} = h/n\}}. \tag{8.2}$$

Figure 3 shows the average values of $\hat{\epsilon}^{(0)}, \hat{\epsilon}^{(1)}, \hat{\epsilon}^{(2)}$ in our five experiments. The decreasing convex nature of these plots demonstrates the deletion sensitivity mentioned in Section 7.

In the bootstrap ANOVA expression (7.8) let

$$\hat{\mu}_i = \hat{\mu}_{X^*_0 = x_i} \quad \text{and} \quad \hat{\mu} = \frac{1}{n}\sum_{i=1}^{n} \hat{\mu}_i. \tag{8.3}$$

Also define

$$\hat{A} = \frac{\sum_{i=1}^{n} \hat{\alpha}_i(i)}{n}\left[-\frac{1}{n-1}\right],$$

$$\hat{B} = \frac{\sum_{i=1}^{n} \hat{\beta}_i(i,i)}{n}\frac{\binom{n}{2}}{n^2}\left[-\frac{1}{n-1}\right],$$

$$\hat{C} = \frac{\sum_{i=1}^{n} \hat{\gamma}_i(i,i,i)}{n}\frac{\binom{n}{3}}{n^3}\left[-\frac{1}{n-1}\right]^3, \tag{8.4}$$

and so on.

Theorem. The repetition error rates $\hat{\epsilon}^{(h)}$ are linear combinations of $\hat{\mu}, \hat{A}, \hat{B}, \hat{C}, \ldots$,

$$\hat{\epsilon}^{(h)} = \hat{\mu} + \lambda_1^{(h)}\hat{A} + \lambda_2^{(h)}\hat{B} + \lambda_3^{(h)}\hat{C} + \cdots, \tag{8.5}$$

where the constants $\lambda_j^{(h)}$ are given by

$$\lambda_j^{(h)} = E[-(n-1)]^{H_j}, \tag{8.6}$$

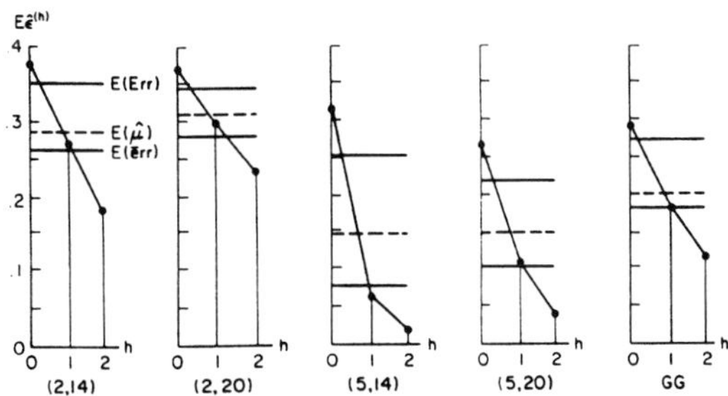

Figure 3. Repetition error rates for the five simulation experiments, h = 0, 1, 2, averaged over all trials in the experiments. Also shown are averages for Err, err, and $\hat{\mu}$.

Table 4. Some of the Coefficients $\lambda_j^{(h)}$ Appearing in (8.5)

		1 A	2 B	3 C	...	j
0	$\hat{\epsilon}^{(0)} - \hat{\mu}$	1	1	1	...	1
1	$\hat{\epsilon}^{(1)} - \hat{\mu}$	0	−1	−2	...	(−j − 1)
2	$\hat{\epsilon}^{(2)} - \hat{\mu}$	−1	$-\frac{n-3}{n-1}$	$\frac{n+5}{n-1}$		
⋮	⋮	⋮				
h	$\hat{\epsilon}^{(h)} - \hat{\mu}$	−(h − 1)				
	$\hat{\omega}^{(\mathrm{BOOT})}$	$\frac{n-1}{n}$	0	0	...	0

H_j being the number of red balls in j draws without replacement from a population of h red balls and $(n - h)$ black balls.

Some of the coefficients $\lambda_j^{(h)}$ are given in Table 4. The coefficients for $\hat{\omega}^{(\mathrm{BOOT})}$ are based on (7.12), which says that $\hat{\omega}^{(\mathrm{BOOT})} = ((n-1)/n)\hat{A}$. The terms $\hat{\mu}, \hat{A}, \hat{B}, \hat{C}, \ldots$, are in declining order of magnitude $O_p(1), O_p(1/n), O_p(1/n^2), O_p(1/n^3), \ldots$. The right side of (8.5) has $n + 1$ terms, and if all of these are included (8.5) is exactly true. Notice that the theorem applies to a single training set $\mathbf{x}$, and not just to expectations over random $\mathbf{X}$.

A wide class of interesting Err estimates can be obtained from the $\hat{\epsilon}^{(h)}$. As a first example consider the bootstrap estimate. Define

$$p_n^{(h)} \equiv \binom{n}{h}\frac{(n-1)^{n-h}}{n^n} = \mathrm{Prob}_*\left\{P_i^* = \frac{h}{n}\right\},$$

$h = 0, 1, 2, \ldots, n$, the last equality following from (2.14). It will turn out that

$$\hat{\omega}^{(\mathrm{BOOT})} = \sum_{h=0}^{n} p_n^{(h)}(1-h)\hat{\epsilon}^{(h)}. \tag{8.7}$$

If $\hat{\epsilon}^{(h)}$ is nearly linear in h, as in experiment (2, 20), say $\hat{\epsilon}^{(h)} \doteq \hat{c}_0 - h\hat{c}_1$, then (8.7) gives $\hat{\omega}^{(\mathrm{BOOT})} \doteq ((n-1)/n)\hat{c}_1$, so $\hat{\omega}^{(\mathrm{BOOT})}$ is the negative slope of the repetition plot, times $(n-1)/n$.

Formula (8.7) gives an improved way to calculate $\hat{\omega}^{(\mathrm{BOOT})}$. First calculate the $\hat{\epsilon}^{(h)}$ as in (8.2), and then combine them as in (8.7). If the number of bootstrap replications B is small this method can be quite a bit more efficient than the obvious Monte Carlo algorithm described in Section 2. The improvement arises from not having to estimate by Monte Carlo the theoretical constants $p_n^{(h)}$.

The quantity $\hat{\mu}$ will be shown to equal

$$\hat{\mu} = \sum_{h=0}^{n} p_n^{(h)}\hat{\epsilon}^{(h)}. \tag{8.8}$$

Then the estimator $\hat{\omega}^{(0)}$ of line 10, Table 2 can be written as

$$\hat{\omega}^{(0)} \equiv \hat{\epsilon}^{(0)} - \hat{\mu} = \sum_{h=0}^{n} [I_{h=0} - p_n^{(h)}]\hat{\epsilon}^{(h)}. \tag{8.9}$$

(Notice that $\hat{\omega}^{(0)}$ is different than the quantity $\hat{\epsilon}^{(0)} - \bar{err}$ appearing in the .632 estimator.) It performed reasonably well in Table 2, having about the same MSE as the bootstrap and about the same bias as cross-validation. Comparing (7.14) with Table 4 shows why $\hat{\omega}^{(0)}$ has smaller bias than does $\hat{\omega}^{(\text{BOOT})}$. The coefficient 1 rather than $(n - 1)/n$ on $\hat{A}$ removes the $E\,\alpha_{X_1}(X_1)/n^2$ term from (7.14). The coefficient 1 rather than 0 on $\hat{B}$ gives an added expectation of $E\,\beta_{X_1}(X_1, X_1)/2n^2 + O(1/n^3)$ thereby removing half the $-E\,\beta_{X_1}(X_1, X_1)/n^2$ term in (7.14). We could remove all of this term, for example with the estimate $(\hat{\epsilon}^{(0)} - \hat{\mu}) - (\hat{\epsilon}^{(1)} - \hat{\mu})$, but then the $E\,\gamma_{X_0}(X_0, X_1, X_1)$ term in (7.14) results in substantial upward biases.

There is an interesting connection between $\hat{\epsilon}^{(0)}$ and cross-validation. For n even we can define the *half sample cross-validation estimate*

$$\hat{\text{Err}}^{(\text{HCV})} = \frac{\sum_i \sum_S Q(x_i, \mathbf{X}^{*S})}{n\binom{n-1}{n/2}} \tag{8.10}$$

the second sum being taken over all subsamples $\mathbf{X}^{*S}$ of $\{x_1, x_2, \ldots, x_n\}$ having $n/2$ elements and not containing the predicted case x_i. We will show that $\hat{\text{Err}}^{(\text{HCV})}$ has the expansion formula

$$\hat{\text{Err}}^{(\text{HCV})} = \hat{\mu} + \hat{A} + \hat{B} + \hat{C} - \frac{n-1}{6n^2(n-3)} \frac{\sum\sum_{i\neq j} \hat{\gamma}_i(j, j, j)}{n(n-1)} - \frac{1}{2n^2(n-3)} \frac{\sum\sum_{i\neq j} \hat{\gamma}_i(i, j, j)}{n(n-1)} + O_p\left(\frac{1}{n^4}\right). \tag{8.11}$$

Compared with Table 4, (8.11) shows that

$$(\hat{\text{Err}}^{(\text{HCV})} - \hat{\mu}) = (\hat{\epsilon}^{(0)} - \hat{\mu}) + c_n + O_p(1/n^{5/2}), \tag{8.12}$$

$c_n = -1/6n^2\,E\,\gamma_{X_0}(X_1, X_1, X_1)$ a constant of order $O(1/n^2)$. This suggests a high correlation between $\hat{\text{Err}}^{(\text{HCV})} - \hat{\mu}$ and $\hat{\epsilon}^{(0)} - \hat{\mu}$, and in fact the observed correlations were .86 experiment (2, 14), .98 experiment (2, 20), .94 experiment (5, 14), and .95 experiment (5, 20).

Expression (7.8), which gives (8.11), applies only when $\mathbf{x}^*$ has n component cases. A half-sample $\mathbf{X}^{*S} = \{x_{i_1}, x_{i_2}, \ldots, x_{i_{n/2}}\}$ can be regarded as a sample of size n by doubling each case, $\{x_{i_1}, x_{i_1}, x_{i_2}, x_{i_2}, \ldots, x_{i_{n/2}}, x_{i_{n/2}}\}$. With this understanding the quantity $Q(x_i, \mathbf{X}^{*S}) = Q[y_i, \eta(t_i, \mathbf{X}^{*S})]$ is well defined and can be evaluated by (7.8).

The cross-validation estimate $\hat{\omega}^{(\text{CV}+)}$ introduced at (7.6), which deletes and adds single cases at a time, turns out to have expansion

$$\hat{\omega}^{(\text{CV}+)} = \hat{A} + 2\hat{B} + 3\frac{n-1}{n-2}\hat{C} + \frac{1}{2n^2} \frac{\sum\sum_{i\neq j} \hat{\gamma}_i(i, j, j)}{n(n-1)} + O_p\left(\frac{1}{n^3}\right). \tag{8.13}$$

As shown at (7.8) this has little bias (cf. the remarks following (8.9)) but high variability, the same as $\hat{\omega}^{(\text{CV})}$.

It seems reasonable for an estimator $\hat{\omega}$ of ω to begin $\hat{A} + O_p(1/n^2)$, as do $\hat{\omega}^{(\text{BOOT})}$, $\hat{\omega}^{(0)}$, $err^{(\text{HCV})} - \hat{\mu}$, and $\hat{\omega}^{(\text{CV}+)}$. This makes $\hat{\omega}$ unbiased for ω order $O(1/n)$, the bias being $O(1/n^2)$. At a more primitive level, $\hat{A} = (-1/(n-1))(\sum \hat{\alpha}_i(i)/n)$ looks like $\omega = (-1/n)(E\,\alpha_{X_1}(X_1))$. One estimator that does not begin this way is connected with the Err estimate E_* (Err*) mentioned in the first paragraph of Section 5. Its use amounts to estimating ω by $\hat{\omega} = E_*(\text{Err}^*) - \bar{err}$. Notice that $E_*\,\text{Err}^* = \hat{\mu}$, by the first line of (7.9). The expansion of $\hat{\omega} = \hat{\mu} - \bar{err}$ turns out to be

$$\hat{\mu} - \bar{err} = \frac{\sum\sum_{i\neq j} \hat{\beta}_i(j, j)}{2n^3} + \left\{ \frac{\sum_i \hat{\beta}_i(i, i)}{2n^3} - \frac{\sum_i \sum_j \hat{\gamma}_i(j, j, j)}{3n^4} - \frac{\sum_i \sum_j \sum_{j'} \hat{\delta}_i(j, j, j', j')}{8n^5} \right\} + O_p\left(\frac{1}{n^3}\right). \tag{8.14}$$

The $O_p(1/n)$ term in (8.14) is $\sum \sum_{i\neq j} \hat{\beta}_i(j, j)/2n^3$, not $\hat{A}$, with expectation $E\,\beta_{X_0}(X_1, X_1)/2n + O(1/n^2)$. There is no theoretical reason for believing that this will be near ω, and the numerical results were terrible, for example $E\,\hat{\omega} = .024$ compared with $\omega = .093$ for experiment (2, 14).

The .632 estimator of Section 6 also has the "wrong" $O_p(1/n)$ term. Table 4 and (8.14) give

$$E\,\hat{\omega}^{(.632)} = \omega\left\{ .632 + \frac{.632}{2} \frac{E\,\beta_{X_0}(X_1, X_1)}{-E\,\alpha_{X_1}(X_1)} \right\} + O\left(\frac{1}{n^2}\right). \tag{8.15}$$

Table 4 gives these values for the bracketed factor,

Experiment:	(5, 14)	(5, 20)	(2, 14)	(2, 20)
{Factor}:	.84	.87	.86	1.02

(8.16)

The rationale for $\hat{\omega}^{(.632)}$ makes it unsurprising that these numbers are near one. On the other hand there is no guarantee that this will always happen, and arbitrarily bad counterexamples can be constructed.

9. REMARKS

Remark A. The sample sizes in our experiments, $n = 14$ or 20, are small. In practice small sample sizes can arise even when n is large, if we are interested in estimating the error rate for only a portion of the population. For example Efron and Gong (1983) consider a medical example with $n = 155$. Of particular interest are the 33 patients who died. Cross-validation, the bootstrap, and so on, can easily be modified to give the prediction rule's estimated error rate for the population of those who die, but the effective sample size is then 33, not 155.

Remark B. The methods of this article can be applied to other problems besides estimating prediction error. Suppose we need to estimate a density function $f(x_0)$ on the basis of a random sample $\mathbf{x} = (x_1, x_2, \ldots, x_n)$ from f, and wish to select among a family of possible density estimators $f_\lambda(x_0, \mathbf{x})$. Here λ might be the window width of a kernel estimator. Let $Q_\lambda(x_0, \mathbf{x}) = -\log f_\lambda(x_0; \mathbf{x})$ and $err_\lambda(\mathbf{x}, f) = E\,Q_\lambda(X_0, \mathbf{x}) = -\int[\log f_\lambda(x_0; \mathbf{x})]f(x_0)dx_0$. In an insightful paper Wong (1983) suggests selecting λ to minimize $E_* \, err_\lambda{}^*$. Using arguments much like those in Section 2, he shows that this is asymptotically equivalent to the older method of "modified likelihood." In light of the remarks in Section 8, we might prefer to define $\bar{err}_\lambda(\mathbf{x}) = -1/n \sum_{i=1}^n \log f_\lambda(x_i, \mathbf{x})$, and select λ minimizing $\bar{err}_\lambda + E_*(err_\lambda - \bar{err}_\lambda)^*$. Wong (1982) has shown that this last approach does in fact lead to a superior asymptotic theory.

Remark C. The statistician may want more than just an estimate of Err. Bootstrap methods are helpful in understanding the variability of all aspects of the prediction problem. As an example, in simulation 1 of experiment GG the forward stepwise logistic regressions selected variables 1 and 2 for inclusion in the fitted prediction rule $\eta(\cdot, \mathbf{x})$. In $B = 100$ bootstrap replications of simulation 1 the following sets of variables were selected:

set selected:	{12}	{123}	{13}	{3}	{2}	all others
# time selected:	53	15	11	8	5	8

(9.1)

Without attempting a quantitative assessment, we see that the "standard error" of the set of variables selected, about the central value {12}, is reasonably small in this case.

Remark D. Stone (1974) and Geisser (1975) emphasize the use of cross-validation to select among competing prediction rules. As a simple example suppose we observe a random sample $x_1, x_2, \ldots, x_n$ from a distribution F on the real line, and wish to choose between two estimators $\hat{\eta}_l = \eta_l(\mathbf{x})$, $l = 1, 2$, perhaps the sample median and the 10 percent trimmed mean. The goal is to minimize the expected squared error of prediction $E[X_0 - \eta(\mathbf{x})]^2$ for a future observation X_0 for F.

We wish to choose $l = 1$ or 2 minimizing $\mathrm{Err}_l(\mathbf{x}, F) = E_F[X_0 - \eta_l(\mathbf{x})]^2$. If F is actually normal, the difference $\mathrm{Err}_2 - \mathrm{Err}_1$ estimated by the bootstrap turns out to be

$$\hat{\mathrm{Err}}_2{}^{(\mathrm{BOOT})} - \hat{\mathrm{Err}}_1{}^{(\mathrm{BOOT})}$$

$$= (\hat{\eta}_2 - \bar{x})^2 - (\hat{\eta}_1 - \bar{x})^2 + O_p\left(\frac{1}{n^{3/2}}\right), \quad (9.2)$$

so asymptotically the bootstrap selects the estimate nearest the sample mean $\bar{x}$. The same can be shown to hold for the cross-validation estimate $\hat{\mathrm{Err}}_2{}^{(\mathrm{CV})} - \hat{\mathrm{Err}}_1{}^{(\mathrm{CV})}$ *if the jackknife estimates of* $\mathrm{var}(\hat{\eta}_l)$ *and* $\mathrm{cov}(\hat{\eta}_l, \bar{x})$ *converge to their correct values*. If not, cross-validation can give strange answers. Stone (1977) shows that for $F \sim N(0, 1)$ the cross-validation method will select $\hat{\eta}_2$ the median as better than $\hat{\eta}_1$ the mean with asymptotic probability 0.5008.

Remark E. Cross-validation is often carried out removing large blocks of observations at a time. If $n = GH$ and $\mathbf{x}_{(g)} = (x_1, x_2, \ldots, x_{(g-1)H}, x_{gH+1}, \ldots, x_n)$, then $\hat{\mathrm{Err}}^{(\mathrm{CVG})} = \sum_{g=1}^G \sum_{h=1}^H Q(x_{(g-1)H+h}, \mathbf{x}_{(g)})/n$ requires only G recomputations of η. There are also theoretical reasons for preferring $\hat{\mathrm{Err}}^{(\mathrm{CVG})}$ to $\hat{\mathrm{Err}}^{(\mathrm{CV})}$. As explained in Section 6.2 of Efron (1982), quadratic approximation formulas like (2.16) tend to be more trustworthy for H large. In other words $\hat{\mathrm{Err}}^{(\mathrm{CVG})}$ is likely to be closer in value to $\hat{\mathrm{Err}}^{(\mathrm{BOOT})}$ than is $\hat{\mathrm{Err}}^{(\mathrm{CV})}$. As an example, grouped cross-validation like the bootstrap, selects the mean in preference to the median with asymptotic probability one in Stone's problem, Remark D, if H is suitably large.

If H is large than $\hat{\mathrm{Err}}^{(\mathrm{CVG})} - \bar{err}$ will have substantial upward bias as an estimate of ω. For example $\hat{\mathrm{Err}}^{(\mathrm{HCV})} - \bar{err}$, (8.10), corresponding to $G = 2$, $H = n/2$, is upwardly biased $O(1/n)$. This will be true for any choice $H = cn$, c fixed as $n \to \infty$. The bias can be removed by estimating Err with $\bar{err} + (\hat{\mathrm{Err}}^{(\mathrm{CVG})} - \hat{\mu})$ instead of $\hat{\mathrm{Err}}^{(\mathrm{CVG})}$, but that involves calculating the bootstrap quantity $\hat{\mu}$. At this point it becomes simpler to estimate ω by $\hat{\omega}^{(0)} = \hat{\epsilon}^{(0)} - \hat{\mu} \doteq \hat{\mathrm{Err}}^{(\mathrm{HCV})} - \hat{\mu}$.

Remark F. Cross-validation behaves more like the bootstrap if $Q[y, \eta]$ is a smooth function, like $(y - \eta)^2$ rather than (1.1), and if $\eta(\cdot, \mathbf{x})$ is also a moderately smooth function of $\mathbf{x}$. Then (2.16) gives more accurate approximations. In this case it is reasonable to estimate Err by $\hat{\mathrm{Err}}^{(\mathrm{CV})}$, though bootstrap calculations may still be helpful for other purposes, as in Remark C.

Remark G. Consider the ordinary least squares (OLS) situation $y_i = t_i\beta + e_i$, $e_i \overset{\mathrm{iid}}{\sim} N(0, \sigma^2)$, $i = 1, \ldots, n$, and $\eta(t_0, \mathbf{x}) = t_0(\mathbf{t}'\mathbf{t})^{-1}\mathbf{t}'\mathbf{y}$. If the predictors t_i are p dimensional then $\omega = (2p/n)\,\sigma^2$. The UMVU estimate of ω is $\hat{\omega} = (2p/n)\,\hat{\sigma}^2$, $\hat{\sigma}^2$ the usual unbiased estimate of σ^2, and it is easily shown that $\mathrm{corr}(\mathrm{op}, \hat{\omega}) = -\sqrt{1 - p/n}$. For $p = 2$, $n = 14$, the correlation is $-.93$.

The .632 estimator had corr(op, $\hat{\omega}$) nearly zero, and this largely accounted for its good performance in the sampling experiments. The OLS example suggests that we cannot always get $\mathrm{corr}(\mathrm{op}, \hat{\omega}) \doteq 0$ for a good estimator of ω.

We can change OLS to be more like the dichotomous models by assuming σ^2 a known function of β, say $\sigma_\beta{}^2 = a_0 + b_0(\beta - \beta_0)$ for β near some fixed vector β_0, a_0 and b_0 given. Then if β is estimated by least squares, the obvious parametric estimate $\hat{\omega} = (2p/n)\,[a_0 + b_0(\hat{\beta} - \beta_0)]$ has $\mathrm{corr}(\mathrm{op}, \hat{\omega}) = 0$, a similar result holding if β is estimated by maximum likelihood. In this case, as in the sampling experiments, we can expect good nonparametric estimators to have corr(op, $\hat{\omega}$) nearly zero.

10. SUMMARY

1. There are a variety of nonparametric methods available for estimating err in the dichotomous prediction problem (1.1), (1.2), all of which are closely related to nonparametric maximum likelihood estimation, that is, to the bootstrap.

2. In practical situations the different methods can give considerably different answers.

3. Cross-validation (1.4) gives a nearly unbiased estimate of err, but often with unacceptably high variability, particularly if n is small.

4. The ordinary bootstrap (2.10) gives an estimate of err with low variability, but with a possibly large downward bias, particularly in highly overfitted situations.

5. The double bootstrap of Section 5 and the $\hat{\omega}^{(0)}$ estimator (8.9) automatically correct the bias of the ordinary bootstrap without increasing the MSE of estimation.

6. The randomized bootstrap, Section 4, requires a modest amount of additional input from the statistician, but results in substantially lower MSE. Overall it performed second best in the sampling experiments.

7. The .632 estimator of Section 6 performed best in the sampling experiments, but has the weakest theoretical justification. It is recommended with caution, pending further numerical and theoretical study.

APPENDIX

Derivation of (5.4), (5.7)

The second-level bootstrap vector $\mathbf{P}^{**}$ has, given $\mathbf{P}^*$, a conditional multinomial distribution, divided by n,

$$\mathbf{P}^{**} \mid \mathbf{P}^* \sim \text{Mult}_n(n, \mathbf{P}^*)/n. \tag{A.1}$$

For instance if $P_i^* = h/n$, then $P_i^{**}|\mathbf{P}^* \sim bi(n, h/n)/n$, the proportion of heads observed in n flips of a coin having probability of heads h/n.

Denote by $E_{**}^{(*)}$ the expectation with respect to probability mechanism (A.1), with $\mathbf{P}^*$ held fixed. Also, let E_{**} indicate expectation with respect to the marginal distribution of $\mathbf{P}^{**}$, obtained from (A.1) and the distribution of $\mathbf{P}^*$,

$$\mathbf{P}^* \sim \text{Mult}_n(n, \mathbf{P}^\circ)/n \quad (\mathbf{P}^\circ = (1, 1, \ldots, 1)/n, \tag{A.2}$$

which agrees with its use in (5.4). Finally, let $E_*^{(**)}$ indicate expectation with respect to the conditional distribution of $\mathbf{P}^*$ given $\mathbf{P}^{**}$. In all these expectations the data $\mathbf{x}$ are fixed.

We need to evaluate $E_*(\hat{\omega}^{(\text{BOOT})})^*$, where $\hat{\omega}^{(\text{BOOT})} = E_* \,\text{op}(\mathbf{X}^*, \hat{F})$. Suppose that $r(\mathbf{x}) \equiv E_* R(\mathbf{X}^*, \hat{F})$ is the bootstrap expectation of a random variable $R(\mathbf{X}, F)$, which is invariant under all permutations of the coordinates of $\mathbf{X}$ (as is op($\mathbf{X}$, F)). A bootstrap replication of r is of the form

$$r(\mathbf{X}^*) = E_{**}^{(*)} R(\mathbf{X}^{**}, \hat{F}^*) = E_{**}^{(*)} R(\mathbf{P}^{**}, \mathbf{P}^*). \tag{A.3}$$

The last expression makes sense because, with data $\mathbf{x}$ fixed $\mathbf{P}^*$ determines F^*, and $\mathbf{P}^{**}$ determines $\mathbf{X}^{**}$ up to permutations of its components.

By carefully following through the various definitions we can apply (A.3) to $r(\mathbf{x}) = \hat{\omega}^{(\text{BOOT})} = E_* \,\text{op}(\mathbf{X}^*, \hat{F})$ and obtain

$$(\hat{\omega}^{(\text{BOOT})})^* = E_{**}^{(*)} \left\{ \sum_{i=1}^{n} (P^* - P^{**})\, Q[y_i, \eta(t_i, \mathbf{X}^{**})] \right\}. \tag{A.4}$$

Since $E_* E_{**}^{(*)} R(\mathbf{P}^{**}, \mathbf{P}^*) = E_{**} E_*^{(**)} R(\mathbf{P}^{**}, \mathbf{P}^*)$ for any function $R(\mathbf{P}^{**}, \mathbf{P}^*)$, (A.4) gives

$$E_*(\hat{\omega}^{(\text{BOOT})})^* = E_{**} \left\{ \sum_{i=1}^{n} e(P_i^{**})\, Q[y_i, \eta(t_i, \mathbf{X}^{**})] \right\}, \tag{A.5}$$

where

$$e(P_i^{**}) = E_*^{(**)} (P_i^* - P_i^{**}). \tag{A.6}$$

This shows that (5.4) holds with $e(P_i^{**})$ given by (A.6).

The conditional expectation $E_*^{(**)} (P_i^* - P_i^{**})$ is actually a function of the entire vector $\mathbf{P}^{**}$ and not just of the ith component P_i^{**}. However, the effect of the other components turns out to be quite small, and will be ignored in what follows. Let $nP_i^{**} \equiv N_i^{**}$ and $nP_i^* \equiv N_i^*$. Then a standard Bayesian calculation gives

$$e(P_i^{**}) = \frac{\displaystyle\sum_{n_i^*=0}^{n} \frac{(n_i^* - N_i^{**})}{n}\, bi\left(n, \frac{1}{n}; n_i^*\right) bi(n, n_i^*/n; N_i^{**})}{\displaystyle\sum_{n_i^*=0}^{n} bi\left(n, \frac{1}{n}; n_i^*\right) bi(n, n_i^*/n; N_i^{**})} \tag{A.7}$$

where $bi(n, p; h)$ is the binomial probability $\binom{n}{h} p^h(1 - p)^{n-h}$.

As $n \to \infty$ the distribution of $N_i^* \to Po(1)$, a Poisson with parameter one, and $N_i^{**} \mid N_i^* \to Po(N_i^{**})$. In this case (A.7) simplifies considerably, and can be rewritten as

$$e(P^{**}) = \frac{1}{n} \left\{ \frac{EZ^{N_i^{**}+1}}{EZ^{N_i^{**}}} - N_i^{**} \right\} \tag{A.8}$$

where $Z \sim Po(e^{-1})$, a Poisson with parameter $\lambda = e^{-1} = .3679$. Formula (A.8) was used to calculate (5.7). These values are within a few percent of (A.7), even for small n. For example with $n = 10$, $P_i^{**} = 0$, (A.7) gives $e(P_i^{**}) = .0359$ compared with $e(P_i^{**}) = .0368$ for (A.8).

Verification of (6.9)

Define the set $T(x_0, \Delta) = \{x : x_0 \in S(x, \Delta)\}$. Then

$$\begin{aligned}\text{Prob}\{\delta(X_0, \mathbf{X}) > \Delta\} &= \text{Prob}\left\{ X_0 \notin \bigcup_{i=1}^{n} S(X_i, \Delta) \right\} \\ &= \text{Prob}\{X_i \notin T(X_0, \Delta),\ i = 1, 2, \ldots, n\} \\ &= E[1 - \text{Prob}\{T(X_0, \Delta)\}]^n. \end{aligned} \tag{A.9}$$

Suppose that $\text{Prob}\{T(X_0, \Delta)\}$ approximately equals Δ. This will be true, at least for small values Δ, if the original neighborhoods $S(x, \Delta)$ are based on a distance function symmetric in x and x_0, as in (6.4). Then (A.9) gives the approximation

$$\text{Prob}\{\delta(X_0, \mathbf{X}) > \Delta\} \doteq (1 - \Delta)^n. \quad (A.10)$$

Let n^* be the number of cases x_i having $P_i^* > 0$ in a given bootstrap sample $\mathbf{X}^*$. The same reasoning as in (A.9) gives

$$\text{Prob}\{\delta(X_0^*, \mathbf{X}^*) > \Delta \mid \delta^* > 0\} = E[1 - \text{Prob}\{T(X_0^*, \Delta)\}]^{n^*}. \quad (A.11)$$

Both the probability and the expectation in (A.11) are marginal over $\mathbf{X}$ and $\mathbf{X}^*$, X_0^*. The approximation $\text{Prob}\{T(X_0^*, \Delta)\} \doteq \Delta$ gives

$$\text{Prob}\{\delta(X_0^*, \mathbf{X}^*) > \Delta \mid \delta^* > 0\} \doteq (1 - \Delta)^{.632n}, \quad (A.12)$$

where we have substituted $E\, n^* \doteq .632n$ for n^*.

If (A.11) and (A.12) can be trusted then $\text{Prob}\{\delta(X_0, \mathbf{X}) > \Delta\} \doteq e^{-n\Delta}$ and $\text{Prob}\{\delta(X_0^*, \mathbf{X}^*) > \Delta/.632\} \doteq e^{-n\Delta}$, verifying (6.9). The insert in Figure 2 compares $\text{Prob}\{\delta(X_0, \mathbf{X}) < \Delta\}$ with

$$\text{Prob}\left\{\delta(X_0^*, \mathbf{X}^*) < \frac{\Delta}{.632} \;\middle|\; \delta^* > 0\right\},$$

showing excellent agreement.

Derivation of (7.13)

There are simple algebraic relationships between the terms in (7.1) and those in (7.8),

$$\hat{\mu}_j = \mu_{x_j} + \alpha_{x_j}(\cdot) + \frac{\binom{n}{2}}{n^2}\beta_{x_j}(\cdot,\cdot) + \frac{\binom{n}{3}}{n^3}\gamma_{x_j}(\cdot,\cdot,\cdot) + \dots,$$

$$\hat{\alpha}_j(j_1) = [\alpha_{x_j}(x_{j_1}) - \alpha_{x_j}(\cdot)] + \frac{\binom{n-1}{1}}{n}[\beta_{x_j}(x_{j_1},\cdot) - \beta_{x_j}(\cdot,\cdot)] + \frac{\binom{n-1}{2}}{n^2}[\gamma_{x_j}(x_{j_1},\cdot,\cdot) - \gamma_{x_j}(\cdot,\cdot,\cdot)] + \dots,$$

$$\hat{\beta}_j(j_1, j_2) = [\beta_{x_j}(x_{j_1}, x_{j_2}) - \beta_{x_j}(x_{j_1},\cdot) - \beta_{x_j}(x_{j_2},\cdot) + \beta_{x_j}(\cdot,\cdot)] + \frac{\binom{n-2}{1}}{n}[\gamma_{x_j}(x_{j_1}, x_{j_2},\cdot) - \gamma_{x_j}(x_{j_1},\cdot,\cdot) + \gamma_{x_j}(x_{j_2},\cdot,\cdot) + \gamma_{x_j}(\cdot,\cdot,\cdot)] + \dots, \quad (A.13)$$

the dot notation indicating averages, $\alpha_{x_j}(\cdot) = \sum_{j_1=1}^n \alpha_{x_j}(x_{j_1})/n$, $\beta_{x_j}(x_{j_1},\cdot) = \sum_{j_2=1}^n \beta_{x_j}(x_{j_1}, x_{j_2})/n$, and so on. (This last average involves terms like $\beta_{x_j}(x_2, x_1)$, which do not seem to exist in (7.1). However, the corresponding terms there are really random variables $\beta_{x_j}(X_i, X_{i'})$, which have well-defined values for $X_i = x_2$, $X_{i'} = x_1$.) Relationships (A.13) are familiar in the comparison of random effects with fixed effects models in ANOVA.

The second line of (A.13) shows that $E\,\hat{\alpha}_j(j) = E\,\alpha_{X_j}(X_j) + O(1/n)$. The $O(1/n)$ term can be computed explicitly giving, from (7.4), (7.12), formula (7.13).

Proofs for Results of Section 8

For any bootstrap random variable $R^* = R(\mathbf{X}^*, \hat{F})$, and for any choice of $i = 1, 2, \dots, n$,

$$E_* R^* = \sum_{h=0}^{n} p_n^{(h)} E_*\{R^* \mid P_i^* = h/n\} \quad (A.14)$$

since $p_n^{(h)} = \text{Prob}_*\{P_i^* = h/n\}$. Applying (A.14) to $R^* = Q(x_i, \mathbf{X}^*)$ gives $\hat{\mu}_i = E_*\, Q(x_i, \mathbf{X}^*) = \sum_{h=0}^n p_n^{(h)} \hat{\epsilon}_i^{(h)}$, where

$$\hat{\epsilon}_i^{(h)} = E_*\{Q(x_i, \mathbf{X}^*) \mid P_i^* = h/n\}. \quad (A.15)$$

It is easy to see that $\sum_{i=1}^n \hat{\epsilon}_i^{(h)}/n = \hat{\epsilon}^{(h)}$, (8.1), so $\hat{\mu} = \sum_{i=1}^n \hat{\mu}_i/n = \sum_{h=0}^n p_n^{(h)} \hat{\epsilon}^{(h)}$, as claimed at (8.8). Likewise letting R^* in (A.14) equal $R_i^* \equiv (1/n - P_i^*)\, Q(x_i, \mathbf{X}^*)$ gives $E_*\, R_i^* = \sum_{h=0}^n p_n^{(h)}((1-h)/n)\,\hat{\epsilon}_i^{(h)}$, and then $\hat{\omega}^{(\text{BOOT})} = E_* \sum_{i=1}^n R_i^* = \sum_{h=0}^n p_n^{(h)}(1-h)\hat{\epsilon}^{(h)}$, verifying (8.7).

Rather than prove the theorem we will prove the stronger result

$$\hat{\epsilon}_i^{(h)} = \hat{\mu}_i + \lambda_1^{(h)} \frac{\hat{\alpha}_i(i)}{[-(n-1)]} + \lambda_2^{(h)} \frac{\hat{\beta}_i(i, i)}{[-(n-1)]^2} \frac{\binom{n}{2}}{n^2} + \dots. \quad (A.16)$$

Averaged over $n = 1, 2, \dots, n$, (A.16) gives (8.5). Because we have assumed that $Q(x_i, \mathbf{X}^*) \equiv Q[y_i, \eta(t_i, \mathbf{X}^*)]$ is unchanged under permutations of $\mathbf{X}^*$'s components, (A.15) can be written as

$$\hat{\epsilon}_i^{(h)} = E_*\{Q(x_i, \mathbf{X}^*) \mid X_1^*, \dots, X_h^* = x_i \text{ and } X_{h+1}^*, \dots, X_n^* \neq x_i\}. \quad (A.17)$$

Now we use (7.8) and (7.11) to evaluate this conditional expectation. For example,

$$E_*\{\hat{\alpha}_i(X_{j_1}^*) \mid X_{j_1}^* \neq x_i\} = \frac{1}{n-1}\sum_{i_1 \neq i} \hat{\alpha}_i(i_1) = \frac{-\hat{\alpha}_i(i)}{n-1} \quad (A.18)$$

and so

$$E_*\left\{\frac{1}{n}\sum_{j=1}^{n}\hat{\alpha}_i(X_j^*) \mid X_1^*,\ \ldots,\ X_h^* = x_i \ \text{ and } \ X_{h-1}^*,\ \ldots,\ X_n^* \neq x_i\right\}$$

$$= \left[\frac{h}{n} + \frac{n-h}{n}\left(-\frac{1}{n-1}\right)\right]\hat{\alpha}_i(i). \qquad \text{(A.19)}$$

Similar calculations for the higher-order terms in (7.8) evaluate (A.17) as

$$\begin{aligned}
\hat{\epsilon}_i^{(h)} = \hat{\mu}_i &+ \frac{1}{n}\left[\binom{h}{1} + \binom{n-h}{1}\left(-\frac{1}{n-1}\right)\right]\hat{\alpha}_i(i) \\
&+ \frac{1}{n^2}\left[\binom{h}{2} + \binom{h}{1}\binom{n-h}{1}\left(-\frac{1}{n-1}\right)\right. \\
&\left. + \binom{n-h}{2}\left(-\frac{1}{n-1}\right)^2\right]\hat{\beta}_i(i,i) \\
&+ \frac{1}{n^3}\left[\binom{h}{3} + \binom{h}{2}\binom{n-h}{1}\left(-\frac{1}{n-1}\right)\right. \\
&+ \binom{h}{1}\binom{n-h}{2}\left(-\frac{1}{n-1}\right)^2 \\
&\left. + \binom{n-h}{3}\left(-\frac{1}{n-1}\right)^3\right]\hat{\gamma}_i(i,i,i) + \ldots .
\end{aligned} \qquad \text{(A.18)}$$

The bracketed terms are obviously related to hypergeometric expectations of the form (8.6), and results (A.16), (8.5) follow easily.

The proof of (8.11) also relies on (7.8) and (7.11). In the notation of (8.10),

$$\begin{aligned}
Q(x_i, \mathbf{X}^{*S}) = \hat{\mu}_i &+ \frac{2}{n}\sum_{j\in S}\hat{\alpha}_i(j) + \frac{4}{n^2}\sum_{j<j'\in S}\sum\hat{\beta}_i(j,j') \\
&+ \frac{1}{n^2}\sum_{j\in S}\hat{\beta}_i(j,j) + \frac{8}{n^3}\sum_{j<j'<j''\in S}\sum\sum\hat{\gamma}_i(j,j',j'') \\
&+ \frac{2}{n^3}\sum_{j<j'\in S}\sum[\hat{\gamma}_i(j,j',j') + \hat{\gamma}_i(j,j,j')] + \ldots ,
\end{aligned} \qquad \text{(A.19)}$$

all sums being only over cases x_j in $\mathbf{X}^{*S}$. Then

$$\begin{aligned}
\hat{\text{Err}}_i^{(\text{HCV})} \equiv \frac{\sum_S Q(x_i, \mathbf{X}^{*S})}{\binom{n-1}{n/2}} &= \hat{\mu}_i + \frac{\sum_{j\neq i}\hat{\alpha}_i(j)}{n-1} \\
&+ \frac{\sum_{j<j'\neq i}\sum\hat{\beta}_i(j,j')}{n(n-1)} + \frac{\sum_{j\neq i}\hat{\beta}_i(j,j)}{2n(n-1)} + \ldots ,
\end{aligned} \qquad \text{(A.20)}$$

or

$$\begin{aligned}
\hat{\text{Err}}_i^{(\text{HCV})} &= \hat{\mu}_i + \hat{\alpha}_i(i\left[-\frac{1}{n-1}\right] + \hat{\beta}_i(i,i)\frac{\binom{n}{2}}{n^2}\left[-\frac{1}{n-1}\right]^2 \\
&+ \hat{\gamma}_i(i,i,i)\frac{\binom{n}{3}}{n^3}\left[-\frac{1}{n-1}\right]^3 \\
&- \frac{\sum_{j\neq 1}\hat{\gamma}_i(j,j,j)}{n-1}\,\frac{n-1}{6n^2(n-3)} \\
&- \frac{1}{2n^2(n-3)}\,\frac{\sum_{j\neq 1}\hat{\gamma}_i(i,j,j)}{n-1} + O_p\left(\frac{1}{n^4}\right),
\end{aligned} \qquad \text{(A.21)}$$

(A.21) following from (A.20) by (7.11). But $\hat{\text{Err}}^{(\text{HCV})} = \sum_{i=1}^{n}\hat{\text{Err}}_i^{(\text{HCV})}/n$ so (A.20) is an improved version of (8.11). Formulas (8.13) and (8.4) follow from similar algebraic manipulations of (7.8) and (7.11).

The first line in (A.21) is the beginning of expansion (A.16) for $\hat{\epsilon}^{(0)}$. The quantity $\sum_{j\neq i}\hat{\gamma}_i(j,j,j)/(n-1)$ appearing in the next term of (A.21) can be written as

$$\begin{aligned}
\frac{\sum_{j\neq i}\hat{\gamma}_i(j,j,j)}{n-1} &= \sum_j\sum_{j'}\sum_{j''} c_i(j,j',j'')\,\gamma_{x_i}(x_j, x_{j'}, x_{j''}) \\
&+ \sum_j\sum_{j'}\sum_{j''}\sum_{j'''} d_i(j,j',j'',j''') \\
&\times \delta_{x_i}(x_j, x_{j'}, x_{j''}, x_{j'''}) + \ldots .
\end{aligned} \qquad \text{(A.22)}$$

The constants $c_i(j,j',j'')$, $d_i(j,j',j'',j''')$, . . . , are calculated from (A.13). For example, $c_i(j,j,j) = 1/(n-1) - 3/n(n-1) + 3/n^2(n-1) - 1/n^3$ if $i \neq j$. The variance of $\sum_j\sum_{j'}\sum_{j''}c_i(j,j',j'')\gamma_{X_i}(X_j, X_{j'}, X_{j''})$ can be calculated exactly and has the limiting form, as $n \to \infty$, $[\text{var}\ \gamma_{X_0}(X_1, X_1, X_1)]/n$. Likewise its expectation approaches $E\ \gamma_{X_0}(X_1, X_1, X_1)$. Ignoring further terms in (A.22) this gives $\sum_{j\neq i}\hat{\gamma}_i(j,j,j)/(n-1) = E\ \gamma_{X_0}(X_1, X_1, X_1) + O_p(1/n^{1/2})$. Returning to (A.21)

$$(\hat{\text{Err}}_i^{(\text{HCV})} - \hat{\mu}_i) = (\hat{\epsilon}_i^{(0)} - \hat{\mu}_i) + \frac{E\ \gamma_{X_0}(X_i, X_1, X_1)}{6n^2} + O_p(1/n^{5/2}), \qquad \text{(A.23)}$$

which is an improved version of (8.12).

[*Received May 1982. Revised October 1982.*]

REFERENCES

EFRON, B. (1975), "The Efficiency of Logistic Regression Compared to Normal Discrimanant Analysis," *Journal of the American Statistical Association*, 70, 892–898.

——— (1979), "Bootstrap Methods: Another Look at the Jackknife," *Annals of Statistics*, 7, 1–26.

——— (1982), "The Jackknife, The Bootstrap, and Other Resampling Plans," SIAM NSF-CBMS, Monograph #38.

EFRON, B., and GONG, G. (1983), "A Leisurely Look at the Bootstrap, the Jackknife, and Cross-Validation," *The American Statistician*, 37, 36–48.

EFRON, B., and STEIN, C. (1981), "The Jackknife Estimate of Variance," *Annals of Statistics*, 9, 586–596.

GEISSER, S. (1975), "The Predictive Sample Reuse Method With Applications," *Journal of the American Statistical Association*, 70, 320–328.

GONG, G. (1982), "Cross-Validation, the Jackknife, and the Bootstrap: Excess Error Estimation in Forward Logistic Regression," Ph.D. dissertation, Stanford University Technical Report No. 80, Dept. of Statistics.

HAMPEL, F. (1974), "The Influence Curve and its Role in Robust Estimation," *Journal of the American Statistical Association*, 64, 1303–1317.

LACHENBRUCH, P., and MICKEY, M. (1968), "Estimation of Error Rates in Discrimant Analysis," *Technometrics*, 10, 1–11.

STONE, M. (1974), "Cross-Validatory Choice and Assessment of Statistical Predictions," *Journal of the Royal Statistical Society*, Ser. B, 36, 111–147.

——— (1977), "Asymptotics For and Against Cross-Validation," *Biometrika*, 64, 29–38.

WONG, W. (1983), "A Note on the Modified Likelihood for Density Estimation," *Journal of the American Statistical Association*, 78, 461–463.

13

Why Isn't Everyone a Bayesian?

Introduction by Larry Wasserman
Carnegie Mellon University

13.1. Introduction

Twenty-one years ago, when this paper first appeared, I was a graduate student asking myself the very question posed in the title of Brad Efron's paper. Bayesians were saviors or crackpots depending on who you asked. Philosophical arguments in favor of and against Bayesian inference were common. Applications using Bayesian methods were not so common. This changed with the advent of Markov chain Monte Carlo methods. Data analyses rooted in Bayesian methods abound. Indeed, Bayesian techniques are used in numerous fields including biology, medicine, physics, ecology, public policy, image recovery and many more. And yet, for all the reasons given in Efron's article, frequentist *thinking*, not subjectivism, is still dominant in the field of statistics. But in related fields, like machine learning, Bayesian thinking does seems to be much more prevalent. So, on the occasion of Efron's 70th birthday, it seems entirely appropriate to reprint and revisit this paper.

13.2. The Reasons

Efron cites four reasons that frequentist theory – which Efron takes to be the union of Fisher, Neyman, Pearson and Wald – has dominated statistics. These are (1) ease of use, (2) model building, (3) division of labor and (4) objectivity. Things haven't changed much. With the possible exception of (2), frequentist methods still have the edge.

Ease of Use. Markov chain Monte Carlo (MCMC) has made it possible to fit very complex models, a feat that would have been unthinkable some years ago. But MCMC has become an entire research industry, a sure sign that it isn't easy. Anyone who has used MCMC in a challenging problem can attest to the difficulties of designing efficient algorithms and the near-impossible task of assessing convergence. In contrast, easy-to-use frequentist methods abound. Since the publication of Efron's paper, even more easy-to-use methods have been discovered such as: the lasso regression method (Tibshirani 1996), the Benjamini–Hochberg multiple testing method (Benjamini and Hochberg 1995), wavelet regression (Donoho and Johnstone 1995), support vector machines (Burges 2004), boosting (Schapire et al. 1998), and generalized additive models (Hastie and Tibshirani 1999), to name a few. In each case, there are also Bayesian versions. But it is telling that the frequentist versions almost always come first.

Model Building. Efron says: "Both Fisherian and NPW theory pay more attention to the preinferential aspects of statistics." Bayesians might disagree with this point. Indeed, much effort in Bayesian analysis is concerned with model building and I think that many would claim this as a place where Bayesian inference is strongest. Ironically, this strength might also be a weakness. We are witnessing a deluge of baroque models thanks to MCMC. These models put a lot of "stuff" between the statistician and the data. For example, multilayered hierarchical models are commonplace in Bayesian inference and it is truly difficult to know if the output is driven by the data or by the numerous modeling assumptions. In contrast, frequentist methods emphasize transparency.

Division of Labor. The idea that statistical problems do not have to be solved as one coherent whole is anathema to Bayesians but is liberating for frequentists. To estimate a quantile, an honest Bayesian needs to put a prior on the space of all distributions and then find the marginal posterior. The frequentist need not care about the rest of the distribution and can focus on much simpler tasks. In a sense, this is another aspect of ease-of-use.

Objectivity. This is the most important, and most divisive, point. Efron claims that "The high ground of scientific objectivity has been seized by the frequentists." Most would agree with this statement, with the exception of diehard subjectivists. To the frequentist, objectivity means that procedures have frequency guarantees such as coverage. A subjectivist will reject that this defines objectivity and, most likely, write off objectivity as a fantasy. The philosophical divide on this issue is probably as insurmountable today as 21 years ago. At any rate, the purely subjective view remains a miniscule part of statistical practice.

Randomization. Although Efron does not state this as a separate point, he does emphasize the role of randomization. I'm referring to ideas like cross-validation, permutation tests and randomized experiments. These are arguably some of the best ideas that statistics has to offer. There is no counterpart in Bayesianism. Quite the opposite. As statistics and machine learning continue to merge and we deal with more and more complicated, high-dimensional problems, tools like cross-validation are becoming ever more important.

13.3. Conclusion

Brad Efron's article challenged Bayesians to make Bayesian inference practical. To some degree, Bayesians rose to the challenge. A Google search yields over 11,000,000 hits for "Bayesian." A casual look at the tables of contents in our leading statistics journals reveals that Bayesian methods are far from rare. There is little doubt that for some applications, Bayesian methods are crucial. For example, in complicated model selection problems, Bayesian methods have been used successfully as a tool for searching through large collections of models. In such cases, Bayesian machinery can be valuable as a search tool even if the priors are somewhat arbitrary.

The field of statistics benefits by having both Bayesian and frequentist methods, but Bayesian methods have not displaced frequentist methods. I suspect that frequentist methods will continue to dominate for precisely the reasons Efron gave in his article.

References

Benjamini, Y. and Hochberg, Y. (1995). Controlling the false discovery rate: A practical and powerful approach to multiple testing. *Journal of the Royal Statistical Society, Series B* **57**, 289–300.

Burges, V. (2004). A tutorial on support vector machines for pattern recognition. *Journal on Data Mining and Knowledge Discovery* **2**, 121–167.

Donoho, D. L. and Johnstone, I. M. (1995). Adapting to unknown smoothness via wavelet shrinkage. *Journal of the American Statistical Association* **90**, 1200–1224.

Hastie, T. and Tibshirani, R. (1999). *Generalized Additive Models*. Chapman & Hall, New York.

Schapire, R., Freund, Y., Bartlett, B. and Lee, W. (1998). Boosting the margin: A new explanation for the effectiveness of voting methods. *Annals of Statistics* **26**, 1651–1686.

Tibshirani, R. (1996). Regression shrinkage and selection via the lasso. *Journal of the Royal Statistical Society, Series B* **58**, 267–288.

Why Isn't Everyone a Bayesian?

B. EFRON*

Originally a talk delivered at a conference on Bayesian statistics, this article attempts to answer the following question: why is most scientific data analysis carried out in a non-Bayesian framework? The argument consists mainly of some practical examples of data analysis, in which the Bayesian approach is difficult but Fisherian/frequentist solutions are relatively easy. There is a brief discussion of objectivity in statistical analyses and of the difficulties of achieving objectivity within a Bayesian framework. The article ends with a list of practical advantages of Fisherian/frequentist methods, which so far seem to have outweighed the philosophical superiority of Bayesianism.

KEY WORDS: Fisherian inference; Frequentist theory; Neyman–Pearson–Wald; Objectivity.

1. INTRODUCTION

The title is a reasonable question to ask on at least two counts. First of all, everyone used to be a Bayesian. Laplace wholeheartedly endorsed Bayes's formulation of the inference problem, and most 19th-century scientists followed suit. This included Gauss, whose statistical work is usually presented in frequentist terms.

A second and more important point is the cogency of the Bayesian argument. Modern statisticians, following the lead of Savage and de Finetti, have advanced powerful theoretical reasons for preferring Bayesian inference. A byproduct of this work is a disturbing catalogue of inconsistencies in the frequentist point of view.

Nevertheless, everyone is not a Bayesian. The current era is the first century in which statistics has been widely used for scientific reporting, and in fact, 20th-century statistics is mainly non-Bayesian. [Lindley (1975) predicts a change for the 21st!] What has happened?

2. TWO POWERFUL COMPETITORS

The first and most obvious fact is the arrival on the scene of two powerful competitors: Fisherian theory and what Jack Kiefer called the Neyman–Pearson–Wald (NPW) school of decision theory, whose constituents are also known as the frequentists. Fisher's theory was invented, and to a remarkable degree completed, by Fisher in the period between 1920 and 1935. NPW began with the famous lemma of 1933, asymptoting in the 1950s, though there have continued to be significant advances such as Stein estimation, empirical Bayes, and robustness theory.

Working together in rather uneasy alliance, Fisher and NPW dominate current theory and practice, with Fisherian ideas particularly prevalent in applied statistics. I am going to try to explain why.

*B. Efron is Professor, Department of Statistics, Stanford University, Stanford, CA 94305.

3. FISHERIAN STATISTICS

In its inferential aspects Fisherian statistics lies closer to Bayes than to NPW in one crucial way: the assumption that there is a *correct* inference in any given situation. For example, if $x_1, x_2, \ldots, x_{20}$ is a random sample from a Cauchy distribution with unknown center θ,

$$f_\theta(x_i) = \frac{1}{\pi[1 + (x_i - \theta)^2]},$$

then in the absence of prior knowledge about θ the correct 95% central confidence interval for θ is, to a good approximation,

$$\hat{\theta} \pm 1.96 \Big/ \sqrt{-\ddot{l}_{\hat{\theta}}},$$

where $\hat{\theta}$ is the maximum likelihood estimator (MLE) and $\ddot{l}_{\hat{\theta}}$ is the second derivative of the log-likelihood function evaluated at $\theta = \hat{\theta}$. The (mathematically) equally good approximation

$$\hat{\theta} \pm 1.96/\sqrt{10}$$

(10 being the expected Fisher information), is not correct (Efron and Hinkley 1978).

Fisher's theory is a theory of archetypes. For any given problem the correct inference is divined by reduction to an archetypal form for which the correct inference is obvious. The first and simplest archetype is that of making inferences about θ from one observation x in the normal model

$$x \sim N(\theta, 1). \tag{1}$$

Fisher was incredibly clever at producing such reductions: sufficiency, ancillarity, permutation distributions, and asymptotic optimality theory are among his inventions, all intended to reduce complicated problems to something like (1). (It is worth noting that Fisher's work superseded an earlier archetypical inference system, Karl Pearson's method of moments and families of frequency curves.)

Why is so much of applied statistics carried out in a Fisherian mode? One big reason is the *automatic nature* of Fisher's theory. Maximum likelihood estimation is the original jackknife, in Tukey's sense of a widely applicable and dependable tool. Faced with a new situation, the working statistician can apply maximum likelihood in an automatic fashion, with little chance (in experienced hands) of going far wrong and considerable chance of providing a nearly optimal inference. In short, he does not have to think a lot about the specific situation in order to get on toward its solution.

Bayesian theory requires a great deal of thought about the given situation to apply sensibly. This is seen clearly in the efforts of Novick (1973), Kadane, Dickey, Winkler, Smith, and Peters (1980), and many others to at least partially automate Bayesian inference. All of this thinking is admirable in principle, but not necessarily in day-to-day practice. The same objection applies to some aspects of

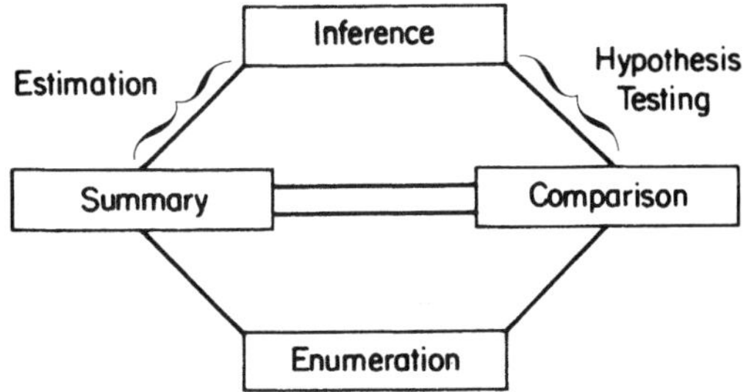

Figure 1. Four Basic Statistical Operations and How They Relate to Estimation. Source: Efron (1982b, fig. 2).

NPW theory, for instance, minimax estimation procedures, and with the same result: they are not used very much.

Not all of statistics is inference. The little diagram of all of statistics in Figure 1 (reprinted from Efron 1982b) starts at the bottom with "enumeration," the collecting and listing of individual datum. The diagram proceeds upward to the reduction of the raw data to more understandable form through the adversarial processes of summary and comparison. This is the part of the analysis where, usually, the statistician decides on a reasonable probabilistic model for the situation. At the top of the diagram is inference. This is the step that takes us from the data actually seen to data that might be expected in the future.

Bayesian theory concentrates on inference, which is the most glamorous part of the statistical world, but not necessarily the most important part. Fisher paid a lot of attention to the earlier steps of the data analysis. Randomization for instance, and experimental design in general, is a statement about how data should be collected, or "enumerated," for best use later in the analysis. Maximum likelihood is a provably efficient way to summarize data, no matter what particular estimation problems are going to be involved in the final inference (Efron 1982b). The NPW school has also contributed to the theory of enumeration, notably in the areas of survey sampling and efficient experimental design.

Fisher's theory culminated in fiducial inference, which to me and most current observers looks like a form of objective (as opposed to subjective) Bayesianism. I will discuss the problems and promise of objective Bayesianism later, but it is interesting to notice that fiducial inference is alone among Fisher's major efforts in its failure to enter common statistical application. In its place, the NPW theory of confidence intervals dominates practice, despite some serious logical problems in its foundations.

4. THE NPW SCHOOL

Unlike Bayes and Fisher, the NPW school does not insist that there is a correct solution for a given inferential situation. Instead, a part of the situation deemed most relevant to the investigator is split off, stated in narrow mathematical fashion, and it is hoped, solved. For example, the correct Bayesian or Fisherian inference for θ in situation (1) leads directly to the correct inference for $\gamma \equiv 1/(1 + \theta)$, but this is not necessarily the case in the NPW formulation. (What is the uniform minimum variance unbiased estimate of γ?)

The NPW piecewise approach to statistical inference has been justly criticized by Bayesians as self-contradictory, inconsistent, and incoherent. The work of Savage, de Finetti, and their successors shows that no logically consistent inference maker can behave in such a non-Bayesian way. The reply of the NPW school is that there is no reason to believe that statistical inference should be logically consistent in the sense of the Bayesians, and that there are good practical reasons for approaching specific inference problems on an individual basis.

As an example consider the following problem: we observe a random sample $x_1, x_2, \ldots, x_{15}$ from a continuous distribution F on the real line and desire an interval estimate for θ, the median of F. The experiment producing the x_i is a new one, so very little is known about F.

A genuine Bayesian solution seems difficult here, since it requires a prior distribution on the space of all distributions on the real line. Frequentist theory produces a simple solution in terms of a confidence interval based on the order statistics of the sample,

$$\theta \in [x_{(3)}, x_{(12)}]$$

with probability .963, no matter what F may be. The fact that this solution, unlike a Bayesian one, does not also solve the corresponding problem for say $\phi \equiv$ 50% trimmed mean of F does not dismay the frequentist, particularly if a satisfactory Bayesian solution is not available.

The Bayesian accusation of incoherency of the frequentist cuts both ways: in order to be coherent Bayesians have to solve all problems at once, an often impossible mental exercise.

As another example consider "rejecting at the .05 level." The inconsistencies of this practice are well documented in the Bayesian literature (see Lindley 1982). On the other hand it is one of the most widely used statistical ideas. Its popularity is founded on a crucial practical observation: it is often easier to compare quantities than to assign them absolute values. In this case the comparison is between the amount of evidence against the null hypothesis provided by different possible outcomes of the data. For testing H_0: $x \sim N(0, 1)$ versus H_1: $x \sim N(2, 1)$, we know that a larger observed x provides greater evidential value against H_0 and in favor of H_1, even if we cannot absolutely quantify "evidence."

A Bayes solution to this problem, "the aposteriori odds ratio is 7 to 1 in favor of H_1," is more satisfactory than "the data are significant at the .05 level," but it also requires more input. In fact, it tacitly implies that we have assigned an *absolute* measure of evidence to every possible outcome. Absolute here means that the meaning of 7 to 1 is the same no matter what experiment it came from. [Good's (1965) Bayes–non-Bayes compromise suggests using Bayesian ideas in a comparative mode, but this is the only example I know.]

The heart attack decision tree (Fig. 2) illustrates another difficult situation for the honest Bayesian. The tree purports to predict coronary patients with high risk of dying (population 2) on the basis of variables observed at hospital admission. A series of dichotomous observations are made, for example, high or low kinase level, which result in a final prediction. The nodes marked "2" on the tree predict death. Of the 389 patients classified by the tree, only 1 out

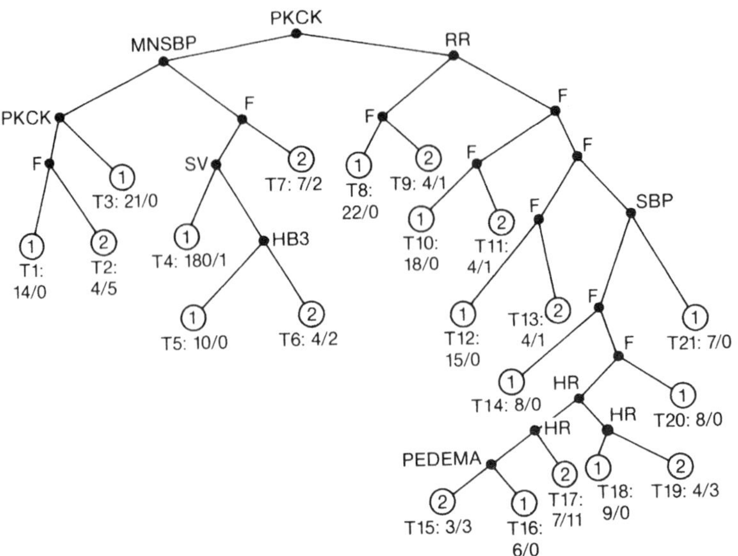

Figure 2. A Decision Tree for Classifying Heart Attack Patients Into Low Risk of Dying (population 1) or High Risk of Dying (population 2). Smaller values of the decision variables go to the left. Circled numbers at terminal nodes indicate population prediction. For example, 6 of the 389 patients in the training set end up at T6, 4 from population 1, 2 from population 2; these patients would all be predicted to be in population 2. Abbreviations: PKCK, peak creatinine kinase level; MNSBP, minimum systolic blood pressure; SBP, systolic blood pressure; FF, respiration rate; HR, average heart rate; SV, superventricular arrhythmia; HB3, heart block 3rd degree; PEDEMA, peripheral edema; F, Fisher linear discriminant function, differing from node to node. Source: Efron (1982a, fig. 7.1).

of 30 deaths was misclassified, that is, predicted to live. Can we believe that the tree has 96.7% probability of successfully predicting deaths?

Because the medical investigators had little prior knowledge of the situation, the tree was constructed by an elaborate data-fitting procedure, which in fact was designed to maximize the apparent success rate. At each stage the dichotomous variable to be used and the splitting point defining "high" or "low" were chosen to give the maximum apparent difference between populations 1 and 2. A bootstrap analysis, much like a cross-validation, gave an unbiased estimate of successful prediction of death of about 70%, rather than 96.7%, for this tree. (Details appear in sec. 7.6 of Efron 1982a.)

The fact that the observed data were used to construct the tree, and how they were used, makes no difference to the Bayesian, since it has no effect on the likelihood function. This is similar in spirit to the fact that the stopping rule used in a sequential procedure has no Bayesian consequence. It makes a world of difference to the frequentist. If exactly the same tree had been constructed by a less flexible rule, the unbiased estimate would move closer to the observed value 96.7%. This is incoherent behavior. The Bayesian estimate, whatever it is, would not change.

"Ad hoc" is a pejorative adjective in Bayesian descriptions of frequentist statistics. On the other hand, ad hoc reasoning produces a reasonable answer here, in a problem that seems far too complicated for a full Bayesian solution. The right to split off the simple part of a complicated inference problem should not be the exclusive property of the frequentists, but I am not aware of much Bayesian activity along these lines. The coherency approach of Savage and de Finetti seems to have discouraged it. (For a counterexample to this statement see Boos and Monahan, in press.)

The NPW school invented decision theory, but it is not decision theory that separates them from the Bayesians. In fact, Bayesians have made good use of decision theory. The parting of the ways occurs on the crucial issue of θ averages, expectations taken with the state of nature θ fixed. In other words, frequentist calculations. Controlling, or at least computing, θ averages is central to the NPW approach and irrelevant to the Bayesians. This brings us to the topic of objectivity, in my opinion the linchpin of non-Bayesian success with statistical practitioners.

5. OBJECTIVITY

So far I have been careful not to define the kind of Bayesian theory under criticism. The dominant Bayesian school, and the one with the legitimate claim to philosophic coherency, is the subjective Bayesianism of de Finetti and Savage. Now by definition one cannot argue with a subjectivist, so I will just state the obvious fact: though subjectivism is undoubtedly useful in situations involving personal decision making, for example, business and legal decisions, it has failed to make much of a dent in scientific statistical practice. The nature of scientific communication makes me doubt that it ever will.

"Scientific objectivity" is more than a catch-phrase. Strict objectivity is one of the crucial factors separating scientific thinking from wishful thinking. Complete objectivity about one's own work is a little much to expect from a human being, even a scientist, but it is not too much to expect from one's colleagues. A prime requirement of any statistical theory intended for scientific use is that it reassures oneself *and others* that the data have been interpreted fairly.

With this in mind, it is not surprising that intuitively "fair" statistical ideas, like unbiasedness, confidence intervals, and .05 significance, are immensely popular with the statistical public. Ideas that do not pass the test of objectivity are not much used. This includes NPW ideas as well as Bayesian ones, for example, James–Stein estimation. (An interesting borderline case, which could go either way depending on how it develops, is robust estimation.)

Of course there is no scientific law that says that objectivity must be interpreted in a frequentist sense, and in fact, there is another line of Bayesian thought that attempts to deal directly with the issue of fairness. I call this "objective Bayes theory" to differentiate it from the Savage–de Finetti approach. Bayes and Laplace were objective Bayesians, and in this century, Jeffreys (1961) wrote a famous book on the subject. The goal of objective Bayesianism is to produce prior distributions that capture the idea of objectivity.

Consider situation (1) again. The obvious prior here is the improper one, spreading probability mass for θ uniformly from $-\infty$ to $+\infty$, often denoted simply by "$d\theta$." A Bayesian using this prior obtains good frequentist results. The central 90% aposteriori interval for θ, for example, agrees exactly with the standard 90% confidence interval.

Next consider the situation where we observe

$$\begin{pmatrix} x_1 \\ x_2 \end{pmatrix} \sim N_2\left(\begin{pmatrix} \theta_1 \\ \theta_2 \end{pmatrix}, I\right), \qquad (2)$$

which is just two independent copies of (1). In some recent work it was necessary for me to make inferences about the parameter $\lambda = \theta_1 \theta_2$. It seemed intuitively reasonable, and objective, to use the improper prior $d\theta_1\, d\theta_2$, which spreads probability mass for (θ_1, θ_2) uniformly over the entire plane.

As Table 1 shows, the aposteriori central 90% probability interval for λ derived from the prior $d\theta_1\, d\theta_2$ does not have good frequentist properties. For values of (θ_1, θ_2) in the first quadrant, it gives overly low probabilities of missing λ on the left and overly high probabilities of missing λ on the right. From a frequentist viewpoint we have not been very objective at all, having biased the interval estimates toward the origin.

Recently Charles Stein has given a method of constructing priors that have better frequentist properties (Stein 1982).

Table 1. Theoretical Versus Actual Probability of Not Covering $\lambda = \theta_1 \theta_2$ Using the Central 90% Aposteriori Interval Based on the Improper Prior $d\theta_1\, d\theta_2$

	Theoretical	
(θ_1, θ_2)	.050	.050
(0, 0)	002 (006)	002 (006)
(2, 2)	023 (036)	087 (065)
(3, 3)	029 (041)	074 (053)
(4, 4)	031 (046)	070 (047)
(5, 5)	036 (048)	064 (043)
(1, 10)	045 (053)	051 (053)

NOTE: Figures in parentheses are, essentially, the same probabilities using the improper prior $(\theta_1^2 + \theta_2^2)^{1/2}\, d\theta_1\, d\theta_2$.

For the parameter λ, Stein's theory suggests using the improper prior $(\theta_1^2 + \theta_2^2)^{1/2}\, d\theta_1\, d\theta_2$, and Table 1 shows that this does indeed give better frequentist coverage probabilities. [In fact, these figures were derived for the bias-corrected percentile intervals, sec. 10.7 of Efron (1982a), and it was then verified that these intervals were almost the same as Stein's.]

The point of the example is that the theory of Bayesian objectivity cannot be a simple one. The correct objective prior seems to depend on which parameter we want to estimate. In higher dimensions when we have several parameters rather than just two, these problems become acute (Efron 1982b). This does not mean that the situation is hopeless. Even a partial solution to the problem of Bayesian objectivity would likely be a valuable contribution to statistical theory and practice. As a hopeful prototype, the Bayesian explanation of the James–Stein estimator has deepened our understanding of this potentially wonderful tool. The whole subject of empirical Bayes can be thought of as an exercise in Bayesian objectivity—trying not to put more information than necessary into the prior—and more progress in this area can be expected.

6. SUMMARY

A summary of the major reasons why Fisherian and NPW ideas have shouldered Bayesian theory aside in statistical practice is as follows:

1. Ease of use: Fisher's theory in particular is well set up to yield answers on an easy and almost automatic basis.
2. Model building: Both Fisherian and NPW theory pay more attention to the preinferential aspects of statistics.
3. Division of labor: The NPW school in particular allows interesting parts of a complicated problem to be broken off and solved separately. These partial solutions often make use of aspects of the situation, for example, the sampling plan, which do not seem to help the Bayesian.
4. Objectivity: The high ground of scientific objectivity has been seized by the frequentists.

None of these points is insurmountable, and in fact, there have been some Bayesian efforts on all four. In my opinion a lot more such effort will be needed to fulfill Lindley's prediction of a Bayesian 21st century.

[*Received July 1985.*]

REFERENCES

Boos, D., and Monahan, J. (in press), "The Bootstrap for Robust Bayesian Analysis: An Adventure in Computing," *14th Symposium on the Interface*.

Efron, B. (1982a), *The Jackknife, the Bootstrap, and Other Resampling Plans*, CBMS-NSF Monograph 39, Philadelphia: SIAM.

——— (1982b), "Maximum Likelihood and Decision Theory," *Annals of Statistics*, 10, 340–356.

Efron, B., and Hinkley, D. V. (1978), "Assessing the Accuracy of the Maximum Likelihood Estimator: Observed Versus Expected Fisher Information," *Biometrika*, 65, 457–487.

Good, I. J. (1965), *The Estimation of Probabilities: An Essay on Modern Bayesian Methods*, Cambridge: MIT Press.

Jeffreys, H. (1961), *Theory of Probability* (3rd ed.), Oxford: Clarendon Press.

Kadane, J. B., Dickey, J. M., Winkler, R. L., Smith, W. S., and Peters, S. C. (1980), "Interactive Elicitation of Opinion for a Normal Linear Model," *Journal of the American Statistical Association*, 75, 845–854.

Lindley, D. V. (1975), "The Future of Statistics—A Bayesian 21st Century," *Supp. Adv. Appl. Prob.*, 7, 106–115.

——— (1982), "Scoring Rules and the Inevitability of Probability," *ISI Review*, 50, 1–26.

Novick, M. (1973), "High School Attainment: An Example of a Computer-Assisted Bayesian Approach to Data Analysis," *ISI Review*, 41, 268–271.

Stein, C. (1982), "The Coverage Probability of Confidence Sets Based on a Prior Distribution," Technical Report 180, Stanford University.

Reply

B. EFRON

This article was actually the text of a talk delivered at a conference on Bayesian statistics held at the Virginia Polytechnic Institute and State University. The question its title asks was not intended to be rhetorical or sarcastic. Despite the considerable philosophical advantages of the Bayesian approach, most scientific data analysis is carried out in a non-Bayesian framework. Why? The talk attempted to answer this question, arguing mainly from a set of examples in which Bayesian analysis is difficult while Fisher/frequentist solutions are relatively easy.

The examples are genuine ones, suggested by my own picaresque adventures as an applied statistician in the Stanford Medical School. For instance, the product-of-normal-means problem arose in the context of comparing two nonnested linear models (Efron 1984). There is nothing particularly striking or unusual about these examples, but of course that was the point of my talk.

Here are a few brief comments on the commentaries. I was surprised at how little support objectivity aroused. Bringing some degree of consensus to the interpretation of noisy data is certainly one of our profession's principal accomplishments. Perhaps, as Smith suggests, we have purchased consensus at a high price in intellectual tyranny, but our scientific clientele seem happy to pay this price.

Understanding the true meaning of objectivity has occupied statistical thinkers from Bayes, Laplace, and Gauss to Fisher, Neyman, and Jeffreys. Fisher's phrase, "the logic of uncertain inference," is particularly evocative of a theory that goes from the data and a family of possible probability models to a consensus agreement of reasonable conclusions.

Such a theory does not yet fully exist, and may never exist, but I hope we have not given up looking for it. In the discussion, only Press had specific good words to say for objectivity. On the other hand, the empirical Bayes approach, very nicely stated by Morris, has strong objective Bayes connections. These connections are made explicit in Good's theory of Type II maximum likelihood (Good 1965).

Lindley makes the serious complaint that I have criticized only a parody of the true Bayesian argument. The heart attack decision tree was particularly offensive. I chose this example because it shows how a simple frequentist technique, cross-validation or the bootstrap in this case, can make headway even in very complicated situations. Devices like cross-validation violate the likelihood principle; they ask what would happen for data sets other than the one actually observed and so are non-Bayesian according to Lindley's strict definitions. What is the "true Bayesian argument" in this example? It is not that I do not think such an argument exists, but Lindley has not given us the slightest idea of what it might be. This is a practical point, not a philosophical one, but practical points are crucial in trying to understand why Bayesian theory is not much used in scientific practice.

Chernoff can be excused a feeling of deja vu, since parts of my talk, in particular the decision tree example, were based on memories of his own lectures. Section 1.b of his discussion is a particularly neat statement of the role of the statistician in scientific practice. It was not my intention to separate decision theory from statistical inference in general (the top box in Fig. 1), as both Chernoff and Press thought I was doing. The relationship between Fisherian and decision-theoretic inferential systems is discussed in Efron (1982b).

My talk was intended as an argument for more Bayesian research, not less. The problems of *practical* Bayesian inference should be given top priority. Morris's discussion is a promising model of how Bayesian theory could be aimed at the actual needs of applied statisticians.

Finally, I must warn Professor Lindley that his brutal, and largely unprovoked, attack has been reported to FOGA (Friends of the Greek Alphabet). He will be in for a *very nasty time indeed* if he wishes to use as much as an epsilon or an iota in any future manuscript.

ADDITIONAL REFERENCE

Efron, B. (1984), "Comparing Non-Nested Linear Models," *Journal of the American Statistical Association*, 79, 791–803.

ADDITIONAL COMMENT BY LINDLEY

Brad Efron asks what *the* true Bayesian argument might be in the heart attack case. *A* (nearly) Bayesian approach was given by David J. Speigelhalter and Robin P. Knill-Jones [*Journal of the Royal Statistical Society*, Ser. A (1984) 147, 35–77, with discussion].

14

Better Bootstrap Confidence Intervals

Introduction by Peter Bickel
University of California, Berkeley

This paper can be viewed as the culmination of Brad's study of the use of the bootstrap in setting confidence intervals, a study he initiated with his introduction of the bootstrap in 1979. The "percentile interval" that he introduced there, although a bona fide asymptotically-correct frequentist confidence interval had a distinct Bayesian flavor, treated the bootstrap distribution of the parameter estimate $\hat{\theta}$ as if it were a posterior distribution of the parameter θ. Although not immediately apparent in 1979, at least to me, his goal all along was to find asymptotic analogues of the known methods developed to obtain exact confidence intervals for a one-dimensional parameter. All of these methods depended, when there were no nuisance parameters present, on having the family of distribution of the estimate be stochastically ordered in the parameters, as Brad mentions at the beginning of Sect. 5. In turn, these intervals depend on inversion of the two families of one-sided tests based on $\hat{\theta}$ for $H: \theta = \theta_0$ vs $K: \theta < \theta_0$ or $K: \theta > \theta_0$, as θ_0 varies.

By construction these intervals are size α and invariant under monotone transformations. That is, if I is the size α interval for θ based on observing $\hat{\theta}$ then $q(I)$ is a size α interval for $q(\theta)$ for q monotone.

If the densities, f_θ, of $\hat{\theta}$ are a monotone likelihood ratio family, then they are also optimal in terms of measures of length – see Lehmann and Romano (2005, pp. 161–168), for instance.

What Brad achieves in this paper, given a sample of size n from a suitably-smooth (in θ) family of densities and a correspondingly well-behaved first-order efficient estimate $\hat{\theta}$ and its nonparametric bootstrap distribution, $\hat{G}$, is to construct intervals $[\underline{\theta}^{(\alpha)}_{\text{BCA}}, \bar{\theta}^{(\alpha)}_{\text{BCA}}]$ which, asymptotically:

(1) Have the correct desired size $1-\alpha$ to order n^{-1}
(2) Have the transformation invariance property mentioned above
(3) Are "second-order" correct in an asymptotic version of the sense described by Lehmann and Romano to order n^{-1} as well

Strictly speaking, Brad does not establish (3) but rather that, if the estimate $\hat{\theta}$ has a formal Edgeworth expansion of the right structure to order n^{-1}, then $[\underline{\theta}^{(\alpha)}_{\text{BCA}}, \bar{\theta}_{\text{BCA}}]$ agree with the corresponding quantities for (the well-behaved) MLE. But, as Pfanzagl (1985) has shown, in regular situations this does imply optimality to order n^{-1}. He then goes on to show that his construction satisfies (1) and (2) even when a nuisance parameter ϕ is present, and conjectures correctly (see Bickel 1987, discussion; Hall 1986) that second-order correctness also holds.

Although other methods achieve the same results, for instance Beran (1987), the elegance and (apparent) simplicity of this solution are, I believe, unsurpassed (although it may have robustness problems).

Its parent is clearly Brad's percentile method, which possesses property (2) and is first-order (to $n^{-\frac{1}{2}}$) efficient. That interval, $[\hat{G}^{-1}(\alpha), \hat{G}^{-1}(1-\alpha)]$ is refined to

$$[\underline{\theta}^{(\alpha)}_{\text{BCA}}, \bar{\theta}^{(\alpha)}_{\text{BCA}}] = [\hat{G}^{-1}\Phi\big(\hat{Z}(\alpha)\big), \hat{G}^{-1}\Phi\big(\hat{Z}(1-\alpha)\big)]\,,$$

where, $\hat{Z}[\alpha]$, $\hat{Z}[1-\alpha]$ are given by a data-dependent perturbation of the Gaussian quantiles, $z^{(\alpha)}$, $z^{(1-\alpha)}$,

$$\hat{Z}[t] = \frac{(\hat{z}_0 + z^{(t)})}{1 - \hat{a}(\hat{z}_0 + z^{(t)})},$$

where $\hat{a}$, $\hat{z}_0$ are computable using $\hat{G}$ and the efficient score function evaluated at $\hat{\theta}$ – Bickel et al. (1993), for instance. Its immediate parent, the BC interval, achieves (1)–(3) but only if a_0, the parameter being estimated by $\hat{a}$, is 0.

Brad's derivation is, as usual, very elegant and deceptively easy to read. I say "deceptively easy" because, from the beginning, Brad works with a 1-parameter subfamily of Gaussian Distributions (3.6) which seems to come from nowhere. This family leads readily to the formulae for $\hat{Z}(\alpha)$, $\hat{Z}(1-\alpha)$. Its origins in Brad's geometrical intuition only become clear in Sect. 3, where he identifies it as the canonical member of the broad family of General Scaled Transformations he introduced in Efron (1982).

The paper, in Brad's usual fashion, jumps from one point of view to another, fruitfully relating all. It is a pleasure to read. The reader should find the discussion (with the possible exception of the writer's contribution) "unusually penetrating," as Brad mentions as well.

References

Beran, R. (1987). Prepivoting to reduce level error of confidence sets. *Biometrika* **74**, 457–468.

Bickel, P. (1987). Comment on "Better bootstrap confidence intervals" by B. Efron. *Journal of the American Statistical Association* **82**, 191.

Bickel, P., Klaassen, C., Ritov, Y. and Wellner, J. (1993). *Efficient and Adaptive Estimation for Semiparametric Models*. Johns Hopkins University Press, Baltimore, MD.

Efron, B. (1982). The jackknife, the bootstrap, and other resampling plans (1982). *SIAM CBMS–NSF Regional Conference Series in Applied Mathematics*, Monograph 38.

Hall, P. (1986). On the bootstrap and confidence intervals. *Annals of Statistics* **14**, 1431–1452.

Lehmann, E. L. and Romano, J. P. (2005). *Testing Statistical Hypotheses*, 3rd edition. Springer, New York.

Pfanzagl, J. (with the assistance of W. Wefelmeyer) (1985). *Asymptotic expansions for general statistical models*. Springer, New York.

Better Bootstrap Confidence Intervals

BRADLEY EFRON*

We consider the problem of setting approximate confidence intervals for a single parameter θ in a multiparameter family. The standard approximate intervals based on maximum likelihood theory, $\hat{\theta} \pm \hat{\sigma}z^{(\alpha)}$, can be quite misleading. In practice, tricks based on transformations, bias corrections, and so forth, are often used to improve their accuracy. The bootstrap confidence intervals discussed in this article automatically incorporate such tricks without requiring the statistician to think them through for each new application, at the price of a considerable increase in computational effort. The new intervals incorporate an improvement over previously suggested methods, which results in second-order correctness in a wide variety of problems. In addition to parametric families, bootstrap intervals are also developed for nonparametric situations.

KEY WORDS: Resampling methods; Approximate confidence intervals; Transformations; Nonparametric intervals; Second-order theory; Skewness corrections.

1. INTRODUCTION

This article concerns setting approximate confidence intervals for a real-valued parameter θ in a multiparameter family. The nonparametric case, where the number of nuisance parameters is infinite, is also considered. The word "approximate" is important, because in only a few special situations can exact confidence intervals be constructed. Table 1 shows one such situation: the data (y_1, y_2) are bivariate normal with unknown mean vector (η_1, η_2), covariance matrix = $\mathbf{I}$ the identity; the parameters of interest are $\theta = \eta_2/\eta_1$ and, in addition, $\xi = 1/\theta$. Fieller's construction (1954) gives central 90% interval (5% error in each tail) of [.29, .76] for θ, having observed $\mathbf{y} = (8, 4)$. The corresponding interval for $\xi = 1/\theta$ is the obvious mapping $\xi \in [1/.76, 1/.29]$.

Table 1 also shows the standard approximate intervals

$$\theta \in [\hat{\theta} + \hat{\sigma}z^{(\alpha)}, \hat{\theta} + \hat{\sigma}z^{(1-\alpha)}], \tag{1.1}$$

where $\hat{\theta}$ is the maximum likelihood estimate (MLE) of θ, $\hat{\sigma}$ is an estimate of its standard deviation, often based on derivatives of the log-likelihood function, and $z^{(\alpha)}$ is the $100 \cdot \alpha$ percentile point of a standard normal variate. In Table 1, $\alpha = .05$ and $z^{(\alpha)} = -z^{(1-\alpha)} = -1.645$.

The standard intervals (1.1) are extremely useful in statistical practice because they can be applied in an automatic way to almost any parametric situation. However, they can be far from perfect, as the results for ξ show. Not only is the standard interval for ξ quite different from the exact interval, it is not even the obvious transformation [1/.73, 1/.27] of the standard interval for θ.

Approximate confidence intervals based on bootstrap computations were introduced by Efron (1981, 1982a). Like the standard intervals, these can be applied automatically to almost any situation, though at greater computational expense than (1.1). Unlike (1.1), the bootstrap intervals transform correctly, so the interval for $\xi = 1/\theta$ in the Fieller example is obtained by inverting the endpoints of the interval for θ. They also tend to be more accurate than the standard intervals. In the situation of Table 1 the bootstrap intervals agree with the exact intervals to three decimal places. Efron (1985) showed that this is no accident; there is a wide class of problems for which the bootstrap intervals are an order of magnitude more accurate than the standard intervals.

In those problems where exact confidence limits exist the endpoints are typically of the form

$$\hat{\theta} + \hat{\sigma}(z^{(\alpha)} + A_n^{(\alpha)}/\sqrt{n} + B_n^{(\alpha)}/n + \cdots), \tag{1.2}$$

where n is the sample size (see Efron 1985). The standard intervals (1.1) are *first-order correct* in the sense that the term $\hat{\theta} + \hat{\sigma}z^{(\alpha)}$ asymptotically dominates (1.2). However, the second-order term $\hat{\sigma}A_n^{(\alpha)}/\sqrt{n}$ can have a major effect in small-sample situations. It is this term that causes the asymmetry of the exact intervals about the MLE as illustrated in Table 1. As a point of comparison the Student-t effect is of third-order magnitude, comparable with $\hat{\sigma}B_n^{(\alpha)}/n$ in (1.2). The bootstrap method described in Efron (1985) was shown to be *second-order correct* in a certain class of problems, automatically producing intervals of correct second-order asymptotic form $\hat{\theta} + \hat{\sigma}(z^{(\alpha)} + A_n^{(\alpha)}/\sqrt{n} + \cdots)$.

This article describes an improved bootstrap method that is second-order correct in a wider class of problems. This wider class includes all of the familiar parametric examples where there are no nuisance parameters and where the data have been reduced to a one-dimensional summary statistic, with asymptotic properties of the usual MLE form (see Sec. 5).

Several authors have developed higher-order correct approximate confidence intervals based on Edgeworth expansions (Abramovitch and Singh 1985; Beran 1984a,b; Hall 1983; Withers 1983), sometimes using bootstrap methods to reduce the theoretical computations. There is a close theoretical relationship between this line of work and the current article (see, e.g., Remark G, Sec. 11). However, the details of the various methods are considerably different, and they can give quite different numerical results. An important point, which will probably have to be settled by extensive simulations, is which method, if any, handles best the practical problems of day-to-day applied statistics.

2. OVERVIEW

The standard interval (1.1) is based on taking literally the asymptotic normal approximation

$$(\hat{\theta} - \theta)/\hat{\sigma} \sim N(0, 1), \tag{2.1}$$

* Bradley Efron is Professor of Statistics and Biostatistics, Department of Statistics, Stanford University, Stanford, CA 94305. The author is grateful to Robert Tibshirani, Timothy Hesterberg, and John Tukey for several useful discussions, suggestions, and references.

Table 1. Central 90% Confidence Intervals for $\theta = \eta_2/\eta_1$ and $\xi = 1/\theta$, Having Observed $(y_1, y_2) = (8, 4)$ From a Bivariate Normal Distribution $\mathbf{y} \sim N_2(\boldsymbol{\eta}, I)$

	For θ	(R/L)	For ξ	(R/L)
Exact interval (also bootstrap)	[.29, .76]	(1.21)	[1.32, 3.50]	(2.20)
Standard approximation (1.1)	[.27, .73]	(1.00)	[1.08, 2.92]	(1.00)
MLE	$\hat{\theta} = .5$		$\hat{\xi} = 2$	

NOTE: The exact intervals are based on Fieller's construction. R/L is the ratio of the right side of the interval, measured from the MLE, to the left side. The exact intervals are markedly asymmetric. The approximate bootstrap intervals of Efron (1982a) agree with the exact intervals to three decimal places in this case.

with the estimated standard error $\hat{\sigma}$ considered to be a fixed constant. In certain cases it is well known that both convergence to normality and constancy of σ can be dramatically improved by considering instead of $\hat{\theta}$ and θ a monotone tranformation $\hat{\phi} = g(\hat{\theta})$ and $\phi = g(\theta)$. The classic example is that of the correlation coefficient from a bivariate normal sample, for which Fisher's inverse hyperbolic tangent transformation works beautifully (see Efron 1982b).

The bias-corrected bootstrap intervals previously introduced by Efron (1981, 1982a), called the BC intervals, assume that normality and constant standard error can be achieved by *some* transformation $\hat{\phi} = g(\hat{\theta})$, $\phi = g(\theta)$, say

$$(\hat{\phi} - \phi)/\tau \sim N(-z_0, 1), \tag{2.2}$$

τ being the constant standard error of $\hat{\phi}$. Allowing the bias constant z_0 in (2.2) considerably improves the approximation in many cases, including that of the normal correlation coefficient. Taking (2.2) literally gives the obvious confidence interval $(\hat{\phi} + \tau z_0) \pm \tau z^{(\alpha)}$ for ϕ, which can then be converted back to a confidence interval for θ by the inverse transformation $\theta = g^{-1}(\phi)$. The advantage of the BC method is that all of this is done automatically from bootstrap calculations, without requiring the statistician to know the correct transformation g.

The improved bootstrap method introduced in this article, called BC_a, makes one further generalization on (2.1): it is assumed that for some monotone transformation g, some bias constant z_0, *and some "acceleration constant"* a, the transformation $\hat{\phi} = g(\hat{\theta})$, $\phi = g(\theta)$ results in

$$(\hat{\phi} - \phi)/\tau \sim N(-z_0\sigma_\phi, \sigma_\phi^2), \qquad \sigma_\phi = 1 + a\phi. \tag{2.3}$$

Notice that (2.2) is the special case of (2.3), with $a = 0$.

Given (2.3), it is not difficult to find the correct confidence interval for ϕ and then convert it back to an interval for θ by $\theta = g^{-1}(\phi)$. The BC_a method produces this interval for θ automatically, without requiring any knowledge of the transformation to form (2.3). This is the gist of Lemma 1 in Section 3.

The difference between (2.2) and (2.3) is greater than it seems. The hypothesized ideal transformation g leading to (2.2) must be both *normalizing* and *variance stabilizing*, whereas in (2.3) g need be only normalizing. Efron (1982b) shows that normalization and stabilization are partially antagonistic goals in familiar families such as the Poisson and the binomial. Schenker's counterexample to the BC method (1985), which helped motivate this article, is based on a family (discussed in Sec. 3) for which (2.2) fails. The main purpose of this article, to produce automatically intervals that are second-order correct, generally requires assumption (2.3) rather than (2.2).

It is not surprising that a theory based on (2.3) is usually more accurate than a theory based on (2.1). In fact, applied statisticians make frequent use of devices like those in (2.3), transformations, bias corrections, and even acceleration adjustments, to improve the performance of the standard intervals. The advantage of the BC_a method is that it automates these improvements, so the statistician does not have to think them through anew for each new application.

The bootstrap was originally introduced as a nonparametric Monte Carlo device for estimating standard errors. The basic idea, however, can be applied to any statistical problem, including parametric ones, and does not necessarily require Monte Carlo simulations. We will begin our discussion of the BC_a method by considering the simplest type of parametric problem: where the data consists only of a single real-valued statistic $\hat{\theta}$ in a one-parameter family of densities $f_\theta(\hat{\theta})$, say $\hat{\theta} \sim f_\theta$, and where we want a confidence interval for θ based on $\hat{\theta}$.

Sections 3, 4, and 5 describe the BC_a method in this simple setting, show how to calculate it from bootstrap computations, and demonstrate that it gives second-order correct intervals for θ under reasonable conditions.

Of course there is no need for the bootstrap in the simple situation $\hat{\theta} \sim f_\theta$, since then it is usually quite easy to calculate exact confidence intervals for θ. There are three reasons for beginning the discussion with the simple situation: (a) it makes clear the logic of the BC_a method; (b) it makes possible the comparison of BC_a intervals with exact intervals, exact intervals usually not existing in complicated problems; (c) it then turns out to be quite easy to extend the BC_a method to complicated situations, where it is more likely to be needed.

The simple situation $\hat{\theta} \sim f_\theta$ can be made more complicated, and more realistic, in two ways: the data can consist of a vector $\mathbf{y}$ instead of a single summary statistic $\hat{\theta}$, and the parameter can be a vector $\boldsymbol{\eta}$ instead of a single unknown scalar θ. Section 6 considers multiparameter families $f_{\boldsymbol{\eta}}(\mathbf{y})$, where we wish to set an approximate confidence interval for a real-valued function $\theta = t(\boldsymbol{\eta})$.

Our approach is to reduce the problem back to the simple situation. The data vector $\mathbf{y}$ is replaced by an efficient estimator $\hat{\theta}$ of θ, perhaps the MLE, and the multiparameter family $f_{\boldsymbol{\eta}}$ is replaced by a *least favorable* one-parameter family. All of the calculations are handled automatically by the BC_a algorithm. For a class of examples, including the Fieller problem of Table 1, the BC_a method automatically produces second-order correct intervals, but a proof of general second-order correctness does not yet exist for multiparameter situations.

Section 7 returns to the original nonparametric setting of the bootstrap: the data $\mathbf{y}$ is assumed to be a random sample $x_1, x_2, \ldots, x_n$ from a completely unknown probability distribution F. We wish to set an approximate confidence interval for $\theta = t(F)$, some real-valued function of F. The BC_a method extends in a natural way to the nonparametric setting. In the case where θ is the expectation, theoretical analysis shows the BC_a intervals performing reasonably. Except for the case of the expectation, not much is proved about nonparametric BC_a intervals, though the empirical results look promising. Section 8 develops a heuristic justification for the nonparametric BC_a method in terms of the geometry of multinomial sampling.

In the simple situation $\hat{\theta} \sim f_\theta$ the parametric bootstrap distribution $\hat{\theta}^* \sim f_{\hat{\theta}}$ can often be written down explicitly, or at least approximated by standard parametric devices such as Edgeworth expansions. The number of bootstrap replications of $\hat{\theta}^*$, to use the terminology of previous papers, is then, effectively, infinity. For more complicated situations like the nonparametric confidence interval problem, Monte Carlo sampling is usually needed to calculate the BC_a intervals. How many bootstrap replications are necessary? The answer, on the order of 1,000, is derived in Section 9. This compares with only about 100 bootstrap replications necessary to adequately calculate a bootstrap standard error. Bootstrap confidence intervals require a lot more computation than bootstrap standard errors, if second-order accuracy is desired.

To get the main ideas across, some important technical points are deferred until Sections 10–12.

3. BOOTSTRAP CONFIDENCE INTERVALS FOR SIMPLE PARAMETRIC SITUATIONS

We first consider the simple situation $\hat{\theta} \sim f_\theta$, where we have a one-parameter family of densities $f_\theta(\hat{\theta})$ for the real-valued statistic $\hat{\theta}$. We wish to set a confidence interval for θ having observed only $\hat{\theta}$. The statistic $\hat{\theta}$ estimates θ. Later we will make specific assumptions about the properties of $\hat{\theta}$ as an estimator of θ—essentially that $\hat{\theta}$ behaves like the MLE asymptotically, though $\hat{\theta}$ may be some first-order efficient estimator other than the MLE.

By definition, the parametric bootstrap distribution in this simple situation is

$$\hat{\theta}^* \sim f_{\hat{\theta}}. \tag{3.1}$$

In other words it is the distribution of the statistic of interest when the unknown parameter θ is set equal to the observed point estimate $\hat{\theta}$. We also need to define the cdf of the bootstrap distribution

$$\hat{G}(s) = \int_{-\infty}^{s} f_{\hat{\theta}}(\hat{\theta}^*)d\hat{\theta}^* = \Pr_{\hat{\theta}}\{\hat{\theta}^* < s\}. \tag{3.2}$$

The integral is replaced by a summation in discrete families. The goal of bootstrap theory is to make inferential statements on the basis of the bootstrap distribution. In this article the inferences are approximate confidence intervals for θ.

Example (chi-squared scale family). Suppose that

$$\hat{\theta} \sim \theta(\chi^2_{19}/19), \tag{3.3}$$

the example considered in Schenker (1985). Then

$$f_\theta(\hat{\theta}) = c(\hat{\theta}/\theta)^{8.5}e^{-9.5(\hat{\theta}/\theta)} \quad \text{for } \hat{\theta} > 0 \quad [c = 9.5^{9.5}/\Gamma(9.5)]. \tag{3.4}$$

Having observed $\hat{\theta}$, the bootstrap distribution $\hat{\theta}^* \sim \hat{\theta}(\chi^2_{19}/19)$ has density $c(\hat{\theta}^*/\hat{\theta})^{8.5}e^{-9.5(\hat{\theta}^*/\hat{\theta})}$ for $\hat{\theta}^* > 0$. The bootstrap cdf is $\hat{G}(s) = I_{9.5}(9.5s/\hat{\theta})$, where $I_{9.5}$ indicates the incomplete gamma function of degree 9.5.

Now suppose that for a family $\hat{\theta} \sim f_\theta$ there exists a monotone-increasing transformation g and constants z_0 and a such that

$$\hat{\phi} = g(\hat{\theta}), \qquad \phi = g(\theta) \tag{3.5}$$

satisfy

$$\hat{\phi} = \phi + \sigma_\phi(Z - z_0), \qquad Z \sim N(0, 1) \tag{3.6}$$

with

$$\sigma_\phi = 1 + a\phi. \tag{3.7}$$

This is of form (2.3), with $\tau = 1$. [Eq. (2.3) can always be reduced to the case $\tau = 1$; see Remark A, Sec. 11.] We will assume that $\phi > -1/a$ if $a > 0$ in (3.7), so $\sigma_\phi > 0$, and likewise $\phi < -1/a$ if $a < 0$. The constant a will typically be in the range $|a| < .2$, as will z_0.

Let Φ denote the standard normal cdf, and let $\hat{G}^{-1}(\alpha)$ denote the $100 \cdot \alpha$ percentile of the bootstrap cdf (3.2).

Lemma 1. Under conditions (3.5)–(3.7), the correct central confidence interval of level 1-2α for θ is

$$\theta \in [\hat{G}^{-1}(\Phi(z[\alpha])), \hat{G}^{-1}(\Phi(z[1 - \alpha]))], \tag{3.8}$$

where

$$z[\alpha] = z_0 + \frac{(z_0 + z^{(\alpha)})}{1 - a(z_0 + z^{(\alpha)})}, \tag{3.9}$$

and likewise for $z[1 - \alpha]$.

The proof of Lemma 1, at the end of this section, makes clear that interval (3.8) is correct in a strong sense: it is equivalent, under assumptions (3.5)–(3.7), to the usual obvious interval for a simple translation problem. Given the bootstrap cdf $\hat{G}(s)$ and values of z_0 and a derived from bootstrap calculations as in the following sections, we can form interval (3.8), (3.9) for θ whether or not assumptions (3.5)–(3.7) apply. This by definition is the BC_a interval for θ.

If z_0 and a equal 0, then $z[\alpha] = z^{(\alpha)}$ and (3.8) becomes $\theta \in [\hat{G}^{-1}(\alpha), \hat{G}^{-1}(1 - \alpha)]$. In this case we just use the obvious percentiles of the bootstrap distribution to form an approximate confidence interval for θ, an approach called the *percentile method* in Efron (1981, 1982a). In general z_0 and a do not equal zero, and formulas (3.8), (3.9) make adjustments to the percentile method that are necessary to achieve second-order correctness.

Continuing example (3.3), the theory of Efron (1982b) shows that for the chi-squared scale family we can find a transformation g very nearly satisfying (3.5)–(3.7). Schenker (1985) proved the same result by a different method. The constants

$$z_0 = .1082, \qquad a = .1077 \tag{3.10}$$

and the transformation g appropriate to family (3.3) are derived in Section 10 and Remark E of Section 11. Simple ways of approximating z_0 and a for general families $\hat{\theta} \sim f_\theta$ are given in Section 4.

Line 2 of Table 2 shows the central 90% BC_a interval, $\alpha = .05$, for family (3.3), with $\hat{G}(s) = I_{9.5}(9.5s/\hat{\theta})$ and z_0 and a as in (3.10). The exact confidence interval is $\theta \in [19\hat{\theta}/\chi_{19}^{2(1-\alpha)}, 19\hat{\theta}/\chi_{19}^{2(\alpha)}]$, where $\chi_{19}^{2(\alpha)}$ is the $100 \cdot \alpha$ percentile point of a χ^2_{19} distribution. Notice how closely the BC_a endpoints match those of the exact interval (see line 1). The standard interval (1.1) is quite inaccurate in this case.

Suppose that we set $a = 0$ in (3.9), so $z[\alpha] = 2z_0 + z^{(\alpha)}$. Interval (3.8) with this definition of $z[\alpha]$ and $z[1 - \alpha]$ is called the *BC interval,* short for bias-corrected bootstrap interval, in Efron (1981, 1982a). In other words, BC = BC_a, with $a = 0$. The constant z_0 is easier to obtain than the constant a, as discussed in the next section, which is why the BC interval might be used. Line 3 of Table 2 shows that for family (3.3) the BC interval is a definite improvement over the standard interval but goes only about half as far as it should toward achieving the asymmetry of the exact interval.

The Fieller situation of Table 1 is an example of a class of multiparameter problems for which $a = 0$, so the BC and BC_a intervals coincide. Efron (1985) showed that the BC intervals are second-order correct for this class, as discussed in Section 6. In general problems the full BC_a method is necessary to get second-order correctness, as shown in Section 5.

Bartlett (1953) and Schenker (1985) discussed problem (3.3). The BC_a method can be thought of as a computer-based way to carry out Bartlett's program of improved approximate confidence intervals without having to do his theoretical calculations.

Proof of Lemma 1. We begin by showing that the BC_a interval for ϕ based on $\hat{\phi}$ is correct in a certain obvious sense: notice that (3.6), (3.7) give

$$\{1 + a\hat{\phi}\} = \{1 + a\phi\}\{1 + a(Z - z_0)\}. \quad (3.11)$$

Taking logarithms puts the problem into standard translation form,

$$\hat{\zeta} = \zeta + W, \quad (3.12)$$

$\hat{\zeta} = \log\{1 + a\hat{\phi}\}$, $\zeta = \log\{1 + a\phi\}$, and $W = \log\{1 + a(Z - z_0)\}$. This example was discussed more carefully in Sections 4 and 8 of Efron (1982b), where the possibility of the bracketed terms in (3.11) being negative was dealt with. Here it will cause no trouble to assume them positive so that it is permissible to take logarithms. In fact the transformation to (3.12) is only for motivational purposes. A quicker but less informative proof of Lemma 1 is possible, working directly on the ϕ scale.

Table 2. Central 90% Confidence Intervals for θ Having Observed $\hat{\theta} \sim \theta\chi^2_{19}/19$

		R/L
1. Exact	$[.631\hat{\theta}, 1.88\hat{\theta}]$	2.38
2. BC_a ($a = .1077$)	$[.630\hat{\theta}, 1.88\hat{\theta}]$	2.37
3. BC ($a = 0$)	$[.580\hat{\theta}, 1.69\hat{\theta}]$	1.64
4. Standard (1.1)	$[.466\hat{\theta}, 1.53\hat{\theta}]$	1.00
5. Nonparametric BC_a	$[.640\hat{\theta}, 1.68\hat{\theta}]$	1.88

NOTE: The BC_a interval, with $a = .1077$, the correct value, is nearly identical to the exact interval. The BC interval, $a = 0$, is only a partial improvement over the standard interval. The nonparametric BC_a interval is discussed in Section 7.

The translation problem (3.12) gives a natural central $1 - 2\alpha$ interval for ζ having observed $\hat{\zeta}$,

$$\zeta \in [\hat{\zeta} - w^{(1-\alpha)}, \hat{\zeta} - w^{(\alpha)}], \quad (3.13)$$

where $w^{(\alpha)}$ is the $100 \cdot \alpha$ percentile point for W, $\Pr\{W < w^{(\alpha)}\} = \alpha$.

We will use the notation $\theta[\alpha]$ for the α-level endpoint of a confidence interval for a parameter θ. For instance (3.13) says that $\zeta[\alpha] = \hat{\zeta} - w^{(1-\alpha)}$, $\zeta[1-\alpha] = \hat{\zeta} - w^{(\alpha)}$. The interval (3.13) can be transformed back to the ϕ scale by the inverse mappings $\hat{\phi} = (e^{\hat{\zeta}} - 1)/a$, $\phi = (e^{\zeta} - 1)/a$, $(Z - z_0) = (e^W - 1)/a$. A little algebraic manipulation shows that the resulting interval for ϕ has α-level endpoint

$$\phi[\alpha] = \hat{\phi} + \sigma_{\hat{\phi}} \frac{(z_0 + z^{(\alpha)})}{1 - a(z_0 + z^{(\alpha)})}. \quad (3.14)$$

The cdf of $\hat{\phi}$ according to (3.6) is $\Phi((s - \phi)/\sigma_\phi + z_0)$, so the bootstrap cdf of $\hat{\phi}^*$, say $\hat{H}$, is $\hat{H}(s) = \Phi((s - \hat{\phi})/\sigma_{\hat{\phi}} + z_0)$. This has inverse $\hat{H}^{-1}(\alpha) = \hat{\phi} + \sigma_{\hat{\phi}}\{\Phi^{-1}(\alpha) - z_0\}$, which shows that $\hat{H}^{-1}(\Phi(z[\alpha]))$ equals (3.14) [see definition (3.9)]. In other words, the BC_a interval for ϕ, based on $\hat{\phi}$, coincides with the correct interval (3.14), "correct" meaning in agreement with the translation interval (3.13).

The BC_a intervals transform in the obvious way: if $\hat{\phi} = g(\hat{\theta})$, $\phi = g(\theta)$, then the BC_a interval endpoints satisfy $\phi[\alpha] = g(\theta[\alpha])$. This follows directly from (3.8), (3.9) and the relationship $\hat{H}(g(s)) = \hat{G}(s)$, equivalently $\hat{H}^{-1}(\alpha) = g(\hat{G}^{-1}(\alpha))$, between the two bootstrap cdf's. Lemma 1 has now been verified: the transformations $\hat{\theta} \to \hat{\phi} \to \hat{\zeta}$ and $\theta \to \phi \to \zeta$ reduce the problem to translation form (3.12); the inverse transformations of the natural interval (3.13) for ζ produce the BC_a interval (3.8), (3.9).

4. THE TWO CONSTANTS

The BC_a intervals require the statistician to calculate the bootstrap distribution $\hat{G}$ and also the two constants z_0 and a. The bootstrap distribution is obtained directly from (3.2). This calculation does not require knowledge of the normalizing transformation g occurring in (3.5). The two constants z_0 and a can also be obtained, or at least approximated, directly from the bootstrap distribution $f_{\hat{\theta}}(\hat{\theta}^*)$. These calculations, which are the subject of this section, assume that a transformation g to form (3.6), (3.7) exists, but do not require g to be known.

In fact the bias-correction constant z_0 is

$$z_0 = \Phi^{-1}(\hat{G}(\hat{\theta})) \quad (4.1)$$

under assumptions (3.5)–(3.7), and so can be computed directly from $\hat{G}$. To verify (4.1) notice that

$$\Pr_\phi\{\hat{\phi} < \phi\} = \Pr\{Z < z_0\} = \Phi(z_0) \quad (4.2)$$

according to (3.6). However, (3.5) gives

$$\Pr_\theta\{\hat{\theta} < \theta\} = \Pr_\phi\{\hat{\phi} < \phi\} = \Phi(z_0) \quad (4.3)$$

for every value of θ. Substituting $\theta = \hat{\theta}$ gives $\hat{G}(\hat{\theta}) = \Pr_{\hat{\theta}}\{\hat{\theta}^* < \hat{\theta}\} = \Phi(z_0)$, which is (4.1).

What about the acceleration constant a? We will show that a good approximation for a is

$$a \doteq \tfrac{1}{6}\,\mathrm{SKEW}_{\theta=\hat{\theta}}(\dot{l}_\theta), \quad (4.4)$$

where $\text{SKEW}_{\theta=\hat\theta}(X)$ indicates the skewness of a random variable X, $\mu_3(X)/\mu_2(X)^{3/2}$, evaluated at parameter point θ equal to $\hat\theta$, and $\dot l_\theta$ is the score function of the family $f_\theta(\hat\theta)$,

$$\dot l_\theta(\hat\theta) = \partial/\partial\theta \log f_\theta(\hat\theta). \tag{4.5}$$

Formula (4.4) allows us to calculate a from the form of the given density f_θ near $\theta = \hat\theta$, without knowing g. Sections 6 and 7 discuss the computation of a in families with nuisance parameters. Section 10 gives a deeper discussion of a and its relationship to other quantities of interest. See also Remark F, Section 11.

Example. For $\hat\theta \sim \theta(\chi^2_{19}/19)$, as in Table 2, standard χ^2 calculations give $\text{SKEW}(\dot l_\theta)/6 = [8/(19 \cdot 36)]^{1/2} = .1081$, which is quite close to the actual value $a = .1077$ derived in Section 10.

Here is a simple heuristic argument that indicates the role of the constant a in setting approximate confidence intervals. Suppose that $z_0 = 0$ and $a > 0$ in (3.6), (3.7). Having observed $\hat\phi = 0$, and noticing $\sigma_{\hat\phi} = 1$, the naive interval for ϕ [which is almost the same as the standard interval (1.1)] is $\phi \in [z^{(\alpha)}, z^{(1-\alpha)}]$. If, however, the statistician checks the situation at the right endpoint $z^{(1-\alpha)}$, he finds that the hypothesized standard deviation of $\hat\phi$ has increased from 1 to $1 + az^{(1-\alpha)}$. This suggests increasing the right endpoint to $z^{(1-\alpha)}(1 + az^{(1-\alpha)})$. Now the hypothesized standard deviation has further increased to $1 + az^{(1-\alpha)}(1 + az^{(1-\alpha)})$, suggesting a still larger right endpoint, and so forth. Continuing on in this way results in formula (3.14), leading to Lemma 1. [Improving the standard interval (1.1) by recomputing $\hat\sigma$ at its endpoints is a useful idea. It was brought to my attention by John Tukey, who pointed out its use by Bartlett (1953); see, e.g., Bartlett's eq. (17). Tukey's (1949) unpublished talk anticipated many of the same points.]

We call a the acceleration constant because of its effect of constantly changing the natural units of measurement as we move along the ϕ (or θ) axis. Notice that we can write (3.7) as

$$\sigma_\phi = \sigma_{\phi_0}[1 + a(\phi - \phi_0)/\sigma_{\phi_0}], \tag{4.6}$$

so

$$a = \frac{d(\sigma_\phi/\sigma_{\phi_0})}{d((\phi - \phi_0)/\sigma_{\phi_0})} \tag{4.7}$$

for any fixed value of ϕ_0. This shows that a is the relative change in σ_ϕ per unit standard deviation change in ϕ, no matter what value ϕ has.

The point $\phi_0 = 0$ is favored in definition (3.7), since σ_0 has been set equal to the convenient value 1. There is no harm in thinking of 0 as the true value of ϕ, the value actually governing the distribution of $\hat\phi$ in (3.8), because in theory we can always choose the transformation g so that this is the case and, in addition, so that $\sigma_0 = 1$ (see Remark A, Sec. 11). The restriction $1 + a\phi > 0$ in (3.7) causes no practical trouble for $|a| \le .2$, since it is then at least 5 standard deviations to the boundary of the permissible ϕ region.

The remainder of this section is devoted to verifying (4.4). The discussion is fairly technical and can be deferred until Section 10 at the reader's preference.

If we make smooth one-to-one transformations $\hat\phi = g(\hat\theta)$, $\phi = h(\theta)$, then $\dot l_\phi(\hat\phi) = \dot l_\theta(\hat\theta)/h'(\theta)$ and $\text{SKEW}(\dot l_\phi) = \text{SKEW}(\dot l_\theta)$. In other words, the right side of (4.4) is invariant under all mappings of this type. Suppose that for some choice of g and h, we can represent the family of distributions of $\hat\phi$ as

$$\hat\phi = \phi + \sigma_\phi q(Z), \qquad Z \sim N(0, 1), \tag{4.8}$$

where σ_ϕ and $q(Z)$ are functions of ϕ and z, having at least one and two derivatives, respectively, $q'(Z) > 0$. Situation (4.8), with the added conditions $q(0) = 0$, $q'(0) = 1$, is called a general scaled transformation family (GSTF) in Efron (1982b). [Please note the corrigenda to Efron (1982b).]

Lemma 2. The family (4.8) has score function $\dot l_\phi(\hat\phi)$ satisfying

$$\sigma_\phi \dot l_\phi(\hat\phi) \sim \left[Z + \frac{q''(Z)}{q'(Z)}\right]\left[\frac{1 + \dot\sigma_\phi q(Z)}{q'(Z)}\right] - \dot\sigma_\phi, \qquad Z \sim N(0, 1). \tag{4.9}$$

Here $\dot\sigma_\phi = d\sigma_\phi/d\phi$ and q' and q'' are the first two derivatives of q.

Before presenting the proof of Lemma 2, we note that it verifies (4.4): in situation (3.6), (3.7), where $\dot\sigma_\phi = a$, $q'(Z) = 1$, $q''(Z) = 0$, the distributional relationship (4.9) becomes

$$\sigma_\phi \dot l_\phi(\hat\phi) \sim (1 - az_0)\left[Z + \frac{a}{1 - az_0}(Z^2 - 1)\right]. \tag{4.10}$$

Let

$$\varepsilon_0 = \frac{a}{1 - az_0}, \tag{4.11}$$

a quantity discussed in Section 10. From the moments of $Z \sim N(0, 1)$, (4.10) gives

$$\frac{\text{SKEW}(\dot l_\phi)}{6} = \varepsilon_0 \frac{1 + \frac{4}{3}\varepsilon_0^2}{(1 + 2\varepsilon_0^2)^{3/2}}. \tag{4.12}$$

We will see in Section 10 that for the usual repeated sampling situation both a and z_0 are order of magnitude $O(n^{-1/2})$ in the sample size n. This means that $\varepsilon_0 = a \cdot [1 + O(n^{-1})]$, (4.11), and that $\text{SKEW}(\dot l_\theta)/6 = \text{SKEW}(\dot l_\phi)/6 = a[1 + O(n^{-1})]$, (4.12), justifying approximation (4.4). The "constant" a actually depends on θ, but substituting $\theta = \hat\theta$ in (4.4) causes errors only at the third-order level, like $\hat\sigma B_n^{(\alpha)}/n$ in (1.2), and so does not affect the second-order properties of the BC_a intervals.

Proof of Lemma 2. Starting from (4.8), the cdf of $\hat\phi$ is $\Phi(q^{-1}((\hat\phi - \phi)/\sigma_\phi))$, so $\hat\phi$ has density $f_\phi(\hat\phi) = \exp(-\frac{1}{2}Z_\phi^2)/(\sqrt{2\pi}\,\sigma_\phi q'(Z_\phi))$, where $Z_\phi \equiv q^{-1}((\hat\phi - \phi)/\sigma_\phi)$. This gives log-likelihood function

$$l_\phi(\hat\phi) = -\tfrac{1}{2}Z_\phi^2 - \log(q'(Z_\phi)) - \log(\sigma_\phi). \tag{4.13}$$

Lemma 2 follows by differentiating (4.13) with respect to ϕ and noting that $Z_\phi \sim N(0, 1)$ when sampling from (4.8).

5. SECOND-ORDER CORRECTNESS OF THE BC_a INTERVALS

The standard intervals are based on approximation (2.1). The BC_a intervals, which improved considerably on the standard intervals in Tables 1 and 2, are based on the more general approximation (2.3). Is it possible to go beyond (2.3), to find still further improvements over the standard intervals? The answer is no, at least not in terms of second-order asymptotics. The theorem of this section states that for simple one-parameter problems the BC_a intervals coincide through second order with the exact intervals. In terms of (1.2), the BC_a intervals have the correct second-order asymptotic form $\hat\theta + \hat\sigma(z^{(\alpha)} + A_n^{(\alpha)}/\sqrt{n} + \cdots)$.

We continue to consider the simple one-parameter problem $\hat\theta \sim f_\theta$. Suppose that the $100 \cdot \alpha$ percentile of $\hat\theta$ as a function of θ, say $\hat\theta_\theta^{(\alpha)}$, is a continuously increasing function of θ for any fixed α. In this case the usual confidence interval construction gives an exact $1 - 2\alpha$ central interval for θ having observed $\hat\theta$, say $[\theta_{Ex}[\alpha], \theta_{Ex}[1 - \alpha]]$, where $\theta_{Ex}[\alpha]$ is the value of θ satisfying $\hat\theta_\theta^{(1-\alpha)} = \hat\theta$. The exact interval in Table 2 is an example of this construction.

It is not necessary that $\hat\theta$ be the MLE of θ. In (3.6), for instance, $\hat\phi$ is not the MLE of ϕ. The BC_a method is quite insensitive to small changes in the form of the estimator (see Remark B, Sec. 11). It will be assumed, however, that $\hat\theta$ behaves asymptotically like the MLE in terms of the orders of magnitude of its bias, standard deviation, skewness, and kurtosis,

$$\hat\theta - \theta \sim (B_\theta/n, C_\theta/\sqrt{n}, D_\theta/\sqrt{n}, E_\theta/n). \quad (5.1)$$

Here n is the sample size upon which the summary statistic $\hat\theta$ is based; B_θ, C_θ, D_θ, and E_θ are bounded functions of θ (and of n, which is suppressed in the notation). Then (5.1) says that the bias of $\hat\theta$, B_θ/n, is $O(n^{-1})$, the standard deviation $C_\theta/\sqrt{n}$ is $O(n^{-1/2})$, skewness $O(n^{-1/2})$, and kurtosis $O(n^{-1})$. Higher cumulants, which are typically of order smaller than $O(n^{-1})$, will be assumed negligible in proving the results that follow (see DiCiccio 1984; Hougaard 1982).

In the simple situation $\hat\theta \sim f_\theta$, $\hat\theta$ is a sufficient statistic for θ. Later when we consider more complicated problems we will take $\hat\theta$ to be the MLE of θ. This guarantees that $\hat\theta$ is first-order efficient and asymptotically sufficient (Efron 1975).

The asymptotics of this article are stated relative to the size of the estimated standard error $\hat\sigma$ of $\hat\theta$, as in (1.2). It is often convenient in what follows to have $\hat\sigma$ be $O_p(1)$. This is easy to accomplish by transforming to $\hat\phi \equiv \sqrt{n}\hat\theta$, $\phi \equiv \sqrt{n}\theta$, so (5.1) becomes

$$\hat\phi - \phi \sim (\beta_\phi, \sigma_\phi, \gamma_\phi, \delta_\phi), \quad (5.2)$$

where $\beta_\phi = B_{\phi/n^{1/2}}/n^{1/2}$, $\sigma_\phi = C_{\phi/n^{1/2}}$, $\gamma_\phi = D_{\phi/n^{1/2}}/n^{1/2}$, and $\delta_\phi = E_{\phi/n^{1/2}}/n$. Notice that $\beta_\phi = O(n^{-1/2})$, $\dot\beta_\phi \equiv d\beta_\phi/d\phi = O(n^{-1})$, and so forth. We can just assume to begin with that $\hat\theta$ and θ are the rescaled quantities previously called $\hat\phi$ and ϕ. Then the following orders of magnitude apply:

$$\begin{array}{cccc} O(1) & O(n^{-1/2}) & O(n^{-1}) & O(n^{-3/2}) \\ \sigma_\theta & \dot\sigma_\theta, \beta_\theta, \gamma_\theta & \ddot\sigma_\theta, \dot\beta_\theta, \dot\gamma_\theta, \delta_\theta & \ddot\beta_\theta, \ddot\gamma_\theta, \dot\delta_\theta \end{array} . \quad (5.3)$$

Theorem 1. If $\hat\theta$ has bias β_θ, standard error σ_θ, skewness γ_θ, and kurtosis δ_θ satisfying (5.3), then the BC_a intervals are second-order correct.

The theorem states that $\theta_{BC_a}[\alpha]$, the α endpoint of the BC_a interval, is asymptotically close to the exact endpoint,

$$(\theta_{BC_a}[\alpha] - \theta_{Ex}[\alpha])/\hat\sigma = O_p(n^{-1}). \quad (5.4)$$

This is not true for the standard intervals (1.1) or the BC intervals, $a = 0$. The proof of Theorem 1, which appears in Section 12, makes it clear that all three of the elements in (2.3), the transformation g, the bias-correction constant z_0, and the acceleration constant a, make necessary corrections of $O_p(n^{-1/2})$ to the standard intervals.

6. NUISANCE PARAMETERS

The discussion so far has centered on the simple case $\hat\theta \sim f_\theta$, where we have only a real-valued parameter θ and a real-valued summary statistic $\hat\theta$ from which we are trying to construct a confidence interval for θ. We have been able to show favorable properties of the BC_a intervals for the simple case, but of course the simple case is where we least need a general method like the bootstrap.

This section discusses the more difficult situation where there are nuisance parameters besides the parameter of interest θ. Section 7 discusses the nonparametric situation, where the number of nuisance parameters is effectively infinite. Because of the inherently simple nature of the bootstrap it will be easy to extend the BC_a method to cover these cases, though we will not be able to provide as strong a justification for the correctness of the resulting intervals.

Suppose then that the data $\mathbf{y}$ comes from a parametric family $\mathcal{F}$ of density functions $f_{\boldsymbol\eta}$, say $\mathbf{y} \sim f_{\boldsymbol\eta}$, where $\boldsymbol\eta$ is an unknown vector of parameters, and we want a confidence interval for the real-valued parameter $\theta = t(\boldsymbol\eta)$. In Efron (1985), the multivariate normal case $\mathbf{y} \sim N_k(\boldsymbol\eta, \mathbf{I})$ is examined in detail.

From $\mathbf{y}$ we obtain $\hat{\boldsymbol\eta}$, the MLE of $\boldsymbol\eta$, and $\hat\theta = t(\hat{\boldsymbol\eta})$, the MLE of θ. The parametric bootstrap distribution of $\mathbf{y}$ is defined to be

$$\mathbf{y}^* \sim f_{\hat{\boldsymbol\eta}}, \quad (6.1)$$

the distribution of the data when $\boldsymbol\eta$ equals $\hat{\boldsymbol\eta}$. From $\mathbf{y}^*$ we obtain $\hat{\boldsymbol\eta}^*$, the bootstrap MLE of $\boldsymbol\eta$, and then $\hat\theta^* = t(\hat{\boldsymbol\eta}^*)$.

The distribution of $\hat\theta^*$ under model (6.1) is the parametric boostrap distribution of $\hat\theta$, generalizing (3.1). This gives the bootstrap cdf

$$\hat G(s) = \Pr_{\hat{\boldsymbol\eta}}\{\hat\theta^* < s\}, \quad (6.2)$$

as in (3.2). The bias-correction constant z_0 equals $\Phi^{-1}(\hat G(\hat\theta))$, as in (4.1).

To compute the BC_a intervals (3.8), (3.9), we also need to know the appropriate value of the acceleration constant a. We will find a by following Stein's (1956) construction, which replaces the multiparameter family $\mathcal{F} = \{f_{\eta}\}$ by a *least favorable* one-parameter family $\hat{\mathcal{F}}$.

Let $\dot{l}_{\eta}$ be the vector with ith coordinate $\partial/\partial\eta_i \log f_{\eta}(\mathbf{y})$, so $\dot{l}_{\hat{\eta}}(\mathbf{y}) = 0$ by definition of the MLE $\hat{\boldsymbol{\eta}}$, and let $\ddot{l}_{\hat{\eta}}$ be the $k \times k$ matrix with ijth entry $\partial^2/(\partial\eta_i\partial\eta_j) \log f_{\eta}(\mathbf{y})|_{\eta=\hat{\eta}}$. In addition, let $\hat{\nabla}$ be the gradient vector of $\theta = t(\boldsymbol{\eta})$ evaluated at the MLE, $\hat{\nabla}_i = (\partial/\partial\eta_i)t(\boldsymbol{\eta})|_{\eta=\hat{\eta}}$. The *least favorable direction* at $\boldsymbol{\eta} = \hat{\boldsymbol{\eta}}$ is defined to be

$$\hat{\boldsymbol{\mu}} \equiv (-\ddot{\mathbf{l}}_{\hat{\eta}})^{-1}\hat{\nabla}. \tag{6.3}$$

Then the least favorable family $\hat{\mathcal{F}}$ is the one-parameter subfamily of $\mathcal{F}$ passing through $\hat{\boldsymbol{\eta}}$ in the direction $\hat{\boldsymbol{\mu}}$,

$$\hat{\mathcal{F}} = \{\hat{f}_{\lambda}(\mathbf{y}^*) \equiv f_{\hat{\eta}+\lambda\hat{\mu}}(\mathbf{y}^*)\}. \tag{6.4}$$

Using $\mathbf{y}^*$ to denote a hypothetical data vector from $\hat{f}_{\lambda}$ is intended to avoid confusion with the actual data vector $\mathbf{y}$ that gave $\hat{\boldsymbol{\eta}}$; $\hat{\boldsymbol{\eta}}$ and $\hat{\boldsymbol{\mu}}$ are fixed in (6.4), only λ being unknown.

Consider the problem of estimating $\theta(\lambda) \equiv t(\hat{\boldsymbol{\eta}} + \lambda\hat{\boldsymbol{\mu}})$ having observed $\mathbf{y}^* \sim \hat{f}_{\lambda}$. The Fisher information bound for an unbiased estimate of θ in this one-parameter family evaluated at $\lambda = 0$ is $\hat{\nabla}'(-\ddot{\mathbf{l}}_{\hat{\eta}})^{-1}\hat{\nabla}$, which is the same as the corresponding bound for estimating $\theta = t(\boldsymbol{\eta})$, at $\boldsymbol{\eta} = \hat{\boldsymbol{\eta}}$, in the multiparameter family $\mathcal{F}$. This is Stein's reason for calling $\hat{\mathcal{F}}$ least favorable.

We will use $\hat{\mathcal{F}}$ to calculate an approximate value for the acceleration constant a,

$$a \doteq \{\text{SKEW}_{\lambda=0}[\partial \log f_{\hat{\eta}+\lambda\hat{\mu}}(\mathbf{y}^*)/\partial\lambda]/6\}. \tag{6.5}$$

This is formula (4.4) applied to $\hat{\mathcal{F}}$, assuming that $\hat{\lambda} = 0$ (which is the MLE of λ in $\hat{\mathcal{F}}$ when $\mathbf{y}^* = \mathbf{y}$, the actual data vector). See Remark F, Section 11.

Formula (6.5) is especially simple in the exponential family case where the densities $f_{\eta}(\mathbf{y})$ are of the form

$$f_{\eta}(\mathbf{y}) = e^{n[\eta'\mathbf{y}-\psi(\eta)]}f_0(\mathbf{y}). \tag{6.6}$$

The factor n in the exponent of (6.6) is not necessary, but it is included to agree with the situation where the data consists of iid observations $\mathbf{x}_1, \mathbf{x}_2, \ldots, \mathbf{x}_n$, each with density $\exp(\boldsymbol{\eta}'\mathbf{x} - \psi(\boldsymbol{\eta}))$, and $\mathbf{y}$ is the sufficient vector $\sum_{i=1}^{n}\mathbf{x}_i/n$.

Lemma 3. For the exponential family (6.6), formula (6.5) gives

$$a = \frac{1}{6\sqrt{n}}\frac{\hat{\psi}^{(3)}(0)}{(\hat{\psi}^{(2)}(0))^{3/2}}, \tag{6.7}$$

where

$$\hat{\psi}^{(j)}(0) = \left.\frac{\partial^j\psi(\hat{\boldsymbol{\eta}} + \lambda\hat{\boldsymbol{\mu}})}{\partial\lambda^j}\right|_{\lambda=0}. \tag{6.8}$$

Proof. We have

$$\left.\frac{\partial \log f_{\hat{\eta}+\lambda\hat{\mu}}(\mathbf{y}^*)}{\partial\lambda}\right|_{\lambda=0} = n\hat{\boldsymbol{\mu}}'(\mathbf{y}^* - \dot{\psi}(\hat{\boldsymbol{\eta}})), \tag{6.9}$$

so $\text{SKEW}_{\lambda=0}[(\partial \log f_{\hat{\eta}+\lambda\hat{\mu}}(\mathbf{y}^*))/\partial\lambda]$ equals the skewness of $\hat{\boldsymbol{\mu}}'\mathbf{y}^*$ for $\mathbf{y}^* \sim f_{\hat{\eta}}$. The fact that $\text{SKEW}(\hat{\boldsymbol{\mu}}'\mathbf{y}^*)$ equals $[\hat{\psi}^{(3)}(0)/(\hat{\psi}^{(2)}(0))^{3/2}]/\sqrt{n}$ is a standard exercise in exponential family theory. Note that Lemma 3 applies to $\mathbf{y} \sim N_k(\boldsymbol{\eta}, \mathbf{I})$, the case considered in Efron (1985), and gives $a = 0$, which is why the unaccelerated BC intervals worked well there.

Table 3 relates to the following example:

$$\mathbf{y} \sim N_4(\boldsymbol{\eta}, \sigma_{\eta}^2\mathbf{I}), \qquad [\sigma_{\eta} = 1 + a(\|\eta\| - 8)], \tag{6.10}$$

where we observe $\mathbf{y} = (8, 0, 0, 0)$ and wish to set confidence intervals for the parameter $\theta = t(\boldsymbol{\eta}) = \|\eta\|$. The case $a = 0$ amounts to finding a confidence interval for the noncentrality parameter of a noncentral χ^2 distribution and can be solved exactly. The theory of Efron (1985) applies to the $a = 0$ case, and we see that the BC_0 interval, that is, the BC interval, well-matches the exact interval.

Table 3 shows the result of varying the constant a from .10 to −.10. This example has a particularly simple geometry: the sphere $C_{\hat{\theta}} = \{\boldsymbol{\eta} : \|\boldsymbol{\eta}\| = \hat{\theta}\}$ is the set of $\boldsymbol{\eta}$ vectors having $t(\boldsymbol{\eta})$ equal to the MLE value $\hat{\theta} = t(\hat{\boldsymbol{\eta}})$; the least favorable direction $\hat{\boldsymbol{\mu}}$ is orthogonal to $C_{\hat{\theta}}$ at $\hat{\boldsymbol{\eta}}$; the distribution of $\hat{\theta}$ is nearly normal (see Efron 1985, Table 2), with standard deviation changing in the least favorable direction at a rate nearly equal to a, as in (4.7). The BC_a intervals alter predictably with a. For instance, comparing the upper endpoint at $a = .10$ with $a = 0$, notice that $(9.70 - 8.00)/(9.44 - 8.00) = 1.18$, closely matching the expansion factor due to acceleration, $1 + .10 \cdot 1.645 = 1.16$.

We could disguise problem (6.10) by making nonlinear transformations

$$\tilde{\mathbf{y}} = g(\mathbf{y}), \qquad \tilde{\boldsymbol{\eta}} = h(\boldsymbol{\eta}), \tag{6.11}$$

in which case the geometry of the BC_a intervals might not be obvious from the form of the parameter $\theta = t(h^{-1}(\tilde{\boldsymbol{\eta}})) = \|h^{-1}(\tilde{\boldsymbol{\eta}})\|$ and the transformed densities $\tilde{f}_{\tilde{\eta}}(\mathbf{y})$. However, the BC_a method is invariant under such transformations (see Remark C, Sec. 11), so the statistician would automatically get the same intervals as if he knew the normalizing transformations $\mathbf{y} = g^{-1}(\tilde{\mathbf{y}})$, $\boldsymbol{\eta} = h^{-1}(\tilde{\boldsymbol{\eta}})$.

Currently we cannot justify the BC_a method as being second-order correct in the multiparameter context of this section, though it seems a likely conjecture that this is so. We know that it is so in the one-parameter case (see Sec. 5) and in the restricted multiparameter case of Efron (1985), where the BC_a and BC methods coincide, and that the BC_a method makes a rather obvious correction to the BC interval in the general multiparameter case.

Table 3. Central 90% Confidence Intervals for $\theta = \|\eta\|$, Having Observed $\|y\| = 8$, From the Parametric Family $y \sim N_4(\eta, \sigma_\eta^2 I)$, With $\sigma_\eta = 1 + a(\|\eta\| - 8)$

	Exact	(R/L)	BC_a	(R/L)	(6.5)
$a = .10$	[6.46, 9.69]	(.96)	[6.47, 9.70]	(.97)	.0984
$a = .05$	[6.32, 9.57]	(.85)	[6.34, 9.56]	(.84)	.0498
$a = 0$	[6.14, 9.47]	(.74)	[6.19, 9.44]	(.75)	0
$a = -.05$	[5.92, 9.38]	(.65)	[6.03, 9.35]	(.66)	−.0498
$a = -.10$	[5.62, 9.30]	(.56)	[5.89, 9.27]	(.60)	−.0984

NOTE: The standard interval (1.1) is [6.36, 9.64] for all values of a. The last column shows that (6.5) nearly equals the constant a in this case. The exact intervals are based on the noncentral χ^2 distribution.

7. THE NONPARAMETRIC CASE

This section concerns the nonparametric case where the data $\mathbf{y} = (x_1, x_2, \ldots, x_n)$ consist of n iid observations x_i that may have come from any probability distribution F on their common sample space $\mathcal{X}$. There is a real-valued parameter $\theta = t(F)$ for which we desire an approximate confidence interval. We will show how the BC_a method can be used to provide such an interval based on the obvious nonparametric estimate $\hat{\theta} = t(\hat{F})$. Here $\hat{F}$ is the empirical probability distribution of the sample, putting mass $1/n$ on each observed value x_i.

A bootstrap sample $\mathbf{y}^* \sim \hat{F}$ consists in this case of an iid sample of size n from $\hat{F}$, say $\mathbf{y}^* = (x_1^*, x_2^*, \ldots, x_n^*)$. In other words, $\mathbf{y}^*$ is a random sample of size n drawn with replacement from $\{x_1, x_2, \ldots, x_n\}$. The bootstrap sample $\mathbf{y}^*$ gives a bootstrap replication of $\hat{\theta}$, $\hat{\theta}^* = t(\hat{F}^*)$, where $\hat{F}^*$ puts mass $1/n$ on each x_i^*. The bootstrap cdf $\hat{G}(s)$ is the probability that a bootstrap replication is less than s,

$$\hat{G}(s) = \Pr_{\hat{F}}\{\hat{\theta}^* < s\}, \tag{7.1}$$

as in (6.2) and (3.2). The bias-correction constant z_0 equals $\Phi^{-1}(\hat{G}(\hat{\theta}))$, as in (4.1).

For most nonparametric problems the bootstrap cdf $\hat{G}$ has to be determined by Monte Carlo sampling. Section 9 discusses how many Monte Carlo replications of $\hat{\theta}^*$ are necessary. Here we will continue to assume that $\hat{G}$ has been computed exactly—in effect, that we have taken an infinite number of bootstrap replications $\hat{\theta}^*$.

At this point we could use $\hat{G}$ to form the BC interval for θ, but to obtain the BC_a interval (3.8), (3.9) we also need the value of the acceleration constant a. We will derive a simple approximation for a, based on Lemma 3. It depends on

$$U_i = \lim_{\Delta \to 0} \frac{t((1-\Delta)\hat{F} + \Delta\delta_i) - t(\hat{F})}{\Delta}, \qquad i = 1, 2, \ldots, n, \tag{7.2}$$

the *empirical influence function* of $\hat{\theta} = t(\hat{F})$. Here δ_i is a point mass at x_i, so U_i is the derivative of the estimate $\hat{\theta}$ with respect to the mass on point x_i. [Jaeckel's infinitesimal jackknife estimate of standard error for $\hat{\theta}$ is $(\Sigma_1^n U_i^2)^{1/2}/n$.] Definition (7.2) assumes that $t(F)$ is smoothly defined for choices of F near $\hat{F}$ [see Efron 1982a, (6.3), or Efron 1979, sec. 5]. Note that $\Sigma_1^n U_i = 0$.

The next section shows that Lemma 3, applied to a family appropriate to the nonparametric situation, gives the following approximation for the constant a,

$$a \doteq \frac{1}{6}\left[\left(\sum_{i=1}^{n} U_i^3\right) \Big/ \left(\sum_{i=1}^{n} U_i^2\right)^{3/2}\right]. \tag{7.3}$$

This is a convenient formula since the U_i can be evaluated easily by using finite differences in definition (7.2).

Example 1: The Law School Data. Table 4 shows two indexes of student excellence, LSAT and GPA, for each of 15 American law schools (see Efron 1982a, sec. 2.2). The Pearson correlation coefficient $\hat{\rho}$ between LSAT and GPA equals .776; we want a confidence interval for the true correlation ρ. Table 4 also shows the values of U_i for the statistic $\hat{\rho}$, from which formula (7.3) produces $a \doteq -.0817$. $B = 100{,}000$ bootstrap replications (about 100 times more than actually needed; see Sec. 9) gave $\hat{G}(\hat{\theta}) = .463$, and so $z_0 = -.0927$. Using these values of a and z_0 in (3.8), (3.9) resulted in the central 90% nonparametric BC_a interval [.43, .92] for ρ. The usual bivariate normal interval, based on Fisher's $\tanh^{-1}$ transformation, is [.49, .90]. This is also the *parametric* BC_a interval based on the simple family $\hat{\rho} \sim f_\rho$, where $f_\rho(\hat{\rho})$ is Fisher's density function for the correlation coefficient from bivariate normal data. The standard interval (1.1), $\hat{\rho} \pm 1.645\hat{\sigma}$, using the bootstrap estimate $\hat{\sigma} = .133$, is [.56, .99].

Table 4. The Law School Data and Values of the Empirical Influence Function for the Correlation Coefficient $\hat{\rho}$

i	(LSAT, GPA)	U_i	i	(LSAT, GPA)	U_i
1	(576, 3.39)	−1.507	9	(651, 3.36)	.310
2	(635, 3.30)	.168	10	(605, 3.13)	.004
3	(558, 2.81)	.273	11	(653, 3.12)	−.526
4	(578, 3.03)	.004	12	(575, 2.74)	−.091
5	(666, 3.44)	.525	13	(545, 2.76)	.434
6	(580, 3.07)	−.049	14	(572, 2.88)	.125
7	(555, 3.00)	−.100	15	(594, 2.96)	−.048
8	(661, 3.43)	.477			

Formula (7.3) is invariant under monotone changes of the parameter of interest. This results in the BC_a intervals having correct transformation properties. Suppose, for example, that we change parameters from ρ to $\phi = g(\rho) \equiv \tanh^{-1}(\rho)$, with corresponding nonparametric estimate $\hat{\phi} = g(\hat{\rho})$. The central 90% BC_a interval for ϕ based on $\hat{\phi}$ is then the obvious transformation of the interval for θ based on $\hat{\theta}$, $[g(.43), g(.92)] = [.46, 1.59]$. This compares with Fisher's $\tanh^{-1}$ interval $[g(.49), g(.90)] = [.54, 1.47]$ and the standard interval $\hat{\phi} \pm 1.645\hat{\sigma}_\phi = [.49, 1.59]$. The standard interval is much more reasonable-looking on the $\tanh^{-1}$ scale, as we might expect from Fisher's transformation theory. As commented before, a major advantage of the BC_a method is that the statistician need not know the correct scale on which to work. In effect the method effectively selects the best (most normal) scale and then transforms the interval back to the scale of interest.

Example 2: The Mean. Suppose that F is a distribution on the real line, and $\theta = t(F)$ equals the expectation $E_F X$. The empirical influence function $U_i = (x_i - \bar{x})$, so (7.3) gives

$$\begin{aligned} a &= \tfrac{1}{6}\{\textstyle\sum (x_i - \bar{x})^3/[\sum (x_i - \bar{x})^2]^{3/2}\} \\ &= (1/6\sqrt{n})(\hat{\mu}_3/\hat{\mu}_2^{3/2}) = \hat{\gamma}/6\sqrt{n}. \end{aligned} \tag{7.4}$$

Here $\hat{\mu}_h = \Sigma (x_i - \bar{x})^h/n$, the hth sample central moment, and $\hat{\gamma} = \hat{\mu}_3/\hat{\mu}_2^{3/2}$, the sample skewness. It turns out also that $z_0 \doteq \hat{\gamma}/6\sqrt{n}$ in this case, by standard Edgeworth arguments. Both a and z_0 are typically of order $n^{-1/2}$.

Because the sample mean is such a simple statistic, we can use Edgeworth methods to get asymptotic expressions for the α-level endpoint of the BC_a interval:

$$\theta_{\mathrm{BC}_a}[\alpha] = \bar{x} + \hat{\sigma}\{z^{(\alpha)} + (\hat{\gamma}/6\sqrt{n})(2z^{(\alpha)^2} + 1) + O_p(n^{-1})\}, \tag{7.5}$$

$\hat{\sigma} \equiv (\hat{\mu}_2/n)^{1/2}$. This compares with

$$\theta_{BC}[\alpha] \doteq \bar{x} + \hat{\sigma}\{z^{(\alpha)} + (\hat{\gamma}/6\sqrt{n})(z^{(\alpha)^2} + 1) + O_p(n^{-1})\}, \tag{7.6}$$

for the BC interval, so the BC_a intervals are shifted approximately $(\hat{\gamma}/6\sqrt{n})z^{(\alpha)^2}$ further right.

Johnson (1978) suggested modifying the usual t statistic $T = (\bar{x} - \theta)/\hat{\sigma}$ to $T_J = T + (\hat{\gamma}/6\sqrt{n})(2T^2 + 1)$ and then considering T_J to have a standard t_{n-1} distribution in order to obtain confidence intervals for $\theta = E_F X$. Efron (1981, sec. 10) showed that this is much like using the bootstrap distribution of $T^* = (\bar{x}^* - \bar{x})/\hat{\sigma}^*$ as a pivotal quantity. Interestingly enough, *the Edgeworth expansion of $\theta_J[\alpha]$, the α endpoint of Johnson's interval, coincides with (7.5).* The BC_a method makes a "t correction" in the case of $\theta = E_F X$, but it is not the familiar Student-t correction, which operates at third order in (1.2), but rather a second-order correction, coming from the correlation between $\bar{x}$ and $\hat{\sigma}$ in nonnormal populations (see Remark D, Sec. 11).

I conjecture that the nonparametric BC_a intervals will be second-order correct for any parameter θ. There is no proof of this, a major difficulty being the definition of second-order correctness in the nonparametric situation. Whether or not it is true, small-sample nonparametric confidence intervals are far from well understood and, as emphasized in Schenker (1985), should be interpreted with some caution.

Example 3: The Variance. Suppose that $\mathfrak{X}$ is the real line and $\theta = \text{var}_F X$, the variance. Line 5 of Table 2 shows the result of applying the nonparametric BC_a method to data sets $x_1, x_2, \ldots, x_{20}$, which were actually iid samples from an $N(0, 1)$ distribution. The number .640, for example, is the average of $\theta_{BC_a}[.05]/\hat{\theta}$ over 40 such data sets, $B = 4{,}000$ bootstrap replications per data set. The upper limit $1.68 \cdot \hat{\theta}$ is noticeably small. The reason is simple: the nonparametric bootstrap distribution of $\hat{\theta}^*$ has a short upper tail compared with the parametric bootstrap distribution, which is a scaled χ^2_{19} random variable. The results of Beran (1984a), Bickel and Freedman (1981), and Singh (1981) show that the nonparametric bootstrap distribution is highly accurate asymptotically, but of course that is not a guarantee of good small-sample behavior. Bootstrapping from a smoothed version of $\hat{F}$, as in Efron (1982a, sec. 5.3), alleviates the problem in this particular example.

8. GEOMETRY OF THE NONPARAMETRIC CASE

Formula (7.3), which allows us to apply the BC_a method nonparametrically, is based on a simple heuristic argument: instead of the actual sample-space $\mathfrak{X}$ of the data points x_i, consider only distributions F supported on $\hat{\mathfrak{X}} = \{x_1, x_2, \ldots, x_n\}$, the observed data set. This is an n-category multinomial family, to which the results of Section 6 can be applied. Because the multinomial is an exponential family, Lemma 3 directly gives (7.3).

We will now examine this argument more carefully, with the help of a simple geometric representation. See Efron (1981, sec. 11) for further discussion of this approach to nonparametric confidence intervals.

A typical distribution supported on $\hat{\mathfrak{X}}$ is

$$F(\mathbf{w}) : \text{mass } w_i \text{ on } x_i, \tag{8.1}$$

where $\mathbf{w} = (w_1, w_2, \ldots, w_n)$ can be any vector in the simplex $\mathcal{S}_n = \{\mathbf{w} : w_i \geq 0\ \forall i, \sum_1^n w_i = 1\}$. The parameter $\theta = t(F)$ is defined on $\mathcal{S}_n$ by $\theta(\mathbf{w}) = t(F(\mathbf{w}))$. The central point of the simplex,

$$\mathbf{w}^O \equiv \mathbf{1}/n = (1/n, 1/n, \ldots, 1/n), \tag{8.2}$$

corresponds to $F(\mathbf{w}^O) = \hat{F}$, the usual empirical distribution; $\theta(\mathbf{w}^O) = \hat{\theta} = t(\hat{F})$, the nonparametric MLE of θ. The curved surface

$$\mathcal{C}_{\hat{\theta}} = \{\mathbf{w} : \theta(\mathbf{w}) = \theta(\mathbf{w}^O) = \hat{\theta}\} \tag{8.3}$$

comprises those distributions $F(\mathbf{w})$ having $\theta(\mathbf{w}) = \hat{\theta}$. *The vector $\mathbf{U}_i$ is orthogonal to $\mathcal{C}_{\hat{\theta}}$ at $\mathbf{w}^O$*, as shown in Figure 1, which follows from definition (7.2) of the empirical influence function. $\mathbf{U}$ is essentially the gradient of $\theta(\mathbf{w})$ at $\mathbf{w}^O$ (see Efron 1982a, sec. 6.3).

With $\mathbf{w}$ unknown, but $\hat{\mathfrak{X}} = \{x_1, \ldots, x_n\}$ considered fixed, one can imagine setting a confidence interval for $\theta(\mathbf{w})$ on the basis of a hypothetical sample $x_1^*, x_2^*, \ldots, x_n^* \overset{\text{iid}}{\sim} F(\mathbf{w})$. A sufficient statistic is the vector of proportions $P_i = \#\{x_j^* = x_i\}/n$, say $\mathbf{P} = (P_1, P_2, \ldots, P_n)$, with distribution

$$\mathbf{P} \sim \text{mult}_n(n, \mathbf{w})/n, \qquad \mathbf{w} \in \mathcal{S}_n. \tag{8.4}$$

The notation here indicates n draws from an n-category multinomial, having probability w_i for category i. We suppose that we have observed $\mathbf{P} = \mathbf{w}^O$ in (8.4), that is, that the hypothetical sample $x_1^*, \ldots, x_n^*$ equals the actual sample $x_1, \ldots, x_n$.

Distributions (8.4) form an n-parameter exponential family (6.6) with $\mathbf{y} = \mathbf{P}$, $\eta_i = \log(nw_i) + c$, and $\psi(\eta) = \log(\sum_1^n \exp(\eta_i)/n)$. Here c can be any constant, since all vectors $\boldsymbol{\eta} + c\mathbf{1}$ correspond to the same probability vector $\mathbf{w}$, namely $w_i = \exp(\eta_i)/\sum_1^n \exp(\eta_j)$.

If one accepts the reduction of the original nonparametric problem to (8.4), with observed value $\mathbf{P} = \mathbf{w}^O$, then it is easy to carry through the least favorable family calculations (6.3)–(6.5): (i) $\hat{\boldsymbol{\eta}} = \mathbf{0}$; (ii) $\hat{\boldsymbol{\mu}} = \mathbf{U}$; (iii) $\hat{f}_\lambda$ is the member of (7.4) corresponding to $\hat{\boldsymbol{\eta}} + \lambda\hat{\boldsymbol{\mu}} = \lambda\mathbf{U}$, namely

$$\mathbf{P}^* \sim \text{mult}(n, \mathbf{w}^\lambda)/n, \; w_i^\lambda = \exp(\lambda U_i) \Big/ \sum_{j=1}^n \exp(\lambda U_j); \tag{8.5}$$

(iv) finally, formula (7.3) follows directly from Lemma 3, by differentiating $\hat{\psi}(\lambda) = \log(\sum_1^n \exp(\lambda' J_j)/n)$ (and remembering that $\sum U_i = 0$).

Only step (ii) is not immediate, but it is a straightforward consequence of definition (6.3) and standard properties of the multinomial. It has already been noted that $\mathbf{U}$ is orthogonal to $\mathcal{C}_{\hat{\theta}}$, so $\mathbf{U}$ is proportional to $\hat{\nabla}$ in (6.3). However, $-\hat{\mathbf{I}}_{\hat{\eta}} = \mathbf{I} - \mathbf{1}\mathbf{1}'/n$, which has pseudo-inverse $\mathbf{I}$. Thus $\hat{\boldsymbol{\mu}}$ is proportional to $\mathbf{U}$. Since (6.7), (6.8) produce the same value of a if $\hat{\boldsymbol{\mu}}$ is multiplied by any constant, this in effect gives $\hat{\boldsymbol{\mu}} = \mathbf{U}$.

An interesting case that provides some support for the

nonparametric BC_a method is that where the sample space is finite to begin with, say $\mathcal{X} = \{1, 2, \ldots, L\}$. A typical distribution on $\mathcal{X}$ is $\mathbf{f} = (f_1, \ldots, f_L)$, where $f_l = \Pr\{x_i = l\}$. The observed sample proportions $\hat{\mathbf{f}} = (\hat{f}_1, \hat{f}_2, \ldots, \hat{f}_L)$, $\hat{f}_l \equiv \#\{x_i = l\}/n$, are sufficient, with distribution $\hat{\mathbf{f}} \sim \text{mult}_L(n, \mathbf{f})/n$. This is an L-parameter exponential family, so the theory of Section 6 applies. It turns out that Lemma 3 agrees with formula (7.3) in this case. *Nonparametric* BC_a *intervals are the same as parametric* BC_a *intervals when* $\mathcal{X}$ *is finite.* See remarks G and H of Efron (1979) for the first-order bootstrap asymptotics of finite sample spaces.

Family (8.4) was used in Section 11 of Efron (1981) to motivate a method called *nonparametric tilting*, a nonparametric analog of the standard hypothesis-testing approach to confidence interval construction. The one-parameter tilting family, (11.12) of Efron (1981), is closely related to the least favorable family $\hat{\mathcal{F}}$ in Figure 1. Efron (1981, table 5) considered samples of size $n = 15$ for the one-sided exponential density $f(x) = \exp[-(x + 1)]$ $(x > -1)$. Central 90% tilting intervals for $\theta = E_F X$ were constructed for each of 10 such samples, averaging $[-.34, .50]$. The corresponding nonparametric BC_a intervals averaged $[-.34, .52]$ and were quite similar to the tilting intervals on a sample-by-sample comparison. The nonparametric BC_a method is computationally simpler than nonparametric tilting and seems likely to give similar results in most problems.

We end this section with a useful approximation formula for the bias-correction constant z_0, developed jointly with Timothy Hesterberg. In addition to (7.2) we need the second-order empirical influence function

$$V_{ij} = \lim_{\Delta \to 0} \{[t((1 - 2\Delta)\hat{F} + \Delta\delta_i + \Delta\delta_j) - t((1 - \Delta)\hat{F} + \Delta\delta_i) - t((1 - \Delta)\hat{F} + \Delta\delta_j) + t(\hat{F})]/\Delta^2\}. \tag{8.6}$$

Define $z_{01} \equiv (\frac{1}{6}) \sum_1^n U_i^3/(\sum_1^n U_i^2)^{3/2}$ [approximation (7.3) for a] and

$$z_{02} \equiv \left[\frac{\mathbf{U}'\mathbf{V}\mathbf{U}}{\|\mathbf{U}\|^2} - \text{tr}\,\mathbf{V}\right]/(2n\|\mathbf{U}\|), \tag{8.7}$$

where $\mathbf{V}$ is the $n \times n$ matrix (V_{ij}).

Lemma 4. The bias-correction constant z_0 approximately equals

$$\Phi^{-1}\{2\Phi(z_{01})\Phi(z_{02})\}. \tag{8.8}$$

For the law school data, Example 1 of Section 7, $z_{01} = -.0817$ and $z_{02} = -.0067$, giving $z_0 = -.0869$ from (8.8), compared with $z_0 = -.0927 \pm .0039$ from $B = 100{,}000$ bootstrap replications.

The term z_{01} relates to skewness in $\hat{\mathcal{F}}$, and z_{02} is a geometric term arising from the curvature of $\mathcal{C}_{\hat{\theta}}$ at $\mathbf{w}^O$. It is analogous to formula (A15) of Efron (1985). Lemma 4 will not be proved here but is important in the sample size considerations of Section 9.

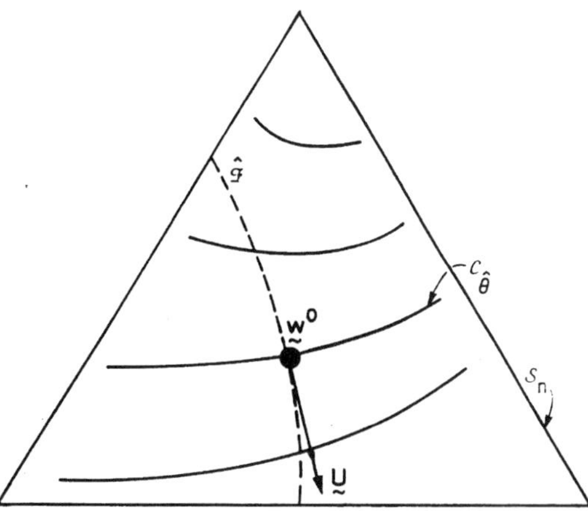

Figure 1. All probability distributions supported on $\{x_1, x_2, \ldots, x_n\}$ are represented as the simplex $\mathcal{S}_n$. The central point $\mathbf{w}^O$ corresponds to the empirical distribution $\hat{F}$. The curves indicate level surfaces of constant value of the parameter θ. In particular $\mathcal{C}_{\hat{\theta}}$ comprises those probability distributions having θ equal to $\theta(\mathbf{w}^O) = \hat{\theta}$, the MLE. The least favorable family $\hat{\mathcal{F}}$ passes through $\mathbf{w}^O$ in the direction $\mathbf{U}$, orthogonal to $\mathcal{C}_{\hat{\theta}}$.

9. BOOTSTRAP SAMPLE SIZES

How many bootstrap replications of $\hat{\theta}^*$ need we take? So far we have pretended that the number of replications $B = \infty$, but if Monte Carlo methods are necessary to obtain the bootstrap cdf $\hat{G}$, then B must be finite, usually the smaller the better. This section gives rough estimates of how small B may be taken in practice. The results are presented without proof, all being standard exercises in error estimation (see, e.g., Kendall and Stuart 1958, chap. 10). They apply to any situation, parametric or nonparametric, where $\hat{G}$ is obtained by Monte Carlo sampling.

First consider the easy problem of estimating the standard error of $\hat{\theta}$ via the bootstrap. The bootstrap estimate based on B replications, $\hat{\sigma}_B = [\sum_{b=1}^B (\hat{\theta}_b^* - \hat{\theta}^{\cdot *})^2/(B - 1)]^{1/2}$, has conditional coefficient of variation (standard deviation divided by expectation)

$$\text{CV}\{\hat{\sigma}_B \mid \mathbf{y}\} \doteq [(\hat{\delta} + 2)/4B]^{1/2}, \tag{9.1}$$

where $\hat{\delta}$ is the kurtosis of the bootstrap distribution $\hat{G}$. The notation indicates that the observed data $\mathbf{y}$ is fixed in this calculation. As $B \to \infty$, then (9.1) $\to 0$ and $\hat{\sigma}_B \to \hat{\sigma}$, the ideal bootstrap estimate of standard error.

Of course $\hat{\sigma}$ itself will usually not estimate the true standard error $\sigma \equiv \text{SD}_\theta\{\hat{\theta}\}$ perfectly. Let $\text{CV}(\hat{\sigma})$ be the coefficient of variation of $\hat{\sigma}$, unconditional now, averaging over the possible realizations of $\mathbf{y}$ [e.g., if $n = 20$, $\hat{\theta} = \bar{x}$, $x_i \overset{\text{iid}}{\sim} N(0, 1)$, then $\text{CV}(\hat{\sigma}) \doteq (1/40)^{1/2} = .16$]. The unconditional CV of $\hat{\sigma}_B$ is then approximated by

$$\text{CV}(\hat{\sigma}_B) \doteq \left[\text{CV}^2(\hat{\sigma}) + \frac{E\hat{\delta} + 2}{4B}\right]^{1/2}. \tag{9.2}$$

Table 5 displays $\text{CV}(\hat{\sigma}_B)$ for various choices of B and $\text{CV}(\hat{\sigma})$, assuming that $E\hat{\delta} = 0$. For values of $\text{CV}(\hat{\sigma}) \geq$

Table 5. Coefficient of Variation of $\hat{\sigma}_B$, the Bootstrap Estimate of Standard Error, as a Function of B, the Number of Bootstrap Replications, and CV($\hat{\sigma}$), the Limiting CV as $B \to \infty$

	$B \to$				
CV($\hat{\sigma}$)	25	50	100	200	∞
.25	.29	.27	.26	.25	.25
.20	.24	.22	.21	.21	.20
.15	.21	.18	.17	.16	.15
.10	.17	.14	.12	.11	.10
.05	.15	.11	.09	.07	.05
0	.14	.10	.07	.05	0

NOTE: These data are based on (9.2), assuming that $E\hat{\delta} = 0$.

.10, typical in practice, *there is little improvement past B = 100*. In fact, B as small as 25 gives reasonable results.

Now we return to bootstrap confidence intervals. In the Monte Carlo situation the bootstrap cdf $\hat{G}$ must be estimated from bootstrap replications $\hat{\theta}_1^*, \hat{\theta}_2^*, \ldots, \hat{\theta}_B^*$, say by

$$\hat{G}_B(s) = \#\{\hat{\theta}_b^* < s\}/B. \tag{9.3}$$

As $B \to \infty$, then $\hat{G}_B \to \hat{G}$, the ideal bootstrap cdf we have been using in the previous sections. Let $\theta_B[\alpha]$ be the level α endpoint of either the BC or BC$_a$ interval obtained from $\hat{G}_B(s)$ by substitution in (3.8), (3.9).

The following formula for the conditional CV of $\theta_B[\alpha] - \hat{\theta}$ assumes that $\hat{G}$ is roughly normal and that z_0 and a are known, for example, from (8.8) and (6.5) or (7.3):

$$\mathrm{CV}\{\theta_B[\alpha] - \hat{\theta} \mid \mathbf{y}\} \doteq \frac{1}{B^{1/2}|z^{(\alpha)}|}\left\{\frac{\alpha(1-\alpha)}{\varphi(z^{(\alpha)})^2}\right\}^{1/2}, \tag{9.4}$$

$\varphi(z) \equiv \exp(-\frac{1}{2}z^2)/\sqrt{2\pi}$. Notice that since we condition on $\mathbf{y}$, the only random quantity on the left side of (9.4) is $\theta_B[\alpha]$. Formula (9.4) measures the variability in $\theta_B[\alpha] - \hat{\theta}$ due to taking only B bootstrap replications, rather than an infinite number.

Here is a brief tabulation of (9.4) $\times B^{1/2}$:

$$\begin{array}{lcccc} \alpha & : .75 & .90 & .95 & .975 \\ (9.4) \times B^{1/2} & : 2.02 & 1.33 & 1.28 & 1.36 \end{array}. \tag{9.5}$$

If $B = 1{,}000$, for instance, then $\mathrm{CV}\{\theta_B[.95] - \hat{\theta} \mid \mathbf{y}\} \doteq 1.28/1000^{1/2} = .040$. Reducing B to 200 increases the conditional CV to .091. This last figure may be too big. The whole purpose of developing a theory better than (1.1) is to capture second-order effects. As the examples have indicated, these become interesting when the asymmetry ratio R/L is larger than say, 1.25, or smaller than .80. In such borderline situations, an extra 9% error in each tail due to inadequate bootstrap sampling may be unacceptable.

If the bias-correction constant z_0 is estimated by Monte Carlo directly from $z_0 = \Phi^{-1}(\hat{G}_B(\hat{\theta}))$, rather than from (8.8), then

$$\mathrm{CV}\{\theta_B[\alpha] - \hat{\theta} \mid \mathbf{y}\} \doteq \frac{1}{B^{1/2}z^{(\alpha)}}\left\{\frac{1}{\varphi(0)^2} - \frac{2(1-\alpha)}{\varphi(0)\varphi(z^{(\alpha)})} + \frac{\alpha(1-\alpha)}{\varphi(z^{(\alpha)})^2}\right\}^{1/2} \tag{9.6}$$

for $\alpha > .50$. This gives larger CV's than (9.4):

$$\begin{array}{lcccc} \alpha & : .75 & .90 & .95 & .975 \\ (9.6) \times B^{1/2} & : 3.04 & 1.97 & 1.75 & 1.71 \end{array}. \tag{9.7}$$

Comparing (9.7) with (9.5) shows that we need B to be about twice as large to get the same CV if z_0 is estimated rather than calculated. Formula (8.8) can be very helpful!

Both (9.4) and (9.6) assume that the bootstrap cdf is estimated by straightforward Monte Carlo sampling, as in (9.3). M. V. Johns (personal communication) has developed importance sampling methods that greatly accelerate the estimation of $\hat{G}$ in some situations.

10. ONE-PARAMETER FAMILIES

We return to the simple situation $\hat{\theta} \sim f_\theta$, where there are no nuisance parameters and where we want a confidence interval for the real-valued parameter θ based on a real-valued summary statistic $\hat{\theta}$. This section gives a more extensive discussion of the acceleration constant a, which has played a basic role in our considerations. Three familiar types of one-parameter families will be investigated: exponential families, translation families, and transformation families.

Efron (1982b) considered the following question: for a given family $\hat{\theta} \sim f_\theta$, do there exist mappings $\hat{\phi} = g(\hat{\theta})$, $\phi = h(\theta)$ such that $\hat{\phi} = \phi + \sigma_\phi q(Z)$, $Z \sim N(0, 1)$, as in (4.8)? This last form, a General Scaled Transformation Family (GSTF), generalizes the concept of the ideal normalization, where $\hat{\phi} = \phi + Z$. [We now add the conditions $q(0) = 0$, $q'(0) = 1$, as in Efron (1982b).]

The question is answered in terms of the diagnostic function $D(z, \theta) \equiv [\varphi(0)/\varphi(z)][\dot{F}_\theta(\hat{\theta}_\theta^{(\alpha)})/\dot{F}_\theta(\mu_\theta)]$. Here $\varphi(z)$ is the standard normal density $(2\pi)^{-1/2}\exp(-z^2/2)$; F_θ is the cdf $F_\theta(s) = \Pr_\theta\{\hat{\theta} \le s\}$; $\dot{F}_\theta(s) = (\partial/\partial\theta)F_\theta(s)$; $\alpha = \Phi(z)$; $\hat{\theta}_\theta^{(\alpha)}$ is the $100 \cdot \alpha$ percentile of $\hat{\theta}$ given θ, $\hat{\theta}_\theta^{(\alpha)} = F_\theta^{-1}(\alpha)$; and μ_θ is the median of $\hat{\theta}$ given θ, $\mu_\theta = \hat{\theta}_\theta^{(.5)} = F_\theta^{-1}(.5)$. It is shown that the form of σ_ϕ and $q(z)$ in (4.8) can be inferred from $D(z, \theta)$, the main advantage being that $D(z, \theta)$ is computed without knowledge of the normalizing transformations g, h.

The connection of transformation family theory with the acceleration constant a is the following: define

$$\varepsilon_\theta \equiv (\partial/\partial z)D(z, \theta)|_{z=0}. \tag{10.1}$$

If $q(z)$ in (4.8) is symmetrically distributed about zero, a situation called a symmetric scaled transformation family (SSTF), then

$$\varepsilon_\theta = d\sigma_\phi/d\phi \tag{10.2}$$

(see Efron 1982b, eq. 4.11). A more complicated relationship holds for the GSTF case.

Notice that (10.2) is quite close to our original description of "a" as the rate of change of standard deviation on the normalized scale. As a matter of fact, we can transform (3.6), (3.7) into an SSTF by considering the statistic

$$\tilde{\phi} = \hat{\phi} + \frac{z_0}{1 - az_0}\sigma_{\hat{\phi}} = \hat{\phi} + \frac{z_0}{1 - az_0}(1 + a\hat{\phi}), \tag{10.3}$$

instead of $\hat{\phi}$ itself. Then it is easy to show that

$$\hat{\phi} = \phi + (1 + \varepsilon_0\phi)Z, \qquad \varepsilon_0 = a/(1 - az_0), \tag{10.4}$$

an SSTF with $\sigma_\phi = 1 + \varepsilon_0\phi$, $\dot{\sigma}_\phi = \varepsilon_0$ for all ϕ. [The quantity ε_0 has the same definition in (10.4) as in (4.11).]

Example. For $\hat{\theta} \sim \theta\chi^2_{19}/19$ as in Table 2, $\varepsilon_\theta = .1090$ for all θ [using Eq. (10.6)]. In addition, $z_0 = \Phi^{-1}\Pr\{\chi^2_{19} < 19\} = .1082$. The relationship $a = \varepsilon_0/(1 + \varepsilon_0 z_0)$ obtained by solving for a in (10.4) gives $a = .1077$, the value used in Table 2. This family is nearly in SSTF (see Remark E, Sec. 11).

We show below that under reasonable asymptotic conditions,

$$\text{SKEW}_\theta(\dot{l}_\theta)/6 \doteq \varepsilon_\theta, \tag{10.5}$$

where $\varepsilon_\theta = (\partial/\partial z)D(z, \theta)|_{z=0}$, as in (10.1). This last definition of ε_θ can be evaluated for any family $\hat{\theta} \sim f_\theta$, assuming only that the necessary derivatives exist. The main point here is that $\text{SKEW}_\theta(\dot{l}_\theta)/6$ always approximates ε_θ (10.1), and in SST families ε_θ has the acceleration interpretation (10.2).

Now to show (10.5). It is possible to reexpress (10.1) as

$$\varepsilon_\theta = -\frac{\varphi(0)}{\dot{\mu}_\theta f_\theta(\mu_\theta)}\,\dot{l}_\theta(\mu_\theta), \tag{10.6}$$

where $\dot{\mu}_\theta = (d/d\theta)\mu_\theta$, the rate of change of the median μ_θ with respect to θ. For notational convenience suppose that $\theta = 0$. Instead of $\hat{\theta}$, consider the statistic $X \equiv \dot{l}_0(\hat{\theta})/i_0$, where i_0 equals the Fisher information $E_0\dot{l}_0(\hat{\theta})^2$. The parameter ε_θ is invariant under one-to-one transformations of $\hat{\theta}$, so we can evaluate the right side of (10.6) in terms of X, $\varepsilon_\theta = -\varphi(0)\dot{l}^X_\theta(\mu^X_\theta)/\dot{\mu}^X_\theta f^X_\theta(\mu^X_\theta)$.

For $\theta = 0$, X has expectation $E_0X = 0$ and standard deviation $\sigma^X_0 = i_0^{-1/2}$; in addition, $\dot{l}^X_0(0) = 0$, since $X = 0$ implies that $\theta = 0$ is a solution of the MLE equation. Assuming the usual asymptotic convergence properties, as in (5.1), (5.3), we have the following approximations: $\dot{\mu}^X_0 \doteq 1$; $\mu^X_0 \doteq -\gamma^X_0 i_0^{-1/2}/6$; $f^X_0(\mu^X_0) \doteq \varphi(0)i_0^{1/2}$; $\dot{l}^X_0(\mu^X_0) \doteq -\sqrt{i_0}\,\gamma^X_0/6$. These are derived from standard Edgeworth and Taylor series arguments, which will not be presented here. Taken together they give $\varepsilon_0 \doteq \text{SKEW}_0(\dot{l}^X_0)/6 = \text{SKEW}_0(\dot{l}_0)/6$, which is (10.5). The quantity $\text{SKEW}_0(\dot{l}_0)/6$ is $O(n^{-1/2})$, and the error of approximation in (10.5) is quite small,

$$\varepsilon_0 = [\text{SKEW}_0(\dot{l}_0)/6][1 + O(n^{-1})]. \tag{10.7}$$

Approximation (10.5) is particularly easy to understand in one-parameter exponential families. Suppose that $x_1, x_2, \ldots, x_n$ are iid observations from such a family, with sufficient statistic $y = \bar{x}$ having density $f_\theta(y) = \exp\{n[\theta y - \psi(\theta)]\}f_0(y)$. In this case formula (10.6) becomes

$$\varepsilon_\theta = \frac{\sigma^Y_\theta\varphi(0)}{\dot{\mu}^Y_\theta f^Y_\theta(\mu^Y_\theta)}\left[\frac{\lambda^Y_\theta - \mu^Y_\theta}{\sigma^Y_\theta}\right], \tag{10.8}$$

where $\lambda^Y_\theta = E_\theta\{y\}$, $\mu^Y_\theta = \text{median}_\theta\{y\}$, $\dot{\mu}^Y_\theta = \partial\mu^Y_\theta/\partial\theta$, and so forth. The term $[(\lambda^Y_\theta - \mu^Y_\theta)/\sigma^Y_\theta] = \gamma^Y_\theta/6[1 + O(n^{-1})]$, and $\sigma^Y_\theta\varphi(0)/\dot{\mu}^Y_\theta f^Y_\theta(\mu^Y_\theta) = 1 + O(n^{-1})$, both of the calculations being quite straightforward. Thus $\varepsilon_\theta = \gamma^Y_\theta/6[1 + O(n^{-1})]$. Since $\dot{l}_\theta(y) = n[y - \lambda_\theta]$, we have $\text{SKEW}_\theta(\dot{l}_\theta(y)) = \text{SKEW}_\theta(y) = \gamma^Y_\theta$, verifying (10.5) for one-parameter exponential families.

Example. If $Y \sim \text{Poisson}(\theta)$, $\theta = 15$, then $\text{SKEW}_\theta(\dot{l}_\theta)/6 = 1/(6 \cdot \theta^{1/2}) = .0430$. For the continued version of the Poisson family used in Efron (1982b; note Corrigenda, p. 1032), $(\partial/\partial z)D(z, \theta)|_{z=0} = .0425$ for $\theta = 15$.

Translation Families. Suppose that we observe a translation family $\hat{\zeta} = \zeta + W$, as in (3.12). Express W as a function $q(Z)$ of $Z \sim N(0, 1)$, for simplicity assuming that $q(0) = 0$ and $q'(0) = 1$, as in Efron (1982b). Then $z_0 = \Phi^{-1}\Pr\{\hat{\zeta} < \zeta\} = 0$. In this case it looks like methods based on the percentiles of the bootstrap distribution must give wrong answers, since if W is long-tailed to the right then the correct interval (3.13) is long-tailed to the left, and vice versa. However, the BC_a method produces at least roughly correct intervals, as we saw in the proof of Lemma 1.

What happens is the following: for any constant A the transformation $g_A(t) \equiv (\exp(At) - 1)/A$ gives $\hat{\phi} = g_A(\hat{\zeta})$, $\phi = g_A(\zeta)$, and $Z_A = g_A(W)$ satisfying

$$\hat{\phi} = \phi + \sigma^A_\phi \cdot Z_A, \qquad \sigma^A_\phi = 1 + A\phi. \tag{10.9}$$

The Taylor series for $W = q(Z)$ begins $W = Z + (\gamma_W/6)Z^2 + \cdots$, where $\gamma_W = \text{SKEW}(W)$. Then $Z_A = Z + (\gamma_W/6)^2Z^2 + (A/2)Z^2 + \cdots$.

The choice $A = a \equiv -\gamma_W/3$ results in $Z_a = Z + cZ^3 + \cdots$, the quadratic term canceling out; Z_a is then approximately normal, so (10.9) is approximately situation (3.6), (3.7), with $z_0 = 0$, $a = -\gamma_W/3$. *But we know that the BC_a intervals are correct if we can transform to situation (3.6), (3.7).* An application of Lemma 2, assuming that $Z_a \sim N(0, 1)$, shows that $a = -\gamma_W/3 \doteq \text{SKEW}(\dot{l}_\zeta(\hat{\zeta}))/6$ for the translation family $\hat{\zeta} = \zeta + W$, reverifying (4.4). [If $Z_a \sim N(0, 1)$ in (10.9), then a must equal ε, the constant value of ε_ζ, (10.1), for the translation family $\hat{\zeta} = \zeta + W$; one can show directly that $\varepsilon \doteq -\gamma_W/3$ for such a family.]

In the example $\hat{\theta} \sim \theta\chi^2_{19}/19$, the two constants z_0 and a are nearly equal. This is no fluke.

Theorem 2. If $\hat{\theta}$ is the MLE of θ in a one-parameter problem having standard asymptotic properties (5.1) or (5.3), then $z_0 \doteq a$,

$$z_0 \equiv \Phi^{-1}\Pr_\theta\{\hat{\theta} < \theta\} = \frac{\text{SKEW}_\theta(\dot{l}_\theta)}{6}[1 + O(n^{-1})]. \tag{10.10}$$

Proof. We follow the notation and results of DiCiccio (1984): thus k_1, k_2, k_3 equal the first three cumulants of $\dot{l}_\theta$ under θ; k_{01}, k_{02}, k_{03} the first three cumulants of $\ddot{l}_\theta$; k_{001}, the first cumulant of $\dddot{l}_\theta$; and $k_{11} = \text{cov}_\theta(\dot{l}_\theta, \ddot{l}_\theta)$. (So $k_2 = i_\theta$, the Fisher information.) All cumulants are assumed to be $O(n)$. Then the relative bias of $\hat{\theta}$ is

$$b \equiv \frac{E_\theta(\hat{\theta} - \theta)}{\text{var}_\theta(\theta)^{1/2}} = \frac{k_{001} - 2k_3}{6k_2^{3/2}} + O(n^{-3/2}), \tag{10.11}$$

and $\hat{\theta}$ has skewness

$$\gamma_\theta = \frac{k_{001} - k_3}{k_2^{3/2}} + O(n^{-3/2}). \qquad (10.12)$$

Both b and γ_θ are $O(n^{-1/2})$.

Standard Edgeworth theory now gives

$$\Pr_\theta\{\hat{\theta} < \theta\} = \Phi(-b) - \frac{\gamma}{6}\varphi(b)(b^2 - 1) + O(n^{-3/2})$$

$$= .5 + \varphi(0)\frac{(2k_3 - k_{001}) + (k_{001} - k_3)}{6k_2^{3/2}} + O(n^{-3/2})$$

$$= .5 + \varphi(0)\frac{k_3}{6k_2^{3/2}} + O(n^{-3/2}).$$

Since $\text{SKEW}_\theta(\dot{l}_\theta) = k_3/k_2^{3/2}$, this verifies (10.10).

In multiparameter problems it is no longer true that $z_0 \doteq a$. The geometry of the level surface $\mathcal{C}_{\hat{\theta}}$ adds another term to z_0, as in (8.8).

11. REMARKS

Remark A. Suppose that instead of (3.6), (3.7) we have $\sigma_\phi = \tau(1 + A\phi)$, so $\sigma_0 = \tau$ ($\tau \neq 1$). The transformations $\hat{\phi}' \equiv \hat{\phi}/\tau$, $\phi' \equiv \phi/\tau$, give $\hat{\phi}' = \phi' + \sigma_{\phi'}'(Z - z_0)$, where $\sigma_{\phi'}' = 1 + a\phi'$ and $a = A\tau$, so we are back in form (3.6), (3.7). Notice that the derivative $d(\sigma_\phi/\sigma_0)/d(\phi/\sigma_0) = a$, as in (4.7). In a similar way we can transform (3.6), (3.7) so that $\sigma_{\phi_0} = 1$ at any point ϕ_0; the resulting value of a satisfies (4.7).

Remark B. Instead of using $\hat{\phi}$ to estimate ϕ in (3.6), (3.7) we might change to the estimator $\hat{\phi}^{(c)} \equiv \hat{\phi} - c\sigma_{\hat{\phi}}$, for some constant c. It turns out that we are still in situation (3.6), (3.7): $\hat{\phi}^{(c)} = \phi + \sigma_\phi^{(c)}(Z - z_0^{(c)})$, where

$$\sigma_\phi^{(c)} = 1 + a^{(c)}(\phi - \phi_0^{(c)}), \qquad \phi_0^{(c)} = c/(1 - ac), \qquad (11.1)$$

and $a^{(c)} = a(1 - ac)$, $z_0^{(c)} = z_0 + \phi_0^{(c)}$. The choice $c = -z_0/(1 - az_0)$ gives $z_0^{(c)} = 0$, as in (10.3), (10.4). The choice $c = a$ gives approximately the MLE of ϕ. Interestingly enough, *the BC_a interval for ϕ based on $\hat{\phi}^{(c)}$ is the same for all choices of c.* Minor changes in the choice of estimator seem to have little effect on the BC_a intervals in general, though for computational reasons it is best not to use very biased estimators having large values of z_0.

Remark C. Section 6 uses the MLE $\hat{\theta} = t(\hat{\eta})$. This has one major advantage: *the BC_a interval for θ, based on $\hat{\theta}$, stays the same under all multivariate transformations* (6.11). Stein (1956) noted that the least favorable direction $\hat{\mu}$ transforms in the obvious way under (6.11), $\hat{\tilde{\mu}} = \hat{\mathbf{D}}\hat{\mu}$, where $\hat{\mathbf{D}}$ is the matrix with ijth element $\partial\tilde{\eta}_j/\partial\eta_i|_{\eta=\hat{\eta}}$, from which it is easy to check that formula (6.5) is invariant: the constant a is assigned the same value no matter what transformations (6.11) are applied. The bootstrap distribution $\hat{G}$ is similarly invariant, as shown in Efron (1985), and so is z_0. This implies that the BC_a intervals are invariant under transformations (6.11).

Remark D. The multiparametric theory of Section 5 gives an interesting result when applied to location-scale families; $y = (x, s)$, $\eta = (\theta, \sigma)$, and family of densities $f_\eta(y)$ of the form

$$f_{\theta,\sigma}(x, s) = (1/\sigma^2)f_{01}((x - \theta)/\sigma, s/\sigma), \qquad (11.2)$$

$f_{01}(x, s)$ being a known bivariate density function.

Suppose that we wish to set a confidence interval for the location parameter θ on the basis of its MLE $\hat{\theta}$. Parametric bootstrap intervals are based on the distribution of $\hat{\theta}^*$ when sampling from $f_{\hat{\theta},\hat{\sigma}}(x^*, s^*)$. The BC interval essentially amounts to pretending that σ is known (and equal to $\hat{\sigma}$) in (11.2) and that we have only a location problem to deal with, rather than a location-scale problem. In contrast, the BC_a interval takes account of the fact that σ is unknown. In particular the least favorable direction $\hat{\mu}$, plotted in the (θ, σ) plane, is *not* parallel to the θ axis. It has a component in the σ direction, whose magnitude is determined by the correlation between x and s. This means that Stein's least favorable family (6.4) does not treat σ as a constant.

Table 6 relates to the following choice of $f_{01}(x, s)$:

$$x \sim \chi_{30}^2/30 - 1, \qquad s \mid x \sim (1 + x)(\chi_{14}^2/14)^{1/2}, \qquad (11.3)$$

the two χ^2 variates being independent. This is a computationally more tractable version of the problem discussed in Efron (1982, tables 4 and 5). Approximate central 90% intervals are given for θ, having observed $(x, s) = (0, 1)$. For any other observed (x, s) the intervals transform in the obvious way, $\theta_{\hat{x}}[\alpha] = x + s\theta_{01}[\alpha]$. Line 3 of Table 6 shows the exact interval, based on inverting the distribution of the pivotal quantity $T = (\hat{\theta} - \theta)/\hat{\sigma}$ for situations (11.2), (11.3).

In this case the BC_a method makes a large "second-order t correction," as in Example 2 of Section 6, shifting the BC interval a considerable way rightward and achieving the correct R/L ratio. The length of the BC_a interval is 90% the length of the T interval. This deficiency is a third-order effect, in the spirit of the familiar Student-t correction. It arises from the variability of $\hat{\sigma}$ as an estimate of σ, rather than the second-order effect due to the correlation of $\hat{\sigma}$ with $\hat{\theta}$.

Remark E. Section 3 says that the family $\hat{\theta} \sim \theta\chi_{19}^2/19$ can be mapped into form (3.6), (3.7). What are the appropriate mappings? It simplifies the problem to consider the equivalent family $\hat{\theta} \sim \theta(\chi_{19}^2/c_0)$, where $c_0 = 18.3337 = \text{median}(\chi_{19}^2)$. Then $\hat{\zeta} \equiv g_1(\hat{\theta})$, $\zeta \equiv g_1(\theta)$, and $W \equiv g_1(\chi_{19}^2/c_0)$ give a translation family (3.12), with median(W)

Table 6. Central 90% Intervals for θ, Having Observed (x, s) = (0, 1) From the Location-Scale Family (11.2), (11.3) so $\hat{\theta}$ = 0 and $\hat{\sigma}$ = .966

		RL	Length
1. BC interval	[−.336, .501]	1.49	.837
2. BC_a interval	[−.303, .603]	1.99	.906
3. T interval	[−.336, .670]	1.99	1.006

NOTE: Line 3 is based on the actual distribution of the pivotal quantity $T = (\hat{\theta} - \theta)/\hat{\sigma}$.

$= 0$, for any mapping $g_1(t) = (\log t)/c_1$. Choosing $c_1 = .3292$ results in $W = q(Z)$ having $q(0) = 0$, $q'(0) = 1$, as in the discussion of translation families in Section 10.

Section 10 suggests normalizing a translation family by $g_A(t) = (\exp(At) - 1)/A$, a good choice for A being the constant ε_θ, (10.1), which equals .1090 for all θ in the family $\hat{\theta} \sim \theta(\chi^2_{19}/c_0)$. The combined transformation $g(t) = g_A(g_1(t))$ is $g(t) = 9.1746[t^{.3311} - 1]$. The transformed family $\hat{\phi} = g(\theta)$, $\phi = g(\theta)$ is of form (3.6), (3.7),

$$\hat{\phi} = \phi + (1 + .1090 \cdot \phi)Z,$$
$$Z = 9.1746[(\chi^2_{19}/c_0)^{.3311} - 1]. \qquad (11.4)$$

Numerical calculations verify that Z as defined in (11.4) is very close to a standard normal variate. In fact we have automatically recovered, nearly, the Wilson–Hilferty cube root transformation (Johnson and Kotz 1970). Using (11.4), it is not difficult to show that $g(t)$, as defined previously, gives approximately (3.6), (3.7) when applied to the family $\hat{\theta} \sim \theta(\chi^2_{19}/19)$ considered in Section 3, with constants z_0 and a as stated. Schenker (1985) gave almost the same result.

Remark F. Suppose that $\mathbf{y} = (x_1, x_2, \ldots, x_n)$, where the x_i are an iid sample from a regular one-parameter family $f_\theta(x_i)$, and that $\hat{\theta}(\mathbf{y})$ is a first-order efficient estimator of θ, like the MLE. The score function $\dot{l}_\theta$ appearing in (4.4) is that based just on $\hat{\theta}$, rather than the score function based on the entire data set $\mathbf{y}$. However, it is easy to show from considerations like those in Efron (1975) that the two score functions are asymptotically identical. Their skewnesses differ only by amount $O_p(n^{-1})$. It is often more convenient to calculate a from the score function for $\mathbf{y}$ rather than for $\hat{\theta}$, as was done, for example, in (6.5).

Remark G. McCullagh (1984) and Cox (1980) gave an interesting approximate confidence interval for θ, which for the simple case $\hat{\theta} \sim f_\theta$ has endpoint

$$\theta_{\text{APP}}[\alpha] = \hat{\theta} + 1/\sqrt{\hat{k}_2} \times \left\{z^{(\alpha)} + \frac{(3\hat{k}_2' + 2\hat{k}_{001}) + \hat{k}_{001}z^{(\alpha)^2}}{6\hat{k}_2^{3/2}}\right\}. \qquad (11.5)$$

Here $\hat{\theta}$ is the MLE of θ; if $k_2(\theta) = E_\theta \dot{l}^2_\theta$, the Fisher information, then $\hat{k}_2 = k_2(\hat{\theta})$ and $\hat{k}_2' = dk_2(\theta)/d\theta|_{\theta=\hat{\theta}}$; and $\hat{k}_{001} = (E_\theta \ddot{l}_\theta)_{\theta=\hat{\theta}}$. Formula (11.5) is based on higher-order asymptotic approximations to the distribution of the MLE (see also Barndorff-Nielsen 1984).

It can be shown, as indicated in Section 12, that $\theta_{\text{BC}_a}[\alpha]$ also closely matches (11.5), $(\theta_{\text{BC}_a}[\alpha] - \theta_{\text{APP}}[\alpha])/\hat{\sigma} = O_p(n^{-1})$. We see again that the BC$_a$ method offers a way to avoid theoretical effort, at the expense of increased numerical computations.

12. PROOF OF THEOREM 1

A monotonic mapping $\hat{\phi} = g(\hat{\theta})$, $\phi = g(\theta)$ transforms the exact confidence interval in the obvious way, $\phi_{\text{EX}}[\alpha] = g(\theta_{\text{EX}}[\alpha])$; likewise for the BC$_a$ interval. By using such a mapping we can always make $\hat{\phi} = 0$ and the distribution of $\hat{\phi}$ given $\phi = 0$ perfectly normal. Because of (5.3), which says that the distributions of $\hat{\theta}$ are approaching normality at the usual $O(n^{-1/2})$ rate, the normalizing transformation g is asymptotically linear, $g(\theta) = \theta + c_2\theta^2 + c_3\theta^3 + \cdots$, $c_2 = O(n^{-1/2})$, $c_3 = O(n^{-1})$.

We will assume that the problem is already in the form $\hat{\theta} = 0$, with the cdf of $\hat{\theta}$ for $\theta = 0$ normal, say

$$G_0 \sim N(-z_0, 1). \qquad (12.1)$$

Here $z_0 = \Phi^{-1}P_0\{\hat{\theta} < 0\}$ must be included because it is not affected by any monotonic transformations; $z_0 \doteq \gamma_\theta/6$ is $O(n^{-1/2})$ by (5.3). A simple exercise, using the mean value theorem of calculus, shows that if (5.4) is true in the transformed problem (12.1), then it is true in the original problem.

Assuming (5.3), $\hat{\theta} = 0$, and (12.1), we will show that the exact interval has endpoint

$$\theta_{\text{Ex}}[\alpha] \doteq \frac{z_0 + z^{(\alpha)}}{1 - \dot{\sigma}_0 z^{(\alpha)} + \dot{\beta}_0 + (\dot{\gamma}_0/6)(z^{(\alpha)^2} - 1)} + (\ddot{\sigma}_0/2)(z_0 + z^{(\alpha)})^3, \qquad (12.2)$$

compared with

$$\theta_{\text{BC}_a}[\alpha] \doteq \frac{z_0 + z^{(\alpha)}}{1 - \dot{\sigma}_0(z_0 + z^{(\alpha)})} \qquad (12.3)$$

for the BC$_a$ interval. In this section the symbol "$\doteq$" indicates accuracy through $O(n^{-1})$ or $O_p(n^{-1})$, with errors $O(n^{-3/2})$ or $O_p(n^{-3/2})$. Then

$$\frac{\theta_{\text{BC}_a}[\alpha] - \theta_{\text{Ex}}[\alpha]}{\sigma_\theta} \doteq \theta_{\text{BC}_a}[\alpha]\{\dot{\sigma}_0 z_0 + \dot{\beta}_0 + (\dot{\gamma}_0/6)(z^{(\alpha)2} - 1)\} - (\ddot{\sigma}_0/2)(z_0 + z^{(\alpha)})^3, \qquad (12.4)$$

which is $O_p(n^{-1})$, as claimed in Theorem 1.

The proof of (12.2) begins by noting that (12.1) implies that $\beta_0 = -z_0$, $\sigma_0 = 1$, $\gamma_0 = 0$, $\delta_0 = 0$. Then (5.3) gives

$$E_\theta\hat{\theta} = \theta + \beta_\theta \doteq (1 + \dot{\beta}_0)\theta - z_0,$$
$$\sigma_\theta \doteq 1 + \dot{\sigma}_0\theta + \ddot{\sigma}_0\theta^2/2,$$
$$\gamma_\theta \doteq \dot{\gamma}_0\theta, \qquad \delta_\theta \doteq 0, \qquad (12.5)$$

for $\theta = O(1)$ [i.e., for θ a bounded function of n, in the sequence of situations referred to in (5.3)]. The $100 \cdot \alpha$ percentile of $\hat{\theta}$ given θ is

$$\hat{\theta}^{(\alpha)}_\theta \doteq (\theta + \beta_\theta) + \sigma_\theta\{z^{(\alpha)} + (\gamma_\theta/6)(z^{(\alpha)2} - 1)\}$$
$$\doteq [(1 + \dot{\beta}_\theta)\theta - z_0] + [1 + \dot{\sigma}_0\theta + (\ddot{\sigma}_0/2)\theta^2] \times [z^{(\alpha)} + (\dot{\gamma}_0\theta/6)(z^{(\alpha)2} - 1)], \qquad (12.6)$$

using a Cornish–Fisher expansion and (12.5). The θ, however, that has $\hat{\theta}^{(\alpha)}_\theta = 0$ is by definition $\theta_{\text{Ex}}[1 - \alpha]$. Solving the lower expression in (12.6) for 0 and substituting $1 - \alpha$ for α gives (12.2).

The proof of (12.3) follows from (3.8), (3.9), and (12.1) [which says that $\hat{G} \sim N(-z_0, 1)$], if we can establish that

$a \doteq \dot{\sigma}_0$. In fact, we show below that

$$\varepsilon_\theta \doteq \dot{\sigma}_0 \quad \text{for} \quad \theta = O(n^{-1/2}), \tag{12.7}$$

which combines with $a = \varepsilon_0/(1 + \varepsilon_0 z_0) \doteq \varepsilon_0$ to give the required result.

Formula (12.7) follows from (12.5), which gives the simpler expressions

$$E_\theta\hat{\theta} \doteq \theta - z_0, \qquad \sigma_\theta \doteq 1 + \dot{\sigma}_0\theta, \qquad \gamma_\theta \doteq 0, \qquad \delta_\theta \doteq 0 \tag{12.8}$$

for $\theta = O(n^{-1/2})$. The cdf of $\hat{\theta}$ given θ is calculated to be

$$G_\theta(\hat{\theta}) \doteq \Phi(z_\theta)\dot{z}_\theta - (\dot{\gamma}_0/6)(\quad - 1), \tag{12.9}$$

$z_\theta \equiv (\hat{\theta} - \theta - \beta_\theta)/\sigma_\theta$, $\dot{z}_\theta = (\partial/\partial\theta)z_\theta$. Straightforward expansions give

$$D(z^{(\alpha)}, \theta) \doteq \frac{1 + \dot{\sigma}_0 z^{(\alpha)} + \dot{\beta}_0 + (\dot{\gamma}_0/6)(z^{(\alpha)2} - 1)}{1 + \dot{\beta}_0 - \dot{\gamma}_0/6}, \tag{12.10}$$

from which $\varepsilon_\theta = (\partial/\partial z)D(z, \theta)\,|_{z=0} \doteq \dot{\sigma}_0/(1 + \dot{\beta}_0 - \dot{\gamma}_0/6)$, verifying (12.7), (12.3), and the main result (12.4).

The proof that $\theta_{BC_a}[\alpha]$ also matches the Cox–McCullagh formula (11.5) is similar to the proof of Theorem 1 and will not be presented here. The main step is an expression for $\theta_{BC_a}[\alpha]$ involving Lemma 5,

$$\theta_{BC_a}[\alpha] \doteq z^{(\alpha)} + (\hat{k}_3/6\hat{k}_2^{3/2})\{z^{(\alpha)2} + 1\} + (\hat{k}_3/6\hat{k}_2^{3/2})^2\{2z^{(\alpha)} + z^{(\alpha)3}\}. \tag{12.11}$$

[*Received November 1984. Revised December 1985.*]

REFERENCES

Abramovitch, L., and Singh, K. (1985), "Edgeworth Corrected Pivotal Statistics and the Bootstrap," *The Annals of Statistics*, 13, 116–132.

Barndorff-Nielsen, O. E. (1984), "Confidence Limits From $c|f|\bar{L}$," Report 104, University of Aarhus, Dept. of Theoretical Statistics.

Bartlett, M. S. (1953), "Approximate Confidence Intervals," *Biometrika*, 40, 12–19.

Beran, R. (1984a), "Bootstrap Methods in Statistics," *Jber. d. Dt. Math-Verein*, 86, 14–30.

——— (1984b), "Jackknife Approximations to Bootstrap Estimates," *The Annals of Statistics*, 12, 101–118.

Bickel, P. J., and Freedman, D. A. (1981), "Some Asymptotic Theory for the Bootstrap," *The Annals of Statistics*, 9, 1196–1217.

Cox, D. R. (1980), "Local Ancillarity," *Biometrika*, 67, 279–286.

DiCiccio, T. J. (1984), "On Parameter Transformations and Interval Estimation," technical report, McMaster University, Dept. of Mathematical Science.

Efron, B. (1975), "Defining the Curvature of a Statistical Problem (With Applications to Second Order Efficiency)" (with discussion), *The Annals of Statistics*, 3, 1189–1242.

——— (1979), "Bootstrap Methods: Another Look at the Jackknife," *The Annals of Statistics*, 7, 1–26.

——— (1981), "Nonparametric Standard Errors and Confidence Intervals" (with discussion), *Canadian Journal of Statistics*, 9, 139–172.

——— (1982a), "The Jackknife, the Bootstrap, and Other Resampling Plans," CBMS 38, SIAM-NSF.

——— (1982b), "Transformation Theory: How Normal Is a Family of Distributions?," *The Annals of Statistics*, 10, 323–339. (NOTE Corrigenda, *The Annals of Statistics*, 10, 1032.)

——— (1984), "Comparing Non-nested Linear Models," *Journal of the American Statistical Association*, 79, 791–803.

——— (1985), "Bootstrap Confidence Intervals for a Class of Parametric Problems," *Biometrika*, 72, 45–58.

Fieller, E. C. (1954), "Some Problems in Interval Estimation," *Journal of the Royal Statistical Society*, Ser. B, 16, 175–183.

Hall, P. (1983), "Inverting an Edgeworth Expansion," *The Annals of Statistics*, 11, 569–576.

Hougaard, P. (1982), "Parameterizations of Non-linear Models," *Journal of the Royal Statistical Society*, Ser. B, 44, 244–252.

Johnson, N. J. (1978), "Modified *t* Tests and Confidence Intervals for Asymmetrical Populations," *Journal of the American Statistical Association*, 73, 536–544.

Johnson, N. L., and Kotz, S. (1970), *Continuous Univariate Distributions—2*, Boston: Houghton-Mifflin.

Kendall, M., and Stuart, A. (1958), *The Advanced Theory of Statistics*, London: Charles W. Griffin.

McCullagh, P. (1984), "Local Sufficiency," *Biometrika*, 71, 233–244.

Schenker, N. (1985), "Qualms About Bootstrap Confidence Intervals," *Journal of the American Statistical Association*, 80, 360–361.

Singh, K. (1981), "On the Asymptotic Accuracy of Efron's Bootstrap," *The Annals of Statistics*, 9, 1187–1195.

Stein, C. (1956), "Efficient Nonparametric Testing and Estimation," in *Proceedings of the Third Berkeley Symposium*, Berkeley: University of California Press, pp. 187–196.

Tukey, J. (1949), "Standard Confidence Points," Memorandum Report 26, unpublished address presented to the Institute of Mathematical Statistics.

Withers, C. S. (1983), "Expansions for the Distribution and Quantiles of a Regular Functional of the Empirical Distribution With Applications to Nonparametric Confidence Intervals," *The Annals of Statistics*, 11, 577–587.

Rejoinder

BRADLEY EFRON

The problem of setting approximate confidence intervals is an important one that is just beginning to receive the attention it deserves. The commentaries here are positive and constructive, whether or not they agree with the approach taken in my article. Other promising approaches are discussed by Beran, Cox, DiCiccio and Tibshirani, Loh and Wu, Schenker, and Schucany. Bickel, Schenker, and Schucany help connect the BC_a method to these other ideas. Freedman and Peters, Loh and Wu, and DiCiccio and Tibshirani provide additional examples of how well or poorly the bootstrap method works in specific cases.

All in all it seems that statisticians are closing in on a workable theory of approximate confidence intervals. To put it another way, after 60 years and a millionfold improvement in computation, we may be able to offer the statistical practitioner a reliable improvement over $\hat{\theta} \pm z^{(\alpha)}\hat{\sigma}$. Here are a few remarks inspired by the commentary, with no attempt to answer all of the points raised.

1. Bootstrap methods enjoy an important practical advantage: only one probability mechanism need be considered, essentially the mechanism that best fits the observed data. More ambitious (and more accurate) methods require the specification of all possible probability distributions for the data. Consider the law school example of Section 7. Should the value $\rho = .50$ be included in a central 90% confidence interval? The classical approach requires us to compute how unusual is the observed value $\hat{\rho} = .776$ for all candidate distributions $f_{\eta}(\mathbf{y})$ having $\rho(\eta) = .50$. Bootstrap methods require only the calculation of $\hat{G}$, the distribution of the statistic of interest under the best-fitting distribution $f_{\hat{\eta}}$. [The standard method (1.1) is a bootstrap in this sense, as discussed in Sec. 2.] This advantage is particularly important in nonparametric settings. See Schenker's discussion for a nice exposition of these points.

2. The likelihood intervals advocated by Cox are non-bootstrap, in the sense of requiring a full specification of the class of distributions f_{η}. When such a specification can be believably made, then likelihood methods should be preferable to the inherently cruder bootstrap procedures.

In fact there is a close relationship between bootstrap and likelihood intervals, as must be the case, since both reduce to the standard interval when (1.1) is appropriate and both transform correctly under monotonic mappings. The connection between likelihood and bootstrap methods is discussed in Section 6 of Efron (1985). It is pointed out there that Cox's region (1) is not second-order correct, and that the "equitailed" correction of Cox's comment 3 is necessary. This correction, which is automatically included in the bootstrap intervals, can be quite complicated, as indicated by the calculations in McCullagh (1984).

3. In their remark (3), DiCiccio and Tibshirani suggest using the likelihood interval method on the least favorable family $\hat{\mathfrak{F}}$ indicated in Figure 1. This is a nice idea, which helps connect the likelihood and BC_a approaches. It is easy to compute that the Fisher information bound for estimating θ in the least favorable family $\hat{\mathfrak{F}}$ is $\{\sum_1^n U_i^2/n^2\}^{1/2}$. This is just the delta method estimate of variance for $\hat{\theta}$ or, equivalently, the infinitesimal jackknife estimate mentioned by Schucany. The delta method can give quite badly biased estimates of $\text{var}(\hat{\theta})$ [see table 5.2 of Efron (1982)], so we might prefer to improve this part of the confidence interval procedure by bootstrapping, which almost gets us back to methods like BC_a.

4. The bootstrap-t method, in which the bootstrap distribution of $t = (\hat{\theta} - \theta)/\hat{\sigma}$ is used to set confidence intervals for θ, now has impressive credentials for general second-order accuracy, thanks to the work of Beran and also Abramovitch and Singh (1985). Bootstrap-t and Beran's more ambitious prepivoting mechanism are genuine bootstrap methods in the sense of remark (1) above, and so are at least potentially applicable to almost any situation. These methods are very promising. Bickel's comments nicely relate the BC_a and bootstrap-t theories.

5. Table A.1 shows 3 different bootstrap-t intervals for ρ, the correlation coefficient in the law school example of Table 4, where $\hat{\rho} = .776$ from a sample of size $n = 15$. Each of the t intervals is based on 1,000 nonparametric bootstrap replications of $t = (\hat{\rho} - \rho)/\hat{\sigma}$, with $\hat{\sigma}$ equal to (a) the delta-method estimate of the standard error of $\hat{\rho}$ [see (6.23) of Efron (1982)], (b) the jackknife estimate of standard error, and (c) the normal-theory estimate $\hat{\sigma} = (1 - \hat{\rho}^2)/\sqrt{12}$. Also shown are the nonparametric BC_a interval and the standard interval based on Fisher's $\tanh^{-1}$ transformation.

It is clear that the bootstrap-t methods do not work well here, especially for the lower endpoint of the intervals. My experience, which is not vast, has been that t methods work better on genuine location-scale problems [see, e.g., table 10.4 of Efron (1982)]. Then the choice of $\hat{\sigma}$ in the statistic $t = (\hat{\theta} - \theta)/\hat{\sigma}$ is more natural (essentially being any reasonably efficient measure of dispersion of the original observations), as is the entire t-statistic approach. [Note that the bootstrap-t method works well for the law school example if we bootstrap *parametrically* rather than nonparametrically, that is, if we draw $x_1^*, x_2^*, \ldots, x_{15}^* \sim N_2(\bar{x}, \hat{\Sigma})$ where $\hat{\Sigma}$ is the sample covariance matrix of the data in Table 4. Then $t^* = (\hat{\rho}^* - \hat{\rho})/\hat{\sigma}^*$, with $\hat{\sigma}^* = (1 - \hat{\rho}^{*2})/\sqrt{12}$, gives [.45, .90] as a central 90% interval for ρ. The nonparametric t methods get into trouble here because of the fat upper tail of $\hat{\rho}^*$ in its nonparametric bootstrap distribution; see figure 5.1 of Efron (1982).]

6. "Second-order accuracy" means that the coverage probabilities on each side of the approximate confidence interval are within $O(n^{-1})$ on their claimed values. "Second-order correctness" means the same thing, plus more:

Journal of the American Statistical Association
March 1987, Vol. 82, No. 397, Theory and Methods

Table A.1. Five Approximate Central 90% Confidence Intervals for the Correlation Coefficient ρ, Law School Data of Table 4

Fisher's $\tanh^{-1}$	[.49, .90]
Nonparametric BC_a	[.43, .92]
Bootstrap-t	
$\hat{\sigma} = \hat{\sigma}_{\text{delta}}$	[−.12, .95]
$\hat{\sigma} = \hat{\sigma}_{\text{jackknife}}$	[−.06, .94]
$\hat{\sigma} = (1 - \hat{\rho}^2)/\sqrt{12}$	[−.01, .91]

NOTE: Each of the bootstrap intervals is based on B = 1,000 nonparametric bootstrap replications. Notice the lower limits of the t intervals.

that the endpoints of the approximate interval agree through order $O_p(n^{-1})$ with those of the *correct* interval, as in Theorem 1. Bickel's final comment, and also some recent results of Peter Hall (private communication), have bolstered my belief that the BC_a intervals are second-order correct in a wide class of problems, including nonparametric situations.

7. Bootstrap-t intervals are second-order accurate, but it has not been clear to me that they are second-order correct when dealing with other than location-scale problems. Bickel's final equation, and some of the comments in Beran (1985) are reassuring on this point. Results like those in Table A.1 are sobering reminders of the limitations of asymptotic optimality theory.

8. Prepivoting requires much more computation than the BC or BC_a intervals, the goal being more dependable pivotal quantities. DiCiccio and Tibshirani's BC_a^0 method requires much less computation, at the expense of a greater reliance on asymptotic approximations. When it was proposed in the 1920s, the standard theory (1.1) was a nearly perfect compromise between theoretical desirability and the computational limitations of mechanical desk calculators. The form of any useful improvement over (1.1) will be shaped by similar compromises, based on how fast, how cheap, and how generally available electronic computation becomes for the majority of applied statisticians. I have tried to keep computational feasibility in mind here. The calculations of Section 9 show that the BC_a intervals are easily computed, even on personal computers, for simple situations like those in my examples. Bigger data sets or more complicated statistics require a bigger computer, though recent work by several authors points in the direction of substantially reduced bootstrap computations.

9. The formulas for the acceleration constant a, like (4.4) and (7.3), were motivated by a desire for computational simplicity. The discussion in Section 10 shows that these are only convenient approximations for the genuine acceleration constant. Loh and Wu, and DiCiccio and Tibshirani, correctly warn against the nonrobustness of formulas (4.4) and (7.3). Both Lemma 4 and Theorem 2 suggest a close connection between "a" and z_0, which might lead to a more robust formula for the acceleration.

10. DiCiccio and Tibshirani's comment (2) exposes an interesting pothole on our primrose path. Notice that the central 95% Fieller interval for $\xi = \eta_1/\eta_2 = 1/\theta$, having observed y = (1, 15), equals [−.064, .200] (which is just [1/(−15.65), 1/5.012]). The BC_a interval is [−.065, .200] in this case [see eq. (10) of Efron (1985)], so everything looks fine.

The trouble starts when we try to go from the parameter ξ to $\theta = 1/\xi$. The mapping for ξ to θ is infinitely *nonmonotonic* in this case, since the relevant values of ξ cross zero. The exact interval for θ turns inside out, as DiCiccio and Tibshirani describe. The BC_a method does not automatically correct for nonmonotonic transformations. A hidden assumption of all of the bootstrap methods [including the standard intervals (1.1)] is that "θ" names the problem in a reasonable way, in the sense that values of θ near $\hat{\theta}$ index the nearby probability mechanisms. This point is made more clearly in Efron (1985). Notice that the DiCiccio–Tibshirani problem also arises in the following simpler situation: we observe $\hat{\xi} = .5$, where $\hat{\xi} \sim N(\xi, 1)$, and we want a 95% confidence interval for $\theta = 1/\xi$.

11. In theory there is no reason why bootstrap calculations cannot be carried out conditionally, rather than unconditionally as in this article. Consider a translation family where the independent observations $x_1, x_2, \ldots, x_n$ each have density $g(x_i - \theta)$, for some known probability function $g(t)$. The MLE $\hat{\theta}$ has an unconditional density, say $f_\theta(\hat{\theta})$, and also a conditional density $f_\theta(\hat{\theta} \mid \mathbf{a})$, where $\mathbf{a}$ is the ancillary statistic $(x_1 - \hat{\theta}, x_2 - \hat{\theta}, \ldots, x_n - \hat{\theta})$. We can apply the BC_a method to $f_\theta(\hat{\theta} \mid \mathbf{a})$, with $\mathbf{a}$ considered fixed, to get an approximate confidence interval for θ based on $\hat{\theta}$. This interval will better agree with the exact Pitman interval for θ than will the BC_a interval based on the unconditional density $f_\theta(\hat{\theta})$. Likewise (11.5) will better agree with McCullagh's and Cox's original result, as mentioned by DiCiccio and Tibshirani, if the bootstrap calculations are done conditionally.

Unfortunately it is difficult to actually carry out conditional bootstrap computations in most situations, especially those where Monte Carlo methods are necessary. A recent article by Hinkley and Schectman (1986) makes interesting progress on this problem. In Efron (1985) and also in the current article I have sidestepped conditionality issues by only considering situations where there are no effective ancillaries.

12. Define

$$\varepsilon(F) \equiv \text{SKEW}_F\{\hat{\theta}(x_1, \ldots, x_n)\},$$

$$\hat{\theta}(x_1, \ldots, x_n) = \sum_1^n (x_i - \bar{x})^2/(n - 1),$$

where $x_1, x_2, \ldots, x_n$ is a random sample from F. If $F = N(\mu, \sigma^2)$ and n = 20, then $\varepsilon(F) = (8/19)^{1/2} = .649$. In this case the nonparametric MLE $\varepsilon(\hat{F})$ is a poor estimator of $\varepsilon(F)$. A small simulation experiment, 100 trials, showed that $\varepsilon(\hat{F})$ has median only .292 in this case and is less than the true value .649 in about 95% of the trials. (These results could be predicted, at least qualitatively, from calculations like those by Peters and Freedman.)

In other words the bootstrap distribution of $\hat{\theta}^* = \sum_{i=1}^n (x_i^* - \bar{x}^*)^2/(n - 1)$ tends not to be long-tailed enough to the right. This deficiency shows up in line 5 of Table 2. It is also at work in the numerical results of Peters and Freedman and of Loh and Wu.

There is no law that bootstrap calculations must start from the empirical distribution $\hat{F}$. Less obvious estimates of F, putting some probability mass outside the range of the observed sample $x_1, \ldots, x_n$, may produce more accurate bootstrap confidence intervals for statistics like $\hat{\theta}$, which depend on higher sample moments.

13. It is worth remembering that the complaint in the case $\theta = \text{var}_F X$ is not that the nonparametric bootstrap intervals are no improvement over the standard intervals, *but rather that they do not depart even further from the standard form (1.1)*. For the situation considered in Example 3 of Section 7, let

$$R = \frac{|\theta_{\text{EXACT}}[.95] - \theta_{\text{BC}_a}[.95]|}{|\theta_{\text{EXACT}}[.95] - \theta_{\text{STANDARD}}[.95]|},$$

so R is the ratio of errors of the two *approximate* nonparametric endpoints, $\theta_{\text{BC}_a}[.95]$ and $\theta_{\text{STANDARD}}[.95]$, from the exact parametric value $\theta_{\text{EXACT}}[.95] = \hat{\theta} \cdot 1.88$. In a simulation experiment involving 500 data sets $x_1, x_2, \ldots, x_{20} \overset{\text{iid}}{\sim} N(0, 1)$, with $B = 2{,}000$ bootstrap replications per trial, the value of R was less than one 492 times; R had mean .68 and median .74. In other words, the BC_a limits were consistently better than the standard limits, but not by an enormous amount. This is consistent with my general impression of the percentile, BC, and BC_a methods as successive cautious improvements over the standard intervals. Less cautious methods, like the bootstrap-t, may be the only way to do *much* better than (1.1) in some situations.

14. Schucany suggests a comparative study of approximate confidence interval methods, perhaps modeled after the Princeton robustness study. A good idea! Here are a few suggestions, culled from painful experience. (a) Simulation studies of bootstrap methods require orders of magnitude more computation than the bootstrap itself, so the investigators have to be computationally well armed. This helps avoid the temptation to take B too small (see Sec. 9) in order to run sufficient numbers of trials. (b) The sensible comparison of confidence intervals involves more than just coverage probabilities or lengths. The shape of the interval, called "R/L" in this article, can tell us whether or not we are on the correct inferential track. (See Table 6 for an example.) (c) The comparison of parametric and nonparametric results, as in Table 2, can help pinpoint the causes of failures and difficulties. Failures can be just as interesting as successes, especially if the reasons for failure suggest improvements. (d) Conventional third-order corrections, for example, dividing by $n - 1$ rather than n in estimating a variance, should be carefully accounted for, since they can substantially affect comparisons when n is small. (e) The standard intervals (1.1) should be included in any comparative study, since they are the current method of choice in most applications.

I am grateful to the discussants for their commentary, which seemed to me unusually penetrating, and to Carl Morris and William Parr for the large amount of editorial effort expended on both the original article and the discussion.

ADDITIONAL REFERENCES

Beran, R. (1985), "Prepivoting to Reduce Level Error of Confidence Sets," technical report, University of California, Berkeley.

Hinkley, D., and Schechtman, E. (1986), "Conditional Bootstrap Methods in the Mean-Shift Model," technical report, University of Texas, Austin.

15

An Introduction to the Bootstrap

Introduction by Rudolf Beran*
University of California, Davis

Brad Efron's (1979) paper on the bootstrap was published at a time when theoretical statisticians, using probability theory to analyze procedures, and applied statisticians, using computer technologies, seemed headed in different directions. The monograph *An Introduction to the Bootstrap* by Efron and Tibshirani (1993) provided a highly-readable account of Efron's bootstrap results, some with co-authors, from 1979 to 1992. The bootstrap concept indicated how modern computing environments may enable intuitive quantification of errors in statistical inference. Research by various other authors clarified the scope of bootstrap methods, in ways that led to the heart of modern estimation theory. This burst of innovative work also brought into focus fundamental distinctions within data, probability models and pseudo-random numbers that had become blurred among statisticians. Thereby it advanced the ongoing transformation of Statistics into an experimentally-based information science.

15.1. Historical Context

Statistical theory was in creative flux in 1979 when Brad Efron's first paper on the bootstrap was published. The asymptotic theory of statistics had just seen major advances that were enabled by sustained developments in probability theory. Prohorov's results on weak convergence of probability measures gave mathematical statisticians powerful tools for investigating the performance of statistical procedures under probability models. Notable statistical achievements included the discovery of super-efficient estimators such as the Hodges and Stein estimators, whose asymptotic risk, as sample size increases, undercuts the information bound on sets of Lebesgue measure zero in the parameter space. Exact risk calculations revealed that the Hodges estimator has poor minimax risk relative to the sample mean while the Stein estimator dominates the sample mean when its dimension exceeds 2. Thereby it became clear that pointwise limiting risks may quietly lose critical information about the behavior of statistical procedures.

One theoretical response to this difficulty, initiated in Le Cam (1953), was to study locally-uniform limits to statistical risks as sample size increases. A second theoretical response, prefigured in the first section of Stein (1956), was to study uniform limiting risks as parameter dimension increases. Both approaches proved powerful in dispelling illusions created by pointwise asymptotics. An early beneficiary was the theory of robust estimation and testing that developed rapidly in the 1970s. More recent instances have included trustworthy theory for regularization techniques such as model selection, penalized least squares, and other estimators of a high-dimensional mean that use multiple shrinkage, whether implicitly or directly. Not surprisingly, both types of asymptotics have proved effective in studying bootstrap procedures. This point will be developed later in the essay.

* This research was supported in part by National Science Foundation Grant DMS 0404547.

Logicians investigating the notion of mathematical proof had refined the concept of algorithm during the first part of the twentieth century. Subsequent realizations of computational algorithms through advances in digital computers, programming languages, numerical linear algebra and graphical output devices opened a new world to statisticians. Number-crunching with computers, no longer laborious or highly restricted, freed analysts to imagine more complex statistical procedures, including procedures not bound to the belief that data is governed by probability models. The performance of such procedures began to be assessed through computational experiments with what-if data. In the new computational environment, the fundamental distinctions within data, probability models for data, pseudo-random numbers and data-analytic procedures rose to greater prominence. For instance, Knuth (1969, Chap. 3) explored the extent to which pseudo-random sequences can imitate properties of random variables. In a book on data analysis that avoided probability models, Tukey (1977) observed, "Today's understanding of how well elementary techniques work owes much to both mathematical analysis and experimentation by computer."

Prior to the computing revolution, theoretical statistics necessarily developed as a mathematical branch of philosophy. This phase became prominent in the 1950s through the influential books by Wald (1950) on decision theory, Savage (1954) on Bayesian reasoning, and Fisher (1956) on proposed alternatives such as the fiducial argument. From a later perspective, these accounts exemplify the grand and simple theories that dominate the early stages of a discipline, before technological advances enable fuller interplay between theory and experiment.

The phenomenon just cited has had repeated parallels in other fields, before the arrival of better tools pushed those disciplines from philosophy towards science. In 1269, Peter Peregrinus wrote *Epistola de Magnete*, a ground-breaking experimental study of magnetism that expressed the principles of scientific method. He noted that an investigator "diligent in the use of his own hands ... will then in a short time be able to correct an error which he would never do in eternity by his knowledge of natural philosophy and mathematics alone." He balanced this opinion with, "There are many things subject to the rule of reason that we cannot completely investigate by hand." His contemporary, Roger Bacon, held Peter Peregrinus in high esteem: "What others strive to see dimly and blindly, like bats in twilight, he gazes at in the full light of day, because he is a master of experiment." The quotations come from Crombie (1953).

Efron's (1979) article, introducing the bootstrap, took a stance midway between those who relied on probability-based asymptotic theory to understand statistical procedures and those who relied on computer experiments with what-if data that is not certifiably random. While invoking a probability model for data, his paper proposed to use Monte Carlo approximations to estimate sampling distributions that depend on the unknown parameters in the model. The idea that computers could replace asymptotic approximations to such sampling distributions was surprising to many at the time and lacked theoretical support.

Efron wrote, "The paper proceeds by a series of examples, with little offered in the way of theory." In fact the paper contained considerable insight, derivation, and correct conjecture. The paper formulated and named the bootstrap idea, pointed to the dual interpretations of a bootstrap distribution as a random probability measure and as a conditional distribution given the sample, described the natural Monte Carlo approach to approximating such conditional distributions, proved for finite sample spaces the first result on consistency of the nonparametric bootstrap distribution, and made an application to bootstrap confidence intervals.

It is not unusual that an important paper goes scarcely noticed for a long time. Efron's (1979) paper sparked immediate interest among theoretical statisticians. It was quickly seen that theoretical tools recently developed for studying asymptotic performance of estimators could be adapted to study the consistency and higher-order asymptotics of bootstrap estimators and confidence sets. Computational algorithms for bootstrapping saw rapid improvements. The reception of the bootstrap contrasts with the long neglect of subsampling, pioneered by Gosset in two 1908 papers. At that time, to do subsampling was laborious and to explain what subsampling did was beyond available statistical theory. It was studies of the bootstrap that ultimately revived subsampling (cf. Politis et al. 1999).

15.2. The Bootstrap World

An Introduction to the Bootstrap, published in 1993 by Efron and Tibshirani, is a monograph shaped by the results in Efron's bootstrap papers, some with co-authors, from 1979 to 1992. The book presents the bootstrap and its links to other methodologies for estimating sampling error as Efron saw them. It is both reader-friendly and highly sophisticated. The first 19 of the 26 chapters are intended to serve audiences who are upper-year undergraduates or first-year graduate students. Each chapter ends with problems complementing the material and with brief bibliographic notes to related work in the literature. The uniqueness of Efron's approaches, for instance, in the development of bootstrap BC_a confidence intervals or in discussing links with cross-validation or the jackknife, gives the Efron–Tibshirani book an intellectual punch that repeatedly provokes a reader to reflection.

In their first chapter, the authors stated, "This book describes the bootstrap and other methods for assessing statistical accuracy. The bootstrap does not work in isolation but rather is applied to a wide variety of statistical procedures." This phrasing expresses the tension between two ideas: (a) the bootstrap as an intuitive general strategy for estimating sampling distributions or functions of sampling distributions; and (b) the bootstrap as a technical object that is carefully defined so as to work in a specific application.

The broad notion of simulation, which underlies the bootstrap, has a distinguished scientific history. "Preserving the appearances" of planetary sightings was achieved in Ptolemaic astronomy by the epicycles that provided accurate mathematical descriptions of planetary orbits relative to the Earth. Subsequent adoption of the Copernican planetary model had more to do with clearer understanding than with better numerical accuracy. (For an account of the issues, see Kuhn 1959.) Of course, statistical models for data in fields other than astronomy may have far less empirical support than the Ptolemaic model. In such instances, statistical model-fitting preserves the appearances of the data to an extent deemed appropriate by the analyst, on grounds that may otherwise remain weak.

A statistical model for a sample X_n of size n is a family of distributions $\{P_{\theta,n}\colon \theta \in \Theta\}$. The parameter space Θ is typically metric and need not be finite-dimensional. The value of θ that identifies the true distribution from which X_n is drawn is not known to the statistician.

The bootstrap applies the simulation concept to formal statistical inference. From the sample X_n, we devise a plausible estimator $\hat{\theta}_n = \hat{\theta}_n(X_n)$. The bootstrap idea is then:

- Create an artificial world in which the true parameter value is $\hat{\theta}_n$ and the sample X_n^* is generated from the fitted model $P_{\hat{\theta}_n,n}$. That is, the conditional distribution of X_n^*, given the data X_n, is $P_{\hat{\theta}_n,n}$.
- Act as if sampling distributions computed in the artificial world are accurate approximations to the corresponding true (but unknown) sampling distributions.

In the *original world* of the statistical model, the distribution of the observable X_n is unknown because the parameter θ is unknown. However, in the *bootstrap world*, the parameter $\hat{\theta}_n$ and the distribution of X_n^* are both known. Hence, in the bootstrap world, the sampling distribution of any measurable function of X_n^* and $\hat{\theta}_n$ can be computed, at least in principle. Monte Carlo or other numerical methods of approximation can be used to this end.

Suppose, for example, we wish to construct a confidence set for the parametric function $\tau(\theta)$, whose range is the set T. By analogy with the classical pivotal method, whose scope is severely limited, let $R_n(X_n, \tau(\theta))$ be a specified real-valued function of the sample and $\tau(\theta)$, to be called a *root*. The sampling distribution of the root under the model is denoted by $H_n(\theta)$. This unknown sampling distribution is plausibly estimated by the plug-in *bootstrap distribution* $H_n(\hat{\theta}_n)$. So defined, the bootstrap distribution is a random probability measure. Alternatively, the bootstrap distribution is the conditional distribution of $R_n(X_n^*, \tau(\hat{\theta}_n))$ given the sample X_n. Most intuitively, the bootstrap distribution is the distribution of the root in the bootstrap world.

Let $H_n^{-1}(\cdot, \hat{\theta}_n)$ denote the quantile function of this bootstrap distribution. The associated *bootstrap confidence set* for $\tau(\theta)$, of nominal coverage probability β, is then

$$C_{n,B} = \{t \in T : R_n(X_n, t) \leq H_n^{-1}(\beta, \hat{\theta}_n)\}. \tag{15.1}$$

The quantile on the right can be approximated, for instance, by Monte Carlo techniques. The intuitive expectation is that the coverage probability of $C_{n,B}$ will be close to β whenever $\hat{\theta}_n$ is close to θ.

This brief account of the general bootstrap strategy omits several important issues. First, for each statistical model, there may be several plausible bootstrap worlds, each corresponding to a different choice of the estimator $\hat{\theta}_n$. Second, theoretical support is needed for the belief that what happens in the bootstrap world closely mimics what happens in the original world of the statistical model. Asymptotic results on the consistency of the bootstrap distribution $H_n(\hat{\theta}_n)$ for the sampling distribution $H_n(\theta)$ have provided a useful starting point, as described in the next section. Third, the basic plug-in strategy in the bootstrap can and sometimes must be generalized. For some problems where a high- or infinite-dimensional θ lacks a consistent estimator, it may be possible to construct a useful bootstrap world by mimicking only certain aspects of the original model. The concept of effective simulation is broader than any specific recipe.

15.3. When does the Bootstrap Work?

Answers to this question depend, of course, on what we mean by "work". Bootstrap samples are perturbations of the data from which they are generated. If the goal is to explore how a statistical procedure performs on data sets similar to the one at hand, then running the statistical procedure on bootstrap samples works by definition. This rationale for the bootstrap is particularly attractive when probability models for the data are dubious.

An Introduction to the Bootstrap uses a technical notion of bootstrap success that appeared in Efron (1979) and, since then, in much of the bootstrap literature. This formulation assumes that the probability model $P_{\theta,n}$ for the data is credible. "Work" is taken to mean that bootstrap distributions, and interesting functionals thereof, converge in probability to the correct limits as sample size increases. The convergence is typically established pointwise for each value of θ in the parameter space.

A broadly useful strategy for verifying correct convergence of bootstrap distributions is based on the following observation: Suppose that the parameter space Θ is metric and that

- $\hat{\theta}_n \to \theta$ in $P_{\theta,n}$-probability as $n \to \infty$;
- for any sequence $\{\theta_n \in \Theta\}$ that converges to θ, $H_n(\theta_n) \Rightarrow H(\theta)$.

Then $H_n(\hat{\theta}_n) \Rightarrow H(\theta)$ in $P_{\theta,n}$-probability. In particular, any weakly continuous functional of the bootstrap distribution converges in probability to the value of that functional at the limit distribution.

This basic equicontinuity reasoning for establishing that the bootstrap works, tweaked or restated in various ways, is widespread in the bootstrap literature (cf. Beran 2003 for references). The second bulleted condition is easier to verify if the metric on Θ is strong, so that the class of convergent sequences $\{\theta_n\}$ is relatively small. The first condition is easier to satisfy if the metric is weak.

The earlier chapters of the Efron–Tibshirani book provide a rich variety of bootstrap examples to which the foregoing convergence reasoning applies. These include: bootstrap standard errors, one- and two-sample problems, fits to linear regression models, bias estimates, bootstrap prediction errors, transformation-respecting confidence intervals versus bootstrap confidence sets based on roots, and bootstrap hypothesis testing. Particularly insightful, motivating further research by others, are chapters that relate successful bootstrap methods to the jackknife, to cross-validation, and to permutation tests, and chapters that address higher-order bootstrap asymptotics.

In general statistical models, skill may be needed to devise a metric on the parameter space under which the two sufficient conditions for bootstrap convergence hold. The book side-steps some of the technical issues by emphasizing nonparametric models with finite support. The examples cited from the book may leave the reader with the impression that bootstrap distributions often work, in the sense of correct pointwise asymptotic convergence over the parameter space. Research by others has strengthened this impression, subject to an important qualification whose significance is discussed in Sect. 4 of this essay.

For instance, Putter (1994) showed: Suppose that the parameter space Θ is complete metric and

- $H_n(\theta) \Rightarrow H(\theta)$ for every $\theta \in \Theta$ as $n \to \infty$;
- $H_n(\theta)$ is continuous in θ, in the topology of weak convergence, for every $n \geq 1$;
- $\hat{\theta}_n \to \theta$ in $P_{\theta,n}$-probability for every $\theta \in \Theta$ as $n \to \infty$.

Then $H_n(\hat{\theta}_n) \Rightarrow H(\theta)$ in $P_{\theta,n}$-probability for "almost all" $\theta \in \Theta$. The technical definition of "almost all" is a set of Baire category II.

While the "almost all" sounds innocuous, it turns out that failure of bootstrap convergence on such a small set in the parameter space is typically a symptom of non-uniform convergence of bootstrap distributions over neighborhoods of that small set. Pointwise limits are deceptive when asymptotics are non-uniform. The difficulty is not just theoretical. As we will see in the next section, it can strike at the heart of bootstrap methods for modern regularized statistical procedures.

15.4. Penalized Least Squares and the Bootstrap

Chapter 18 of the Efron–Tibshirani book treats nonparametric regression by penalized least squares (PLS), the penalty weight being chosen to minimize a bootstrap estimate of prediction error. The statistical model used treats the observed response–covariate pairs as an independent, identically-distributed sample of size n drawn from an unknown distribution. Because PLS is also used with other statistical models, this example provokes further questions, including:

- When can intuitive bootstrap methods select a good data-based penalty weight?
- When can we trust an intuitive bootstrap distribution for the PLS estimator defined by this data-based penalty weight?

The following simple example of PLS estimation in a fixed effects model illustrates the force of these questions and exhibits naturally arising modes of bootstrap failure.

Suppose that $X_n = (x_1, x_2, \ldots, x_n)$ consists of n independent, identically-distributed random p-vectors, each having a $N_p(\theta, I_p)$ distribution, where θ is an unknown p-vector. Let $\bar{X}_n$ denote the sample mean vector and let $|\cdot|$ denote Euclidean norm. To estimate θ, consider the candidate estimator that minimizes, over $\theta \in R^p$, the PLS criterion

$$\sum_{i=1}^{n} |x_i - \theta|^2 + \lambda|\theta|^2, \qquad \lambda \geq 0. \tag{15.2}$$

If we allow λ to range over $[0, \infty]$, then this class of candidate PLS estimators coincides with the class of shrinkage estimators

$$\hat{\theta}_n(c) = c\bar{X}_n, \qquad 0 \leq c \leq 1. \tag{15.3}$$

The normalized quadratic risk of $\hat{\theta}_n(c)$ is

$$r_n(c, \theta) = p^{-1}\mathrm{E}|\hat{\theta}_n(c) - \theta|^2 = c^2/n + (1 - c)^2|\theta|^2/p. \tag{15.4}$$

Because $\bar{X}_n$ is consistent for θ, a natural parametric bootstrap estimator of the risk is

$$r_n(c, \bar{X}_n) = c^2/n + (1-c)^2|\bar{X}_n|^2/p. \tag{15.5}$$

The value of c that minimizes the bootstrap risk (5) is

$$\hat{c}_{n,B} = n|\bar{X}_n|^2/(p + n|\bar{X}_n|^2). \tag{15.6}$$

Unfortunately, the corresponding adaptive PLS estimator $\hat{\theta}_n(\hat{c}_{n,B})$ is known to be unsatisfactory. It is dominated for every $p \geq 3$, substantially so for large p, by the James–Stein estimator $\hat{\theta}_n(\hat{c}_{n,S})$, where

$$\hat{c}_{n,S} = 1 - (p-2)/(n|\bar{X}_n|^2). \tag{15.7}$$

From the work of Mallows (1973) on C_p or the more general results of Stein (1981), it is known the bootstrap risk estimator $r_n(c, \bar{X}_n)$ is biased upwards. Bias-correction of the bootstrap risk estimator before minimizing over c is a good idea in this example. To do so, replace $|\bar{X}_n|^2$ in (5) by $|\bar{X}_n|^2 - p/n$. The value of c that minimizes the bias-corrected bootstrap risk estimator differs from the James–Stein value (7) only in using p instead of the better $p-2$.

The bootstrap algorithm we used in this simple PLS or James–Stein example can be patched so as to generate an unbiased bootstrap estimator for the risk of $\hat{\theta}_n(c)$ and, thereby, a satisfactory minimizing value of the shrinkage parameter c. This does not change the main points:

- The need for such a bootstrap patch is not obvious without detailed mathematical analysis of the example.
- Intuitive bootstrapping can mislead seriously when selecting penalty weights for PLS in fixed effects models.

What about bootstrap distributions for PLS estimators that use a good data-based choice of penalty weight? To continue with the James–Stein version of the example, let $H_n(\theta)$ be the distribution of $n^{1/2}(\hat{\theta}_n(\hat{c}_{n,S}) - \theta)$ under the $N_p(\theta, I_p)$ model for the data vectors. Let Z be a random p-vector with the standard $N_p(0, I_p)$ distribution. For every p-vector h, consider the probability measure

$$\pi(h) = \mathcal{L}[Z - \{(p-2)/|Z+h|^2\}(Z+h)]. \tag{15.8}$$

Then, from Beran (1997),

- As n increases, $H_n(\theta) \Rightarrow N_p(0, I_p)$ when $\theta \neq 0$ but $H_n(0) \Rightarrow \pi(0)$.
- The intuitive bootstrap distribution $H_n(\bar{X}_n)$ converges weakly in probability to the correct $N_p(0, I_p)$ distribution when $\theta \neq 0$. However, if $\theta = 0$, then $H_n(\bar{X}_n)$ converges *in distribution*, as a random probability measure, to the random probability measure $\pi(Z)$ rather than to the desired constant probability measure $\pi(0)$.

These and other technical findings from the last century bring out several points about the distribution and bootstrap distribution of the James–Stein estimator:

- The bootstrap distribution in this example converges correctly almost everywhere on the parameter space, except at $\theta = 0$, in accordance with results mentioned at the end of Sect. 3.
- Unfortunately, the weak convergence of the sampling distribution and of the corresponding bootstrap distribution is not uniform over neighborhoods of the point of bootstrap failure, $\theta = 0$.
- Through Stein's exact analysis for $p \geq 3$, the James–Stein estimator dominates $\bar{X}_n$ in quadratic risk at *every* θ, especially at $\theta = 0$. If the dimension p is fixed, the region of substantial dominance shrinks towards $\theta = 0$ as n increases. In asymptotic risk, the James–Stein estimator dominates the sample mean only at $\theta = 0$. This non-uniform limit loses important information.
- The bootstrap "works" concept used in Sect. 3 and in much of the bootstrap literature to date does not suffice for trustworthy understanding of the bootstrap in the James–Stein example or in the linked PLS example.

The James–Stein example is a prototype of modern estimators that dominate their classical counterparts in risk and fail to bootstrap naively. In fixed effects models, these include regularized estimators obtained from adaptive penalized least squares, adaptive submodel selection, or adaptive multiple shrinkage. Since the publication of the Efron–Tibshirani book, one line of research has sought to characterize those estimators that can be bootstrapped naively, using asymptotics where n tends to infinity (cf. Beran 1997; van Zwet and van Zwet 1999). The results supplement the earlier findings sketched in this essay. Another line of research has studied bootstrap procedures under asymptotics in which the dimension of the parameter space increases while n is held fixed. Such bootstrap asymptotics turn out to be uniform over usefully-large subsets of the parameter space and have generated promising bootstrap confidence sets around the James–Stein estimator and around other regularization estimators (cf. Beran 1995; Beran and Dümbgen 1998).

This essay outlines the writer's retrospective reflections on selected chapters in *An Introduction to the Bootstrap*. Considerably more could be discussed, the book containing a wealth of provocative examples. The success of the book in describing Efron's unique approaches to the bootstrap and in stimulating research by his co-authors and others should be evident from the little I have said. How to bootstrap successfully when the "dimension" of the parameter space is comparable to or larger than sample size remains a major topic only lightly explored. Developments in empirical process theory play an important role in these investigations. The Efron–Tibshirani book continues to provide a baseline for exploration of bootstrap or bootstrap-like methodologies. Thinking about the bootstrap raises fundamental questions about the nature of statistical theory, with and without probability models. The magnitude of the research challenge is fortunate for the long-term intellectual future of Statistics.

References

Beran, R. (1995). Stein confidence sets and the bootstrap. *Statistica Sinica* **5**, 109–127.

Beran, R. (1997). Diagnosing bootstrap success. *Annals of the Institute of Statistical Mathematics* **49**, 1–24.

Beran, R. (2003). The impact of the bootstrap on statistical theory and algorithms. *Statistical Science* **18**, 175–184.

Beran, R. and Dümbgen, L. (1998). Modulation of estimators and confidence sets. *Annals of Statistics* **26**, 1826–1856.

Crombie, A. C. (1953). *Robert Grosseteste and the Origins of Experimental Science 1100–1700*. Clarendon, Oxford.

Efron, B. (1979). Bootstrap methods: Another look at the jackknife. *Annals of Statistics* **7**, 1–26.

Efron, B. and Tibshirani, R. (1993). *An Introduction to the Bootstrap*. Chapman and Hall, New York.

Fisher, R. A. (1956). *Statistical Methods and Scientific Inference*. Oliver and Boyd, Edinburgh.

Knuth, D. E. (1969). *The Art of Computer Programming* vol. 2. Addison-Wesley, Reading, MA.

Kuhn, T. S. (1959). *The Copernican Revolution* (second corrected printing). Vintage, New York.

Le Cam, L. (1953). On some asymptotic properties of maximum likelihood estimators and related Bayes estimates. *University of California Publications in Statistics* **1**, 277–330.

Mallows, C. (1973). Some comments on C_p. *Technometrics* **15**, 661–675.

Politis, D. N., Romano, J. P. and Wolf, M. (1999). *Subsampling*. Springer, New York.

Putter, H. (1994). *Consistency of Resampling Methods*. Ph.D. dissertation, Leiden University.

Savage, L. J. (1954). *The Foundations of Statistics*. Wiley, New York.

Stein, C. (1956). Inadmissibility of the usual estimator for the mean of a multivariate normal distribution. In *Proceedings of the Third Berkeley Symposium on Mathematical Statistics and Probability*, 197–206. Edited by J. Neyman. University of California Press, Berkeley.

Stein, C. (1981). Estimation of the mean of a multivariate normal distribution. *Annals of Statistics* **9**, 1135–1151.

Tukey, J. W. (1977). *Exploratory Data Analysis*. Addison-Wesley, Reading, MA.

van Zwet, E. W. and van Zwet, W. R. (1999). A remark on consistent estimation. *Mathematical Methods of Statistics* **8**, 277–284.

Wald, A. (1950). *Statistical Decision Functions*. Wiley, New York.

Contents

16

Using Specially Designed Exponential Families for Density Estimation

Introduction by Nancy Reid
University of Toronto

This paper grew out of a talk given at the Statistical Society of Canada's Annual Meeting in Banff, Alberta, in 1994; Rob Tibshirani was the invited discussant. Brad has been a good friend to statistics in Canada over the years, and I am very pleased to have the opportunity to comment on a paper with strong Canadian connections.

The paper is ostensibly concerned with density estimation, but as with so much of Brad's work, it touches on a wide variety of related topics. The method introduced combines a kernel density estimate with a parametric model; this model is the "specially designed exponential family" of the title. The result is that the density estimation problem becomes semi-parametric instead of non-parametric, and the resulting estimator is automatically smoothed by the introduction of the parametric components.

The model for the data is

$$g(y) = g_0(y)\exp\{\beta_0 + t(y)\beta_1\}; \tag{16.1}$$

the parametric part of the model is a constructed exponential family for a set of statistics $t(y)$. The term $\exp(\beta_0)$ normalizes the density estimate to integrate to 1; so β_1 is the only free parameter in the model. The form of the density estimator is

$$\hat{g}(y) = \hat{g}_0(y)\exp\{\hat{\beta}_0 + t(y)\hat{\beta}_1\}$$

where $\hat{g}_0(\cdot)$ is a kernel density estimate, and $\exp(\hat{\beta}_0)$ is the normalizing constant for the estimator. The effect of this construction is to ensure that the mean of $t(y)$ under $\hat{g}$ is the same as the observed sample average of $t(y_i)$ based on the sample $y_1, \ldots, y_n$.

After illustrating the use of the new method on a univariate example, the fitting method and construction of the exponential family is detailed for a discretized version of the problem, using a Poisson/multinomial formulation. This has computational advantages, as needed integrals become finite sums, but it also makes theoretical analysis of the resulting estimator more transparent. In Section 2 the computational aspects are illustrated, and in Section 3 the properties of the method are investigated, with emphasis on estimation of the covariance matrix of the exponential family parameters. These estimated parameters smooth the density estimate, so having an estimate of their standard errors provides an assessment of how much smoothing is needed. The approach is illustrated on a bivariate example in Section 4, and further theoretical analysis in Sections 5 and 6 outline how ideas from the theory of smoothing, such as estimated degrees of freedom, and ideas from the theory of exponential families, such as deviance, can be extended to this density estimation problem. In Section 7 the method is extended to multi-sample problems, and illustrated on a clinical trial in which the treatment and control groups serve as the two samples. The methodology developed in this paper is, in Section 8, related to several ideas current in the literature at the time, and the paper closes with the familiar series of "Remarks".

The work in this paper is, as far as I know, the only excursion of Brad's research into density estimation, but it combines his long-standing interests in nonparametrics and exponential families in a typically ingenious way. The parameter β_1 is estimated by the usual maximum likelihood formulation for fixed $g_0(\cdot)$, but its properties, such as its mean and covariance matrix, do need to take into account the fact that $\hat{g}_0(\cdot)$ was estimated from the data. The discretization described in Section 2 enables both easy computation of $\hat{\beta}_1$, using generalized linear models, and a fairly accessible way to estimate its covariance matrix. The use of multinomial models to simplify analysis and computation is a familiar theme in Brad's work. Another hallmark is the incorporation of applications from collaborative research; Brad's writing always seems to make a convincing case that the new theory being discussed was genuinely motivated by the science. Geometric arguments, persuasive graphical displays (although a bit dated here), and a rapid-fire string of new concepts and statistical pyrotechnics combine in quintessential 'Brad-style': often imitated but rarely equalled.

The multi-sample section is not developed in detail, but has echoes in his work on empirical Bayes methods, and ideas related to it have appeared in the literature in various guises. Qin (1999) connects (1) to Anderson's (1979) density ratio model: if the sufficient statistic $t(x)$ indexes the populations, then (1) becomes

$$f_2(x) = f_1(x)\exp\{\alpha + \beta t(x)\} \tag{16.2}$$

which Qin showed can can be interpreted as a weighted sampling model, as well as a semi-parametric density estimation specification. One version of this is retrospective sampling in case-control models. This model and a mixture generalization of it is analyzed by empirical likelihood methods in Qin (1999), with emphasis on efficient estimation of the mixing proportion. Zhang (2000) uses the density ratio model in an M-estimation problem. An m-sample version of the problem is discussed in Fokianos (2004). Qin and Zhang (2005) use (2) with $t(x) = (x, x^2)$ and emphasize improved estimation of the underlying density $f_x(\cdot)$.

The special exponential family version was used in Efron and Petrosian (1999) to analyse the quasar data, and again in Efron (2004), where he showed that with a large number of test statistics arising from the same construction, such as in microarray analysis, the smooth density estimate can be used to establish an empirical null hypothesis. These techniques would seem to be potentially useful in high energy physics problems as well, where bumpy histograms are often the basic unit of analysis. A Bayesian version of the problem is discussed in Walker, Damien and Lenk (2004).

As discussed in Sections 1 and 8, Hjort and Glad (1995) proposed a very similar approach, in which the parametric part is fitted to the data first, and the nonparametric smoother is fit to the residuals. Looking at the two approaches simultaneously leads to the connections to backfitting in various types of semiparametric and nonparametric models. Related work is summarized in Glad, Hjort and Ushakov (2003).

References

Efron, B. (2004). Large-scale simultaneous hypothesis testing: The choice of a null hypothesis. *Journal of the American Statistical Association* **99**, 96–104.

Efron, B. and Petrosian, V. (1999). Nonparametric methods for doubly truncated data. *Journal of the American Statistical Association* **94**, 824–834.

Glad, I. K., Hjort, N. L. and Ushakov, N. G. (2003). Correction of density estimators that are not densities. *Scandinavian Journal of Statistics* **30**, 415–427.

Fokianos, K. (2004). Merging information for semi-parametric density estimation. *Journal of the Royal Statistical Society, Series B* **66**, 941–958.

Hjort, N. and Glad, I. (1995). Nonparametric density estimation with a parametric start. *Annals of Statistics* **23**, 882–904.

Qin, J. (1999). Empirical likelihood ratio based confidence intervals for mixture proportions. *Annals of Statistics* **27**, 1368–1384.

Qin, J. and Zhang, B. (2005). Density estimation under a two-sample semiparametric model. *Journal of Nonparametric Statistics* **17**, 665–683.

Walker, S., Damien, P. and Lenk, P. (2004). On priors with a Kullback–Leibler property. *Journal of the American Statistical Association* **99**, 404–408.

Zhang, B. (2000). M-estimation under a two-sample semiparametric model. *Scandinavian Journal of Statistics* **27**, 263–280.

The Annals of Statistics
1996, Vol. 24, No. 6, 2431–2461

USING SPECIALLY DESIGNED EXPONENTIAL FAMILIES FOR DENSITY ESTIMATION

BY BRADLEY EFRON[1] AND ROBERT TIBSHIRANI

Stanford University

We wish to estimate the probability density $g(y)$ that produced an observed random sample of vectors $y_1, y_2, \ldots, y_n$. Estimates of $g(y)$ are traditionally constructed in two quite different ways: by maximum likelihood fitting within some parametric family such as the normal or by nonparametric methods such as kernel density estimation. These two methods can be combined by putting an exponential family "through" a kernel estimator. These are the specially designed exponential families mentioned in the title. Poisson regression methods play a major role in calculations concerning such families.

1. Introduction. Suppose that we wish to estimate the probability density $g(y)$ that produced an observed random sample of vectors $y_1, y_2, \ldots, y_n$,

$$y_i \overset{\text{i.i.d.}}{\sim} g(y) \quad \text{for } i = 1, 2, \ldots, n. \tag{1.1}$$

The vectors y_i take values in a sample space $\mathscr{Y}$. The numerical examples in this paper have $\mathscr{Y}$ being portions of the real line or of the plane, but the methodology applies just as well to higher dimensionalities and to more complicated spaces.

Estimates of $g(y)$ are traditionally constructed in two quite different ways: by maximum likelihood fitting within some parametric family such as the normal or by nonparametric methods such as kernel density estimation. These two methods can be combined by putting an exponential family "through" a nonparametric estimator. The resulting hybrid estimators are the specially designed exponential families of the title.

Figure 1 shows a simple example of this methodology. The y_i are pain scores for $n = 67$ women, each obtained by averaging the results from a questionnaire administered after an operation. The scale runs from $y = 0 =$ no pain to $y = 4 =$ worst pain, so the sample space $\mathscr{Y}$ is the interval $[0, 4]$. The 67 scores y_i, indicated by the histogram, run from 0.02 to 3.08. The dashed curve $\hat{g}_0(y)$ is a normal kernel density estimator with window width $\lambda = 1$, described more carefully in Section 2. Also shown are two special exponential family estimates, $\hat{g}_1(y)$ and $\hat{g}_2(y)$, described below.

Received January 1995; revised October 1995.
[1]Supported by NSF Grant DMS-95-04379 and Public Health Service Grant 5 ROI CA59039-20.
AMS 1991 *subject classifications*. 62F05, 62G05.
Key words and phrases. Poisson regression, degrees of freedom, expected deviance, local and global smoothing, moment-matching.

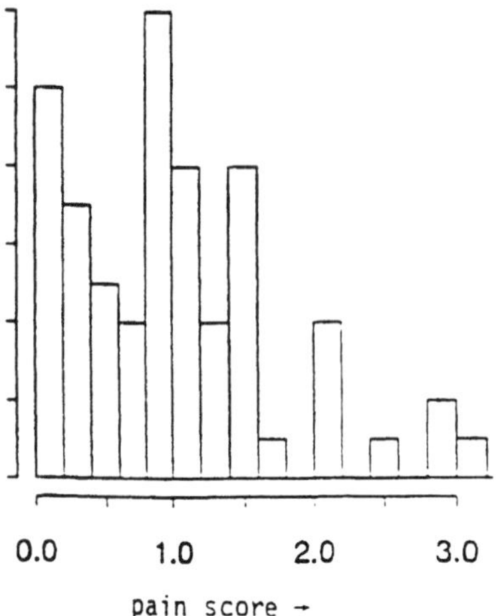

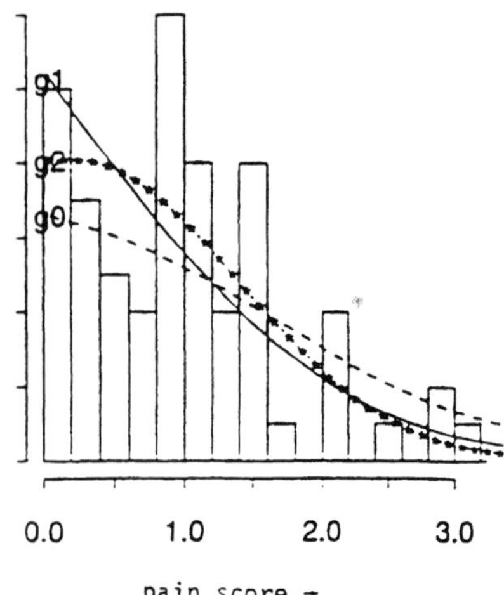

FIG. 1. *Pain-score data. Left: Histogram of pain scores for* 67 *women following an operation:* 0 = *no pain;* 4 = *worst pain. Right:* $\hat{g}_0(y)$ *is normal kernel density estimator, window width* 1; $\hat{g}_1(y)$ *is the special exponential family through* $\hat{g}_0(y)$ *with sufficient statistic* $t(y) = y$; $\hat{g}_2(y)$ *uses sufficient statistics* $t(y) = (y, y^2)$. *Density* $\hat{g}_2(y)$ *matches the empirical mean and variance of the* 67 *data points.*

An exponential family of densities on $\mathscr{Y}$, $\{g_\beta(y)\}$ is given by

$$g_\beta(y) = g_0(y)\exp(\beta_0 + t(y)\beta_1), \tag{1.2}$$

which is called the exponential family through $g_0(y)$ with sufficient statistics $t(y)$. Here $g_0(y)$ is a *carrier density*, $t(y)$ is a $1 \times p$ vector of sufficient statistics, β_1 is a $p \times 1$ parameter vector and β_0 is a normalizing parameter that makes $g_\beta(y)$ integrate to 1 over $\mathscr{Y}$. For example, the one-dimensional normal family, with all possible choices of expectation and variance, can be obtained using the standard normal carrier $g_0(y) = \varphi(y) = \exp\{-0.5y^2\}/\sqrt{2\pi}$, with the sufficient statistics $t(y) = (y, y^2)$. The densities in (1.2) are defined with respect to some background measure, which we will take to be Lebesgue measure. See Section 1.4 of Lehmann (1983).

The estimates $\hat{g}_1(y)$ and $\hat{g}_2(y)$ in Figure 1 are of the form

$$g_{\hat{\beta}}(y) = \hat{g}_0(y)\exp(\hat{\beta}_0 + t(y)\hat{\beta}_1), \tag{1.3}$$

where $\hat{g}_0(y)$ is the kernel density estimate indicated by the dashed curve. Estimate $\hat{g}_1(y)$ uses the single sufficient statistic $t(y) = y$, so $p = 1$, while $\hat{g}_2(y)$ uses $t(y) = (y, y^2)$, $p = 2$. The parameter values $\hat{\beta} = (\hat{\beta}_0, \hat{\beta}_1)$ were chosen by maximum likelihood, that is, by maximizing $\prod_{i=1}^n g_{\hat{\beta}}(y_i)$, ignoring the fact that the carrier $\hat{g}_0(y)$ is itself data-dependent. This choice of β matches the $t(y)$ moments of $g_{\hat{\beta}}(y)$ to their empirical averages:

$$\int_{\mathscr{Y}} t(y) g_{\hat{\beta}}(y)\, dy = \frac{1}{n} \sum_{i=1}^{n} t(y_i). \tag{1.4}$$

Thus $\hat{g}_2$ matches the first two empirical moments of the 67 pain scores or, equivalently, it matches the empirical mean and variance. Property (1.4) implies that the special exponential family estimate of $g(y)$ is unbiased for the moments of $t(y)$. Linear transformations to "post-repair" the mean and variance of a kernel estimate are familiar in the density-estimation literature; see, for example, Jones (1993). The two-dimensional example of Section 4 shows a more ambitious example of moment-matching.

We can think of a special exponential family estimator in two complementary ways: (1) as being a standard exponential family estimator, except one that is preceded by an adaptive choice of the carrier, or (2) as being a standard nonparametric smoothing estimator, except one that is followed by a correction to match certain sample moments.

Either way, we will argue that special exponential families can be a favorable compromise between parametric and nonparametric density estimation. From the first point of view, we will be able to use exponential family theory more flexibly than when restricted to the usual small catalogue of normals, gammas, betas and so forth. An approach more in the spirit of a regression analysis than a density estimate is possible, including exploratory choices of the sufficient statistics $t(y)$.

From the second point of view, the moment-matching correction will usually reduce the bias of a nonparametric smoother, since the moments $t(y)$ are estimated unbiasedly. This has an important practical consequence shown in the numerical examples: *it allows the nonparametric smoother to use a substantially greater window width without badly degrading the overall fit to the data*. The result is a substantially smoother estimate of the density $g(y)$. (This phenomenon is illustrated in the bivariate example of Figures 3 and 4.)

To put things another way, the special exponential family estimate (1.3) works at two different scales. The nonparametric smoother allows local adaptation to the data, while the exponential term matches some of the data's global properties.

Sections 2–5 describe how to compute and interpret special exponential family (SEF) estimates such as those in Figure 1. Most of our computations are done using a Poisson regression model for density estimation (Section 2), originally introduced in Lindsey (1974a, b). Section 3 gives a delta-method formula for the covariance of $\hat{\beta}_1$ in (1.3), which takes into account the data-based choice of the carrier $\hat{g}_0(y)$. We can use this formula in the usual way to select among possible choices of the sufficient statistic. Section 6 concerns formulas for choosing the window width of the smoother that produces $\hat{g}_0(y)$. This choice involves "degrees of freedom" calculations like those introduced by Hastie and Tibshirani (1990). Multisample SEF estimates are discussed in Section 7. Remarks appear in Section 9.

Special exponential families are an example of what Green and Silverman (1994) call *semiparametric methods*. Many other semiparametric methods have been proposed for density estimation. Hjort and Glad (1995) propose reversing the SEF order: first fit a parametric family to the data and then fit a nonparametric smoother to the residuals from the parametric estimator.

Hastie and Tibshirani's *backfitting* approach can also be applied to semiparametric density estimation. It amounts to an iteration to convergence between the smoother and the exponential family. The principal advantage of the SEF estimator is its theoretical tractability. The relationship between these ideas is discussed in Section 8.

Stone (1994) and Kooperberg and Stone (1991) use adaptive exponential families based on spline functions for density estimation. The carrier $g_0(y)$ is effectively the uniform density in this work, but adaptation is still possible because of the local character of the spline basis. Olkin and Spiegelman (1987) combine parametric and nonparametric density estimates by mixtures, rather than by exponential family methods as in this paper.

2. Poisson regression for density estimation. Density estimation problems can be rephrased in terms of Poisson regression models. This technique was introduced by Lindsey (1974a, b) as a way of using generalized linear model software to fit difficult exponential family models. See also Aitken (1993), Lindsey and Mersch (1992) and Efron (1988). Our results here will be phrased in Poisson regression terms, mainly for reasons of conceptual clarity, though there are also some computational advantages.

Lindsey's construction begins with a discretization of the problem: the sample space $\mathscr{Y}$ is partitioned into K disjoint cells $\mathscr{Y}_k$,

$$\mathscr{Y} = \bigcup_{k=1}^{K} \mathscr{Y}_k, \tag{2.1}$$

and the data $\mathbf{y} = (y_1, y_2, \ldots, y_n)$ are reduced to the cell counts

$$s_k = \#\{y_i \in \mathscr{Y}_k\} \quad \text{for } k = 1, 2, \ldots, K. \tag{2.2}$$

The vector of counts $\mathbf{s} = (s_1, s_2, \ldots, s_K)$ has sum $s_+ = n$. Table 1 shows the counts for the pain-score data of Figure 1, where $\mathscr{Y} = [0, 4]$ has been partitioned into $K = 40$ cells $\mathscr{Y}_k$ each of length 0.1.

Suppose that $\{g_\theta(y),\ \theta \in \Theta\}$ is a family of probability densities on $\mathscr{Y}$. The probability of observing y in the kth cell is

$$\pi_k(\theta) = \int_{\mathscr{Y}_k} g_\theta(y)\, dy, \tag{2.3}$$

TABLE 1

Discretized version of the pain-score data of Figure 1; $\mathscr{Y} = [0, 4]$ *partitioned into* $K = 40$ *cells of length* 0.1; $s_1 = 3$ *of the* 67 *scores occurred in* $\mathscr{Y}_1 = [0, 0.1)$, 7 *in* $\mathscr{Y}_2 = [0.1, 0.2)$ *and so forth*

3	7	6	1	2	3	3	1	7	5
4	4	1	3	3	5	0	1	0	0
2	2	0	0	0	1	0	0	1	1
1	0	0	0	0	0	0	0	0	0

with sum $\pi_+(\theta) = 1$. Then $\mathbf{s}$ has a multinomial distribution on K categories, with n draws and probability vector $\boldsymbol{\pi}(\theta) = (\pi_1(\theta), \pi_2(\theta), \dots, \pi_K(\theta))$,

$$\mathbf{s} \sim \mathrm{Mult}_K(n, \boldsymbol{\pi}(\theta)). \tag{2.4}$$

We could find the maximum likelihood estimate (MLE) of θ, based on $\mathbf{s}$, by maximizing the multinomial probability of $\mathbf{s}$.

Instead we consider the s_k to be *independent* Poisson observations

$$s_k \overset{\mathrm{ind}}{\sim} \mathrm{Po}(\mu_k(\gamma, \theta)), \qquad k = 1, 2, \dots, K, \tag{2.5}$$

with expectations

$$\mu_k(\gamma, \theta) = \gamma \pi_k(\theta). \tag{2.6}$$

Here γ is a free parameter, restricted only to be positive. Standard Poisson properties allow (2.5) and (2.6) to be expressed as

$$s_+ \sim \mathrm{Po}(\gamma) \quad \text{and} \quad s | s_+ \sim \mathrm{Mult}_K(s_+, \boldsymbol{\pi}(\theta)). \tag{2.7}$$

This means that the maximum likelihood estimates from (2.5) and (2.6) are

$$\hat{\gamma} = s_+ = n, \tag{2.8}$$

and $\hat{\theta}$ is equal to the MLE for θ in (2.4). Lindsey's method is to (approximately) maximize the original likelihood $\prod_{i=1}^n g_\theta(y_i)$ by finding the Poisson MLE in (2.5) and (2.6). *Note*: The parameters γ and θ are orthogonal in Cox and Reid's (1987) sense, so that the information for estimating θ is the same in (2.4) and (2.5).

This method of finding the MLE is particularly convenient when the original densities are of the exponential family form (1.2). We will consider the density estimation problem (1.1) and (1.2) in the Poisson regression form (2.5) and (2.6):

$$s_k \overset{\mathrm{ind}}{\sim} \mathrm{Po}(\mu_k(\beta)) \quad \text{for } k = 1, 2, \dots, K, \tag{2.9a}$$

with

$$\mu_k(\beta) = \mu_k^o \exp(\beta_0 + t_k \beta_1). \tag{2.9b}$$

Here μ_k^o is proportional to $\pi_k^o = \int_{\mathscr{Y}_k} g_0(y)\, dy$, a discretized version of the carrier, and

$$t_k = t(y_{(k)}), \tag{2.10}$$

the sufficient vector $t(y)$ evaluated at a convenient point $y_{(k)}$ in $\mathscr{Y}_k$. The free parameter β_0 corresponds to $\gamma = e^{\beta_0}$ in (2.6).

Define X to be the $K \times (p+1)$ matrix whose kth row equals

$$x_k = (1, t_k). \tag{2.11}$$

The maximum likelihood equations for $\beta = (\beta_0, \beta_1)$ in the generalized linear model (2.9) are

$$X'\left[\mathbf{s} - \boldsymbol{\mu}(\hat{\beta})\right] = 0, \tag{2.12}$$

where $\boldsymbol{\mu}(\hat{\beta})$ indicates the vector with kth component $\mu_k^o \exp(x_k \hat{\beta})$. Standard generalized linear model software easily solves for $\hat{\beta}$ in (2.12), even for difficult nonstandard forms of the exponential family (1.2). This was Lindsey's principal point.

Here is how Figure 1 was constructed. The problem was discretized into $K = 40$ cells as in Table 1. The carrier vector $\boldsymbol{\mu}^o = (\mu_1^o, \mu_2^o, \ldots, \mu_K^o)$ in (2.9b) was estimated using a $K \times K$ smoothing matrix $M(\lambda)$,

$$\hat{\boldsymbol{\mu}}^o = M(\lambda)\mathbf{s}. \tag{2.13}$$

Matrix $M(\lambda)$ was taken to be a normal kernel smoother, having kjth element

$$M_{kj}(\lambda) = \frac{c_k}{\lambda}\varphi\left(\frac{y_{(k)} - y_{(j)}}{\lambda}\right), \tag{2.14}$$

with $y_{(k)} = (k - 0.5)/10$, the midpoint of cell $\mathscr{Y}_k$. The constants c_k were chosen to make $M_{k+} = 1$. The starred curve labelled $\hat{g}_0$ in Figure 1 is actually $\hat{\mu}_k^o$ plotted as a function of $y_{(k)}$, with the window width λ in (2.14) set equal to 1.

The curves labelled $\hat{g}_1(y)$ and $\hat{g}_2(y)$ are really the discrete analogs of the special exponential family (1.3), say $\hat{\boldsymbol{\mu}} = (\hat{\mu}_1, \hat{\mu}_2, \ldots, \hat{\mu}_K)$,

$$\hat{\mu}_k = \hat{\mu}_k^o \exp\left(\hat{\beta}_0 + t_k \hat{\beta}_1\right), \tag{2.15}$$

plotted versus $y_{(k)}$: $\hat{g}_1$ uses $t_k = y_{(k)}$, while $\hat{g}_2$ is based on the quadratic vector $t_k = (y_{(k)}, y_{(k)}^2)$. The MLE estimates $\hat{\beta}$ were obtained by iterative solution of (2.12), so

$$X'[\mathbf{s} - \hat{\boldsymbol{\mu}}] = 0, \tag{2.16}$$

where X is a 40×2 matrix for $\hat{g}_1$ and a 40×3 matrix for $\hat{g}_2$. [The estimates were actually computed using a centered version of y, $\tilde{y}_{(k)} = (y_{(k)} - 2)/4$.] To fit this model in the GLIM language or the `glm` function in SPlus, one simply includes log $\hat{\mu}^o$ as an offset in a Poisson generalized linear model.

Equation (2.16) shows that

$$\hat{\mu}_+ = s_+ = n \tag{2.17a}$$

and

$$\sum_{k=1}^{K} \frac{\hat{\mu}_k}{n} t_k = \sum_{k=1}^{K} \frac{s_k}{n} t_k, \tag{2.17b}$$

the discrete analog of the moment-matching property (1.4). Notice that because of (2.17a) and the fact that the cells are of length 0.1, the curves in Figure 1 integrate over $\mathscr{Y}$ to 6.7 rather than to 1.

Changing K to 20 or to 80 made very little difference in Figure 1. The numerical calculations in this paper were insensitive to the form of discretization. In fact, discretization is not really necessary for any of our results, as discussed in Remark E of Section 9.

Nevertheless, it is conceptually easier to discuss special exponential family density estimation in terms of the discrete Poisson model (2.9). The problem becomes one of fitting a smooth regression curve to the independent observations s_k, and this lets us make use of the arsenal of regression tools. It also emphasizes the important point that density estimation is equivalent to Poisson regression, and not to ordinary least squares regression. The Poisson nature of the problem will be evident in the formulas developed below.

3. Estimating the covariance of $\hat{\boldsymbol{\beta}}$. This section derives an approximate covariance matrix for $\hat{\beta} = (\hat{\beta}_0, \hat{\beta}_1)$, the estimated parameters in a special exponential family such as (1.3) or (2.15). The formula for the covariance takes into account the data-based choice of the carrier. We will use the estimated standard errors of the components of $\hat{\beta}$ for model building, checking the significance of the corresponding components of the sufficient statistic $t(y)$ in the usual way.

We consider the Poisson form (2.9) of the SEF model,

$$\mathbf{s} \sim \mathrm{Po}_K(\boldsymbol{\mu}(\beta)) \quad \text{with } \boldsymbol{\mu}(\beta) = \boldsymbol{\mu}^o e^{X\beta}, \tag{3.1}$$

where $\mathrm{Po}_K(\boldsymbol{\mu})$ indicates a vector of K independent Poisson variates having expectations $(\mu_1, \mu_2, \ldots, \mu_K) = \boldsymbol{\mu}$. The notation $\boldsymbol{\mu}^o e^{X\beta}$ indicates the vector with kth component $\mu_k^o e^{x_k \beta}$, $x_k = (1, t_k)$, as in (2.11). Generalizing (2.13), we first estimate $\boldsymbol{\mu}^o$ by some function of $\mathbf{s}$, say $m(\mathbf{s})$, and then solve for $\hat{\beta}$ in the MLE equations (2.16):

$$\hat{\boldsymbol{\mu}}^o = m(\mathbf{s}) \quad \text{and} \quad \hat{\beta}\colon X'\left[\mathbf{s} - \hat{\boldsymbol{\mu}}^o e^{X\hat{\beta}}\right] = 0. \tag{3.2a}$$

The special exponential family estimate of $\boldsymbol{\mu}$ is

$$\hat{\boldsymbol{\mu}} = \hat{\boldsymbol{\mu}}^o e^{X\hat{\beta}}. \tag{3.2b}$$

LEMMA 1. *Let $\hat{D}$ be the $K \times K$ diagonal matrix with kth diagonal element $\hat{\mu}_k = \hat{\mu}_k^o e^{x_k \hat{\beta}}$ and let $\hat{H}$ be the $K \times K$ derivative matrix of* $\log(\hat{\boldsymbol{\mu}}^o) = (\log(\hat{\mu}_1^o), \log(\hat{\mu}_2^o), \ldots, \log(\hat{\mu}_K^o))$ *with respect to* $\mathbf{s}$, *with "j" indexing columns,*

$$\hat{H} = \frac{d \log \hat{\boldsymbol{\mu}}^o}{d\mathbf{s}} = \left(\frac{\partial \log(\hat{\mu}_k^o)}{\partial s_j}\right). \tag{3.3}$$

Then the $(p + 1) \times K$ derivative matrix of $\hat{\beta}$ with respect to $\mathbf{s}$ *is*

$$\frac{d\hat{\beta}}{d\mathbf{s}} = [X'\hat{D}X]^{-1} Z', \tag{3.4a}$$

where

$$Z' = X'(I - \hat{D}\hat{H}). \tag{3.4b}$$

In case (2.13), $\hat{\boldsymbol{\mu}}^o = M\mathbf{s}$, this becomes

$$Z' = X'\left(I - D(e^{X\hat{\beta}})M\right), \tag{3.4c}$$

where $D(e^{X\hat{\beta}})$ is the diagonal matrix with kth diagonal element $e^{x_k \hat{\beta}}$. (*The proof appears below.*)

By the usual delta-method argument, an approximate covariance matrix estimate for $\hat{\beta}$ is given by

$$(3.5)\qquad \left(\frac{d\hat{\beta}}{d\mathbf{s}}\right)\widehat{\mathrm{Cov}}(\mathbf{s})\left(\frac{d\hat{\beta}}{d\mathbf{s}}\right)'.$$

COROLLARY 1. *The* SEF *vector* $\hat{\beta}$ *obtained from* (3.1) *and* (3.2) *has approximate covariance matrix*

$$(3.6a)\qquad \overline{\mathrm{Cov}}(\hat{\beta}) = [X'\hat{D}X]^{-1}[Z'\overline{D}Z][X'\hat{D}X]^{-1},$$

where $\overline{D}$ *is the diagonal matrix with kth diagonal element* s_k. *An alternative estimate is*

$$(3.6b)\qquad \widehat{\mathrm{Cov}}(\hat{\beta}) = [X'\hat{D}X]^{-1}[Z'\hat{D}Z][X'\hat{D}X]^{-1}.$$

For any K-vector $\mathbf{v}$, we let $D(\mathbf{v})$ be the $K \times K$ diagonal matrix with kth diagonal element v_k. The true covariance of $\mathbf{s} \sim \mathrm{Po}_k(\boldsymbol{\mu})$ is $D(\boldsymbol{\mu})$. Approximation (3.6a) estimates Cov(**s**) by $D(\mathbf{s}) = \overline{D}$ in (3.5), while (3.6b) uses $D(\hat{\boldsymbol{\mu}}) = \hat{D}$. The former may be preferred if model (3.1) is suspect. In our numerical examples the two formulas gave nearly the same results.

Table 2 applies the corollary to the pain-score data. The quadratic model, described in (2.13)–(2.16) has $\hat{\beta}_1 = (-2.74, -3.80)$, with standard errors $(0.93, 2.45)$ according to (3.6a). [$\hat{\beta}_0$ serves only to normalize $\hat{\boldsymbol{\mu}}$ to $\hat{\mu}_+ = n$, (2.17a), so its value is of no statistical interest.] The coefficient of $\tilde{y} = (y - 2)/4$, the centered version of y, is nearly 3 standard errors below zero. The coefficient of $\tilde{y}^2$ is -1.55 standard errors below zero, so it is not so clear that the quadratic term is significant. The cubic model, in which $t(y) = (\tilde{y}, \tilde{y}^2, \tilde{y}^3)$, has the cubic coefficient only 0.07 standard errors above zero. Either the linear or the quadratic model seem reasonable here; the cubic model is definitely excessive. Section 6 discusses model selection in more detail.

TABLE 2

Parameter estimates $\hat{\beta}$ *and estimated standard errors for the pain-score data of Table* 1. *The quadratic model is described in* (2.13)–(2.16); $\tilde{y} = (y - 2)/4$, *centered version of* y; $\overline{\mathrm{se}}$ *square root of diagonal elements* (3.6a); $\widehat{\mathrm{se}}$ *from* (3.6b); *"jack" is jackknife standard error*; *"naive" is the usual exponential family* sterr *estimated ignoring data-based choice of carrier. The cubic model uses the same* M, *but* $t(y) = (\tilde{y}, \tilde{y}^2, \tilde{y}^3)$; *the cubic term is not at all significant*

	Quadratic model						**Cubic model**	
	$\hat{\beta}_1$	$\overline{\mathrm{se}}$	**(ratio)**	$\widehat{\mathrm{se}}$	**jack**	**naive**	$\hat{\beta}_1$	$\overline{\mathrm{se}}$
$\tilde{y}$	−2.74	0.93	(−2.95)	0.96	1.07	1.18	−2.78	1.24
$\tilde{y}^2$	−3.80	2.45	(−1.55)	2.50	2.66	2.85	−3.59	2.70
$\tilde{y}^3$							0.59	8.30

The jackknife standard errors for the components of $\hat{\beta}_1$ [formula (11.5) of Efron and Tibshirani (1993)] were computed as a check on the corollary. They came out a little larger than the delta-method estimates from (3.6a) or (3.6b), as is often the case; see Section 2 of Efron (1992).

Suppose that we ignore the fact that the carrier $\hat{\boldsymbol{\mu}}^o$ in (3.2) is a function of the data $\mathbf{s}$. This amounts to taking $\hat{H} = 0$ in (3.3), so $Z' = X'$ in (3.4b). Then (3.6b) reduces to the usual covariance estimate for exponential families,

$$\widehat{\mathrm{Cov}}(\hat{\beta}) = (X'\hat{D}X)^{-1}, \tag{3.7}$$

while (3.6a) becomes the "sandwich" estimate $(X'\hat{D}X)^{-1}(X'\bar{D}X)(X'\hat{D}X)^{-1}$. The standard error estimates from (3.7), labelled "naive" in Table 2, are considerably larger than the standard errors that take into account the adaptive choice of the carrier.

The naive standard errors will usually exceed those from (3.6). The reason is that the adaptive choice of $\hat{\boldsymbol{\mu}}^o$ absorbs some of the variability in $\hat{\beta}$. Here is a simple normal-theory version of the same phenomenon: suppose we observe $z \sim N(\mu^o + \beta, 1)$ and we wish to estimate β, the distance of the expectation $\mu = \mu^o + \beta$ from some origin of measurement μ^o. Then $\hat{\beta} = z - \mu^o$ has standard error 1. However, if the origin is chosen adaptively, say by $\hat{\mu}^o = 0.75 \cdot z$, then $\hat{\beta} = z - \hat{\mu}^o$ has standard error 0.25. Observing $z = 4$, for example, gives $\hat{\beta} = 1$, which in the adaptive case is 4 standard errors above 0. See Remark B in Section 9.

Proof of Lemma 1. Using the D notation for diagonal matrices, (3.2) for $\hat{\beta}$ can be expressed as

$$X'\left[\mathbf{s} - D(\hat{\boldsymbol{\mu}}^o)e^{X\hat{\beta}}\right] = 0, \tag{3.8}$$

where $e^{x\hat{\beta}}$ is the vector with components $e^{x_k\hat{\beta}}$. A small change $d\mathbf{s}$ in $\mathbf{s}$ produces change $d\hat{\beta}$ in the MLE vector and change

$$d\hat{\boldsymbol{\mu}}^o \doteq D(\hat{\boldsymbol{\mu}}^o)\hat{H}\,d\mathbf{s} \tag{3.9}$$

in $\hat{\boldsymbol{\mu}}^o$. Then (3.8) gives

$$\begin{aligned} \mathbf{0} &= X'\left[\mathbf{s} + d\mathbf{s} - D(\hat{\boldsymbol{\mu}}^o + d\hat{\boldsymbol{\mu}}^o)\exp\left(X(\hat{\beta} + d\hat{\beta})\right)\right] \\ &\doteq X'\left[\mathbf{s} + D(\hat{\boldsymbol{\mu}}^o)\exp(X\hat{\beta}) + d\mathbf{s} - D(d\hat{\boldsymbol{\mu}}^o)\exp(X\hat{\beta}) \right. \\ &\qquad \left. - D\left(\hat{\boldsymbol{\mu}}^o \exp(X\hat{\beta})\right)X\,d\hat{\beta}\right]. \end{aligned} \tag{3.10}$$

Using (3.8), (3.9) and the fact that $D(d\hat{\boldsymbol{\mu}}^o)e^{X\hat{\beta}} = D(e^{X\hat{\beta}})\,d\hat{\boldsymbol{\mu}}^o$, (3.10) becomes

$$\begin{aligned} 0 &= X'\,d\mathbf{s} - X'D(\hat{\boldsymbol{\mu}}^o e^{X\hat{\beta}})\hat{H}\,d\mathbf{s} - X'D(\hat{\boldsymbol{\mu}}^o e^{X\hat{\beta}})X\,d\hat{\beta} \\ &= X'\left[I - \hat{D}\hat{H}\right]d\mathbf{s} - X'\hat{D}X\,d\hat{\beta}. \end{aligned} \tag{3.11}$$

This verifies (3.4). □

4. A bivariate example. Density estimation becomes more interesting, and more difficult, when the sample space $\mathscr{Y}$ is of higher dimension. This section introduces an example when $\mathscr{Y}$ is a portion of the plane. We will use this example to illustrate some of the advantages of special exponential family density estimation.

Figure 2 shows $l = \log(\text{redshift})$ and m equal to the apparent magnitude for 486 galaxies taken from Loh and Spillar's (1988) redshift survey,

$$y_i = (l_i, m_i) \quad \text{for } i = 1, 2, 3, \ldots, n = 486. \tag{4.1}$$

Hubble's law, that larger redshift implies greater distance from Earth, is apparent in the figure. The galaxies with larger values of l tend to appear dimmer, that is, to have larger apparent magnitudes, leaving the lower right corner nearly empty.

The data in Figure 2 are a truncated subsample of the 879 galaxies in the Loh–Spillar catalog. It is all of the catalog entries falling into the rectangle

$$\log(0.2) \le l \le \log(1.2) \quad \text{and} \quad 17.2 \le m \le 21.5. \tag{4.2}$$

We will take this rectangle to be the sample space $\mathscr{Y}$. Some of the scientific reasons for truncation are discussed in Efron and Petrosian (1992).

Figure 2 shows $\mathscr{Y}$ partitioned into $K = 285$ rectangular cells $\mathscr{Y}_k$, by dividing the l axis into 15 equal strips and dividing the m axis into 19 equal strips. The corresponding counts s_k, (3.2), are shown on the right side of the figure. It is convenient to index the cells by

$$k = (i, j), \quad i = 1, 2, \ldots, 15, \; j = 1, 2, \ldots, 19. \tag{4.3}$$

The midpoint of rectangle $\mathscr{Y}_k$ is $y_{(k)} \equiv (l_{(i)}, m_{(j)})$.

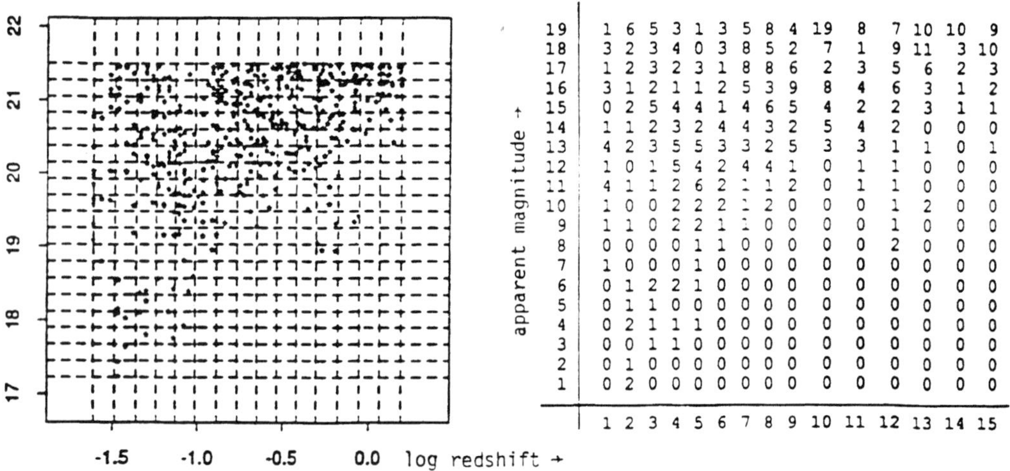

	1	2	3	4	5	6	7	8	9	10	11	12	13	14	15
19	1	6	5	3	1	3	5	8	4	19	8	7	10	10	9
18	3	2	3	4	0	3	8	5	2	7	1	9	11	3	10
17	1	2	3	2	3	1	8	8	6	2	3	5	6	2	3
16	3	1	2	1	1	2	5	3	9	8	4	6	3	1	2
15	0	2	5	4	4	1	4	6	5	4	2	2	3	1	1
14	1	1	2	3	2	4	4	3	2	5	4	2	0	0	0
13	4	2	3	5	5	3	3	2	5	3	3	1	1	0	1
12	1	0	1	5	4	2	4	4	1	0	1	1	0	0	0
11	4	1	1	2	6	2	1	1	2	0	1	1	0	0	0
10	1	0	0	2	2	2	1	2	0	0	0	1	2	0	0
9	1	1	0	2	2	1	1	0	0	0	0	1	0	0	0
8	0	0	0	0	1	1	0	0	0	0	0	2	0	0	0
7	1	0	0	0	1	0	0	0	0	0	0	0	0	0	0
6	0	1	2	2	1	0	0	0	0	0	0	0	0	0	0
5	0	1	1	0	0	0	0	0	0	0	0	0	0	0	0
4	0	2	1	1	1	0	0	0	0	0	0	0	0	0	0
3	0	0	1	1	0	0	0	0	0	0	0	0	0	0	0
2	0	1	0	0	0	0	0	0	0	0	0	0	0	0	0
1	0	2	0	0	0	0	0	0	0	0	0	0	0	0	0

FIG. 2. *The galaxy data*: 486 *galaxies from Loh and Spillar's* (1988) *redshift survey* (log *redshift l and apparent magnitude m*) *discretized into* $285 = 15 \times 19$ *equal cells. The counts* **s** *for the* 285 *cells are shown at right.*

The right side of Figure 3 shows the results of applying a linear smoother $\hat{\boldsymbol{\mu}} = M(\lambda)\mathbf{s}$ to the 285-dimensional count vector $\mathbf{s}$ given in Figure 2. The $K \times K$ matrix $M(\lambda)$ is a two-dimensional version of (2.14), with kk'th element

$$M_{kk'}(\lambda) = \frac{c_k}{\lambda^2}\exp\left(-\frac{1}{2\lambda^2}\left[(i-i')^2 + (j-j')^2\right]\right) \tag{4.4}$$

for $k = (i, j)$ and $k' = (i', j')$. The c_k are chosen to make $\sum_{k'} M_{kk'}(\lambda) = 1$. The choice $\lambda = 1.5$, suggested by the expected deviance calculations of Section 6, gave the smoothed estimate $\hat{\boldsymbol{\mu}}^o = M(1.5)\cdot\mathbf{s}$ plotted versus $y_{(k)}$ on the right side of Figure 3.

The left side of Figure 3 shows an SEF estimate $\hat{\boldsymbol{\mu}}$ of form(3.2), with

$$\hat{\boldsymbol{\mu}}^o = M(2)\mathbf{s} \quad \text{and} \quad X = (\mathbf{1}, \mathbf{i}, \mathbf{j}, \mathbf{i}^2, \mathbf{j}^2, \mathbf{ij}). \tag{4.5}$$

Here $\mathbf{i}$ is the K vector with kth element $i - 8$, and $\mathbf{j}$ is the K vector with kth element $j - 9$. This choice amounts to using the usual sufficient statistics for a bivariate normal in (1.2), $t(y) = (l, m, l^2, m^2, lm)$. The fitted density matches the empirical means, variances and correlation of the galaxy data.

The variance calculations of Section 5 and the expected deviance calculations of Section 6 suggest that these two estimates are roughly equal in their overall ability to predict the true density. However, the SEF estimate is much smoother than the smoothing-only choice. This is obvious in Figure 4, which shows contour plots of the two density estimates.

Suppose that the carrier $\hat{\boldsymbol{\mu}}^o$ in (4.5) was taken to be the constant vector $\hat{\mu}_k^o \equiv 1$ instead of $M(2)\mathbf{s}$. Then the SEF $\hat{\boldsymbol{\mu}}$ would be the (discretized) truncated bivariate normal MLE for the galaxy data, the truncation being to the rectangle (4.2). In fact, the SEF in Figure 3 looks like the lower corner of a bivariate normal, though there are some discrepancies due to the adaptation of $\hat{\boldsymbol{\mu}}^o$ to the galaxy data.

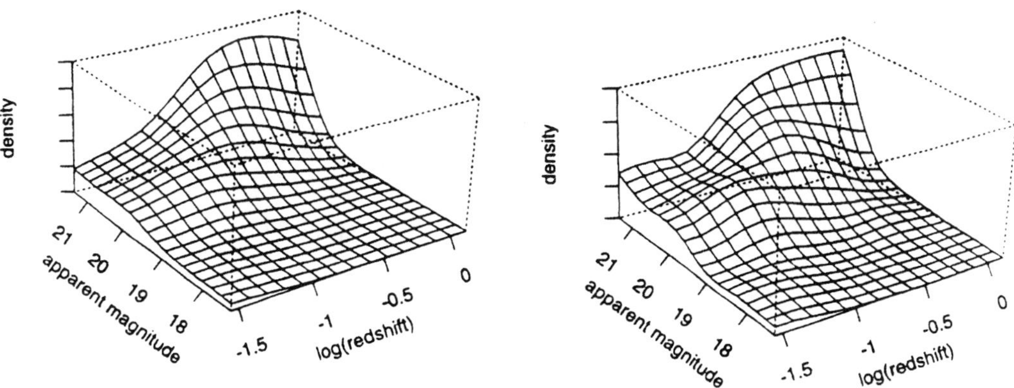

FIG. 3. *Left panel: SEF estimate* (3.2) *for the galaxy data as discretized in Figure* 2; $\hat{\mu}^o = M(2)\cdot\mathbf{s}$, *quadratic matrix* X, (4.5). *Right panel: The smoothing-only estimate* $\hat{\mu}^o = M(1.5)\cdot\mathbf{s}$. *The calculations of Sections* 5 *and* 6 *suggest that the two estimates are roughly equal in accuracy. However, the SEF estimate is much smoother.*

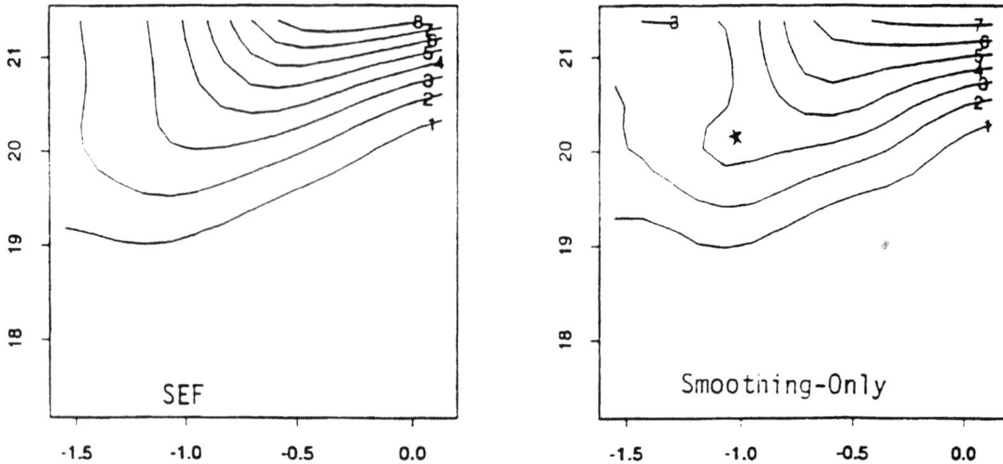

FIG. 4. *Contour plots of the density estimates in Figure 3. The SEF contours (left panel) are much smoother than those from the smoothing-only estimate (right panel). The smoothing-only estimate has a weak second mode at the starred point.*

Parametric models such as the bivariate normal are fierce data smoothers. This can be a big advantage if the statistician is interested in global properties of the density, like the general shape of its contours, and especially if the data-sampling process is suspect. In Figure 2 we can see that the galaxy data are quite patchy and clumpy, which is not surprising since the Loh–Spillar catalog was a census of the brighter galaxies in a few degrees of sky, and not a random sample of the full sky.

If those few degrees of sky are of great individual interest, then a narrow-band smoother like that in the right side of Figure 3 may be appropriate. [Notice that it estimates a weak second mode near $(l, m) = (-1, 20)$.] However if we really want to know the density for the whole sky, then it pays to *oversmooth* the estimate from a flawed sample like that in Figure 2. The SEF methodology allows us to oversmooth without losing much estimating efficiency compared to smoothing-only estimates and without making the drastic assumptions of the usual parametric models.

Various SEF models besides the quadratic choice of X in (4.5) were tried. One of these added a further cross-term to (4.5),

$$\hat{\boldsymbol{\mu}}^o = M(2)\mathbf{s} \quad \text{and} \quad X = \left(\mathbf{1}, \mathbf{i}, \mathbf{j}, \mathbf{i}^2, \mathbf{j}^2, \mathbf{ij}, (\mathbf{i}^2\mathbf{j}^2)_{\text{orthog}}\right), \tag{4.6}$$

where $(\mathbf{i}^2\mathbf{j}^2)_{\text{orthog}}$ is the component of $\mathbf{i}^2\mathbf{j}^2$ orthogonal to $(\mathbf{1}, \mathbf{i}, \mathbf{j}, \mathbf{i}^2, \mathbf{j}^2, \mathbf{ij})$. Orthogonalization makes the first six components of $\hat{\beta}$ have roughly the same values and standard errors as in (4.5). The term $(\mathbf{i}^2\mathbf{j}^2)_{\text{orthog}}$ in (4.6) allows the regression surface to turn more quickly near the corners of the rectangle $\mathscr{Y}$.

The MLE vector $\hat{\beta}_1$ for model (4.6) appears in Table 3, along with the standard error estimates $\overline{\text{se}}$ from (3.6a) and the t-values $\hat{\beta}/\overline{\text{se}}$. All of the coefficients except the one for m^2 are significantly different than zero [the

TABLE 3

Parameter estimates and standard errors for the components of $\hat{\beta}_1$ in the SEF *model* (4.6). *All of the components of $\hat{\beta}_1$ are significantly nonzero, except for the coefficient of m^2. l indicates the coefficient corresponding to* **i**, *lm to* **ij** *and so forth. Standard errors are from* (3.6a). *Model* (4.5) *gave similar estimates, standard errors and t-values for l, m, l^2, m^2, lm*

	l	m	l^2	m^2	lm	$(l^2m^2)_{\text{orthog}}$
$\hat{\beta}_1$	−0.105	0.084	−0.0115	−0.0018	0.0158	0.000267
$\overline{\text{se}}$	0.019	0.015	0.0021	0.0014	0.0027	0.000071
t-Value	−5.6	5.8	−5.5	−1.4	5.8	3.75

same is true for model (4.5)]. This includes the coefficient for the term $(l^2m^2)_{\text{orthog}}$, which takes us beyond the normal-theory SEF analogue (4.5). Compared to the left side of Figure 3, the SEF density estimate from (4.6) has its highest point at the upper right corner of the rectangle $\mathcal{Y}$.

The next two sections discuss several criteria for choosing among possible SEF models: observed deviance, expected deviance, degrees of freedom and so forth. However, none of these criteria is sharp enough to entirely free the statistician from model-choice quandries. A considerable amount of subjectivity must still go into the model-building process, just as in ordinary regression situations.

5. Total relative variance. A variety of diagnostic tools is available to assist model selection in standard regression situations. Similar tools are available for model selection in special exponential families. These ideas are developed in the next two sections, beginning here with the total relative variance, a simple measure of overall variability for an SEF density estimate. For example, looking ahead to Table 4 the reader can compare the total relative variances for the two galaxy-data SEF estimates in Figure 3: 6.8 for the quadratic model in the left panel versus 8.8 for the smoothing-only model in the right panel.

Let

$$\hat{\boldsymbol{\mu}} = \text{SEF}(\mathbf{s}; m, X) \tag{5.1}$$

indicate $\hat{\boldsymbol{\mu}} = \hat{\boldsymbol{\mu}}^o e^{X\hat{\beta}}$, the special exponential family estimate (3.2). First we will compute the $K \times K$ derivative matrix of $\hat{\boldsymbol{\mu}}$ with respect to $\mathbf{s}$, which leads immediately to a delta-method estimate of $\text{Cov}(\hat{\boldsymbol{\mu}})$. This derivative matrix involves the projection matrices

$$\hat{P} = X(X'\hat{D}X)^{-1}X' \quad \text{and} \quad \hat{Q} = \hat{D}^{-1} - \hat{P}, \tag{5.2}$$

where $\hat{D} = D(\hat{\boldsymbol{\mu}})$ as before, $\hat{P}$ is the symmetric projection matrix into the linear space spanned by the columns of X, in the inner product $\hat{D}$, the projection of vector $\mathbf{v}$ being $\hat{P}\hat{D}\mathbf{v}$, and $\hat{Q}$ is the projection orthogonal to X's column space. Because they represent orthogonal projections, we have

$$\hat{P}\hat{D}\hat{P} = \hat{P}, \qquad \hat{Q}\hat{D}\hat{Q} = \hat{Q} \quad \text{and} \quad \hat{P}\hat{D}\hat{Q} = 0. \tag{5.3}$$

LEMMA 2. *The derivative matrix of* $\hat{\boldsymbol{\mu}} = \mathrm{SEF}(\mathbf{s}; m, X)$ *with respect to* $\mathbf{s}$ *is*

$$\frac{d\hat{\boldsymbol{\mu}}}{d\mathbf{s}} = \hat{D}\hat{O} \tag{5.4a}$$

where, in terms of $\hat{H} = d \log \hat{\boldsymbol{\mu}}^o / d\mathbf{s}$, (3.3),

$$\hat{O} = \hat{P} + \hat{Q}\hat{D}\hat{H} = \hat{P} + \hat{H} - \hat{P}\hat{D}\hat{H}. \tag{5.4b}$$

If $\hat{\boldsymbol{\mu}}^o = M\mathbf{s}$, *then* $\hat{H} = D(1/\hat{\boldsymbol{\mu}}^o)M$ *and*

$$\hat{O} = \hat{P} + \hat{Q}D(e^{X'\hat{\beta}})M. \tag{5.4c}$$

(The proof appears below.)

The canonical parameter vector for the Poisson family (3.1) is

$$\boldsymbol{\eta} = \log(\boldsymbol{\mu}) = (\log(\mu_1), \log(\mu_2), \dots, \log(\mu_k)). \tag{5.5}$$

Likewise, define $\hat{\boldsymbol{\eta}} = \log(\hat{\boldsymbol{\mu}})$ and $\hat{\boldsymbol{\eta}}^o = \log(\hat{\boldsymbol{\mu}}^o)$. Then we can write Lemma 2 as

$$\frac{d\hat{\boldsymbol{\eta}}}{d\mathbf{s}} = \hat{O} = \hat{P} + \hat{Q}\hat{D}\frac{d\hat{\eta}^o}{d\mathbf{s}}. \tag{5.6}$$

This decomposes $d\hat{\boldsymbol{\eta}}/d\mathbf{s}$ into a part $\hat{P}$ coming from the exponential family factor $e^{X\hat{\beta}}$ and an orthogonal part $\hat{Q}\hat{D}(d\hat{\boldsymbol{\eta}}^o/d\mathbf{s})$ coming from the adaptive choice of the carrier.

Lemma 2 leads directly to delta-method estimates of the covariance matrix of $\hat{\boldsymbol{\mu}}$, as in (3.6),

$$\overline{\mathrm{Cov}}(\hat{\boldsymbol{\mu}}) = \hat{D}\hat{O}\overline{D}\hat{O}'\hat{D} \quad \text{or} \quad \widehat{\mathrm{Cov}}(\hat{\boldsymbol{\mu}}) = \hat{D}\hat{O}\hat{D}\hat{O}'\hat{D} \tag{5.7}$$

with $\overline{D} = D(s)$. The diagonal elements give variance estimates $\overline{\mathrm{var}}(\hat{\mu}_k)$ or $\widehat{\mathrm{var}}(\hat{\mu}_k)$ for the individual components. For Poisson variables it is natural to measure variance relative to the etsimate $\hat{\mu}_k$. We define the *total relative variance* (TRV) estimate for $\hat{\boldsymbol{\mu}}$ to be

$$\overline{\mathrm{TRV}} = \sum_{k=1}^{K} \frac{\overline{\mathrm{var}}(\hat{\mu}_k)}{\hat{\mu}_k} \quad \text{or} \quad \widehat{\mathrm{TRV}} = \sum_{k=1}^{K} \frac{\widehat{\mathrm{var}}(\hat{\mu}_k)}{\hat{\mu}_k}. \tag{5.8}$$

COROLLARY 2. *For* $\hat{\boldsymbol{\mu}} = \mathrm{SEF}(\mathbf{s}; m, X)$ *the total relative variance estimates are*

$$\begin{aligned} \overline{\mathrm{TRV}} &= \mathrm{tr}\Big[\overline{D}\hat{P} + \big(\hat{H}\overline{D}\hat{H}'\big)\big(\hat{D}\hat{Q}\hat{D}\big)\Big] \quad \textit{or} \\ \widehat{\mathrm{TRV}} &= (p+1) + \mathrm{tr}\big(\hat{H}\hat{D}\hat{H}'\big)\big(\hat{D}\hat{Q}\hat{D}\big), \end{aligned} \tag{5.9}$$

where $p + 1$ *is the number of columns of* X, $\overline{D} = D(\mathbf{s})$, $\hat{D} = D(\hat{\boldsymbol{\mu}})$ *and* $\hat{P}, \hat{Q}$ *are as in* (5.2).

The proof of Corollary 2 follows directly from (5.7) by writing $\overline{\text{TRV}}$ in the trace form $\operatorname{tr} \hat{D}^{-1} \overline{\text{Cov}}(\hat{\mu})$ and using the orthogonality relationship (5.3), and similarly for $\widetilde{\text{TRV}}$. Computationally more efficient expressions for $\overline{\text{TRV}}$ and $\widetilde{\text{TRV}}$ appear in Remark I, Section 9.

Table 4 shows $\overline{\text{TRV}}$ for various SEF estimates for the galaxy data, as discretized in Figure 2. The three estimates correspond to three choices of X in (5.1): sef 2 is for X as in (4.5), sef 3 for X as in (4.6), and sef 0 for $X = \mathbf{1}$. This last choice is the smoothing-only estimate $\hat{\boldsymbol{\mu}}^o$ rescaled to $(n/\hat{\mu}_+^o)\hat{\boldsymbol{\mu}}^o$, so that it sums to $n = s_+$. The carrier $\hat{\boldsymbol{\mu}}^o$ for the SEF is

$$\hat{\boldsymbol{\mu}}^o = M(\lambda)\mathbf{s}, \tag{5.10}$$

where $M(\lambda)$ is the matrix (4.4).

All three TRV estimates decrease as the smoothing parameter λ increases because greater window width λ decreases the variability of $\hat{\boldsymbol{\mu}}^o$. If the carrier $\hat{\boldsymbol{\mu}}^o$ were prechosen instead of adaptive, then sef 2 would exceed sef 0 by about 5, this being the increased number of free parameters, and likewise sef 3 would exceed sef 2 by about 1. This is seen in (5.9) for the case $\hat{H} = 0$. In fact, this is nearly the case at the right side of Table 4, where λ is so large that $\hat{\boldsymbol{\mu}}^o$ has nearly constant entries. The difference between the three estimates decreases at smaller values of λ because the adaptability of the common carrier $\hat{\boldsymbol{\mu}}^o = M(\lambda) \cdot \mathbf{s}$ absorbs some of the difference in the exponential family fits $e^{X\hat{\beta}}$.

Small variability is a good property of course, but we also want $\hat{\boldsymbol{\mu}}$ to have small bias for estimating the true density vector $\boldsymbol{\mu}$. The next section puts TRV into the context of bias–variance tradeoffs for Poisson regression estimates.

Proof of Lemma 2. In the notation following (5.5), $\hat{\boldsymbol{\eta}} = \hat{\boldsymbol{\eta}}^o + X\hat{\beta}$. Differentiating this with respect to $\mathbf{s}$ and using (3.4) gives

$$\frac{d\hat{\boldsymbol{\eta}}}{d\mathbf{s}} = \hat{H} + X(X'\hat{D}X)^{-1}X'\left[I - \hat{D}\hat{H}\right] = \hat{P} + \hat{H} - \hat{P}\hat{D}\hat{H}, \tag{5.11}$$

which is (5.6), the equivalent of Lemma 2. □

Table 4

Total relative variance $\overline{\text{TRV}}$ *for three* SEF *estimates, galaxy data, at increasing values of smoothing parameter* λ

	$\lambda = 1.5$	$\lambda = 2$	$\lambda = 4$	$\lambda = 6$
sef 0 (smoothing only)	8.8[a]	5.2	1.2	0.4
sef 2 (4.5)	9.6	6.8[b]	5.2	5.2
sef 3 (4.6)	10.2	7.6	6.2	6.2

[a] Right panel of Figure 3.
[b] Left panel of Figure 3.

6. Degrees of freedom and estimated deviance. Selecting a good SEF estimate $\hat{\boldsymbol{\mu}}$ for a particular application involves making the usual tradeoffs between variance and bias. This section concerns estimating the total expected deviance of $\hat{\boldsymbol{\mu}}$ from the expectation vector $\boldsymbol{\mu}$. This is a measure of accuracy for $\hat{\boldsymbol{\mu}}$ that involves both variance and bias. An important role is played by the *degrees of freedom* of the estimator $\hat{\boldsymbol{\mu}}$, an idea related to the total relative variance of Section 5. The ideas here are an extension of those in Section 6.8 of Hastie and Tibshirani (1990).

Figure 5 relates to the *expected deviance* measure $\widehat{\text{EDEV}}$ developed below, a diagnostic measure for comparing the goodness-of-fit of different SEF models. It shows $\widehat{\text{EDEV}}$ for both the pain-score and galaxy examples. $\widehat{\text{EDEV}}$ is plotted versus the smoothing parameter λ for the normal kernel estimates (2.13) and (4.4) used to obtain $\hat{\boldsymbol{\mu}}^o = M(\lambda) \cdot \mathbf{s}$. The different curves correspond to different choices of X in the SEF formula (3.2). "Smoothing-only" refers to $X = \mathbf{1}$, which gives the estimate $\hat{\boldsymbol{\mu}}^o$ renormalized to sum to n. For the pain-score data, "linear" and "quadratic" are the cases referred to in (2.15). For the galaxy data, (4.5) and (4.6) are as in Table 4.

Notice that $\widehat{\text{EDEV}}(\lambda)$ has a sharp minimum as a function of λ for both smoothing-only cases. (At $\lambda = 0.63$ for the pain-score data and at $\lambda = 1.5$ for the galaxy data.) This is not true for the genuine SEF estimates. They allow the smoothing parameter λ to be chosen much larger without incurring too much EDEV penalty. In other words, *the SEF methodology allows us to oversmooth the density estimate*, with the advantages seen in Figure 4.

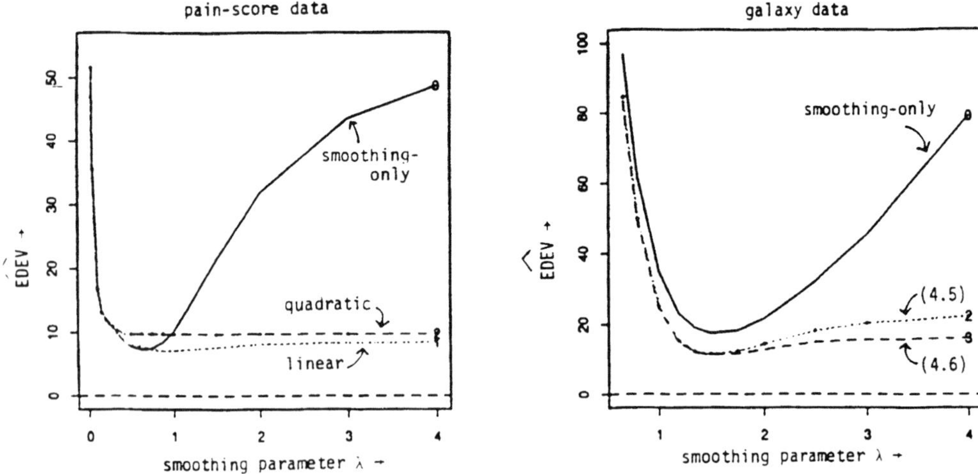

FIG. 5. *Expected deviance estimates* $\widehat{\text{EDev}}$, (6.22). *Left panel*: *SEF estimates for pain-score data* (2.13)–(2.15). *Right panel*: *SEF estimates for galaxy data as in Figure* 5. *In both cases the SEF estimates permit the use of larger smoothing parameters* λ, *compared to the smoothing-only estimates.*

In order to motivate $\widehat{\text{EDEV}}$, it helps to begin the discussion with the normal case. Suppose that the statistician observes a K-dimensional normal vector

$$\mathbf{z} \sim N_K(\mu, I) \tag{6.1}$$

and wishes to estimate μ using some linear estimator

$$\hat{\boldsymbol{\mu}} = S\mathbf{z}, \tag{6.2}$$

S being a $K \times K$ matrix. Model selection amounts to making a good choice of S.

Let err be the observed total residual squared error of $\hat{\boldsymbol{\mu}}$:

$$\text{err} = \|\mathbf{z} - \hat{\boldsymbol{\mu}}\|^2. \tag{6.3}$$

Also define

$$\boldsymbol{\nu} = E\{\hat{\boldsymbol{\mu}}\}, \qquad \text{DF} = \text{tr}\, S \quad \text{and} \quad \text{TV} = \text{tr}\, SS', \tag{6.4}$$

where DF stands for *degrees of freedom* and TV stands for *total variance*. Total variance TV equals $\Sigma_k \text{var}(\hat{\mu}_k)$, and if S is a projection matrix, then $\text{tr}\, S$ is the usual degrees of freedom. It is easy to prove the following two relationships:

$$E\{\text{err} - (K - 2\text{DF})\} = E\{\|\hat{\boldsymbol{\mu}} - \boldsymbol{\mu}\|^2\} \tag{6.5a}$$

and

$$E\{\text{err} - (K - 2\text{DF} + \text{TV})\} = \|\boldsymbol{\nu} - \boldsymbol{\mu}\|^2. \tag{6.5b}$$

Both (6.5a) and (6.5b) play an important role in normal-theory model selection. The first of these is essentially the C_p or AIC criterion. Its extension to nonlinear estimators is Stein's unbiased risk estimate. Extensions of formula (6.5b) are used in hypothesis testing. The quantity $K - 2\text{DF} + \text{TV}$ equals the *residual degrees of freedom*:

$$\text{RDF} = \text{tr}(I - S)(I - S)'. \tag{6.6}$$

If S is a projection matrix, then (6.5b) can be improved to

$$\text{err} \sim \chi^2_{\text{RDF}}(\|\boldsymbol{\nu} - \boldsymbol{\mu}\|^2), \tag{6.7}$$

where the notation indicates a noncentral chi-square variate with noncentrality parameter $\|\boldsymbol{\nu} - \boldsymbol{\mu}\|^2$. We test the adequacy of the estimate $\hat{\boldsymbol{\mu}} = S\mathbf{z}$ by comparing err to a central chi-square distribution χ^2_{RDF}.

We now return to the Poisson situation where we observe

$$\mathbf{s} \sim \text{Po}_K(\boldsymbol{\mu}) \tag{6.8}$$

and estimate $\boldsymbol{\mu}$ by some estimator $\hat{\boldsymbol{\mu}} = \hat{\boldsymbol{\mu}}(\mathbf{s})$, not necessarily of the SEF form (3.2), having expectation

$$\boldsymbol{\nu} \equiv E\{\hat{\boldsymbol{\mu}}(\mathbf{s})\}. \tag{6.9}$$

The *observed error* err = err($\mathbf{s}$) in the Poisson context is

$$\text{err}(\mathbf{s}) = \text{Dev}(\mathbf{s}, \hat{\boldsymbol{\mu}}), \tag{6.10}$$

where Dev indicates the total Poisson deviance

$$\mathrm{Dev}(\boldsymbol{\mu}, \boldsymbol{\nu}) = 2 \textstyle\sum_k \{\log(\mu_k/\nu_k) - (\mu_k - \nu_k)\}.$$

The Poisson equivalents of K, DF and TV in (6.5) are

$$K(\boldsymbol{\mu}) = E\{\mathrm{Dev}(\mathbf{s}, \boldsymbol{\mu})\}, \qquad \mathrm{DF}(\boldsymbol{\mu}) = E\left\{\sum_k (s_k - \mu_k)\log(\hat{\mu}_k)\right\}, \tag{6.11}$$

$$\mathrm{TRV}(\boldsymbol{\mu}) = 2\sum_k E\{\mu_k \log(\nu_k/\hat{\mu}_k)\},$$

all expectations being with respect to $\mathbf{s} \sim \mathrm{Po}_K(\boldsymbol{\mu})$.

LEMMA 3. *In the Poisson situation* (6.8)–(6.11),

$$E\{\mathrm{err}(\mathbf{s}) - [K(\boldsymbol{\mu}) - 2\mathrm{DF}(\boldsymbol{\mu})]\} = E\{\mathrm{Dev}(\boldsymbol{\mu}, \hat{\boldsymbol{\mu}})\} \tag{6.12a}$$

and

$$E\{\mathrm{err}(\mathbf{s}) - [K(\boldsymbol{\mu}) - 2\mathrm{DF}(\boldsymbol{\mu}) + \mathrm{TRV}(\boldsymbol{\mu})]\} = \mathrm{Dev}(\boldsymbol{\mu}, \boldsymbol{\nu}). \tag{6.12b}$$

(The proof is given below.) Formulas (6.12a, b) are the Poisson versions of (6.5a, b).

In order to use Lemma 3, we can approximate $K(\boldsymbol{\mu})$, $\mathrm{DF}(\boldsymbol{\mu})$ and $\mathrm{TRV}(\boldsymbol{\mu})$ by their plug-in estimates $K(\hat{\boldsymbol{\mu}})$, $\mathrm{DF}(\hat{\boldsymbol{\mu}})$ and $\mathrm{TRV}(\hat{\boldsymbol{\mu}})$. Using bootstrap notation, let $\mathbf{s}^*$ given $\mathbf{s}$ have Poisson distribution

$$\mathbf{s}^* | \mathbf{s} \sim \mathrm{Po}_K(\hat{\boldsymbol{\mu}}(\mathbf{s})) \tag{6.13}$$

and let E_* indicate expectations with respect to (6.13), with $\mathbf{s}$ and $\hat{\boldsymbol{\mu}}(\mathbf{s})$ fixed. Then

$$\begin{aligned} K(\hat{\boldsymbol{\mu}}) &= \sum_k E_*\{\mathrm{Dev}(s_k^*, \hat{\mu}_k)\} \\ &= \sum_k E_*\{2[s_k^* \log(s_k^*/\hat{\mu}_k) - (s_k^* - \hat{\mu}_k)]\}. \end{aligned} \tag{6.14}$$

It is easy to evaluate (6.14) numerically since it is the sum of univariate Poisson deviances, the $\hat{\mu}_k$ being just fixed constants in the E_* expectations.

Using this same notation we can write the degrees of freedom estimate $\mathrm{DF}(\hat{\boldsymbol{\mu}})$ as

$$\mathrm{DF}(\hat{\boldsymbol{\mu}}) = E_*\left\{\sum_k (s_k^* - \hat{\mu}_k)\log \hat{\mu}_k^*\right\} = E_*(\mathbf{s}^* - \hat{\boldsymbol{\mu}})'\hat{\boldsymbol{\eta}}^*, \tag{6.15}$$

where $\hat{\boldsymbol{\eta}}^* = \log(\hat{\boldsymbol{\mu}}^*) = \log(\hat{\boldsymbol{\mu}}(\mathbf{s}^*))$, the vector with kth component $\log \hat{\mu}_k^*$. Since by (5.6), $d\hat{\boldsymbol{\eta}}/d\mathbf{s} = \hat{O}$, (6.15) has the Taylor series approximation

$$\mathrm{DF}(\hat{\boldsymbol{\mu}}) \doteq E_*(\mathbf{s}^* - \hat{\boldsymbol{\mu}})'\hat{O}(s^* - \hat{\boldsymbol{\mu}}) = \mathrm{tr}\,\hat{D}\hat{O} \equiv \widehat{\mathrm{DF}}. \tag{6.16}$$

See Remark K. An alternative estimate is $\overline{\mathrm{DF}} = \mathrm{tr}\,\overline{D}\hat{O}$ as in (5.8).

The quadratic expansion $\log(\hat{\mu}/\nu) \doteq (\hat{\mu}/\nu - 1) - \frac{1}{2}(\hat{\mu}/\nu - 1)^2$ gives

$$E\left\{\mu_k \log\left(\frac{\nu_k}{\hat{\mu}_k}\right)\right\} \doteq \frac{\mu_k}{2}\frac{\operatorname{var}(\hat{\mu}_k)}{\nu_k^2} \doteq \frac{\operatorname{var}(\hat{\mu}_k)}{2\mu_k}, \tag{6.17}$$

permitting us to approximate (6.11) by

$$\operatorname{TRV}(\boldsymbol{\mu}) \doteq \sum_k \frac{\operatorname{var}(\hat{\mu}_k)}{\mu_k}. \tag{6.18}$$

The quantities $\overline{\text{TRV}}$ and $\widehat{\text{TRV}}$ in (5.8) are the obvious plug-in estimates for (6.18). The substitution of μ_k for ν_k in the denominator of (6.17) looks worrisome, but Remark K of Section 9 shows that the resulting error is asymptotically negligible.

The TRV estimates (5.8) can be written as

$$\overline{\text{TRV}} = \operatorname{tr} \bar{D}\hat{O}'\hat{D}\hat{O} \quad \text{or} \quad \widehat{\text{TRV}} = \operatorname{tr} \hat{D}\hat{O}'\hat{D}\hat{O}, \tag{6.19}$$

using (5.7), compared to the DF estimates

$$\overline{\text{DF}} = \operatorname{tr} \bar{D}\hat{O} \quad \text{or} \quad \widehat{\text{DF}} = \operatorname{tr} \hat{D}\hat{O}. \tag{6.20}$$

Notice the similarity to the normal-theory definitions in (6.4). Comparing (6.5b) with (6.12b), we can also define residual degrees of freedom RDF for the Poisson situation, estimated by

$$\overline{\text{RDF}} = K(\hat{\boldsymbol{\mu}}) - 2\overline{\text{DF}} + \overline{\text{TRV}} \quad \text{or} \quad \widehat{\text{RDF}} = K(\hat{\boldsymbol{\mu}}) - 2\widehat{\text{DF}} + \widehat{\text{TRV}}. \tag{6.21}$$

See Remark L. The quantity graphed in Figure 5 was the expected deviance estimate obtained from (6.12a), (6.14) and (6.20):

$$\widehat{\text{EDEV}} = \operatorname{err}(\mathbf{s}) - K(\hat{\boldsymbol{\mu}}) + 2\widehat{\text{DF}}. \tag{6.22}$$

The vertical scale in Figure 5 is misleading. For the galaxy data, reducing λ from 4 to 1.5 reduces $\widehat{\text{EDEV}}$ for the quadratic SEF (4.5) from 22.2 to 11.5, a considerable amount. Is this significant? The observed deviance error (6.10), err, decreases by 45.3, while the residual degrees of freedom $\widehat{\text{RDF}}$, (6.21), decreases by 29.3. This gives the naive chi-square significance value $\text{prob}\{\chi^2_{29.3} > 45.3\} = 0.03$. A more trustworthy significance level could be obtained by Monte Carlo methods, bootstrapping with $\mathbf{s}^* \sim \text{Prob}(\hat{\boldsymbol{\mu}})$, with $\hat{\boldsymbol{\mu}}$ the (1.5, 2) SEF vector.

The quantitative aspects of Figure 5 cannot be taken too literally. For instance, using $\overline{\text{EDEV}}$ instead of $\widehat{\text{EDEV}}$ moved the smoothing-only curve below the others for $\lambda < 2$ in the galaxy data. In general the careted (hat) formulas gave less erratic answers than the bar formulas, but there are really no strong reasons for preferring $\widehat{\text{EDEV}}$. The fact is that it is usually difficult to estimate the performance of competing decision rules, and the SEF density estimates are no exception. Formulas like (5.9) and (6.12) help with model selection, but considerable subjectivity remains. The main point of Figure 5 is the qualitative one that the SEF estimates permit extensive oversmoothing.

Proof of Lemma 3. Define $\mathscr{E}(\boldsymbol{\mu}) = E\{\text{Dev}(\mathbf{s}^\dagger, \hat{\boldsymbol{\mu}})\}$, where $\mathbf{s}^\dagger \sim \text{Po}_K(\boldsymbol{\mu})$ independent of $\mathbf{s}$, so $\mathscr{E}(\boldsymbol{\mu})$ is the expected prediction error of $\hat{\boldsymbol{\mu}}(\mathbf{s})$. Efron (1986) shows that

$$(6.23\text{a}) \qquad \mathscr{E}(\boldsymbol{\mu}) = E\{\text{err}(\mathbf{s}) + 2\text{DF}(\mu)\}.$$

We also have the identities

$$(6.23\text{b}) \qquad \mathscr{E}(\boldsymbol{\mu}) = K(\boldsymbol{\mu}) + E\{\text{Dev}(\boldsymbol{\mu}, \hat{\boldsymbol{\mu}})\}$$

and

$$(6.23\text{c}) \qquad E(\text{Dev}(\boldsymbol{\mu}, \hat{\boldsymbol{\mu}})\} = \text{Dev}(\boldsymbol{\mu}, \boldsymbol{\nu}) + \text{TRV}(\boldsymbol{\mu}).$$

All three of these relationships are easy to prove directly from the definition of the total Poisson deviance. Substituting (6.23b) into (6.23a) gives (6.12a). Substituting (6.23c) on the right side of (6.12a) gives (6.12b). □

7. Multisample problems. So far we have only considered one-sample problems. SEF estimates are particularly useful for investigating density differences in multisample situations. We use the exponential family model (1.2) for the different densities, with a shared carrier g_0 estimated nonparametrically, but with possibly different values of the exponential β parameters. An example will precede the theory.

Figure 6 concerns a two-sample application of SEF modelling. The data are the compliances of men in the Stanford arm of a randomized trial of the cholesterol-lowering drug Cholostyramine; see Efron and Feldman (1991). There were $n_1 = 172$ men in the control group and $n_2 = 165$ men in the treatment group. Compliance ran from 0 to 100%, so $\mathscr{Y} = [0, 100]$. A discretization $\mathscr{Y} = \bigcup_k \mathscr{Y}_k$ partitioned $\mathscr{Y}$ into $K = 46$ intervals of equal length. The left panel of Figure 6 shows the counts in the two groups. Compliance is significantly worse in the treatment group, as shown by standard two-sample tests.

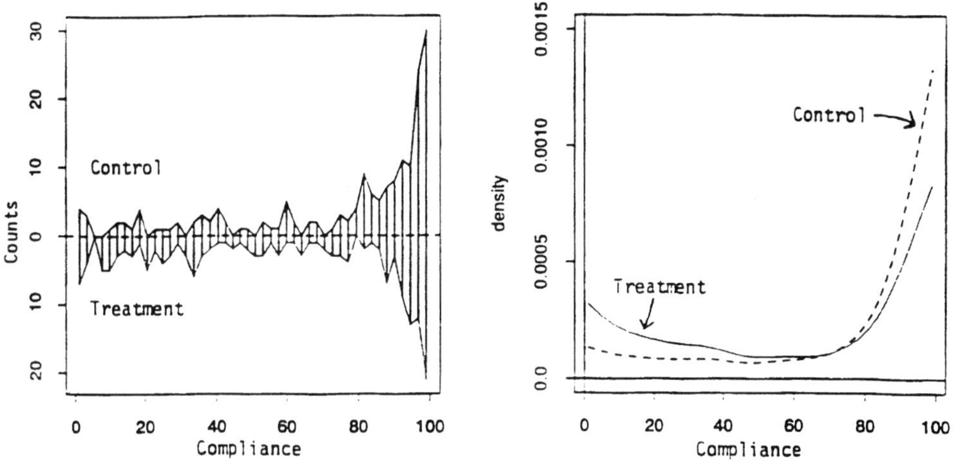

FIG. 6. *An application of SEF modelling to the Cholostyramine trial compliance data of Efron and Feldman* (1991). *Left panel: The count vectors for the control group and the treatment group;* $K = 46$ *equal divisions of* $\mathscr{Y} = [0, 100]$. *Right panel: SEF density estimates* (7.4) *for the two groups;* m *and* X *as in* (7.11). *The poorer compliance in the treatment group is graphically evident.*

The right panel of Figure 6 is the output of a two-sample SEF analysis. It neatly displays the compliance differences between the two groups in terms of their estimated densities. The ratio $\hat{g}_{\text{treat}}(y)/\hat{g}_{\text{cont}}(y)$ decreases almost linearly as y goes from 0 to 100%, but both densities are greatest at $y = 100\%$.

These densities were derived from a simple extension of the previous theory. In the multisample situation we observe independent random samples from L possibly different densities $g_1, g_2, \ldots, g_L$ on the same sample space $\mathscr{Y}$,

$$y_i(l) \overset{\text{i.i.d.}}{\sim} g_l(y) \quad \text{for } i = 1,2,\ldots,n_l \text{ and } l = 1,2,\ldots,L. \tag{7.1}$$

We discretize the problem as in Section 2, obtaining count vector $\mathbf{s}(l)$ for the lth sample,

$$s_k(l) = \#\{y_i(l) \in \mathscr{Y}_k\} \quad \text{for } k = 1,2,\ldots,K \text{ and } l = 1,2,\ldots,L. \tag{7.2}$$

The many-sample version of the Poisson regression model (3.1) is

$$\mathbf{s}(l) \overset{\text{ind.}}{\sim} \text{Po}_K(\boldsymbol{\mu}(l)) \quad \text{for } l = 1,2,\ldots,L, \text{ with } \boldsymbol{\mu}(l) = \boldsymbol{\mu}^o e^{X\beta(l)}. \tag{7.3}$$

The SEF estimates corresponding to (3.2) are

$$\hat{\boldsymbol{\mu}}(l) = \hat{\boldsymbol{\mu}}^o e^{X\hat{\beta}(l)}, \tag{7.4a}$$

where

$$\hat{\boldsymbol{\mu}}^o = m(\mathbf{s}(1), \mathbf{s}(2), \ldots, \mathbf{s}(L)) \quad \text{and} \quad \hat{\beta}(l)\colon X'\big[\mathbf{s}(l) - \hat{\boldsymbol{\mu}}^o e^{X\hat{\beta}(l)}\big] = 0. \tag{7.4b}$$

In this model, a common carrier $\hat{\boldsymbol{\mu}}^o$, estimated from all of the data $(\mathbf{s}(1), \mathbf{s}(2), \ldots, \mathbf{s}(L))$ is modified by an exponential factor $e^{X\hat{\beta}(L)}$ which can vary with l.

The many-sample version of Lemma 1 in Section 3 is the following lemma:

LEMMA 4. *Let*

$$\hat{H}(l) = \frac{\partial \log(\hat{\boldsymbol{\mu}}^o)}{\partial \mathbf{s}(l)} \tag{7.5}$$

be the $K \times K$ matrix with kjth entry $\partial \log(\hat{\mu}_k^o)/ds_j(l)$. Then the $(p+1) \times K$ derivative matrix of $\hat{\beta}(l)$ with respect to $\mathbf{s}(j)$ is

$$\frac{\partial \hat{\beta}(l)}{\partial \mathbf{s}(j)} = \hat{G}(l)^{-1} Z'_{lj}, \tag{7.6a}$$

where

$$\hat{G}(l) = X'\hat{D}(l)X \quad \text{and} \quad Z'_{lj} = X'_{lj}\big[\delta_{lj} I - \hat{D}(l)\hat{H}(j)\big], \tag{7.6b}$$

$\hat{D}(l) \equiv D(\hat{\boldsymbol{\mu}}(l))$, δ_{lj} equalling 1 or 0 as l does or does not equal j. If

$$\hat{\boldsymbol{\mu}}^o = M\mathbf{s}_+ \quad \Big(\mathbf{s}_+ \equiv \sum_l \mathbf{s}_l\Big), \tag{7.7}$$

then

$$Z'_{lj} = X'_{lj}\left[\delta_{lj} I - D(e^{X\hat{\beta}(l)})M\right]. \tag{7.8}$$

(The proof is nearly the same as for Lemma 1 and will not be given here.)

Lemma 4 leads to delta-method estimates of the covariance matrix for $\hat{\beta}(l)$, just as in (3.6),

$$\begin{aligned} \overline{\operatorname{Cov}}(\hat{\beta}(l)) &= \sum_{j=1}^{L} \hat{G}(l)^{-1}\left[Z'_{lj}\overline{D}(j)Z_{lj}\right]\hat{G}(l)^{-1} \quad \text{or} \\ \widehat{\operatorname{Cov}}(\hat{\beta}(l)) &= \sum_{j=1}^{L} \hat{G}(l)^{-1}\left[Z'_{lj}\hat{D}(j)Z_{jl}\right]\hat{G}(l)^{-1}, \end{aligned} \tag{7.9}$$

$\overline{D}(j) \equiv D(\mathbf{s}_j)$. We can also obtain covariance estimates for functions of the $\hat{\beta}(l)$. For example, if $\hat{\gamma} = \hat{\beta}(2) - \hat{\beta}(1)$, then

$$\begin{aligned} \frac{\partial\hat{\gamma}}{\partial\mathbf{s}(1)} &= \hat{G}(2)^{-1}Z'_{21} - \hat{G}(1)^{-1}Z'_{11}, \\ \frac{\partial\hat{\gamma}}{\partial\mathbf{s}(2)} &= \hat{G}(2)^{-1}Z'_{22} - \hat{G}(1)^{-1}Z'_{12} \end{aligned} \tag{7.10a}$$

and

$$\begin{aligned} \overline{\operatorname{Cov}}(\hat{\gamma}) &= \sum_{j=1}^{2}\left(\frac{\partial\hat{\gamma}}{\partial\mathbf{s}(j)}\right)\overline{D}(j)\left(\frac{\partial\hat{\gamma}}{\partial\mathbf{s}(j)}\right)', \\ \widehat{\operatorname{Cov}}(\hat{\gamma}) &= \sum_{j=1}^{2}\left(\frac{\partial\hat{\gamma}}{\partial\mathbf{s}(j)}\right)\hat{D}(j)\left(\frac{\partial\hat{\gamma}}{\partial\mathbf{s}(j)}\right)'. \end{aligned} \tag{7.10b}$$

SEF model (7.4) was used to estimate the two compliance densities in Figure 6. The kth row of X was quadratic in compliance,

$$x_k = \left(1, \tilde{y}_{(k)}, \tilde{y}_{(k)}^2\right), \tag{7.11a}$$

where $\tilde{y}_{(k)} = y_{(k)} - 50/100$, $y_{(k)}$ being the midpoint of $\mathscr{Y}_k$; $\hat{\boldsymbol{\mu}}^o = M\mathbf{s}_1$ as in (7.7), with

$$M = M(7), \tag{7.11b}$$

as in (2.14). The estimated difference $\hat{\gamma} = \hat{\beta}(\text{treatment}) - \hat{\beta}(\text{control})$ was

$$(0.26, -1.36, -0.44) \pm (0.22, 0.38, 1.51), \tag{7.12}$$

with the standard errors taken from $\widehat{\operatorname{Cov}}(\hat{\gamma})$, (7.10b). We see that the linear coefficient is significantly negative, t-value $= -3.6$, but the quadratic coefficient is not.

8. Other types of estimators. In the SEF methodology the application of an initial nonparametric smoother is followed by the fitting of an exponential family parametric model. Hjort and Glad's (1995) semiparametric density estimator reverses this order, with the parametric model coming before the smoother. *Backfitting*, as applied to generalized linear models like (3.1), repeatedly iterates between the parametric and nonparametric fitting methods. This section discusses the SEF methodology in the context of these other possibilities.

As in Section 6, we begin with the normal case where the statistician wishes to estimate $\boldsymbol{\mu}$ having observed

$$\mathbf{z} \sim N_K(\boldsymbol{\mu}, I). \tag{8.1}$$

The linear model $\boldsymbol{\mu} = \boldsymbol{\xi} + X\beta$, with $\boldsymbol{\xi}$ a known *origin* vector and X a known $K \times (p+1)$ structure matrix, gives the estimate

$$\hat{\boldsymbol{\mu}} = \mathscr{P}(\mathbf{z}; \boldsymbol{\xi}) \equiv \boldsymbol{\xi} + P(\mathbf{z} - \boldsymbol{\xi}). \tag{8.2}$$

Here $P = X(X'X)^{-1}X'$ is the $K \times K$ projection matrix into $\mathscr{L}(X)$, the column space of X. In the normal case, $\mathscr{P}(\mathbf{z}; \boldsymbol{\xi})$ plays the role of the parametric exponential family model estimator. The role of the nonparametric smoother is played by

$$\mathscr{M}(\mathbf{z}; \boldsymbol{\xi}) = \boldsymbol{\xi} + M(\mathbf{z} - \boldsymbol{\xi}), \tag{8.3}$$

where M is some fixed $K \times K$ smoothing matrix such as $M(\lambda)$ in (2.13). Once again $\boldsymbol{\xi}$ represents a fixed and known origin. Often we take $\boldsymbol{\xi} = \mathbf{0}$, but it will be important here to consider more general choices of the origin.

The normal-theory analog of the SEF estimate (3.2) is

$$\begin{aligned} \hat{\mu}_{\text{sef}} = \mathscr{P}(\mathbf{z}; \mathscr{M}(\mathbf{z}; \mathbf{0})) &= (P + M - PM)\mathbf{z} \\ &= (P + QM)\mathbf{z}. \end{aligned} \tag{8.4}$$

Here $Q = I - P$ is the projection matrix into $\overset{\perp}{\mathscr{L}}(X)$, the orthocomplement to $\mathscr{L}(X)$. In other words, we begin with $\mathbf{0}$ as the origin, apply $\mathscr{M}$ to get an updated origin $\hat{\boldsymbol{\mu}}^o = \mathscr{M}(\mathbf{z}, \mathbf{0}) = M\mathbf{z}$ and finally take the estimate of $\boldsymbol{\mu}$ to be $\hat{\boldsymbol{\mu}} = \mathscr{P}(\mathbf{z}; \hat{\boldsymbol{\mu}}^o) = \hat{\boldsymbol{\mu}}^o + P(\mathbf{z} - \hat{\boldsymbol{\mu}}^o)$. Notice that

$$X'\mathbf{z} = X'\hat{\boldsymbol{\mu}} \tag{8.5}$$

according to (8.4), which says that $\mathbf{z}$ and $\hat{\boldsymbol{\mu}}$ have the same projection into $\mathscr{L}(X)$, namely, $P\mathbf{z}$. Equality (8.5) is the normal-theory analog of the moment-matching property (2.17).

Reversing the order of $\mathscr{P}$ and $\mathscr{M}$, as Hjort and Glad do in the density estimation problem, gives the estimator

$$\hat{\boldsymbol{\mu}}_{\text{HG}} = \mathscr{M}(\mathbf{z}; \mathscr{P}(\mathbf{z}; \mathbf{0})) = (P + MQ)\mathbf{z}. \tag{8.6}$$

This no longer enjoys the moment-matching property (8.5), but we can restore it with one further application of $\mathscr{P}$. This defines the *symmetrized estimator*

$$\hat{\boldsymbol{\mu}}_{\text{sym}} = \mathscr{P}(\mathbf{z}; \hat{\boldsymbol{\mu}}_{\text{HG}}) = (P + QMQ)\mathbf{z}. \tag{8.7}$$

If M is a symmetric matrix with eigenvalues between 0 and 1, then so is $P + QMQ$. This makes $\hat{\mu}_{\text{sym}}$ a formal Bayes estimator for $\boldsymbol{\mu}$ in situation (8.1), versus a normal prior distribution on $\boldsymbol{\mu}$ possibly having infinite variance in some directions.

The backfitting estimator $\hat{\boldsymbol{\mu}}_{\text{back}}$ is defined as follows in Chapter 5 of Hastie and Tibshirani (1990): suppose we can find vectors $\hat{\boldsymbol{\mu}}^o$ and $\hat{\boldsymbol{\mu}}^1$ such that

$$\hat{\boldsymbol{\mu}}^o = \mathscr{M}(\mathbf{z}; \hat{\boldsymbol{\mu}}^1) - \hat{\boldsymbol{\mu}}^1 \quad \text{and} \quad \hat{\boldsymbol{\mu}}^1 = \mathscr{P}(\mathbf{z}; \hat{\boldsymbol{\mu}}^o) - \hat{\boldsymbol{\mu}}^o. \tag{8.8a}$$

Then

$$\hat{\boldsymbol{\mu}}_{\text{back}} = \hat{\boldsymbol{\mu}}^o + \hat{\boldsymbol{\mu}}^1 \tag{8.8b}$$

[so $\hat{\boldsymbol{\mu}}_{\text{back}} = \mathscr{M}(\mathbf{z}; \hat{\boldsymbol{\mu}}^1) = P(\mathbf{z}; \hat{\boldsymbol{\mu}}^o)$.] Letting $\mathbf{v} = \mathbf{z} - \hat{\boldsymbol{\mu}}^1$, it is easy to show that

$$\mathbf{v} - M\mathbf{v} = (\mathbf{z} - \hat{\boldsymbol{\mu}}^1) - \hat{\boldsymbol{\mu}}^o \in \overset{\perp}{\mathscr{L}}(X), \tag{8.9}$$

which implies the moment-matching property (8.5). Hastie and Tibshirani show that $\hat{\boldsymbol{\mu}}_{\text{back}} = S_{\text{back}}\mathbf{z}$ for a matrix S_{back} that is symmetric and with eigenvalues in $[0, 1]$ if M has these same properties.

Further insight into the backfitting estimator can be gained by expressing it in an explicit form. The backfitting estimator satisfies

$$\hat{\beta} = (X'X)^{-1}X'(\mathbf{z} - \hat{\boldsymbol{\mu}}^0),$$
$$\hat{\boldsymbol{\mu}}^1 = M(\mathbf{z} - X\hat{\beta}).$$

Solving these equations yields the explicit expression

$$\hat{\beta} = [X'(I - M)X]^{-1}X'(I - M)\mathbf{z}.$$

This looks like a weighted least squares estimate with weight matrix $I - M$ or, equivalently, variance matrix $(I - M)^{-1}$. It has the same form as Hjort and Glad's estimator except that they use an unweighted least squares estimator. The weights $I - M$ can be justified from a mixed effects model in which $\boldsymbol{\mu}^1$ is a random effect.

We have discussed four linear estimators $\hat{\boldsymbol{\mu}} = S\mathbf{z}$, three of which have the moment-matching property (8.5). These three can be described as follows: the $\mathscr{L}(X)$ component of $\hat{\boldsymbol{\mu}}$ matches the $\mathscr{L}(X)$ component of $\mathbf{z}$; the $\overset{\perp}{\mathscr{L}}(X)$ component of $\hat{\boldsymbol{\mu}}$ equals the $\overset{\perp}{\mathscr{L}}(X)$ component of a point $\mathbf{v}$ in the flat space $\mathbf{z} \oplus \mathscr{L}(X)$; $\mathbf{v} = \mathbf{z}$ for $\hat{\mu}_{\text{sef}}$, $\mathbf{v} = Q\mathbf{z}$ for $\hat{\mu}_{\text{sym}}$ and $\mathbf{v}$ equals the point or points satisfying the orthogonality condition (8.9) for $\hat{\boldsymbol{\mu}}_{\text{back}}$. *Note*: All four estimators are the same if M and P commute, $MP = PM$.

We calculated the "equivalent kernels" for each of the four estimators, as in Figure 2.5 of Hastie and Tibshirani (1990). The calculation was done for the situation that produced the quadratic estimator in Figure 1: 40 equally spaced x values, P based on a matrix X with rows $(1, x, x^2)$ and M of the form (2.14). The smoothing parameter λ was chosen to give $\operatorname{tr} S = 5$ in each case. The SEF and backfitting kernels were remarkably similar, with both the HG and symmetrical kernels being slightly different.

We can define analogs to $\hat{\boldsymbol{\mu}}_{\text{sef}}$, $\hat{\boldsymbol{\mu}}_{\text{HG}}$, $\hat{\boldsymbol{\mu}}_{\text{sym}}$ and $\hat{\boldsymbol{\mu}}_{\text{back}}$ for the Poisson situation $\mathbf{s} \sim \text{Po}_K(\boldsymbol{\mu})$. Given a positive origin vector $\boldsymbol{\xi}$ and structure matrix X, we let the analog of (8.2) be

$$\mathscr{P}(\mathbf{s}; \boldsymbol{\xi}) \equiv D(\boldsymbol{\xi})e^{X\hat{\beta}(\xi)}, \quad \text{where } X'\left[\mathbf{s} - D(\boldsymbol{\xi})e^{X\hat{\beta}(\xi)}\right] = 0; \tag{8.10}$$

$\hat{\mu} = \mathscr{P}(\mathbf{s}; \boldsymbol{\xi})$ is the MLE estimate of $\boldsymbol{\mu}$ in the "offset" generalized linear model $\boldsymbol{\mu} = D(\boldsymbol{\xi})e^{X\beta}$. The analog of (8.3) is

$$\mathscr{M}(\mathbf{s}; \boldsymbol{\xi}) = D(\boldsymbol{\xi})MD(1/\boldsymbol{\xi})\mathbf{s} = D(\boldsymbol{\xi})M(\mathbf{s}/\boldsymbol{\xi}); \tag{8.11}$$

$\hat{\mu} = \mathscr{M}(\mathbf{s}; \boldsymbol{\xi})$ is a discretized version of what Hjort and Glad (1994) call a *nonparametric density estimate with a parametric start*, the "start" being the choice of $\boldsymbol{\xi}$.

The Poisson analogs of $\hat{\boldsymbol{\mu}}_{\text{sef}}$, $\hat{\boldsymbol{\mu}}_{\text{HG}}$ and $\hat{\boldsymbol{\mu}}_{\text{sym}}$ [(8.4)–(8.7)] are

$$\hat{\boldsymbol{\mu}}_{\text{sef}} = \mathscr{P}(\mathbf{s}; \mathscr{M}(\mathbf{s}; \mathbf{1})), \qquad \hat{\boldsymbol{\mu}}_{\text{HG}} = \mathscr{M}(\mathbf{s}; \mathscr{P}(\mathbf{s}; \mathbf{1})), \qquad \hat{\boldsymbol{\mu}}_{\text{sym}} = \mathscr{P}(\mathbf{s}; \hat{\boldsymbol{\mu}}_{\text{HG}}). \tag{8.12}$$

The backfitting estimate (8.8) is now defined by $\hat{\boldsymbol{\mu}}_{\text{back}} = \hat{\boldsymbol{\mu}}^o + \hat{\boldsymbol{\mu}}^1$, where $\hat{\boldsymbol{\mu}}^o$ and $\hat{\boldsymbol{\mu}}^1$ satisfy the fixed-point relationships

$$\hat{\boldsymbol{\mu}}^o = D(1/\hat{\boldsymbol{\mu}}^1)\mathscr{M}(\mathbf{s}; \hat{\boldsymbol{\mu}}^1) \quad \text{and} \quad \hat{\boldsymbol{\mu}}^1 = D(1/\hat{\boldsymbol{\mu}}^o)\mathscr{P}(\mathbf{s}; \hat{\boldsymbol{\mu}}^o). \tag{8.13}$$

Chapter 6 of Hastie and Tibshirani (1990) discusses an iterative algorithm for computing $\hat{\boldsymbol{\mu}}_{\text{back}}$.

The moment-matching property (2.17) is satisfied by $\hat{\boldsymbol{\mu}}_{\text{sef}}$, $\hat{\boldsymbol{\mu}}_{\text{sym}}$ and $\hat{\boldsymbol{\mu}}_{\text{back}}$. Each of these estimators can be written in the form $\hat{\boldsymbol{\mu}}^o e^{X\hat{\beta}}$, (3.26), with

$$\hat{\boldsymbol{\mu}}^o_{\text{sef}} = M\mathbf{s}, \qquad \hat{\boldsymbol{\mu}}^o_{\text{sym}} = \hat{\boldsymbol{\mu}}_{\text{HG}} \quad \text{and} \quad \hat{\boldsymbol{\mu}}^o_{\text{back}} = M(\mathbf{s}/\hat{\boldsymbol{\mu}}^1). \tag{8.14}$$

We have used $\hat{\boldsymbol{\mu}}_{\text{sef}}$ in all of our examples because it makes the computation of $\hat{H} = d\log(\hat{\boldsymbol{\mu}}^o)/d\mathbf{s}$, a crucial part of the formulas in Sections 2–7, so simple. [An even simpler choice, $\log(\hat{\boldsymbol{\mu}}^o) = H\mathbf{s}$ for some fixed matrix H, does not satisfy (5.14) and seems to have undesirable small-sample properties.]

In theory at least we can compute $\hat{H}$ for any choice of $\hat{\boldsymbol{\mu}}^o = m(\mathbf{s})$. Here, without proof, is $\hat{H}$ for $\hat{\boldsymbol{\mu}}^o_{\text{sym}}$:

LEMMA 5. *For* $\hat{\boldsymbol{\mu}}^o = \hat{\boldsymbol{\mu}}^o_{\text{sym}} = \mathscr{M}(\mathbf{s}; \mathscr{P}(\mathbf{s}, \mathbf{1}))$, *the derivative matrix* $\hat{H} = d\log(\hat{\boldsymbol{\mu}}^o)/d\mathbf{s}$ *is*

$$\hat{H} = D(\hat{\boldsymbol{\mu}}^{oo}/\hat{\boldsymbol{\mu}}^o)MD(1/\hat{\boldsymbol{\mu}}^{oo}) + \left[I - D(\hat{\boldsymbol{\mu}}^{oo}/\hat{\boldsymbol{\mu}}^o)MD(\mathbf{s}/\hat{\boldsymbol{\mu}}^{oo})\right]\hat{P}^{oo}, \tag{8.15a}$$

where

$$\hat{\boldsymbol{\mu}}^{oo} = \mathscr{P}(\mathbf{s}; \mathbf{1}) \quad \textit{and} \quad \hat{P}^{oo} = X[X'D(\hat{\boldsymbol{\mu}}^{oo})X]^{-1}X'. \tag{8.15b}$$

The symmetrized version of the $\hat{\boldsymbol{\mu}}_{\text{sef}}$ in Figure 3, $\hat{\mu}_{\text{sym}} = \mathscr{P}(\mathbf{s}; \mathscr{M}(\mathbf{s}; \hat{\boldsymbol{\mu}}_{\text{sef}}))$, gave similar but somewhat rougher contours than those in the left panel of Figure 4. There is no compelling theoretical reason for preferring $\hat{\boldsymbol{\mu}}_{\text{sym}}$ or $\hat{\boldsymbol{\mu}}_{\text{back}}$ to $\hat{\boldsymbol{\mu}}_{\text{sef}}$, though they seem closer in structure to Bayes and maximum likelihood estimators. In practice, the specific choices of M and X seem more crucial to successful estimation than does the choice between $\hat{\boldsymbol{\mu}}_{\text{sef}}$, $\hat{\boldsymbol{\mu}}_{\text{sym}}$ or $\hat{\boldsymbol{\mu}}_{\text{back}}$.

9. Remarks.

The following remarks apply to the indicated sections.

REMARK A (Section 2). There is an interesting connection between moment-matching and function-preserving properties of smoothers. For a smoother $\hat{\mu} = M\mathbf{s}$, the condition $\Sigma\hat{\mu}_i = \Sigma s_i$ requires $\mathbf{1}'M = \mathbf{1}'$ (where $\mathbf{1}$ is a column of ones), or equivalently $M'\mathbf{1} = \mathbf{1}$. Now most smoothers preserve constants so that $M\mathbf{1} = \mathbf{1}$. For symmetric M we see that matching the zeroth moment is the same as preserving the constant vector. However, most smoothers such as kernels are not symmetric, and hence will not match the zeroth moment exactly. Similarly, a smoother may preserve a vector $\mathbf{t}$ (so $M\mathbf{t} = \mathbf{t}$) without satisfying the moment-matching property $M'\mathbf{t} = \mathbf{t}$. In general, moment-matching is equivalent to function-preserving for the transpose of the smoother matrix.

REMARK B (Section 3). What is the true parameter β being estimated by $\hat{\beta}$? Suppose that $\mathbf{s} \sim \mathrm{Po}_k(\boldsymbol{\mu})$, but that $\boldsymbol{\mu}$ is not necessarily of the form $\boldsymbol{\mu}(\beta)$ in (3.1). From $\boldsymbol{\mu}$ we determine $\boldsymbol{\mu}^o = m(\boldsymbol{\mu})$ and then β the solution vector to the equations $X'[\boldsymbol{\mu} - \boldsymbol{\mu}^o e^{X\beta}] = 0$. Under reasonable conditions $\hat{\beta}$ will be an asymptotically normal estimate for β, in the usual manner of an MLE. For instance if $\boldsymbol{\mu} = n\boldsymbol{\pi}$, with $\boldsymbol{\pi}$ fixed and $n \to \infty$, and if the $\boldsymbol{\mu}^o$ estimate is homogeneous, $m(c\mathbf{s}) = cm(\mathbf{s})$, then it is easy to show that

$$\sqrt{n}\,(\hat{\beta} - \beta) \to N_{p+1}(\mathbf{0}, A^{-1}BA^{-1}), \tag{9.1}$$

where, with $Z' = X'[I - D(e^{X\beta})\,d\boldsymbol{\mu}^o/d\boldsymbol{\mu}]$,

$$A = \lim_{n\to\infty} X'D\left(\frac{\boldsymbol{\mu}^o e^{X\beta}}{n}\right)X \quad \text{and} \quad B = \lim_{n\to\infty} Z'D\left(\frac{\boldsymbol{\mu}}{n}\right)Z. \tag{9.2}$$

The bias of $\hat{\beta}$ as an estimate of β is of order $1/n$.

REMARK C (Section 3). In case (2.13), $\hat{\boldsymbol{\mu}}^o = M\mathbf{s}$, the matrix Z in (3.4a) is a function of $\hat{\beta}$ but not of $\hat{\boldsymbol{\mu}}^o$, say $Z = Z(\hat{\beta})$. If $d\mathbf{s}$ is a K vector orthogonal to the columns of $Z(\hat{\beta})$, then

$$d\hat{\beta} = [X'\hat{D}X]^{-1} Z'(\hat{\beta})\,d\mathbf{s} = 0. \tag{9.3}$$

This implies that the level surfaces of constant $\hat{\beta}$ value are $K - (p+1)$-dimensional flat subspaces in the K-dimensional $\mathbf{s}$ space. Flat level surfaces tend to make delta-method covariance estimates such as (3.6) more accurate.

REMARK D (Section 4). Here are the total Poisson deviances:

$$\mathrm{Dev}(\mathbf{s}, \hat{\boldsymbol{\mu}}) = \sum_k 2\left\{ s_k \log\left(\frac{s_k}{\hat{\mu}_k}\right) - (s_k - \hat{\mu}_k)\right\} \tag{9.4}$$

for four choices of $\hat{\boldsymbol{\mu}}$: $\hat{\boldsymbol{\mu}}^o = M(2)\mathbf{s}$; $\hat{\boldsymbol{\mu}}^1$ the SEF based on this $\hat{\boldsymbol{\mu}}^o$ and $X = (\mathbf{1}, \mathbf{i}, \mathbf{j})$; $\hat{\boldsymbol{\mu}}^2$ the SEF (4.5); $\hat{\boldsymbol{\mu}}^3$ the SEF (9.3):

$$\begin{matrix} \hat{\boldsymbol{\mu}}^0 & \hat{\boldsymbol{\mu}}^1 & \hat{\boldsymbol{\mu}}^2 & \hat{\boldsymbol{\mu}}^3 \\ 247.1 & 240.3 & 226.1 & 220.3. \end{matrix} \tag{9.5}$$

The deviance decrease $D(\mathbf{s}, \hat{\boldsymbol{\mu}}^2) - D(\mathbf{s}\hat{\boldsymbol{\mu}}^3) = 5.77$ is much smaller than 3.75^2, the square of the corresponding t-value in Table 3. The naive t-value, calculated using the standard error estimate from (3.7), is the right one for approximating the deviance decrease. In this case the naive t-value is $0.000267/0.000111 = 2.74$, predicting $2.74^2 = 5.76$ for the deviance decrease.

REMARK E (Section 4). Results like (3.6) can be directly derived for continuous special exponential families (1.3) without going through the Poisson discretization argument. Suppose we estimate $g_0(y)$ in (1.2) by a continuous version of (2.13),

$$\hat{g}_0(y) = \frac{1}{n}\sum_{i=1}^{n} M(y|y_i), \tag{9.6}$$

where, for any y_i, $M(y|y_i)$ is a distribution over $\mathscr{Y}$ [not necessarily satisfying $\int_{\mathscr{Y}} M(y|y_i)\,dy = 1$]. We define the SEF density estimate to be $g_{\hat{\beta}}(y) = \hat{g}_0(y)\exp(\hat{\beta}_0 + t(y)\hat{\beta}_1)$, where $\hat{\beta} = (\hat{\beta}_0, \hat{\beta}_1)$ satisfies the "maximum likelihood" equations

$$\int_{\mathscr{Y}} t(y) g_{\hat{\beta}}(y)\,dy = \bar{t} = \frac{1}{n}\sum_{i=1}^{n} t(y_i), \tag{9.7}$$

as well as the constraint $\int_{\mathscr{Y}} g_{\hat{\beta}}(y)\,dy = 1$.

Define

$$z_i = (t(y_i) - \bar{t}) - \int_{\mathscr{Y}} (t(y) - \bar{t})\exp\big(\hat{\beta}_0 + (t(y) - \bar{t})\hat{\beta}_1\big) M(y|y_i)\,dy \tag{9.8}$$

and let $\widehat{\mathrm{Cov}}(t)$ indicate the covariance matrix of $t(y)$ for the distribution on $\mathscr{Y}$ corresponding to $g_{\hat{\beta}}(y)$. Then the continuous analog of (3.6a) is

$$\overline{\mathrm{Cov}}\big(\hat{\beta}_1\big) = \frac{1}{n}\big[\widehat{\mathrm{Cov}}(t)\big]^{-1}\left[\sum_{i=1}^{n} \frac{z_i' z_i}{n}\right]\big[\widehat{\mathrm{Cov}}(t)\big]^{-1}. \tag{9.9}$$

(9.9) is the limit of (3.6a) as the discretization (2.1) becomes infinitely fine, after the superfluous parameter $\hat{\beta}_0$ is removed.

In order to use (9.9), we need to evaluate the integrals over $\mathscr{Y}$ involved in (9.7), (9.8) and $\widehat{\mathrm{Cov}}(t)$. The discretization argument effectively does such integrals by summation over $k = 1, 2, \ldots, K$. If $\mathscr{Y}$ is high dimensional, we might prefer to work in the continuous mode, doing the integrals by some more efficient algorithm such as componentwise Simpson rules.

REMARK F (Section 5). We will often be interested in the probability vector

$$\hat{\boldsymbol{\pi}} = \hat{\boldsymbol{\mu}}/\hat{\mu}_+ = \hat{\boldsymbol{\mu}}/n, \tag{9.10}$$

rather than in $\hat{\boldsymbol{\mu}}$ itself. Suppose that $\hat{\boldsymbol{\mu}}^o = m(\mathbf{s})$ is scale homogeneous,

$$m(c\mathbf{s}) = cm(\mathbf{s}) \qquad (c > 0). \tag{9.11}$$

The $\hat{\boldsymbol{\mu}} = \mathrm{SEF}(\mathbf{s}; m, X)$ is also scale homogeneous and $\hat{\pi}(c\mathbf{s}) = \hat{\pi}(\mathbf{s})$. Familiar homogeneity properties give

$$\mathbf{1}'\frac{d\hat{\boldsymbol{\mu}}}{d\mathbf{s}} = \mathbf{1}', \qquad \frac{d\hat{\boldsymbol{\mu}}}{d\mathbf{s}}\mathbf{s} = \hat{\boldsymbol{\mu}} \quad \text{and} \quad \frac{d\hat{\boldsymbol{\pi}}}{d\mathbf{s}} = \frac{1}{\hat{\mu}_+}(I - \hat{\boldsymbol{\pi}}\mathbf{1}')\frac{d\hat{\boldsymbol{\mu}}}{d\mathbf{s}}. \tag{9.12}$$

Using (9.12) it is not difficult to show that

$$\overline{\text{Cov}}(\hat{\boldsymbol{\pi}}) \equiv \left(\frac{d\hat{\boldsymbol{\pi}}}{d\mathbf{s}}\right)\overline{D}\left(\frac{d\hat{\boldsymbol{\pi}}}{d\mathbf{s}}\right)' = \frac{1}{n^2}\left[\overline{\text{Cov}}(\hat{\boldsymbol{\mu}}) - \frac{\hat{\boldsymbol{\mu}}\hat{\boldsymbol{\mu}}'}{n}\right] \tag{9.13}$$

and that the diagonal elements of $\overline{\text{Cov}}(\hat{\boldsymbol{\pi}})$ satisfy

$$\overline{\text{trv}} \equiv n \sum_k \frac{\overline{\text{var}}(\hat{\pi}_i)}{\hat{\pi}_k} \equiv \overline{\text{TRV}} - 1. \tag{9.14}$$

Thus the comparisons in Table 4 remain valid for $\overline{\text{trv}}$. The results for the equivalent of $\widehat{\text{TRV}}$ are a little less neat.

REMARK G (Section 5). Let $\mathbf{p} = \mathbf{s}/n$, the vector of empirical probabilities, and define

$$\begin{aligned} &\bar{d} \equiv D(\mathbf{p}) = \overline{D}/n, \qquad \hat{d} \equiv D(\hat{\boldsymbol{\pi}}) = \hat{D}/n, \\ &\hat{p} \equiv X(X'\hat{d}X)^{-1}X' = n\hat{P}, \qquad \hat{q} \equiv \hat{d}^{-1} - \hat{p} = n\hat{Q}, \qquad \hat{h} \equiv n\hat{H}. \end{aligned} \tag{9.15}$$

[Under the homogeneity condition (5.14), $\hat{h} = d\log(\hat{\boldsymbol{\pi}}^o)/d\mathbf{p}$, for $\hat{\boldsymbol{\pi}}^o \equiv \hat{\boldsymbol{\mu}}^o/n$.] Then (5.9) can be written as

$$\overline{\text{TRV}} = \text{tr}\left[\bar{d}\hat{p} + (\hat{h}\bar{d}\hat{h})'(\hat{d}\hat{q}\hat{d}\right] \quad \text{or} \quad \widehat{\text{TRV}} = (p+1) + \text{tr}(\hat{h}\bar{d}\hat{h})(\hat{d}\hat{q}\hat{d}) \tag{9.16}$$

not depending on n. This shows that $\overline{\text{TRV}}$ and $\widehat{\text{TRV}}$ are $O_p(1)$ as $n \to \infty$. See Remark K.

REMARK H (Section 5). Suppose that the discretization of $\mathscr{Y}$ becomes infinitely fine, with K going to infinity and the kth cell $\mathscr{Y}_k$ having volume Δ_k going to zero. Under sufficient regularity conditions, $\hat{\pi}(y_{(k)})/\Delta_k$ will approach $\hat{g}(y_{(k)})$, an estimate of the original continuous density with approximate variance

$$\overline{\text{var}}\{\hat{g}(y_{(k)})\} = \frac{\overline{\text{var}}\{\hat{\pi}(y_{(k)})\}}{\Delta_k^2}. \tag{9.17}$$

Then $\overline{\text{trv}}$, (9.14), will approach

$$n\int_{\mathscr{Y}} \frac{\overline{\text{var}}\{\hat{g}(y)\}}{\hat{g}(y)}\,dy = n\int_{\mathscr{Y}} \overline{\text{CV}}(y)^2 \hat{g}(y)\,dy, \tag{9.18}$$

with $\overline{\text{CV}}(y) \equiv [\overline{\text{Var}}(\hat{g}(y))]^{1/2}/\hat{g}(y)$, a coefficient of variation measure for $\hat{g}(y)$. From Remark G we see that $\overline{\text{CV}}(y)$ is typically $O_p(1/\sqrt{n})$ as $n \to \infty$. The estimated average value for the coefficient of variation of SEF (4.5) was 0.15.

REMARK I (Section 5). A computationally more efficient expression than (5.9) is

$$\overline{\text{TRV}} = \text{tr}\,\overline{D}\left[\hat{P} + \hat{H}'\hat{D}\hat{H} - \hat{H}'\hat{D}\hat{P}\hat{D}\hat{H}\right]. \tag{9.19}$$

Each of the three terms in (9.19) can be evaluated using $O(K^2)$ multiplications rather than $O(K^3)$. This makes values of K as large as 1000 practical. Letting $\hat{G} = X'\hat{D}X$, the $O(K^2)$ computing formula is

$$\overline{\text{TRV}} = \operatorname{tr} \hat{G}^{-1}(X'\overline{D}X) + \|\overline{D}^{1/2}\hat{H}\hat{D}^{1/2}\|^2 + \operatorname{tr} \hat{G}^{-1}(X'\hat{D}\hat{H}\overline{D}^{1/2})(\overline{D}^{1/2}\hat{H}'\hat{D}X) \tag{9.20}$$

($\|a\|^2 \equiv \sum_i \sum_j a_{ij}^2$ for matrix a), and similarly for $\widehat{\text{TRV}}$, substituting $\hat{D}$ for $\overline{D}$.

When $\hat{\boldsymbol{\mu}}^o = \boldsymbol{M}\mathbf{s}$ the middle term in (9.20) equals $\sum_k \hat{\mu}_k A_k$, where

$$A_k = \sum_j \hat{M}_{kj}^2 \tilde{s}_j \Big/ \Big(\sum_j \tilde{M}_{kj} s_j\Big)^2 \qquad \left[\tilde{M}_{kj} \equiv (\hat{\mu}_k/\hat{\mu}_k^o) M_{kj}\right], \tag{9.21a}$$

and $\tilde{s}_j$ equals s_j for $\overline{\text{TRV}}$ or $\hat{\mu}_j$ for $\widehat{\text{TRV}}$. For the $\overline{\text{TRV}}$ case $A_k \leq 1$, but A_k can blow up for $\widehat{\text{TRV}}$. The calculations of Section 6 replace A_k with

$$\min(A_k, 1) \tag{9.21b}$$

in the $\widehat{\text{TRV}}$ case.

Remark J (Section 6). The degrees of freedom formula (6.20) can be rewritten as

$$\overline{\text{DF}} = \operatorname{tr} \overline{D}\left[\hat{P} + \hat{H} - \hat{P}\hat{D}\hat{H}\right] \quad \text{or} \quad \widehat{\text{DF}} = (p+1) + \operatorname{tr} \hat{D}(\hat{H} - \hat{P}\hat{D}\hat{H}). \tag{9.22}$$

The $\hat{P}$ term corresponds to the $e^{X\hat{\beta}}$ part of the SEF definition (3.2), while the $\hat{H}$ term corresponds to the choice of the carrier $\hat{\boldsymbol{\mu}}^o$. The $\hat{P}\hat{D}\hat{H}$ subtraction term corrects for collinearity between the carrier and the exponential family terms. The degrees of freedom definition for $\hat{\boldsymbol{\pi}}$ rather than $\hat{\boldsymbol{\mu}}$, as in Remark F, subtracts 1 from formula (6.20) or (9.22). It takes $O(K^2)$ multiplications to compute $\overline{\text{DF}}$ or $\widehat{\text{DF}}$ from either formula.

Remark K (Section 6). Write $\mathbf{s} = n\mathbf{p}$ as in Remark G and consider expression (6.15) for $\text{DF}(\hat{\boldsymbol{\mu}})$ as $n \to \infty$ with $\mathbf{p}$ fixed. Assuming $m(\mathbf{s}) = cm(\mathbf{s})$, (9.11), the approximation $\widehat{\text{DF}} = \hat{D}\hat{O}$ does not depend on n and so is $O_p(1)$ just as in (9.16). Using higher-order expansions it is easy to show that

$$\text{DF}(\hat{\boldsymbol{\mu}}) - \widehat{\text{DF}} = O_p(1/n), \tag{9.23}$$

so at least in this sense $\widehat{\text{DF}}$ is a good approximation to $\text{DF}(\hat{\boldsymbol{\mu}})$. The corresponding results hold for $\overline{\text{DF}}$, $\overline{\text{TRV}}$, and $\widehat{\text{TRV}}$.

Remark L (Section 6). The Poisson residual degrees of freedom formula

$$\text{RDF}(\boldsymbol{\mu}) = K(\mu) - 2\text{DF}(\mu) + \text{TRV}(\boldsymbol{\mu}) \tag{9.24}$$

is always nonnegative. To prove this we use (6.12b) to write

$$\text{RDF}(\boldsymbol{\mu}) = E\{\text{Dev}(\mathbf{s}, \hat{\boldsymbol{\mu}})\} - \text{Dev}(\boldsymbol{\mu}, \boldsymbol{\nu}), \tag{9.25}$$

and note that $E\{(\mathbf{s}, \hat{\boldsymbol{\mu}})\} = (\boldsymbol{\mu}, \boldsymbol{\nu})$. The proof is completed with the fact that the Poisson deviance $\text{Dev}(\boldsymbol{\mu}, \boldsymbol{\nu})$ is a jointly convex function of $(\boldsymbol{\mu}, \boldsymbol{\nu})$.

The RDF estimates (6.21) are not necessarily nonnegative. They become so if we replace $K(\boldsymbol{\mu})$ in (6.21) with the Taylor series estimates

$$\bar{K} = \operatorname{tr} \bar{D}\hat{D}^{-1} \quad \text{or} \quad \hat{K} = \operatorname{tr} \hat{D}\hat{D}^{-1} = (p+1), \tag{9.26}$$

but these were poor approximations in our examples.

REMARK M (Section 8). We tried a "loess" smoother [Cleveland and Gross (1992)] for the galaxy data, in the form of a generalized additive Poisson model [Hastie and Tibshirani (1990)]. We used a degree 2 loess model, that is, one that fits second degree polynomials locally. This gave an expected deviance picture similar to Figure 5. This is not surprising given the similarity between the moment-matching and function-preserving properties mentioned in Remark A.

REFERENCES

AITKEN, M. (1993). Model choice in single samples from the exponential and double exponential families using the posterior Bayes factor. Technical report, Tel-Aviv Univ., Israel.

CLEVELAND, W. C. and GROSS, E. (1992). Locally weighted regression. In *Statistical Models in S* (J. Chambers and T. Hastie, eds.). Wadsworth, Belmont, CA.

COX, D. and REID, N. (1987). Parametric orthogonality and approximate conditional inference. *J. Roy. Statist. Soc. Ser. B* **49** 1–39.

EFRON, B. (1986). How biased is the apparent error rate of a logistic regression? *J. Amer. Statist. Assoc.* **81** 461–470.

EFRON, B. (1988). Logistic regression, survival analysis, and the Kaplan–Meier curve. *J. Amer. Statist. Assoc.* **83** 414–425.

EFRON, B. (1992). Six questions raised by the bootstrap. In *Bootstrap Proceedings Volume* (R. LaPage and L. Billard, eds.). Wiley, New York.

EFRON, B. and FELDMAN, D. (1991). Compliance as an explanatory variable in clinical trials (with comments and rejoinder). *J. Amer. Statist. Assoc.* **86** 9–25.

EFRON, B. and PETROSIAN, V. (1992). A simple test of independence for truncated data with applications to redshift surveys. *Astrophysical Journal* **399** 345–352.

EFRON, B. and TIBSHIRANI, R. J. (1993). *An Introduction to the Bootstrap*. Chapman and Hall, New York.

GREEN, P. and SILVERMAN, B. (1994). *Nonparametric Regression and Generalized Linear Models*. Chapman and Hall, New York.

HASTIE, T. and TIBSHIRANI, R. (1990). *Generalized Additive Models*. Chapman and Hall, New York.

HJORT, N. and GLAD, I. (1995). Nonparametric density estimation with a parametric start. *Ann. Statist.* **23** 882–904.

JONES, M. C. (1993). Kernel density estimation when the bandwidth is large. *Australian J. Statist.* **35** 319–326.

KOOPERBERG, C. and STONE, C. (1991). A study of logspline density estimation. *Comput. Statist. Data Anal.* **12** 327–347.

LEHMANN, E. (1983). *Theory of Point Estimation*. Wiley, New York.

LINDSEY, J. (1974a). Comparison of probability distributions. *J. Roy. Statist. Soc. Ser. B* **36** 38–47.

LINDSEY, J. (1974b). Construction and comparison of statistical models. *J. Roy. Statist. Soc. Ser. B* **36** 418–425.

LINDSEY, J. and MERSCH, G. (1992). Fitting and comparing probability distributions with log linear models. *Comput. Statist. Data Anal.* **13** 373–384.

LOH, E. and SPILLAR, E. (1986). Photometric redshift of galaxies. *Astrophysical Journal* **303** 154–161.

MARDIA, K. V., KENT, J. T. AND BIBBY, J. M. (1979). *Multivariate Analysis*. Academic Press, New York.

OLKIN, I. and SPIEGELMAN, C. (1987). A semiparametric approach to density estimation. *J. Amer. Statist. Assoc.* **82** 858–865.

PIERCE, D. AND PETERS, D. (1992). Practical use of higher order asymptotics for multiparameter exponential families (with discussion). *J. Roy. Statist. Soc. Ser. B* **54** 701–737.

STONE, C. (1994). The use of polynomial splines and their tensor products in multivariate function estimation. *Ann. Statist.* **22** 118–170.

DEPARTMENT OF STATISTICS
STANFORD UNIVERSITY
STANFORD, CALIFORNIA 94305

DEPARTMENT OF STATISTICS
UNIVERSITY OF TORONTO
TORONTO, ONTARIO
CANADA

17

Bootstrap Confidence Levels for Phylogenetic Trees

Introduction by Joe Felsenstein
University of Washington, Seattle

In the early 1980s, biologists were increasingly interested in reconstructing phylogenies (evolutionary trees), often using DNA sequences. It was becoming clear that this should be treated statistically. But although point estimates could be made by maximum likelihood or least squares methods, the space of possible trees was large enough, and strange enough, that it was not clear how to know how much confidence to have in features of the tree. In 1985 I suggested that Efron's nonparametric bootstrap could be applied to the problem (Felsenstein 1985). If we have a table of data with rows being different species and columns being different sites in the DNA molecule, we would bootstrap by drawing columns (keeping the rows in the same order within each column), until we had a data set with the same species and the same number of sites. For each of these bootstrap sample data sets, we would infer a tree.

An important question was whether a given branch of the tree (such as the branch that leads to humans and chimpanzees but not to gorillas or orangutans) could be validated. I naïvely thought of the presence or absence of the branch as a discrete 0/1 variable and assumed that we could use the percentile method with the bootstrap to see whether (say) 95% of the mass of the distribution of this variable was in the one atom. We should declare the branch significantly supported if it appeared in more than 95% of the trees inferred from bootstrap samples. The method met a need and was very widely used – Ryan and Woodall (2005) list it as number 7 in a list of most-cited statistical papers. I was astonished to see that it outranked Efron's original paper on the bootstrap, until I realized that the list was not of citations by statisticians. It thus tended to emphasize applications rather than the theory underlying them. Efron's paper introducing the bootstrap was far more influential, but there were a great many people inferring phylogenies.

Doubts soon arose. I kept hearing from users that the confidence the method assigned to their groups seemed to be too low. A careful analysis of a simple model with four-species trees was made by Zharkikh and Li (1992). They found that the P-value assigned to tree topologies was too low. Hillis and Bull (1993) verified this by extensive simulations in larger cases. In response, I was forced (Felsenstein and Kishino 1993) to concede the point and show in a simple example that the use of P-values, rather than the bootstrap itself, was the culprit.

Efron, Halloran and Holmes show in this paper that there is a way of correcting the P-value assigned to a group, and provide a practical method for making the correction. Their method competes with an iterated bootstrap method by Rodrigo (1993), a complete-and-partial bootstrap approach by Zharkikh and Li (1995), and Shimodaira's (2002) "Almost Unbiased" method. While the best approach is not yet clear, I hope that there will be further work following up on these methods. Not only is the application to phylogenies important, but the issues raised affect all uses of the bootstrap to address "the problem of regions" (Efron and Tibshirani 1998). These issues presumably arise in all cases where there is a highly-structured family of hypotheses, including those being investigated under the slogan of "algebraic statistics". The bootstrap must have some role here, and the methods of Efron, Halloran and Holmes will be important in investigating it.

References

Efron, B. and R. Tibshirani (1998). The problem of regions. *Annals of Statistics* **26**, 1687–1718.

Felsenstein, J. (1985). Confidence limits on phylogenies: An approach using the bootstrap. *Evolution* **39**, 783–791.

Felsenstein, J. and H. Kishino (1993). Is there something wrong with the bootstrap on phylogenies? A reply to Hillis and Bull. *Systematic Biology* **42**, 193–200.

Hillis, D. and J. J. Bull (1993). An empirical test of bootstrapping as a method for assessing confidence in phylogenetic analysis. *Systematic Biology* **42**, 182–192.

Rodrigo, A. G. (1993). Calibrating the bootstrap test of monophyly. *International Journal for Parasitology* **23**, 507–514.

Ryan, T. P. and W. H. Woodall (2005). The most-cited statistical papers. *Journal of Applied Statistics* **32**, 461–474.

Shimodaira, H. (2002). An approximately unbiased test of phylogenetic tree selection. *Systematic Biology* **51**, 492–508.

Zharkikh, A. and W.-H. Li (1992). Statistical properties of bootstrap estimation of phylogenetic variability from nucleotide sequences. I. Four taxa with a molecular clock. *Molecular Biology and Evolution* **9**, 1119–1147.

Zharkikh, A. and W.-H. Li (1995). Estimation of confidence in phylogeny: The complete-and-partial bootstrap technique. *Molecular Phylogenetics and Evolution* **4**, 44–63.

Evolution. The following article, which appeared in number 14, July 1996, of *Proc. Natl. Acad. Sci. USA* (**93,** 7085–7090) is reprinted in its entirety with the author's corrections incorporated.

Bootstrap confidence levels for phylogenetic trees

BRADLEY EFRON, ELIZABETH HALLORAN‡, AND SUSAN HOLMES†§

†Department of Statistics, Stanford University, Stanford, CA 94305; and ‡Department of Biostatistics, Rollins School of Public Health, Emory University, Atlanta, GA 30322

Contributed by Bradley Efron, January 26, 1996

ABSTRACT **Evolutionary trees are often estimated from DNA or RNA sequence data. How much confidence should we have in the estimated trees? In 1985, Felsenstein [Felsenstein, J. (1985) *Evolution* 39, 783–791] suggested the use of the bootstrap to answer this question. Felsenstein's method, which in concept is a straightforward application of the bootstrap, is widely used, but has been criticized as biased in the genetics literature. This paper concerns the use of the bootstrap in the tree problem. We show that Felsenstein's method is not biased, but that it can be corrected to better agree with standard ideas of confidence levels and hypothesis testing. These corrections can be made by using the more elaborate bootstrap method presented here, at the expense of considerably more computation.**

The bootstrap, as described in ref. 1, is a computer-based technique for assessing the accuracy of almost any statistical estimate. It is particularly useful in complicated nonparametric estimation problems, where analytic methods are impractical. Felsenstein (2) introduced the use of the bootstrap in the estimation of phylogenetic trees. His technique, which has been widely used, provides assessments of "confidence" for each clade of an observed tree, based on the proportion of bootstrap trees showing that same clade. However Felsenstein's method has been criticized as biased. Hillis and Bull's paper (3), for example, says that the bootstrap confidence values are consistently too conservative (i.e., biased downward) as an assessment of the tree's accuracy.

Is the bootstrap biased for the assessment of phylogenetic trees? We will show that the answer is no, at least to a first order of statistical accuracy. Felsenstein's method provides a reasonable first approximation to the actual confidence levels of the observed clades. More ambitious bootstrap methods can be fashioned to give still better assessments of confidence. We will describe one such method and apply it to the estimation of a phylogenetic tree for the malaria parasite *Plasmodium*.

Bootstrapping Trees

Fig. 1 shows part of a data set used to construct phylogenetic trees for malaria. The data are the aligned sequences of small subunit RNA genes from 11 malaria species of the genus *Plasmodium*. The 11 × 221 data matrix we will first consider is composed of the 221 polytypic sites. Fig. 1 shows the first 20 columns of **x**. There are another 1399 monotypic sites, where the 11 species are identical.

Fig. 2 shows a phylogenetic tree constructed from **x**. The tree-building algorithm proceeds in two main steps: (*i*) an 11 × 11 distance matrix $\hat{D}$ is constructed for the 11 species, measuring differences between the row vectors of **x**; and (*ii*) $\hat{D}$ is converted into a tree by a connection algorithm that connects the closest two entries (species 9 and 10 here), reduces $\hat{D}$ to a 10 × 10 matrix according to some merging rule, connects the two closest entries of the new D matrix, etc.

We can indicate the tree-building process schematically as

$$\mathbf{x} \rightarrow \hat{D} \rightarrow \widehat{\text{TREE}},$$

the hats indicating that we are dealing with estimated quantities. A deliberately simple choice of algorithms was made in constructing Fig. 2: $\hat{D}$ was the matrix of the Euclidean distances between the rows of **x**, with (A, G, C, T) interpreted numerically as (1, 2, 5, 6), while the connection algorithm merged nodes by maximization. Other, better, tree-building algorithms are available, as mentioned later in the paper. Some of these, such as the maximum parsimony method, do not involve a distance matrix, and some use all of the sites, including the monotypical ones. The discussion here applies just as well to all such tree-building algorithms.

Felsenstein's method proceeds as follows. A bootstrap data matrix $\mathbf{x}^*$ is formed by randomly selecting 221 columns from the original matrix **x** *with replacement*. For example the first column of $\mathbf{x}^*$ might be the 17th column of **x**, the second might be the 209th column of **x**, the third the 17th column of **x**, etc. Then the original tree-building algorithm is applied to $\mathbf{x}^*$, giving a bootstrap tree $\widehat{\text{TREE}}^*$,

$$x^* \rightarrow \hat{D}^* \rightarrow \widehat{\text{TREE}}^*,$$

This whole process is independently repeated some large number B times, $B = 200$ in Fig. 2, and the proportions of bootstrap trees agreeing with the original tree are calculated. "Agreeing" here refers to the topology of the tree and not to the length of its arms.

These proportions are the bootstrap confidence values. For example the 9-10 clade seen in Fig. 2 appeared in 193 of the 200 bootstrap trees, for an estimated confidence value of 0.965. Species 7-8-9-10 occurred as a clade in 199 of the 200 bootstrap trees, giving 0.995 confidence. (Not all of these 199 trees had the configuration shown in Fig. 2; some instead first having 8 joined to 9-10 and then 7 joined to 8-9-10, as well as other variations.)[2]

Felsenstein's method is, nearly, a standard application of the *nonparametric bootstrap*. The basic assumption, further discussed in the next section, is that the columns of the data matrix **x** are independent of each other and drawn from the same probability distribution. Of course, if this assumption is a bad one, then Felsenstein's method goes wrong, but that is not the point of concern here nor in the references, and we will take the independence assumption as a given truth.

The bootstrap is more typically applied to statistics $\hat{\theta}$ that estimate a parameter of interest θ, both $\hat{\theta}$ and θ being single numbers. For example, $\hat{\theta}$ could be the sample correlation

The publication costs of this article were defrayed in part by page charge payment. This article must therefore be hereby marked "*advertisement*" in accordance with 18 U.S.C. §1734 solely to indicate this fact.

§Permanent address: Biometrie–Institut National de la Recherche Agronomique, Montpellier, France.

Species	Site: 1	2	3	4	5	6	7	8	9	10	11	12	13	14	15	16	17	18	19	20
1 Pre (Chimp)	C	T	T	G	A	G	A	A	A	A	T	T	C	T	T	A	G	A	T	A
2 Pme (Lizard)	T	C	T	A	A	A	A	G	A	T	T	A	T	A	T	A	G	A	T	A
3 Pma (Human)	T	T	T	A	A	G	G	A	A	A	T	T	C	T	T	A	A	A	T	T
4 Pfa (Human)	T	T	T	G	A	G	A	A	A	A	T	T	C	T	T	A	G	A	T	A
5 Pbe (Rodent)	T	T	T	A	A	G	A	A	A	A	T	T	T	A	T	A	A	A	T	A
6 Plo (Bird)	T	T	T	A	A	G	A	A	A	A	C	T	C	A	C	A	A	A	T	C
7 Pfr (Monkey)	C	T	T	A	A	G	A	A	G	A	T	T	C	T	T	A	G	G	A	A
8 Pkn (Monkey)	C	T	T	A	A	G	A	A	A	G	T	T	C	T	T	A	G	A	T	A
9 Pcy (Monkey)	C	T	C	A	T	G	A	A	A	A	T	T	C	T	T	A	G	A	T	A
10 Pv (Human)	C	T	T	A	T	G	A	A	A	A	T	T	C	T	C	G	G	A	T	A
11 Pga (Bird)	T	T	T	A	A	G	A	A	A	A	T	T	T	T	C	A	A	A	T	C

FIG. 1. Part of the data matrix of aligned nucleotide sequences for the malaria parasite *Plasmodium*. Shown are the first 20 columns of the 11 × 221 matrix **x** of polytypic sites used in most of the analyses below. The final analysis of the last section also uses the data from 1399 monotypic sites.

coefficient between the first two malaria species, Pre and Pme, at the 221 sites, with (A, G, C, T) interpreted as (1, 2, 5, 6): $\hat{\theta} = 0.616$. How accurate is $\hat{\theta}$ as an estimate of the true correlation θ? The nonparametric bootstrap answers such questions without making distributional assumptions.

Each bootstrap data set $\mathbf{x}^*$ gives a bootstrap estimate $\hat{\theta}^*$, in this case the sample correlation between the first two rows of $\mathbf{x}^*$. The central idea of the bootstrap is to use the observed distribution of the differences $\hat{\theta}^* - \hat{\theta}$ to infer the unobservable distribution of $\hat{\theta} - \theta$; in other words to learn about the accuracy of $\hat{\theta}$. In our example, the 200 bootstrap replications of $\hat{\theta}^* - \hat{\theta}$ were observed to have expectation 0.622 and standard deviation 0.052. The inference is that $\hat{\theta}$ is nearly unbiased for estimating θ, with a standard error of about 0.052. We can also calculate bootstrap confidence intervals for θ. A well-developed theory supports the validity of these inferences [see Efron and Tibshirani (1)].

Felsenstein's application of the bootstrap is nonstandard in one important way: the statistic $\widehat{\text{TREE}}$, unlike the correlation coefficient, does not change smoothly as a function of the data set **x**. Rather, $\widehat{\text{TREE}}$ is constant within large regions of the **x**-space, and then changes discontinuously as certain boundaries are crossed. This behavior raises questions about the bootstrap inferences, questions that are investigated in the sections that follow.

A Model For The Bootstrap

The rationale underlying the bootstrap confidence values depends on a simple multinomial probability model. There are $K = 4^{11} - 4$ possible column vectors for **x**, the number of vectors of length 11 based on a 4-letter alphabet, not counting the 4 monotypic ones. Call these vectors $X_1, X_2, \ldots, X_K$, and suppose that each observed column of **x** is an independent

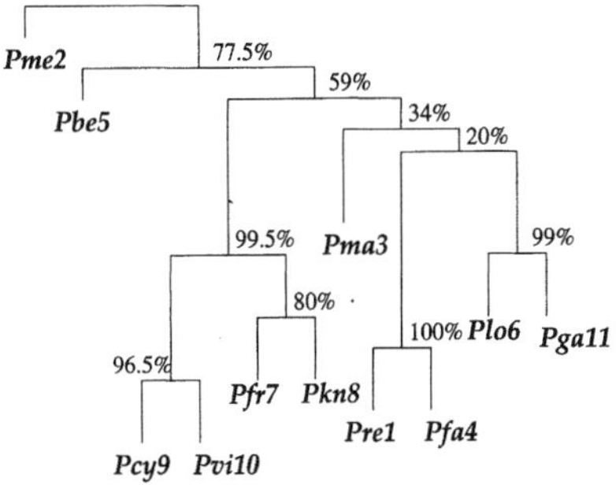

FIG. 2. Phylogenetic tree based on the malaria data matrix; species are numbered as in Fig. 1. The numbers at the branches are confidence values based on Felsenstein's bootstrap method. $B = 200$ bootstrap replications.

selection from $X_1, X_2, \ldots, X_K$, equaling X_k with probability π_k. This is the *multinomial model* for the generation of **x**.

Denote $\underset{\sim}{\pi} = (\pi_1, \pi_2, \ldots, \pi_K)$, so the sum of $\underset{\sim}{\pi}$'s coordinates is 1. The data matrix **x** can be characterized by the proportion of its $n = 221$ columns equalling each possible X_k, say

$$\hat{\pi}_k = \#\{\text{columns of } \mathbf{x} \text{ equalling } X_k\}/n,$$

with $\hat{\underset{\sim}{\pi}} = (\hat{\pi}_1, \hat{\pi}_2, \ldots, \hat{\pi}_K)$. This is a very inefficient way to represent the data, since $4^{11} - 4$ is so much bigger than 221, but it is useful for understanding the bootstrap. Later we will see that only the vectors X_k that actually occur in **x** need be considered, at most n of them.

Almost always the distance matrix $\hat{D}$ is a function of the observed proportions $\hat{\underset{\sim}{\pi}}$, so we can write the tree-building algorithm as

$$\hat{\underset{\sim}{\pi}} \rightarrow \hat{D} \rightarrow \widehat{\text{TREE}}.$$

In a similar way the vector of true probabilities $\underset{\sim}{\pi}$ gives a true distance matrix and a true tree,

$$\underset{\sim}{\pi} \rightarrow D \rightarrow \text{TREE}.$$

D would be the matrix with ijth element $\{\Sigma_k \pi_k (X_{ki} - X_{kj})^2\}^{1/2}$ in our example, and TREE the tree obtained by applying the maximizing connection algorithm to D.

Fig. 3 is a schematic picture of the space of possible $\underset{\sim}{\pi}$ vectors, divided into regions $\mathcal{R}_1, \mathcal{R}_2, \ldots$. The regions correspond to different possible trees, so if $\underset{\sim}{\pi} \in \mathcal{R}_j$ the jth possible tree results. We hope that $\widehat{\text{TREE}} = \text{TREE}$, which is to say

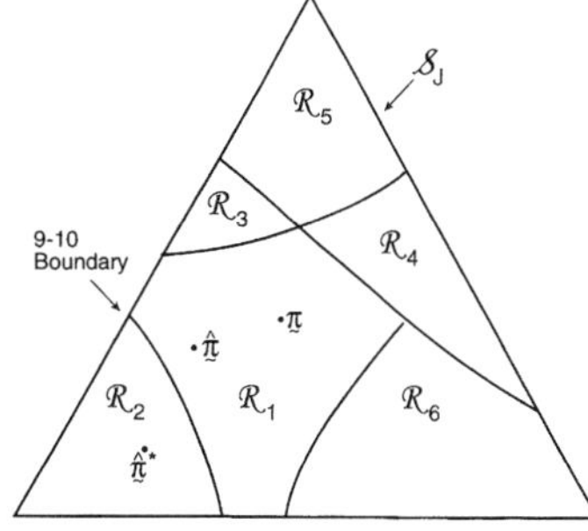

FIG. 3. Schematic diagram of tree estimation; triangle represents the space of all possible $\underset{\sim}{\pi}$ vectors in the multinomial probability model; regions $\mathcal{R}_1, \mathcal{R}_2 \ldots$ correspond to the different possible trees. In the case shown $\underset{\sim}{\pi}$ and $\hat{\underset{\sim}{\pi}}$ lie in the same region so TREE = $\widehat{\text{TREE}}$, but $\hat{\underset{\sim}{\pi}}^*$ lies in a region where TREE* does not have the 9-10 clade.

that π and $\hat{\pi}$ lie in the same region, or at least that $\widehat{\text{TREE}}$ and TREE agree in their most important aspects.

The bootstrap data matrix $\mathbf{x}^*$ has proportions of columns say

$$\hat{\pi}_k^* = \#\{\text{columns of } \mathbf{x}^* \text{ equalling } X_k\}/n,$$

$\hat{\pi}^* = (\hat{\pi}_1^*, \hat{\pi}_2^*, \ldots, \hat{\pi}_K^*)$. We can indicate the bootstrap tree-building

$$\hat{\pi}^* \to \hat{D}^* \to \widehat{\text{TREE}}^*,$$

The hypothetical example of Fig. 3 puts π and $\hat{\pi}$ in the same region, so that the estimate $\widehat{\text{TREE}}$ exactly equals the true TREE. However $\hat{\pi}^*$ lies in a different region, with $\widehat{\text{TREE}}^*$ not having the 9-10 clade. This actually happened in 7 out of the 200 bootstrap replications for Fig. 2.

What the critics of Felsenstein's method call its bias is the fact that the probability $\widehat{\text{TREE}}^*$ = TREE is usually less than the probability $\widehat{\text{TREE}}$ = TREE. In terms of Fig. 3, this means that $\hat{\pi}^*$ has less probability than $\hat{\pi}$ of lying in the same region as π. Hillis and Bull (3) give specific simulation examples. The discussion below is intended to show that this property is not a bias, and that to a first order of approximation the bootstrap confidence values provide a correct assessment of $\widehat{\text{TREE}}$'s accuracy. A more valid criticism of Felsenstein's method, discussed later, involves its relationship with the standard theory of statistical confidence levels based on hypothesis tests.

Returning to the correlation example of the previous section, it is *not* true that $\hat{\theta}^* - \theta$ (as opposed to $\hat{\theta}^* - \hat{\theta}$) has the same distribution as $\hat{\theta} - \theta$, even approximately. In fact $\hat{\theta}^* - \theta$ will have nearly twice the variance of $\hat{\theta} - \theta$, the sum of the variances of $\hat{\theta}$ around θ and of $\hat{\theta}^*$ around $\hat{\theta}$. Similarly in Fig. 3 the average distance from $\hat{\pi}^*$ to π will be greater than the average distance from $\hat{\pi}$ to π. This is the underlying reason for results like those of Hillis and Bull, that $\hat{\pi}^*$ has less probability than $\hat{\pi}$ of lying in the same region as π. However, to make valid bootstrap inferences we need to use the observed differences between $\widehat{\text{TREE}}^*$ and $\widehat{\text{TREE}}$ (not between $\widehat{\text{TREE}}^*$ and TREE) to infer the differences between $\widehat{\text{TREE}}$ and TREE. Just how this can be done is discussed using a simplified model in the next two sections.

A Simpler Model

The meaning of the bootstrap confidence values can be more easily explained using a simple normal model rather than the multinomial model. This same tactic is used in Felsenstein and Kishino (4). Now we assume that the data $\mathbf{x} = (x_1, x_2)$ is a two dimensional normal vector with expectation vector $\boldsymbol{\mu} = (\mu_1, \mu_2)$ and identity covariance matrix, written

$$\mathbf{x} \sim N_2(\boldsymbol{\mu}, I).$$

In other words x_1 and x_2 are independent normal variates with expectations μ_1 and μ_2, and variances 1. The obvious estimate of $\boldsymbol{\mu}$ is $\hat{\boldsymbol{\mu}} = \mathbf{x}$, and we will use this notation in what follows. The $\boldsymbol{\mu}$-plane is partitioned into regions $\mathcal{R}_1, \mathcal{R}_2, \mathcal{R}_3, \ldots$ similarly to Fig. 3. We observe that $\hat{\boldsymbol{\mu}}$ lies in one of these regions, say $\mathcal{R}_1$, and we wish to assign a confidence value to the event that $\boldsymbol{\mu}$ itself lies in $\mathcal{R}_1$.

Two examples are illustrated in Fig. 4. In both of them $\mathbf{x} = \hat{\boldsymbol{\mu}} = (4.5,0)$ lies in $\mathcal{R}_1$, one of two possible regions. Case I has $\mathcal{R}_2 = \{\boldsymbol{\mu} : \mu_1 \le 3\}$, a half-plane, while case II has $\mathcal{R}_2 = \{\boldsymbol{\mu} : \|\boldsymbol{\mu}\| \le 3\}$, a disk of radius 3.

Bootstrap sampling in our simplified problem can be taken to be

$$\mathbf{x}^* \sim N_2(\hat{\boldsymbol{\mu}}, I).$$

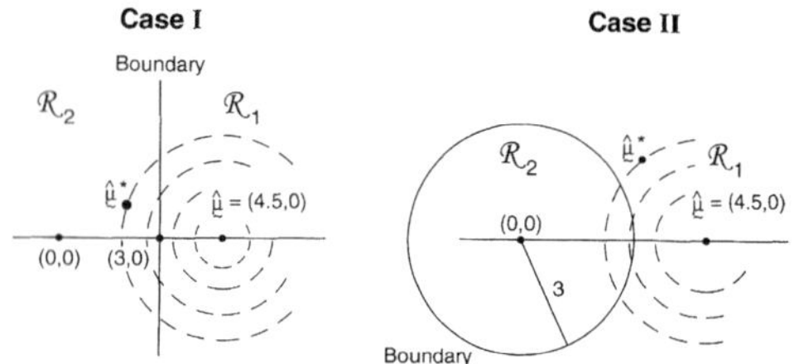

FIG. 4. Two cases of the simple normal model; in both we observe $\hat{\boldsymbol{\mu}} = (4.5, 0) \in \mathcal{R}_1$, and wish to assign a confidence value to $\boldsymbol{\mu} \in \mathcal{R}_1$. Case I, $\mathcal{R}_2$ is the region $\{\mu_1 \le 3\}$. Case II, $\mathcal{R}_2$ is the region $\{\|\boldsymbol{\mu}\| < 3\}$. The dashed circles indicate bootstrap sampling $\hat{\boldsymbol{\mu}}^* \sim N_2(\hat{\boldsymbol{\mu}}, I)$.

This is a parametric version of the bootstrap, as in section 6.5 of Efron and Tibshirani (1), rather than the more familiar nonparametric bootstrap considered previously, but it provides the proper analogy with the multinomial model. The dashed circles in Fig. 4 indicate the bootstrap density of $\hat{\boldsymbol{\mu}}^* = \mathbf{x}^*$, centered at $\hat{\boldsymbol{\mu}}$. Felsenstein's confidence value is the bootstrap probability that $\hat{\boldsymbol{\mu}}^*$ lies in $\mathcal{R}_1$, say

$$\tilde{\alpha} = \text{Prob}_{\hat{\mu}}\{\hat{\boldsymbol{\mu}}^* \in \mathcal{R}_1\}.$$

The notation $\text{Prob}_{\hat{\mu}}$ emphasizes that the bootstrap probability is computed with $\hat{\boldsymbol{\mu}}$ fixed and only $\hat{\boldsymbol{\mu}}^*$ random. The bivariate normal model of this section is simple enough to allow the $\tilde{\alpha}$ values to be calculated theoretically, without doing simulations,

$$\tilde{\alpha}_{\text{I}} = 0.933 \quad \text{and} \quad \tilde{\alpha}_{\text{II}} = 0.949.$$

Notice that $\tilde{\alpha}_{\text{II}}$ is bigger than $\tilde{\alpha}_{\text{I}}$ because $\mathcal{R}_1$ is bigger in case II.

In our normal model, $\hat{\boldsymbol{\mu}}^* - \hat{\boldsymbol{\mu}}$ has the same distribution as $\hat{\boldsymbol{\mu}} - \boldsymbol{\mu}$, both distributions being the standard bivariate normal $N_2(\mathbf{0}, I)$. The general idea of the bootstrap is to use the observable bootstrap distribution of $\hat{\boldsymbol{\mu}}^* - \hat{\boldsymbol{\mu}}$ to say something about the unobservable distribution of the error $\hat{\boldsymbol{\mu}} - \boldsymbol{\mu}$. Notice, however, that the marginal distribution of $\hat{\boldsymbol{\mu}}^* - \boldsymbol{\mu}$ has *twice* as much variance,

$$\hat{\boldsymbol{\mu}}^* - \boldsymbol{\mu} \sim N_2(\mathbf{O}, 2I).$$

This generates the "bias" discussed previously, that $\hat{\boldsymbol{\mu}}^*$ has less probability than $\hat{\boldsymbol{\mu}}$ of being in the same region as $\boldsymbol{\mu}$. But this kind of interpretation of bootstrap results cannot give correct inferences. Newton (5) makes a similar point, as do Zharkikh and Li (6) and Felsenstein and Kishino (4).

We can use a Bayesian model to show that $\tilde{\alpha}$ is a reasonable assessment of the probability that $\mathcal{R}_1$ contains μ. Suppose we believe *apriori* that $\boldsymbol{\mu}$ could lie anywhere in the plane with equal probability. Then having observed $\hat{\boldsymbol{\mu}}$, the *aposteriori* distribution of $\boldsymbol{\mu}$ given $\hat{\boldsymbol{\mu}}$ is $N_2(\hat{\boldsymbol{\mu}}, I)$, exactly the same as the bootstrap distribution of $\hat{\boldsymbol{\mu}}^*$. In other words, $\tilde{\alpha}$ is the *aposteriori* probability of the event $\boldsymbol{\mu} \in \mathcal{R}_1$, if we begin with an "uninformative" prior density for $\boldsymbol{\mu}$.

Almost the same thing happens in the multinomial model. The bootstrap probability that $\widehat{\text{TREE}}^* = \widehat{\text{TREE}}$ is almost the same as the *aposteriori* probability that TREE = $\widehat{\text{TREE}}$ starting from an uninformative prior density on π [see section 10.6 of Efron (7)]. The same statement holds for any part of the tree, for example the existence of the 9-10 clade in Fig. 2. There are reasons for being skeptical about the Bayesian argument, as discussed in the next section. However, the argument shows that Felsenstein's bootstrap confidence values are at least reasonable and certainly cannot be universally biased downward.

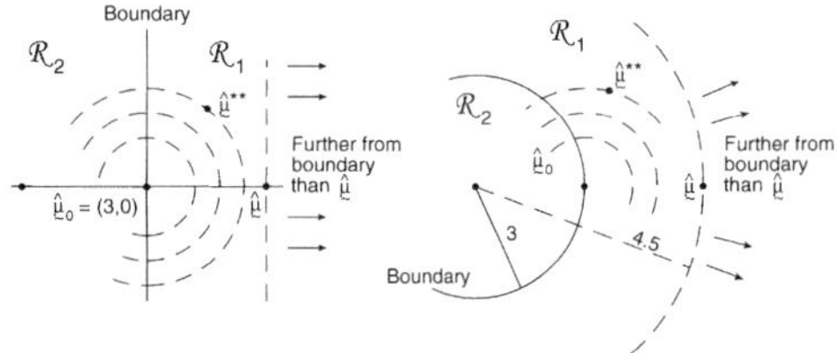

FIG. 5. Confidence levels of the two cases in Fig. 4; $\hat{\boldsymbol{\mu}}_0 = (3, 0)$ is the closest point to $\hat{\boldsymbol{\mu}} = (4.5, 0)$ on the boundary separating $\mathcal{R}_1$ from $\mathcal{R}_2$; bootstrap vector $\hat{\boldsymbol{\mu}}^{**} \sim N_2(\hat{\boldsymbol{\mu}}_0, I)$. The confidence level $\hat{\alpha}$ is the probability that $\hat{\boldsymbol{\mu}}^{**}$ is closer than $\hat{\boldsymbol{\mu}}$ to the boundary.

Hypothesis-Testing Confidence Levels

Fig. 5 illustrates another more customary way of assigning a confidence level to the event $\boldsymbol{\mu} \in \mathcal{R}_1$. In both case I and case II, $\hat{\boldsymbol{\mu}}_0 = (3, 0)$ is the closest point to $\hat{\boldsymbol{\mu}}$ on the boundary separating $\mathcal{R}_1$ from $\mathcal{R}_2$. We now bootstrap from $\hat{\boldsymbol{\mu}}_0$ rather than from $\hat{\boldsymbol{\mu}}$, obtaining bootstrap data vectors

$$\mathbf{x}^{**} \sim N_2(\hat{\boldsymbol{\mu}}_0, I).$$

[The double star notation is intended to avoid confusion with the previous bootstrap vectors $\mathbf{x}^* \sim N_2(\hat{\boldsymbol{\mu}}, I)$.] The *confidence level* $\hat{\alpha}$ for the event $\boldsymbol{\mu} \in \mathcal{R}_1$ is the probability that the bootstrap vector $\hat{\boldsymbol{\mu}}^{**} = \mathbf{x}^{**}$ lies closer than $\hat{\boldsymbol{\mu}}$ to the boundary. This has a familiar interpretation: $1 - \hat{\alpha}$ is the rejection level, one-sided, of the usual likelihood ratio test of the null hypothesis that $\boldsymbol{\mu}$ does *not* lie in $\mathcal{R}_1$. Here we are computing $\hat{\alpha}$ numerically, rather than relying on an asymptotic χ^2 approximation. In a one-dimensional testing problem, $1 - \hat{\alpha}$ would exactly equal the usual p value obtained from the test of the null hypothesis that the true parameter value lies in $\mathcal{R}_2$.

Once again it is simple to compute the confidence level for our two cases, at least numerically,

$$\hat{\alpha}_{\text{I}} = 0.933 \quad \text{and} \quad \hat{\alpha}_{\text{II}} = 0.914.$$

In the first case $\hat{\alpha}_{\text{I}}$ equals $\tilde{\alpha}_{\text{I}}$, the Felsenstein bootstrap confidence value. However $\hat{\alpha}_{\text{II}} = 0.914$ is less than $\tilde{\alpha}_{\text{II}} = 0.949$.

Why do the answers differ? Comparing Figs. 4 and 5, we see that, roughly speaking, the confidence value $\tilde{\alpha}$ is a probabilistic measure of the distance from $\hat{\boldsymbol{\mu}}$ to the boundary, while the confidence level $\hat{\alpha}$ measures distance from the boundary to $\hat{\boldsymbol{\mu}}$. The two ways of measuring distance agree for the straight boundary, but not for the curved boundary of case II. Because the boundary curves *away* from $\hat{\boldsymbol{\mu}}$, the confidence value $\tilde{\alpha}$ is increased from the straight-line case. However the set of vectors further than $\hat{\boldsymbol{\mu}}$ from the boundary curves *toward* $\hat{\boldsymbol{\mu}}_0$, which decreases $\hat{\alpha}$. We would get the opposite results if the boundary between $\mathcal{R}_1$ and $\mathcal{R}_2$ curved in the other direction.

The confidence level $\hat{\alpha}$, rather than $\tilde{\alpha}$, provides the more usual assessment of statistical belief. For example in case II let $\theta = \|\boldsymbol{\mu}\|$ be the length of the expectation vector $\boldsymbol{\mu}$. Then $\hat{\alpha} = 0.914$ is the usual confidence level attained for the event $\{\theta \geq 3\}$, based on observing $\hat{\theta} = \|\hat{\boldsymbol{\mu}}\| = 4.5$. And $\{\theta \geq 3\}$ is the same as the event $\{\boldsymbol{\mu} \in \mathcal{R}_1\}$.

Using the confidence value $\tilde{\alpha}$ is equivalent to assuming a flat Bayesian prior for $\boldsymbol{\mu}$. It can be shown that using $\hat{\alpha}$ amounts, approximately, to assuming a different prior density for $\boldsymbol{\mu}$, one that depends on the shape of the boundary. In case II this prior is uniform on polar coordinates for $\boldsymbol{\mu}$, rather than uniform on the original rectangular coordinates [see Tibshirani (8)].

The Relationship Between the Two Measures of Confidence

There is a simple approximation formula for converting a Felsenstein confidence value $\tilde{\alpha}$ to a hypothesis-testing confidence level $\hat{\alpha}$. This formula is conveniently expressed in terms of the cumulative distribution function $\Phi(z)$ of a standard one-dimensional normal variate, and its inverse function Φ^{-1}: $\Phi(1.645) = 0.95$, $\Phi^{-1}(0.95) = 1.645$, etc. We define the "z values" corresponding to $\tilde{\alpha}$ and $\hat{\alpha}$,

$$\tilde{z} = \Phi^{-1}(\tilde{\alpha}) \quad \text{and} \quad \hat{z} = \Phi^{-1}(\hat{\alpha}).$$

In case II, $\tilde{z} = \Phi^{-1}(0.949) = 1.64$ and $\hat{z} = \Phi^{-1}(0.914) = 1.37$.

Now let $\hat{\boldsymbol{\mu}}^{**} \sim N_2(\hat{\boldsymbol{\mu}}_0, I)$ as in Fig. 5, and define

$$z_0 = \Phi^{-1}(\text{Prob}_{\hat{\boldsymbol{\mu}}_0}\{\hat{\boldsymbol{\mu}}^{**} \in R_1\}).$$

For case I it is easy to see that $z_0 = \Phi^{-1}(0.50) = 0$. For case II, standard calculations show that $z_0 = \Phi^{-1}(0.567) = 0.17$.

In normal problems of the sort shown in Figs. 4 and 5 we can approximate $\hat{z}$ in terms of $\tilde{z}$ and z_0:

$$\hat{z} \doteq \tilde{z} - 2z_0. \quad [1]$$

Formula **1** is developed in Efron (9), where it is shown to have "second order accuracy." This means that in repeated sampling situations [where we observe independent data vectors $x_1, x_2, \ldots, x_n \sim N_2(\boldsymbol{\mu}, I)$ and estimate $\boldsymbol{\mu}$ by $\hat{\boldsymbol{\mu}} = \Sigma_i^n = x_i/n$] z_0 is of order $1/\sqrt{n}$, and formula **1** estimates $\hat{z}$ with an error of order only $1/n$.

Second-order accuracy is a large sample property, but it usually indicates good performance in actual problems. For case I, Eq. **1** correctly predicts $\hat{z} = \tilde{z}$, both equalling $\Phi^{-1}(0.933) = 1.50$. For case II the prediction is $\hat{z} = 1.64 - 0.34 = 1.30$, compared with the actual value $\hat{z} = 1.37$.

Formula **1** allows us to compute the confidence level $\hat{\alpha}$ for the event $\{\boldsymbol{\mu} \in \mathcal{R}_1\}$ solely in terms of bootstrap calculations, no matter how complicated the boundary may be. A first level of bootstrap replications with $\hat{\boldsymbol{\mu}}^* \sim N_2(\hat{\boldsymbol{\mu}}, I)$ gives bootstrap data vectors $\hat{\boldsymbol{\mu}}^*(1), \hat{\boldsymbol{\mu}}^*(2), \ldots, \hat{\boldsymbol{\mu}}^*(B)$, from which we calculate

$$\tilde{z} = \Phi^{-1}\left(\frac{\#\{\hat{\boldsymbol{\mu}}^* \text{ vectors in } \mathcal{R}_1\}}{B}\right).$$

A second level of bootstrap replications with $\hat{\boldsymbol{\mu}}^{**} \sim N_2(\hat{\boldsymbol{\mu}}_0, I)$, giving say $\hat{\boldsymbol{\mu}}^{**}(1), \hat{\boldsymbol{\mu}}^{**}(2), \ldots, \hat{\boldsymbol{\mu}}^{**}(B_2)$, allows us to calculate

$$\hat{z}_0 = \Phi^{-1}\left(\frac{\#\{\hat{\boldsymbol{\mu}}^{**} \text{ vectors in } \mathcal{R}_1\}}{B_2}\right).$$

Then formula **1** gives $\hat{z} = \tilde{z} - 2z_0$.

As few as $B = 100$, or even 50, replications $\hat{\boldsymbol{\mu}}^*$ are enough to provide a rough but useful estimate of the confidence value $\tilde{\alpha}$. However, because the difference between $\tilde{z} = \Phi^{-1}(\tilde{\alpha})$ and $\hat{z} = \Phi^{-1}(\hat{\alpha})$ is relatively small, considerably larger bootstrap samples are necessary to make formula **1** worthwhile. The calculations in section 9 of Efron (9) suggest both B and B_2 must be on the order of at least 1000. This point did not arise in cases I and II where we were able to do the calculations by direct numerical integration, but it is important in the kind of complicated tree-construction problems we are actually considering.

We now return to the problem of trees, as seen in Fig. 2. The version of formula **1** that applies to the multinomial model of Fig. 3 is

$$\hat{z} = \frac{\tilde{z} - z_0}{1 + a(\tilde{z} - z_0)} - z_0. \quad [2]$$

Here "a" is the *acceleration constant* introduced in ref. 9. It is quite a bit easier to calculate than z_0, as shown in the next section. Formula **2** is based on the bootstrap confidence intervals called "BC_a" in ref. 9.

If we tried to draw Fig. 3 accurately we would find that the multi-dimensional boundaries were hopelessly complicated. Nevertheless, formula **2** allows us to obtain a good approximation to the hypothesis-testing confidence level $\hat{\alpha} = \Phi(\hat{z})$ solely in terms of bootstrap computations. How to do so is illustrated in the next section.

An Example Concerning the Malaria Data

Fig. 2 shows an estimated confidence value of

$$\tilde{\alpha} = 0.965$$

for the existence of the 9-10 clade on the malaria evolutionary tree. This value was based on $B = 200$ bootstrap replications, but (with some luck) it agrees very closely with the value $\tilde{\alpha} = 0.962$ obtained from $B = 2000$ replications. How does $\tilde{\alpha}$ compare with $\hat{\alpha}$, the hypothesis-testing confidence level for the 9-10 clade? We will show that

$$\hat{\alpha} = 0.942$$

(or $\hat{\alpha} = 0.938$ if we begin with $\tilde{\alpha} = 0.962$ instead of 0.965). To put it another way, our nonconfidence in the 9-10 clade goes from $1 - \tilde{\alpha} = 0.035$ to $1 - \hat{\alpha} = 0.058$, a substantial change.

We will describe, briefly, the computational steps necessary to compute $\hat{\alpha}$. To do so we need notation for multinomial sampling. Let $\mathbf{P} = (P_1, P_2, \ldots, P_n)$ indicate a probability vector on $n = 221$ components, so the entries of the vector $\mathbf{P}$ are nonnegative numbers summing to 1. The notation

$$\mathbf{P}^* \sim \text{Mult}(\mathbf{P})$$

will indicate that $\mathbf{P}^* = (P_1^*, P_2^*, \ldots, P_n^*)$ is the vector of proportions obtained in a multinomial sample of size n from $\mathbf{P}$. In other words we independently draw integers $I_1^*, I_2^*, \ldots, I_n^*$ from $\{1, 2, \ldots, n\}$ with probability P_k on k, and record the proportions $P_k^* = \#\{I_i^* = k\}/n$. This is the kind of multinomial sampling pictured in Fig. 3, expressed more efficiently in terms of $n = 221$ coordinates instead of $K = 4^{11} - 4$.

Each vector $\mathbf{P}^*$ is associated with a data matrix $\mathbf{x}^*$ that has proportion P_k^* of its columns equal to the kth column of the original data matrix $\mathbf{x}$. Then $\mathbf{P}^*$ determines a distance matrix and a tree according to the original tree-building algorithm,

$$\mathbf{P}^* \rightarrow \hat{D}^* \rightarrow \widehat{\text{TREE}}^*.$$

The "central" vector

$$\mathbf{P}^{(\text{cent})} = (1/n, 1/n, \ldots, 1/n)$$

corresponds to the original data matrix $\mathbf{x}$ and the original tree $\widehat{\text{TREE}}$. Notice that taking $\mathbf{P}^* \sim \text{Mult}(\mathbf{P}^{(\text{cent})})$ amounts to doing ordinary bootstrap sampling, since then $\mathbf{x}^*$ has its columns chosen independently and with equal probability from the columns of $\mathbf{x}$.

Resampling from $\mathbf{P}^{(\text{cent})}$ means that each of the 221 columns is equally likely, but this is not the same as all possible 11 vectors being equally likely. There were only 149 *distinct* 11 vectors among the columns of $\mathbf{x}$, and these are the only ones that can appear in $\mathbf{x}^*$. The vector *TTTTCTTTTTT* appeared seven times among the columns of $\mathbf{x}$, so it shows up seven times as frequently in the columns of $\mathbf{x}^*$, compared with *ATAAAAAAAAA* which appeared only once in $\mathbf{x}$.

Here are the steps in the computation of $\hat{\alpha}$.

Step 1. $B = 2000$ first-level bootstrap vectors $\mathbf{P}^*(1), \mathbf{P}^*(2), \ldots, \mathbf{P}^*(B)$ were obtained as independent multinomials $\mathbf{P}^* \sim \text{Mult}(\mathbf{P}^{(\text{cent})})$. Some 1923 of the corresponding bootstrap trees had the 9-10 clade, giving the estimate $\tilde{\alpha} = 0.962 = 1923/2000$.

Step 2. The first 200 of these included seven cases without the 9-10 clade. Call the seven $\mathbf{P}^*$ vectors $\mathbf{P}^{(1)}, \mathbf{P}^{(2)}, \ldots, \mathbf{P}^{(7)}$. For each of them, a value of w between 0 and 1 was found such that the vector

$$\mathbf{p}^{(j)} = w \cdot \mathbf{P}^{(j)} + (1 - w)\mathbf{P}^{(\text{cent})}$$

was right on the 9-10 boundary. The vectors $\mathbf{p}^{(j)}$ play the role of $\hat{\boldsymbol{\mu}}_0$ in Fig. 5.

Finding w is easy using a one-dimensional binary search program, as on page 90 of ref. 10. At each step of the search it is only necessary to check whether or not the current value of $w\mathbf{P}^{(j)} + (1 - w)\mathbf{P}^{(\text{cent})}$ gives a tree having the 9-10 clade. Twelve steps of the binary search, the number used here, locates the boundary value of w within $1/2^{12}$. The vectors $\mathbf{p}^{(j)}$ play the role of $\hat{\boldsymbol{\mu}}_0$ in Fig. 5.

Step 3. For each of the boundary vectors $\mathbf{p}^{(j)}$ we generated $B_2 = 400$ second-level bootstrap vectors

$$\mathbf{P}^{**} \sim \text{Mult}(\mathbf{p}^{(j)}),$$

computed the corresponding tree, and counted the number of trees having the 9-10 clade. The numbers were as follows for the seven cases:

Case	No.	B_2
1	218	400
2	204	400
3	223	400
4	214	400
5	213	400
6	216	400
7	223	400
Total	1151	2800

From the total we calculated an estimate of the correction term z_0 in formula **2**,

$$z_0 = \Phi^{-1}\left(\frac{1511}{2800}\right) = 0.0995.$$

Binomial calculations indicate that $z_0 = 0.0995$ has a standard error of about 0.02 due to the bootstrap sampling (that is, due to taking 2800 instead of all possible bootstrap replications), so 2800 is not lavishly excessive. Notice that we could have started with the 77 out of the 2000 $\mathbf{P}^*$ vectors not having the 9-10 clade, rather than the 7 out of the first 200, and taken $B_2 = 40$ for each $\mathbf{p}^{(j)}$, giving about the same total second-level sample.

Step 4. The acceleration constant "a" appearing in formula **2** depends on the direction from $\mathbf{P}^{(\text{cent})}$ to the boundary, as explained in section 8 of ref. 9. For a given direction vector $\mathbf{U}$,

$$a(\mathbf{U}) = \frac{1}{6}\sum_1^n U_k^3 / \left(\sum_1^n U_k^2\right)^{3/2}.$$

Taking $\mathbf{U} = \mathbf{p}^{(j)} - \mathbf{P}^{(\text{cent})}$ for each of the seven cases gave

Case	a
1	0.014
2	0.009
3	0.014
4	0.012
5	0.014
6	0.012
7	0.014
Average	0.0129

Step 5. Finally we applied formula **2** with $\tilde{z} = \Phi^{-1}(0.962) = 1.77$, $z_0 = 0.0995$, and $a = 0.0129$, to get $\hat{z} = 1.54$, or $\hat{\alpha} = \Phi(\hat{z}) = 0.938$. If we begin with $\tilde{z} = \Phi^{-1}(0.965)$ then $\hat{\alpha} = 0.942$.

Notice that in this example we could say that Felsenstein's bootstrap confidence value $\tilde{\alpha}$ was biased *upward*, not downward, at least compared with the hypothesis-testing level $\hat{\alpha}$. This happened because z_0 was positive, indicating that the 9-10 boundary was curving away from $\mathbf{P}^{(\text{cent})}$, just as in case 2 of Fig. 5. The opposite can also occur, and in fact did for other clades. For example the clade at the top of Fig. 2 that includes all of the species except lizard (species 2) had $\tilde{\alpha} = 0.775$ compared with $\hat{\alpha} = 0.875$.

We carried out these same calculations using the more efficient tree-building algorithm employed in Escalante and Ayala (11); that is we used Felsenstein's PHYLIP package (12) on the complete RNA sequences, neighbor-joining trees based on Kimura's (13) two-parameter distances.

In order to vary our problem slightly, we looked at the clade 7-8 (Pfr-Pkn), which is more questionable than the 9-10 clade. The tree produced from the original set is:

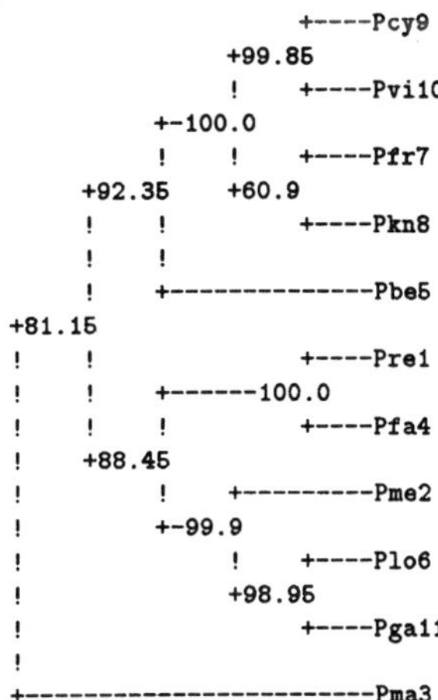

Step 1. $B = 2000$ first-level bootstrap vectors. Some 1218 of the corresponding bootstrap trees had the 7-8 clade, giving the estimate $\tilde{\alpha} = 0.609 = 1218/2000$.

Step 2. We took, as before, seven cases without the 7-8 clade, and for each one found a multinomial vector near the 7-8 boundary.

Step 3. For each of the boundary vectors $\mathbf{p}^{(j)}$ we generated $B_2 = 400$ second-level bootstrap vectors

$$\mathbf{P}^{**} \sim \text{Mult}(\mathbf{p}^{(j)}),$$

computed the corresponding tree, and counted the number of trees having the 7-8 clade. The numbers were as follows for the seven cases:

Case	No.	B_2
1	120	400
2	184	400
3	145	400
4	187	400
5	176	400
6	197	400
7	240	400
Total	1249	2800

From the total we calculated an estimate of the correction term z_0 in formula **2**,

$$z_0 = \Phi^{-1}\left(\frac{1249}{2800}\right) = -0.136$$

Step 4. The acceleration constant "a" appearing in formula **2** was computed as before giving:

Case	a
1	−0.118
2	−0.0176
3	0.0172
4	−0.0256
5	0.00981
6	−0.0540
7	−0.0198
Average	−0.0296

Step 5. Finally we applied formula 2 with $\tilde{z} = \Phi^{-1}(0.609) = 0.277$ to get $\hat{z} = 0.417$, or $\hat{\alpha} = \Phi(\hat{z}) = 0.662$. In this case $\hat{\alpha}$ is bigger than $\tilde{\alpha}$, reflecting the fact that the 7-8 boundary curves toward the central point, at least in a global sense.

Computing $\hat{\alpha}$ is about 20 times as much work as $\tilde{\alpha}$, but it is work for the computer and not for the investigator. Once the tree-building algorithm is available, all of the computations require no more than applying this algorithm to resampled versions of the original data set.

Discussion and Summary

The discussion in this paper, which has gone lightly over many technical details of statistical inference, makes the following main points about the bootstrapping of phylogenetic trees.

(*i*) The confidence values $\tilde{\alpha}$ obtained by Felsenstein's bootstrap method are not biased systematically downward.

(*ii*) In a Bayesian sense, the $\tilde{\alpha}$ can be thought of as reasonable assessments of error for the estimated tree.

(*iii*) More familiar non-Bayesian confidence levels $\hat{\alpha}$ can also be defined. Typically $\hat{\alpha}$ and $\tilde{\alpha}$ will converge as the number n of independent sites grows large, at rate $1/\sqrt{n}$.

(*iv*) The $\hat{\alpha}$ can be estimated by a two-level bootstrap algorithm.

(*v*) As few as 100 or even 50 bootstrap replications can give useful estimates of $\tilde{\alpha}$, while $\hat{\alpha}$ estimates require at least 2000 total replications. None of the computations requires more than applying the original tree-building algorithm to resampled data sets.

We are grateful to A. Escalante and F. Ayala (14) for providing us with these data. B.E. is grateful for support from Public Health Service Grant 5 R01 CA59039-20 and National Science Foundation Grant DMS95-04379. E.H. is supported by National Science Foundation Grant DMS94-10138 and National Institutes of Health Grant NIAID R29-A131057.

1. Efron, B. & Tibshirani, R. (1993) *An Introduction to the Bootstrap* (Chapman & Hall, London).
2. Felsenstein, J. (1985) *Evolution* **39,** 783–791.
3. Hillis, D. & Bull, J. (1993) *Syst. Biol.* **42,** 182–192.
4. Felsenstein, J. & Kishino, H. (1993) *Syst. Biol.* **42,** 193–200.
5. Newton, M. A. (1996) *Biometrika* **83,** 315–328.
6. Zharkikh, A. & Li, W. H. (1992) *Mol. Biol. Evol.* **9,** 1119–1147.
7. Efron, B. (1982) *SIAM CBMS-NSF Monogr.* **38.**
8. Tibshirani, R. J. (1989) *Biometrika* **76,** 604–608.
9. Efron, B. (1987) *J. Am. Stat. Assoc.* **82,** 171–185.
10. Press, W., Flannery, B., Teukolsky, S. & Vetterling, W. (1987) *Numerical Recipes* (Cambridge Univ. Press, New York).
11. Escalante, A. & Ayala, F. (1995) *Proc. Natl. Acad. Sci. USA* **92,** 5793–5797.
12. Felsenstein, J. (1993) PHYLIP (Univ. of Washington, Seattle).
13. Kimura, M. (1980) *J. Mol. Evol.* **16,** 111–120.
14. Escalante, A. & Ayala, F. (1994) *Proc. Natl. Acad. Sci. USA* **91,** 11371–11377.

18

R. A. Fisher in the 21st Century

Introduction by Stephen M. Stigler
University of Chicago

Despite R. A. Fisher's pervasive influence in statistics and genetics, he had very few students in either field. Bradley Efron could, I think, be properly called a student of Fisher, and he was one of the best in the class. Efron never met Fisher, as he tells us in this far-ranging and deeply informative essay, based upon the 1996 Fisher Memorial Lecture to the Joint Statistical Meetings in Chicago. And it is just as well they never met: Fisher was immensely creative and independent-minded, but when he met others of that same type with similar interests, the result tended to be permanent mutual alienation.

The depth of Efron's understanding of Fisher is palpable in some of his earlier work and in this essay in particular. Still, there is a sense in which discussing Fisher is like a Rorschach test for later statisticians: when they write of Fisher's work and influence, they tell us even more about themselves and their own work than they do about Fisher. This is the case for the published discussants of Efron's paper, each of whom sees something different in Fisher, but it is also true of the essay itself. This is not to deny that what Efron (and the others) say about Fisher sheds important light on Fisher; rather it is the choice of what parts of Fisher to comment on and how to represent them that is quite revealing.

In his earlier work Efron gave particular attention to geometric thinking in Fisher; here the emphasis is more upon tactics and logic. We find singled out as Fisherian characteristics some of the hallmarks of Efron's own approach: the logical reduction of complicated inference problems to a simple form where all would agree the answer is obvious, the arrival at a compromise between Bayesian and frequentist approaches, and a philosophy always expressed in very practical terms. As a bonus we are given a personal guided tour showing how Efron's own bootstrap is based upon deep Fisherian ideas of statistical information; it is not the simplistic tool it can be misunderstood to be by lesser practitioners.

The clarity of the exposition not only informs, it also invites the reader into the discussion, to weigh whether or not the Fisher–Efron tactics will work for them: a continuing Rorschach test, as it were. For example, Efron nicely identifies as one of Fisher's tactics what he calls the "Plug-in Principle": "All possible inferential questions are answered by simply plugging in estimates, usually maximum likelihood estimates, for unknown parameters." I wonder, though, if this can be properly raised to the level of a "Principle", except when skillfully wielded by a statistician such as Fisher or Efron. When Lagrange tried this in 1776 he ended up with the now-notoriously inefficient method of moments, even though starting with maximum likelihood!

Eager young statistical adventurers heading out into research might take caution from Efron's Statistical Triangle, a barycentric picture of modern research with Fisherian, Bayesian, and frequentist as the three vertices. It bears some resemblance to the dreaded Bermuda Triangle: relative safety near the edges, and Fiducial probability lurking in the center. But no such adventurer with this essay firmly in hand is likely to perish, and it would not surprise me if, a century or so from now, some one of them were to write a self-revealing essay entitled "Brad Efron in the 22nd century."

Statistical Science
1998, Vol. 13, No. 2, 95–122

R. A. Fisher in the 21st Century

Invited Paper Presented at the 1996 R. A. Fisher Lecture

Bradley Efron

Abstract. Fisher is the single most important figure in 20th century statistics. This talk examines his influence on modern statistical thinking, trying to predict how Fisherian we can expect the 21st century to be. Fisher's philosophy is characterized as a series of shrewd compromises between the Bayesian and frequentist viewpoints, augmented by some unique characteristics that are particularly useful in applied problems. Several current research topics are examined with an eye toward Fisherian influence, or the lack of it, and what this portends for future statistical developments. Based on the 1996 Fisher lecture, the article closely follows the text of that talk.

Key words and phrases: Statistical inference, Bayes, frequentist, fiducial, empirical Bayes, model selection, bootstrap, confidence intervals.

1. INTRODUCTION

Even scientists need their heroes, and R. A. Fisher was certainly the hero of 20th century statistics. His ideas dominated and transformed our field to an extent a Caesar or an Alexander might have envied. Most of this happened in the second quarter of the century, but by the time of my own education Fisher had been reduced to a somewhat minor figure in American academic statistics, with the influence of Neyman and Wald rising to their high water mark.

There has been a late 20th century resurgence of interest in Fisherian statistics, in England where his influence never much waned, but also in America and the rest of the statistical world. Much of this revival has gone unnoticed because it is hidden behind the dazzle of modern computational methods. One of my main goals here will be to clarify Fisher's influence on modern statistics. Both the strengths and limitations of Fisherian thinking will be described, mainly by example, finally leading up to some speculations on Fisher's role in the statistical world of the 21st century.

What follows is basically the text of the Fisher lecture presented to the August 1966 Joint Statistical meetings in Chicago. The talk format has certain advantages over a standard journal article. First and foremost, it is meant to be absorbed quickly, in an hour, forcing the presentation to concentrate on main points rather than technical details. Spoken language tends to be livelier than the gray prose of a journal paper. A talk encourages bolder distinctions and personal opinions, which are dangerously vulnerable in a written article but appropriate I believe for speculations about the future. In other words, this will be a broad-brush painting, long on color but short on detail.

These advantages may be viewed in a less favorable light by the careful reader. Fisher's mathematical arguments are beautiful in their power and economy, and most of that is missing here. The broad brush strokes sometimes conceal important areas of controversy. Most of the argumentation is by example rather than theory, with examples from my own work playing an exaggerated role. References are minimal, and not indicated in the usual author–year format but rather collected in annotated form at the end of the text. Most seriously, the one-hour limit required a somewhat arbitrary selection of topics, and in doing so I concentrated on

Bradley Efron is Max H. Stein Professor of Humanities and Sciences and Professor of Statistics and Biostatistics, Department of Statistics, Stanford University, Stanford, California 94305-4065 (e-mail: brad@stat.stanford.edu).

those parts of Fisher's work that have been most important to me, omitting whole areas of Fisherian influence such as randomization and experimental design. The result is more a personal essay than a systematic survey.

This is a talk (as I will now refer to it) on Fisher's influence, not mainly on Fisher himself or even his intellectual history. A much more thorough study of the work itself appears in L. J. Savage's famous talk and essay, "On rereading R. A. Fisher," the 1971 Fisher lecture, a brilliant account of Fisher's statistical ideas as sympathetically viewed by a leading Bayesian (Savage, 1976). Thanks to John Pratt's editorial efforts, Savage's talk appeared, posthumously, in the 1976 *Annals of Statistics*. In the article's discussion, Oscar Kempthorne called it the best statistics talk he had ever heard, and Churchill Eisenhart said the same. Another fine reference is Yates and Mather's introduction to the 1971 five-volume set of Fisher's collected works. The definitive Fisher reference in Joan Fisher Box's 1978 biography, *The Life of a Scientist*.

It is a good rule never to meet your heroes. I inadvertently followed this rule when Fisher spoke at the Stanford Medical School in 1961, without notice to the Statistics Department. The strength of Fisher's powerful personality is missing from this talk, but not I hope the strength of his ideas. Heroic is a good word for Fisher's attempts to change statistical thinking, attempts that had a profound influence on this century's development of statistics into a major force on the scientific landscape. "What about the next century?" is the implicit question asked in the title, but I won't try to address that question until later.

2. THE STATISTICAL CENTURY

Despite its title, the greater portion of the talk concerns the past and the present. I am going to begin by looking back on statistics in the 20th century, which has been a time of great advancement for our profession. During the 20th century statistical thinking and methodology have become the scientific framework for literally dozens of fields, including education, agriculture, economics, biology and medicine, and with increasing influence recently on the hard sciences such as astronomy, geology and physics.

In other words, we have grown from a small obscure field into a big obscure field. Most people and even most scientists still don't know much about statistics except that there is something good about the number ".05" and perhaps something bad about the bell curve. But I believe that this will change in the 21st century and that statistical methods will be widely recognized as a central element of scientific thinking.

The 20th century began on an auspicious statistical note with the appearance of Karl Pearson's famous χ^2 paper in the spring of 1900. The groundwork for statistics's growth was laid by a pre–World War II collection of intellectual giants: Neyman, the Pearsons, Student, Kolmogorov, Hotelling and Wald, with Neyman's work being especially influential. But from our viewpoint at the century's end, or at least from my viewpoint, the dominant figure has been R. A. Fisher. Fisher's influence is especially pervasive in statistical applications, but it also runs through the pages of our theoretical journals. With the end of the century in view this seemed like a good occasion for taking stock of the vitality of Fisher's legacy and its potential for future development.

A more accurate but less provocative title for this talk would have been "Fisher's influence on modern statistics." What I will mostly do is examine some topics of current interest and assess how much Fisher's ideas have or have not influenced them. The central part of the talk concerns six research areas of current interest that I think will be important during the next couple of decades. This will also give me a chance to say something about the kinds of applied problems we might be dealing with soon, and whether or not Fisherian statistics is going to be of much help with them.

First though I want to give a brief review of Fisher's ideas and the ideas he was reacting to. One difficulty in assessing the importance of Fisherian statistics is that it's hard to say just what it is. Fisher had an amazing number of important ideas and some of them, like randomization inference and conditionality, are contradictory. It's a little as if in economics Marx, Adam Smith and Keynes turned out to be the same person. So I am just going to outline some of the main Fisherian themes, with no attempt at completeness or philosophical reconciliation. This and the rest of the talk will be very short on references and details, especially technical details, which I will try to avoid entirely.

In 1910, two years before the 20-year-old Fisher published his first paper, an inventory of the statistics world's great ideas would have included the following impressive list: Bayes theorem, least squares, the normal distribution and the central limit theorem, binomial and Poisson methods for count data, Galton's correlation and regression, multivariate distributions, Pearson's χ^2 and Student's t. What was missing was a core for these ideas. The list existed as an ingenious collection of ad hoc devices. The situation for statistics was similar to the one now faced by computer science.

In Joan Fisher Box's words, "The whole field was like an unexplored archaeological site, its structure hardly perceptible above the accretions of rubble, its treasures scattered throughout the literature."

There were two obvious candidates to provide a statistical core: "objective" Bayesian statistics in the Laplace tradition of using uniform priors for unknown parameters, and a rough frequentism exemplified by Pearson's χ^2 test. In fact, Pearson was working on a core program of his own through his system of Pearson distributions and the method of moments.

By 1925, Fisher had provided a central core for statistics—one that was quite different and more compelling than either the Laplacian or Pearsonian schemes. The great 1925 paper already contains most of the main elements of Fisherian estimation theory: consistency; sufficiency; likelihood; Fisher information; efficiency; and the asymptotic optimality of the maximum likelihood estimator. Partly missing is ancillarity, which is mentioned but not fully developed until the 1934 paper.

The 1925 paper even contains a fascinating and still controversial section on what Rao has called the second order efficiency of the maximum likelihood estimate (MLE). Fisher, never really satisfied with asymptotic results, says that in small samples the MLE loses less information than competing asymptotically efficient estimators, and implies that this helps solve the problem of small-sample inference (at which point Savage wonders why one should care about the amount of information in a point estimator).

Fisher's great accomplishment was to provide an optimality standard for statistical estimation—a yardstick of the best it's possible to do in any given estimation problem. Moreover, he provided a practical method, maximum likelihood, that quite reliably produces estimators coming close to the ideal optimum even in small samples.

Optimality results are a mark of scientific maturity. I mark 1925 as the year statistical theory came of age, the year statistics went from an ad hoc collection of ingenious techniques to a coherent discipline. Statistics was lucky to get a Fisher at the beginning of the 20th century. We badly need another one to begin the 21st, as will be discussed near the end of the talk.

3. THE LOGIC OF STATISTICAL INFERENCE

Fisher believed that there must exist a logic of inductive inference that would yield a correct answer to any statistical problem, in the same way that ordinary logic solves deductive problems. By using such an inductive logic the statistician would be freed from the a priori assumptions of the Bayesian school.

Fisher's main tactic was to logically reduce a given inference problem, sometimes a very complicated one, to a simple form where everyone should agree that the answer is obvious. His favorite target for the "obvious" was the situation where we observe a single normally distributed quantity x with unknown expectation θ,

$$x \sim N(\theta, \sigma^2), \tag{1}$$

the variance σ^2 being known. Everyone agrees, says Fisher, that in this case, the best estimate is $\hat{\theta} = x$ and the correct 90% confidence interval for 0 (to use terminology Fisher hated) is

$$\hat{\theta} \pm 1.645\sigma. \tag{2}$$

Fisher's inductive logic might be called a theory of types, in which problems are reduced to a small catalogue of obvious situations. This had been tried before in statistics, the Pearson system being a good example, but never so forcefully nor successfully. Fisher was astoundingly resourceful at reducing problems to simple forms like (1). Some of the devices he invented for this purpose were sufficiency, ancillarity and conditionality, transformations, pivotal methods, geometric arguments, randomization inference and asymptotic maximum likelihood theory. Only one major reduction principle has been added to this list since Fisher's time, invariance, and that one is not in universal favor these days.

Fisher always preferred exact small-sample results but the asymptotic optimality of the MLE has been by far the most influential, or at least the most popular, of his reduction principles. The 1925 paper shows that in large samples the MLE $\hat{\theta}$ of an unknown parameter θ approaches the ideal form (1),

$$\hat{\theta} \rightarrow N(\theta, \sigma^2),$$

with the variance σ^2 determined by the Fisher information and the sample size. Moreover, no other "reasonable" estimator of θ has a smaller asymptotic variance. In other words, the maximum likelihood method automatically produces an estimator that can reasonably be termed "optimal," without ever invoking the Bayes theorem.

Fisher's great accomplishment triggered a burst of interest in optimality results. The most spectacular product of this burst was the Neyman–Pearson lemma for optimal hypothesis testing, followed soon by Neyman's theory of confidence intervals. The Neyman–Pearson lemma did for hypothesis testing what Fisher's MLE theory did for estimation, by pointing the way toward optimality.

Philosophically, the Neyman–Pearson lemma fits in well with Fisher's program: using mathematical logic it reduces a complicated problem to an obvious solution without invoking Bayesian priors. Moreover, it is a tremendously useful idea in applications, so that Neyman's ideas on hypotheses testing and confidence intervals now play a major role in day-to-day applied statistics.

However, the success of the Neyman–Pearson lemma triggered new developments, leading to a more extreme form of statistical optimality that Fisher deeply distrusted. Even though Fisher's personal motives are suspect here, his philosophical qualms were far from groundless. Neyman's ideas, as later developed by Wald into decision theory, brought a qualitatively different spirit into statistics.

Fisher's maximum likelihood theory was launched in reaction to the rather shallow Laplacian Bayesianism of the previous century. Fisher's work demonstrated a more stringent approach to statistical inference. The Neyman–Wald decision theoretic school carried this spirit of astringency much further. A strict mathematical statement of the problem at hand, often phrased quite narrowly, followed by an optimal solution became the ideal. The practical result was a more sophisticated form of frequentist inference having enormous mathematical appeal.

Fisher, caught I think by surprise by this flanking attack from his right, complained that the Neyman–Wald decision theorists could be *accurate* without being *correct*. A favorite example of his concerned a Cauchy distribution with unknown center

$$(3) \qquad f_\theta(x) = \frac{1}{\pi[1 + (x - \theta)^2]}.$$

Given a random sample $\mathbf{x} = (x_1, x_2, \ldots, x_n)$ from (3), decision theorists might try to provide the shortest interval of the form $\hat{\theta} \pm c$ that covers the true θ with probability 0.90. Fisher's objection, spelled out in his 1934 paper on ancillarity, was that c should be different for different samples $\mathbf{x}$ depending upon the correct amount of information in $\mathbf{x}$.

The decision theory movement eventually spawned its own counter-reformation. The neo-Bayesians, led by Savage and de Finetti, produced a more logical and persuasive Bayesianism, emphasizing subjective probabilities and personal decision making. In its most extreme form the Savage–de Finetti theory directly denies Fisher's claim of an impersonal logic of statistical inference. There has also been a postwar revival of interest in objectivist Bayesian theory, Laplacian in intent but based on Jeffreys's more sophisticated methods for choosing objective priors, which I shall talk more about later on.

Very briefly then, this is the way we arrived at the end of the 20th century with three competing philosophies of statistical inference: Bayesian; Neyman–Wald frequentist; and Fisherian. In many ways the Bayesian and frequentist philosophies stand at opposite poles from each other, with Fisher's ideas being somewhat of a compromise. I want to talk about that compromise next because it has a lot to do with the popularity of Fisher's methods.

4. THREE COMPETING PHILOSOPHIES

The chart in Figure 1 shows four major areas of disagreement between the Bayesians and the frequentists. These are not just philosophical disagreements. I chose the four categories because they lead to different behavior at the data-analytic level. For each category I have given a rough indication of Fisher's preferred position.

4.1 Individual Decision Making versus Scientific Inference

Bayes theory, and in particular Savage–de Finetti Bayesianism (the kind I'm focusing on here, though later I'll also talk about the Jeffreys brand of objec-

BAYES	FISHER	FREQUENTIST
1. Individual (personal decisions)		*** Universal (world of science)
2. Coherent (correct)	*************	Optimal (accurate)
3. Synthetic (combination)	****	Analytic (separation)
4. Optimistic (aggressive)	*****	Pessimistic (defensive)

FIG. 1. *Four major areas of disagreement between Bayesian and frequentist methods. For each one I have inserted a row of stars to indicate, very roughly, the preferred location of Fisherian inference.*

tive Bayesianism), emphasizes the individual decision maker, and it has been most successful in fields like business where individual decisions are paramount. Frequentists aim for universal acceptance of their inferences. Fisher felt that the proper realm of statistics was scientific inference, where it is necessary to persuade all or at least most of the world of science that you have reached the correct conclusion. Here Fisher is far over to the frequentist side of the chart (which is philosophically accurate but anachronistic, since Fisher's position predates both the Savage–de Finetti and Neyman–Wald schools).

4.2 Coherence versus Optimality

Bayesian theory emphasizes the coherence of its judgments, in various technical ways but also in the wider sense of enforcing consistency relationships between different aspects of a decision-making situation. Optimality in the frequentist sense is frequently incoherent. For example, the uniform minimum variance unbiased (UMVU) estimate of $\exp\{\theta\}$ does not have to equal exp{the UMVU of θ}, and more seriously there is no simple calculus relating the two different estimates. Fisher wanted to have things both ways, coherent and optimal, and in fact maximum likelihood estimation does satisfy

$$\exp\{\hat{\theta}\} = \widehat{\exp\{\theta\}}.$$

The tension between coherence and optimality is like the correctness–accuracy disagreement concerning the Cauchy example (3), where Fisher argued strongly for correctness. The emphasis on correctness, and a belief in the existence of a logic of statistical inference, moves Fisherian philosophy toward the Bayesian side of Figure 1. Fisherian practice is a less clear story. Different parts of the Fisherian program don't cohere with each other and in practice Fisher seemed quite willing to sacrifice logical consistency for a neat solution to a particular problem, for example, switching back and forth between frequentist and nonfrequentist justifications of the Fisher information. This kind of case-to-case expediency, which is a common attribute of modern data analysis has a frequentist flavor. I have located the Fisherian stars for this category a little closer to the Bayesian side of Figure 1, but spreading over a wide range.

4.3 Synthesis versus Analysis

Bayesian decision making emphasizes the collection of information across all possible sources, and the synthesis of that information into the final inference. Frequentists tend to break problems into separate small pieces that can be analyzed separately (and optimally). Fisher emphasized the use of all available information as a hallmark of correct inference, and in this way he is more in sympathy with the Bayesian position.

In this case Fisher tended toward the Bayesian position both in theory and in methodology: maximum likelihood estimation and its attendant theory of approximate confidence intervals based on Fisher information are superbly suited to the combination of information from different sources. (On the other hand, we have this quote from Yates and Mather: "In his own work Fisher was at his best when confronted with small self-contained sets of data. ... He was never much interested in the assembly and analysis of large amounts of data from varied sources bearing on a given issue." They blame this for his stubbornness on the smoking–cancer controversy. Here as elsewhere we will have to view Fisher as a lapsed Fisherian.)

4.4 Optimism versus Pessimism

This last category is more psychological than philosophical, but it is psychology rooted in the basic nature of the two competing philosophies. Bayesians tend to be more aggressive and risk-taking in their data analyses. There couldn't be a more pessimistic and defensive theory than minimax, to choose an extreme example of frequentist philosophy. It says that if anything can go wrong it will. Of course a minimax person might characterize the Bayesian position as "If anything can go right it will."

Fisher took a middle ground here. He scorns the finer mathematical concerns of the decision theorists ("Not only does it take a cannon to shoot a sparrow, but it misses the sparrow!"), but he fears averaging over the states of nature in a Bayesian way. One of the really appealing features of Fisher's work is its spirit of reasonable compromise, cautious but not overly concerned with pathological situations. This has always struck me as the right attitude toward most real-life problems, and it's certainly a large part of Fisher's dominance in statistical applications.

Looking at Figure 1, I think it is a mistake trying too hard to make a coherent philosophy out of Fisher's theories. From our current point of view they are easier to understand as a collection of extremely shrewd compromises between Bayesian and frequentist ideas. Fisher usually wrote as if he had a complete logic of statistical inference in hand, but that didn't stop him from changing his system when he thought up another landmark idea.

De Finetti, as quoted by Cifarelli and Regazzini, puts it this way: "Fisher's rich and manifold personality shows a few contradictions. His common sense in applications on one hand and his lofty conception of scientific research on the other lead him to disdain the narrowness of a genuinely objectivist formulation, which he regarded as a *wooden attitude*. He professes his adherence to the objectivist point of view by rejecting the errors of the Bayes–Laplace formulation. What is not so good here is his mathematics, which he handles with mastery in individual problems but rather cavalierly in conceptual matters, thus exposing himself to clear and sometimes heavy criticism. From our point of view it appears probable that many of Fisher's observations and ideas are valid provided we go back to the intuitions from which they spring and free them from the arguments by which he thought to justify them."

Figure 1 describes Fisherian statistics as a compromise between the Bayesian and frequentist schools, but in one crucial way it is not a compromise: in its ease of use. Fisher's philosophy was always expressed in very practical terms. He seemed to think naturally in terms of computational algorithms, as with maximum likelihood estimation, analysis of variance and permutation tests. If anything is going to replace Fisher in the 21st century it will have to be a methodology that is equally easy to apply in day-to-day practice.

5. FISHER'S INFLUENCE ON CURRENT RESEARCH

There are three parts to this talk: past, present and future. The past part, which you have just seen, didn't do justice to Fisher's ideas, but the subject here is more one of influence than ideas, admitting of course that the influence is founded on the ideas's strengths. So now I am going to discuss Fisher's influence on current research.

What follows are several (actually six) examples of current research topics that have attracted a lot of attention recently. No claim of completeness is being made here. The main point I'm trying to make with these examples is that Fisher's ideas are still exerting a powerful influence on developments in statistical theory, and that this is an important indication of their future relevance. The examples will gradually get more speculative and futuristic, and will include some areas of development *not* satisfactorily handled by Fisher–holes in the Fisherian fabric—where we might expect future work to be more frequentist or Bayesian in motivation.

The examples will also allow me to talk about the new breed of applied problems statisticians are starting to see, the bigger, messier, more complicated data sets that we will have to deal with in the coming decades. Fisherian methods were fashioned to deal with the problems of the 1920s and 1930s. It is not a certainty that they will be equally applicable to the problems of the 21st century—a question I hope to shed at least a little light upon.

5.1 Fisher Information and the Bootstrap

This first example is intended to show how Fisher's ideas can pop up in current work, but be difficult to recognize because of computational advances. First, here is a very brief review of Fisher information. Suppose we observe a random sample $x_1, x_2, \ldots, x_n$ from a density function $f_\theta(x)$ depending on a single unknown parameter θ,

$$f_\theta(x) \to x_1, x_2, \ldots, x_n.$$

The Fisher information in any one x is the expected value of minus the second derivative of the log density,

$$i_\theta = \mathbf{E}_\theta\left\{-\frac{\partial^2}{\partial\theta^2}\log f_\theta(x)\right\},$$

and the total Fisher information in the whole sample is ni_θ.

Fisher showed that the asymptotic standard error of the MLE is inversely proportional to the square root of the total information,

$$\text{se}_\theta(\hat\theta) \doteq \frac{1}{\sqrt{ni_\theta}}, \tag{4}$$

and that no other consistent and sufficiently regular estimation of θ—essentially no other asymptotically, unbiased estimator—can do better.

A tremendous amount of philosophical interpretation has been attached to i_θ, concerning the meaning of statistical information, but in practice Fisher's formula (4) is most often used simply as a handy estimate of the standard error of the MLE. Of course, (4) by itself cannot be used directly because i_θ involves the unknown parameter θ. Fisher's tactic, which seems obvious but in fact is quite central to Fisherian methodology, is to *plug in* the MLE $\hat\theta$ for θ in (4), giving a usable estimate of standard error,

$$\widehat{\text{se}} = \frac{1}{\sqrt{ni_{\hat\theta}}}. \tag{5}$$

Here is an example of formula (5) in action. Figure 2 shows the results of a small study designed to test the efficacy of an experimental antiviral drug.

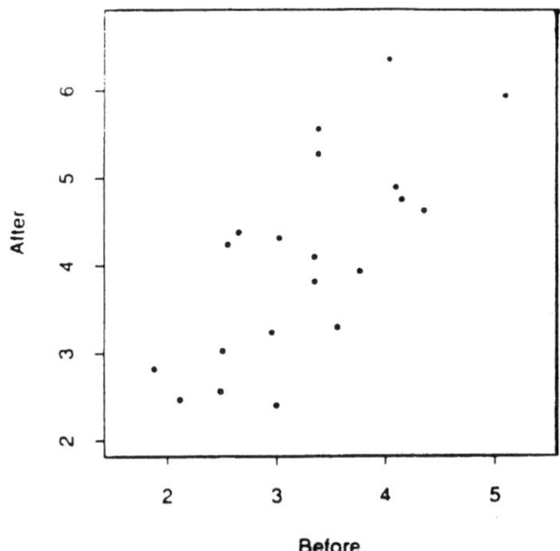

FIG. 2. *The cd4 data; 20 AIDS patients had their cd4 counts measured before and after taking an experimental drug; correlation coefficient* $\hat{\theta} = 0.723$.

A total of $n = 20$ AIDS patients had their cd4 counts measured before and after taking the drug, yielding data

$$x_i = (\text{before}_i, \text{after}_i) \quad \text{for } i = 1, 2, \ldots, 20.$$

The Pearson sample correlation coefficient was $\hat{\theta} = 0.723$. How accurate is this estimate?

If we assume a bivariate normal model for the data,

$$N_2(\mu, \Sigma) \to x_1, x_2, x_3, \ldots, x_{20}, \tag{6}$$

the notation indicating a random sample of 20 pairs from a bivariate normal distribution with expectation vector μ and covariance matrix Σ, then $\hat{\theta}$ is the MLE for the true correlation coefficient θ. The Fisher information for estimating θ turns out to be $i_\theta = 1/(1 - \theta^2)^2$ (after taking proper account of the "nuisance parameters" in (6)—one of those technical points I am avoiding in this talk) so (5) gives estimated standard error

$$\widehat{\text{se}} = \frac{(1 - \hat{\theta}^2)}{\sqrt{20}} = 0.107.$$

Here is a bootstrap estimate of standard error for the same problem, also assuming that the bivariate normal model is correct. In this context the bootstrap samples are generated from model (6), but with estimates $\hat{\mu}$ and $\hat{\Sigma}$ substituted for the unknown parameters μ and Σ:

$$N(\hat{\mu}, \hat{\Sigma}) \to x_1^*, x_2^*, x_3^*, \ldots, x_{20}^* \to \hat{\theta}^*,$$

where $\hat{\theta}^*$ is the sample correlation coefficient for the bootstrap data set $x_1^*, x_2^*, x_3^*, \ldots, x_{20}^*$.

This whole process was independently repeated 2,000 times, giving 2,000 bootstrap correlation coefficients $\hat{\theta}^*$. Figure 3 shows their histogram.

The empirical standard deviation of the 2,000 $\hat{\theta}^*$ values is

$$\widehat{\text{se}}_{\text{boot}} = 0.112,$$

which is the normal-theory bootstrap estimate of standard error for $\hat{\theta}$; 2,000 is 10 times more than needed for a standard error, but we will need all 2,000 later for the discussion of approximate confidence intervals.

5.2 The Plug-in Principle

The Fisher information and bootstrap standard error estimates, 0.107 and 0.112, are quite close to each other. This is no accident. Despite the fact that they look completely different, the two methods are doing very similar calculations. Both are using the "plug-in principle" as a crucial step in getting the answer.

Here is a plug-in description of the two methods:

- Fisher information—(i) compute an (approximate) formula for the standard error of the sample correlation coefficient as a function of the unknown parameters (μ, Σ); (ii) plug in estimates $(\hat{\mu}, \hat{\Sigma})$ for the unknown parameters (μ, Σ) in the formula;
- bootstrap—(i) plug in $(\hat{\mu}, \hat{\Sigma})$ for the unknown parameters (μ, Σ) in the mechanism generating the data; (ii) compute the standard error of the sample correlation coefficient, for the plugged-in mechanism, by Monte Carlo simulation.

The two methods proceed in reverse order, "compute and then plug in" versus "plug in and then compute," but this is a relatively minor technical difference. The crucial step in both methods, and the only statistical inference going on, is the substitution of the estimates $(\hat{\mu}, \hat{\Sigma})$ for the unknown parameters (μ, Σ), in other words the plug-in principle. Fisherian inference makes frequent use of the plug-in principle, and this is one of the main reasons that Fisher's methods are so convenient to use in practice. All possible inferential questions are answered by simply plugging in estimates, usually maximum likelihood estimates, for unknown parameters.

The Fisher information method involves cleverer mathematics than the bootstrap, but it has to because we enjoy a 10^7 computational advantage over Fisher. A year's combined computational effort by

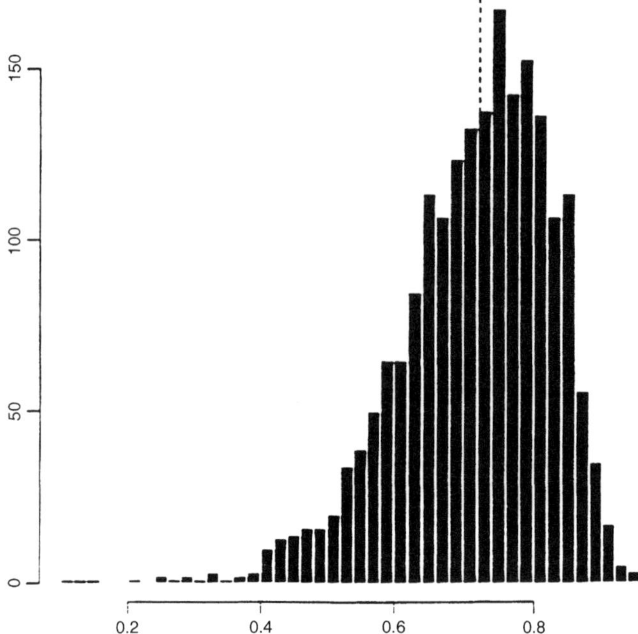

FIG. 3. *Histogram of* 2,000 *bootstrap correlation coefficients; bivariate normal sampling model.*

all the statisticians of 1925 wouldn't equal a minute of modern computer time. The bootstrap exploits this advantage to numerically extend Fisher's calculations to situations where the mathematics becomes hopelessly complicated. One of the less attractive aspects of Fisherian statistics is its overreliance on a small catalog of simple parametric models like the normal, understandable enough given the limitations of the mechanical calculators Fisher had to work with.

Modern computation has given us the opportunity to extend Fisher's methods to a much wider class of models, including nonparametric ones (the more usual arena of the bootstrap). We are beginning to see many such extensions, for example, the extension of discriminant analysis to CART, and the extension of linear regression to generalized additive models.

6. THE STANDARD INTERVALS

I want to continue the cd4 example, but proceeding from standard errors to confidence intervals. The confidence interval story illustrates how computer-based inference can be used to extend Fisher's ideas in a more ambitious way.

The MLE and its estimated standard error were used by Fisher to form approximate confidence intervals, which I like to call the *standard intervals* because of their ubiquity in day-to-day practice,

$$\hat{\theta} \pm 1.645\,\widehat{\mathrm{se}}. \tag{7}$$

The constant, 1.645, gives intervals of approximate 90% coverage for the unknown parameter θ, with 5% noncoverage probabilities at each end of the interval. We could use 1.96 instead of 1.645 for 95% coverage, and so on, but here I'll stick to 90%.

The standard intervals follow from Fisher's result that $\hat{\theta}$ is asymptotically normal, unbiased and with standard error fixed by the sample size and the Fisher information,

$$\hat{\theta} \to N(\theta, \mathrm{se}^2), \tag{8}$$

as in (4). We recognize (8) as one of Fisher's ideal "obvious" forms.

If usage determines importance then the standard intervals were Fisher's most important invention. Their popularity is due to a combination of optimality, or at least asymptotic optimality, with computation tractability. The standard intervals are:

- *accurate*—their noncoverage probabilities, which are supposed to be 0.05 at each end of the interval, are actually

$$0.05 + c/\sqrt{n}, \tag{9}$$

  where c depends on the situation, so as the sample size n gets large we approach the nominal value 0.05 at rate $n^{-1/2}$;
- *correct*—the estimated standard error based on the Fisher information is the minimum possible for any asymptotically unbiased estimate of θ so interval (7) doesn't waste any information nor is it misleadingly optimistic;
- *automatic*—$\hat{\theta}$ and $\widehat{\text{se}}$ are computed from the same basic algorithm no matter how complicated the problem may be.

Despite these advantages, applied statisticians know that the standard intervals can be quite inaccurate in small samples. This is illustrated in the left panel of Figure 4 for the cd4 correlation example, where we see that the standard interval endpoints lie far to the right of the endpoints for the normal-theory exact 90% central confidence interval. In fact, we can see from the bootstrap histogram (reproduced from Figure 3) that in this case the asymptotic normality of the MLE hasn't taken hold at $n = 20$, so that there is every reason to doubt the standard interval. Being able to look at the histogram, which has a lot of information in it, is a luxury Fisher did not have.

Fisher suggested a fix for this specific situation: transform the correlation coefficient to $\hat{\phi} = \tanh^{-1}(\hat{\theta})$, that is, to

$$\hat{\phi} = \frac{1}{2}\log\frac{1+\hat{\theta}}{1-\hat{\theta}}, \tag{10}$$

apply the standard method on this scale and then transform the standard interval back to the θ scale. This was another one of Fisher's ingenious reduction methods. The $\tanh^{-1}$ transformation greatly accelerates convergence to normality, as we can see from the histogram of the 2,000 values of $\hat{\theta}^* = \tanh^{-1}(\hat{\theta}^*)$ in the right panel of Figure 4, and makes the standard intervals far more accurate. However, we have now lost the "automatic" property of the standard intervals. The $\tanh^{-1}$ transformation works only for the normal correlation coefficient and not for most other problems.

The standard intervals take literally the large sample approximation $\hat{\theta} \sim N(\theta, \text{se}^2)$, which says that $\hat{\theta}$ is normally distributed, is unbiased for θ and has a constant standard error. A more careful look at the asymptotics shows that each of these three assumptions can fail in a substantial way: the sampling distribution of $\hat{\theta}$ can be skewed; $\hat{\theta}$ can be biased as an estimate of θ; and its standard error can change with θ. Modern computation makes it practical to correct all three errors. I am going to mention two methods of doing so, the first using the bootstrap histogram, the second based on likelihood methods.

It turns out that there is enough information in the bootstrap histogram to correct all three errors

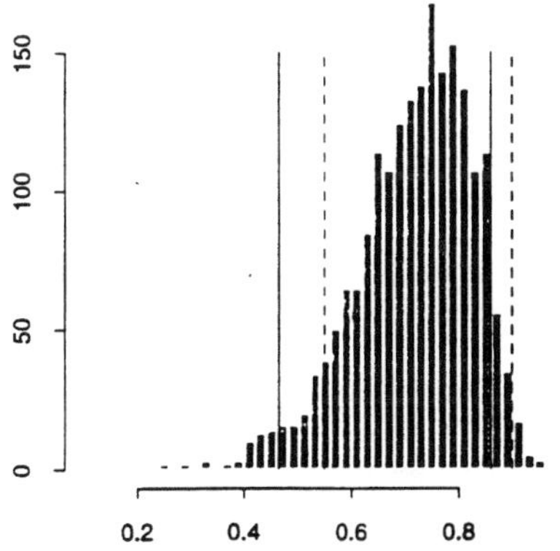

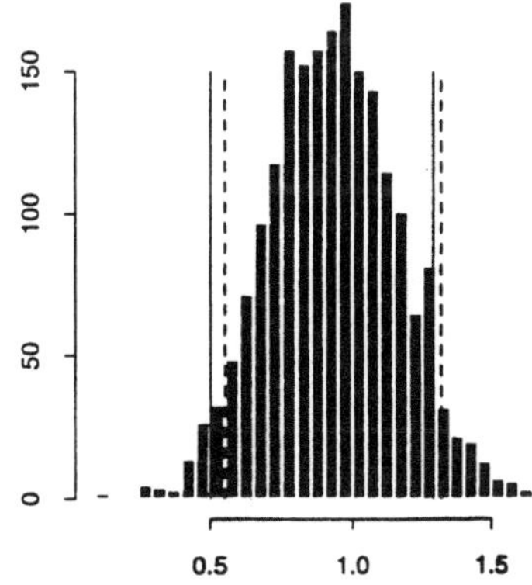

FIG. 4. *(Left panel) Endpoints of exact 90% confidence interval for cd4 correlation coefficient (solid lines) are much different than standard interval endpoints (dashed lines), as suggested by the nonnormality of the bootstrap histogram. (Right panel) Fisher's transformation normalizes the bootstrap histogram and makes the standard interval more accurate.*

of the standard intervals. The result is a system of approximate confidence intervals an order of magnitude more accurate, with noncoverage probabilities

$$0.05 + c/n$$

compared to (9), achieving what is called *second order accuracy*. Table 1 demonstrates the practical advantages of second order accuracy. In most situations we would not have exact endpoints as a "gold standard" for comparison, but second order accuracy would still point to the superiority of the bootstrap intervals.

The bootstrap method, and also the likelihood-based methods of the next section, are *transformation invariant*; that is, they give the same interval for the correlation coefficient whether or not you go through the $\tanh^{-1}$ transformation. In this sense they automate Fisher's wonderful transformation trick.

I like this example because it shows how a basic Fisherian construction, the standard intervals, can be extended by modern computation. The extension lets us deal easily with very complicated probability models, even nonparametric ones, and also with complicated statistics such as a coefficient in a stepwise robust regression.

Moreover, the extension is not just to a wider set of applications. Some progress in understanding the theoretical basis of approximate confidence intervals is made along the way. Other topics are springing up in the same fashion. For example, Fisher's 1925 work on the information loss for insufficient estimators has transmuted into our modern theories of the EM algorithm and Gibbs sampling.

7. CONDITIONAL INFERENCE, ANCILLARITY AND THE MAGIC FORMULA

Table 2 shows the occurrence of a very undesirable side effect in a randomized experiment that will be described more fully later. The treatment produces a smaller ratio of these undesirable effects than does the control, the sample log odds ratio being

$$\hat{\theta} = \log\left(\frac{1}{15}\bigg/\frac{13}{3}\right) = -4.2.$$

TABLE 1

Endpoints of exact and approximate 90% confidence intervals for the cd4 correlation coefficient assuming bivariate normality

	Exact	Bootstrap	Standard
0.05	0.464	0.468	0.547
0.95	0.859	0.856	0.899

TABLE 2

The occurrence of adverse events in a randomized experiment; sample log odds ratio $\hat{\theta} = -4.2$

	Yes	No	
Treatment	1	15	16
Control	13	3	16
	14	18	

Fisher wondered how one might make appropriate inferences for θ, the true log odds ratio. The trouble here is nuisance parameters. A multinomial model for the 2×2 table has three free parameters, representing four cell probabilities constrained to add up to 1, and in some sense two of the three parameters have to be eliminated in order to get at θ. To do this Fisher came up with another device for reducing a complicated situation to a simple form.

Fisher showed that if we condition on the marginals of the table, then the conditional density of $\hat{\theta}$ given the marginals depends only θ. The nuisance parameters disappear. This conditioning is "correct" he argued because the marginals are acting as what might be called *approximate ancillary statistics*. That is, they do not carry much direct information concerning the value of θ, but they have something to say about how accurately $\hat{\theta}$ estimates θ. Later Neyman gave a much more specific frequentist justification for conditioning on the marginals, through what is now called *Neyman structure*.

For the data in Table 2, the conditional distribution of $\hat{\theta}$ given the marginals yields $[-6.3, -2.4]$ as a 90% confidence interval for θ, ruling out the null hypothesis value $\theta = 0$ where Treatment equals Control. However, the conditional distribution is not easy to calculate, even in this simple case, and it becomes prohibitive in more complicated situations.

In his 1934 paper, which was the capstone of Fisher's work on efficient estimation, he solved the conditioning problem for translation families. Suppose that $\mathbf{x} = (x_1, x_2, \ldots, x_n)$ is a random sample from a Cauchy distribution (3) and that we wish to use $\mathbf{x}$ to make inferences about θ, the unknown center point of the distribution. In this case there is a genuine ancillary statistic $\mathbf{A}$, the vector of spacings between the ordered values of $\mathbf{x}$. Again Fisher argued that correct inferences about θ should be based on $f_\theta(\hat{\theta}|\mathbf{A})$, the conditional density of the MLE $\hat{\theta}$ given the ancillary $\mathbf{A}$, not on the unconditional density $f_\theta(\hat{\theta})$.

Fisher also provided a wonderful trick for calculating $f_\theta(\hat{\theta}|\mathbf{A})$. Let $L(\theta)$ be the likelihood function:

the unconditional density of the whole sample, considered as a function of θ with $\mathbf{x}$ fixed. Then it turns out that

$$f_\theta(\hat{\theta}|\mathbf{A}) = c\frac{L(\theta)}{L(\hat{\theta})}, \tag{11}$$

where c is a constant. Formula (11) allows us to compute the conditional density $f_\theta(\hat{\theta}|\mathbf{A})$ from the likelihood, which is easy to calculate. It also hints at a deep connection between likelihood-based inference, a Fisherian trademark, and frequentist methods.

Despite this promising start, the promise went unfulfilled in the years following 1934. The trouble was that formula (11) applies only in very special circumstances, not including the 2×2 table example, for instance. Recently, though, there has been a revival of interest in likelihood-based conditional inference. Durbin, Barndorff-Nielsen, Hinkley and others have developed a wonderful generalization of (11) that applies to a wide variety of problems having approximate ancillaries, the so-called magic formula

$$f_\theta(\hat{\theta}|\mathbf{A}) = c\frac{L(\theta)}{L(\hat{\theta})}\left\{-\frac{d^2}{d\theta^2}\log L(\theta)|_{\theta=\hat{\theta}}\right\}^{1/2}. \tag{12}$$

The bracketed factor is constant in the Cauchy situation, reducing (12) back to (11).

Likelihood-based conditional inference has been pushed forward in current work by Fraser, Cox and Reid, McCullagh, Barndorff-Nielson, Pierce, DiCiccio and many others. It represents a major effort to perfect and extend Fisher's goal of an inferential system based directly on likelihoods.

In particular the magic formula can be used to generate approximate confidence intervals that are more accurate than the standard intervals, at least second order accurate. These intervals agree to second order with the bootstrap intervals. If this were not true, then one or both of them would not be second order correct. Right now it looks like attempts to improve upon the standard intervals are converging from two directions: likelihood and bootstrap.

Results like (12) have enormous potential. Likelihood inference is the great unfulfilled promise of Fisherian statistics—the promise of a theory that directly interprets likelihood functions in a way that simultaneously satisfies Bayesians and frequentists. Fulfilling that promise, even partially, would greatly influence the shape of 21st century statistics.

8. FISHER'S BIGGEST BLUNDER

Now I'll start edging gingerly into the 21st century by discussing some topics where Fisher's ideas have not been dominant, but where they might or might not be important in future developments. I am going to begin with the fiducial distribution, generally considered to be Fisher's biggest blunder. But in Arthur Koestler's words "The history of ideas is filled with barren truths and fertile errors." If fiducial inference is an error it certainly has been a fertile one.

In terms of Figure 1, the Bayesian–frequentist comparison chart, fiducial inference was Fisher's closest approach to the Bayesian side of the ledger. Fisher was trying to codify an objective Bayesianism in the Laplace tradition but without using Laplace's ad hoc uniform prior distributions. I believe that Fisher's continuing devotion to fiducial inference had two major influences, a negative reaction against Neyman's ideas and a positive attraction to Jeffreys's point of view.

The solid line in Figure 5 is the fiducial density for a binomial parameter θ having observed 3 successes in 10 trials,

$$s \sim \text{Binomial}(n, \theta), \quad s = 3 \text{ and } n = 10.$$

Also shown is an approximate fiducial density that I will refer to later. Fisher's fiducial theory at its boldest treated the solid curve as a genuine a posteriori density for θ even though, or perhaps because, no prior assumptions had been made.

8.1 The Confidence Density

We could also call the fiducial distribution the "confidence density" because this is an easy way to motivate the fiducial construction. As I said earlier, Fisher would have hated this name.

Suppose that for every value of α between 0 and 1 we have an upper 100 αth confidence limit $\hat{\theta}[\alpha]$ for θ, so that by definition

$$\text{prob}\{\theta < \hat{\theta}[\alpha]\} = \alpha.$$

We can interpret this as a probability distribution for θ given the data if we are willing to accept the classic *wrong* interpretation of confidence,

> θ is in the interval $(\hat{\theta}[0.90], \hat{\theta}[0.91])$
> with probability 0.01, and so on.

Going to the continuous limit gives the "confidence density," a name Neyman would have hated.

The confidence density *is* the fiducial distribution, at least in those cases where Fisher would have considered the confidence limits to be inferen-

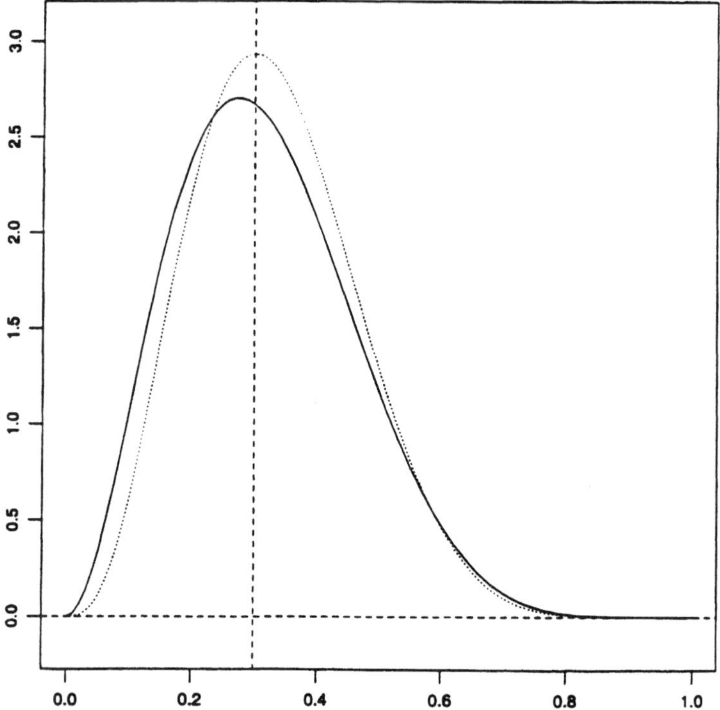

FIG. 5. *Fiducial density for a binomial parameter* θ *having observed* 3 *successes out of* 10 *trials. The dashed line is an approximation that is useful in complicated situations.*

tially correct. The fiducial distribution in Figure 5 is the confidence density based on the usual confidence limits for θ (taking into account the discrete nature of the binomial distribution): $\hat{\theta}[\alpha]$ is the value of θ such that $S \sim \text{Binomial}(10, \theta)$ satisfies

$$\text{prob}\,\{S > 3\} + \tfrac{1}{2}\,\text{prob}\{S = 3\} = \alpha.$$

Fisher was uncomfortable applying fiducial arguments to discrete distributions because of the ad hoc continuity corrections required, but the difficulties caused are more theoretical than practical.

The advantage of stating fiducial ideas in terms of the confidence density is that they then can be applied to a wider class of problems. We can use the approximate confidence intervals mentioned earlier, either the bootstrap or the likelihood ones, to get approximate fiducial distribution even in very complicated situations having lots of nuisance parameters. (The dashed curve in Figure 5 is the confidence density based on approximate bootstrap intervals.) And there are practical reasons why it would be very convenient to have good approximate fiducial distributions, reasons connected with our profession's 250-year search for a dependable objective Bayes theory.

8.2 Objective Bayes

By "objective Bayes" I mean a Bayesian theory in which the subjective element is removed from the choice of prior distribution; in practical terms a universal recipe for applying Bayes theorem in the absence of prior information. A widely accepted objective Bayes theory, which fiducial inference was intended to be, would be of immense theoretical and practical importance.

I have in mind here dealing with messy, complicated problems where we are trying to combine information from disparate sources—doing a metaanalysis, for example. Bayesian methods are particularly well-suited to such problems. This is particularly true now that techniques like the Gibbs sampler and Markov chain Monte Carlo are available for integrating the nuisance parameters out of high-dimensional posterior distributions.

The trouble of course is that the statistician still has to choose a prior distribution in order to use

Bayes's theorem. An unthinking use of uniform priors is no better now than it was in Laplace's day. A lot of recent effort has been put into the development of uninformative or objective prior distributions, priors that eliminate nuisance parameters safely while remaining neutral with respect to the parameter of interest. Kass and Wasserman's 1996 *JASA* article reviews current developments by Berger, Bernardo and many others, but the task of finding genuinely objective priors for high-dimensional problems remains daunting.

Fiducial distributions, or confidence densities, offer a way to finesse this difficulty. A good argument can be made that the confidence density is the posterior density for the parameter of interest, after all of the nuisance parameters have been integrated out in an objective way. If this argument turns out to be valid, then our progress in constructing approximate confidence intervals, and approximate confidence densities, could lead to an easier use of Bayesian thinking in practical problems.

This is all quite speculative, but here is a safe prediction for the 21st century: statisticians will be asked to solve bigger and more complicated problems. I believe that there is a good chance that objective Bayes methods will be developed for such problems, and that something like fiducial inference will play an important role in this development. Maybe Fisher's biggest blunder will become a big hit in the 21st century!

9. MODEL SELECTION

Model selection is another area of statistical research where important developments seem to be building up, but without a definitive breakthrough. The question asked here is how to select the model itself, not just the continuous parameters of a given model, from the observed data. *F*-tests, and "*F*" stands for Fisher, help with this task, and are certainly the most widely used model selection techniques. However, even in relatively simple problems things can get complicated fast, as anyone who has gotten lost in a tangle of forward and backward stepwise regression programs can testify.

The fact is that classic Fisherian estimation and testing theory are a good start, but not much more than that, on model selection. In particular, maximum likelihood estimation theory and model fitting do not account for the number of free parameters being fit, and that is why frequentist methods like Mallow's C_p, the Akaike information criterion and cross-validation have evolved. Model selection seems to be moving away from its Fisherian roots.

Now statisticians are starting to see really complicated model selection problems, with thousands and even millions of data points and hundreds of candidate models. A thriving area called machine learning has developed to handle such problems, in ways that are not yet very well connected to statistical theory.

Table 3, taken from Gail Gong's 1982 thesis, shows part of the data from a model selection problem that is only moderately complicated by today's standards, though hopelessly difficult from a prewar viewpoint. A "training set" of 155 chronic hepatitis patients were measured on 19 diagnostic prediction variables. The outcome variable y was whether or not the patient died from liver failure (122 lived, 33 died), the goal of the study being to develop a prediction rule for y in terms of the diagnostic variables.

In order to predict the outcome, a logistic regression model was built up in three steps:

- Individual logistic regressions were run for each of the 19 predictors, yielding 13 that were significant at the 0.05 level.
- A forward stepwise logistic regression program, including only those patients with none of the 13 predictors missing, retained 5 of the 13 predictors at significance level 0.10.
- A second forward stepwise logistic regression program, including those patients with none of the 5 predictors missing, retained 4 of the 5 at significance level 0.05.

These last four variables,

(13) ascites, (15) bilirubin,
(7) malaise, (20) histology,

were deemed the "important predictors." The logistic regression based on them misclassified 16% of the 155 patients, with cross-validation suggesting a true error rate of about 20%.

A crucial question concerns the validity of the selected model. Should we take the four "important predictors" very seriously in a medical sense? The bootstrap answer seems to be "probably not," even though it was natural for the medical investigator to do so given the impressive amount of statistical machinery involved in their selection.

Gail Gong resampled the 155 patients, taking as a unit each patient's entire record of 19 predictors and response. For each bootstrap data set of 155 resampled records, she reran the three-stage logistic regression model, yielding a bootstrap set of "important predictors." This was done 500 times. Figure 6 shows the important predictors for the final 25 bootstrap data sets. The first of these is (13, 7, 20, 15), agreeing except for order with the set (13, 15, 7, 20) from the original data. This didn't happen in any other of the 499 bootstrap cases. In

TABLE 3

155 chronic hepatitis patients were measured on 19 diagnostic variables; data shown for the last 11 patients; outcome $y = 0$ or 1 as patient lived or died; negative numbers indicate missing data

y	Cons-tant 1	Age 2	Sex 3	Ster-oid 4	Anti-viral 5	Fa-tigue 6	Mal-aise 7	Anor-exia 8	Liver Big 9	Liver Firm 10	Spleen Palp 11	Spi-ders 12	As-cites 13	Var-ices 14	Bili-rubin 15	Alk Phos 16	SGOT 17	Albu-min 18	Pro-tein 19	Histo-logy 20	#
1	1	45	1	2	2	1	1	1	2	2	2	1	1	2	1.90	−1	114	2.4	−1	−3	145
0	1	31	1	1	2	1	2	2	2	2	2	2	2	2	1.20	75	193	4.2	54	2	146
1	1	41	1	2	2	1	2	2	2	1	1	1	2	1	4.20	65	120	3.4	−1	−3	147
1	1	70	1	1	2	1	1	1	−3	−3	−3	−3	−3	−3	1.70	109	528	2.8	35	2	148
0	1	20	1	1	2	2	2	2	2	−3	2	2	2	2	0.90	89	152	4.0	−1	2	149
0	1	36	1	2	2	2	2	2	2	2	2	2	2	2	0.60	120	30	4.0	−1	2	150
1	1	46	1	2	2	1	1	1	2	2	2	1	1	1	7.60	−1	242	3.3	50	−3	151
0	1	44	1	2	2	1	2	2	2	1	2	2	2	2	0.90	126	142	4.3	−1	2	152
0	1	61	1	1	2	1	1	2	1	1	2	1	2	2	0.80	95	20	4.1	−1	2	153
0	1	53	2	1	2	1	2	2	2	2	1	1	2	1	1.50	84	19	4.1	48	−3	154
1	1	43	1	2	2	1	2	2	2	2	1	1	1	2	1.20	100	19	3.1	42	2	155

```
13  7 20 15
13 19  6
20 16 19
20 19
14 18  7 16  2
18 20  7 11
20 19 15
20
13 12 15  8 18  7 19
15 13 19
13  4
12 15  3
15 16  3
15 20  4
16 13  2 19
18 20  3
13 15 20
15 13
15 20  7
13
15
13 14
12 20 18
 2 20 15  7 19 12
13 20 15 19
```

FIG. 6. *The set of "important predictors" selected in the last* 25 *of* 500 *bootstrap replications of the three-step logistic regression model selection program; original choices were* (13, 15, 7, 20).

all 500 bootstrap replications only variable 20, histology, which appeared 295 times, was "important" more than half of the time. These results certainly discourage confidence in the causal nature of the predictor variables (13, 15, 7, 20).

Or do they? It seems like we should be able to use the bootstrap results to quantitatively assess the validity of the various predictors. Perhaps they could also help in selecting a better prediction model. Questions like these are being asked these days, but the answers so far are more intriguing than conclusive.

It is not clear to me whether Fisherian methods will play much of a role in the further progress of model selection theory. Figure 6 makes model selection look like an exercise in discrete estimation, while Fisher's MLE theory was always aimed at continuous situations. Direct frequentist methods like cross-validation seem more promising right now, and there have been some recent developments in Bayesian model selection, but in fact our best efforts so far are inadequate for problems like the hepatitis data. We could badly use a clever Fisherian trick for reducing complicated model selection problems to simple obvious ones.

10. EMPIRICAL BAYES METHODS

As a final example, I wanted to say a few words about empirical Bayes methods. Empirical Bayes seems like the wave of the future to me, but it seemed that way 25 years ago and the wave still hasn't washed in, despite the fact that it is an area of enormous potential importance. It is not a topic that has had much Fisherian input.

Table 4 shows the data for an empirical Bayes situation: independent clinical trials were run in 41 cities, comparing the occurrence of recurrent bleeding, an undesirable side effect, for two stomach ulcer surgical techniques, a new treatment and an older control. Each trial yielded an estimate of the true log odds ratio for recurrent bleeding, Treatment versus Control,

$$\theta_i = \text{log odds ratio in city } i, \quad i = 1, 2, \ldots, 41.$$

In city 8, for example, we have the estimate seen earlier in Table 2,

$$\hat{\theta} = \log\left(\frac{1}{15} \Big/ \frac{13}{3}\right) = -4.2,$$

indicating that the new surgery was very effective in reducing recurrent bleeding, at least in city 8.

Figure 7 shows the likelihoods for θ_i in 10 of the 41 cities. These are conditional likelihoods, using Fisher's trick of conditioning on the marginals to get rid of the nuisance parameters in each city. It seems clear that the log odds ratios θ_i are not all the same. For instance, the likelihoods for cities 8 and 13 barely overlap. On the other hand, the θ_i values are not wildly discrepant, most of the 41 likelihood functions concentrating themselves on the range $(-6, 3)$. (This is the kind of complicated inferential situation I was worrying about in the discussion of fiducial inference, confidence densities and objective Bayes methods.)

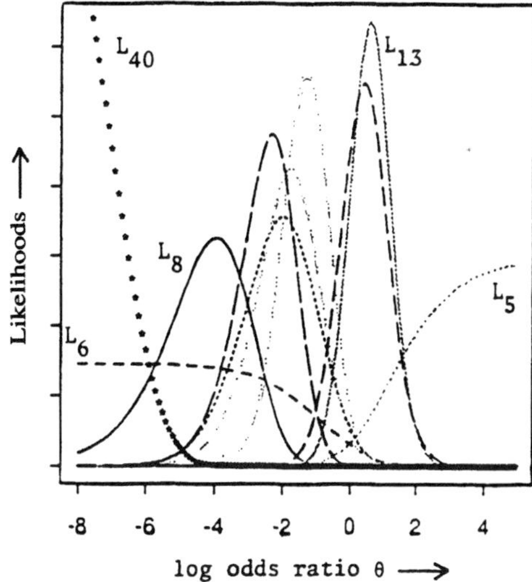

FIG. 7. *Individual likelihood functions for θ_i, for 10 of the 41 experiments in Table 4; L_8, the likelihood for the log odds ratio in city 8, lies to the left of most of the others.*

TABLE 4

Ulcer data: 41 independent experiments concerning the number of occurrences of recurrent bleeding following ulcer surgery; (a, b) = (# bleeding; # nonbleeding) for Treatment, a new surgical technique; (c, d) is the same for Control, an older surgery; $\hat{\theta}$ is the sample log odds ratio, with estimated standard deviation $\widehat{\text{SD}}$; stars indicate cases shown in Figure 7

Experiment	a	b	c	d	$\hat{\theta}$	$\widehat{\text{SD}}$	Experiment	a	b	c	d	$\hat{\theta}$	$\widehat{\text{SD}}$
1*	7	8	11	2	−1.84	0.86	21	6	34	13	8	−2.22	0.61
2	8	11	8	8	−0.32	0.66	22	4	14	5	34	66	0.71
3*	5	29	4	35	0.41	0.68	23	14	54	13	61	20	0.42
4	7	29	4	27	0.49	0.65	24	6	15	8	13	−43	0.64
5*	3	9	0	12	inf	1.57	25	0	6	6	0	−inf	2.08
6*	4	3	4	0	−inf	1.65	26	1	9	5	10	−1.50	1.02
7*	4	13	13	11	−1.35	0.68	27	5	12	5	10	−0.18	0.73
8*	1	15	13	3	−4.17	1.04	28	0	10	12	2	−inf	1.60
9	3	11	7	15	−0.54	0.76	29	0	22	8	16	−inf	1.49
10*	2	36	12	20	−2.38	0.75	30	2	16	10	11	−1.98	0.80
11	6	6	8	0	−inf	1.56	31	1	14	7	6	−2.79	1.01
12*	2	5	7	2	−2.17	1.06	32	8	16	15	12	−0.92	0.57
13*	9	12	7	17	0.60	0.61	33	6	6	7	2	−1.25	0.92
14	7	14	5	20	0.69	0.66	34	0	20	5	18	−inf	1.51
15	3	22	11	21	−1.35	0.68	35	4	13	2	14	0.77	0.87
16	4	7	6	4	−0.97	0.86	36	10	30	12	8	−1.50	0.57
17	2	8	8	2	−2.77	1.02	37	3	13	2	14	0.48	0.91
18	1	30	4	23	−1.65	0.98	38	4	30	5	14	−0.99	0.71
19	4	24	15	16	−1.73	0.62	39	7	31	15	22	−1.11	0.52
20	7	36	16	27	−1.11	0.51	40*	0	34	34	0	−inf	2.01
							41	0	9	0	16	NA	2.04

Notice that L_8, the likelihood for θ_8, lies to the left of most of the other curves. This would still be true if we could see all 41 curves instead of just 10 of them. In other words, θ_8 appears to be more negative than the log odds ratios in most of the other cities.

What is a good estimate or confidence interval for θ_8? Answering this question depends on how much the results in other cities influence our thinking about city 8. That is where empirical Bayes theory comes in, giving us a systematic framework for combining the direct information for θ_8 from city 8's experiment with the indirect information from the experiments in the other 40 cities.

The ordinary 90% confidence interval for θ_8, based only on the data $(1, 15, 13, 3)$ from its own experiment, is

$$\theta_8 \in [-6.3, -2.4]. \tag{13}$$

Empirical Bayes methods give a considerably different result. The empirical Bayes analysis uses the data in the other 40 cities to estimate a prior density for log odds ratios. This prior density can be combined with the likelihood L_8 for city 8, using Bayes theorem, to get a central 90% a posteriori interval for θ_8,

$$\theta_8 \in [-5.1, -1.8]. \tag{14}$$

The fact that most of the cities had less negatively tending results than city 8 plays an important role in the empirical Bayes analysis. The Bayesian prior estimated from the other 40 cities says that θ_8 is unlikely to be as negative as its own data by itself would indicate.

The empirical Bayes analysis implies that there is a lot of information in the other 40 cities's data for estimating θ_8, as a matter of fact, just about as much as in city 8's own data. This kind of "other" information does not have a clear Fisherian interpretation. The whole empirical Bayes analysis is heavily Bayesian, as if we had begun with a genuinely informative prior for θ_8 and yet it still has some claims to frequentist objectivity.

Perhaps we are verging here on a new compromise between Bayesian and frequentist methods, one that is fundamentally different from Fisher's proposals. If so, the 21st century could look a lot less Fisherian, at least for problems with parallel structure like the ulcer data. Right now there aren't many such problems. This could change quickly if the statistics community became more confident about analyzing empirical Bayes problems. There weren't many factorial design problems before Fisher provided an effective methodology for handling them. Scientists tend to bring us the problems we can solve. The current attention to metaanalysis and hierarchical models certainly suggests a growing interest in the empirical Bayes kind of situation.

11. THE STATISTICAL TRIANGLE

The development of modern statistical theory has been a three-sided tug of war between the Bayesian, frequentist and Fisherian viewpoints. What I have been trying to do with my examples is apportion the influence of the three philosophies on several topics of current interest: standard error estimation; approximate confidence intervals; conditional inference; objective Bayes theories and fiducial inference; model selection; and empirical Bayes techniques.

Figure 8, the statistical triangle, does this more concisely. It uses barycentric coordinates to indicate the influence of Bayesian, frequentist and Fisherian thinking upon a variety of active research areas. The Fisherian pole of the triangle is located between the Bayesian and frequentist poles, as in Figure 1, but here I have allocated Fisherian philosophy its own dimension to take account of its distinctive operational features: reduction to "obvious" types; the plug-in principle; an emphasis on inferential correctness; the direct interpretation of likelihoods; and the use of automatic computational algorithms.

Of course, a picture like this cannot be more than roughly accurate, even if one accepts the author's prejudices, but many of the locations are difficult to argue with. I had no trouble placing conditional inference and partial likelihood near the Fisherian pole, robustness at the frequentist pole and multiple imputation near the Bayesian pole. Empirical Bayes is clearly a mixture of Bayesian and frequentist ideas. Bootstrap methods combine the convenience of the plug-in principle with a strong frequentist desire for accurate operating characteristics, particularly for approximate confidence intervals, while the jackknife's development has been more purely frequentistic.

Some of the other locations in Figure 8 are more problematical. Fisher provided the original idea behind the EM algorithm, and in fact the self-consistency of maximum likelihood estimation (when missing data is filled in by the statistician) is a classic Fisherian correctness argument. On the other hand EM's modern development has had a strong Bayesian component, seen more clearly in the related topic of Gibbs sampling. Similarly, Fisher's method for combining independent p-values is an early form of metaanalysis, but the subject's recent growth has been strongly frequentist.

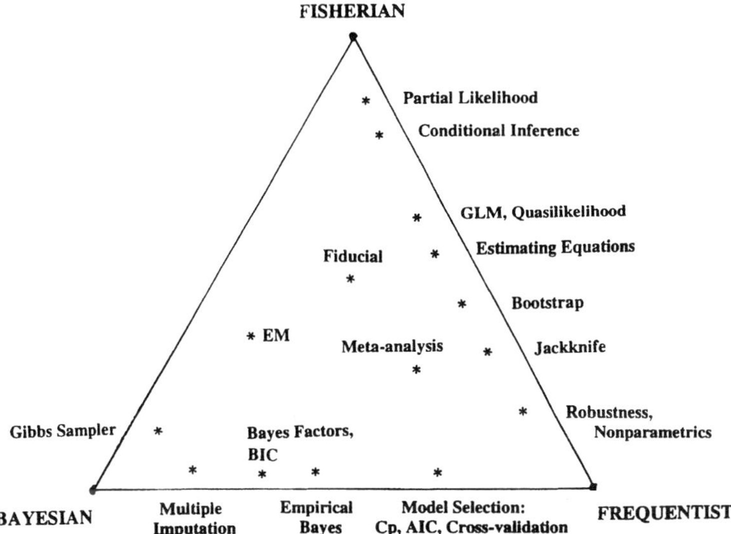

FIG. 8. *A barycentric picture of modern statistical research, showing the relative influence of the Bayesian, frequentist and Fisherian philosophies upon various topics of current interest.*

The trouble here is that Fisher wasn't always a Fisherian, so it is easy to confuse parentage with development.

The most difficult and embarrassing case concerns what I have been calling "objective Bayes" methods, among which I included fiducial inference. One definition of frequentism is the desire to do well, or at least not to do poorly, against every possible prior distribution. The Bayesian spirit, as epitomized by Savage and de Finetti, is to do very well against one prior distribution, presumably the right one.

There have been a variety of objective Bayes compromises between these two poles. Working near the frequentist end of the spectrum, Welch and Peers showed how to calculate priors whose a posteriori credibility intervals coincide closely with standard confidence intervals. Jeffreys's work, which has led to vigorous modern development of Bayesian model selection, is less frequentistic. In a bivariate normal situation Jeffreys would recommend the same prior distribution for estimating the correlation coefficient or for the ratio of expectations, while the Welch–Peers theory would use two different priors in order to separately match each of the frequentist solutions.

Nevertheless Jeffreys's Bayesianism has an undeniable objectivist flavor. Erich Lehmann (personal communication) had this to say: "If one separates the two Bayesian concepts [Savage–de Finetti and Jeffreys] and puts only the subjective version in your Bayesian corner, it seems to me that something interesting happens: the Jeffreys concept moves to the right and winds up much closer to the frequency corner than to the Bayesian one. For example, you contrasted Bayes as optimistic and risk-taking with frequentist as pessimistic and playing it safe. On both of these scales Jeffreys is much closer to the frequentist end of the spectrum. In fact, the concept of uninformative prior is philosophically close to Wald's least favorable distribution, and the two often coincide."

Lehmann's advice is followed a bit in Figure 8, where the Bayesian model selection (BIC) point, a direct legacy of Jeffreys's work, has been moved a little ways toward the frequentist pole. However, I have located fiducial inference, Fisher's form of objective Bayesianism, near the center of the triangle. There isn't much work in that area right now but there is a lot of demand coming from all three directions.

The point of my examples, and the main point of this talk, was to show that Fisherian statistics, is not a dead language and that it continues to inspire new research. I think this is clear in Figure 8, even allowing for its inaccuracies. But Fisher's language is not the only language in town, and it is not even the dominant language of our research journals. That prize would have to go to a rather casual frequentism, not usually as hard-edged as pure decision theory these days. We might ask what Figure 8 will look like 20 or 30 years from now, and whether there will be many points of active research interest lying near the Fisherian pole of the triangle.

12. R. A. FISHER IN THE 21ST CENTURY

Most talks about the future are really about the present, and this one has certainly been no exception. But here at the end of the talk, and nearly at the end of the 20th century, we can peek cautiously ahead and speculate at least a little bit about the future of Fisherian statistics.

Of course Fisher's fundamental discoveries like sufficiency, Fisher information, the asymptotic efficiency of the MLE, experimental design and randomization inference are not going to disappear. They might become less visible though. Right now we use those ideas almost exactly as Fisher coined them, but modern computing equipment could change that.

For example, maximum likelihood estimates can be badly biased in certain situations involving a great many nuisance parameters (as in the Neyman–Scott paradox.) A computer-modified version of the MLE that was less less biased could become the default estimator of choice in applied problems. REML estimation of variance components offers a current example. Likewise, with the universal spread of high-powered computers statisticians might automatically use some form of the more accurate confidence intervals I mentioned earlier instead of the standard intervals.

Changes like these would conceal Fishers's influence, but not really diminish it. There are a couple of good reasons though that one might expect more dramatic changes in the statistical world, the first of these being the miraculous improvement in our computational equipment, by orders of magnitude every decade. Equipment is destiny in science, and statistics is no exception to that rule. Second, statisticians are being asked to solve bigger, harder, more complicated problems, under such names as pattern recognition, DNA screening, neural networks, imaging and machine learning. New problems have always evoked new solutions in statistics, but this time the solutions might have to be quite radical ones.

Almost by definition it's hard to predict radical change, but I thought I would finish with a few speculative possibilities about a statistical future that might, or might not, be a good deal less Fisherian.

12.1 A Bayesian World

In 1974 Dennis Lindley predicted that the 21st century would be Bayesian. (I notice that his recent *Statistical Science* interview now predicts the year 2020.) He could be right. Bayesian methods are attractive for complicated problems like the ones just mentioned, but unless the scientific world changes the way it thinks I can't imagine subjective Bayes methods taking over. What I called objective Bayes, the use of neutral or uninformative priors, seems a lot more promising and is certainly in the air these days.

A successful objective Bayes theory would have to provide good frequentist properties in familiar situations, for instance, reasonable coverage probabilities for whatever replaces confidence intervals. Such a Bayesian world might not seem much different than the current situation except for more straightforward analyses of complicated problems like multiple comparisons. One can imagine the statistician of the year 2020 hunched over his or her supercomputer terminal, trying to get Proc Prior to run successfully, and we can only wish that future colleague "good luck."

12.2 Nonparametrics

As part of our Fisherian legacy we tend to overuse simple parametric models like the normal. A nonparametric world, where parametric models were a last resort instead of the first, would favor the frequentist vertex of the triangle picture.

12.3 A New Synthesis

The postwar years and especially the last few decades have been more notable for methodological progress than the development of fundamental new ideas in the theory of statistical inference. This doesn't mean that such developments are finished forever. Fisher's work came out of the blue in the 1920s, and maybe our field is due for another bolt of lightening.

It's easy for us to imagine that Fisher, Neyman and the others were lucky to live in a time when all the good ideas hadn't been plucked from the trees. In fact, we are the ones living in the golden age of statistics—the time when computation has become fast and easy. In this sense we are overdue for a new statistical paradigm, to consolidate the methodological gains of the postwar period. The rubble is building up again, to use Joan Fisher Box's simile, and we could badly use a new Fisher to put our world in order.

My actual guess is that the old Fisher will have a very good 21st century. The world of applied statistics seems to need an effective compromise between Bayesian and frequentist ideas, and right now there is no substitute in sight for the Fisherian synthesis. Moreover, Fisher's theories are well suited to life in the computer age. Fisher seemed naturally to think in algorithmic terms. Maximum likelihood estimates, the standard intervals, ANOVA tables, permutation tests are all expressed algorithmically and are easy to extend with modern computation.

Let me say finally that Fisher was a genius of the first rank, who has a solid claim to being the most important applied mathematician of the 20th century. His work has a unique quality of daring mathematical synthesis combined with the utmost practicality. The stamp of his work is very much upon our field and shows no sign of fading. It is the stamp of a great thinker, and statistics—and science in general—is much in his debt.

REFERENCES

Section 1

SAVAGE, L. J. H. (1976). On rereading R. A. Fisher (with discussion). *Ann. Statist.* **4** 441–500. (Savage says that Fisher's work greatly influenced his seminal book on subjective Bayesianism. Fisher's great ideas are examined lovingly here, but not uncritically.)

YATES, F. and MATHER K. (1971). Ronald Aylmer Fisher. In *Collected Papers of R. A. Fisher* (K. Mather, ed.) **1** 23–52. Univ. Adelaide Press. (Reprinted from a 1963 Royal Statistical Society memoir. Gives a nontechnical assessment of Fisher's ideas, personality and attitudes toward science.)

BOX, J. F. (1978). *The Life of a Scientist.* Wiley, New York. (This is both a personal and an intellectual biography by Fisher's daughter, a scientist in her own right and also an historian of science, containing some unforgettable vignettes of precocious mathematical genius mixed with a difficulty in ordinary human interaction. The sparrow quote in Section 4 is put in context on page 130.)

Section 2

FISHER, R. A. (1925). Theory of statistical estimation. *Proc. Cambridge Philos. Soc.* **22** 200–225. (Reprinted in the Mather collection, and also in the 1950 Wiley Fisher collection *Contributions to Mathematical Statistics.* This is my choice for the most important single paper in statistical theory. A competitor might be Fisher's 1922 Philosophical Society paper, but as Fisher himself points out in the Wiley collection, the 1925 paper is more compact and businesslike than was possible in 1922, and more sophisticated as well.)

EFRON B. (1995). The statistical century. *Royal Statistical Society News* **22** (5) 1–2. (This is mostly about the postwar boom in statistical methodology and uses a different statistical triangle than Figure 8.)

Section 3

FISHER, R. A. (1934). Two new properties of mathematical likelihood. *Proc. Roy. Soc. Ser. A* **144** 285–307. (Concerns two situations when fully efficient estimation is possible in finite samples: one-parameter exponential families, where the MLE is a sufficient statistic, and location–scale families, where there are exhaustive ancillary statistics. Reprinted in the Mather and the Wiley collections.)

Section 4

EFRON, B. (1978). Controversies in the foundations of statistics. *Amer. Math. Monthly* **85** 231–246. (The Bayes–Frequentist–Fisherian argument in terms of what kinds of averages should the statistician take. Includes Fisher's famous circle example of ancillarity.)

EFRON, B. (1982). Maximum likelihood and decision theory. *Ann. Statist.* **10** 240–356. (Examines five questions concerning maximum likelihood estimation: What kind of theory is it? How is it used in practice? How does it look from a frequentistic decision-theory point of view? What are its principal virtues and defects? What improvements have been suggested by decision theory?)

CIFARELLI, D. and REGAZZINI, E. (1996). De Finetti's contribution to probability and statistics. *Statist. Sci.* **11** 253–282. [The second half of the quote in my Section 4, their Section 3.2.2, goes on to criticize the Neyman–Pearson school. De Finetti is less kind to Fisher in the discussion following Savage's (1976) article.]

Sections 5 and 6

DICICCIO, T. and EFRON, B. (1996). Bootstrap confidence intervals (with discussion). *Statist. Sci.* **11** 189–228. (Presents and discusses the cd4 data of Figure 2. The bootstrap confidence limits in Table 1 were obtained by the BC_a method.)

Section 7

REID, N. (1995). The roles of conditioning in inference. *Statist. Sci.* **10** 138–157. [This is a survey of the p^* formula, what I called the magic formula following Ghosh's terminology, and many other topics in conditional inference; see also the discussion (following the companion article) on pages 173–199, in particular McCullagh's commentary. Gives an extensive bibliography.]

EFRON, B. and HINKLEY, D. (1978). Assessing the accuracy of the maximum likelihood estimator: observed versus expected Fisher information. *Biometrika* **65** 457–487. (Concerns ancillarity, approximate ancillarity and the assessment of accuracy for a MLE.)

Section 8

EFRON, B. (1993). Bayes and likelihood calculations from confidence intervals. *Biometrika* **80** 3–26. (Shows how approximate confidence intervals can be used to get good approximate confidence densities, even in complicated problems with a great many nuisance parameters.)

Section 9

EFRON, B. and GONG, G. (1983). A leisurely look at the bootstrap, the jackknife, and cross-validation. *Amer. Statist.* **37** 36–48. (The chronic hepatitis example is discussed in Section 10 of this bootstrap–jackknife survey article.)

O'HAGAN, A. (1995). Fractional Bayes factors for model comparison (with discussion). *J. Roy. Statist. Soc. Ser. B* **57** 99–138. (This paper and the ensuing discussion, occasionally rather heated, give a nice sense of Bayesian model selection in the Jeffreys tradition.)

KASS, R. and RAFTERY, A. (1995). Bayes factors. *J. Amer. Statist. Soc.* **90** 773–795. (This review of Bayesian model selection features five specific applications and an enormous bibliography.)

Section 10

EFRON, B. (1996). Empirical Bayes methods for combining likelihoods. *J. Amer. Statist. Assoc.* **91** 538–565.

Section 11

KASS, R. and WASSERMAN, L. (1996). The selection of prior distributions by formal rules. *J. Amer. Statist. Assoc.* **91** 1343–1370. (Begins "Subjectivism has become the dominant philosophical tradition for Bayesian inference. Yet in practice, most Bayesian analyses are performed with so-called noninformative priors. ...")

Section 12

LINDLEY, D. V. (1974). The future of statistics—a Bayesian 21st century. In *Proceedings of the Conference on Directions for Mathematical Statistics*. Univ. College, London.

SMITH, A. (1995). A conversation with Dennis Lindley. *Statist. Sci.* **10** 305–319. (This is a nice view of Bayesians and Bayesianism. The 2020 prediction is attributed to de Finetti.)

Comment

D. R. Cox

I very much enjoyed Professor Efron's eloquent and perceptive assessment of R. A. Fisher's contributions and of their current relevance. I am sure that Professor Efron is right to attach outstanding importance to Fisher's ideas.

As Professor Efron emphasizes, Fisher's ideas are so wide ranging that it is not feasible to cover them all in a single paper. The following outline notes are in supplementation of rather than in disagreement with the paper.

(1) Fisher stressed the need for different modes of attack on different types of inferential problem.

(2) While his formal ideas deal with fully parametric problems he gave the "exact" randomization test based on the procedure used in design. The status to him of such tests is not entirely clear. Did he regard them as reassurance for the faint of heart, timid about working assumptions of normality, or are they the preferred method of analysis to which normal theory based results are often a convenient approximation? Yates vigorously rejected the second interpretation. The more important point is probably that Fisher recognized that randomization indicated the appropriate analysis of variance, that is, appropriate estimate of error. This replaced the need for special ad hoc assumptions of a new linear model for each design.

(3) A key to understanding some of the distinctions between Fisherian and Neyman–Pearson approaches lies in Fisher's special notion of the meaning of the probability p of a "unique" event as set out, for example, in Fisher (1956, pages 31–36). There are two aspects, one that the individual belongs to an ensemble or population in a proportion p of which which the event holds. The other is that it is not possible to recognize the individual as lying in a subpopulation with a different proportion. Fisher considered, it seems to me correctly, that this enabled probability statements to be attached to an unknown parameter on the basis of a random sample, no other information being available, without invoking a prior distribution. The snag is that such distributions cannot be manipulated or combined by the ordinary laws of probability.

(4) The development of Fisher's ideas on Bayesian inference can be traced by comparing the polemical remarks at the start of *Design of Experiments* (Fisher, 1935) with the more measured comments in Fisher (1956).

(5) Fisher was of course an extremely powerful mathematician especially with distributional and combinatorial calculations. It helps us to understand his attitude to mathematical rigor to note the remarks of Mahalanobis (1938), who wrote:

> The explicit statement of a rigorous argument interested him but only on the important condition that such explicit demonstration of rigor was needed. Mechanical drill in the technique of rigorous statement was abhorrent to him, partly for its pedantry, and partly as an inhibition to the active use of the mind. He felt it was more important to think actively, even at the expense of occasional errors from which an alert intelligence would soon recover, than to proceed with perfect safety at a snail's pace along well known paths with the aid of a most perfectly designed mechanical crutch.

D. R. Cox is an Honorary Fellow, Nuffield College, Oxford OX1 1NF, United Kingdom (e-mail: david.cox@nuf.ox.ac.uk).

This is a comment on Fisher's attitude when an undergraduate at Cambridge; it is tempting to think that Mahalanobis was using Fisher's own words.

(6) In some ways Fisher's greatest methodological contributions, the analysis of variance including discriminant analysis, and ideas about experimental design, appear to owe relatively little directly to his ideas on general statistical theory. There are of course connections to be perceived subsequently, for example, to sufficiency and transformation models, and in the case of randomization somewhat underdeveloped connections to conditioning and ancillary statistics. Fisher's mastery of distribution theory was, however, obviously relevant, perhaps most strikingly in his approach to the connection between the distribution theory of multiple regression to that of linear discriminant analysis.

(7) In the normal process of scientific development important notions get simplified and absorbed into the general ethos of the subject and reference to original sources becomes unnecessary except for the historian of ideas. It is a measure of the range and subtlety of Fisher's ideas that it is still fruitful to read parts at least of the two books mentioned above as well as the papers that Professor Efron mentions in his list of references.

(8) I agree with Professor Efron that one key current issue is the synthesis of ideas on modified likelihood functions, empirical Bayes methods and some notion of reference priors.

(9) I like Professor Efron's triangle although on balance I would prefer a square, labeled by one axis that represents mathematical formulation and the other that represents conceptual objective. For example Fisher and Jeffreys were virtually identical in their objective although of course different in their mathematics.

(10) Finally, in contemplating Fisher's contributions, one must not forget that he was as eminent as a geneticist as he was as a statistician.

Comment

Rob Kass

Very nice, very provocative—as I read Efron's version of Fisher's profound influence on our discipline, I found myself wondering whether statistics could have succeeded to this point so spectacularly, across so many disciplines, if it had been based primarily on Bayesian logic rather than largely Fisherian frequentist logic. Without the historical counterfactual, the question might be stated this way: from a Bayesian point of view, when, if ever, is frequentist logic necessary?

I believe there are three places where frequentist reasoning is essential to the success of our enterprise. First, we need goodness-of-fit assessments, or what Dempster in his Fisherian ruminations has called "postdictive inference" (see Gelman, Meng and Stern, 1996, and the discussion of it). Second, although many principles have been formulated for defining noninformative, or reference, priors, it seems that good behavior under repeated sampling should play a role somehow. It is possible, as Cox implied in his 1994 *Statistical Science* interview, that we have not yet recognized how, and I have the sense that Efron shares this view. But the third and perhaps most important place even we Bayesians currently find frequentist methods useful is that they offer highly desirable shortcuts. This is related to Efron's point at the end of Section 4. The Fisherian frequentist methods for what are now relatively simple situations, such as analysis of variance, are certainly easy to use; it remains unclear whether standardized Bayesian methods could entirely replace them. Equally crucial, however, are the more sophisticated data analytic methods, such as modern nonparametric regression, which share a big relative advantage in ease of use over their Bayesian counterparts. In short, despite my strong preference for Bayesian thinking, based on our current understanding of inference, I cannot see how the next century or any other could be exclusively Bayesian. I have felt for a long time that the Bayesian versus frequentist contrast, while strictly a matter of logic, might more usefully be considered, metaphorically, a matter of language—that is, they are two alternative languages that are used to grapple with uncertainty, with fluent speakers of

Rob Kass is Professor and Head, Department of Statistics, Carnegie Mellon University, Pittsburgh, Pennsylvania 15213 (e-mail: kass@stat.cmu.edu).

one being capable of arriving at an understanding of essentially any phenomenon that is understood by fluent speakers of the other, despite there being no good translation of certain phrases.

I suspect we all agree on Fisher's greatness. I like to say Fisher was to statisics what Newton was to physics. Continuing the analogy, Efron suggests that we need a statistical Einstein. But the real question is whether it is possible to obtain a new framework that achieves the goal of fiducial inference. The situation in statistical inference is beautifully peaceful and compelling for one-parameter problems: reference Bayesian and Fisherian roads converge to second-order, via the magic formula. When we go to the multiparameter world, however, the hope dims not only for a reconciliation of Bayesian and frequentist paradigms, but for any satisfactory, unified approach in either a frequentist or Bayesian framework, and we must wonder whether the world is simply depressingly messy. Indeed, some of the cautionary notes sounded in the 1996 *JASA* review paper I wrote with Larry Wasserman implicitly suggest that (as pure subjectivists are quick to argue) there may be no way around the fundamental difficulties.

It is clear that statistical problems are becoming much more complicated. I got the possibly erroneous sense (from his comment at the end of Section 8) that Efron connects this to his unease with the current situation and the need for a new paradigm—to be furnished perhaps by our Einstein Messiah. I would instead look toward a different big new theoretical development. Bayesian and frequentist analyses based on parametric models are, from a consumer's point of view, really quite similar. Bayesian nonparametrics, however, is in its infancy and its connection with frequentist nonparametrics almost nonexistent. I hope for a much more thorough and successful Bayesian nonparametric theory and a resulting deeper understanding of infinite-dimensional problems. Perhaps entirely new principles would have to be invoked to supplement those of Fisher, Jeffreys, Neyman and de Finetti–Savage. If it happens, an increasingly nonparametric future would not, in principle, move us toward the frequentist vertex of Efron's triangle. Rather, Bayesian inference would continue to play its illuminating foundational role, important new methodology would be developed, and it might even turn out that there is a genuine, deep and detailed sense in which frequentist methods could be considered shortcut substitutes for full-fledged Bayesian alternatives.

Comment

Ole E. Barndorff-Nielsen

It has been a pleasure reading Professor Efron's far-ranging, thoughtful and valiant paper. I agree with most of the views presented there and this contribution to the discussion of the paper consists of a number of disperse comments, mostly adding to what has been mentioned in the paper.

(1) One, potentially major, omission from the paper's vision of Fisherian influence in the next century is the lack of discussion of the role that ideas and methods of statistical inference may have in quantum mechanics. Such ideas and methods are likely to become of increasing importance, particularly in connection with the developments in experimental techniques that allow the study of very small quantum systems.

Moreover, there is already now, in the physical literature, a substantial body of results on quantum analogues of Fisher information and on associated results of statistical differential geometry, in the vein of Amari.

(2) It also seems pertinent to stress the importance of "pseudolikelihood," that is, functions of part or all of the data and part or all of the parameters that to a large extent can be treated as genuine likelihoods. Many of the most fruitful advances in the second half of the 20th century centers around such functions.

(3) My own view on optimality is, perhaps, somewhat different from that of Bradley Efron in that I find that the focussing on optimality has to a considerable extent been to the detriment of statistical development. The problems and danger stem from too narrow definitions of what is meant by optimal-

Ole E. Barndorff-Nielsen is Professor, Theoretical Statistics, Institute of Mathematics, Aarhus University, 8000 AARHUS C, Denmark (e-mail: oebn@mi.aau.dk).

ity. One prominent instance of how too much emphasis on a restricted concept of optimality may lead astray from general principles of scientific enquiry is provided by the upper storeys of the Neyman–Person test theory edifice.

In general, what good is it to have a procedure that is "optimal" if this optimality" does not fit in naturally with a comprehensive and integrated methodological approach to scientific studies.

However, if I read correctly, Professor Efron and I do not disagree strongly here. I am thinking in particular of his reference to the "spirit of reasonable compromise."

(4) In relation to the statement (at the end of Section 8.1, on "The Confidence Density") concerning approximate confidence intervals and approximate fiducial distributions in situations with many nuisance parameters, I wonder how this is to be reconciled with the many counterexamples to Fisher's idea(s) for fiducial inference under multiparameter distributions.

(5) Concerning model selection I suspect that techniques in Fisherian spirit can be developed although this largely remains to be done.

(6) On a historical note, neither Bayesian statistics nor Neyman–Pearson–Wald type theory has ever taken serious foothold in statistics in Denmark, and thinking along lines close to those of Fisher has been prominent throughout.

The tradition does, in fact, go back to the latter part of the 19th century, in particular to Thiele, who, one might add, essentially worked with analysis of variance long before Fisher (cf. Hald, 1981).

Comment

D. V. Hinkley

Using this paper as an excuse, I revisited the *Collected Papers* (Bennett, 1972) after a gap of 15 years and was struck again by how clear and concise the writing was—usually firm and irreverent, but often good-humored.

Fisher's work certainly colored my own view of the principles and practices of statistics, especially in the early 1980s. But I think that the biological context of much of Fisher's work made him wrongly dismissive of Bayes's theorem as a potential tool. It is instructive to read the two-page note by Fisher (1929), which answered Student's suggestion that normal-theory methods such as analysis of variance be extended to nonnormal data. Fisher says "I have never known difficulty to arise in biological work from imperfect normality of variation, often though I have examined data for this particular cause of difficulty.... This is not to say that the deviation from [normal-theory methods] may not have a real application in some technological work..." (what would Fisher have thought of mathematical finance, one wonders!). I think that Fisher saw sampling distributions as concrete, logically distinct from the weaker uncertainty characterizing prior distributions, and so incompatible with the application of Bayes's theorem. Fisher's view about sampling models is still widely held, especially among those using non-Bayesian methods, and so model uncertainty is a nonissue in nearly all of statistical education (not so in some sciences, however). But it may yet turn out to be one of the most important topics in statistics. Note that model uncertainty is the antithesis of model selection, which we are beginning to understand fairly well using non-Fisherian ideas.

Other discussants may comment on Efron's interpretation of Fisher, so I shall note only that he may overstate the contradictions. For example, randomization inference could be conditioned, when appropriate, by appropriate design—as happened with the Knight's Move and Diagonal Squares; see Yates (1970, page 58) and Savage et al. (1976, page 464). Indeed the combination of Efron (1971) and Cox (1982) seems quintessentially Fisherian! (Similar ideas extend into resampling–bootstrap–cross-validation analysis, but are not yet common.)

But what of Fisher's possible influence on the future of statistics? With the bootstrap I am not sure. Certainly the parametric bootstrap involves a harmless use of computers to do Fisher's calculations. The ingenious theoretical basis for the parametric BC_a bootstrap method (Efron, 1987) could almost have been written by Fisher himself. The idea of going beyond a simple standard error and

D. V. Hinkley is Professor, Department of Statistics and Applied Probability, University of California, Santa Barbara, California 93106-3110 (e-mail: hinkley@pstat.ucsb.edu).

standardized estimate (for confidence intervals or tests) follows Fisher; see Fisher (1928), where in reference to the conventional standard error of the sample correlation Fisher quips "among the things 'which students have to know, but only a fool would use'." Nevertheless, the alternative developments based on likelihood seem closer to continuation of the Fisher approach. (All of these methods do need further development to become easily and widely useable.)

A major question is whether or not the nonparametric bootstrap methods have or will have Fisherian influence. I think not. Certainly when we get into model selection, highly nonparametric regression such as regression trees (Ripley, 1996), we are more in the arena of Tukey than Fisher. Purely empirical validation and assessment does not seem to appear in Fisher's work, possibly due to impossibility of practical calculations.

The topic of empirical Bayes estimation, or random-effect modelling, has become of increasing importance with the spread of metaanalysis ideas and methods. To paraphrase Efron (Section 10), the information about one subpopulation in another subpopulation "does not have a clear Fisherian interpretation." But surely Fisher would have addressed this problem in a sensible way, so we need to figure out how by going back to read Fisher.

Fisher's major contributions may be the ideas about designs to avoid bias in experimental results and to enable calculation of reliable measures of uncertainty. Compared to these, the fine points about whether confidence intervals are Bayesian or not seem relatively unimportant. Fisher did make valuable contributions to the discussion of the roles of probability in statistics, much of it usefully surveyed by Lane (1980). As far as Fisher's theoretical work goes, particularly the 1925 and 1934 masterpieces, it has long seemed to me that Fisher may have inadvertently prepared us for an acceptable future of Bayesian methodology.

Comment

D. A. S. Fraser

It is a delight to read Brad's thoughtful and insightful overview giving deep credits to Fisher's contributions to current and future statistics. Brad mentions large areas not covered by his review and it is our loss that these areas such as randomization and experimental design have been omitted; they may also be among Fisher's major contributions although much neglected in current practice. I strongly endorse Brad's positive approval of Fisher's contributions and add just a few further approving remarks.

Brad notes that at "the time of (Brad's) education Fisher had been reduced to a somewhat minor figure in American academic statistics." This seems to be a substantial understatement particularly at Stanford in 1961–1962 where, and when, Brad began his graduate work in statistics. In the fall of 1961 a psychologist–statistician Sidney Siegel spoke at the Stanford statistics seminar describing how he had graphed the likelihood function for an applied problem to obtain insight on a parameter of interest, something that now would be viewed as a sensible step in a statistical analysis. Siegel was widely challenged by the Statistics faculty then, that this cannot and should not be done, a frontal rejection of a Fisherian idea, leaving Siegel feeling dispossessed of his credibility as a statistician. Also that same fall "Fisher spoke at the Stanford Medical School" as noted by Brad. I was visiting the Stanford Statistics Department that fall and did have notice of Fisher's talk. The absence of Statistics faculty from Fisher's talk was a measure of Fisher's influence at that time, perhaps the nadir of his influence. Brad does much to correct this and give credit for the wealth we have inherited from Fisher.

I prefer a somewhat different interpretation for "frequentist" than that indicated by "competing philosophies . . . : Bayesian; Neyman–Wald frequentist; and Fisherian." "Frequentist" seems to refer to the interpretation for the probabilities given by the statistical model alone, thus embracing both Fishe-

D. A. S. Fraser is Professor, Department of Statistics, Sidney Smith Hall, University of Toronto, Toronto, Canada M5S 3G3 (e-mail: dfraser@utstat.toronto.edu).

rian and decision theoretic philosophies. Bayesian statistics adds prior probabilities as the distinguishing feature. But even here the distinction may now be blurring when we view such additions on a "what if" basis of temporarily enlarging the model, in a way similar to the "what if" basis we often apply to the model itself.

Brad refers to empirical Bayes as "not a topic that has had much Fisherian input." It does seem, however, to be quite consonant with Fisher's view and indeed Fisher should be treated as the initiator of what was latter called empirical Bayes. In his 1956 book *Statistical Methods and Scientific Inference* (page 18f) Fisher considers the genetic origins of an animal under investigation and appends to the initial statistical model the theoretical–empirical probabilities concerning the origins of the animal. This is pure what-we-now-call empirical Bayes and Fisher should be credited. From another viewpoint it is just enlarging the statistical model to provide in some sense proper modeling.

In referring to Fisher's "devices" for solving problems, Brad sometimes uses the seemingly pejorative term "trick." Most of these devices did seem to be tricks when introduced but hardly now. Perhaps we both long for more of these "tricks" as guides to future devices.

Two formulas, (10) and (11), are recorded for "the conditional density" $f_\theta(\hat\theta|A)$ "of the MLE $\hat\theta$ given [an] ancillary" A. These formulas use cryptic notation that makes them technically wrong without clarification to indicate additional dependencies on $\hat\theta$. The context assumes that $\mathbf{x}$ is equivalent to $(\hat\theta, A)$ so that density and likelihood have the form $f_\theta(\hat\theta, A)$ and $L(\theta; \hat\theta, A)$; the amplified formulas would then appear as

$$f_\theta(\hat\theta|A) = c\frac{L(\theta;\hat\theta,A)}{L(\hat\theta;\hat\theta,A)},$$

$$(1)\quad f_\theta(\hat\theta|A) = c\frac{L(\theta;\hat\theta,A)}{L(\hat\theta;\hat\theta,A)} \cdot\left\{-\frac{d^2}{d\theta^2}\log L(\theta;\hat\theta,A)|_{\theta=\hat\theta}\right\}^{1/2}$$

for the location and general cases (with appropriate accuracies). In the location model case with

$$f_\theta(x;\theta) = g(\hat\theta-\theta|A)h(A)$$

the amplified version of (10) then gives

$$c\frac{g(\hat\theta-\theta|A)}{g(0|A)},$$

which reproduces the conditional density except for a norming constant. By contrast the given formula (10) taken literally with the preamble describing $L(\theta)$ "as a function of θ with $\mathbf{x}$ fixed" at an observed value $(\hat\theta^0, A)$ gives

$$c\frac{g(\hat\theta^0-\theta|A)}{g(\hat\theta^0-\hat\theta|A)},$$

which is the MLE density approximation only for the observed data point $(\hat\theta, A) = (\hat\theta^0, A)$. This may seem like a very technical point but the formulas are often cited as a way "for calculating $f_\theta(\hat\theta|A)$." In practice they can rarely be used for such a calculation as it would require the likelihood function to be available at points $(\hat\theta, A)$ with fixed A and varying $\hat\theta$, and this would require an explicit ancillary and much computation.

For the location model case there is a formula from Fisher that gives the conditional density of $\hat\theta$ from just the observed likelihood $L^0(\theta) = L(\theta; \hat\theta^0, A)$:

$$f_\theta(\hat\theta|A;\theta) = cL^0(\theta-\hat\theta+\hat\theta^0).$$

An extension of this is available for transformation models.

Brad mentions that "the magic formula can be used to generate approximate confidence intervals [with] at least second order accura[cy]." In fact for continuous models third order confidence intervals are available in wide generality but require additional theory for approximate ancillaries.

I particularly welcome Brad's positive views on the prospects for fiducial methods. As one who has worked extensively on contexts where fiducial calculations have good conventional properties (the transformation group context) or are useful for increasing the accuracy of a significance value (using fiducial to eliminate a nuisance parameter), I find it reassuring to see optimism elsewhere. Certainly the typical statistician feels fiducial is wrong but in most cases he is also unfamiliar with the details or the overlap with confidence methods.

For the case where the dimension of the variable has been reduced to the dimension of the parameter both fiducial and confidence make use quite generally of a pivotal quantity. The confidence procedure chooses a 90% region on the pivot space and then inverts to the parameter space to get the confidence region; by contrast the fiducial inverts to the parameter space and then chooses the 90% region. The fiducial issues arise then from the increased generality in the second procedure. Of course any observed 90% fiducial region has a corresponding 90% pivot region and is then of course the 90% confidence region for that pivot region. The controversial issues arise with simulations and repetitions. Should these repetitions always be with the same value of the parameter? This is not the reality of applications! And there are alternatives. I do applaud Brad's boldness in inverting the pivotal distribution and getting a confidence density; the fiducial whisper campaign has been too constraining on the profession.

Comment

A. P. Dempster

Bradley Efron's colorful lecture is fun to read. Brad is generous and accurate in crediting Fisher's important mathematical foundations for sparking the "frequentist" school whose framework is evidently deeply embedded in Brad's psyche and work. At the same time, I question whether he is sufficiently in touch with Fisher's thinking to do justice to Fisher's also remarkable contributions to the logic of inference. A balanced reading of the Fisher–Neyman disputes suggests that the history of 20th century statistics is not a linear path from Fisher, to Neyman and ultimately to a modern "frequentist" statistics whose main challenger is "Bayesianism" of either "subjective" or "objective" varieties. The hackneyed terms in quotes need fundamental clarification and definition, laying out roles in science as actually practiced.

Fisher's conception of inference is built around the interpretation of sampling distributions in relation to sample data. Sample data are frequency data, and sampling distributions have natural frequency interpretations. But these roles for frequency are basic to any view of statistics as a discipline and are far from making Fisher "frequentist" in Neyman's sense. Fisher aimed to characterize the information in data, whereas Neyman settled on a theory that guides a statistician's behavior in choosing among procedures seen as competing on the basis of long run operating characteristics. "Neyman–Wald" theory provides useful insights, but creates a sterile view of practice. After a procedure is chosen and applied, how does one then think postanalysis about uncertainties? Fisher gave interpretations of significance tests, likelihood and fiducial intervals that whatever their strengths and weaknesses meet the question head-on.

A. P. Dempster is Professor, Department of Statistics, Harvard University, Cambridge, Massachusetts 02138 (e-mail: dempster@stat.harvard.edu).

Understanding Fisher requires understanding that a probability in practice determines a "well-specified state of uncertainty" (Fisher, 1958) about a specific situation in hand, a position identifiable as describing a kind of formal subjectivity that complements rather than contradicts the conventional view of science as objective. Interpreting the p-value of a significance test Fisher's way (Fisher, 1956, page 39), after the result is reported, requires "postdictive" assessment of such a formal subjective probability (Dempster, 1971). As early as 1935, Fisher recognized that "confidence intervals" are "only another way of saying that, by a certain test of significance, some kinds of hypothetical possibilities are to be rejected, while others are not" (Bennett, 1990, page 187). It may be, as Brad says, "generally considered" that fiducial probability was Fisher's "biggest blunder," but should not be on the basis of the simplistic arguments accompanying Neyman's exasperated polemics, for example, "a conglomeration of mutually inconsistent assertions, not a mathematical theory" (Neyman, 1977, page 100). Interpreting a fiducial probability shares with interpreting a Bayesian posterior probability an intended postanalysis predictive interpretation of formal subjective probability that depends on accepting detailed prior assumptions, including specifically in the case of the fiducial argument that the distribution of an identified pivotal quantity is independent of, and hence unaffected by, the observations. Both fiducial inferences and Bayesian inferences stand or fall on case-by-case judgments of both models and independences assumed.

Was Fisher a "lapsed Fisherian" in the smoking and lung cancer debate of the late 1950s? He was certainly correct to draw attention to the potential for misleading selection biases affecting causal inferences from observational data. Are Fisher's innovative ideas "like randomization inference and conditionality... contradictory"? Not if one accepts that postdictive reasoning and predictive reasoning are complementary activities. Conditionality is important for estimation, but irrelevant for interpreting the outcome of a randomization test. Also, in Fisher's view of estimation, conditionality is not the blanket principle of Bayesian statistics. Indeed, selective conditioning was for Fisher a key to avoiding universal submission to Bayes. Fisher was unsuccessful in obvious ways, but he addresses deeper and harder questions than Neymanian theory attempts to solve. I believe we should downplay sound-bite criticisms coming from self-limiting theoretical perspectives, either frequentist or Bayesian. I do not wish to protest too much, however. Brad and I agree in seeking to resurrect Fisher from his relative obscurity in American academic statistics.

Rejoinder

Bradley Efron

One could scarcely ask for a clearer set of discussions, or a more qualified set of discussants, leaving me with very little to rejoin. The lecture itself was written quickly, in a few weeks, and barely revised for publication. My fear was that a careful revision would become just that, careful, and lose its force in an attempt to cover all Fisherian bases. The commentaries each add important ideas that I omitted, while scarcely overlapping with each other. Fisher's world must be a very high-dimensional one! Let me end here with just a few brief reactions to the commentaries.

(1) Paragraph (3) of Professor Cox's comments concerns Fisher's definition of probability, hinging on the absence of recognizable subsets. This is a crucial point for the relevance of statistical arguments to actual decision making, but it is not an easy criterion to apply. As an analog I like to imagine a field of intermixed orange and white poppies, with the proportion of orange poppies representing the probability of an event of interest. Perhaps the orange proportion increases steadily from east to west or from north to south or diagonally from corner to corner. A model-building program based on logistic regression or CART amounts to a systematic search for the field's inhomogeneities.

We are all used to basing statistical inferences on the outputs of such programs, saying perhaps that the estimated probability of an orange poppy at the field's center point is 0.20, but how does this "probability" relate to Fisher's definition? A different model, amounting to a different partitioning of the field, might give a different probability. My problem here is not with recognizability, which is obviously an important idea, but with its application to contexts where the statistician must choose which subsets might be recognizable.

(2) Professor Kass discusses one of the deepest problems of statistical philosophy: why are frequentist computations often useful and compelling, even for those who prefer the Bayesian paradigm? Fisher's work provides the best answers so far to that question but we are still a long way from having a satisfactory resolution. Kass offers a different hope, via Bayesian nonparametrics, for bridging the chasm. Going this route requires us to solve another deep problem, how to put uninformative prior distributions on high- (or infinite-) dimensional parameter spaces.

(3) On the same point, Professor Barndorff-Nielsen asks how we can reconcile fiducial inference with high-dimensional estimation results like the James–Stein theorem. For a multivariate version of formula (1), $x \sim N_K(\theta, \sigma^2 I)$, the original fiducial argument seems to say that $\theta | x \sim N_K(x, \sigma^2 I)$, a terrible answer if we are interested in estimating $\|\theta\|^2$, for example. The confidence density approach gives sensible results in such situations. My 1993 *Biometrika* paper argues, a la Kass, that this is a way of using frequentist methods to aid Bayesian calculations.

(4) Professor Hinkley notes the continued vitality of Tukey-style data analysis. In its purest form this line of work is statistics without probability theory (see, e.g., Mosteller and Tukey's 1977 book "*Data Analysis and Regression*") and as such I could not place it anywhere in the statistical triangle of Section 11. This is my picture's fault of course, not Tukey's. Problem-driven areas like neural networks often begin with a healthy burst of pure data analysis before settling down to an accommodation with statistical theory.

(5) I am grateful to Professor Fraser for presenting a more intelligible version of the magic formula. This was the spot in the talk where "avoiding technicalities" almost avoided coherency. "Trick" is a positive word in my vocabulary, reflecting a Caltech education, and I only wish I could think of some more Fisher-level tricks. Fisherian statistics was

not entirely missing from early 1960s Stanford: it flourished in the medical school, where Lincoln Moses and Rupert Miller ran the biostatistics training program.

(6) I resisted the urge to locate the discussants in the statistical triangle, perhaps because Professor Dempster has placed me closer to the frequentist corner than I feel comfortable with. Dempster emphasizes an important point: that Fisher's theory, more so than Neyman's, aimed at postdata interpretations of uncertainty. This appeared in almost pure Bayesian form with fiducial inference, but usually was expressed less formally, as in his interpretations of significance test p-values. (See Section 20 of the 1954 edition of *Statistical Methods for Research Workers*.)

One cannot consider Fisher without also discussing Neyman, and he is likely to come out of such discussions sounding like the bad guy. This is most unfair. To invert Kass's metaphor, Neyman played Niels Bohr to Fisher's Einstein. Nobody could match Fisher's intuition for statistical inference, but even the strongest intuition can sometimes go astray. Neyman–Wald decision theory was an heroic and largely successful attempt to put statistical inference on a sound mathematical foundation. Too much mathematical soundness can be stultifying, as Barndorff-Nielsen points out, but too much intuition can cross over into mysticism, and one can sympathize with Neyman's frustration over Fisher's sometimes Delphic fiducial pronouncements.

(7) Hinkley and Dempster (and others at the lecture) questioned whether randomization inference really contradicts the conditionality principle. I have to admit to using both ends of the contradiction myself without feeling any great amount of guilt, but the logical inconsistency still seems to be there. Isn't a randomly chosen experimental design an ancillary, and shouldn't we condition on it rather than basing inference on its random properties? Dempster's statement "Conditionality is important for estimation, but irrelevant for interpreting the outcome of a randomization test" seems to just restate the problem. We condition on a randomly selected sample size n (to borrow Professor Cox's famous example) whether the data is used for estimation or testing.

(8) Lumping subjective and objective Bayesianism together simplified my presentation, but maybe to a dangerous degree. It elicited Professor Lehman's objection quoted in Section 11. Perhaps it would be better to follow Professor Cox's preference for a square instead of a triangle.

Statistical philosophy is best taken in small doses. The discussants have followed this rule (even if I have not) with six excellent brief essays. I am grateful to them, to COPSS for the invitation to give the Fisher lecture, and to the Editors of *Statistical Science* for arranging this discussion.

ADDITIONAL REFERENCES

BENNETT, J. H. (1972). *Collected Papers of R. A. Fisher*. Univ. Adelaide Press.

BENNETT, J. H., ed. (1990). *Statistical Inference and Analysis. Selected Correspondence of R. A. Fisher*. Oxford Univ. Press.

COX, D. R. (1982). A remark on randomization in clinical trials. *Utilitas Math.* **21A** 245–252. (Birthday volume for F. Yates.)

DEMPSTER, A. P. (1971). Model searching and estimation in the logic of inference. In *Foundations of Statistical Inference* (V. P. Godambe and D. A. Sprott, eds.) 56–78. Holt, Rinehart and Winston, Toronto.

EFRON, B. (1971). Forcing a sequential experiment to be balanced. *Biometrika* **58** 403–417.

EFRON, B. (1987). Better bootstrap confidence intervals (with discussion). *J. Amer. Statist. Assoc.* **82** 171–200.

FISHER, R. A. (1928). Correlation coefficients in meteorology. *Nature* **121** 712.

FISHER, R. A. (1929). Statistics and biological research. *Nature* **124** 266–267.

FISHER, R. A. (1935). *Design of Experiments*. Oliver and Boyd, Edinburgh.

FISHER, R. A. (1956). *Statistical Methods and Scientific Inference*. Oliver and Boyd, Edinburgh. (Slightly revised versions appeared in 1958 and 1960.)

FISHER, R. A. (1958). The nature of probability. *Centennial Review* **2** 261–274.

GELMAN, A., MENG, X.-L. and STERN, H. (1996). Posterior predictive assessments of model fitness (with discussion). *Statist. Sinica* **6** 773–807.

HALD, A. (1981). T. N. Thiele's contributions to statistics. *Internat. Statist. Rev.* **49** 1–20.

LANE, D. A. (1980). Fisher, Jeffreys and the nature of probability. *R.A. Fisher: An Appreciation Lecture Notes in Statist.* **1**. Springer, New York.

MAHALANOBIS, P. C. (1938). Professor Ronald Aylmer Fisher. *Sankhyā* **4** 265–272.

NEYMAN, J. (1977). Frequentist probability and frequentist statistics. *Synthese* **36** 97–131.

RIPLEY, B. D. (1996). *Pattern Recognition and Neural Networks*. Cambridge Univ. Press.

SAVAGE, L. J. (1976). On rereading R. A. Fisher. *Ann. Statist.* **4** 441–483.

YATES, F. (1970). *Experimental Design. Selected Papers of Frank Yates, C.B.E., F.R.S.* Griffin, London.

19

Empirical Bayes Analysis of a Microarray Experiment

Introduction by Rafael Irizarry
Johns Hopkins University

Microarrays are an example of the powerful high-throughput genomics tools that revolutionized the measurement of biological systems. This technology measures the quantity of nucleic acid molecules present in a sample. The most popular example is the measurement of gene expression. To do this we take advantage of hybridization properties of nucleic acid and use complementary molecules attached to a solid surface, referred to as *probes*. The molecules in the target are labeled and a specialized scanner is used to measure the amount of hybridization at each probe, reported as a *feature* intensity. A defining characteristic of microarray technology is that it includes thousands of probes in a relatively-small surface, such as a glass slide. The raw or *feature-level* data are the intensities read from the different probes on the array (CF 1999, 2002). A typical dataset consists of a large table with rows representing genes and columns representing samples. Statisticians have made great contributions to the analysis of microarray data. In fact, these days it is rare to find a competitive high-throughput biology lab that does not collaborate with or employ a statistician. This was not true five years ago and this paper played an important role in this change.

Although the first microarray papers started to appear in 1995 (Schena et al. 1995), a CIS search shows only two methodology papers published in the Statistics literature before 2002. Efron et al.'s paper is one of them (the other is by Kerr and Churchill 2001). This paper deals with the problem of finding genes that are differentially expressed in different populations (e.g., tumor versus normal, 2×2 factorial design, etc.). For the last few years this has been the most common application in academia. Before this seminal paper was published, the state of the art among biologists was to find groups of genes that behaved differently across the populations using clustering algorithms. Others would decide on arbitrary cut-offs, make all pair-wise comparison across populations, and identify as interesting genes those with most comparisons surpassing the cut-off. Professor Efron and colleagues quickly realized that by focusing on one gene at a time, the problem was reduced to a simple regression/ANOVA problem. In microarray papers comparing two populations, gene-specific t-tests were an obvious first approach that most seemed to have missed. Today, t-test type statistics, such as the one described in this paper, are standard practice for identifying differentially-expressed genes in two populations.

An important insight described in this paper is that one can greatly improve gene-specific test statistics by incorporating across gene information. In this paper a simple, yet very useful, idea was first proposed in the statistics literature: add a constant to the denominator of the t-test to reduce its dependence on the standard error estimate. A PNAS paper presenting this idea appeared in the same year (Tusher et al. 2001). In Efron's paper, the constant was defined as the 90th percentile of the genewise standard error estimates. Many of the approaches that are widely used today shrink the standard error as first done by Efron et al. (Tusher et al. 2001; Smyth 2004).

This paper also pioneered the implementation of useful inference in an application where it was desperately needed. This was not an easy problem due to the fact that thousands of simultaneous tests are performed. A nonparametric empirical Bayes analysis, which provided a posterior probability of differential expression for each gene, was proposed. A clever implementation of a two-group model to describe the behavior of interesting (differentially-expressed) and uninteresting genes was perhaps the most insightful contribution of this work. In fact, two papers (Storey 2002), including this one, laid the groundwork for the development of the multiple comparison correction methods currently used by most practitioners.

Finally, I would like to point out that at the time this paper was published, the field of microarray data analysis was dominated by overly-complicated methods. In many cases the developers were simply rediscovering the wheel. The approach followed by Professor Efron and colleagues was different: they demonstrated that the clever use and extension of existing statistical methodology could result in extremely-useful data analysis tools. Most of the currently-successful developers of data analysis tools for this field seemed to have followed their lead.

References

Chipping Forecast (1999). *Nature Genetics* **21** (supplement).

Chipping Forecast (2002). *Nature Genetics* **32** (supplement).

Schena, M., Shalon, D., Davis, R. W. and Brown, P. O. (1995). Quantitative monitoring of gene expression patterns with a complementary DNA microarray. *Science* **270**, 467–470.

Kerr, M. Kathleen and Churchill, Gary A. (2001). Experimental design for gene expression microarrays. *Biostatistics* **2**, 183–201.

Tusher, V. G., Tibshirani, R. and Chu, G. (2001). Significance analysis of microarrays applied to the ionizing radiation response. In *Proceedings of the National Academy of Sciences* **98**, 5116–5121.

Smyth, Gordon K. (2004). Linear models and empirical Bayes methods for assessing differential expression in microarray experiments. *Statistical Applications in Genetics and Molecular Biology* **3**, 1–27. [Available at: `http://www.bepress.com/sagmb/vol3/iss1/art3`.]

Storey, John D. (2002). A direct approach to false discovery rates *Journal of the Royal Statistical Society, Series B: Statistical Methodology* **64**, 479–498.

Empirical Bayes Analysis of a Microarray Experiment

Bradley EFRON, Robert TIBSHIRANI, John D. STOREY, and Virginia TUSHER

Microarrays are a novel technology that facilitates the simultaneous measurement of thousands of gene expression levels. A typical microarray experiment can produce millions of data points, raising serious problems of data reduction, and simultaneous inference. We consider one such experiment in which oligonucleotide arrays were employed to assess the genetic effects of ionizing radiation on seven thousand human genes. A simple nonparametric empirical Bayes model is introduced, which is used to guide the efficient reduction of the data to a single summary statistic per gene, and also to make simultaneous inferences concerning which genes were affected by the radiation. Although our focus is on one specific experiment, the proposed methods can be applied quite generally. The empirical Bayes inferences are closely related to the frequentist false discovery rate (FDR) criterion.

1. INTRODUCTION

Through the use of DNA microarrays, a novel technology, it is now possible to obtain quantitative measurements for the expression of thousands of genes present in a biological sample. DNA microarrays have been used to monitor changes in gene expression during important biological processes (e.g., cellular replication and the response to changes in the environment), and to study variation in gene expression across collections of related samples (e.g., tumor samples from patients with cancer). A major statistical task is to understand the structure of the data from such studies, which often consist of measurements on thousands of genes in dozens of conditions.

This article concerns the use of microarrays in a comparative experiment, where it is desired to compare gene expression under Treatment versus Control conditions. We wish to identify which of several thousand candidate genes have had their expression levels changed, either positively or negatively, by the Treatment. Answering this question requires an efficient data reduction strategy, because microarrays deliver megabytes of information, and also statistical inference techniques that deal with the difficulties of simultaneous inference on thousands of genes. We discuss both problems here, working in the context of an experiment on radiation sensitivity discussed later.

The statistics literature for microarrays, still in its infancy and with much of it unpublished, has tended to focus on frequentist data-analytic devices such as cluster analysis, bootstrapping, and linear models (see Li and Wong 2000; Kerr and Churchill 2000; Black and Doerge 2000; Van del Laan and Bryan 2000; Eisen, Spellman, Brown, and Botslein 1998). Parametric Bayesian modeling was featured in Newton, Kendziorski, Richmond, Blatter, and Tsui (2000) and to a less extent in Lee, Kuo, Whitmore, and Sklar (2000). Multiple comparison techniques, designed to control error rates in thousands of simultaneous hypotheses tests, were explored in Dudoit, Yang, Callow, and Speed (2000). Tusher, Tibshirani, and Chu (2001) approached the simultaneity problem through the method of false discovery rates (FDRs), as discussed later.

Our inferences here will be based on a simple nonparametric empirical Bayes model. The model produces useful *a posteriori* probabilities of effect for the individual genes, with a minimum of prior assumptions. It also connects well with Benjamini and Hochberg's (1995) frequentist theory of FDRs, as discussed in Section 5. Besides being useful in its own right, the empirical Bayes model helps to select from among competing data reduction schemes, a crucial point in dealing with the massive datasets microarrays produce.

Here is some background on microarrays in general and the specific experiment analyzed in this article. Virtually all living cells contain chromosomes, large pieces of DNA containing hundreds or thousands of genes, each of which specifies the composition and structure of a protein (or sometimes several related proteins). Protein polymers of amino acids are the workhorse molecules of the cell, responsible, for example, for cellular structure, producing energy and important biomolecules like DNA and proteins, and for reproducing the cell chromosomes. Every cell in an organism has nearly the same set of chromosomes, and thus contains the same repertoire of proteins. However, cells have remarkably distinct properties, such as the differences between human eye cells, hair cells and liver cells, distinctions that are the result of differences in the abundance, distribution, and state of the cell proteins. One of the seminal discoveries of molecular biology was that these changes in protein abundance are determined in part by changes in the levels of messenger RNA (mRNA), small and relatively unstable nucleic acid polymers that shuttle information from chromosomes to the cellular machines that synthesize new proteins. Thus, there is a logical connection between the state of a cell and the details of its protein and mRNA composition.

Whereas it remains difficult to measure the abundances of a cell's proteins, the recently developed DNA microarray makes it possible to quickly and efficiently measure the relative representation of each mRNA species in the total cellular mRNA population, or in more familiar terms to measure gene expression levels.

There are two major kinds of microarrays. In an oligonucleotide array, the kind featured in this article, there are 20 probe pairs (pm, mm) for each gene. The perfect match (pm)

Bradley Efron, Department of Statistics and Division of Biostatistics, Stanford University, Stanford, CA 94305 (E-mail: *brad@stat.stanford.edu*). Robert Tibshirani, Division of Biostatistics, Department of Health Research and Policy and Department of Statistics, Stanford University, Stanford CA 94305 (E-mail: *tibs@stat.stanford.edu*). John D. Storey, Department of Statistics, Stanford University, Stanford CA 94305 (E-mail: *jstorey@stat.stanford.edu*). Virginia Tusher, Department of Biochemistry, Stanford University, Stanford CA 94305 (E-mail: *tusher@cmgm.stanford.edu*). We are grateful to Dr. Gilbert Chu of the Stanford Biochemistry Department for sharing his ideas and data with us, and thank the Editor and Associate Editor for suggestions that improved the manuscript.

Journal of the American Statistical Association
December 2001, Vol. 96, No. 456, Applications and Case Studies

probe is designed to match a small subsequence of the gene about 25 bases long. The mismatch (mm) probe is a control, being identical to pm except with the middle base flipped to its complement. An experimental sample is hybridized on the microarray, and the RNA expression of the gene is estimated by the difference in signal pm–mm averaged over the 20 probe pairs. There is some concern that subtracting mismatch numbers may actually degrade the inferences, a question we consider in this article.

In a spotted cDNA microarray, the other major variety, one base sequence matching all or part of a gene is printed on a glass slide. The experimental sample is labeled with red dye and hybridized on the slide. As a control, a reference sample is labeled with green dye and hybridized on the same slide. Using a fluorescent microscope, the log (red–green) intensities of RNA hybridization at each site are measured. The red–green microarray is featured in much of the recent literature, see Newton et al. (2001), Dudoit et al. (2000), and Lee et al. (2001). Our discussion, like that in Li and Wong (2000) centers in the Affymetrix oligonucleotide microarray, but similar analysis problems arise for both types of array. However, our Empirical Bayes procedure, summarized in Algorithm 1, can be applied quite generally. An example extending the empirical Bayes analysis to a cDNA microarray experiment appears in Remark D of Section 6, showing how our methods can be applied to other experimental situations.

From either type of microarray, we obtain several thousand expression values, one or many for each gene. Microarrays in current use measure anywhere from 1,000 to 25,000 genes; larger ones will soon be available. In a typical study, a number of experimental samples are each hybridized to a different microarray to learn about gene expression differences across different conditions. For example Alizadeh et al. (2000) studied gene expression patterns from tissue samples from a number of lymphoma patients and related gene expression to patient survival. Clustering methods (Eisen et al. 1998) were the main tool used in that article, and in a number of other similar studies. Here we will be interested in the more familiar statistical task of comparing Treatment and Control arrays, though carried out in a novel setting.

Our particular dataset comes from a set of eight oligonucleotide microarrays in an experiment designed by Professor Gilbert Chu of the Stanford Biochemistry Department to study transcriptional responses to ionizing radiation. Some cancer patents have severe life-threatening reactions to radiation treatment. It is important to understand the genetic basis of this sensitivity, so that such patients can be identified before the treatment is given. The eight microarrays were labeled

$$(U1A, U1B, I1A, I1B, U2A, U2B, I2A, I2B), \quad (1.1)$$

the labels indicating the following experimental design: RNA was harvested from two wild-type human lymphoblastoid cell lines, designated "1" and "2," growing in an unirradiated state "U," or in an irradiated state "I." RNA samples were labeled and divided into two identical aliquots for independent hybridizations, "A" and "B." Each microarray provided expression estimates for 6,810 genes. Further experimental details appear in Remark A of Section 6.

Here is the article's plan: the data structure of the radiation experiment is described in Section 2. This sets up the main thrust of the article, the efficient reduction of microarray data (320 numbers per gene in this case) to a single summary statistic "Z_i" for each gene, followed by an appropriate simultaneous inference for the activity of each gene based on all the Z scores. Section 3 presents the simple nonparametric empirical Bayes model used to make our simultaneous inferences. The model is presented in algorithmic form, suggesting how it can be applied to other microarray comparative experiments, both oligoneucliotide and cDNA types. (Another such experiment is briefly discussed in Section 6.)

Section 4 concerns the efficient reduction of the data to a single score Z_i per gene. The reduction makes use of the empirical Bayes model, essentially selecting mappings that maximize the amount of Bayesian information preserved in Z_i. Frequentist justification of the empirical Bayes approach appears in Section 5, where it is related to Benjamini and Hochberg's (1995) theory of FDR. Section 6 closes with a summary and some detailed remarks, including a comparison of our analysis with a "gold standard" assay of some of the genes.

2. THE DATA

Microarray experiments produce enormous amounts of data, more than two million feature numbers in the relatively small experiment we are discussing here. The statistical task is to efficiently reduce these numbers to simple summaries of the genes' activities. One goal in this article is to provide a method for comparing the statistical efficiency of different data reduction strategies.

Here is a description of the data in the radiation experiment, and the notation we will use to describe it. Expression levels were recorded for 6,810 different genes,

$$\text{genes:} \quad i = 1, 2, \ldots, n = 6{,}810. \quad (2.1)$$

(There were actually 7,129 genes, 319 of which had some missing data. For convenience, this article considers only the 6,810 genes having complete data. The various analyses were also carried out on all 7,129 genes, with nearly identical results.) Each gene on each plate was represented by 20 oligonucleotide "probes,"

$$\text{probes:} \quad j = 1, 2, \ldots, J = 20. \quad (2.2)$$

Finally there were eight plates, (the individual microarrays) representing the eight experimental conditions of the experiment described in section 1, (U1A, U1B, I1A, I1B, U2A, U2B, I2A, I2B),

$$\text{plates:} \quad k = 1, 2, \ldots, K = 8. \quad (2.3)$$

Two features were recorded for each probe of each gene on each plate, a "perfect match number" pm_{ijk} and a "mismatch number" mm_{ijk}, the latter referring to a deliberately distorted version of the oligonucleotide included as a control. Table 1 shows the 20 pairs of numbers for gene $i = 2{,}715$ on plate $k = 1$.

We will investigate three separate stages of data reduction: "probe reduction", the mapping that takes the 20 probe pair

Table 1. The 20 Pairs of Perfect Match and Mismatch Feature Numbers for Gene i = 2,715 on Plate k = 1 (U1A)

Probe	1	2	3	4	5	6	7	8	9	10
pm	1,054	3,242	1,470	4,050	1,356	1,476	561	606	1,307	1,057
mm	793	2,333	826	1,912	561	558	942	526	699	1,060
Probe	11	12	13	14	15	16	17	18	19	20
pm	974	1,584	802	1,399	1,670	2,514	2,096	6,592	5,662	2,244
mm	829	1,771	601	569	840	950	700	8,717	1,484	668

numbers into a single expression value "M_{ik}" for gene i on plate k,

$$\text{probe reduction:}\quad \{(\text{pm}_{ijk}, \text{mm}_{ijk}), j = 1, 2, \ldots, 20\} \to M_{ik}, \tag{2.4}$$

"gene reduction," the mapping that takes the $K = 8$ expression values M_{ik} for gene i into a single expression score "Z_i,"

$$\text{gene reduction:}\quad \{M_{ik}, k = 1, 2, \ldots, 8\} \to Z_i, \tag{2.5}$$

and finally an *inference mapping* that re-expresses Z_i in terms of a statistical inference concerning i activity of gene. The nonparametric empirical Bayes analysis of Section 3 will provide inferences of the form Prob $\{\text{Event}_i | Z_i\}$, where Event$_i$ is an event of interest such as "gene i's activity was affected by radiation." Section 5 connects these probabilities with the frequentist FDR criteria of Benjamini and Hochberg (1995).

There are of course an unlimited selection of possible data reductions from the original data, 320 numbers per gene in the radiation experiment, to the expression scores Z_i. For reasons explained in Section 4, the empirical Bayes analysis will lead us to prefer the following choices: For the probe reduction let

$$M_{ik} = \text{mean}\{\log(\text{pm}_{ijk}) - .5 \cdot \log(\text{mm}_{ijk}), j = 1, 2, \ldots, 20\}. \tag{2.6}$$

For the gene reduction, first compute the four differences $(D_{i1}, D_{i2}, D_{i3}, D_{i4})$ between the irradiated and unirradiated values within the same wild-type sample and aliquot, for example

$$D_{i1} = M_{i3} - M_{i1}, \tag{2.7}$$

the difference between the I1A and U1A values M_{ik}. Then take

$$Z_i = \overline{D}_i/(a_0 + S_i), \tag{2.8}$$

where $\overline{D}_i$ is the average of the four differences, S_i is their sample standard deviation, and a_0 is the 90th percentile of the 6,810 S values. Specifications (2.6)–(2.8) will be used as a comparison point in all of our numerical examples. They will be compared with other choices in Section 4, including the current one included in the Affymetrix software.

3. EMPIRICAL BAYES INFERENCES

Besides analyzing the radiation data, our goal here is to provide data analytic techniques useful in a variety of microarray situations. With generality in mind we will avoid highly specified models, relying instead on a simple inference model that is likely to apply to most comparative experiments: that a gene is either affected on unaffected by the treatment of interest, radiation in our case, giving two possible distributions for the expression score "Z," (2.5). Lee et al. (2000) used a normal theory version of this idea, as, less directly, did Li and Wong (2000). Newton et al. (2000) focused on Gamma models. Here we will avoid parametric assumptions. The resulting nonparametric empirical Bayes analysis, which provides a posteriori probabilities of effect for the various genes, is further justified in Sections 4–6.

Let

$$p_1 = \text{probability that a gene is affected,}$$
$$p_0 = 1 - p_1 = \text{probability unaffected,} \tag{3.1}$$

and

$$f_1(z) = \text{the density of } Z \text{ for affected genes}$$
$$f_0(z) = \text{the density of } Z \text{ for unaffected genes.} \tag{3.2}$$

Then

$$f(z) = p_0 f_0(z) + p_1 f_1(z) \tag{3.3}$$

is the mixture density of the two populations. In our situation, we can estimate $f(z)$ directly from the 6,810 expression scores Z_i obtained from the data reduction (2.4) and (2.5).

In the absence of strong parametric assumptions such as normality, model (3.3) is useless without an estimate of the "null density" $f_0(z)$. Fortunately, it is easy to obtain such estimates. What follows is the method we used to estimate $f_0(z)$ in the radiation experiment. Section 6 discusses variants of this method that are applicable more generally.

The $6,810 \times 8$ matrix $\mathbf{M}$ of expression values (2.4), one value for each gene on each plate, gives a $6,810 \times 4$ matrix $\mathbf{D}$ of differences between the irradiated and unirradiated expression values, as in (2.7). Let $\mathbf{M}_k$ indicate the kth column of $\mathbf{M}$, a 6,810 vector. With the plates ordered as before, (U1A, U1B, I1A, I1B, U2A, U2B, I2A, I2B), the "difference matrix" $\mathbf{D}$ is

$$\mathbf{D} = (\mathbf{M}_3 - \mathbf{M}_1, \mathbf{M}_4 - \mathbf{M}_2, \mathbf{M}_7 - \mathbf{M}_5, \mathbf{M}_8 - \mathbf{M}_6). \tag{3.4}$$

Symbolically, the vector $\mathbf{Z}$ of expression scores (2.5) is obtained via

$$\underset{6,810\times 20\times 2\times 8}{\{\text{original data}\}} \to \underset{6,810\times 8}{\mathbf{M}} \to \underset{6,810\times 4}{\mathbf{D}} \to \underset{6,810}{\mathbf{Z}}. \tag{3.5}$$

Now let the "null difference matrix" $\mathbf{d}$ be the $6,810 \times 4$ matrix obtained by differencing within the aliquot splits,

$$\mathbf{d} = (\mathbf{M}_2 - \mathbf{M}_1, \mathbf{M}_4 - \mathbf{M}_3, \mathbf{M}_6 - \mathbf{M}_5, \mathbf{M}_8 - \mathbf{M}_7), \tag{3.6}$$

so for example the first column of $\mathbf{d}$ records differences between the B and A splits of the unirradiated wild-type 1 experiments. We define "null scores" $\mathbf{z} = (z_1, z_2, \ldots, z_{6,810})'$ by

$$\{\text{original data}\} \rightarrow \mathbf{M} \rightarrow \mathbf{d} \rightarrow \mathbf{z}, \tag{3.7}$$

with the understanding that except for the substitution of $\mathbf{d}$ for $\mathbf{D}$, the arrows in (3.7) indicate the same mappings as in (3.5).

We will use the empirical distribution of the null scores $\{z_i\}$ to estimate the null density $f_0(z)$ in (3.3). One could just as well take $\mathbf{M}_1 - \mathbf{M}_2$ as $\mathbf{M}_2 - \mathbf{M}_1$ in (3.6), etc., and in fact our numerical algorithm employs random sign permutations of the columns of $\mathbf{d}$ to improve the estimation of f_0. The basic idea here, that we can recover the "null hypothesis" from differences that negate treatment effects, shows up in one form or another in many of the microarray references, being essentially unavoidable in a comparative experiment. Further discussion appears in Section 6, which describes strategies that might be used for estimating f_0 in situations less intricate than the radiation experiment.

An application of Bayes' rule to the mixture model (3.3) gives the a posteriori probabilities $p_1(Z)$ and $p_0(Z)$ that a gene with score Z was affected or unaffected by the treatment:

$$\text{Bayes' Rule:} \quad p_1(Z) = 1 - p_0 f_0(Z)/f(Z)$$

and

$$p_0(Z) = p_0 f_0(Z)/f(Z). \tag{3.8}$$

The ratio $f_0(Z)/f(Z)$ can be estimated directly from the $\{Z_i\}$ and $\{z_i\}$ empirical distributions. The probabilities p_0 and $p_1 = 1 - p_0$ are unidentifiable without strong parametric assumptions, but this will turn out to be less problematic than it might seem. The constraint that $p_1(Z)$ be nonnegative for all Z does restrict p_0 and p_1,

$$p_1 \geq 1 - \min_Z \{f(Z)/f_0(Z)\}$$

and

$$p_0 \leq \min_Z \{f(Z)/f_0(Z)\} \tag{3.9}$$

A more stable bound for p_1 and p_o is given in Remark F of Section 6.

Figure 1 displays the Bayesian inference curve $p_1(Z) = \text{Prob}\{\text{Event}|Z\}$ obtained from the probe and gene data reductions (2.6), (2.8). It was constructed as follows (skipping some technical details that appear in Section 6):

Algorithm 1: Empirical Bayes analysis for microarrays

(a) Compute the scores $\{Z_i\}$ according to (3.5), using probe reduction (2.6) and gene reduction (2.8).
(b) Compute the null scores $\{z_i\}$ in the same way, beginning with (3.7). Generate 20 versions of the $\{z_i\}$, based on 20 independent row-wise sign permutations of $\mathbf{d}$ (see Remark D).
(c) Use logistic regression to estimate the ratio $f_0(z)/f(z)$ based on the relative densities of the $\{Z_i\}$ to the $\{z_i\}$ (see Remark C).
(d) Use relationship (3.9) to obtain an estimated upper bound for p_0: here $p_0 = .811$ (See Remark F.)
(e) For each gene compute Prob{Event|Z} from (3.8), with f_0/f estimated from the logistic regression, and p_0 equaling its estimated maximum value (or more conservatively with $p_0 = 1$.)

We have focussed on our particular experimental setup, but Algorithm 1 is quite general. It can be applied to any two-class situation, for example two sets of unpaired samples. All that changes is the generation of null scores z_i in step (b). For instance, for unpaired samples the values of z_i would be generated by random permutations of the column labels "1" and "2." Remark E of Section 6 gives another example.

Our Bayesian analysis is actually "empirical Bayes" in the sense that the crucial ratio $f_0(Z)/f(Z)$ in (3.8) is estimated

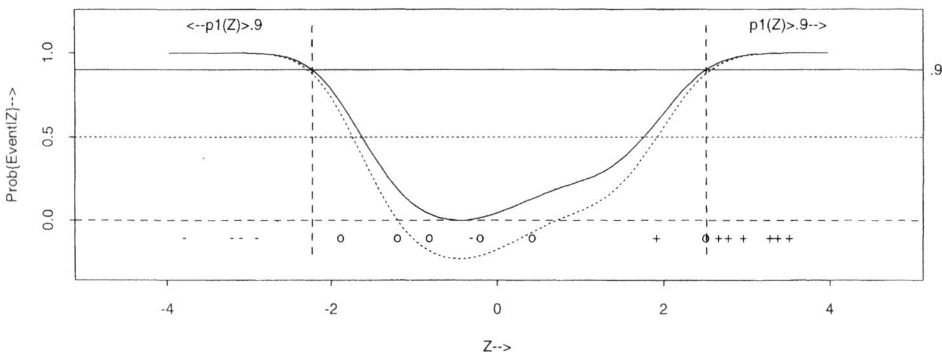

Figure 1. Solid curve: Bayesian Inference Mapping Prob{Event$_i$|Z$_i$} From Data Reductions (2.6), (2.8); Event$_i$ is "Gene i Affected by Radiation." Symbols show Z values for 18 genes separately analyzed by Northern Blot: "+" positively affected; "−" negatively affected"; "o" not affected. Dotted curve is lower bound (3.10).

from the data rather than from *a priori* assumptions. Newton et al. (2000) carried out a similar analysis, but using specific Bayesian modeling assumptions beyond (3.1)–(3.3).

In Figure 1, the *a posteriori* probability of being affected is seen to increase as Z or $-Z$ grows large. The positive end of the Z-axis corresponds to genes turned *on* by the radiation, with their expression values increased, while negative values of Z indicate decreased expression under radiation. Out of the 6,810 genes 127 genes had $p_1(Z)$ exceeding .90, more on the negative than positive end of the Z scales.

Eighteen of the 6,810 genes were independently assessed by a Northern Blot analysis, a pre microarray assay that serves here as a gold standard for gene expression. Seven of these, indicated by "+" in Figure 1, were deemed "affected positively by radiation," five indicated by "−" were "affected negatively," and six indicated by "o" were "not affected." There is good agreement between the Northern Blot assessments and the probabilities assigned in Figure 1. The full results, given in Section 6, show a high correlation between the gold standard and our results.

In comparing different data reductions, it is convenient to always have the same marginal distribution for Z. To this end, the raw scores $\{Z_i\}$ from (2.8) were monotonically transformed to have a nearly perfect $N(0, 1)$ distribution, say by transformation $m(Z)$, and then the null scores were transformed according to the same $m(z)$. Notice that the crucial ratio $f_0(z)/f(z)$ remains the same under such transformations, so that $p_1(Z)$ and $p_o(Z)$ in (3.8) are transformation invariant. We will always make the empirical distribution of the $\{Z_i\}$ almost perfectly $N(0, 1)$, using a normal scores transformation, implying for example that $42 = 6{,}810 \cdot (1 - \Phi(2.5))$ of the 6,810 genes have $Z_i > 2.5$, with Φ the standard normal cumulative distribution function.

Figure 2 shows the estimates of f_0, f_1, and f contributing to Figure 1; $f(Z)$ is a standard $N(0, 1)$ density, by construction, while $f_0(z)$ is a less dispersed density. This is what we hoped for of course: the values of Z should be more dispersed than the values of z because they reflect the disturbing effects of the radiation treatment. The large values of Prob{Event|Z} in the tails of Figure 1 come from (3.8), and the small ratio of $f_0(z)$ to $f(Z)$. A good choice of data reductions makes $f_0(z)/f(z)$ small for $|z|$ large, and we will use this criterion to guide our choices of the probe and gene reductions in Section 4.

Looking again at (3.8),

$$p_1(Z) \geq 1 - f_0(Z)/f(Z), \tag{3.10}$$

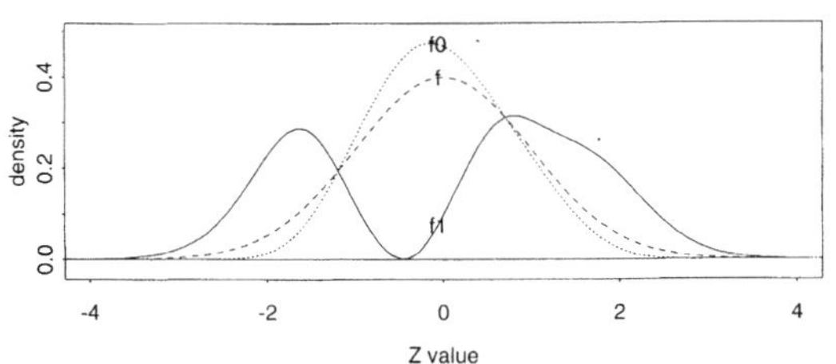

Figure 2. Estimates of $f(\cdot)$, $f_0(\cdot)$ and $f_1(\cdot)$ for the Situation of Figure 1, Model (3.1)–(3.3); $p_0 = .811$, its Estimated Maximum From (3.9), Used in the Construction of f_1.

because this corresponds to $p_0 = 1$, the largest possible value. The dotted curve in Figure 1 is $1 - f_0(Z)/f(Z)$. This is not much less than the solid curve for large values of $|Z|$, giving 106 genes with $p_1(Z) \geq .90$.

4. EFFICIENT DATA REDUCTIONS FOR MICROARRAY EXPERIMENTS

The empirical Bayes analysis of Section 3 depends on a drastic data reduction: from the full vector $\mathbf{v}_i$ of data for gene i, a 320-vector in the radiation experiment, to a single number Z_i (and its null counterpart z_i.) Information is bound to be lost in the mapping from $\mathbf{v}_i$ to Z_i, but the less we lose the more powerful will be the analysis, and better our chance of detecting genuinely affected genes.

To state things more exactly, we can imagine applying model (3.1)–(3.2) to the 320-dimensional densities of $\mathbf{v}$,

$$f_1^{\mathbf{v}}(\mathbf{v}) = \text{density of } \mathbf{v} \text{ for affected genes}, \tag{4.1}$$
$$f_0^{\mathbf{v}}(\mathbf{v}) = \text{density of } \mathbf{v} \text{ for unaffected genes},$$

and

$$f^{\mathbf{v}}(\mathbf{v}) = p_0 f_0^{\mathbf{v}}(\mathbf{v}) + p_1 f_1^{\mathbf{v}}(\mathbf{v}), \tag{4.2}$$

the mixture density; p_0 and p_1 have the same meaning here as in (3.1). Defining the likelihood ratio statistic $R^{\mathbf{v}}(\mathbf{v}) = f^{\mathbf{v}}(\mathbf{v})/f_0^{\mathbf{v}}(\mathbf{v})$, Bayes theorem gives

$$p_1^{\mathbf{v}}(\mathbf{v}_i) = 1 - p_0/R^{\mathbf{v}}(\mathbf{v}_i) = \text{Prob}\{\text{gene } i \text{ affected}|\mathbf{v}_i\}, \tag{4.3}$$

compared to $p_1(Z_i) = 1 - p_0/R(Z_i)$ in (3.8), where $R(Z_i) = f(Z_i)/f_0(Z_i)$.

In our situation it is not practical to estimate the high-dimensional densities $f^{\mathbf{v}}(\mathbf{v})$ and $f_0^{\mathbf{v}}(\mathbf{v})$, at least not without extensive modeling. However, we can easily estimate the corresponding densities $f(Z)$ and $f_0(Z)$ for a one-dimensional statistic Z. The goal is to choose a mapping $Z = s(\mathbf{v})$ that does not lose much information. Information loss manifests itself by reductions in the likelihood ratio $R(Z_i)$, compared to $R^{\mathbf{v}}(\mathbf{v}_i)$, which reduces the number of genes having convincingly large values at $p_1(Z_i)$.

4.1 Estimation of a_0

With this background in mind we searched for mappings $Z = s(\mathbf{v})$ that produced large values of $R(Z)$, i.e., good separation between $f(Z)$ and $f_0(Z)$ as in Figure 2. Figure 3 shows the part of the search relating to the choice of a_0 in the denominator of (2.8). The curve marked "90" is equivalent to the dashed curve in Figure 1, the difference here being that the vertical axis is plotted on the logit scale, $\log p_1(Z)/(1 - p_1(Z))$, to emphasize differences in the tails. Keeping probe reduction (2.6) fixed, Figure 3 compares five different choices of a_0 in the gene reduction (2.8): a_0 equal to the 90th percentile of the 6,810 S_i values; the 50th percentile; the 5th percentile; $a_0 = 0$; and $a_0 \to \infty$. The choice $a_0 = 0$ makes Z_i in (2.8) proportional to the one-sample t-statistic for the four differences $(D_{i1}, D_{i2}, D_{i3}, D_{i4})$, whereas $a_0 \to \infty$ makes Z_i equivalent to the numerator $\overline{D}_i$. The plotted curves

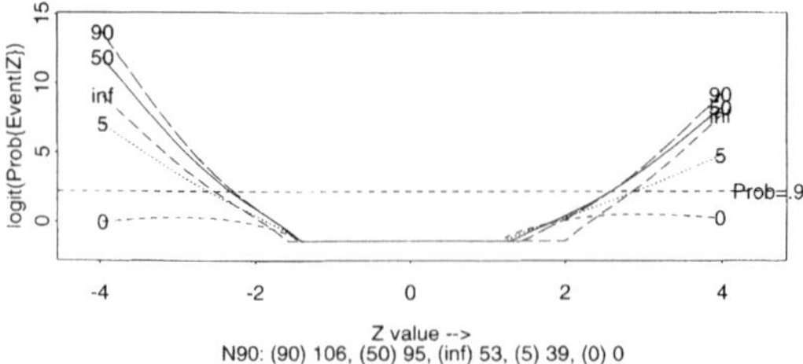

Figure 3. Choice of a_0 in the Gene Mapping $Z_i = \bar{D}_i/(a_0 + S_i)$, (2.8); Vertical Axis is Logit of Prob{Event|Z}, Estimated as in (3.8) With $p_0 = 1$; "90" Indicates a_0 Equaling 90th Percentile of the 6,810 S_i Values, etc.; "inf" is Limit as $a_0 \to \infty$. We see that 90 is the best choice in terms of maximizing Prob{Event|Z} for large |Z|; $a_0 = 0$ is worst. All choices used probe reduction (2.6). The vertical axis is truncated at lower bound Prob{Event|Z} = .20. ***N90*** *is the number genes having Prob{Event|Z} ≥ .90.*

are the logits of (3.10), the conservative lower bound for $p_1(Z)$, (taking $p_0 = 1$ in (3.8).)

Figure 3 shows that the best choice for a_0 is the one we used before, a_0 the 90th percentile. This manifests itself as higher values of Prob{Event|Z} at both ends of the Z scale. The density $f_0(z)$ in Figure 2 is more concentrated around zero than it is say for the disastrous choice $a_0 = 0$, raising $f(Z)/f_0(Z)$ in the tails and thus $p_1(Z)$, (3.10). The numbers N90 in Figure 3 indicate the number of genes having lower bound (3.10) for $p_1(Z_i)$ greater than .90. These range downward from 106 for $a_0 = 90$ to 0 for $a_0 = 0$. Larger values of $N90$ indicate less information loss in going from the full data vector $\mathbf{v}_i$ to the summary statistic Z_i. (Efron, Tibshirani, Goss, and Chu (2001) also use Kulback–Liebler distance to measure information loss.)

4.2 Choosing the Probe Reduction

The GeneChip software distributed by Affymetrix uses a simple average difference, (with some outlier rejection) to estimate what we called the probe reduction in (2.4), the expression for gene i on plate k : $M_{ik} = \text{mean}_j\{\text{pm}_{ijk} - \text{mm}_{ijk}\}$. However, this choice is controversial, and some researchers have suggested that ignoring the mismatch entirely might produce better expression estimates. We investigate the issue here.

Keeping the gene reduction fixed as in (2.6), $a_0 = .90$, Figure 4 compares probe reductions of the form

$$M_{ik} = \text{mean}_j\{s(\text{pm}_{ijk}) - c \cdot s(\text{mm}_{ijk})\}, \tag{4.4}$$

with s either the log function or the identity function. For example curve 2 in the left panel uses $M_{ik} = \text{mean}_j\{\text{pm}_{ijk} - \text{mm}_{ijk}\}$ whereas the dotted curve in the right panel uses $M_{ik} = \text{mean}_j\{\log(\text{pm}_{ijk})\}$. Our preferred choice (2.6)–(2.8) is curve 1, "$c = .5$ & logs." The "Affy" curve in the left panel was based on the algorithm provided by Affymetrix, which is similar to the "$c = 1$ no logs" choice, but with a provision for removing apparent outliers among the 20 $\text{pm}_{ijk} - \text{mm}_{ijk}$ differences before averaging.

Figure 4 indicates a substantial advantage to taking logs, and a mild advantage to using $c = .5$ rather than $c = 1$ or $c = 0$. The comparison between $c = .5$ and $c = 1$ is close on the log scale, but other comparisons, reported in Efron et al. (2000), reinforce the superiority of $c = .5$. We also tried using various L-estimators in (4.7), including trimmed means. When applied on the log scale, this form of robustification made almost no difference to our results.

Some comments are in order, which apply to the whole section:

- There is no claim that the mapping $Z = s(\mathbf{v})$ described by (2.6), (2.8) is "correct," only that it is relatively efficient in preserving the information in $\mathbf{v}$. The estimated curve

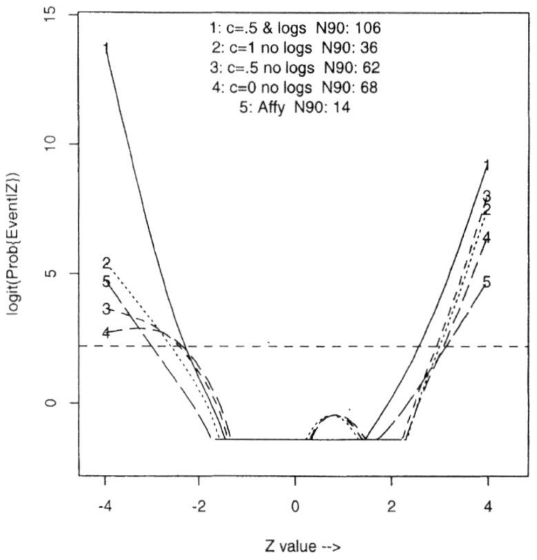

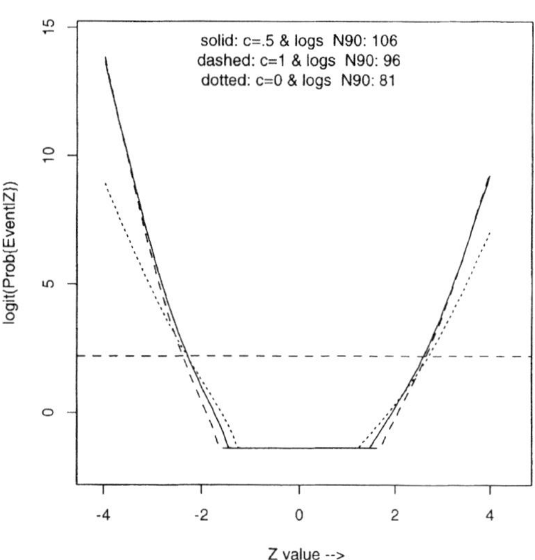

Figure 4. Comparison of Various Probe Reductions (gene reduction fixed as in (2.8), a_0 = 90th percentile). The solid curve in both panels is the choice (2.6) used previously; constant "c" is multiple of mm level subtracted from pm level, for example. "c = 1 no logs" uses M_{ik} = mean{$\text{pm}_{ijk} - \text{mm}_{ijk}$}. "Affy" based on the probe reduction software provided with the Affymetrix Genechip.

$p_1(Z)$ still is meaningful as the *a posteriori* probability of effect given the insufficient statistic Z, and also has the FDR interpretation of Section 5. (Efron et al. (2000) showed that in fact a better "$s(\mathbf{v})$" can be obtained in the radiation experiment by removing plates U1A and I1A from the dataset (1.1); a processing error appears to have degraded the results from I1A.) Our tactic of choosing the Z mapping to maximize $p_1(Z)$ is nearly equivalent to minimizing FDR, which was the approach taken in Tusher et al. (2000).

- There is also no claim that mappings (2.6) and (2.8) enjoy general superiority. The equivalent of Figures 3 and 4 might point to a different choice of $s(\mathbf{v})$ in another dataset. Section 6 discusses how our methodology can be applied to other comparative microarray experiments.
- Overfitting is not a threat in a genuine Bayesian framework, where results like those from Figures 3 and 4 can be thought of as just computer-based attempts to numerically solve a probabilistic maximization problem. However, in our empirical Bayes framework, too much data-based maximization could in fact lead to overfitting. Two forms of bootstrapping were employed as a check on our results: "gene resampling," in which the rows of the $6{,}810 \times 8$ matrix $\mathbf{M}$ were resampled to give $\mathbf{M}^*$; and "row resampling" in which row i of $\mathbf{M}^*$ was obtained as the average of 20 resampled rows from the 20×8 matrix $\mathbf{x}_i$ having entries

$$x_{ijk} = \log(\text{pm}_{ijk}) - .5 \cdot \log(\text{mm}_{ijk}). \tag{4.5}$$

The bootstrap results indicated that the differences seen in Figures 3 and 4 were much greater than the standard errors of the curves, so that overfitting was not a threat. For example, row resampling showed that the difference between the $a_0 = .90$ and $a_0 = .50$ curves at $Z = -3$, which looks suspiciously small in Figure 3, had point estimate and standard error $.68 \pm .13$.

5. FALSE DISCOVERY RATES

The empirical Bayes analysis of Section 3 is closely related to Benjamini and Hochberg's FDR criterion. For a collection of simultaneous hypothesis tests, FDR is the expected proportion of type I errors made using a given rejection rule. Define the *local false discovery rate* to be

$$\text{fdr}(Z) = p_0 f_0(Z)/f(Z), \tag{5.1}$$

so fdr(Z) is the *a posteriori* probability $p_0(Z)$, (3.8), that a gene with score Z is unaffected. It will be shown that (6.1) has a natural FDR interpretation. We begin with a numerical example.

In the calculations for Figure 1, $N = 74$ of the 6810 genes had Z scores in the interval $Z \in [1.9, 2.1]$, whereas the twenty permuted null score datasets $\{z_i\}$ had 676 falling into $[1.9, 2.1]$, an average of $33.8 = 676/20$ per set. Taking p_0 to be its estimated maximum .811, this suggests that among the $N = 74$ binned Z values, the expected number of "unaffecteds" is $27.4 = .811 \cdot 33.8$. If we now declare all genes with Z in $[1.9, 2.1]$ to be affected, our expected proportion of false discoveries is

$$\frac{27.4}{74} = 37\%. \tag{5.2}$$

Notice that (5.2) is an obvious estimator of (5.1) for $Z = 2$,

$$\widehat{\text{fdr}}(2) = \hat{p}_0 \hat{f}_0(2)/\hat{f}(2)$$
$$\left[\hat{p}_0 = .811, \quad \hat{f}_0(2) = \frac{33.8}{6810}, \quad \hat{f}(2) = \frac{74}{6810}\right]. \tag{5.3}$$

In general, if we bin the genes into small intervals on the Z scale, then a bin declared "affected" will have a FDR of about fdr(Z), (5.1), the equality becoming exact as the number of genes goes to infinity. This last statement can be rigorously verified under modest ergodic conditions that preclude extremely high correlations among the values fo Z or the z.

Figure 5 reports on a simulation experiment used to check the accuracy of fdr(Z) as an estimate of FDR. A 6810×8

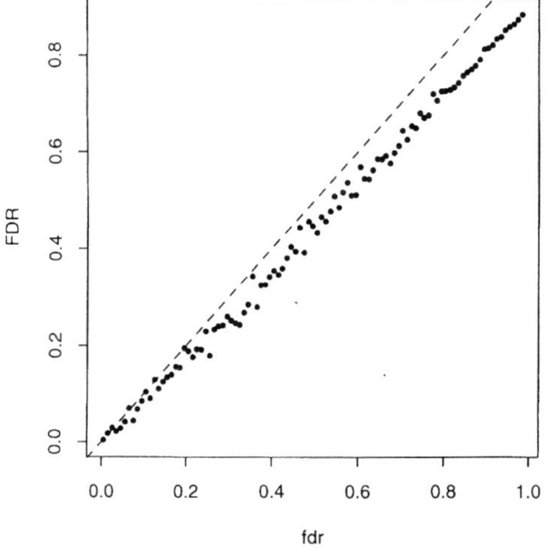

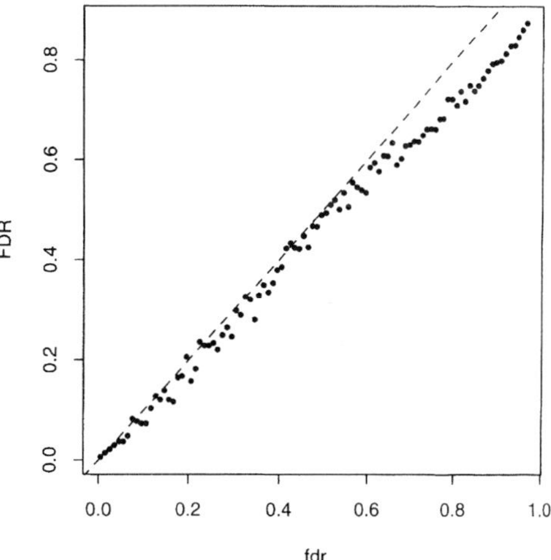

Figure 5. Simulations Comparing Empirical Bayes Formula fdr, (5.1) With Actual FDR, as Explained in Text. Left panel: $a_0 = .90$ in mapping (2.8). Right panel: $a_0 \to \infty$.

matrix $\mathbf{M}$ was constructed in a way that mimicked the radiation experiment,

$$M_{ik} = \theta_i t_k + \epsilon_{ik} \quad [\epsilon_{ik} \overset{\text{ind}}{\sim} N(0, 2)], \tag{5.4}$$

$(t_1, t_2, \ldots, t_8) = (0, 0, 1, 1, 0, 0, 1, 1)$; 681 of the "gene effects" θ_i were chosen from an $N(-1.5, 1)$ distribution, 681 from $N(1.5, 1)$, and the remaining 5448 set at zero. In other words 80% of the genes were unaffected and 20% were affected, 10% in each direction.

The matrix $\mathbf{M}$ was processed into a $\mathbf{Z}$ vector according to (3.5), and also into twenty $\mathbf{z}$ vectors according to (3.7), using (2.8) for the mappings from $\mathbf{D} \to \mathbf{Z}$ and $\mathbf{d} \to \mathbf{z}$. Following the same algorithm that lead to Figure 1, these gave an estimated $\text{fdr}(Z)$ curve, (5.1), that in fact looked much like the one for the actual experiment.

Figure 5 reports on two different choices of a_0 in (2.8), $a_0 = .90$ in the left panel and $a_0 \to \infty$ on the right. There are 100 points in each panel, corresponding to a binning of the fdr axis in units of .01 from 0 to 1, with the jth point plotted at $\text{fdr}_j = j/100$ and

$$\text{FDR}_j = \text{proportion of } j\text{th bin with } \theta_i = 0, \tag{5.5}$$

the empirical FDR for that bin. The FDR_j values are averages over 50 simulations, each of the individual simulations being noisy versions of the same picture. If formula (5.1) is actually the local fdr, then the points should lie near the main diagonal FDR = fdr as they do. The slight conservative bias $\text{FDR}_j \leq \text{fdr}_j$, came from the fact that the upper bound for p_0 used in (5.1) (calculated as in Remark F of Section 6) substantially overestimated the true value $p_0 = .80$.

In the artificial situation (5.4), taking $a_0 \to \infty$ in (2.8) gives a more efficient choice of Z_i than $a_0 = .90$, doubling $N90$. Nevertheless, the inefficient choice $a_0 = .90$ is still accurately calibrated: $\text{FDR}_j \doteq \text{fdr}_j$.

FDRs are usually defined for an entire rejection region, for example for

$$\mathcal{R} = \{Z\colon R(Z) \geq r_0\} \quad [R(Z) = f(Z)/f_0(Z)] \tag{5.6}$$

rather than locally as in (5.1). We can think of this as replacing our original choice of summary statistic $Z = s(\mathbf{v})$ with

$$\tilde{Z} = \begin{cases} 1 & \text{if } Z \in \mathcal{R}, \\ 0 & Z \notin \mathcal{R} \end{cases} \tag{5.7}$$

Assuming that genes having $\tilde{Z} = 1$ are declared affected, the empirical Bayes formula (5.1) now becomes

$$\widetilde{\text{fdr}}(1) = p_0 \tilde{f}_0(1)/\tilde{f}(1), \tag{5.8}$$

with straightforward estimate

$$\hat{p}_0 \frac{\text{proportion}\{z_i \in \mathcal{R}\}}{\text{proportion}\{Z_i \in \mathcal{R}\}}. \tag{5.9}$$

The heuristic argument proceeding (5.2) is more obvious here: (5.9) estimates the proportion of genes in the "affected" region $\mathcal{R}$, which is actually unaffected, and in this sense estimates Benjamini and Hochberg's FDR. (Current work by the authors strengthens this connection: choosing $\mathcal{R}$ in (5.6) as large as possible subject to keeping (5.9) below some fixed limit exactly matches the Benjamini–Hochberg choice of rejection region.)

The global definition of FDR has the advantage of being easier to estimate. We can use the totally nonparametric estimator (5.9) rather than having to estimate the local ratio $f_0(Z)/f(Z)$ in (5.1). On the other hand (5.8) is a composite measure that assigns the same FDR to all the genes in $\mathcal{R}$ even though some of them have $\text{Prob}\{\text{Event}|Z\}$ much greater than others. Comparing (5.1) with (5.8) it is easy to see that $\widetilde{\text{fdr}}(1)$ is the conditional expectation of $\text{fdr}(Z)$ given $Z \in \mathcal{R}$,

$$\widetilde{\text{fdr}}(1) = E_f\{\text{fdr}(Z)|\mathcal{R}\} \tag{5.10}$$

so that $\text{fdr}(Z)$ is more precise than $\widetilde{\text{fdr}}(1)$. More on the connection between FDR and fdr appears in Storey (2001).

6. SUMMARY AND REMARKS

The Empirical Bayes procedure described in this article provides an effective framework for studying the relative changes in gene expression for a large number of genes. It uses a simple nonparametric mixture prior to model the population of affected and unaffected genes, thereby avoiding parametric assumptions about gene expression. We establish a close connection between the estimated posterior probabilities and a local version of the FDR, thereby allowing for the analyst to handle multiple testing issues that arise when dealing with a large number of simultaneous tests. As we have detailed in Algorithm 1 and Remark D, the proposed procedure can be applied quiet generally to other kinds of comparative microarray experiments.

We conclude with a number of remarks, giving important practical details for the proposed methods.

(A) The Experiment. Lymphoblastoid cell lines GM14660 and GM08925, (Coriell Cell Repositories, Camden, New Jersey) were seeded at 2.5 × 105 cells/ml. The treatment consisted of 5 Gy of ionizing radiation. After 24 hours, RNA was isolated, labeled, and divided into two aliquots that were independently hybridized to the HuGeneFL Genechip microarray, Affymetrix Corporation.

(B) Northern Blot Analysis. Northern Blot Analysis produced a quantitative score "G_i" for each of the 18 genes indicated in Figure 1, G standing for gold standard. Following previous biomedical convention, G scores exceeding 1.30 were taken to indicate a positive effect of radiation on gene activity, the "+" symbols in Figure 1; likewise "−" for $G_i < .70$ and "o" for $.70 \leq G_i \leq 1.30$. Figure 6 compares the Z_i scores from Figure 1 with $\log G_i$ for the 18 test genes. We see a strong monotone relationship, correlation .87.

The agreement in Figure 6 is impressive, especially considering the magnitude of the sampling errors in the individual expression values. Our gold standard, the Northern Blot score, is not pure gold, itself being subject to experimental error. There is only one flagrant disagreement in Figure 6, the "−" gene at $Z_i = -.31$. The vector of differences (3.4) was $\mathbf{D}_i = (-1.59, .55, , 88, -.83)$ for this gene, so that both wild types yielded aliquots of opposing signs. In contrast the "o"

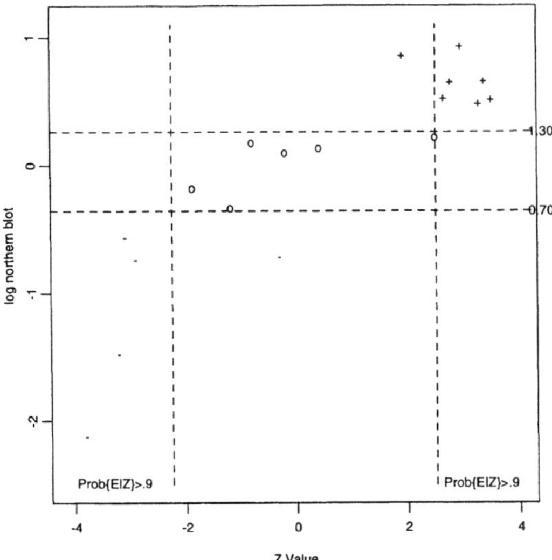

Figure 6. Comparison of Z Scores From the Analysis in Figure 1 With the Logarithm of the Northern Blot Results. Correlation .87. Z values outside the two vertical lines have Prob{Event$_i$ | Z_i} ≥ .90.

point at $Z_i = 2.51$, lying just below the "+" cutoff value $G = 1.30$, was consistently positive, $\mathbf{D}_i = (4.54, 2.81, 1.64, 2.65)$, strengthening our belief that this gene was positively affected by the radiation.

(C) Debrightening and Desumming. Some microarray plates are "brighter" than others in that they produce systematically larger expression levels. Following probe reduction (2.4) we debrightened the data by separately standardizing the columns of **M**. That is, each column of **M** was linearly transformed to have mean 0 and empirical standard deviation 1.

"Desumming" corrects for another type of data inhomogeneity. Corresponding to **D** (3.4), let

$$\mathbf{S} = (\mathbf{M}_3 + \mathbf{M}_1, \mathbf{M}_4 + \mathbf{M}_2, \mathbf{M}_7 + \mathbf{M}_5, \mathbf{M}_8 + \mathbf{M}_6). \qquad (6.1)$$

A gene with larger **S** values tended to have larger values of **D**, which undercut the exchangeability across genes implicit in our empirical Bayes analyses. (Newton et al. 2001 adjusted their data for a similar problem.) After debrightening, the individual columns of **D** were desummed as follows: a linear regression $|D_{ik}| = a_0 + a_1|S_{ik}| + \text{error}$ was fit individually to each column, and then each D_{ik} was transformed to

$$D_{ik}/(\hat{a}_0 + \hat{a}_1|S_{ik}|). \qquad (6.2)$$

Similar transformations were made on the columns of **d**, (3.6). It was the transformed **D** and **d** matrices that were used to compute the scores **Z** and **z** via (2.8). Desumming made almost no difference to the results in Figure 1, but the exchangeability issue is an important general point of concern for the empirical Bayes analysis, see Remark F.

(D) Logistic Regression Estimate of $f_0(z)/f(z)$. The ratio $f_0(z)/f(z)$ in (3.8) was estimated by logistic regression. Given $B = 20$ replications of **z**, all $n \cdot (1 + B) = 6,810 \cdot 21$ scores Z_i and z_i were plotted on a line, with values of Z_i considered as "successes" and values of **z**, as "failures". The probability $\pi(z)$ of a success at point z is given in terms of the densities (3.2),

$$\pi(z) = f(z)/(f(z) + Bf_0(z)), \qquad (6.3)$$

so that (3.8) becomes

$$p_1(Z) = 1 - p_0 \frac{1 - \pi(Z)}{B\pi(Z)}. \qquad (6.4)$$

With $n = 6,810$ genes, the normal scores transformation resulted in $\max\{Z_i\} = -\min\{Z_i\} = 3.80$, whereas the null scores $\{z_i\}$ were confined to a smaller range, as in Figure 2. Our algorithm divided the range $[-4, 4]$ into 139 equal intervals, counted the number of values of Z_i and z_i in each interval, and estimated $\pi(z)$ by logistic regression, for use in (6.4). The regression function was a natural spline with 5 degrees of freedom, called by the Splus command $ns(x, df = 5)$, x being the 139 center points of the intervals. The choice of $B = 20$ **z** replications was based on an analysis like that in Figure 3, which showed considerable improvement for B increasing from 1 to 10, but little gain past 20. Other methods of estimating $f_0(z)/f(z)$ are possible, and in fact the details of the logistic regression method made little difference to our results. The "global" estimate (5.9) avoids regression entirely, at the expense of providing less-specific results.

(E) Estimating the Null Distribution. The null density $f_0(z)$ is supposed to describe the distribution of expression scores for genes unaffected by the treatment of interest. Basing f_0 on **d** in (3.6) seems natural for the radiation experiment, but other choices are possible and may be necessary for other experimental designs. Table 2 shows a small portion (5 of 2,638 genes) of the data from a microarray study comparing two different types of liver cancer, 36 Type I patients versus 36 Type II, with Type II having worse prognosis. Spotted cDNA arrays were used, the "red–green" variety, the tabled values being $10^3 \cdot \log$ (red/green) intensity ratio. The full table corresponds to matrix **M** in (3.5), now $2,638 \times 72$. Probe reduction (2.8) is very simple here, (red, green) → log(red/green), though more efficient reductions may be possible as shown in Dudoit et al. (2000). The analog of Figure 3 indicated a preference for $a_0 = 0$, i.e. for taking Z to be the two-sample t-statistic between the Types.

We obtained the null scores z_i and the null density $f_0(z)$ by randomly splitting the Type I patients into two groups of 18 each, say "A" and "B," and likewise "C" and "D" for the Type II patients, and defining the values of z as t-statistics between groups A ∪ C versus B ∪ D. In other words, we used balanced permutations that put equal numbers of those of Type I and Type II into each of the two permuted groups; using unbalanced permutations would add an unwanted component of variance to the null scores. The empirical Bayes analysis produced results similar to those in Figure 2, with Type II playing the role of the Treatment group.

As a simple, but informative, model for the radiation experiment, suppose that M_{ik} in (3.5) can be expressed as

$$M_{ik} = \mu_i + \alpha_i w_k + \theta_i t_k + \epsilon_{ik}, \qquad (6.5)$$

Table 2. Some Data From a Microarray Study Comparing two Types of Liver Cancer

	TYPE I						TYPE II				
	pat1	pat2	pat3	pat4	pat5	...	pat37	pat38	pat39	pat40	pat41
GENE1	230.0	−1,350	−1,580.0	−400	−760	...	970	110	−50	−190.0	−200
GENE2	470.0	−850	−.8	−280	120	...	390	−1,730	−1,360	−.8	−330
GENE3	−920.0	−1,070	1,360.0	−510	−1,120	...	70	−1,150	340	−400.0	580
GENE4	.1	380	730.0	180	−90	...	1,040	180	1,070	250.0	1880
GENE5	390.0	−1,960	−210.0	200	230	...	530	−1,170	670	.7	890

where $\mathbf{t}' = (0, 0, 1, 1, 0, 0, 1, 1)$, $\mathbf{w}' = (-1, -1, -1, -1, 1, 1, 1, 1)$, and ϵ_{ik} is an independent noise term. Here θ_i represents the treatment effect whereas α_i is the differential response for gene i between the first and second wild types. Then Z_i in (2.8) is

$$Z_i = (\theta_i + e_i)/(a_0 + S_i), \quad S_i = \left[\sum_{\ell=1}^{4}(e_{i\ell} - e_i)^{2/3}\right]^{1/2}, \quad (6.6)$$

where each $e_{i\ell}$ is the difference of two values of ϵ_{ik} and e_i is the average of the four values of $e_{i\ell}$. The values of z_i have the same expression except that $\theta_i = 0$ in (6.6). We can see that $f_0(z)$ is a legitimate null hypothesis comparator for $f(Z)$.

Suppose we had defined null scores by differencing across wild types instead of across aliquots: $\mathbf{d} = (\mathbf{M}_5 - \mathbf{M}_1, \mathbf{M}_6 - \mathbf{M}_2, \mathbf{M}_7 - \mathbf{M}_3, \mathbf{M}_8 - \mathbf{M}_4)$ replacing (3.6). Then z_i would pick up an additional term due to the gene/wild-type interaction α_i in (6.5), adding a component of variance to $f_0(z)$, and decreasing the likelihood ratio $f(z)/f_0(z)$. Models like (6.5) are helpful in guiding the choice of the Z and z mappings, even if we do not need them for the data-based estimation of f and f_0.

The additive model (6.5) gives every column of $\mathbf{d}$ the same distribution, but we might not trust the Treatment differences to really have the same distribution as the Control, I2B–I2A compared to U2B–U2A for example. Empirically this turned out not to be a problem for the radiation experiment, but if it had we might have used only the first and third columns of $\mathbf{d}$ in (3.6).

(F) Better Upper Bound Estimates for "p_0." The upper bound (3.9), $p_0 \le \min\{f(Z)/f_0(Z)\}$, can be poorly estimated by the choice $\min\{\hat{f}(Z)/\hat{f}_0(Z)\}$ used in Figure 1. More stable upper bounds can be constructed by integrating over an interval "$\mathcal{A}$" near $Z = 0$,

$$p_0 \le \frac{\int_{\mathcal{A}}[f(Z)/f_0(Z)]f_0(Z)}{\int_{\mathcal{A}} f_0(Z)} = \frac{\int_{\mathcal{A}} f(Z)}{\int_{\mathcal{A}} f_0(Z)}. \quad (6.7)$$

Simulation showed that the choice $\mathcal{A} = [-.5, .5]$ performed better than $\min\{f(Z)/f_0(Z)\}$, particularly when the true p_0 was near 1. The upper bound (6.7) is directly estimated by proportion{values of Z_i in $\mathcal{A}$}/proportion{values of z_i in $\mathcal{A}$}, avoiding the logistic regression estimate for $f(Z)/f_0(Z)$. This gave $p_0 \le .825$ in the context of Figure 1, not much different than the previous estimate $p_0 \le .811$. *[Received 8 November, 2000. Revised 4 September 2001.]*

REFERENCES

Alizadeh, A., Eisen, M., Davis, R. E., Ma, C., Lossos, I., Rosenwal, A., Boldrick, J., Sabet, H., Tran, T., Yu, X. J. P., Marti, G., Moore, T., Hudsom, J., Lu, L., Lewis, D., Tibshirani, R., Sherlock, G., Chan, W., Greiner, T., Weisenburger, D., Armitage, K., Levy, R., Wilson, W., Greve, M., Byrd, J., Botstein, D., Brown, P., and Staudt, L. (2000), "Identification of Molecularly and Clinically Distinct Substypes of Diffuse Large b Cell Lymphoma by Gene Expression Profiling," *Nature*, 403, 503–511.

Benjamini, Y., and Hochberg, Y. (1995), "Controlling the False Discovery Rate: A Practical and Powerful Approach to Multiple Testing," *Journal of The Royal Statistical Society*, Ser. B 57, 289–300.

Black, M., and Doerge, R. (2000), "Calculation of the Minimum Number of Replicate Spots Required for Detection of Significant Gene Expression Fold Change for cDNA Microarrays," Purdue University, Dept. of Statistics.

Dudoit, S., Yang, Y., Callow, M., and Speed, T. (2000), "Statistical Methods for Identifying Differentially Expressed Genes in Replicated cDNA Microarray Experiments," Technical Report, University of California, Berkeley, Dept. Statistics.

Efron, B., Tibshirani, R., Goss, V., and Chu, G. (2000), "Microarrays and Their use in a Comparative Experiment," Stanford Technical Report 213.

Eisen, M., Spellman, P., Brown, P., and Botstein, D. (1998), "Cluster Analysis and Display of Genomewide Expression Patterns," *Proceedings of the National Academy of Science*, 95, 14863–14868.

Kerr, K., and Churchill, G. (2000), "Bootstrapping Cluster Analysis: Assessing the Reliability of Conclusions From Microarray Experiments," *Proceedings of the National of Academy of Science*, to appear.

Lee, M., Kuo, F., Whitmore, G., and Sklar, J. (2000), "Importance of Replication in Microarray Gene Expression Studies: Statistical Methods and Evidence from a Single cDNA Array Experiment," *Proceedings National Academy of Science*, 97, 9834–9.

Li, C., and Wong, W. H. (2000), "Model-based Analysis of Oligonucleotide Arrays: Expression Index Computation and Outlier Detection," unpublished.

Newton, M., Kendziorski, C., Richmond, C., Blatter, F., and Tsui, K. (2001), "On Differential Variability of Expression Ratios: Improving Statistical Inference About Gene Expression Changes From Microarray Data," *Journal of Computational Biology*, 8, 37–52.

Storey, J. (2001), "The False Discovery Rate: A Bayesian Interpretation and the q-value." Stanford Technical Report. Available at jstorey@stat.stanford.edu.

Tusher, V., Tibshirani, R., and Chu, C. (2001), "Significance Analysis of Microarrays Applied to Transcriptional Responses to Ionizing Radiation," *Proceedings National Academy Science*, 98, 5116–21.

Van del Laan, M., and Bryan, J. (2000), "Gene Expression Analysis With the Parametric Bootstrap," Report #81, University of California, Berkeley, Biostatistics Group.

20

Least Angle Regression

Introduction by David Madigan
Columbia University

Classification and regression tools play a central role in the modern statistical arsenal. Over the last few decades, researchers have developed an extraordinary variety of associated methods and algorithms ranging from non- and semi-parametric methods to computer-intensive boosting and bagging approaches to neural networks to Bayesian approaches. Despite all this activity, the linear regression model (and its close relative, the logistic regression model) continues to enjoy widespread use. In addition to being more interpretable, linear models often provide reasonable out-of-sample predictive performance. Indeed, Hand (2006) provides a cogent argument that the incremental improvement in predictive performance that more complex methods sometimes enjoy proves illusory in the face of real problems.

Notwithstanding the extensive knowledge base surrounding the linear model, some gnarly challenges remain, chief amongst them the problem of variable selection. Standard practice (and standard software) focuses on stepwise methods for variable selection that implement somewhat crude and unsatisfactory search algorithms. The size of the search space does present a significant hurdle; with p predictor variables, the space contains 2^p models. Many important applications now exist with p in the hundreds of thousands or even millions.

Efron et al. (2004, "the LAR paper") significantly advanced the state of the art in variable selection. The LAR paper has its roots in the closely related "homotopy method" of Osborne et al. (2000a,b) and in "forward stagewise regression" (which in turn builds on ideas in boosting). Forward stagewise regression starts out like a standard forward stepwise algorithm, first selecting the predictor variable that is most correlated with the response. Next, forward stagewise adds to the model just a small fraction of the least squares estimate of the corresponding coefficient. Then it picks the variable having the highest correlation with the current residuals, and so on. Coupled with a stopping rule, forward stagewise performs variable selection and provides impressive predictive performance. The LAR paper overcomes the principal drawback to forward stagewise, namely computational cost. Rather than taking many small steps, least angle regression (LAR) uses an ingenious geometric argument to combine many small steps into a sequence of much larger steps. LAR starts out with the predictor that makes the smallest angle with the response. Then, it computes a coefficient for this predictor such that adding this coefficient to the model results in residuals with which some other predictor begins to have a smaller angle, and so on. Remarkably, the computational requirements for LAR are of the same order as a regular least squares fit.

The $L1$-regularized "lasso" model of Tibshirani (1996) also provides an elegant approach to variable selection. The lasso simultaneously selects variables and provides shrinkage estimates of regression coefficients. Chen et al. (1998) discuss similar ideas in a signal processing context. Surprisingly, as pointed out in Hastie et al. (2001, p. 329), forward stagewise yields solutions that are very similar to the lasso. Prior to the LAR paper, this similarity was poorly understood. The LAR paper brilliantly resolved this mystery via a modified version of the LAR algorithm that yields the lasso solution.

Nontrivial theoretical and computational challenges still remain open. For example, lasso is "consistent for variable selection" only under rather restrictive conditions (see, for example, Meinshausen and Bühlmann 2006; Zhao and Yu 2006). A variable selection procedure is said to be consistent if the probability that the procedure correctly identifies the set of important covariates approaches 1 when the sample size goes to infinity (Leng et al. 2006). Even in the case of an orthogonal design, Leng et al. (2006) show that lasso will select the wrong model with a positive probability that does not depend on the sample size. On the other hand, Meinshausen and Bühlmann (2006) show that with high probability lasso will select a superset of all the "substantial" predictors. A number of authors have proposed variations on the lasso theme that are consistent for variable selection. For example, both Zou (2006) and Lu and Zhang (2006) propose versions of the lasso that involve separate data-adaptive penalties for each predictor. Bühlmann (2007) suggests a simple implementation of the adaptive lasso that first generates initial regression coefficient estimates using the lasso (trained in a predictively-optimal fashion) and then incorporates these estimates as penalization weights in a second iteration of the lasso.

Statistical challenges continue to grow rapidly in scale, and variable selection approaches will need to avoid combinatorial search. Except for extremely sparse datasets, applications involving tens of millions of predictor variables remain beyond the reach of current algorithms. Text mining, pharmacovigilance, and risk adjustment, for example, can require models on this scale. It seems that the size of important applications like these will always outstrip available computing power, and simply waiting for faster computers does not represent a viable strategy. Parallel computing may hold the key to rapid progress.

I fully expect that LAR, lasso, and related ideas will dominate linear modeling practice in years to come.

References

Bühlmann, P. (2007). Variable selection for high-dimensional data with applications in molecular biology. *ISI Proceedings*, 56th Session, Lisbon.

Chen, S. S., Donoho, D. L., and Saunders, M. A. (1998). Atomic Decomposition by Basis Pursuit. *SIAM Journal on Scientific Computing* **20**, 33–61.

Efron, B., Hastie, T., Johnstone, I. and Tibshirani, R. (2004). Least angle regression. *Annals of Statistics* **32**, 407–499 with discussion and rejoinder.

Hand, D. J. (2006). Classifier technology and the illusion of progress. *Statistical Science* **21**, 1–15.

Hastie, T., Tibshirani, R. and Friedman, J. (2001). *The Elements of Statistical Learning: Data Mining, Inference and Prediction*. Springer, New York.

Leng, C., Lin, Y. and Wahba, G. (2006). A note on the lasso and related procedures in model selection. *Statistica Sinica* **16**, 1273–1284.

Lu, W. and Zhang, H. (2006). Variable selection via penalized likelihood with adaptive penalty. *Institute of Statistics Mimeo Series, No. 2594*, Department of Statistics, North Carolina State University. [Available from `www.stat.ncsu.edu/library/mimeo.html`]

Meinshausen, N. and Bühlmann, P. (2006). High dimensional graphs and variable selection with the Lasso. *Annals of Statistics* **34**, 1436–1462.

Osborne, M. R., Presnell, B., and Turlach, B. A. (2000a). A new approach to variable selection in least squares problems. *IMA Journal of Numerical Analysis* **20**, 389–404.

Osborne, M. R., Presnell, B., and Turlach, B. A. (2000b). On the LASSO and its dual. *Journal of Computational and Graphical Statistics* **9**, 319–337.

Tibshirani, R. (1996). Regression shrinkage and selection via the lasso. *Journal of the Royal Statistical Society, Series B* **58**, 267–288.

Zou, H. (2006). The adaptive lasso and its oracle properties. *Journal of the American Statistical Association* **101**, 1418–1429.

Zhao, P. and Yu, B. (2006). On model selection consistency of LASSO. Technical Report 702, Department of Statistics, University of California, Berkeley.

The Annals of Statistics
2004, Vol. 32, No. 2, 407–499

LEAST ANGLE REGRESSION

BY BRADLEY EFRON,[1] TREVOR HASTIE,[2] IAIN JOHNSTONE[3] AND ROBERT TIBSHIRANI[4]

Stanford University

The purpose of model selection algorithms such as *All Subsets*, *Forward Selection* and *Backward Elimination* is to choose a linear model on the basis of the same set of data to which the model will be applied. Typically we have available a large collection of possible covariates from which we hope to select a parsimonious set for the efficient prediction of a response variable. *Least Angle Regression* (LARS), a new model selection algorithm, is a useful and less greedy version of traditional forward selection methods. Three main properties are derived: (1) A simple modification of the LARS algorithm implements the Lasso, an attractive version of ordinary least squares that constrains the sum of the absolute regression coefficients; the LARS modification calculates all possible Lasso estimates for a given problem, using an order of magnitude less computer time than previous methods. (2) A different LARS modification efficiently implements Forward Stagewise linear regression, another promising new model selection method; this connection explains the similar numerical results previously observed for the Lasso and Stagewise, and helps us understand the properties of both methods, which are seen as constrained versions of the simpler LARS algorithm. (3) A simple approximation for the degrees of freedom of a LARS estimate is available, from which we derive a C_p estimate of prediction error; this allows a principled choice among the range of possible LARS estimates. LARS and its variants are computationally efficient: the paper describes a publicly available algorithm that requires only the same order of magnitude of computational effort as ordinary least squares applied to the full set of covariates.

1. Introduction. Automatic model-building algorithms are familiar, and sometimes notorious, in the linear model literature: Forward Selection, Backward Elimination, All Subsets regression and various combinations are used to automatically produce "good" linear models for predicting a response y on the basis of some measured covariates $x_1, x_2, \dots, x_m$. Goodness is often defined in terms of prediction accuracy, but parsimony is another important criterion: simpler models are preferred for the sake of scientific insight into the $x - y$ relationship. Two promising recent model-building algorithms, the Lasso and Forward Stagewise lin-

Received March 2002; revised January 2003.
[1]Supported in part by NSF Grant DMS-00-72360 and NIH Grant 8R01-EB002784.
[2]Supported in part by NSF Grant DMS-02-04162 and NIH Grant R01-EB0011988-08.
[3]Supported in part by NSF Grant DMS-00-72661 and NIH Grant R01-EB001988-08.
[4]Supported in part by NSF Grant DMS-99-71405 and NIH Grant 2R01-CA72028.
AMS 2000 subject classification. 62J07.
Key words and phrases. Lasso, boosting, linear regression, coefficient paths, variable selection.

ear regression, will be discussed here, and motivated in terms of a computationally simpler method called Least Angle Regression.

Least Angle Regression (LARS) relates to the classic model-selection method known as Forward Selection, or "forward stepwise regression," described in Weisberg [(1980), Section 8.5]: given a collection of possible predictors, we select the one having largest absolute correlation with the response y, say x_{j_1}, and perform simple linear regression of y on x_{j_1}. This leaves a residual vector orthogonal to x_{j_1}, now considered to be the response. We project the other predictors orthogonally to x_{j_1} and repeat the selection process. After k steps this results in a set of predictors $x_{j_1}, x_{j_2}, \dots, x_{j_k}$ that are then used in the usual way to construct a k-parameter linear model. Forward Selection is an aggressive fitting technique that can be overly greedy, perhaps eliminating at the second step useful predictors that happen to be correlated with x_{j_1}.

Forward Stagewise, as described below, is a much more cautious version of Forward Selection, which may take thousands of tiny steps as it moves toward a final model. It turns out, and this was the original motivation for the LARS algorithm, that a simple formula allows Forward Stagewise to be implemented using fairly large steps, though not as large as a classic Forward Selection, greatly reducing the computational burden. The geometry of the algorithm, described in Section 2, suggests the name "Least Angle Regression." It then happens that this same geometry applies to another, seemingly quite different, selection method called the Lasso [Tibshirani (1996)]. The LARS–Lasso–Stagewise connection is conceptually as well as computationally useful. The Lasso is described next, in terms of the main example used in this paper.

Table 1 shows a small part of the data for our main example.

Ten baseline variables, age, sex, body mass index, average blood pressure and six blood serum measurements, were obtained for each of $n = 442$ diabetes

TABLE 1

Diabetes study: 442 *diabetes patients were measured on* 10 *baseline variables*; *a prediction model was desired for the response variable, a measure of disease progression one year after baseline*

	AGE	SEX	BMI	BP	Serum measurements						Response
Patient	x_1	x_2	x_3	x_4	x_5	x_6	x_7	x_8	x_9	x_{10}	y
1	59	2	32.1	101	157	93.2	38	4	4.9	87	151
2	48	1	21.6	87	183	103.2	70	3	3.9	69	75
3	72	2	30.5	93	156	93.6	41	4	4.7	85	141
4	24	1	25.3	84	198	131.4	40	5	4.9	89	206
5	50	1	23.0	101	192	125.4	52	4	4.3	80	135
6	23	1	22.6	89	139	64.8	61	2	4.2	68	97
⋮	⋮	⋮	⋮	⋮	⋮	⋮	⋮	⋮	⋮	⋮	⋮
441	36	1	30.0	95	201	125.2	42	5	5.1	85	220
442	36	1	19.6	71	250	133.2	97	3	4.6	92	57

patients, as well as the response of interest, a quantitative measure of disease progression one year after baseline. The statisticians were asked to construct a model that predicted response y from covariates $x_1, x_2, \ldots, x_{10}$. Two hopes were evident here, that the model would produce accurate baseline predictions of response for future patients and that the form of the model would suggest which covariates were important factors in disease progression.

The Lasso is a constrained version of ordinary least squares (OLS). Let $\mathbf{x}_1, \mathbf{x}_2, \ldots, \mathbf{x}_m$ be n-vectors representing the covariates, $m = 10$ and $n = 442$ in the diabetes study, and let $\mathbf{y}$ be the vector of responses for the n cases. By location and scale transformations we can always assume that the covariates have been standardized to have mean 0 and unit length, and that the response has mean 0,

$$\sum_{i=1}^{n} y_i = 0, \qquad \sum_{i=1}^{n} x_{ij} = 0, \qquad \sum_{i=1}^{n} x_{ij}^2 = 1 \qquad \text{for } j = 1, 2, \ldots, m. \tag{1.1}$$

This is assumed to be the case in the theory which follows, except that numerical results are expressed in the original units of the diabetes example.

A candidate vector of regression coefficients $\widehat{\boldsymbol{\beta}} = (\widehat{\beta}_1, \widehat{\beta}_2, \ldots, \widehat{\beta}_m)'$ gives prediction vector $\widehat{\boldsymbol{\mu}}$,

$$\widehat{\boldsymbol{\mu}} = \sum_{j=1}^{m} \mathbf{x}_j \widehat{\beta}_j = X\widehat{\boldsymbol{\beta}} \qquad [X_{n\times m} = (\mathbf{x}_1, \mathbf{x}_2, \ldots, \mathbf{x}_m)] \tag{1.2}$$

with total squared error

$$S(\widehat{\boldsymbol{\beta}}) = \|\mathbf{y} - \widehat{\boldsymbol{\mu}}\|^2 = \sum_{i=1}^{n} (y_i - \widehat{\mu}_i)^2. \tag{1.3}$$

Let $T(\widehat{\boldsymbol{\beta}})$ be the absolute norm of $\widehat{\boldsymbol{\beta}}$,

$$T(\widehat{\boldsymbol{\beta}}) = \sum_{j=1}^{m} |\widehat{\beta}_j|. \tag{1.4}$$

The Lasso chooses $\widehat{\boldsymbol{\beta}}$ by minimizing $S(\widehat{\boldsymbol{\beta}})$ subject to a bound t on $T(\widehat{\boldsymbol{\beta}})$,

$$\textit{Lasso}: \quad \text{minimize} \quad S(\widehat{\boldsymbol{\beta}}) \qquad \text{subject to} \quad T(\widehat{\boldsymbol{\beta}}) \le t. \tag{1.5}$$

Quadratic programming techniques can be used to solve (1.5) though we will present an easier method here, closely related to the "homotopy method" of Osborne, Presnell and Turlach (2000a).

The left panel of Figure 1 shows all Lasso solutions $\widehat{\boldsymbol{\beta}}(t)$ for the diabetes study, as t increases from 0, where $\widehat{\boldsymbol{\beta}} = 0$, to $t = 3460.00$, where $\widehat{\boldsymbol{\beta}}$ equals the OLS regression vector, the constraint in (1.5) no longer binding. We see that the Lasso tends to shrink the OLS coefficients toward 0, more so for small values of t. Shrinkage often improves prediction accuracy, trading off decreased variance for increased bias as discussed in Hastie, Tibshirani and Friedman (2001).

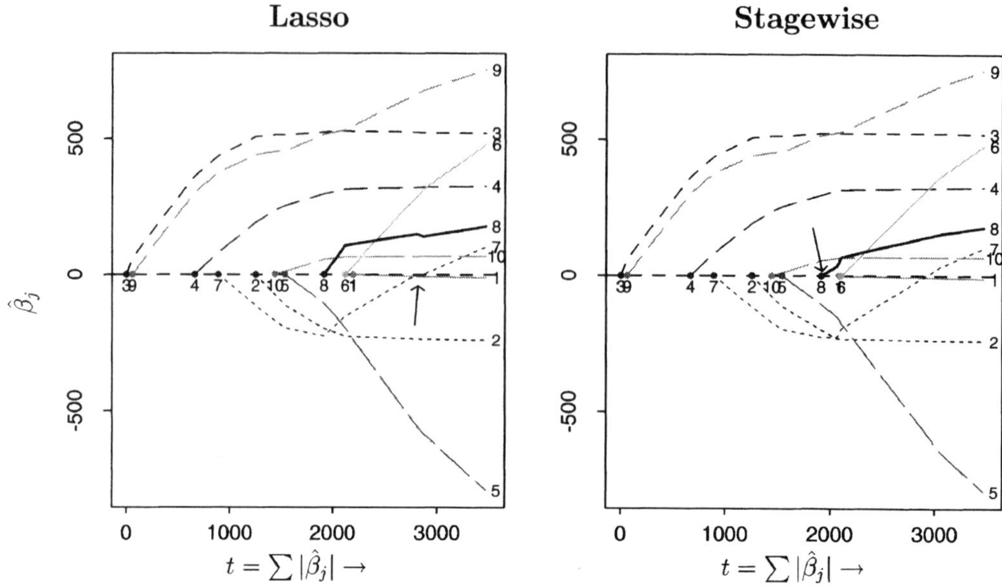

FIG. 1. *Estimates of regression coefficients* $\widehat{\beta}_j$, $j = 1, 2, \ldots, 10$, *for the diabetes study.* (Left panel) *Lasso estimates, as a function of* $t = \sum_j |\widehat{\beta}_j|$. *The covariates enter the regression equation sequentially as* t *increases, in order* $j = 3, 9, 4, 7, \ldots, 1$. (Right panel) *The same plot for Forward Stagewise Linear Regression. The two plots are nearly identical, but differ slightly for large* t *as shown in the track of covariate* 8.

The Lasso also has a parsimony property: for any given constraint value t, only a subset of the covariates have nonzero values of $\widehat{\beta}_j$. At $t = 1000$, for example, only variables 3, 9, 4 and 7 enter the Lasso regression model (1.2). If this model provides adequate predictions, a crucial question considered in Section 4, the statisticians could report these four variables as the important ones.

Forward Stagewise Linear Regression, henceforth called *Stagewise*, is an iterative technique that begins with $\widehat{\boldsymbol{\mu}} = 0$ and builds up the regression function in successive small steps. If $\widehat{\boldsymbol{\mu}}$ is the current Stagewise estimate, let $\mathbf{c}(\widehat{\boldsymbol{\mu}})$ be the vector of *current correlations*

$$\widehat{\mathbf{c}} = \mathbf{c}(\widehat{\boldsymbol{\mu}}) = X'(\mathbf{y} - \widehat{\boldsymbol{\mu}}), \tag{1.6}$$

so that $\widehat{c}_j$ is proportional to the correlation between covariate x_j and the current residual vector. The next step of the Stagewise algorithm is taken in the direction of the greatest current correlation,

$$\widehat{j} = \arg\max |\widehat{c}_j| \quad \text{and} \quad \widehat{\boldsymbol{\mu}} \rightarrow \widehat{\boldsymbol{\mu}} + \varepsilon \cdot \operatorname{sign}(\widehat{c}_{\widehat{j}}) \cdot \mathbf{x}_{\widehat{j}}, \tag{1.7}$$

with ε some small constant. "Small" is important here: the "big" choice $\varepsilon = |\widehat{c}_{\widehat{j}}|$ leads to the classic Forward Selection technique, which can be overly greedy, impulsively eliminating covariates which are correlated with $x_{\widehat{j}}$. The Stagewise procedure is related to boosting and also to Friedman's MART algorithm

[Friedman (2001)]; see Section 8, as well as Hastie, Tibshirani and Friedman [(2001), Chapter 10 and Algorithm 10.4].

The right panel of Figure 1 shows the coefficient plot for Stagewise applied to the diabetes data. The estimates were built up in 6000 Stagewise steps [making ε in (1.7) small enough to conceal the "Etch-a-Sketch" staircase seen in Figure 2, Section 2]. The striking fact is the similarity between the Lasso and Stagewise estimates. Although their definitions look completely different, the results are nearly, *but not exactly*, identical.

The main point of this paper is that both Lasso and Stagewise are variants of a basic procedure called Least Angle Regression, abbreviated LARS (the "S" suggesting "Lasso" and "Stagewise"). Section 2 describes the LARS algorithm while Section 3 discusses modifications that turn LARS into Lasso or Stagewise, reducing the computational burden by at least an order of magnitude for either one. Sections 5 and 6 verify the connections stated in Section 3.

Least Angle Regression is interesting in its own right, its simple structure lending itself to inferential analysis. Section 4 analyzes the "degrees of freedom" of a LARS regression estimate. This leads to a C_p type statistic that suggests which estimate we should prefer among a collection of possibilities like those in Figure 1. A particularly simple C_p approximation, requiring no additional computation beyond that for the $\widehat{\boldsymbol{\beta}}$ vectors, is available for LARS.

Section 7 briefly discusses computational questions. An efficient *S* program for all three methods, LARS, Lasso and Stagewise, is available. Section 8 elaborates on the connections with boosting.

2. The LARS algorithm. Least Angle Regression is a stylized version of the Stagewise procedure that uses a simple mathematical formula to accelerate the computations. Only m steps are required for the full set of solutions, where m is the number of covariates: $m = 10$ in the diabetes example compared to the 6000 steps used in the right panel of Figure 1. This section describes the LARS algorithm. Modifications of LARS that produce Lasso and Stagewise solutions are discussed in Section 3, and verified in Sections 5 and 6. Section 4 uses the simple structure of LARS to help analyze its estimation properties.

The LARS procedure works roughly as follows. As with classic Forward Selection, we start with all coefficients equal to zero, and find the predictor most correlated with the response, say x_{j_1}. We take the largest step possible in the direction of this predictor until some other predictor, say x_{j_2}, has as much correlation with the current residual. At this point LARS parts company with Forward Selection. Instead of continuing along x_{j_1}, LARS proceeds in a direction equiangular between the two predictors until a third variable x_{j_3} earns its way into the "most correlated" set. LARS then proceeds equiangularly between x_{j_1}, x_{j_2} and x_{j_3}, that is, along the "least angle direction," until a fourth variable enters, and so on.

The remainder of this section describes the algebra necessary to execute the equiangular strategy. As usual the algebraic details look more complicated than the simple underlying geometry, but they lead to the highly efficient computational algorithm described in Section 7.

LARS builds up estimates $\widehat{\boldsymbol{\mu}} = X\widehat{\boldsymbol{\beta}}$, (1.2), in successive steps, each step adding one covariate to the model, so that after k steps just k of the $\widehat{\beta}_j$'s are nonzero. Figure 2 illustrates the algorithm in the situation with $m = 2$ covariates, $X = (\mathbf{x}_1, \mathbf{x}_2)$. In this case the current correlations (1.6) depend only on the projection $\bar{\mathbf{y}}_2$ of $\mathbf{y}$ into the linear space $\mathcal{L}(X)$ spanned by $\mathbf{x}_1$ and $\mathbf{x}_2$,

$$\mathbf{c}(\widehat{\boldsymbol{\mu}}) = X'(\mathbf{y} - \widehat{\boldsymbol{\mu}}) = X'(\bar{\mathbf{y}}_2 - \widehat{\boldsymbol{\mu}}). \tag{2.1}$$

The algorithm begins at $\widehat{\boldsymbol{\mu}}_0 = \mathbf{0}$ [remembering that the response has had its mean subtracted off, as in (1.1)]. Figure 2 has $\bar{\mathbf{y}}_2 - \widehat{\boldsymbol{\mu}}_0$ making a smaller angle with $\mathbf{x}_1$ than $\mathbf{x}_2$, that is, $c_1(\widehat{\boldsymbol{\mu}}_0) > c_2(\widehat{\boldsymbol{\mu}}_0)$. LARS then augments $\widehat{\boldsymbol{\mu}}_0$ in the direction of $\mathbf{x}_1$, to

$$\widehat{\boldsymbol{\mu}}_1 = \widehat{\boldsymbol{\mu}}_0 + \widehat{\gamma}_1 \mathbf{x}_1. \tag{2.2}$$

Stagewise would choose $\widehat{\gamma}_1$ equal to some small value ε, and then repeat the process many times. Classic Forward Selection would take $\widehat{\gamma}_1$ large enough to make $\widehat{\boldsymbol{\mu}}_1$ equal $\bar{\mathbf{y}}_1$, the projection of $\mathbf{y}$ into $\mathcal{L}(\mathbf{x}_1)$. LARS uses an intermediate value of $\widehat{\gamma}_1$, the value that makes $\bar{\mathbf{y}}_2 - \widehat{\boldsymbol{\mu}}$, *equally* correlated with $\mathbf{x}_1$ and $\mathbf{x}_2$; that is, $\bar{\mathbf{y}}_2 - \widehat{\boldsymbol{\mu}}_1$ bisects the angle between $\mathbf{x}_1$ and $\mathbf{x}_2$, so $c_1(\widehat{\boldsymbol{\mu}}_1) = c_2(\widehat{\boldsymbol{\mu}}_1)$.

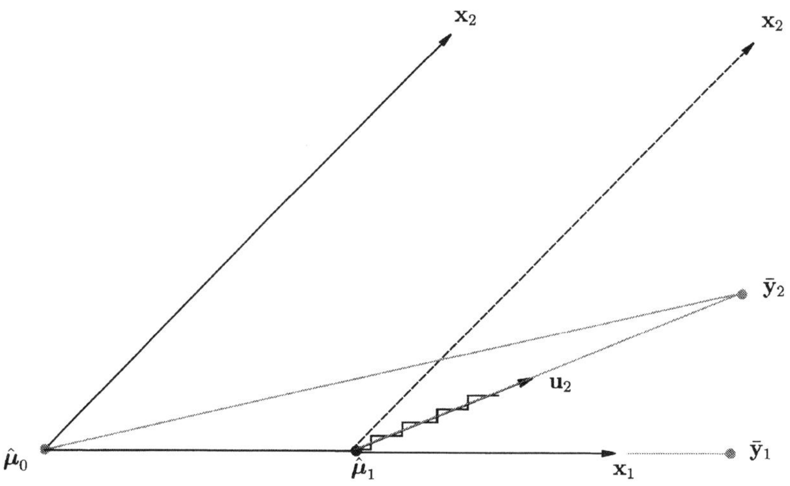

FIG. 2. *The LARS algorithm in the case of $m = 2$ covariates; $\bar{\mathbf{y}}_2$ is the projection of $\mathbf{y}$ into $\mathcal{L}(\mathbf{x}_1, \mathbf{x}_2)$. Beginning at $\widehat{\boldsymbol{\mu}}_0 = \mathbf{0}$, the residual vector $\bar{\mathbf{y}}_2 - \widehat{\boldsymbol{\mu}}_0$ has greater correlation with $\mathbf{x}_1$ than $\mathbf{x}_2$; the next LARS estimate is $\widehat{\boldsymbol{\mu}}_1 = \widehat{\boldsymbol{\mu}}_0 + \widehat{\gamma}_1 \mathbf{x}_1$, where $\widehat{\gamma}_1$ is chosen such that $\bar{\mathbf{y}}_2 - \widehat{\boldsymbol{\mu}}_1$ bisects the angle between $\mathbf{x}_1$ and $\mathbf{x}_2$; then $\widehat{\boldsymbol{\mu}}_2 = \widehat{\boldsymbol{\mu}}_1 + \widehat{\gamma}_2 \mathbf{u}_2$, where $\mathbf{u}_2$ is the unit bisector; $\widehat{\boldsymbol{\mu}}_2 = \bar{\mathbf{y}}_2$ in the case $m = 2$, but not for the case $m > 2$; see Figure 4. The staircase indicates a typical Stagewise path. Here LARS gives the Stagewise track as $\varepsilon \to 0$, but a modification is necessary to guarantee agreement in higher dimensions; see Section 3.2.*

Let $\mathbf{u}_2$ be the unit vector lying along the bisector. The next LARS estimate is

$$\widehat{\boldsymbol{\mu}}_2 = \widehat{\boldsymbol{\mu}}_1 + \widehat{\gamma}_2 \mathbf{u}_2, \tag{2.3}$$

with $\widehat{\gamma}_2$ chosen to make $\widehat{\boldsymbol{\mu}}_2 = \bar{\mathbf{y}}_2$ in the case $m = 2$. With $m > 2$ covariates, $\widehat{\gamma}_2$ would be smaller, leading to another change of direction, as illustrated in Figure 4. The "staircase" in Figure 2 indicates a typical Stagewise path. LARS is motivated by the fact that it is easy to calculate the step sizes $\widehat{\gamma}_1, \widehat{\gamma}_2, \ldots$ theoretically, short-circuiting the small Stagewise steps.

Subsequent LARS steps, beyond two covariates, are taken along *equiangular vectors*, generalizing the bisector $\mathbf{u}_2$ in Figure 2. *We assume that the covariate vectors* $\mathbf{x}_1, \mathbf{x}_2, \ldots, \mathbf{x}_m$ *are linearly independent*. For $\mathcal{A}$ a subset of the indices $\{1, 2, \ldots, m\}$, define the matrix

$$X_{\mathcal{A}} = (\cdots s_j \mathbf{x}_j \cdots)_{j \in \mathcal{A}}, \tag{2.4}$$

where the signs s_j equal ± 1. Let

$$\mathcal{G}_{\mathcal{A}} = X'_{\mathcal{A}} X_{\mathcal{A}} \quad \text{and} \quad A_{\mathcal{A}} = (\mathbf{1}'_{\mathcal{A}} \mathcal{G}_{\mathcal{A}}^{-1} \mathbf{1}_{\mathcal{A}})^{-1/2}, \tag{2.5}$$

$\mathbf{1}_{\mathcal{A}}$ being a vector of 1's of length equaling $|\mathcal{A}|$, the size of $\mathcal{A}$. The

$$\textit{equiangular vector} \quad \mathbf{u}_{\mathcal{A}} = X_{\mathcal{A}} w_{\mathcal{A}} \qquad \text{where } w_{\mathcal{A}} = A_{\mathcal{A}} G_{\mathcal{A}}^{-1} \mathbf{1}_{\mathcal{A}}, \tag{2.6}$$

is the unit vector making equal angles, less than 90°, with the columns of $X_{\mathcal{A}}$,

$$X'_{\mathcal{A}} \mathbf{u}_{\mathcal{A}} = A_{\mathcal{A}} \mathbf{1}_{\mathcal{A}} \quad \text{and} \quad \|\mathbf{u}_{\mathcal{A}}\|^2 = 1. \tag{2.7}$$

We can now fully describe the LARS algorithm. As with the Stagewise procedure we begin at $\widehat{\boldsymbol{\mu}}_0 = \mathbf{0}$ and build up $\widehat{\boldsymbol{\mu}}$ by steps, larger steps in the LARS case. Suppose that $\widehat{\boldsymbol{\mu}}_{\mathcal{A}}$ is the current LARS estimate and that

$$\widehat{\mathbf{c}} = X'(\mathbf{y} - \widehat{\boldsymbol{\mu}}_{\mathcal{A}}) \tag{2.8}$$

is the vector of current correlations (1.6). The *active set* $\mathcal{A}$ is the set of indices corresponding to covariates with the greatest absolute current correlations,

$$\widehat{C} = \max_j \{|\widehat{c}_j|\} \quad \text{and} \quad \mathcal{A} = \{j : |\widehat{c}_j| = \widehat{C}\}. \tag{2.9}$$

Letting

$$s_j = \text{sign}\{\widehat{c}_j\} \qquad \text{for } j \in \mathcal{A}, \tag{2.10}$$

we compute $X_{\mathcal{A}}$, $A_{\mathcal{A}}$ and $\mathbf{u}_{\mathcal{A}}$ as in (2.4)–(2.6), and also the inner product vector

$$\mathbf{a} \equiv X' \mathbf{u}_{\mathcal{A}}. \tag{2.11}$$

Then the next step of the LARS algorithm updates $\widehat{\boldsymbol{\mu}}_{\mathcal{A}}$, say to

$$\widehat{\boldsymbol{\mu}}_{\mathcal{A}_+} = \widehat{\boldsymbol{\mu}}_{\mathcal{A}} + \widehat{\gamma} \mathbf{u}_{\mathcal{A}}, \tag{2.12}$$

where

$$\widehat{\gamma} = \min_{j \in \mathcal{A}^c}{}^{+} \left\{ \frac{\widehat{C} - \widehat{c}_j}{A_{\mathcal{A}} - a_j}, \frac{\widehat{C} + \widehat{c}_j}{A_{\mathcal{A}} + a_j} \right\}; \tag{2.13}$$

"$\min^+$" indicates that the minimum is taken over only positive components within each choice of j in (2.13).

Formulas (2.12) and (2.13) have the following interpretation: define

$$\boldsymbol{\mu}(\gamma) = \widehat{\boldsymbol{\mu}}_{\mathcal{A}} + \gamma \mathbf{u}_{\mathcal{A}}, \tag{2.14}$$

for $\gamma > 0$, so that the current correlation

$$c_j(\gamma) = \mathbf{x}_j'(\mathbf{y} - \boldsymbol{\mu}(\gamma)) = \widehat{c}_j - \gamma a_j. \tag{2.15}$$

For $j \in \mathcal{A}$, (2.7)–(2.9) yield

$$|c_j(\gamma)| = \widehat{C} - \gamma A_{\mathcal{A}}, \tag{2.16}$$

showing that all of the maximal absolute current correlations decline equally. For $j \in \mathcal{A}^c$, equating (2.15) with (2.16) shows that $c_j(\gamma)$ equals the maximal value at $\gamma = (\widehat{C} - \widehat{c}_j)/(A_{\mathcal{A}} - a_j)$. Likewise $-c_j(\gamma)$, the current correlation for the reversed covariate $-\mathbf{x}_j$, achieves maximality at $(\widehat{C} + \widehat{c}_j)/(A_{\mathcal{A}} + a_j)$. Therefore $\widehat{\gamma}$ *in* (2.13) *is the smallest positive value of* γ *such that some new index* $\widehat{j}$ *joins the active set*; $\widehat{j}$ is the minimizing index in (2.13), and the new active set $\mathcal{A}_+$ is $\mathcal{A} \cup \{\widehat{j}\}$; the new maximum absolute correlation is $\widehat{C}_+ = \widehat{C} - \widehat{\gamma} A_{\mathcal{A}}$.

Figure 3 concerns the LARS analysis of the diabetes data. The complete algorithm required only $m = 10$ steps of procedure (2.8)–(2.13), with the variables

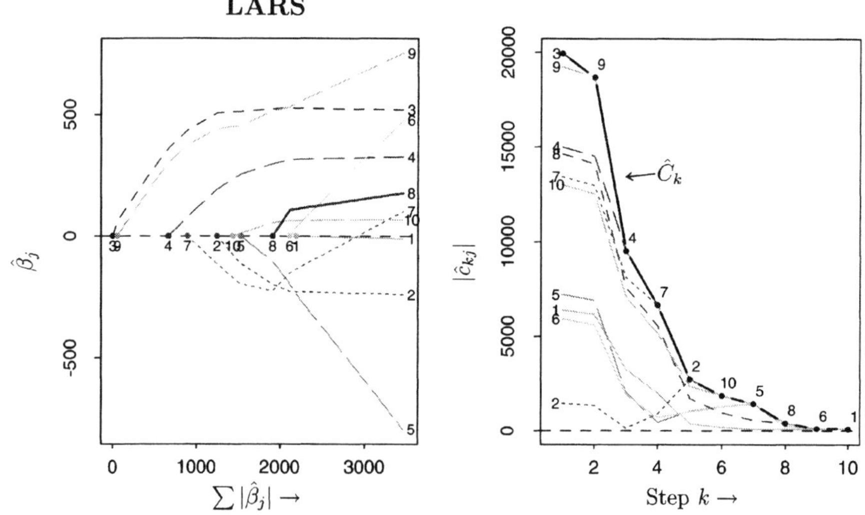

FIG. 3. *LARS analysis of the diabetes study*: (left) *estimates of regression coefficients* $\widehat{\beta}_j$, $j = 1, 2, \ldots, 10$; *plotted versus* $\sum |\widehat{\beta}_j|$; *plot is slightly different than either Lasso or Stagewise, Figure* 1; (right) *absolute current correlations as function of LARS step*; *variables enter active set* (2.9) *in order* $3, 9, 4, 7, \ldots, 1$; *heavy curve shows maximum current correlation* $\widehat{C}_k$ *declining with* k.

joining the active set $\mathcal{A}$ in the same order as for the Lasso: $3, 9, 4, 7, \ldots, 1$. Tracks of the regression coefficients $\widehat{\beta}_j$ are nearly but not exactly the same as either the Lasso or Stagewise tracks of Figure 1.

The right panel shows the absolute current correlations

$$|\widehat{c}_{kj}| = |\mathbf{x}'_j(\mathbf{y} - \widehat{\boldsymbol{\mu}}_{k-1})| \tag{2.17}$$

for variables $j = 1, 2, \ldots, 10$, as a function of the LARS step k. The maximum correlation

$$\widehat{C}_k = \max\{|\widehat{c}_{kj}|\} = \widehat{C}_{k-1} - \widehat{\gamma}_{k-1} A_{k-1} \tag{2.18}$$

declines with k, as it must. At each step a new variable j joins the active set, henceforth having $|\widehat{c}_{kj}| = \widehat{C}_k$. The sign s_j of each $\mathbf{x}_j$ in (2.4) stays constant as the active set increases.

Section 4 makes use of the relationship between Least Angle Regression and Ordinary Least Squares illustrated in Figure 4. Suppose LARS has just completed step $k-1$, giving $\widehat{\boldsymbol{\mu}}_{k-1}$, and is embarking upon step k. The active set $\mathcal{A}_k$, (2.9), will have k members, giving $X_k, \mathcal{G}_k, A_k$ and $\mathbf{u}_k$ as in (2.4)–(2.6) (here replacing subscript $\mathcal{A}$ with "k"). Let $\bar{\mathbf{y}}_k$ indicate the projection of $\mathbf{y}$ into $\mathcal{L}(X_k)$, which, since $\widehat{\boldsymbol{\mu}}_{k-1} \in \mathcal{L}(X_{k-1})$, is

$$\bar{\mathbf{y}}_k = \widehat{\boldsymbol{\mu}}_{k-1} + X_k \mathcal{G}_k^{-1} X'_k(\mathbf{y} - \widehat{\boldsymbol{\mu}}_{k-1}) = \widehat{\boldsymbol{\mu}}_{k-1} + \frac{\widehat{C}_k}{A_k}\mathbf{u}_k, \tag{2.19}$$

the last equality following from (2.6) and the fact that the signed current correlations in $\mathcal{A}_k$ all equal $\widehat{C}_k$,

$$X'_k(\mathbf{y} - \widehat{\boldsymbol{\mu}}_{k-1}) = \widehat{C}_k \mathbf{1}_{\mathcal{A}}. \tag{2.20}$$

Since $\mathbf{u}_k$ is a unit vector, (2.19) says that $\bar{\mathbf{y}}_k - \widehat{\boldsymbol{\mu}}_{k-1}$ has length

$$\bar{\gamma}_k \equiv \frac{\widehat{C}_k}{A_k}. \tag{2.21}$$

Comparison with (2.12) shows that the LARS estimate $\widehat{\boldsymbol{\mu}}_k$ lies on the line

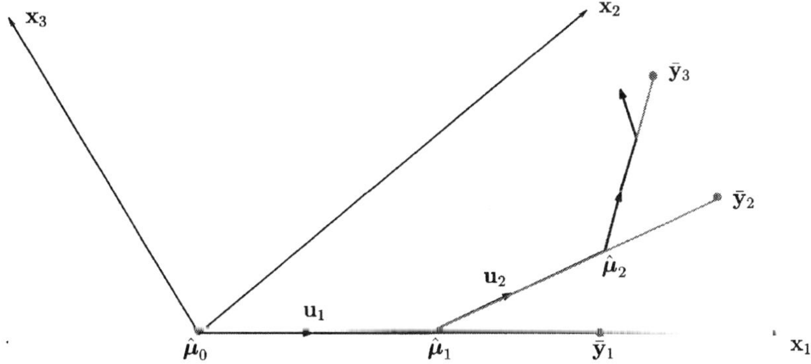

FIG. 4. *At each stage the LARS estimate $\widehat{\boldsymbol{\mu}}_k$ approaches, but does not reach, the corresponding OLS estimate $\bar{\mathbf{y}}_k$.*

from $\widehat{\boldsymbol{\mu}}_{k-1}$ to $\bar{\mathbf{y}}_k$,

$$\widehat{\boldsymbol{\mu}}_k - \widehat{\boldsymbol{\mu}}_{k-1} = \frac{\widehat{\gamma}_k}{\bar{\gamma}_k}(\bar{\mathbf{y}}_k - \widehat{\boldsymbol{\mu}}_{k-1}). \tag{2.22}$$

It is easy to see that $\widehat{\gamma}_k$, (2.12), is always less than $\bar{\gamma}_k$, so that $\widehat{\boldsymbol{\mu}}_k$ lies closer than $\bar{\mathbf{y}}_k$ to $\widehat{\boldsymbol{\mu}}_{k-1}$. Figure 4 shows the successive LARS estimates $\widehat{\boldsymbol{\mu}}_k$ always approaching but never reaching the OLS estimates $\bar{\mathbf{y}}_k$.

The exception is at the last stage: since $\mathcal{A}_m$ contains all covariates, (2.13) is not defined. By convention the algorithm takes $\widehat{\gamma}_m = \bar{\gamma}_m = \widehat{C}_m / A_m$, making $\widehat{\boldsymbol{\mu}}_m = \bar{\mathbf{y}}_m$ and $\widehat{\boldsymbol{\beta}}_m$ equal the OLS estimate for the full set of m covariates.

The LARS algorithm is computationally thrifty. Organizing the calculations correctly, the computational cost for the entire m steps is of the same order as that required for the usual Least Squares solution for the full set of m covariates. Section 7 describes an efficient LARS program available from the authors. With the modifications described in the next section, this program also provides economical Lasso and Stagewise solutions.

3. Modified versions of Least Angle Regression. Figures 1 and 3 show Lasso, Stagewise and LARS yielding remarkably similar estimates for the diabetes data. The similarity is no coincidence. This section describes simple modifications of the LARS algorithm that produce Lasso or Stagewise estimates. Besides improved computational efficiency, these relationships elucidate the methods' rationale: all three algorithms can be viewed as moderately greedy forward stepwise procedures whose forward progress is determined by compromise among the currently most correlated covariates. LARS moves along the most obvious compromise direction, the equiangular vector (2.6), while Lasso and Stagewise put some restrictions on the equiangular strategy.

3.1. *The LARS–Lasso relationship.* The full set of Lasso solutions, as shown for the diabetes study in Figure 1, can be generated by a minor modification of the LARS algorithm (2.8)–(2.13). Our main result is described here and verified in Section 5. It closely parallels the homotopy method in the papers by Osborne, Presnell and Turlach (2000a, b), though the LARS approach is somewhat more direct.

Let $\widehat{\boldsymbol{\beta}}$ be a Lasso solution (1.5), with $\widehat{\boldsymbol{\mu}} = X\widehat{\boldsymbol{\beta}}$. Then it is easy to show that the sign of any nonzero coordinate $\widehat{\beta}_j$ must agree with the sign s_j of the current correlation $\widehat{c}_j = \mathbf{x}'_j(\mathbf{y} - \widehat{\boldsymbol{\mu}})$,

$$\operatorname{sign}(\widehat{\beta}_j) = \operatorname{sign}(\widehat{c}_j) = s_j; \tag{3.1}$$

see Lemma 8 of Section 5. The LARS algorithm does not enforce restriction (3.1), but it can easily be modified to do so.

Suppose we have just completed a LARS step, giving a new active set $\mathcal{A}$ as in (2.9), and that the corresponding LARS estimate $\widehat{\boldsymbol{\mu}}_{\mathcal{A}}$ corresponds to a Lasso solution $\widehat{\boldsymbol{\mu}} = X\widehat{\boldsymbol{\beta}}$. Let

$$w_{\mathcal{A}} = A_{\mathcal{A}} \mathcal{G}_{\mathcal{A}}^{-1} \mathbf{1}_{\mathcal{A}}, \tag{3.2}$$

a vector of length the size of $\mathcal{A}$, and (somewhat abusing subscript notation) define $\widehat{\mathbf{d}}$ to be the m-vector equaling $s_j w_{\mathcal{A}j}$ for $j \in \mathcal{A}$ and zero elsewhere. Moving in the positive γ direction along the LARS line (2.14), we see that

$$\boldsymbol{\mu}(\gamma) = X\boldsymbol{\beta}(\gamma), \qquad \text{where } \beta_j(\gamma) = \widehat{\beta}_j + \gamma \widehat{d}_j \tag{3.3}$$

for $j \in \mathcal{A}$. Therefore $\beta_j(\gamma)$ will change sign at

$$\gamma_j = -\widehat{\beta}_j / \widehat{d}_j, \tag{3.4}$$

the first such change occurring at

$$\widetilde{\gamma} = \min_{\gamma_j > 0} \{\gamma_j\}, \tag{3.5}$$

say for covariate $x_{\widetilde{j}}$; $\widetilde{\gamma}$ equals infinity by definition if there is no $\gamma_j > 0$.

If $\widetilde{\gamma}$ is less than $\widehat{\gamma}$, (2.13), then $\beta_j(\gamma)$ cannot be a Lasso solution for $\gamma > \widetilde{\gamma}$ since the sign restriction (3.1) must be violated: $\beta_{\widetilde{j}}(\gamma)$ has changed sign while $c_{\widetilde{j}}(\gamma)$ has not. [The continuous function $c_{\widetilde{j}}(\gamma)$ cannot change sign within a single LARS step since $|c_{\widetilde{j}}(\gamma)| = \widehat{C} - \gamma A_{\mathcal{A}} > 0$, (2.16).]

LASSO MODIFICATION. If $\widetilde{\gamma} < \widehat{\gamma}$, stop the ongoing LARS step at $\gamma = \widetilde{\gamma}$ and remove $\widetilde{j}$ from the calculation of the next equiangular direction. That is,

$$\widehat{\boldsymbol{\mu}}_{\mathcal{A}_+} = \widehat{\boldsymbol{\mu}}_{\mathcal{A}} + \widetilde{\gamma}\mathbf{u}_{\mathcal{A}} \quad \text{and} \quad \mathcal{A}_+ = \mathcal{A} - \{\widetilde{j}\} \tag{3.6}$$

rather than (2.12).

THEOREM 1. *Under the Lasso modification, and assuming the "one at a time" condition discussed below, the LARS algorithm yields all Lasso solutions.*

The active sets $\mathcal{A}$ grow monotonically larger as the original LARS algorithm progresses, but the Lasso modification allows $\mathcal{A}$ to decrease. "One at a time" means that the increases and decreases never involve more than a single index j. This is the usual case for quantitative data and can always be realized by adding a little jitter to the y values. Section 5 discusses tied situations.

The Lasso diagram in Figure 1 was actually calculated using the modified LARS algorithm. Modification (3.6) came into play only once, at the arrowed point in the left panel. There $\mathcal{A}$ contained all 10 indices while $\mathcal{A}_+ = \mathcal{A} - \{7\}$. Variable 7 was restored to the active set one LARS step later, the next and last step then taking $\widehat{\boldsymbol{\beta}}$ all the way to the full OLS solution. The brief absence of variable 7 had an effect on the tracks of the others, noticeably $\widehat{\beta}_8$. The price of using Lasso instead of unmodified LARS comes in the form of added steps, 12 instead of 10 in this example. For the more complicated "quadratic model" of Section 4, the comparison was 103 Lasso steps versus 64 for LARS.

3.2. *The LARS–Stagewise relationship.* The staircase in Figure 2 indicates how the Stagewise algorithm might proceed forward from $\hat{\boldsymbol{\mu}}_1$, a point of equal current correlations $\hat{c}_1 = \hat{c}_2$, (2.8). The first small step has (randomly) selected index $j = 1$, taking us to $\hat{\boldsymbol{\mu}}_1 + \varepsilon \mathbf{x}_1$. Now variable 2 is more correlated,

$$\mathbf{x}_2'(\mathbf{y} - \hat{\boldsymbol{\mu}}_1 - \varepsilon \mathbf{x}_1) > \mathbf{x}_1'(\mathbf{y} - \hat{\boldsymbol{\mu}}_1 - \varepsilon \mathbf{x}_1), \tag{3.7}$$

forcing $j = 2$ to be the next Stagewise choice and so on.

We will consider an idealized Stagewise procedure in which the step size ε goes to zero. This collapses the staircase along the direction of the bisector $\mathbf{u}_2$ in Figure 2, making the Stagewise and LARS estimates agree. They always agree for $m = 2$ covariates, but another modification is necessary for LARS to produce Stagewise estimates in general. Section 6 verifies the main result described next.

Suppose that the Stagewise procedure has taken N steps of infinitesimal size ε from some previous estimate $\hat{\boldsymbol{\mu}}$, with

$$N_j \equiv \#\{\text{steps with selected index } j\}, \qquad j = 1, 2, \ldots, m. \tag{3.8}$$

It is easy to show, as in Lemma 11 of Section 6, that $N_j = 0$ for j not in the active set $\mathcal{A}$ defined by the current correlations $\mathbf{x}_j'(\mathbf{y} - \hat{\boldsymbol{\mu}})$, (2.9). Letting

$$P \equiv (N_1, N_2, \ldots, N_m)/N, \tag{3.9}$$

with $P_{\mathcal{A}}$ indicating the coordinates of P for $j \in \mathcal{A}$, the new estimate is

$$\boldsymbol{\mu} = \hat{\boldsymbol{\mu}} + N\varepsilon X_{\mathcal{A}} P_{\mathcal{A}} \qquad [(2.4)]. \tag{3.10}$$

(Notice that the Stagewise steps are taken along the directions $s_j \mathbf{x}_j$.)

The LARS algorithm (2.14) progresses along

$$\boldsymbol{\mu}_{\mathcal{A}} + \gamma X_{\mathcal{A}} w_{\mathcal{A}}, \qquad \text{where } w_{\mathcal{A}} = A_{\mathcal{A}} \mathcal{G}_{\mathcal{A}}^{-1} \mathbf{1}_{\mathcal{A}} \qquad [(2.6)\text{–}(3.2)]. \tag{3.11}$$

Comparing (3.10) with (3.11) shows that LARS cannot agree with Stagewise if $w_{\mathcal{A}}$ has negative components, since $P_{\mathcal{A}}$ is nonnegative. To put it another way, the direction of Stagewise progress $X_{\mathcal{A}} P_{\mathcal{A}}$ must lie in the convex cone generated by the columns of $X_{\mathcal{A}}$,

$$\mathcal{C}_{\mathcal{A}} = \left\{ \mathbf{v} = \sum_{j \in \mathcal{A}} s_j \mathbf{x}_j P_j,\ P_j \geq 0 \right\}. \tag{3.12}$$

If $\mathbf{u}_{\mathcal{A}} \in \mathcal{C}_{\mathcal{A}}$ then there is no contradiction between (3.12) and (3.13). If not it seems natural to replace $\mathbf{u}_{\mathcal{A}}$ with its projection into $\mathcal{C}_{\mathcal{A}}$, that is, the nearest point in the convex cone.

STAGEWISE MODIFICATION. Proceed as in (2.8)–(2.13), except with $\mathbf{u}_{\mathcal{A}}$ replaced by $\mathbf{u}_{\hat{\mathcal{B}}}$, the unit vector lying along the projection of $\mathbf{u}_{\mathcal{A}}$ into $\mathcal{C}_{\mathcal{A}}$. (See Figure 9 in Section 6.)

THEOREM 2. *Under the Stagewise modification, the LARS algorithm yields all Stagewise solutions.*

The vector $\mathbf{u}_{\hat{\mathcal{B}}}$ in the Stagewise modification is the equiangular vector (2.6) for the subset $\widehat{\mathcal{B}} \subseteq \mathcal{A}$ corresponding to the face of $\mathcal{C}_{\mathcal{A}}$ into which the projection falls. Stagewise is a LARS type algorithm that allows the active set to decrease by one or more indices. This happened at the arrowed point in the right panel of Figure 1: there the set $\mathcal{A} = \{3, 9, 4, 7, 2, 10, 5, 8\}$ was decreased to $\widehat{\mathcal{B}} = \mathcal{A} - \{3, 7\}$. It took a total of 13 modified LARS steps to reach the full OLS solution $\bar{\boldsymbol{\beta}}_m = (X'X)^{-1}X'\mathbf{y}$. The three methods, LARS, Lasso and Stagewise, always reach OLS eventually, but LARS does so in only m steps while Lasso and, especially, Stagewise can take longer. For the $m = 64$ quadratic model of Section 4, Stagewise took 255 steps.

According to Theorem 2 the difference between successive Stagewise–modified LARS estimates is

$$\widehat{\boldsymbol{\mu}}_{\mathcal{A}_+} - \widehat{\boldsymbol{\mu}}_{\mathcal{A}} = \widehat{\gamma}\mathbf{u}_{\hat{\mathcal{B}}} = \widehat{\gamma} X_{\hat{\mathcal{B}}} w_{\hat{\mathcal{B}}}, \tag{3.13}$$

as in (3.13). Since $\mathbf{u}_{\hat{\mathcal{B}}}$ exists in the convex cone $\mathcal{C}_{\mathcal{A}}$, $w_{\hat{\mathcal{B}}}$ must have nonnegative components. This says that the difference of successive coefficient estimates for coordinate $j \in \widehat{\mathcal{B}}$ satisfies

$$\operatorname{sign}(\widehat{\beta}_{+j} - \widehat{\beta}_j) = s_j, \tag{3.14}$$

where $s_j = \operatorname{sign}\{\mathbf{x}_j'(\mathbf{y} - \widehat{\boldsymbol{\mu}})\}$.

We can now make a useful comparison of the three methods:

1. *Stagewise*—successive differences of $\widehat{\beta}_j$ agree in sign with the current correlation $\widehat{c}_j = \mathbf{x}_j'(\mathbf{y} - \widehat{\boldsymbol{\mu}})$;
2. *Lasso*—$\widehat{\beta}_j$ agrees in sign with $\widehat{c}_j$;
3. *LARS*—no sign restrictions (but see Lemma 4 of Section 5).

From this point of view, Lasso is intermediate between the LARS and Stagewise methods.

The successive difference property (3.14) makes the Stagewise $\widehat{\beta}_j$ estimates move monotonically away from 0. Reversals are possible only if $\widehat{c}_j$ changes sign while $\widehat{\beta}_j$ is "resting" between two periods of change. This happened to variable 7 in Figure 1 between the 8th and 10th Stagewise-modified LARS steps.

3.3. *Simulation study.* A small simulation study was carried out comparing the LARS, Lasso and Stagewise algorithms. The X matrix for the simulation was based on the diabetes example of Table 1, but now using a "Quadratic Model" having $m = 64$ predictors, including interactions and squares of the 10 original covariates:

(3.15) *Quadratic Model* 10 main effects, 45 interactions, 9 squares, the last being the squares of each $\mathbf{x}_j$ except the dichotomous variable $\mathbf{x}_2$. The true mean vector $\boldsymbol{\mu}$ for the simulation was $\boldsymbol{\mu} = X\boldsymbol{\beta}$, where $\boldsymbol{\beta}$ was obtained by running LARS for 10 steps on the original $(X, \mathbf{y})$ diabetes data (agreeing in this case with the 10-step Lasso or Stagewise analysis). Subtracting $\boldsymbol{\mu}$ from a centered version of the original $\mathbf{y}$ vector of Table 1 gave a vector $\boldsymbol{\varepsilon} = \mathbf{y} - \boldsymbol{\mu}$ of $n = 442$ residuals. The "true R^2" for this model, $\|\boldsymbol{\mu}\|^2/(\|\boldsymbol{\mu}\|^2 + \|\boldsymbol{\varepsilon}\|^2)$, equaled 0.416.

100 simulated response vectors $\mathbf{y}^*$ were generated from the model

$$\mathbf{y}^* = \boldsymbol{\mu} + \boldsymbol{\varepsilon}^*, \tag{3.16}$$

with $\boldsymbol{\varepsilon}^* = (\varepsilon_1^*, \varepsilon_2^*, \ldots, \varepsilon_n^*)$ a random sample, with replacement, from the components of $\boldsymbol{\varepsilon}$. The LARS algorithm with $K = 40$ steps was run for each simulated data set $(X, \mathbf{y}^*)$, yielding a sequence of estimates $\widehat{\boldsymbol{\mu}}^{(k)*}$, $k = 1, 2, \ldots, 40$, and likewise using the Lasso and Stagewise algorithms.

Figure 5 compares the LARS, Lasso and Stagewise estimates. For a given estimate $\widehat{\boldsymbol{\mu}}$ define the *proportion explained* $\mathrm{pe}(\widehat{\boldsymbol{\mu}})$ to be

$$\mathrm{pe}(\widehat{\boldsymbol{\mu}}) = 1 - \|\widehat{\boldsymbol{\mu}} - \boldsymbol{\mu}\|^2/\|\boldsymbol{\mu}\|^2, \tag{3.17}$$

so $\mathrm{pe}(\mathbf{0}) = 0$ and $\mathrm{pe}(\boldsymbol{\mu}) = 1$. The solid curve graphs the average of $\mathrm{pe}(\widehat{\boldsymbol{\mu}}^{(k)*})$ over the 100 simulations, versus step number k for LARS, $k = 1, 2, \ldots, 40$. The corresponding curves are graphed for Lasso and Stagewise, except that the horizontal axis is now the average number of nonzero $\widehat{\beta}_j^*$ terms composing $\widehat{\boldsymbol{\mu}}^{(k)*}$. For example, $\widehat{\boldsymbol{\mu}}^{(40)*}$ averaged 33.23 nonzero terms with Stagewise, compared to 35.83 for Lasso and 40 for LARS.

Figure 5's most striking message is that the three algorithms performed almost identically, and rather well. The average proportion explained rises quickly,

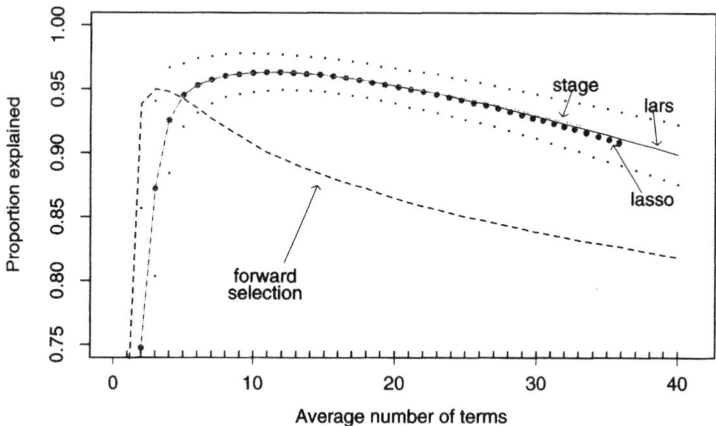

FIG. 5. *Simulation study comparing LARS, Lasso and Stagewise algorithms*; 100 *replications of model* (3.15)–(3.16). *Solid curve shows average proportion explained,* (3.17), *for LARS estimates as function of number of steps* $k = 1, 2, \ldots, 40$; *Lasso and Stagewise give nearly identical results*; *small dots indicate plus or minus one standard deviation over the* 100 *simulations. Classic Forward Selection* (*heavy dashed curve*) *rises and falls more abruptly.*

reaching a maximum of 0.963 at $k = 10$, and then declines slowly as k grows to 40. The light dots display the small standard deviation of $\mathrm{pe}(\widehat{\boldsymbol{\mu}}^{(k)*})$ over the 100 simulations, roughly ± 0.02. Stopping at any point between $k = 5$ and 25 typically gave a $\widehat{\boldsymbol{\mu}}^{(k)*}$ with true predictive R^2 about 0.40, compared to the ideal value 0.416 for $\boldsymbol{\mu}$.

The dashed curve in Figure 5 tracks the average proportion explained by classic Forward Selection. It rises very quickly, to a maximum of 0.950 after $k = 3$ steps, and then falls back more abruptly than the LARS–Lasso–Stagewise curves. This behavior agrees with the characterization of Forward Selection as a dangerously greedy algorithm.

3.4. *Other LARS modifications.* Here are a few more examples of LARS type model-building algorithms.

POSITIVE LASSO. Constraint (1.5) can be strengthened to

$$\text{minimize} \quad S(\widehat{\boldsymbol{\beta}}) \qquad \text{subject to} \quad T(\widehat{\boldsymbol{\beta}}) \le t \text{ and all } \widehat{\beta}_j \ge 0. \tag{3.18}$$

This would be appropriate if the statisticians or scientists believed that the variables x_j *must* enter the prediction equation in their defined directions. Situation (3.18) is a more difficult quadratic programming problem than (1.5), but it can be solved by a further modification of the Lasso-modified LARS algorithm: change $|\widehat{c}_j|$ to $\widehat{c}_j$ at both places in (2.9), set $s_j = 1$ instead of (2.10) and change (2.13) to

$$\widehat{\gamma} = \min_{j \in \mathcal{A}^c}{}^{+} \left\{ \frac{\widehat{C} - \widehat{c}_j}{A_{\mathcal{A}} - a_j} \right\}. \tag{3.19}$$

The positive Lasso usually does *not* converge to the full OLS solution $\bar{\beta}_m$, even for very large choices of t.

The changes above amount to considering the $\mathbf{x}_j$ as generating half-lines rather than full one-dimensional spaces. A positive Stagewise version can be developed in the same way, and has the property that the $\widehat{\beta}_j$ tracks are always monotone.

LARS–OLS hybrid. After k steps the LARS algorithm has identified a set $\mathcal{A}_k$ of covariates, for example, $\mathcal{A}_4 = \{3, 9, 4, 7\}$ in the diabetes study. Instead of $\widehat{\boldsymbol{\beta}}_k$ we might prefer $\bar{\boldsymbol{\beta}}_k$, the OLS coefficients based on the linear model with covariates in $\mathcal{A}_k$—using LARS to find the model but not to estimate the coefficients. Besides looking more familiar, this will always increase the usual empirical R^2 measure of fit (though not necessarily the true fitting accuracy),

$$R^2(\bar{\boldsymbol{\beta}}_k) - R^2(\widehat{\boldsymbol{\beta}}_k) = \frac{1 - \rho_k^2}{\rho_k(2 - \rho_k)}[R^2(\widehat{\boldsymbol{\beta}}_k) - R^2(\widehat{\boldsymbol{\beta}}_{k-1})], \tag{3.20}$$

where $\rho_k = \widehat{\gamma}_k / \bar{\gamma}_k$ as in (2.22).

The increases in R^2 were small in the diabetes example, on the order of 0.01 for $k \geq 4$ compared with $R^2 \doteq 0.50$, which is expected from (3.20) since we would usually continue LARS until $R^2(\widehat{\boldsymbol{\beta}}_k) - R^2(\widehat{\boldsymbol{\beta}}_{k-1})$ was small. For the same reason $\bar{\boldsymbol{\beta}}_k$ and $\widehat{\boldsymbol{\beta}}_k$ are likely to lie near each other as they did in the diabetes example.

Main effects first. It is straightforward to restrict the order in which variables are allowed to enter the LARS algorithm. For example, having obtained $\mathcal{A}_4 = \{3, 9, 4, 7\}$ for the diabetes study, we might *then* wish to check for interactions. To do this we begin LARS again, replacing $\mathbf{y}$ with $\mathbf{y} - \widehat{\boldsymbol{\mu}}_4$ and $\mathbf{x}$ with the $n \times 6$ matrix whose columns represent the interactions $\mathbf{x}_{3:9}, \mathbf{x}_{3:4}, \ldots, \mathbf{x}_{4:7}$.

Backward Lasso. The Lasso–modified LARS algorithm can be run backward, starting from the full OLS solution $\bar{\boldsymbol{\beta}}_m$. Assuming that all the coordinates of $\bar{\boldsymbol{\beta}}_m$ are nonzero, their signs must agree with the signs s_j that the current correlations had during the final LARS step. This allows us to calculate the last equiangular direction $\mathbf{u}_{\mathcal{A}}$, (2.4)–(2.6). Moving backward from $\widehat{\boldsymbol{\mu}}_m = X\bar{\boldsymbol{\beta}}_m$ along the line $\boldsymbol{\mu}(\gamma) = \widehat{\boldsymbol{\mu}}_m - \gamma \mathbf{u}_{\mathcal{A}}$, we eliminate from the active set the index of the first $\widehat{\beta}_j$ that becomes zero. Continuing backward, we keep track of all coefficients $\widehat{\beta}_j$ and current correlations $\widehat{c}_j$, following essentially the same rules for changing $\mathcal{A}$ as in Section 3.1. As in (2.3), (3.5) the calculation of $\widetilde{\gamma}$ and $\widehat{\gamma}$ is easy.

The crucial property of the Lasso that makes backward navigation possible is (3.1), which permits calculation of the correct equiangular direction $\mathbf{u}_{\mathcal{A}}$ at each step. In this sense Lasso can be just as well thought of as a backward-moving algorithm. This is not the case for LARS or Stagewise, both of which are inherently forward-moving algorithms.

4. Degrees of freedom and C_p estimates. Figures 1 and 3 show all possible Lasso, Stagewise or LARS estimates of the vector $\boldsymbol{\beta}$ for the diabetes data. The scientists want just a single $\widehat{\boldsymbol{\beta}}$ of course, so we need some rule for selecting among the possibilities. This section concerns a C_p-type selection criterion, especially as it applies to the choice of LARS estimate.

Let $\widehat{\boldsymbol{\mu}} = g(\mathbf{y})$ represent a formula for estimating $\boldsymbol{\mu}$ from the data vector $\mathbf{y}$. Here, as usual in regression situations, we are considering the covariate vectors $\mathbf{x}_1, \mathbf{x}_2, \ldots, \mathbf{x}_m$ fixed at their observed values. We assume that given the $\mathbf{x}$'s, $\mathbf{y}$ is generated according to an homoskedastic model

$$\mathbf{y} \sim (\boldsymbol{\mu}, \sigma^2 \mathbf{I}), \tag{4.1}$$

meaning that the components y_i are uncorrelated, with mean μ_i and variance σ^2. Taking expectations in the identity

$$(\widehat{\mu}_i - \mu_i)^2 = (y_i - \widehat{\mu}_i)^2 - (y_i - \mu_i)^2 + 2(\widehat{\mu}_i - \mu_i)(y_i - \mu_i), \tag{4.2}$$

and summing over i, yields

$$E\left\{\frac{\|\widehat{\boldsymbol{\mu}}-\boldsymbol{\mu}\|^2}{\sigma^2}\right\}=E\left\{\frac{\|\mathbf{y}-\widehat{\boldsymbol{\mu}}\|^2}{\sigma^2}-n\right\}+2\sum_{i=1}^{n}\frac{\operatorname{cov}(\widehat{\mu}_i,y_i)}{\sigma^2}. \tag{4.3}$$

The last term of (4.3) leads to a convenient definition of the *degrees of freedom* for an estimator $\widehat{\boldsymbol{\mu}}=g(\mathbf{y})$,

$$df_{\mu,\sigma^2}=\sum_{i=1}^{n}\operatorname{cov}(\widehat{\mu}_i,y_i)/\sigma^2, \tag{4.4}$$

and a C_p-type risk estimation formula,

$$C_p(\widehat{\boldsymbol{\mu}})=\frac{\|\mathbf{y}-\widehat{\boldsymbol{\mu}}\|^2}{\sigma^2}-n+2df_{\mu,\sigma^2}. \tag{4.5}$$

If σ^2 and df_{μ,σ^2} are known, $C_p(\widehat{\boldsymbol{\mu}})$ is an unbiased estimator of the true risk $E\{\|\widehat{\boldsymbol{\mu}}-\boldsymbol{\mu}\|^2/\sigma^2\}$. For linear estimators $\widehat{\boldsymbol{\mu}}=M\mathbf{y}$, model (4.1) makes $df_{\mu,\sigma^2}=\operatorname{trace}(M)$, equaling the usual definition of degrees of freedom for OLS, and coinciding with the proposal of Mallows (1973). Section 6 of Efron and Tibshirani (1997) and Section 7 of Efron (1986) discuss formulas (4.4) and (4.5) and their role in C_p, Akaike information criterion (AIC) and Stein's unbiased risk estimated (SURE) estimation theory, a more recent reference being Ye (1998).

Practical use of C_p formula (4.5) requires preliminary estimates of $\boldsymbol{\mu},\sigma^2$ and df_{μ,σ^2}. In the numerical results below, the usual OLS estimates $\bar{\boldsymbol{\mu}}$ and $\bar{\sigma}^2$ from the full OLS model were used to calculate bootstrap estimates of df_{μ,σ^2}; bootstrap samples $\mathbf{y}^*$ and replications $\widehat{\boldsymbol{\mu}}^*$ were then generated according to

$$\mathbf{y}^*\sim N(\bar{\boldsymbol{\mu}},\bar{\sigma}^2)\quad\text{and}\quad\widehat{\boldsymbol{\mu}}^*=g(\mathbf{y}^*). \tag{4.6}$$

Independently repeating (4.6) say B times gives straightforward estimates for the covariances in (4.4),

$$\widehat{\operatorname{cov}}_i=\frac{\sum_{b=1}^{B}\widehat{\boldsymbol{\mu}}_i^*(b)[\mathbf{y}_i^*(b)-\mathbf{y}_i^*(\cdot)]}{B-1},\qquad\text{where }\mathbf{y}^*(\cdot)=\frac{\sum_{b=1}^{B}\mathbf{y}^*(b)}{B}, \tag{4.7}$$

and then

$$\widehat{df}=\sum_{i=1}^{n}\widehat{\operatorname{cov}}_i/\bar{\sigma}^2. \tag{4.8}$$

Normality is not crucial in (4.6). Nearly the same results were obtained using $\mathbf{y}^*=\bar{\boldsymbol{\mu}}^*+\mathbf{e}^*$, where the components of $\mathbf{e}^*$ were resampled from $\mathbf{e}=\mathbf{y}-\bar{\boldsymbol{\mu}}$.

The left panel of Figure 6 shows $\widehat{df}_k$ for the diabetes data LARS estimates $\widehat{\boldsymbol{\mu}}_k, k=1,2,\dots,m=10$. It portrays a startlingly simple situation that we will call the "simple approximation,"

$$df(\widehat{\boldsymbol{\mu}}_k)\doteq k. \tag{4.9}$$

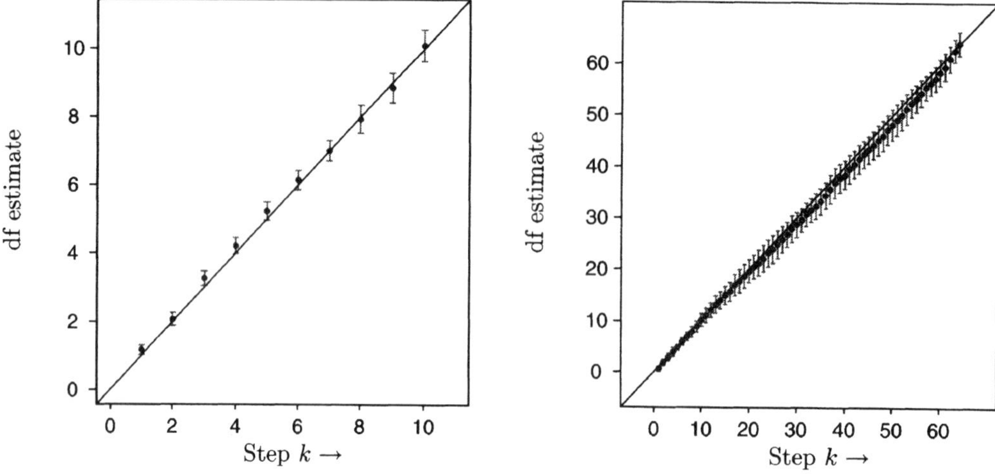

FIG. 6. *Degrees of freedom for LARS estimates* $\widehat{\mu}_k$: (left) *diabetes study, Table* 1, $k = 1, 2, \ldots, m = 10$; (right) *quadratic model* (3.15) *for the diabetes data,* $m = 64$. *Solid line is simple approximation* $df_k = k$. *Dashed lines are approximate* 95% *confidence intervals for the bootstrap estimates. Each panel based on* $B = 500$ *bootstrap replications.*

The right panel also applies to the diabetes data, but this time with the quadratic model (3.15), having $m = 64$ predictors. We see that the simple approximation (4.9) is again accurate within the limits of the bootstrap computation (4.8), where $B = 500$ replications were divided into 10 groups of 50 each in order to calculate Student-t confidence intervals.

If (4.9) can be believed, and we will offer some evidence in its behalf, we can estimate the risk of a k-step LARS estimator $\widehat{\mu}_k$ by

$$C_p(\widehat{\mu}_k) \doteq \|\mathbf{y} - \widehat{\mu}_k\|^2/\bar{\sigma}^2 - n + 2k. \tag{4.10}$$

The formula, which is the same as the C_p estimate of risk for an OLS estimator based on a subset of k preselected predictor vectors, has the great advantage of not requiring any further calculations beyond those for the original LARS estimates. The formula applies only to LARS, and not to Lasso or Stagewise.

Figure 7 displays $C_p(\widehat{\mu}_k)$ as a function of k for the two situations of Figure 6. Minimum C_p was achieved at steps $k = 7$ and $k = 16$, respectively. Both of the minimum C_p models looked sensible, their first several selections of "important" covariates agreeing with an earlier model based on a detailed inspection of the data assisted by medical expertise.

The simple approximation becomes a theorem in two cases.

THEOREM 3. *If the covariate vectors* $\mathbf{x}_1, \mathbf{x}_2, \ldots, \mathbf{x}_m$ *are mutually orthogonal, then the k-step LARS estimate* $\hat{\mu}_k$ *has* $df(\hat{\mu}_k) = k$.

To state the second more general setting we introduce the following condition.

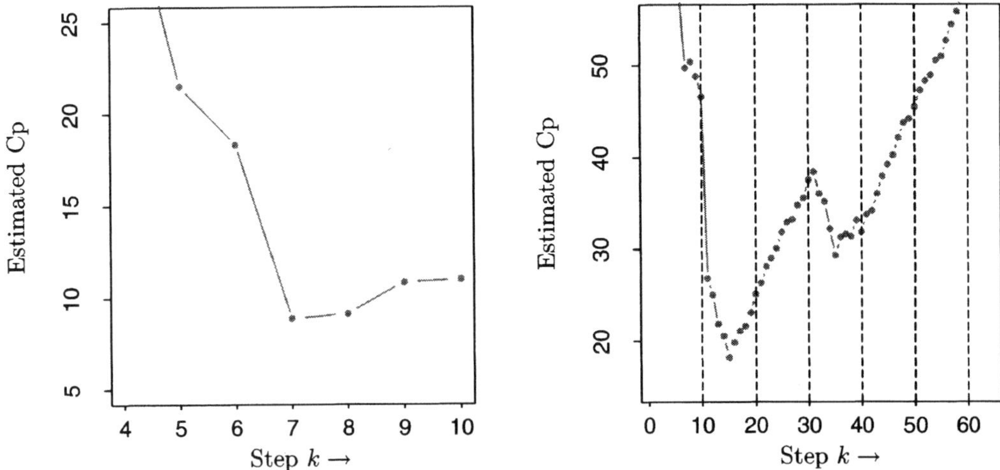

FIG. 7. *C_p estimates of risk* (4.10) *for the two situations of Figure* 6: (left) $m = 10$ *model has smallest* C_p *at* $k = 7$; (right) $m = 64$ *model has smallest* C_p *at* $k = 16$.

POSITIVE CONE CONDITION. For all possible subsets $X_{\mathcal{A}}$ of the full design matrix X,

$$G_{\mathcal{A}}^{-1}\mathbf{1}_{\mathcal{A}} > 0, \tag{4.11}$$

where the inequality is taken element-wise.

The positive cone condition holds if X is orthogonal. It is strictly more general than orthogonality, but counterexamples (such as the diabetes data) show that not all design matrices X satisfy it.

It is also easy to show that LARS, Lasso and Stagewise all coincide under the positive cone condition, so the degrees-of-freedom formula applies to them too in this case.

THEOREM 4. *Under the positive cone condition,* $df(\hat{\boldsymbol{\mu}}_k) = k$.

The proof, which appears later in this section, is an application of Stein's unbiased risk estimate (SURE) [Stein (1981)]. Suppose that $g : \mathbb{R}^n \rightarrow \mathbb{R}^n$ is almost differentiable (see Remark A.1 in the Appendix) and set $\nabla \cdot g = \sum_{i=1}^n \partial g_i / \partial x_i$. If $\mathbf{y} \sim N_n(\boldsymbol{\mu}, \sigma^2\mathbf{I})$, then Stein's formula states that

$$\sum_{i=1}^n \operatorname{cov}(g_i, y_i)/\sigma^2 = E[\nabla \cdot g(\mathbf{y})]. \tag{4.12}$$

The left-hand side is $df(g)$ for the general estimator $g(\mathbf{y})$. Focusing specifically on LARS, it will turn out that $\nabla \cdot \hat{\boldsymbol{\mu}}_k(\mathbf{y}) = k$ in *all* situations with probability 1, but that the continuity assumptions underlying (4.12) and SURE can fail in certain nonorthogonal cases where the positive cone condition does not hold.

A range of simulations suggested that the simple approximation is quite accurate even when the $\mathbf{x}_j$'s are highly correlated and that it requires concerted effort at pathology to make $df(\widehat{\boldsymbol{\mu}}_k)$ much different than k.

Stein's formula assumes normality, $\mathbf{y} \sim N(\boldsymbol{\mu}, \sigma^2\mathbf{I})$. A cruder "delta method" rationale for the simple approximation requires only homoskedasticity, (4.1). The geometry of Figure 4 implies

$$\widehat{\boldsymbol{\mu}}_k = \bar{\mathbf{y}}_k - \cot_k \cdot \|\bar{\mathbf{y}}_{k+1} - \bar{\mathbf{y}}_k\|, \tag{4.13}$$

where $\cot_k$ is the cotangent of the angle between $\mathbf{u}_k$ and $\mathbf{u}_{k+1}$,

$$\cot_k = \frac{\mathbf{u}_k'\mathbf{u}_{k+1}}{[1 - (\mathbf{u}_k'\mathbf{u}_{k+1})^2]^{1/2}}. \tag{4.14}$$

Let $\mathbf{v}_k$ be the unit vector orthogonal to $\mathcal{L}(X_b)$, the linear space spanned by the first k covariates selected by LARS, and pointing into $\mathcal{L}(X_{k+1})$ along the direction of $\bar{\mathbf{y}}_{k+1} - \bar{\mathbf{y}}_k$. For $\mathbf{y}^*$ near $\mathbf{y}$ we can reexpress (4.13) as a locally linear transformation,

$$\widehat{\boldsymbol{\mu}}_k^* = \widehat{\boldsymbol{\mu}}_k + M_k(\mathbf{y}^* - \mathbf{y}) \qquad \text{with } M_k = P_k - \cot_k \cdot \mathbf{u}_k\mathbf{v}_k', \tag{4.15}$$

P_k being the usual projection matrix from $\mathbb{R}^n$ into $\mathcal{L}(X_k)$; (4.15) holds within a neighborhood of $\mathbf{y}$ such that the LARS choices $\mathcal{L}(X_k)$ and $\mathbf{v}_k$ remain the same.

The matrix M_k has $\text{trace}(M_k) = k$. Since the trace equals the degrees of freedom for linear estimators, the simple approximation (4.9) is seen to be a delta method approximation to the bootstrap estimates (4.6) and (4.7).

It is clear that (4.9) $df(\widehat{\boldsymbol{\mu}}_k) \doteq k$ cannot hold for the Lasso, since the degrees of freedom is m for the full model but the total number of steps taken can exceed m. However, we have found empirically that an intuitively plausible result holds: the degrees of freedom is well approximated by the number of nonzero predictors in the model. Specifically, starting at step 0, let $\ell(k)$ be the index of the last model in the Lasso sequence containing k predictors. Then $df(\widehat{\boldsymbol{\mu}}_{\ell(k)}) \doteq k$. We do not yet have any mathematical support for this claim.

4.1. *Orthogonal designs.* In the orthogonal case, we assume that $\mathbf{x}_j = \mathbf{e}_j$ for $j = 1, \dots, m$. The LARS algorithm then has a particularly simple form, reducing to soft thresholding at the order statistics of the data.

To be specific, define the soft thresholding operation on a scalar y_1 at threshold t by

$$\eta(y_1; t) = \begin{cases} y_1 - t, & \text{if } y_1 > t, \\ 0, & \text{if } |y_1| \le t, \\ y_1 + t, & \text{if } y_1 < -t. \end{cases}$$

The order statistics of the absolute values of the data are denoted by

$$|y|_{(1)} \ge |y|_{(2)} \ge \cdots \ge |y|_{(n)} \ge |y|_{(n+1)} := 0. \tag{4.16}$$

We note that $y_{m+1}, \dots, y_n$ do not enter into the estimation procedure, and so we may as well assume that $m = n$.

LEMMA 1. *For an orthogonal design with* $\mathbf{x}_j = \mathbf{e}_j$, $j = 1, \dots, n$, *the kth LARS estimate* $(0 \le k \le n)$ *is given by*

$$\hat{\mu}_{k,i}(\mathbf{y}) = \begin{cases} y_i - |y|_{(k+1)}, & \text{if } y_i > |y|_{(k+1)}, \\ 0, & \text{if } |y_i| \le |y|_{(k+1)}, \\ y_i + |y|_{(k+1)}, & \text{if } y_i < -|y|_{(k+1)}, \end{cases} \tag{4.17}$$

$$= \eta\big(y_i; |y|_{(k+1)}\big). \tag{4.18}$$

PROOF. The proof is by induction, stepping through the LARS sequence. First note that the LARS parameters take a simple form in the orthogonal setting:

$$G_{\mathcal{A}} = I_{\mathcal{A}}, \qquad A_{\mathcal{A}} = |\mathcal{A}|^{-1/2}, \qquad \mathbf{u}_{\mathcal{A}} = |\mathcal{A}|^{-1/2}\mathbf{1}_{\mathcal{A}}, \qquad a_{k,j} = 0, \qquad j \notin \mathcal{A}_k.$$

We assume for the moment that there are no ties in the order statistics (4.16), so that the variables enter one at a time. Let $j(l)$ be the index corresponding to the lth order statistic, $|y|_{(l)} = s_l y_{j(l)}$: we will see that $\mathcal{A}_k = \{j(1), \dots, j(k)\}$.

We have $\mathbf{x}_j'\mathbf{y} = y_j$, and so at the first step LARS picks variable $j(1)$ and sets $\hat{C}_1 = |y|_{(1)}$. It is easily seen that

$$\hat{\gamma}_1 = \min_{j \ne j(1)} \{|y|_{(1)} - |y_j|\} = |y|_{(1)} - |y|_{(2)}$$

and so

$$\hat{\boldsymbol{\mu}}_1 = [|y|_{(1)} - |y|_{(2)}]\mathbf{e}_{j(1)},$$

which is precisely (4.17) for $k = 1$.

Suppose now that step $k - 1$ has been completed, so that $\mathcal{A}_k = \{j(1), \dots, j(k)\}$ and (4.17) holds for $\hat{\boldsymbol{\mu}}_{k-1}$. The current correlations $\hat{C}_k = |y|_{(k)}$ and $\hat{c}_{k,j} = y_j$ for $j \notin \mathcal{A}_k$. Since $A_k - a_{k,j} = k^{-1/2}$, we have

$$\hat{\gamma}_k = \min_{j \notin \mathcal{A}_k} k^{1/2}\{|y|_{(k)} - |y_j|\}$$

and

$$\hat{\gamma}_k \mathbf{u}_k = [|y|_{(k)} - |y|_{(k+1)}]\mathbf{1}\{j \in \mathcal{A}_k\}.$$

Adding this term to $\hat{\boldsymbol{\mu}}_{k-1}$ yields (4.17) for step k.

The argument clearly extends to the case in which there are ties in the order statistics (4.16): if $|y|_{(k+1)} = \cdots = |y|_{(k+r)}$, then $\mathcal{A}_k(\mathbf{y})$ expands by r variables at step $k + 1$ and $\hat{\boldsymbol{\mu}}_{k+\nu}(\mathbf{y})$, $\nu = 1, \dots, r$, are all determined at the same time and are equal to $\hat{\boldsymbol{\mu}}_{k+1}(\mathbf{y})$. □

PROOF OF THEOREM 4 (Orthogonal case). The argument is particularly simple in this setting, and so worth giving separately. First we note from (4.17) that $\hat{\boldsymbol{\mu}}_k$ is continuous and Lipschitz(1) and so certainly almost differentiable.

Hence (4.12) shows that we simply have to calculate $\nabla \cdot \hat{\boldsymbol{\mu}}_k$. Inspection of (4.17) shows that

$$\nabla \cdot \hat{\boldsymbol{\mu}}_k = \sum_i \frac{\partial \hat{\mu}_{k,i}}{\partial y_i}(\mathbf{y}) = \sum_i I\{|y_i| > |y|_{(k+1)}\} = k$$

almost surely, that is, except for ties. This completes the proof. □

4.2. *The divergence formula.* While for the most general design matrices X, it can happen that $\hat{\boldsymbol{\mu}}_k$ fails to be almost differentiable, we will see that the divergence formula

$$\nabla \cdot \hat{\boldsymbol{\mu}}_k(\mathbf{y}) = k \tag{4.19}$$

does hold almost everywhere. Indeed, certain authors [e.g., Meyer and Woodroofe (2000)] have argued that the divergence $\nabla \cdot \hat{\boldsymbol{\mu}}$ of an estimator provides itself a useful measure of the effective dimension of a model.

Turning to LARS, we shall say that $\hat{\boldsymbol{\mu}}(\mathbf{y})$ is locally linear at a data point y_0 if there is some small open neighborhood of y_0 on which $\hat{\boldsymbol{\mu}}(\mathbf{y}) = M\mathbf{y}$ is exactly linear. Of course, the matrix $M = M(y_0)$ can depend on y_0—in the case of LARS, it will be seen to be constant on the interior of polygonal regions, with jumps across the boundaries. We say that a set G has full measure if its complement has Lebesgue measure zero.

LEMMA 2. *There is an open set G_k of full measure such that, at all $\mathbf{y} \in G_k$, $\hat{\boldsymbol{\mu}}_k(\mathbf{y})$ is locally linear and $\nabla \cdot \hat{\boldsymbol{\mu}}_k(\mathbf{y}) = k$.*

PROOF. We give here only the part of the proof that relates to actual calculation of the divergence in (4.19). The arguments establishing continuity and local linearity are delayed to the Appendix.

So, let us fix a point $\mathbf{y}$ in the interior of G_k. From Lemma 13 in the Appendix, this means that near $\mathbf{y}$ the active set $\mathcal{A}_k(\mathbf{y})$ is locally constant, that a single variable enters at the next step, this variable being the same near $\mathbf{y}$. In addition, $\hat{\boldsymbol{\mu}}_k(\mathbf{y})$ is locally linear, and hence in particular differentiable. Since $G_k \subset G_l$ for $l < k$, the same story applies at all previous steps and we have

$$\hat{\boldsymbol{\mu}}_k(\mathbf{y}) = \sum_{l=1}^{k} \gamma_l(\mathbf{y})\mathbf{u}_l. \tag{4.20}$$

Differentiating the jth component of vector $\hat{\boldsymbol{\mu}}_k(\mathbf{y})$ yields

$$\frac{\partial \hat{\mu}_{k,j}}{\partial y_i}(\mathbf{y}) = \sum_{l=1}^{k} \frac{\partial \gamma_l(\mathbf{y})}{\partial y_i} u_{l,j}.$$

In particular, for the divergence

$$\nabla \cdot \hat{\boldsymbol{\mu}}_k(\mathbf{y}) = \sum_{i=1}^{n} \frac{\partial \hat{\mu}_{k,i}}{\partial y_i} = \sum_{l=1}^{k} \langle \nabla \gamma_l, \mathbf{u}_l \rangle, \tag{4.21}$$

the brackets indicating inner product.

The active set is $\mathcal{A}_k = \{1, 2, \ldots, k\}$ and $\mathbf{x}_{k+1}$ is the variable to enter next. For $k \geq 2$, write $\boldsymbol{\delta}_k = \mathbf{x}_l - \mathbf{x}_k$ for any choice $l < k$—as remarked in the Conventions in the Appendix, the choice of l is immaterial (e.g., $l = 1$ for definiteness). Let $b_{k+1} = \langle \boldsymbol{\delta}_{k+1}, \mathbf{u}_k \rangle$, which is nonzero, as argued in the proof of Lemma 13. As shown in (A.4) in the Appendix, (2.13) can be rewritten

$$\gamma_k(\mathbf{y}) = b_{k+1}^{-1} \langle \boldsymbol{\delta}_{k+1}, \mathbf{y} - \hat{\boldsymbol{\mu}}_{k-1} \rangle. \tag{4.22}$$

For $k \geq 2$, define the linear space of vectors equiangular with the active set

$$\mathcal{L}_k = \mathcal{L}_k(\mathbf{y}) = \{\mathbf{u} : \langle \mathbf{x}_1, \mathbf{u} \rangle = \cdots = \langle \mathbf{x}_k, \mathbf{u} \rangle \text{ for } \mathbf{x}_l \text{ with } l \in \mathcal{A}_k(\mathbf{y})\}.$$

[We may drop the dependence on $\mathbf{y}$ since $\mathcal{A}_k(\mathbf{y})$ is locally fixed.] Clearly $\dim \mathcal{L}_k = n - k + 1$ and

$$\mathbf{u}_k \in \mathcal{L}_k, \qquad \mathcal{L}_{k+1} \subset \mathcal{L}_k. \tag{4.23}$$

We shall now verify that, for each $k \geq 1$,

$$\langle \nabla \gamma_k, \mathbf{u}_k \rangle = 1 \quad \text{and} \quad \langle \nabla \gamma_k, \mathbf{u} \rangle = 0 \qquad \text{for } \mathbf{u} \in \mathcal{L}_{k+1}. \tag{4.24}$$

Formula (4.21) shows that this suffices to prove Lemma 2.

First, for $k = 1$ we have $\gamma_1(\mathbf{y}) = b_2^{-1} \langle \boldsymbol{\delta}_2, \mathbf{y} \rangle$ and $\langle \nabla \gamma_1, \mathbf{u} \rangle = b_2^{-1} \langle \boldsymbol{\delta}_2, \mathbf{u} \rangle$, and that

$$\langle \boldsymbol{\delta}_2, \mathbf{u} \rangle = \langle \mathbf{x}_1 - \mathbf{x}_2, \mathbf{u} \rangle = \begin{cases} b_2, & \text{if } \mathbf{u} = \mathbf{u}_1, \\ 0, & \text{if } \mathbf{u} \in \mathcal{L}_2. \end{cases}$$

Now, for general k, combine (4.22) and (4.20):

$$b_{k+1} \gamma_k(\mathbf{y}) = \langle \boldsymbol{\delta}_{k+1}, \mathbf{y} \rangle - \sum_{l=1}^{k-1} \langle \boldsymbol{\delta}_{k+1}, \mathbf{u}_l \rangle \gamma_l(\mathbf{y}),$$

and hence

$$b_{k+1} \langle \nabla \gamma_k, \mathbf{u} \rangle = \langle \boldsymbol{\delta}_{k+1}, \mathbf{u} \rangle - \sum_{l=1}^{k-1} \langle \boldsymbol{\delta}_{k+1}, \mathbf{u}_l \rangle \langle \nabla \gamma_l, \mathbf{u} \rangle.$$

From the definitions of b_{k+1} and $\mathcal{L}_{k+1}$ we have

$$\langle \boldsymbol{\delta}_{k+1}, \mathbf{u} \rangle = \langle \mathbf{x}_l - \mathbf{x}_{k+1} \rangle = \begin{cases} b_{k+1}, & \text{if } \mathbf{u} = \mathbf{u}_k, \\ 0, & \text{if } \mathbf{u} \in \mathcal{L}_{k+1}. \end{cases}$$

Hence the truth of (4.24) for step k follows from its truth at step $k - 1$ because of the containment properties (4.23). □

4.3. *Proof of Theorem* 4. To complete the proof of Theorem 4, we state the following regularity result, proved in the Appendix.

LEMMA 3. *Under the positive cone condition, $\hat{\boldsymbol{\mu}}_k(\mathbf{y})$ is continuous and almost differentiable.*

This guarantees that Stein's formula (4.12) is valid for $\hat{\boldsymbol{\mu}}_k$ under the positive cone condition, so the divergence formula of Lemma 2 then immediately yields Theorem 4.

5. LARS and Lasso properties. The LARS and Lasso algorithms are described more carefully in this section, with an eye toward fully understanding their relationship. Theorem 1 of Section 3 will be verified. The latter material overlaps results in Osborne, Presnell and Turlach (2000a), particularly in their Section 4. Our point of view here allows the Lasso to be described as a quite simple modification of LARS, itself a variation of traditional Forward Selection methodology, and in this sense should be more accessible to statistical audiences. In any case we will stick to the language of regression and correlation rather than convex optimization, though some of the techniques are familiar from the optimization literature.

The results will be developed in a series of lemmas, eventually lending to a proof of Theorem 1 and its generalizations. The first three lemmas refer to attributes of the LARS procedure that are not specific to its Lasso modification.

Using notation as in (2.17)–(2.20), suppose LARS has completed step $k-1$, giving estimate $\widehat{\boldsymbol{\mu}}_{k-1}$ and active set $\mathcal{A}_k$ for step k, with covariate $\mathbf{x}_k$ the newest addition to the active set.

LEMMA 4. *If $\mathbf{x}_k$ is the only addition to the active set at the end of step $k-1$, then the coefficient vector $w_k = A_k \mathcal{G}_k^{-1} \mathbf{1}_k$ for the equiangular vector $\mathbf{u}_k = X_k w_k$, (2.6), has its kth component w_{kk} agreeing in sign with the current correlation $c_{kk} = \mathbf{x}_k'(\mathbf{y} - \widehat{\boldsymbol{\mu}}_{k-1})$. Moreover, the regression vector $\widehat{\boldsymbol{\beta}}_k$ for $\widehat{\boldsymbol{\mu}}_k = X\widehat{\boldsymbol{\beta}}_k$ has its kth component $\widehat{\beta}_{kk}$ agreeing in sign with c_{kk}.*

Lemma 4 says that new variables *enter* the LARS active set in the "correct" direction, a weakened version of the Lasso requirement (3.1). This will turn out to be a crucial connection for the LARS–Lasso relationship.

PROOF OF LEMMA 4. The case $k = 1$ is apparent. Note that since

$$X_k'(\mathbf{y} - \widehat{\boldsymbol{\mu}}_{k-1}) = \widehat{C}_k \mathbf{1}_k,$$

(2.20), from (2.6) we have

$$w_k = A_k \widehat{C}_k^{-1}[(X_k' X_k)^{-1} X_k'(\mathbf{y} - \widehat{\boldsymbol{\mu}}_{k-1})] := A_k \widehat{C}_k^{-1} w_k^*. \tag{5.1}$$

The term in square braces is the least squares coefficient vector in the regression of the current residual on X_k, and the term preceding it is positive.

Note also that

$$X_k'(\mathbf{y} - \bar{\mathbf{y}}_{k-1}) = (\mathbf{0}, \delta)' \qquad \text{with } \delta > 0, \tag{5.2}$$

since $X_{k-1}'(\mathbf{y} - \bar{\mathbf{y}}_{k-1}) = \mathbf{0}$ by definition (this $\mathbf{0}$ has $k-1$ elements), and $c_k(\gamma) = \mathbf{x}_k'(\mathbf{y} - \gamma \mathbf{u}_{k-1})$ decreases more slowly in γ than $c_j(\gamma)$ for $j \in \mathcal{A}_{k-1}$:

$$c_k(\gamma) \begin{cases} < c_j(\gamma), & \text{for } \gamma < \widehat{\gamma}_{k-1}, \\ = c_j(\gamma) = \widehat{C}_k, & \text{for } \gamma = \widehat{\gamma}_{k-1}, \\ > c_j(\gamma), & \text{for } \widehat{\gamma}_{k-1} < \gamma < \bar{\gamma}_{k-1}. \end{cases} \tag{5.3}$$

Thus

$$\widehat{w}_k^* = (X_k'X_k)^{-1}X_k'(\mathbf{y} - \bar{\mathbf{y}}_{k-1} + \bar{\mathbf{y}}_{k-1} - \widehat{\boldsymbol{\mu}}_{k-1}) \tag{5.4}$$

$$= (X_k'X_k)^{-1}\begin{pmatrix}\mathbf{0}\\ \delta\end{pmatrix} + (X_k'X_k)^{-1}X_k'[(\bar{\gamma}_{k-1} - \widehat{\gamma}_{k-1})\mathbf{u}_{k-1}]. \tag{5.5}$$

The kth element of $\widehat{w}_k^*$ is positive, because it is in the first term in (5.5) [$(X_k'X_k)$ is positive definite], and in the second term it is 0 since $\mathbf{u}_{k-1} \in \mathcal{L}(X_{k-1})$.

This proves the first statement in Lemma 4. The second follows from

$$\widehat{\beta}_{kk} = \widehat{\beta}_{k-1,k} + \widehat{\gamma}_k w_{kk}, \tag{5.6}$$

and $\widehat{\beta}_{k-1,k} = 0$, $\mathbf{x}_k$ not being active before step k. □

Our second lemma interprets the quantity $A_{\mathcal{A}} = (\mathbf{1}'\mathcal{G}_{\mathcal{A}}^{-1}\mathbf{1})^{-1/2}$, (2.4) and (2.5). Let $\mathcal{S}_{\mathcal{A}}$ indicate the extended simplex generated by the columns of $X_{\mathcal{A}}$,

$$\mathcal{S}_{\mathcal{A}} = \left\{\mathbf{v} = \sum_{j\in\mathcal{A}} s_j \mathbf{x}_j P_j : \sum_{j\in\mathcal{A}} P_j = 1\right\}, \tag{5.7}$$

"extended" meaning that the coefficients P_j are allowed to be negative.

LEMMA 5. *The point in $\mathcal{S}_{\mathcal{A}}$ nearest the origin is*

$$\mathbf{v}_{\mathcal{A}} = A_{\mathcal{A}}\mathbf{u}_{\mathcal{A}} = A_{\mathcal{A}}X_{\mathcal{A}}w_{\mathcal{A}} \qquad \text{where } w_{\mathcal{A}} = A_{\mathcal{A}}\mathcal{G}_{\mathcal{A}}^{-1}\mathbf{1}_{\mathcal{A}}, \tag{5.8}$$

with length $\|\mathbf{v}_{\mathcal{A}}\| = A_{\mathcal{A}}$. If $\mathcal{A} \subseteq \mathcal{B}$, then $A_{\mathcal{A}} \geq A_{\mathcal{B}}$, the largest possible value being $A_{\mathcal{A}} = 1$ for $\mathcal{A}$ a singleton.

PROOF. For any $\mathbf{v} \in \mathcal{S}_{\mathcal{A}}$, the squared distance to the origin is $\|X_{\mathcal{A}}P\|^2 = P'\mathcal{G}_{\mathcal{A}}P$. Introducing a Lagrange multiplier to enforce the summation constraint, we differentiate

$$P'\mathcal{G}_{\mathcal{A}}P - \lambda(\mathbf{1}_{\mathcal{A}}'P - 1), \tag{5.9}$$

and find that the minimizing $P_{\mathcal{A}} = \lambda \mathcal{G}_{\mathcal{A}}^{-1} \mathbf{1}_{\mathcal{A}}$. Summing, we get $\lambda \mathbf{1}'_{\mathcal{A}} \mathcal{G}_{\mathcal{A}}^{-1} \mathbf{1}_{\mathcal{A}} = 1$, and hence

$$P_{\mathcal{A}} = A_{\mathcal{A}}^2 \mathcal{G}_{\mathcal{A}}^{-1} \mathbf{1}_{\mathcal{A}} = A_{\mathcal{A}} w_{\mathcal{A}}. \tag{5.10}$$

Hence $\mathbf{v}_{\mathcal{A}} = X_{\mathcal{A}} P_{\mathcal{A}} \in \mathcal{S}_{\mathcal{A}}$ and

$$\|\mathbf{v}_{\mathcal{A}}\|^2 = P'_{\mathcal{A}} \mathcal{G}_{\mathcal{A}}^{-1} P_{\mathcal{A}} = A_{\mathcal{A}}^4 \mathbf{1}'_{\mathcal{A}} \mathcal{G}_{\mathcal{A}}^{-1} \mathbf{1}_{\mathcal{A}} = A_{\mathcal{A}}^2, \tag{5.11}$$

verifying (5.8). If $\mathcal{A} \subseteq \mathcal{B}$, then $\mathcal{S}_{\mathcal{A}} \subseteq \mathcal{S}_{\mathcal{B}}$, so the nearest distance $A_{\mathcal{B}}$ must be equal to or less than the nearest distance $A_{\mathcal{A}}$. $A_{\mathcal{A}}$ obviously equals 1 if and only if $\mathcal{A}$ has only one member. □

The LARS algorithm and its various modifications proceed in piecewise linear steps. For m-vectors $\widehat{\boldsymbol{\beta}}$ and $\mathbf{d}$, let

$$\boldsymbol{\beta}(\gamma) = \widehat{\boldsymbol{\beta}} + \gamma \mathbf{d} \quad \text{and} \quad S(\gamma) = \|\mathbf{y} - X\boldsymbol{\beta}(\gamma)\|^2. \tag{5.12}$$

Lemma 6. *Letting $\widehat{\mathbf{c}} = X'(\mathbf{y} - X\widehat{\boldsymbol{\beta}})$ be the current correlation vector at $\widehat{\boldsymbol{\mu}} = X\widehat{\boldsymbol{\beta}}$,*

$$S(\gamma) - S(0) = -2\widehat{\mathbf{c}}'\mathbf{d}\gamma + \mathbf{d}'X'X\mathbf{d}\gamma^2. \tag{5.13}$$

Proof. $\mathcal{S}(\gamma)$ is a quadratic function of γ, with first two derivatives at $\gamma = 0$,

$$\dot{S}(0) = -2\widehat{\mathbf{c}}'\mathbf{d} \quad \text{and} \quad \ddot{S}(0) = 2\mathbf{d}'X'X\mathbf{d}. \tag{5.14}$$

□

The remainder of this section concerns the LARS–Lasso relationship. Now $\widehat{\boldsymbol{\beta}} = \widehat{\boldsymbol{\beta}}(t)$ will indicate a Lasso solution (1.5), and likewise $\widehat{\boldsymbol{\mu}} = \widehat{\boldsymbol{\mu}}(t) = X\widehat{\boldsymbol{\beta}}(t)$. Because $S(\widehat{\boldsymbol{\beta}})$ and $T(\widehat{\boldsymbol{\beta}})$ are both convex functions of $\widehat{\boldsymbol{\beta}}$, with S strictly convex, standard results show that $\widehat{\boldsymbol{\beta}}(t)$ and $\widehat{\boldsymbol{\mu}}(t)$ are unique and continuous functions of t.

For a given value of t let

$$\mathcal{A} = \{j : \widehat{\beta}_j(t) \neq 0\}. \tag{5.15}$$

We will show later that $\mathcal{A}$ is also the active set that determines the equiangular direction $\mathbf{u}_{\mathcal{A}}$, (2.6), for the LARS–Lasso computations.

We wish to characterize the track of the Lasso solutions $\widehat{\boldsymbol{\beta}}(t)$ or equivalently of $\widehat{\boldsymbol{\mu}}(t)$ as t increases from 0 to its maximum effective value. Let $\mathcal{T}$ be an open interval of the t axis, with infimum t_0, within which the set $\mathcal{A}$ of nonzero Lasso coefficients $\widehat{\beta}_j(t)$ remains constant.

Lemma 7. *The Lasso estimates $\widehat{\boldsymbol{\mu}}(t)$ satisfy*

$$\widehat{\boldsymbol{\mu}}(t) = \widehat{\boldsymbol{\mu}}(t_0) + A_{\mathcal{A}}(t - t_0)\mathbf{u}_{\mathcal{A}} \tag{5.16}$$

for $t \in \mathcal{T}$, where $\mathbf{u}_{\mathcal{A}}$ is the equiangular vector $X_{\mathcal{A}} w_{\mathcal{A}}$, $w_{\mathcal{A}} = A_{\mathcal{A}} \mathcal{G}_{\mathcal{A}}^{-1} \mathbf{1}_{\mathcal{A}}$, (2.7).

PROOF. The lemma says that, for t in $\mathcal{T}$, $\widehat{\boldsymbol{\mu}}(t)$ moves linearly along the equiangular vector $\mathbf{u}_{\mathcal{A}}$ determined by $\mathcal{A}$. We can also state this in terms of the nonzero regression coefficients $\widehat{\beta}_{\mathcal{A}}(t)$,

$$\widehat{\beta}_{\mathcal{A}}(t) = \widehat{\beta}_{\mathcal{A}}(t_0) + S_{\mathcal{A}} A_{\mathcal{A}} (t - t_0) w_{\mathcal{A}}, \tag{5.17}$$

where $S_{\mathcal{A}}$ is the diagonal matrix with diagonal elements s_j, $j \in \mathcal{A}$. [$S_{\mathcal{A}}$ is needed in (5.17) because definitions (2.4), (2.10) require $\widehat{\boldsymbol{\mu}}(t) = X\widehat{\boldsymbol{\beta}}(t) = X_{\mathcal{A}} S_{\mathcal{A}} \widehat{\beta}_{\mathcal{A}}(t)$.]

Since $\widehat{\boldsymbol{\beta}}(t)$ satisfies (1.5) and has nonzero set $\mathcal{A}$, it also minimizes

$$S(\widehat{\beta}_{\mathcal{A}}) = \|\mathbf{y} - X_{\mathcal{A}} S_{\mathcal{A}} \widehat{\beta}_{\mathcal{A}}\|^2 \tag{5.18}$$

subject to

$$\sum_{\mathcal{A}} s_j \widehat{\beta}_j = t \quad \text{and} \quad \operatorname{sign}(\widehat{\beta}_j) = s_j \qquad \text{for } j \in \mathcal{A}. \tag{5.19}$$

[The inequality in (1.5) can be replaced by $T(\widehat{\boldsymbol{\beta}}) = t$ as long as t is less than $\sum |\bar{\beta}_j|$ for the full m-variable OLS solution $\bar{\boldsymbol{\beta}}_m$.] Moreover, the fact that the minimizing point $\widehat{\beta}_{\mathcal{A}}(t)$ occurs strictly *inside* the simplex (5.19), combined with the strict convexity of $S(\widehat{\beta}_{\mathcal{A}})$, implies we can drop the second condition in (5.19) so that $\widehat{\beta}_{\mathcal{A}}(t)$ solves

$$\text{minimize} \quad \{S(\widehat{\beta}_{\mathcal{A}})\} \qquad \text{subject to} \quad \sum_{\mathcal{A}} s_j \widehat{\beta}_j = t. \tag{5.20}$$

Introducing a Lagrange multiplier, (5.20) becomes

$$\text{minimize} \quad \tfrac{1}{2}\|\mathbf{y} - X_{\mathcal{A}} S_{\mathcal{A}} \widehat{\beta}_{\mathcal{A}}\|^2 + \lambda \sum_{\mathcal{A}} s_j \widehat{\beta}_j. \tag{5.21}$$

Differentiating we get

$$-S_{\mathcal{A}} X'_{\mathcal{A}} (\mathbf{y} - X_{\mathcal{A}} S_{\mathcal{A}} \widehat{\beta}_{\mathcal{A}}) + \lambda S_{\mathcal{A}} \mathbf{1}_{\mathcal{A}} = 0. \tag{5.22}$$

Consider two values t_1 and t_2 in $\mathcal{T}$ with $t_0 < t_1 < t_2$. Corresponding to each of these are values for the Lagrange multiplier λ such that $\lambda_1 > \lambda_2$, and solutions $\widehat{\beta}_{\mathcal{A}}(t_1)$ and $\widehat{\beta}_{\mathcal{A}}(t_2)$. Inserting these into (5.22), differencing and premultiplying by $S_{\mathcal{A}}$ we get

$$X'_{\mathcal{A}} X_{\mathcal{A}} S_{\mathcal{A}} (\widehat{\beta}_{\mathcal{A}}(t_2) - \widehat{\beta}_{\mathcal{A}}(t_1)) = (\lambda_1 - \lambda_2) \mathbf{1}_{\mathcal{A}}. \tag{5.23}$$

Hence

$$\widehat{\beta}_{\mathcal{A}}(t_2) - \widehat{\beta}_{\mathcal{A}}(t_1) = (\lambda_1 - \lambda_2) S_{\mathcal{A}} \mathcal{G}_{\mathcal{A}}^{-1} \mathbf{1}_{\mathcal{A}}. \tag{5.24}$$

However, $s'_{\mathcal{A}}[(\widehat{\beta}_{\mathcal{A}}(t_2) - \widehat{\beta}_{\mathcal{A}}(t_1)] = t_2 - t_1$ according to the Lasso definition, so

$$t_2 - t_1 = (\lambda_1 - \lambda_2) s'_{\mathcal{A}} S_{\mathcal{A}} \mathcal{G}_{\mathcal{A}}^{-1} \mathbf{1}_{\mathcal{A}} = (\lambda_1 - \lambda_2) \mathbf{1}'_{\mathcal{A}} \mathcal{G}_{\mathcal{A}}^{-1} \mathbf{1}_{\mathcal{A}} = (\lambda_1 - \lambda_2) A_{\mathcal{A}}^{-2} \tag{5.25}$$

and

$$\widehat{\beta}_{\mathcal{A}}(t_2) - \widehat{\beta}_{\mathcal{A}}(t_1) = S_{\mathcal{A}} A_{\mathcal{A}}^2 (t_2 - t_1) \mathcal{G}_{\mathcal{A}}^{-1} \mathbf{1}_{\mathcal{A}} = S_{\mathcal{A}} A_{\mathcal{A}} (t - t_1) w_{\mathcal{A}}. \tag{5.26}$$

Letting $t_2 = t$ and $t_1 \to t_0$ gives (5.17) by the continuity of $\widehat{\boldsymbol{\beta}}(t)$, and finally (5.16). Note that (5.16) implies that the maximum absolute correlation $\widehat{C}(t)$ equals $\widehat{C}(t_0) - A^2_{\mathcal{A}}(t - t_0)$, so that $\widehat{C}(t)$ is a piecewise linear decreasing function of the Lasso parameter t. □

The Lasso solution $\widehat{\boldsymbol{\beta}}(t)$ occurs on the surface of the diamond-shaped convex polytope

$$\mathcal{D}(t) = \left\{\boldsymbol{\beta} : \sum |\beta_j| \leq t\right\}, \tag{5.27}$$

$\mathcal{D}(t)$ increasing with t. Lemma 7 says that, for $t \in \mathcal{T}$, $\widehat{\boldsymbol{\beta}}(t)$ moves linearly along edge $\mathcal{A}$ of the polytope, the edge having $\beta_j = 0$ for $j \notin \mathcal{A}$. Moreover the regression estimates $\widehat{\boldsymbol{\mu}}(t)$ move in the LARS equiangular direction $\mathbf{u}_{\mathcal{A}}$, (2.6). It remains to show that "$\mathcal{A}$" changes according to the rules of Theorem 1, which is the purpose of the next three lemmas.

LEMMA 8. *A Lasso solution $\widehat{\boldsymbol{\beta}}$ has*

$$\widehat{c}_j = \widehat{C} \cdot \operatorname{sign}(\widehat{\beta}_j) \qquad \textit{for } j \in \mathcal{A}, \tag{5.28}$$

where $\widehat{c}_j$ equals the current correlation $\mathbf{x}'_j(\mathbf{y} - \widehat{\boldsymbol{\mu}}) = \mathbf{x}'_j(\mathbf{y} - X\widehat{\boldsymbol{\beta}})$. In particular, this implies that

$$\operatorname{sign}(\widehat{\beta}_j) = \operatorname{sign}(\widehat{c}_j) \qquad \textit{for } j \in \mathcal{A}. \tag{5.29}$$

PROOF. This follows immediately from (5.22) by noting that the jth element of the left-hand side is $\widehat{c}_j$, and the right-hand side is $\lambda \cdot \operatorname{sign}(\widehat{\beta}_j)$ for $j \in \mathcal{A}$. Likewise $\lambda = |\widehat{c}_j| = \widehat{C}$. □

LEMMA 9. *Within an interval $\mathcal{T}$ of constant nonzero set $\mathcal{A}$, and also at $t_0 = \inf(\mathcal{T})$, the Lasso current correlations $c_j(t) = \mathbf{x}'_j(\mathbf{y} - \widehat{\boldsymbol{\mu}}(t))$ satisfy*

$$|c_j(t)| = \widehat{C}(t) \equiv \max\{|c_\ell(t)|\} \qquad \textit{for } j \in \mathcal{A}$$

and

$$|c_j(t)| \leq \widehat{C}(t) \qquad \textit{for } j \notin \mathcal{A}. \tag{5.30}$$

PROOF. Equation (5.28) says that the $|c_j(t)|$ have identical values, say $\widehat{C}_t$, for $j \in \mathcal{A}$. It remains to show that $\widehat{C}_t$ has the extremum properties indicated in (5.30). For an m-vector $\mathbf{d}$ we define $\boldsymbol{\beta}(\gamma) = \widehat{\boldsymbol{\beta}}(t) + \gamma\mathbf{d}$ and $S(\gamma)$ as in (5.12), likewise $T(\gamma) = \sum |\beta_j(\gamma)|$, and

$$R_t(d) = -\dot{S}(0)/\dot{T}(0). \tag{5.31}$$

Again assuming $\widehat{\beta}_j > 0$ for $j \in \mathcal{A}$, by redefinition of $\mathbf{x}_j$ if necessary, (5.14) and (5.28) yield

$$R_t(\mathbf{d}) = 2\left[\widehat{C}_t \sum_{\mathcal{A}} d_j + \sum_{\mathcal{A}^c} c_j(t) d_j\right] \Big/ \left[\sum_{\mathcal{A}} d_j + \sum_{\mathcal{A}^c} |d_j|\right]. \tag{5.32}$$

If $d_j = 0$ for $j \notin \mathcal{A}$, and $\sum d_j \neq 0$,

$$R_t(\mathbf{d}) = 2\widehat{C}_t, \tag{5.33}$$

while if $\mathbf{d}$ has only component j nonzero we can make

$$R_t(\mathbf{d}) = 2|c_j(t)|. \tag{5.34}$$

According to Lemma 7 the Lasso solutions for $t \in \mathcal{T}$ use $d_{\mathcal{A}}$ proportional to $w_{\mathcal{A}}$ with $d_j = 0$ for $j \notin \mathcal{A}$, so

$$R_t \equiv R_t(w_{\mathcal{A}}) \tag{5.35}$$

is the downward slope of the curve $(T, S(T))$ at $T = t$, and by the definition of the Lasso must maximize $R_t(\mathbf{d})$. This shows that $\widehat{C}_t = \widehat{C}(t)$, and verifies (5.30), which also holds at $t_0 = \inf(\mathcal{T})$ by the continuity of the current correlations. □

We note that Lemmas 7–9 follow relatively easily from the Karush–Kuhn–Tucker conditions for optimality for the quadratic programming Lasso problem [Osborne, Presnell and Turlach (2000a)]; we have chosen a more geometrical argument here to demonstrate the nature of the Lasso path.

Figure 8 shows the (T, S) curve corresponding to the Lasso estimates in Figure 1. The arrow indicates the tangent to the curve at $t = 1000$, which has

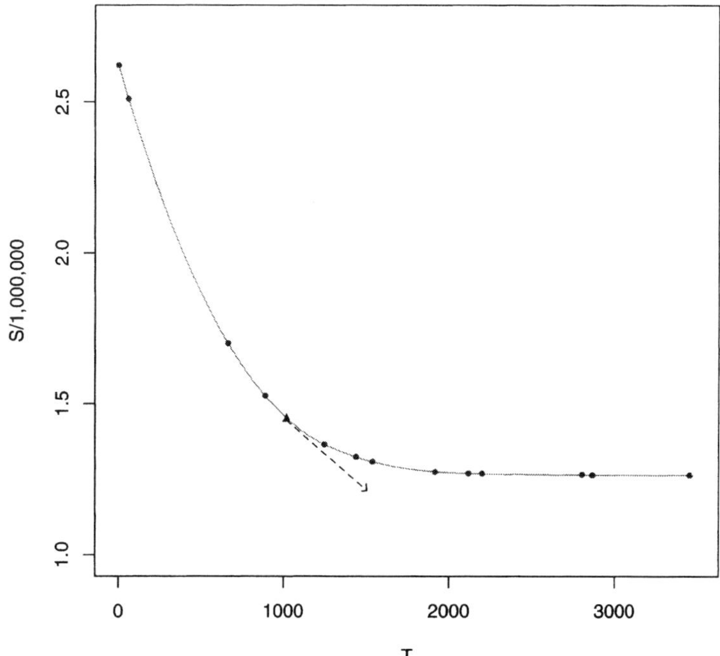

FIG. 8. *Plot of S versus T for Lasso applied to diabetes data; points indicate the* 12 *modified LARS steps of Figure* 1; *triangle is* (T, S) *boundary point at* $t = 1000$; *dashed arrow is tangent at* $t = 1000$, *negative slope* R_t, (5.31). *The* (T, S) *curve is a decreasing, convex, quadratic spline.*

downward slope R_{1000}. The argument above relies on the fact that $R_t(\mathbf{d})$ cannot be greater than R_t, or else there would be (T, S) values lying below the optimal curve. Using Lemmas 3 and 4 it can be shown that the (T, S) curve is always convex, as in Figure 8, being a quadratic spline with $\dot{S}(T) = -2\widehat{C}(T)$ and $\ddot{S}(T) = 2A_{\mathcal{A}}^2$.

We now consider in detail the choice of active set at a breakpoint of the piecewise linear Lasso path. Let $t = t_0$ indicate such a point, $t_0 = \inf(\mathcal{T})$ as in Lemma 9, with Lasso regression vector $\widehat{\boldsymbol{\beta}}$, prediction estimate $\widehat{\boldsymbol{\mu}} = X\widehat{\boldsymbol{\beta}}$, current correlations $\widehat{\mathbf{c}} = X'(\mathbf{y} - \widehat{\boldsymbol{\mu}})$, $s_j = \text{sign}(\widehat{c}_j)$ and maximum absolute correlation $\widehat{C}$. Define

$$\mathcal{A}_1 = \{j : \widehat{\beta}_j \neq 0\}, \qquad \mathcal{A}_0 = \{j : \widehat{\beta}_j = 0 \text{ and } |\widehat{c}_j| = \widehat{C}\}, \tag{5.36}$$

$\mathcal{A}_{10} = \mathcal{A}_1 \cup \mathcal{A}_0$ and $\mathcal{A}_2 = \mathcal{A}_{10}^c$, and take $\boldsymbol{\beta}(\gamma) = \widehat{\boldsymbol{\beta}} + \gamma\mathbf{d}$ for some m-vector $\mathbf{d}$; also $S(\gamma) = \|\mathbf{y} - X\boldsymbol{\beta}(\gamma)\|^2$ and $T(\gamma) = \sum |\beta_j(\gamma)|$.

LEMMA 10. *The negative slope* (5.31) *at* t_0 *is bounded by* $2\widehat{C}$,

$$R(\mathbf{d}) = -\dot{S}(0)/\dot{T}(0) \leq 2\widehat{C}, \tag{5.37}$$

with equality only if $d_j = 0$ *for* $j \in \mathcal{A}_2$. *If so, the differences* $\Delta S = S(\gamma) - S(0)$ *and* $\Delta T = T(\gamma) - T(0)$ *satisfy*

$$\Delta S = -2\widehat{C}\Delta T + L(\mathbf{d})^2 \cdot (\Delta T)^2, \tag{5.38}$$

where

$$L(\mathbf{d}) = \|X\mathbf{d}/d_+\|. \tag{5.39}$$

PROOF. We can assume $\widehat{c}_j \geq 0$ for all j, by redefinition if necessary, so $\widehat{\beta}_j \geq 0$ according to Lemma 8. Proceeding as in (5.32),

$$R(\mathbf{d}) = 2\widehat{C}\left[\sum_{\mathcal{A}_{10}} d_j + \sum_{\mathcal{A}_2} (\widehat{c}_j/\widehat{C}) d_j\right] \bigg/ \left[\sum_{\mathcal{A}_1} d_j + \sum_{\mathcal{A}_0 \cup \mathcal{A}_2} |d_j|\right]. \tag{5.40}$$

We need $d_j \geq 0$ for $j \in \mathcal{A}_0 \cup \mathcal{A}_2$ in order to maximize (5.40), in which case

$$R(\mathbf{d}) = 2\widehat{C}\left[\sum_{\mathcal{A}_{10}} d_j + \sum_{\mathcal{A}_2} (\widehat{c}_j/\widehat{C}) d_j\right] \bigg/ \left[\sum_{\mathcal{A}_{10}} d_j + \sum_{\mathcal{A}_2} d_j\right]. \tag{5.41}$$

This is $< 2\widehat{C}$ unless $d_j = 0$ for $j \in \mathcal{A}_2$, verifying (5.37), and also implying

$$T(\gamma) = T(0) + \gamma \sum_{\mathcal{A}_{10}} d_j. \tag{5.42}$$

The first term on the right-hand side of (5.13) is then $-2\widehat{C}(\Delta T)$, while the second term equals $(\mathbf{d}/d_+)'X'X(\mathbf{d}/d_+)(\Delta T)^2 = L(\mathbf{d})^2$. □

Lemma 10 has an important consequence. Suppose that $\mathcal{A}$ is the current active set for the Lasso, as in (5.17), and that $\mathcal{A} \subseteq \mathcal{A}_{10}$. Then Lemma 5 says that $L(\mathbf{d})$ is $\geq A_{\mathcal{A}}$, and (5.38) gives

$$\Delta S \geq -2\widehat{C} \cdot \Delta T + A_{\mathcal{A}}^2 \cdot (\Delta T)^2, \tag{5.43}$$

with equality if $\mathbf{d}$ is chosen to give the equiangular vector $\mathbf{u}_{\mathcal{A}}$, $d_{\mathcal{A}} = S_{\mathcal{A}} w_{\mathcal{A}}$, $d_{\mathcal{A}^c} = 0$. The Lasso operates to minimize $S(T)$ so we want ΔS to be as negative as possible. Lemma 10 says that if the support of $\mathbf{d}$ is not confined to $\mathcal{A}_{10}$, then $\dot{S}(0)$ exceeds the optimum value $-2\widehat{C}$; if it is confined, then $\dot{S}(0) = -2\widehat{C}$ but $\ddot{S}(0)$ exceeds the minimum value $2A_{\mathcal{A}}$ unless $d_{\mathcal{A}}$ is proportional to $S_{\mathcal{A}} w_{\mathcal{A}}$ as in (5.17).

Suppose that $\widehat{\boldsymbol{\beta}}$, a Lasso solution, exactly equals a $\widehat{\boldsymbol{\beta}}$ obtained from the Lasso-modified LARS algorithm, henceforth called LARS–Lasso, as at $t = 1000$ in Figures 1 and 3. We know from Lemma 7 that subsequent Lasso estimates will follow a linear track determined by some subset $\mathcal{A}$, $\boldsymbol{\mu}(\gamma) = \widehat{\boldsymbol{\mu}} + \gamma \mathbf{u}_{\mathcal{A}}$, and so will the LARS–Lasso estimates, but to verify Theorem 1 we need to show that "$\mathcal{A}$" is the same set in both cases.

Lemmas 4–7 put four constraints on the Lasso choice of $\mathcal{A}$. Define $\mathcal{A}_1$, $\mathcal{A}_0$ and $\mathcal{A}_{10}$ as at (5.36).

Constraint 1. $\mathcal{A}_1 \subseteq \mathcal{A}$. This follows from Lemma 7 since for sufficiently small γ the subsequent Lasso coefficients (5.17),

$$\widehat{\beta}_{\mathcal{A}}(\gamma) = \widehat{\beta}_{\mathcal{A}} + \gamma S_{\mathcal{A}} w_{\mathcal{A}}, \tag{5.44}$$

will have $\widehat{\beta}_j(\gamma) \neq 0$, $j \in \mathcal{A}_1$.

Constraint 2. $\mathcal{A} \subseteq \mathcal{A}_{10}$. Lemma 10, (5.37) shows that the Lasso choice $\widehat{\mathbf{d}}$ in $\boldsymbol{\beta}(\gamma) = \widehat{\boldsymbol{\beta}} + \gamma \widehat{\mathbf{d}}$ must have its nonzero support in $\mathcal{A}_{10}$, or equivalently that $\widehat{\boldsymbol{\mu}}(\gamma) = \widehat{\boldsymbol{\mu}} + \gamma \mathbf{u}_{\mathcal{A}}$ must have $\mathbf{u}_{\mathcal{A}} \in \mathcal{L}(X_{\mathcal{A}_{10}})$. (It is possible that $\mathbf{u}_{\mathcal{A}}$ happens to equal $\mathbf{u}_{\mathcal{B}}$ for some $\mathcal{B} \supset \mathcal{A}_{10}$, but that does not affect the argument below.)

Constraint 3. $w_{\mathcal{A}} = A_{\mathcal{A}} \mathcal{G}_{\mathcal{A}}^{-1} \mathbf{1}_{\mathcal{A}}$ cannot have $\operatorname{sign}(w_j) \neq \operatorname{sign}(\widehat{c}_j)$ for any coordinate $j \in \mathcal{A}_0$. If it does, then $\operatorname{sign}(\widehat{\beta}_j(\gamma)) \neq \operatorname{sign}(\widehat{c}_j(\gamma))$ for sufficiently small γ, violating Lemma 8.

Constraint 4. Subject to Constraints 1–3, $\mathcal{A}$ must minimize $A_{\mathcal{A}}$. This follows from Lemma 10 as in (5.43), and the requirement that the Lasso curve $S(T)$ declines at the fastest possible rate.

Theorem 1 follows by induction: beginning at $\widehat{\boldsymbol{\beta}}_0 = 0$, we follow the LARS–Lasso algorithm and show that at every succeeding step it must continue to agree with the Lasso definition (1.5). First of all, suppose that $\widehat{\boldsymbol{\beta}}$, our hypothesized Lasso and LARS–Lasso solution, has occurred strictly *within* a LARS–Lasso step. Then

$\mathcal{A}_0$ is empty so that Constraints 1 and 2 imply that $\mathcal{A}$ cannot change its current value: the equivalence between Lasso and LARS–Lasso must continue at least to the end of the step.

The one-at-a-time assumption of Theorem 1 says that at a LARS–Lasso breakpoint, $\mathcal{A}_0$ has exactly one member, say j_0, so $\mathcal{A}$ must equal $\mathcal{A}_1$ or $\mathcal{A}_{10}$. There are two cases: if j_0 has just been *added* to the set $\{|\widehat{c}_j| = \widehat{C}\}$, then Lemma 4 says that $\operatorname{sign}(w_{j_0}) = \operatorname{sign}(\widehat{c}_{j_0})$, so that Constraint 3 is not violated; the other three constraints and Lemma 5 imply that the Lasso choice $\mathcal{A} = \mathcal{A}_{10}$ agrees with the LARS–Lasso algorithm. The other case has j_0 *deleted* from the active set as in (3.6). Now the choice $\mathcal{A} = \mathcal{A}_{10}$ is ruled out by Constraint 3: it would keep $w_{\mathcal{A}}$ the same as in the previous LARS–Lasso step, and we know that that was stopped in (3.6) to prevent a sign contradiction at coordinate j_0. In other words, $\mathcal{A} = \mathcal{A}_1$, in accordance with the Lasso modification of LARS. This completes the proof of Theorem 1.

A LARS–Lasso algorithm is available even if the one-at-a-time condition does not hold, but at the expense of additional computation. Suppose, for example, *two* new members j_1 and j_2 are added to the set $\{|\widehat{c}_j| = \widehat{C}\}$, so $\mathcal{A}_0 = \{j_1, j_2\}$. It is possible but not certain that $\mathcal{A}_{10}$ does not violate Constraint 3, in which case $\mathcal{A} = \mathcal{A}_{10}$. However, if it does violate Constraint 3, then both possibilities $\mathcal{A} = \mathcal{A}_1 \cup \{j_1\}$ and $\mathcal{A} = \mathcal{A}_1 \cup \{j_2\}$ must be examined to see which one gives the smaller value of $A_{\mathcal{A}}$. Since one-at-a-time computations, perhaps with some added $\mathbf{y}$ jitter, apply to all practical situations, the LARS algorithm described in Section 7 is not equipped to handle many-at-a-time problems.

6. Stagewise properties. The main goal of this section is to verify Theorem 2. Doing so also gives us a chance to make a more detailed comparison of the LARS and Stagewise procedures. Assume that $\widehat{\boldsymbol{\beta}}$ is a Stagewise estimate of the regression coefficients, for example, as indicated at $\sum|\widehat{\beta}_j| = 2000$ in the right panel of Figure 1, with prediction vector $\widehat{\boldsymbol{\mu}} = X\widehat{\boldsymbol{\beta}}$, current correlations $\widehat{\mathbf{c}} = X'(\mathbf{y} - \widehat{\boldsymbol{\mu}})$, $\widehat{C} = \max\{|\widehat{c}_j|\}$ and maximal set $\mathcal{A} = \{j : |\widehat{c}_j| = \widehat{C}\}$. We must show that successive Stagewise estimates of $\boldsymbol{\beta}$ develop according to the modified LARS algorithm of Theorem 2, henceforth called LARS–Stagewise. For convenience we can assume, by redefinition of $\mathbf{x}_j$ as $-\mathbf{x}_j$, if necessary, that the signs $s_j = \operatorname{sign}(\widehat{c}_j)$ are all non-negative.

As in (3.8)–(3.10) we suppose that the Stagewise procedure (1.7) has taken N additional ε-steps forward from $\widehat{\boldsymbol{\mu}} = X\widehat{\boldsymbol{\beta}}$, giving new prediction vector $\widehat{\boldsymbol{\mu}}(N)$.

LEMMA 11. *For sufficiently small ε, only $j \in \mathcal{A}$ can have $P_j = N_j/N > 0$.*

PROOF. Letting $N\varepsilon \equiv \gamma$, $\|\widehat{\boldsymbol{\mu}}(N) - \widehat{\boldsymbol{\mu}}\| \le \gamma$ so that $\widehat{\mathbf{c}}(N) = X'(\mathbf{y} - \widehat{\boldsymbol{\mu}}(N))$ satisfies

$$|\widehat{c}_j(N) - \widehat{c}_j| = |\mathbf{x}_j'(\widehat{\boldsymbol{\mu}}(N) - \widehat{\boldsymbol{\mu}})| \le \|\mathbf{x}_j\| \cdot \|\widehat{\boldsymbol{\mu}}(N) - \widehat{\boldsymbol{\mu}}\| \le \gamma. \tag{6.1}$$

For $\gamma < \frac{1}{2}[\widehat{C} - \max_{\mathcal{A}^c}\{\widehat{c}_j\}]$, j in $\mathcal{A}^c$ cannot have maximal current correlation and can never be involved in the N steps. □

Lemma 11 says that we can write the developing Stagewise prediction vector as

$$\widehat{\boldsymbol{\mu}}(\gamma) = \widehat{\boldsymbol{\mu}} + \gamma \mathbf{v}, \qquad \text{where } \mathbf{v} = X_{\mathcal{A}} P_{\mathcal{A}}, \tag{6.2}$$

$P_{\mathcal{A}}$ a vector of length $|\mathcal{A}|$, with components N_j/N for $j \in \mathcal{A}$. The nature of the Stagewise procedure puts three constraints on $\mathbf{v}$, the most obvious of which is the following.

CONSTRAINT I. The vector $\mathbf{v} \in \mathcal{S}_{\mathcal{A}}^{+}$, the nonnegative simplex

$$\mathcal{S}_{\mathcal{A}}^{+} = \left\{ \mathbf{v} : \mathbf{v} = \sum_{j \in \mathcal{A}} \mathbf{x}_j P_j, \ P_j \geq 0, \ \sum_{j \in \mathcal{A}} P_j = 1 \right\}. \tag{6.3}$$

Equivalently, $\gamma \mathbf{v} \in \mathcal{C}_{\mathcal{A}}$, the convex cone (3.12).

The Stagewise procedure, unlike LARS, is not required to use all of the maximal set $\mathcal{A}$ as the active set, and can instead restrict the nonzero coordinates P_j to a subset $\mathcal{B} \subseteq \mathcal{A}$. Then $\mathbf{v} \in \mathcal{L}(X_{\mathcal{B}})$, the linear space spanned by the columns of $X_{\mathcal{B}}$, but not all such vectors $\mathbf{v}$ are allowable Stagewise forward directions.

CONSTRAINT II. The vector $\mathbf{v}$ must be proportional to the equiangular vector $\mathbf{u}_{\mathcal{B}}$, (2.6), that is, $\mathbf{v} = \mathbf{v}_{\mathcal{B}}$, (5.8),

$$\mathbf{v}_{\mathcal{B}} = A_{\mathcal{B}}^2 X_{\mathcal{B}} \mathcal{G}_{\mathcal{B}}^{-1} \mathbf{1}_{\mathcal{B}} = A_{\mathcal{B}} \mathbf{u}_{\mathcal{B}}. \tag{6.4}$$

Constraint II amounts to requiring that the current correlations in $\mathcal{B}$ decline at an equal rate: since

$$\widehat{c}_j(\gamma) = \mathbf{x}_j'(\mathbf{y} - \widehat{\boldsymbol{\mu}} - \gamma \mathbf{v}) = \widehat{c}_j - \gamma \mathbf{x}_j' \mathbf{v}, \tag{6.5}$$

we need $X_{\mathcal{B}}' \mathbf{v} = \lambda \mathbf{1}_{\mathcal{B}}$ for some $\lambda > 0$, implying $\mathbf{v} = \lambda \mathcal{G}_{\mathcal{B}}^{-1} \mathbf{1}_{\mathcal{B}}$; choosing $\lambda = A_{\mathcal{B}}^2$ satisfies Constraint II. Violating Constraint II makes the current correlations $\widehat{c}_j(\gamma)$ unequal so that the Stagewise algorithm as defined at (1.7) could not proceed in direction $\mathbf{v}$.

Equation (6.4) gives $X_{\mathcal{B}}' \mathbf{v}_{\mathcal{B}} = A_{\mathcal{B}}^2 \mathbf{1}_{\mathcal{B}}$, or

$$\mathbf{x}_j' \mathbf{v}_{\mathcal{B}} = A_{\mathcal{B}}^2 \qquad \text{for } j \in \mathcal{B}. \tag{6.6}$$

CONSTRAINT III. The vector $\mathbf{v} = \mathbf{v}_{\mathcal{B}}$ must satisfy

$$\mathbf{x}_j' \mathbf{v}_{\mathcal{B}} \geq A_{\mathcal{B}}^2 \qquad \text{for } j \in \mathcal{A} - \mathcal{B}. \tag{6.7}$$

Constraint III follows from (6.5). It says that the current correlations for members of $\mathcal{A} = \{j : |\widehat{c}_j| = \widehat{C}\}$ *not* in $\mathcal{B}$ must decline at least as quickly as those in $\mathcal{B}$. If this were not true, then $\mathbf{v}_{\mathcal{B}}$ would not be an allowable direction for Stagewise development since variables in $\mathcal{A} - \mathcal{B}$ would immediately reenter (1.7).

To obtain strict inequality in (6.7), let $\mathcal{B}_0 \subset \mathcal{A} - \mathcal{B}$ be the set of indices for which $\mathbf{x}_j'\mathbf{v}_{\mathcal{B}} = A_{\mathcal{B}}^2$. It is easy to show that $\mathbf{v}_{\mathcal{B} \cup \mathcal{B}_o} = \mathbf{v}_{\mathcal{B}}$. In other words, if we take $\mathcal{B}$ to be the *largest* set having a given $\mathbf{v}_{\mathcal{B}}$ proportional to its equiangular vector, then $\mathbf{x}_j'\mathbf{v}_{\mathcal{B}} > A_{\mathcal{B}}^2$ for $j \in \mathcal{A} - \mathcal{B}$.

Writing $\widehat{\boldsymbol{\mu}}(\gamma) = \widehat{\boldsymbol{\mu}} + \gamma\mathbf{v}$ as in (6.2) presupposes that the Stagewise solutions follow a piecewise linear track. However, the presupposition can be reduced to one of piecewise differentiability by taking γ infinitesimally small. We can always express the family of Stagewise solutions as $\widehat{\boldsymbol{\beta}}(z)$, where the real-valued parameter Z plays the role of T for the Lasso, increasing from 0 to some maximum value as $\widehat{\boldsymbol{\beta}}(z)$ goes from $\mathbf{0}$ to the full OLS estimate. [The choice $Z = T$ used in Figure 1 may not necessarily yield a one-to-one mapping; $Z = S(\mathbf{0}) - S(\widehat{\boldsymbol{\beta}})$, the reduction in residual squared error, always does.] We suppose that the Stagewise estimate $\widehat{\boldsymbol{\beta}}(z)$ is everywhere right differentiable with respect to z. Then the right derivative

$$\widehat{\mathbf{v}} = d\widehat{\boldsymbol{\beta}}(z)/dz \tag{6.8}$$

must obey the three constraints.

The definition of the idealized Stagewise procedure in Section 3.2, in which $\varepsilon \to 0$ in rule (1.7), is somewhat vague but the three constraints apply to any reasonable interpretation. It turns out that the LARS–Stagewise algorithm satisfies the constraints and is unique in doing so. This is the meaning of Theorem 2. [Of course the LARS–Stagewise algorithm is also supported by direct numerical comparisons with (1.7), as in Figure 1's right panel.]

If $\mathbf{u}_{\mathcal{A}} \in \mathcal{C}_{\mathcal{A}}$, then $\mathbf{v} = \mathbf{v}_{\mathcal{A}}$ obviously satisfies the three constraints. The interesting situation for Theorem 2 is $\mathbf{u}_{\mathcal{A}} \notin \mathcal{C}_{\mathcal{A}}$, which we now assume to be the case. Any subset $\mathcal{B} \subset \mathcal{A}$ determines a face of the convex cone of dimension $|\mathcal{B}|$, the face having $P_j > 0$ in (3.12) for $j \in \mathcal{B}$ and $P_j = 0$ for $j \in \mathcal{A} - \mathcal{B}$. The orthogonal projection of $\mathbf{u}_{\mathcal{A}}$ into the linear subspace $\mathcal{L}(X_{\mathcal{B}})$, say $\mathrm{Proj}_{\mathcal{B}}(\mathbf{u}_{\mathcal{A}})$, is proportional to $\mathcal{B}$'s equiangular vector $\mathbf{u}_{\mathcal{B}}$: using (2.7),

$$\mathrm{Proj}_{\mathcal{B}}(\mathbf{u}_{\mathcal{A}}) = X_{\mathcal{B}}\mathcal{G}_{\mathcal{B}}^{-1}X_{\mathcal{B}}'\mathbf{u}_{\mathcal{A}} = X_{\mathcal{B}}\mathcal{G}_{\mathcal{B}}^{-1}A_{\mathcal{A}}\mathbf{1}_{\mathcal{B}} = (A_{\mathcal{A}}/A_{\mathcal{B}}) \cdot \mathbf{u}_{\mathcal{B}}, \tag{6.9}$$

or equivalently

$$\mathrm{Proj}_{\mathcal{B}}(\mathbf{v}_{\mathcal{A}}) = (A_{\mathcal{A}}/A_{\mathcal{B}})^2\mathbf{v}_{\mathcal{B}}. \tag{6.10}$$

The nearest point to $\mathbf{u}_{\mathcal{A}}$ in $\mathcal{C}_{\mathcal{A}}$, say $\widehat{\mathbf{u}}_{\mathcal{A}}$, is of the form $\Sigma_{\mathcal{A}}\mathbf{x}_j\widehat{P}_j$ with $\widehat{P}_j \geq 0$. Therefore $\widehat{\mathbf{u}}_{\mathcal{A}}$ exists strictly within face $\widehat{\mathcal{B}}$, where $\widehat{\mathcal{B}} = \{j : \widehat{P}_j > 0\}$, and must equal $\mathrm{Proj}_{\widehat{\mathcal{B}}}(\mathbf{u}_{\mathcal{A}})$. According to (6.9), $\widehat{\mathbf{u}}_{\mathcal{A}}$ is proportional to $\widehat{\mathcal{B}}$'s equiangular vector $\mathbf{u}_{\widehat{\mathcal{B}}}$, and also to $\mathbf{v}_{\widehat{\mathcal{B}}} = A_{\mathcal{B}}\mathbf{u}_{\mathcal{B}}$. In other words $\mathbf{v}_{\widehat{\mathcal{B}}}$ satisfies Constraint II, and it obviously also satisfies Constraint I. Figure 9 schematically illustrates the geometry.

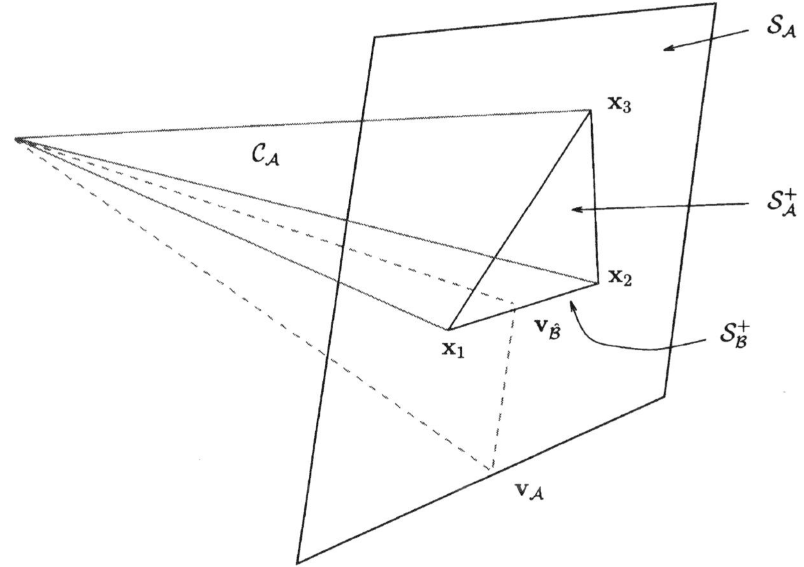

FIG. 9. *The geometry of the LARS–Stagewise modification.*

LEMMA 12. *The vector* $\mathbf{v}_{\hat{\mathcal{B}}}$ *satisfies Constraints* I–III, *and conversely if* $\mathbf{v}$ *satisfies the three constraints, then* $\mathbf{v} = \mathbf{v}_{\hat{\mathcal{B}}}$.

PROOF. Let $\mathrm{Cos} \equiv A_{\mathcal{A}}/A_{\mathcal{B}}$ and $\mathrm{Sin} = [1 - \mathrm{Cos}^2]^{1/2}$, the latter being greater than zero by Lemma 5. For any face $\mathcal{B} \subset \mathcal{A}$, (6.9) implies

$$\mathbf{u}_{\mathcal{A}} = \mathrm{Cos} \cdot \mathbf{u}_{\mathcal{B}} + \mathrm{Sin} \cdot \mathbf{z}_{\mathcal{B}}, \tag{6.11}$$

where $\mathbf{z}_{\mathcal{B}}$ is a unit vector orthogonal to $\mathcal{L}(X_{\mathcal{B}})$, pointing away from $\mathcal{C}_{\mathcal{A}}$. By an n-dimensional coordinate rotation we can make $\mathcal{L}(X_{\mathcal{B}}) = \mathcal{L}(\mathbf{c}_1, \mathbf{c}_2, \dots, \mathbf{c}_J)$, $J = |\mathcal{B}|$, the space of n-vectors with last $n - J$ coordinates zero, and also

$$\mathbf{u}_{\mathcal{B}} = (1, \mathbf{0}, 0, \mathbf{0}), \qquad \mathbf{u}_{\mathcal{A}} = (\mathrm{Cos}, \mathbf{0}, \mathrm{Sin}, \mathbf{0}), \tag{6.12}$$

the first $\mathbf{0}$ having length $J - 1$, the second $\mathbf{0}$ length $n - J - 1$. Then we can write

$$\mathbf{x}_j = (A_{\mathcal{B}}, \mathbf{x}_{j_2}, 0, \mathbf{0}) \qquad \text{for } j \in \mathcal{B}, \tag{6.13}$$

the first coordinate $A_{\mathcal{B}}$ being required since $\mathbf{x}'_j \mathbf{u}_{\mathcal{B}} = A_{\mathcal{B}}$, (2.7). Notice that $\mathbf{x}'_j \mathbf{u}_{\mathcal{A}} = \mathrm{Cos} \cdot A_{\mathcal{B}} = A_{\mathcal{A}}$, as also required by (2.7).

For $\ell \in \mathcal{A} - \mathcal{B}$ denote $\mathbf{x}_\ell$ as

$$\mathbf{x}_\ell = (x_{\ell_1}, \mathbf{x}_{\ell_2}, x_{\ell_3}, \mathbf{x}_{\ell_4}), \tag{6.14}$$

so (2.7) yields

$$A_{\mathcal{A}} = \mathbf{x}'_\ell \mathbf{u}_{\mathcal{A}} = \mathrm{Cos} \cdot x_{\ell_1} + \mathrm{Sin} \cdot x_{\ell_3}. \tag{6.15}$$

Now assume $\mathcal{B} = \widehat{\mathcal{B}}$. In this case a separating hyperplane $\mathcal{H}$ orthogonal to $\mathbf{z}_{\hat{\mathcal{B}}}$ in (6.11) passes between the convex cone $\mathcal{C}_{\mathcal{A}}$ and $\mathbf{u}_{\mathcal{A}}$, through $\widehat{\mathbf{u}}_{\mathcal{A}} = \text{Cos} \cdot \mathbf{u}_{\hat{\mathcal{B}}}$, implying $x_{\ell_3} \le 0$ [i.e., $\mathbf{x}_\ell$ and $\mathbf{u}_{\mathcal{A}}$ are on opposite sides of $\mathcal{H}$, x_{ℓ_3} being negative since the corresponding coordinate of $\mathbf{u}_{\mathcal{A}}$, "Sin" in (6.12), is positive]. Equation (6.15) gives $\text{Cos} \cdot x_{\ell_1} \ge A_{\mathcal{A}} = \text{Cos} \cdot A_{\hat{\mathcal{B}}}$ or

$$\mathbf{x}_\ell' \mathbf{v}_{\hat{\mathcal{B}}} = x_\ell'(A_{\hat{\mathcal{B}}} \mathbf{u}_{\hat{\mathcal{B}}}) = A_{\hat{\mathcal{B}}} x_{\ell_1} \ge A_{\hat{\mathcal{B}}}^2, \tag{6.16}$$

verifying that Constraint III is satisfied.

Conversely suppose that $\mathbf{v}$ satisfies Constraints I–III so that $\mathbf{v} \in \mathcal{S}_{\mathcal{A}}^+$ and $\mathbf{v} = \mathbf{v}_{\mathcal{B}}$ for the nonzero coefficient set $\mathcal{B}$: $\mathbf{v}_{\mathcal{B}} = \Sigma_{\mathcal{B}} \mathbf{x}_j P_j$, $P_j > 0$. Let $\mathcal{H}$ be the hyperplane passing through $\text{Cos} \cdot \mathbf{u}_{\mathcal{B}}$ orthogonally to $\mathbf{z}_{\mathcal{B}}$, (6.9), (6.11). If $\mathbf{v}_{\mathcal{B}} \ne \mathbf{v}_{\hat{\mathcal{B}}}$, then at least one of the vectors $\mathbf{x}_\ell$, $\ell \in \mathcal{A} - \mathcal{B}$, must lie on the same side of $\mathcal{H}$ as $\mathbf{u}_{\mathcal{A}}$, so that $x_{\ell_3} > 0$ (or else $\mathcal{H}$ would be a separating hyperplane between $\mathbf{u}_{\mathcal{A}}$ and $\mathcal{C}_{\mathcal{A}}$, and $\mathbf{v}_{\mathcal{B}}$ would be proportional to $\widehat{\mathbf{u}}_{\mathcal{A}}$, the nearest point to $\mathbf{u}_{\mathcal{A}}$ in $\mathcal{C}_{\mathcal{A}}$, implying $\mathbf{v}_{\mathcal{B}} = \mathbf{v}_{\hat{\mathcal{B}}}$). Now (6.15) gives $\text{Cos} \cdot x_{\ell_1} < A_{\mathcal{A}} = \text{Cos} \cdot A_{\mathcal{B}}$, or

$$\mathbf{x}_\ell' \mathbf{v}_{\mathcal{B}} = \mathbf{x}_\ell'(A_{\mathcal{B}} \mathbf{u}_{\mathcal{B}}) = A_{\mathcal{B}} x_{\ell_1} < A_{\mathcal{B}}^2. \tag{6.17}$$

This violates Constraint III, showing that $\mathbf{v}$ must equal $\mathbf{v}_{\hat{\mathcal{B}}}$. $\square$

Notice that the direction of advance $\widehat{\mathbf{v}} = \mathbf{v}_{\hat{\mathcal{B}}}$ of the idealized Stagewise procedure is a function only of the current maximal set $\widehat{\mathcal{A}} = \{j : |\widehat{c}_j| = \widehat{C}\}$, say $\widehat{\mathbf{v}} = \phi(\widehat{\mathcal{A}})$. In the language of (6.7),

$$\frac{d\widehat{\boldsymbol{\beta}}(z)}{dz} = \phi(\widehat{\mathcal{A}}). \tag{6.18}$$

The LARS–Stagewise algorithm of Theorem 2 produces an evolving family of estimates $\widehat{\boldsymbol{\beta}}$ that everywhere satisfies (6.18). This is true at every LARS–Stagewise breakpoint by the definition of the Stagewise modification. It is also true between breakpoints. Let $\widehat{\mathcal{A}}$ be the maximal set at the breakpoint, giving $\widehat{\mathbf{v}} = \mathbf{v}_{\hat{\mathcal{B}}} = \phi(\widehat{\mathcal{A}})$. In the succeeding LARS–Stagewise interval $\widehat{\boldsymbol{\mu}}(\gamma) = \widehat{\boldsymbol{\mu}} + \gamma \mathbf{v}_{\hat{\mathcal{B}}}$, the maximal set is immediately reduced to $\widehat{\mathcal{B}}$, according to properties (6.6), (6.7) of $\mathbf{v}_{\hat{\mathcal{B}}}$, at which it stays during the entire interval. However, $\phi(\widehat{\mathcal{B}}) = \phi(\widehat{\mathcal{A}}) = \mathbf{v}_{\hat{\mathcal{B}}}$ since $\mathbf{v}_{\hat{\mathcal{B}}} \in \mathcal{C}_{\hat{\mathcal{B}}}$, so the LARS–Stagewise procedure, which continues in the direction $\widehat{\mathbf{v}}$ until a new member is added to the active set, continues to obey the idealized Stagewise equation (6.18).

All of this shows that the LARS–Stagewise algorithm produces a legitimate version of the idealized Stagewise track. The converse of Lemma 12 says that there are no other versions, verifying Theorem 2.

The Stagewise procedure has its potential generality as an advantage over LARS and Lasso: it is easy to define forward Stagewise methods for a wide variety of nonlinear fitting problems, as in Hastie, Tibshirani and Friedman [(2001), Chapter 10, which begins with a Stagewise analysis of "boosting"]. Comparisons with LARS and Lasso within the linear model framework, as at the end of Section 3.2, help us better understand Stagewise methodology. This section's results permit further comparisons.

Consider proceeding forward from $\widehat{\boldsymbol{\mu}}$ along unit vector $\mathbf{u}$, $\widehat{\boldsymbol{\mu}}(\gamma) = \widehat{\boldsymbol{\mu}} + \gamma \mathbf{u}$, two interesting choices being the LARS direction $\mathbf{u}_{\hat{\mathcal{A}}}$ and the Stagewise direction $\widehat{\boldsymbol{\mu}}_{\hat{\mathcal{B}}}$. For $\mathbf{u} \in \mathcal{L}(X_{\hat{\mathcal{A}}})$, the rate of change of $S(\gamma) = \|\mathbf{y} - \widehat{\boldsymbol{\mu}}(\gamma)\|^2$ is

$$-\frac{\partial S(\gamma)}{\partial \gamma}\bigg|_0 = 2\widehat{C} \cdot \frac{\mathbf{u}'_{\mathcal{A}} \cdot \mathbf{u}}{A_{\hat{\mathcal{A}}}}, \tag{6.19}$$

(6.19) following quickly from (5.14). This shows that the LARS direction $\mathbf{u}_{\hat{\mathcal{A}}}$ maximizes the instantaneous decrease in S. The ratio

$$\frac{\partial S_{\text{Stage}}(\gamma)}{\partial \gamma}\bigg|_0 \bigg/ \frac{\partial S_{\text{LARS}}(\gamma)}{\partial \gamma}\bigg|_0 = \frac{A_{\hat{\mathcal{A}}}}{A_{\hat{\mathcal{B}}}}, \tag{6.20}$$

equaling the quantity "Cos" in (6.15).

The comparison goes the other way for the maximum absolute correlation $\widehat{C}(\gamma)$. Proceeding as in (2.15),

$$-\frac{\partial \widehat{C}(\gamma)}{\partial \gamma}\bigg|_0 = \min_{\hat{\mathcal{A}}}\{|x'_j \mathbf{u}|\}. \tag{6.21}$$

The argument for Lemma 12, using Constraints II and III, shows that $\mathbf{u}_{\hat{\mathcal{B}}}$ maximizes (6.21) at $A_{\hat{\mathcal{B}}}$, and that

$$\frac{\partial \widehat{C}_{\text{LARS}}(\gamma)}{\partial \gamma}\bigg|_0 \bigg/ \frac{\partial \widehat{C}_{\text{Stage}}(\gamma)}{\partial \gamma}\bigg|_0 = \frac{A_{\hat{\mathcal{A}}}}{A_{\hat{\mathcal{B}}}}. \tag{6.22}$$

The original motivation for the Stagewise procedure was to minimize residual squared error within a framework of parsimonious forward search. However, (6.20) shows that Stagewise is less greedy than LARS in this regard, it being more accurate to describe Stagewise as striving to minimize the maximum absolute residual correlation.

7. Computations. The entire sequence of steps in the LARS algorithm with $m < n$ variables requires $O(m^3 + nm^2)$ computations—the cost of a least squares fit on m variables.

In detail, at the kth of m steps, we compute $m - k$ inner products c_{jk} of the nonactive $\mathbf{x}_j$ with the current residuals to identify the next active variable, and then invert the $k \times k$ matrix $\mathcal{G}_k = X'_k X_k$ to find the next LARS direction. We do this by updating the Cholesky factorization R_{k-1} of $\mathcal{G}_{k-1}$ found at the previous step [Golub and Van Loan (1983)]. At the final step m, we have computed the Cholesky $R = R_m$ for the full cross-product matrix, which is the dominant calculation for a least squares fit. Hence the LARS sequence can be seen as a Cholesky factorization with a guided ordering of the variables.

The computations can be reduced further by recognizing that the inner products above can be updated at each iteration using the cross-product matrix $X'X$ and the current directions. For $m \gg n$, this strategy is counterproductive and is not used.

For the *lasso* modification, the computations are similar, except that occasionally one has to drop a variable, and hence *downdate* R_k [costing at most $O(m^2)$ operations per downdate]. For the *stagewise* modification of LARS, we need to check at each iteration that the components of w are all positive. If not, one or more variables are dropped [using the *inner loop* of the NNLS algorithm described in Lawson and Hanson (1974)], again requiring downdating of R_k. With many correlated variables, the stagewise version can take many more steps than LARS because of frequent dropping and adding of variables, increasing the computations by a factor up to 5 or more in extreme cases.

The LARS algorithm (in any of the three states above) works gracefully for the case where there are many more variables than observations: $m \gg n$. In this case LARS terminates at the saturated least squares fit after $n-1$ variables have entered the active set [at a cost of $O(n^3)$ operations]. (This number is $n-1$ rather than n, because the columns of X have been mean centered, and hence it has row-rank $n-1$.) We make a few more remarks about the $m \gg n$ case in the *lasso* state:

1. The LARS algorithm continues to provide Lasso solutions along the way, and the final solution highlights the fact that a Lasso fit can have no more than $n-1$ (mean centered) variables with nonzero coefficients.
2. Although the model involves no more than $n-1$ variables at any time, the number of *different* variables ever to have entered the model during the entire sequence can be—and typically is—greater than $n-1$.
3. The model sequence, particularly near the saturated end, tends to be quite variable with respect to small changes in $\mathbf{y}$.
4. The estimation of σ^2 may have to depend on an auxiliary method such as nearest neighbors (since the final model is saturated). We have not investigated the accuracy of the simple approximation formula (4.12) for the case $m > n$.

Documented S-PLUS implementations of LARS and associated functions are available from www-stat.stanford.edu/~hastie/Papers/; the diabetes data also appears there.

8. Boosting procedures. One motivation for studying the Forward Stagewise algorithm is its usefulness in adaptive fitting for data mining. In particular, Forward Stagewise ideas are used in "boosting," an important class of fitting methods for data mining introduced by Freund and Schapire (1997). These methods are one of the hottest topics in the area of machine learning, and one of the most effective prediction methods in current use. Boosting can use any adaptive fitting procedure as its "base learner" (model fitter): trees are a popular choice, as implemented in CART [Breiman, Friedman, Olshen and Stone (1984)].

Friedman, Hastie and Tibshirani (2000) and Friedman (2001) studied boosting and proposed a number of procedures, the most relevant to this discussion being *least squares boosting*. This procedure works by successive fitting of regression trees to the current residuals. Specifically we start with the residual $\mathbf{r} = \mathbf{y}$ and the fit $\hat{\mathbf{y}} = 0$. We fit a tree in $\mathbf{x}_1, \mathbf{x}_2, \ldots, \mathbf{x}_m$ to the response $\mathbf{y}$ giving a fitted tree $\mathbf{t}_1$ (an n-vector of fitted values). Then we update $\hat{\mathbf{y}}$ to $\hat{\mathbf{y}} + \varepsilon \cdot \mathbf{t}_1$, $\mathbf{r}$ to $\mathbf{y} - \hat{\mathbf{y}}$ and continue for many iterations. Here ε is a small positive constant. Empirical studies show that small values of ε work better than $\varepsilon = 1$: in fact, for prediction accuracy "the smaller the better." The only drawback in taking very small values of ε is computational slowness.

A major research question has been why boosting works so well, and specifically why is ε-shrinkage so important? To understand boosted trees in the present context, we think of our predictors not as our original variables $\mathbf{x}_1, \mathbf{x}_2, \ldots, \mathbf{x}_m$, but instead as the set of all trees $\mathbf{t}_k$ that could be fitted to our data. There is a strong similarity between least squares boosting and Forward Stagewise regression as defined earlier. Fitting a tree to the current residual is a numerical way of finding the "predictor" most correlated with the residual. Note, however, that the greedy algorithms used in CART do not search among all possible trees, but only a subset of them. In addition the set of all trees, including a parametrization for the predicted values in the terminal nodes, is infinite. Nevertheless one can define idealized versions of least-squares boosting that look much like Forward Stagewise regression.

Hastie, Tibshirani and Friedman (2001) noted the the striking similarity between Forward Stagewise regression and the Lasso, and conjectured that this may help explain the success of the Forward Stagewise process used in least squares boosting. That is, in some sense least squares boosting may be carrying out a Lasso fit on the infinite set of tree predictors. Note that direct computation of the Lasso via the LARS procedure would not be feasible in this setting because the number of trees is infinite and one could not compute the optimal step length. However, Forward Stagewise regression is feasible because it only need find the the most correlated predictor among the infinite set, where it approximates by numerical search.

In this paper we have established the connection between the Lasso and Forward Stagewise regression. We are now thinking about how these results can help to understand and improve boosting procedures. One such idea is a modified form of Forward Stagewise: we find the best tree as usual, but rather than taking a small step in only that tree, we take a small least squares step in all trees currently in our model. One can show that for small step sizes this procedure approximates LARS; its advantage is that it can be carried out on an infinite set of predictors such as trees.

APPENDIX

A.1. Local linearity and Lemma 2.

CONVENTIONS. We write $\mathbf{x}_l$ with subscript l for members of the active set $\mathcal{A}_k$. Thus $\mathbf{x}_l$ denotes the lth variable to enter, being an abuse of notation for $s_l \mathbf{x}_{j(l)} = \operatorname{sgn}(\hat{c}_{j(l)})\mathbf{x}_{j(l)}$. Expressions $\mathbf{x}_l'(\mathbf{y} - \hat{\boldsymbol{\mu}}_{k-1}(\mathbf{y})) = \hat{C}_k(\mathbf{y})$ and $\mathbf{x}_l'\mathbf{u}_k = A_k$ clearly do not depend on which $\mathbf{x}_l \in \mathcal{A}_k$ we choose.

By writing $j \notin \mathcal{A}_k$, we intend that both $\mathbf{x}_j$ and $-\mathbf{x}_j$ are candidates for inclusion at the next step. One could think of negative indices $-j$ corresponding to "new" variables $\mathbf{x}_{-j} = -\mathbf{x}_j$.

The active set $\mathcal{A}_k(\mathbf{y})$ depends on the data $\mathbf{y}$. When $\mathcal{A}_k(\mathbf{y})$ is the same for all $\mathbf{y}$ in a neighborhood of $\mathbf{y}_0$, we say that $\mathcal{A}_k(\mathbf{y})$ is locally fixed [at $\mathcal{A}_k = \mathcal{A}_k(\mathbf{y}_0)$].

A function $g(\mathbf{y})$ is locally Lipschitz at $\mathbf{y}$ if for all sufficiently small vectors $\Delta\mathbf{y}$,

$$\|\Delta g\| = \|g(\mathbf{y} + \Delta\mathbf{y}) - g(\mathbf{y})\| \le L\|\Delta\mathbf{y}\|. \tag{A.1}$$

If the constant L applies for all $\mathbf{y}$, we say that g is uniformly locally Lipschitz (L), and the word "locally" may be dropped.

LEMMA 13. *For each k, $0 \le k \le m$, there is an open set G_k of full measure on which $\mathcal{A}_k(\mathbf{y})$ and $\mathcal{A}_{k+1}(\mathbf{y})$ are locally fixed and differ by 1, and $\hat{\boldsymbol{\mu}}_k(\mathbf{y})$ is locally linear. The sets G_k are decreasing as k increases.*

PROOF. The argument is by induction. The induction hypothesis states that for each $\mathbf{y}_0 \in G_{k-1}$ there is a small ball $B(\mathbf{y}_0)$ on which (a) the active sets $\mathcal{A}_{k-1}(\mathbf{y})$ and $\mathcal{A}_k(\mathbf{y})$ are fixed and equal to $\mathcal{A}_{k-1}$ and $\mathcal{A}_k$, respectively, (b) $|\mathcal{A}_k \setminus \mathcal{A}_{k-1}| = 1$ so that the same single variable enters locally at stage $k-1$ and (c) $\hat{\boldsymbol{\mu}}_{k-1}(\mathbf{y}) = M\mathbf{y}$ is linear. We construct a set G_k with the same property.

Fix a point $\mathbf{y}_0$ and the corresponding ball $B(\mathbf{y}_0) \subset G_{k-1}$, on which $\mathbf{y} - \hat{\boldsymbol{\mu}}_{k-1}(\mathbf{y}) = \mathbf{y} - M\mathbf{y} = R\mathbf{y}$, say. For indices $j_1, j_2 \notin \mathcal{A}$, let $N(j_1, j_2)$ be the set of $\mathbf{y}$ for which there exists a γ such that

$$w'(R\mathbf{y} - \gamma\mathbf{u}_k) = \mathbf{x}_{j_1}'(R\mathbf{y} - \gamma\mathbf{u}_k) = \mathbf{x}_{j_2}'(R\mathbf{y} - \gamma\mathbf{u}_k). \tag{A.2}$$

Setting $\boldsymbol{\delta}_1 = \mathbf{x}_l - \mathbf{x}_{j_1}$, the first equality may be written $\boldsymbol{\delta}_1' R\mathbf{y} = \gamma\boldsymbol{\delta}_1'\mathbf{u}_k$ and so when $\boldsymbol{\delta}_1'\mathbf{u}_k \neq 0$ determines

$$\gamma = \boldsymbol{\delta}_1' R\mathbf{y}/\boldsymbol{\delta}_1'\mathbf{u}_k =: \boldsymbol{\eta}_1'\mathbf{y}.$$

[If $\boldsymbol{\delta}_1'\mathbf{u}_k = 0$, there are no qualifying $\mathbf{y}$, and $N(j_1, j_2)$ is empty.] Now using the second equality and setting $\boldsymbol{\delta}_2 = \mathbf{x}_l - \mathbf{x}_{j_2}$, we see that $N(j_1, j_2)$ is contained in the set of $\mathbf{y}$ for which

$$\boldsymbol{\delta}_2' R\mathbf{y} = \boldsymbol{\eta}_1'\mathbf{y}\,\boldsymbol{\delta}_2'\mathbf{u}_k.$$

In other words, setting $\boldsymbol{\eta}_2 = R'\boldsymbol{\delta}_2 - (\boldsymbol{\delta}_2'\mathbf{u}_k)\boldsymbol{\eta}_1$, we have

$$N(j_1, j_2) \subset \{\mathbf{y} : \boldsymbol{\eta}_2'\mathbf{y} = 0\}.$$

If we define

$$N(\mathbf{y}_0) = \bigcup\{N(j_1, j_2) : j_1, j_2 \notin \mathcal{A}, j_1 \neq j_2\},$$

it is evident that $N(\mathbf{y}_0)$ is a finite union of hyperplanes and hence closed. For $\mathbf{y} \in B(\mathbf{y}_0) \setminus N(\mathbf{y}_0)$, a unique new variable joins the active set at step k. Near each such $\mathbf{y}$ the "joining" variable is locally the same and $\gamma_k(\mathbf{y})\mathbf{u}_k$ is locally linear.

We then define $G_k \subset G_{k-1}$ as the union of such sets $B(\mathbf{y}) \setminus N(\mathbf{y})$ over $\mathbf{y} \in G_{k-1}$. Thus G_k is open and, on G_k, $\mathcal{A}_{k+1}(\mathbf{y})$ is locally constant and $\hat{\boldsymbol{\mu}}_k(\mathbf{y})$ is locally linear. Thus properties (a)–(c) hold for G_k.

The same argument works for the initial case $k = 0$: since $\hat{\boldsymbol{\mu}}_0 = 0$, there is no circularity.

Finally, since the intersection of G_k with any compact set is covered by a finite number of $B(y_i) \setminus N(y_i)$, it is clear that G_k has full measure. □

LEMMA 14. *Suppose that, for* $\mathbf{y}$ *near* $\mathbf{y}_0$, $\hat{\boldsymbol{\mu}}_{k-1}(\mathbf{y})$ *is continuous* (*resp. linear*) *and that* $\mathcal{A}_k(\mathbf{y}) = \mathcal{A}_k$. *Suppose also that, at* $\mathbf{y}_0$, $\mathcal{A}_{k+1}(\mathbf{y}_0) = \mathcal{A} \cup \{k+1\}$.

Then for $\mathbf{y}$ *near* $\mathbf{y}_0$, $\mathcal{A}_{k+1}(\mathbf{y}) = \mathcal{A}_k \cup \{k+1\}$ *and* $\hat{\gamma}_k(\mathbf{y})$ *and hence* $\hat{\boldsymbol{\mu}}_k(\mathbf{y})$ *are continuous* (*resp. linear*) *and uniformly Lipschitz.*

PROOF. Consider first the situation at $\mathbf{y}_0$, with $\widehat{C}_k$ and $\widehat{c}_{kj}$ defined in (2.18) and (2.17), respectively. Since $k+1 \notin \mathcal{A}_k$, we have $|\widehat{C}_k(\mathbf{y}_0)| > \hat{c}_{k,k+1}(\mathbf{y}_0)$, and $\hat{\gamma}_k(\mathbf{y}_0) > 0$ satisfies

$$\hat{C}_k(\mathbf{y}_0) - \hat{\gamma}_k(\mathbf{y}_0)A_k \begin{Bmatrix} = \\ > \end{Bmatrix} \hat{c}_{k,j}(\mathbf{y}_0) - \hat{\gamma}_k(\mathbf{y}_0)a_{k,j} \qquad \text{as} \begin{Bmatrix} j = k+1 \\ j > k+1 \end{Bmatrix}. \tag{A.3}$$

In particular, it must be that $A_k \neq a_{k,k+1}$, and hence

$$\hat{\gamma}_k(\mathbf{y}_0) = \frac{\hat{C}_k(\mathbf{y}_0) - \hat{c}_{k,k+1}(\mathbf{y}_0)}{A_k - a_{k,k+1}} > 0.$$

Call an index j admissible if $j \notin \mathcal{A}_k$ and $a_{k,j} \neq A_k$. For $\mathbf{y}$ near $\mathbf{y}_0$, this property is independent of $\mathbf{y}$. For admissible j, define

$$R_{k,j}(\mathbf{y}) = \frac{\hat{C}_k(\mathbf{y}) - \hat{c}_{k,j}(\mathbf{y})}{A_k - a_{k,j}},$$

which is continuous (resp. linear) near $\mathbf{y}_0$ from the assumption on $\hat{\boldsymbol{\mu}}_{k-1}$. By definition,

$$\hat{\gamma}_k(\mathbf{y}) = \min_{j \in \mathcal{P}_k(\mathbf{y})} R_{k,j}(\mathbf{y}),$$

where

$$\mathcal{P}_k(\mathbf{y}) = \{j \text{ admissible and } R_{k,j}(\mathbf{y}) > 0\}.$$

For admissible j, $R_{k,j}(\mathbf{y}_0) \neq 0$, and near $\mathbf{y}_0$ the functions $\mathbf{y} \to R_{k,j}(\mathbf{y})$ are continuous and of fixed sign. Thus, near $\mathbf{y}_0$ the set $\mathcal{P}_k(\mathbf{y})$ stays fixed at $\mathcal{P}_k(\mathbf{y}_0)$ and (A.3) implies that

$$R_{k,k+1}(\mathbf{y}) < R_{k,j}(\mathbf{y}), \qquad j > k+1, j \in \mathcal{P}_k(\mathbf{y}).$$

Consequently, for $\mathbf{y}$ near $\mathbf{y}_0$, only variable $k+1$ joins the active set, and so $\mathcal{A}_{k+1}(\mathbf{y}) = \mathcal{A}_k \cup \{k+1\}$, and

$$\hat{\gamma}_k(\mathbf{y}) = R_{k,k+1}(\mathbf{y}) = \frac{(\mathbf{x}_l - \mathbf{x}_{k+1})'(\mathbf{y} - \hat{\boldsymbol{\mu}}_{k-1}(\mathbf{y}))}{(\mathbf{x}_l - \mathbf{x}_{k+1})'\mathbf{u}_k}. \tag{A.4}$$

This representation shows that both $\hat{\gamma}_k(\mathbf{y})$ and hence $\hat{\boldsymbol{\mu}}_k(\mathbf{y}) = \hat{\boldsymbol{\mu}}_{k-1}(\mathbf{y}) + \hat{\gamma}_k(\mathbf{y})\mathbf{u}_k$ are continuous (resp. linear) near $\mathbf{y}_0$.

To show that $\hat{\gamma}_k$ is locally Lipschitz at $\mathbf{y}$, we set $\boldsymbol{\delta} = \mathbf{w} - \mathbf{x}_{k+1}$ and write, using notation from (A.1),

$$\Delta\hat{\gamma}_k = \frac{\boldsymbol{\delta}'(\Delta\mathbf{y} - \Delta\hat{\boldsymbol{\mu}}_{k-1})}{\boldsymbol{\delta}'\mathbf{u}_k}.$$

As $\mathbf{y}$ varies, there is a finite list of vectors $(\mathbf{x}_l, \mathbf{x}_{k+1}, \mathbf{u}_k)$ that can occur in the denominator term $\boldsymbol{\delta}'\mathbf{u}_k$, and since all such terms are positive [as observed below (A.3)], they have a uniform positive lower bound, $a_{\min}$ say. Since $\|\boldsymbol{\delta}\| \leq 2$ and $\hat{\boldsymbol{\mu}}_{k-1}$ is Lipschitz (L_{k-1}) by assumption, we conclude that

$$\frac{|\Delta\hat{\gamma}_k|}{\|\Delta\mathbf{y}\|} \leq 2a_{\min}^{-1}(1 + L_{k-1}) =: L_k. \qquad \square$$

A.2. Consequences of the positive cone condition.

LEMMA 15. *Suppose that* $|\mathcal{A}_+| = |\mathcal{A}| + 1$ *and that* $X_{\mathcal{A}+} = [X_{\mathcal{A}} \ \mathbf{x}_+]$ (*where* $\mathbf{x}_+ = s_j\mathbf{x}_j$ *for some* $j \notin \mathcal{A}$). *Let* $P_{\mathcal{A}} = X_{\mathcal{A}}G_{\mathcal{A}}^{-1}X_{\mathcal{A}}'$ *denote projection on* span$(X_{\mathcal{A}})$, *so that* $a = \mathbf{x}_+'P_{\mathcal{A}}\mathbf{x}_+ < 1$. *The* $+$-*component of* $G_{\mathcal{A}+}^{-1}\mathbf{1}_{\mathcal{A}+}$ *is*

$$(G_{\mathcal{A}+}^{-1}\mathbf{1}_{\mathcal{A}+})_+ = (1-a)^{-1}\left(1 - \frac{\mathbf{x}_+'\mathbf{u}_{\mathcal{A}}}{A_{\mathcal{A}}}\right). \tag{A.5}$$

Consequently, under the positive cone condition (4.11),

$$\mathbf{x}_+'\mathbf{u}_{\mathcal{A}} < A_{\mathcal{A}}. \tag{A.6}$$

PROOF. Write $G_{\mathcal{A}+}$ as a partitioned matrix

$$G_{\mathcal{A}+} = \begin{pmatrix} X'X & X'\mathbf{x}_+ \\ \mathbf{x}_+'X & \mathbf{x}_+'\mathbf{x}_+ \end{pmatrix} = \begin{pmatrix} A & B \\ B' & D \end{pmatrix}.$$

Applying the formula for the inverse of a partitioned matrix [e.g., Rao (1973), page 33],

$$(G_{\mathcal{A}+}^{-1}\mathbf{1}_{\mathcal{A}+})_+ = -E^{-1}F'\mathbf{1} + E^{-1},$$

where

$$E = D - B'A^{-1}B = 1 - \mathbf{x}_+' P_{\mathcal{A}}\mathbf{x}_+,$$

$$F = A^{-1}B = G_{\mathcal{A}}^{-1}X'\mathbf{x}_+,$$

from which (A.5) follows. The positive cone condition implies that $G_{\mathcal{A}+}^{-1}\mathbf{1}_{\mathcal{A}+} > 0$, and so (A.6) is immediate. □

A.3. Global continuity and Lemma 3. We shall call $\mathbf{y}_0$ a multiple point at step k if two or more variables enter at the same time. Lemma 14 shows that such points form a set of measure zero, but they can and do cause discontinuities in $\hat{\boldsymbol{\mu}}_{k+1}$ at $\mathbf{y}_0$ in general. We will see, however, that the positive cone condition prevents such discontinuities.

We confine our discussion to double points, hoping that these arguments will be sufficient to establish the same pattern of behavior at points of multiplicity 3 or higher. In addition, by renumbering, we shall suppose that indices $k+1$ and $k+2$ are those that are added at double point $\mathbf{y}_0$. Similarly, for convenience only, we assume that $\mathcal{A}_k(\mathbf{y})$ is constant near $\mathbf{y}_0$. Our task then is to show that, for $\mathbf{y}$ near a double point $\mathbf{y}_0$, both $\hat{\boldsymbol{\mu}}_k(\mathbf{y})$ and $\hat{\boldsymbol{\mu}}_{k+1}(\mathbf{y})$ are continuous and uniformly locally Lipschitz.

LEMMA 16. *Suppose that $\mathcal{A}_k(\mathbf{y}) = \mathcal{A}_k$ is constant near $\mathbf{y}_0$ and that $\mathcal{A}_{k+}(\mathbf{y}_0) = \mathcal{A}_k \cup \{k+1, k+2\}$. Then for $\mathbf{y}$ near $\mathbf{y}_0$, $\mathcal{A}_{k+}(\mathbf{y}) \setminus \mathcal{A}_k$ can only be one of three possibilities, namely $\{k+1\}, \{k+2\}$ or $\{k+1, k+2\}$. In all cases $\hat{\boldsymbol{\mu}}_k(\mathbf{y}) = \hat{\boldsymbol{\mu}}_{k-1}(\mathbf{y}) + \hat{\gamma}_k(\mathbf{y})\mathbf{u}_k$ as usual, and both $\gamma_k(\mathbf{y})$ and $\hat{\boldsymbol{\mu}}_k(\mathbf{y})$ are continuous and locally Lipschitz.*

PROOF. We use notation and tools from the proof of Lemma 14. Since $\mathbf{y}_0$ is a double point and the positivity set $\mathcal{P}_k(\mathbf{y}) = \mathcal{P}_k$ near $\mathbf{y}_0$, we have

$$0 < R_{k,k+1}(\mathbf{y}_0) = R_{k,k+2}(\mathbf{y}_0) < R_{k,j}(\mathbf{y}_0) \qquad \text{for } j \in \mathcal{P}_k \setminus \{k+1, k+2\}.$$

Continuity of $R_{k,j}$ implies that near $\mathbf{y}_0$ we still have

$$0 < R_{k,k+1}(\mathbf{y}), R_{k,k+2}(\mathbf{y}) < \min\{R_{k,j}(\mathbf{y}); j \in \mathcal{P}_k \setminus \{k+1, k+2\}\}.$$

Hence $\mathcal{A}_{k+} \setminus \mathcal{A}_k$ must equal $\{k+1\}$ or $\{k+2\}$ or $\{k+1, k+2\}$ according as $R_{k,k+1}(\mathbf{y})$ is less than, greater than or equal to $R_{k,k+2}(\mathbf{y})$. The continuity of

$$\hat{\gamma}_k(\mathbf{y}) = \min\{R_{k,k+1}(\mathbf{y}), R_{k,k+2}(\mathbf{y})\}$$

is immediate, and the local Lipschitz property follows from the arguments of Lemma 14. □

LEMMA 17. *Assume the conditions of Lemma* 16 *and in addition that the positive cone condition* (4.11) *holds. Then* $\hat{\boldsymbol{\mu}}_{k+1}(\mathbf{y})$ *is continuous and locally Lipschitz near* $\mathbf{y}_0$.

PROOF. Since $\mathbf{y}_0$ is a double point, property (A.3) holds, but now with equality when $j = k+1$ or $k+2$ and strict inequality otherwise. In other words, there exists $\delta_0 > 0$ for which

$$\hat{C}_{k+1}(\mathbf{y}_0) - \hat{c}_{k+1,j}(\mathbf{y}_0) \begin{cases} = 0, & \text{if } j = k+2, \\ \geq \delta_0, & \text{if } j > k+2. \end{cases}$$

Consider a neighborhood $B(\mathbf{y}_0)$ of $\mathbf{y}_0$ and let $N(\mathbf{y}_0)$ be the set of double points in $B(\mathbf{y}_0)$, that is, those for which $\mathcal{A}_{k+1}(\mathbf{y}) \setminus \mathcal{A}_k = \{k+1, k+2\}$. We establish the convention that at such double points $\hat{\boldsymbol{\mu}}_{k+1}(\mathbf{y}) = \hat{\boldsymbol{\mu}}_k(\mathbf{y})$; at other points $\mathbf{y}$ in $B(\mathbf{y}_0)$, $\hat{\boldsymbol{\mu}}_{k+1}(\mathbf{y})$ is defined by $\hat{\boldsymbol{\mu}}_k(\mathbf{y}) + \hat{\gamma}_{k+1}(\mathbf{y})\mathbf{u}_{k+1}$ as usual.

Now consider those $\mathbf{y}$ near $\mathbf{y}_0$ for which $\mathcal{A}_{k+1}(\mathbf{y}) \setminus \mathcal{A}_k = \{k+1\}$, and so, from the previous lemma, $\mathcal{A}_{k+2}(\mathbf{y}) \setminus \mathcal{A}_{k+1} = \{k+2\}$. For such $\mathbf{y}$, continuity and the local Lipschitz property for $\hat{\boldsymbol{\mu}}_k$ imply that

$$\hat{C}_{k+1}(\mathbf{y}) - \hat{c}_{k+1,j}(\mathbf{y}) \begin{cases} = O(\|\mathbf{y} - \mathbf{y}_0\|), & \text{if } j = k+2, \\ > \delta_0/2, & \text{if } j > k+2. \end{cases}$$

It is at this point that we use the positive cone condition (via Lemma 15) to guarantee that $A_{k+1} > a_{k+1,k+2}$. Also, since $\mathcal{A}_{k+1}(\mathbf{y}) \setminus \mathcal{A}_k = \{k+1\}$, we have

$$\hat{C}_{k+1}(\mathbf{y}) > \hat{c}_{k+1,k+2}(\mathbf{y}).$$

These two facts together show that $k+2 \in \mathcal{P}_{k+1}(\mathbf{y})$ and hence that

$$\hat{\gamma}_{k+1}(\mathbf{y}) = \frac{\hat{C}_{k+1}(\mathbf{y}) - \hat{c}_{k+1,k+2}(\mathbf{y})}{A_{k+1} - a_{k+1,k+2}} = O(\|\mathbf{y} - \mathbf{y}_0\|)$$

is continuous and locally Lipschitz. In particular, as $\mathbf{y}$ approaches $N(\mathbf{y}_0)$, we have $\hat{\gamma}_{k+1}(\mathbf{y}) \to 0$. $\square$

REMARK A.1. We say that a function $g : \mathbb{R}^n \to \mathbb{R}$ is *almost differentiable* if it is absolutely continuous on almost all line segments parallel to the coordinate axes, and its partial derivatives (which consequently exist a.e.) are locally integrable. This definition of almost differentiability appears superficially to be weaker than that given by Stein, but it is in fact precisely the property used in his proof. Furthermore, this definition is equivalent to the standard definition of weak differentiability used in analysis.

PROOF OF LEMMA 3. We have shown explicitly that $\hat{\boldsymbol{\mu}}_k(\mathbf{y})$ is continuous and uniformly locally Lipschitz near single and double points. Similar arguments extend the property to points of multiplicity 3 and higher, and so all points $\mathbf{y}$ are covered. Finally, absolute continuity of $\mathbf{y} \to \hat{\boldsymbol{\mu}}_k(\mathbf{y})$ on line segments is a simple consequence of the uniform Lipschitz property, and so $\hat{\boldsymbol{\mu}}_k$ is almost differentiable. □

Acknowledgments. The authors thank Jerome Friedman, Bogdan Popescu, Saharon Rosset and Ji Zhu for helpful discussions.

REFERENCES

BREIMAN, L., FRIEDMAN, J., OLSHEN, R. and STONE, C. (1984). *Classification and Regression Trees*. Wadsworth, Belmont, CA.

EFRON, B. (1986). How biased is the apparent error rate of a prediction rule? *J. Amer. Statist. Assoc.* **81** 461–470.

EFRON, B. and TIBSHIRANI, R. (1997). Improvements on cross-validation: The .632+ bootstrap method. *J. Amer. Statist. Assoc.* **92** 548–560.

FREUND, Y. and SCHAPIRE, R. (1997). A decision-theoretic generalization of online learning and an application to boosting. *J. Comput. System Sci.* **55** 119–139.

FRIEDMAN, J. (2001). Greedy function approximation: A gradient boosting machine. *Ann. Statist.* **29** 1189–1232.

FRIEDMAN, J., HASTIE, T. and TIBSHIRANI, R. (2000). Additive logistic regression: A statistical view of boosting (with discussion). *Ann. Statist.* **28** 337–407.

GOLUB, G. and VAN LOAN, C. (1983). *Matrix Computations*. Johns Hopkins Univ. Press, Baltimore, MD.

HASTIE, T., TIBSHIRANI, R. and FRIEDMAN, J. (2001). *The Elements of Statistical Learning: Data Mining, Inference and Prediction*. Springer, New York.

LAWSON, C. and HANSON, R. (1974). *Solving Least Squares Problems*. Prentice-Hall, Englewood Cliffs, NJ.

MALLOWS, C. (1973). Some comments on C_p. *Technometrics* **15** 661–675.

MEYER, M. and WOODROOFE, M. (2000). On the degrees of freedom in shape-restricted regression. *Ann. Statist.* **28** 1083–1104.

OSBORNE, M., PRESNELL, B. and TURLACH, B. (2000a). A new approach to variable selection in least squares problems. *IMA J. Numer. Anal.* **20** 389–403.

OSBORNE, M. R., PRESNELL, B. and TURLACH, B. (2000b). On the LASSO and its dual. *J. Comput. Graph. Statist.* **9** 319–337.

RAO, C. R. (1973). *Linear Statistical Inference and Its Applications*, 2nd ed. Wiley, New York.

STEIN, C. (1981). Estimation of the mean of a multivariate normal distribution. *Ann. Statist.* **9** 1135–1151.

TIBSHIRANI, R. (1996). Regression shrinkage and selection via the lasso. *J. Roy. Statist. Soc. Ser. B.* **58** 267–288.

WEISBERG, S. (1980). *Applied Linear Regression*. Wiley, New York.

YE, J. (1998). On measuring and correcting the effects of data mining and model selection. *J. Amer. Statist. Assoc.* **93** 120–131.

DEPARTMENT OF STATISTICS
STANFORD UNIVERSITY
SEQUOIA HALL
STANFORD, CALIFORNIA 94305-4065
USA
E-MAIL: brad@stat.stanford.edu

DISCUSSION

BY HEMANT ISHWARAN

Cleveland Clinic Foundation

Being able to reliably, and automatically, select variables in linear regression models is a notoriously difficult problem. This research attacks this question head on, introducing not only a computationally efficient algorithm and method, LARS (and its derivatives), but at the same time introducing comprehensive theory explaining the intricate details of the procedure as well as theory to guide its practical implementation. This is a fascinating paper and I commend the authors for this important work.

Automatic variable selection, the main theme of this paper, has many goals. So before embarking upon a discussion of the paper it is important to first sit down and clearly identify what the objectives are. The authors make it clear in their introduction that, while often the goal in variable selection is to select a "good" linear model, where goodness is measured in terms of prediction accuracy performance, it is also important at the same time to choose models which lean toward the parsimonious side. So here the goals are pretty clear: we want good prediction error performance but also simpler models. These are certainly reasonable objectives and quite justifiable in many scientific settings. At the same, however, one should recognize the difficulty of the task, as the two goals, low prediction error and smaller models, can be diametrically opposed. By this I mean that certainly from an oracle point of view it is true that minimizing prediction error will identify the true model, and thus, by going after prediction error (in a perfect world), we will also get smaller models by default. However, in practice, what happens is that small gains in prediction error often translate into larger models and less dimension reduction. So as procedures get better at reducing prediction error, they can also get worse at picking out variables accurately.

Unfortunately, I have some misgivings that LARS might be falling into this trap. Mostly my concern is fueled by the fact that Mallows' C_p is the criterion used for determining the optimal LARS model. The use of C_p often leads to overfitting, and this coupled with the fact that LARS is a forward optimization procedure, which is often found to be greedy, raises some potential flags. This, by the way, does not necessarily mean that LARS per se is overfitting, but rather that I think C_p may be an inappropriate model selection criterion for LARS. It is this point that will be the focus of my discussion. I will offer some evidence that C_p can sometimes be used effectively if *model uncertainty* is accounted for, thus pointing to ways for its more appropriate use within LARS. Mostly I will make my arguments by way of high-dimensional simulations. My focus on high dimensions is motivated in part by the increasing interest in such problems, but also because it is in such problems that performance breakdowns become magnified and are more easily identified.

Note that throughout my discussion I will talk only about LARS, but, given the connections outlined in the paper, the results should also naturally apply to the Lasso and Stagewise derivatives.

1. Is C_p the correct stopping rule for LARS? The C_p criterion was introduced by Mallows (1973) to be used with the OLS as an unbiased estimator for the model error. However, it is important to keep in mind that it was not intended to be used *when the model is selected by the data* as this can lead to selection bias and in some cases poor subset selection [Breiman (1992)]. Thus, choosing the model with lowest C_p value is only a heuristic technique with sometimes bad performance. Indeed, ultimately, this leads to an inconsistent procedure for the OLS [Shao (1993)]. Therefore, while I think it is reasonable to assume that the C_p formula (4.10) is correct [i.e., that it is reasonable to expect that $df(\widehat{\mu}_k) \approx k$ under a wide variety of settings], there is really no reason to expect that minimizing the C_p value will lead to an optimal procedure for LARS.

In fact, using C_p in a Forward Stagewise procedure of any kind seems to me to be a risky thing to do given that C_p often overfits and that Stagewise procedures are typically greedy. Figure 5 of the paper is introduced (partly) to dispel these types of concerns about LARS being greedy. The message there is that $\mathrm{pe}(\widehat{\mu})$, a performance measurement related to prediction error, declines slowly from its maximum value for LARS compared to the quick drop seen with standard forward stepwise regression. Thus, LARS acts differently than well-known greedy algorithms and so we should not be worried. However, I see the message quite differently. If the maximum proportion explained for LARS is roughly the same over a large range of steps, and hence models of different dimension, then this implies that there is not much to distinguish between higher- and lower-dimensional models. Combine this with the use of C_p which could provide poor estimates for the prediction error due to selection bias and there is real concern for estimating models that are too large.

To study this issue, let me start by reanalyzing the diabetes data (which was the basis for generating Figure 5). In this analysis I will compare LARS to a Bayesian method developed in Ishwaran and Rao (2000), referred to as SVS (short for Stochastic Variable Selection). The SVS procedure is a hybrid of the spike-and-slab model approach pioneered by Mitchell and Beauchamp (1988) and later developed in George and McCulloch (1993). Details for SVS can be found in Ishwaran and Rao (2000, 2003). My reason for using SVS as a comparison procedure is that, like LARS, its coefficient estimates are derived via shrinkage. However, unlike LARS, these estimates are based on model averaging *in combination* with shrinkage. The use of model averaging is a way of accounting for model uncertainty, and my argument will be that models selected via C_p based on SVS coefficients will be more stable than those found using LARS thanks to the extra benefit of model averaging.

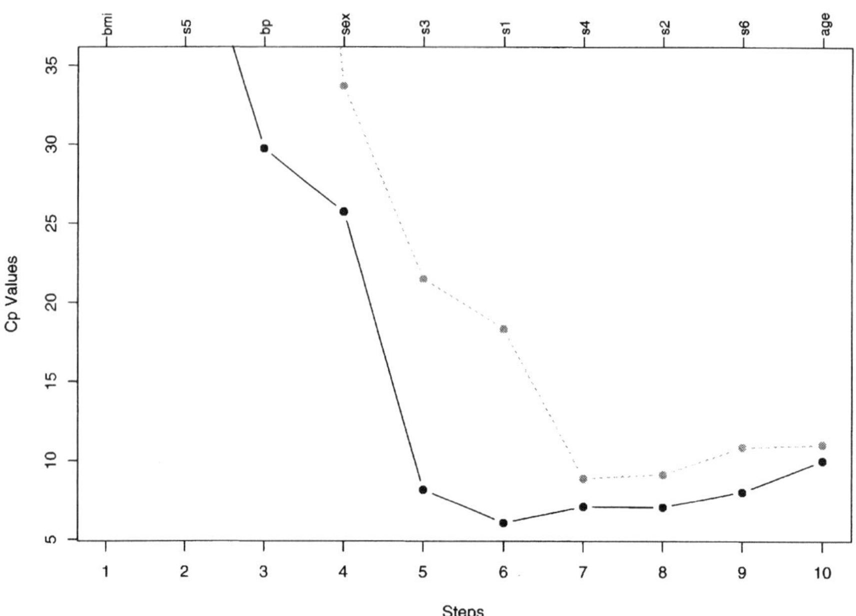

FIG. 1. *C_p values from main effects model for diabetes data: thick line is values from SVS; thin dashed line is from LARS. Covariates listed at the top of the graph are ordered by importance as measured by their absolute posterior mean.*

Figures 1 and 2 present the C_p values for the main effects model and the quadratic model from both procedures (the analysis for LARS was based on S-PLUS code kindly provided by Trevor Hastie). The C_p values for SVS were computed by (a) finding the posterior mean values for coefficients, (b) ranking covariates by the size of their absolute posterior mean coefficient values (with the top rank going to the largest absolute mean) and (c) computing the C_p value $C_p(\widetilde{\boldsymbol{\mu}}_k) = \|\mathbf{y} - \widetilde{\boldsymbol{\mu}}_k\|/\overline{\sigma}^2 - n + 2k$, where $\widetilde{\boldsymbol{\mu}}_k$ is the OLS estimate based on the k top ranked covariates. All covariates were standardized. This technique of using C_p with SVS was discussed in Ishwaran and Rao (2000).

We immediately see some differences in the figures. In Figure 1, the final model selected by SVS had $k = 6$ variables, while LARS had $k = 7$ variables. More interesting, though, are the discrepancies for the quadratic model seen in Figure 2. Here the optimal SVS model had $k = 8$ variables in contrast to the much higher $k = 15$ variables found by LARS. The top eight variables from SVS (some of these can be read off the top of the plot) are bmi, ltg, map, hdl, sex, age.sex, bmi.map and glu.2. The last three variables are interaction effects and a squared main effects term. The top eight variables from LARS are bmi, ltg, map, hdl, bmi.map, age.sex, glu.2 and bmi.2. Although there is a reasonable overlap in variables, there is still enough of a discrepancy to be concerned. The different model sizes are also cause for concern. Another worrisome aspect for LARS seen in Figure 2 is that its C_p values remain bounded away from zero. This should be compared to the C_p values for SVS, which attain a near-zero mininum value, as we would hope for.

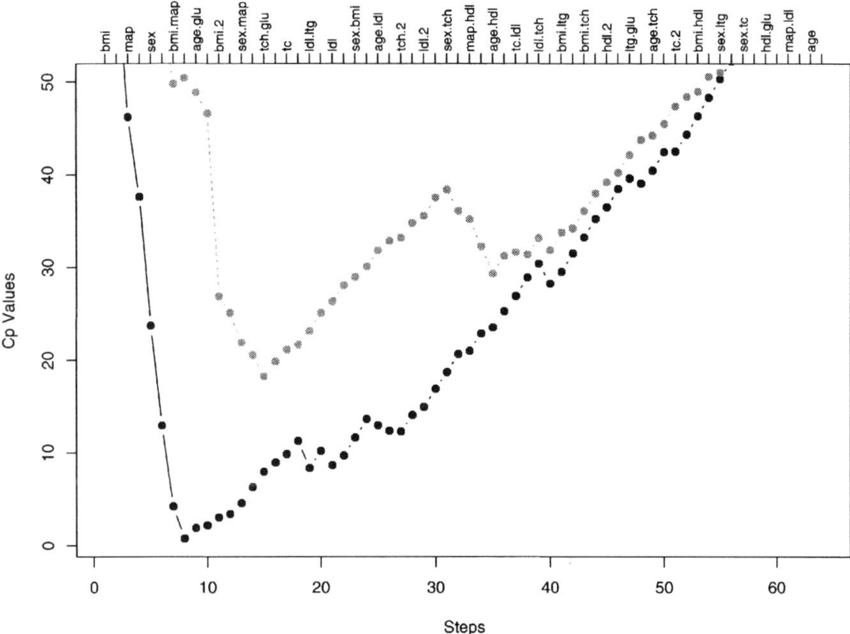

FIG. 2. *C_p values from quadratic model: best model from SVS is $k = 8$ (thick line) compared with $k = 15$ from LARS (thin dashed line). Note how the minimum value for SVS is nearly zero.*

2. High-dimensional simulations. Of course, since we do not know the true answer in the diabetes example, we cannot definitively assess if the LARS models are too large. Instead, it will be helpful to look at some simulations for a more systematic study. The simulations I used were designed following the recipe given in Breiman (1992). Data was simulated in all cases by using i.i.d. N(0, 1) variables for ε_i. Covariates x_i, for $i = 1, \dots, n$, were generated independently from a multivariate normal distribution with zero mean and with covariance satisfying $E(x_{i,j}x_{i,k}) = \rho^{|j-k|}$. I considered two settings for ρ: (i) $\rho = 0$ (uncorrelated); (ii) $\rho = 0.90$ (correlated). In all simulations, $n = 800$ and $m = 400$. Nonzero β_j coefficients were in 15 clusters of 7 adjacent variables centered at every 25th variable. For example, for the variables clustered around the 25th variable, the coefficient values were given by $\beta_{25+j} = |h - j|^{1.25}$ for $|j| < h$, where $h = 4$. The other 14 clusters were defined similarly. All other coefficients were set to zero. This gave a total of 105 nonzero values and 295 zero values. Coefficient values were adjusted by multiplying by a common constant to make the theoretical R^2 value equal to 0.75 [see Breiman (1992) for a discussion of this point]. Please note that, while the various parameters chosen for the simulations might appear specific, I also experimented with other simulations (not reported) by considering different configurations for the dimension m, sample size n, correlation ρ and the number of nonzero coefficients. What I found was consistent with the results presented here.

For each ρ correlation setting, simulations were repeated 100 times independently. Results are recorded in Table 1. There I have recorded what I call TotalMiss,

TABLE 1
Breiman simulation: $m = 400$, $n = 800$ *and* 105 *nonzero* β_j

	$\rho = 0$ (uncorrelated X)					$\rho = 0.9$ (correlated X)				
	$\widehat{m}$	pe($\widehat{\mu}$)	TotalMiss	FDR	FNR	$\widehat{m}$	pe($\widehat{\mu}$)	TotalMiss	FDR	FNR
LARS	210.69	0.907	126.63	0.547	0.055	99.51	0.962	75.77	0.347	0.135
svsCp	126.66	0.887	61.14	0.323	0.072	58.86	0.952	66.38	0.153	0.164
svsBMA	400.00	0.918	295.00	0.737	0.000	400.00	0.966	295.00	0.737	0.000
Step	135.53	0.876	70.35	0.367	0.075	129.24	0.884	137.10	0.552	0.208

FDR and FNR. TotalMiss is the total number of misclassified variables, that is, the total number of falsely identified nonzero β_j coefficients and falsely identified zero coefficients; FDR and FNR are the false discovery and false nondiscovery rates defined as the false positive and false negative rates for those coefficients identified as nonzero and zero, respectively. The TotalMiss, FDR and FNR values reported are the averaged values from the 100 simulations. Also recorded in the table is $\widehat{m}$, the average number of variables selected by a procedure, as well as the performance value pe($\widehat{\mu}$) [cf. (3.17)], again averaged over the 100 simulations.

Table 1 records the results from various procedures. The entry "svsCp" refers to the C_p-based SVS method used earlier; "Step" is standard forward stepwise regression using the C_p criterion; "svsBMA" is the Bayesian model averaged estimator from SVS. My only reason for including svsBMA is to gauge the prediction error performance of the other procedures. Its variable selection performance is not of interest. Pure Bayesian model averaging leads to improved prediction, but because it does no dimension reduction at all it cannot be considered as a serious candidate for selecting variables.

The overall conclusions from Table 1 are summarized as follows:

1. The total number of misclassified coefficients and FDR values is high in the uncorrelated case for LARS and high in the correlated case for stepwise regression. Their estimated models are just too large. In comparison, svsCp does well in both cases. Overall it does the best in terms of selecting variables by maintaining low FDR and TotalMiss values. It also maintains good performance values.
2. LARS's performance values are good, second only to svsBMA. However, low prediction error does not necessarily imply good variable selection.

3. LARS C_p values in orthogonal models. Figure 3 shows the C_p values for LARS from the two sets of simulations. It is immediately apparent that the C_p curve in the uncorrelated case is too flat, leading to models which are too large. These simulations were designed to reflect an orthogonal design setting (at least asymptotically), so what is it about the orthogonal case that is adversely affecting LARS?

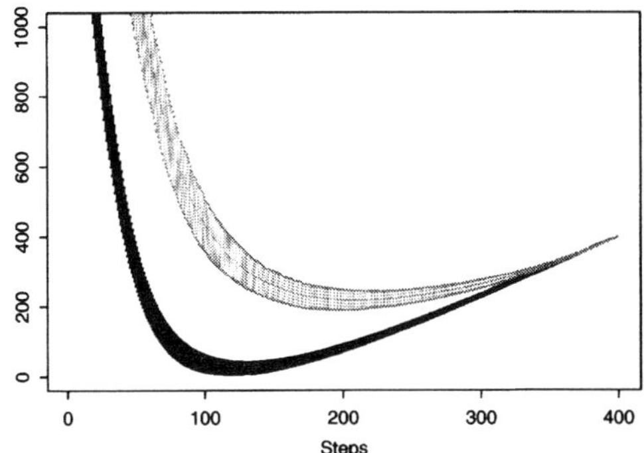

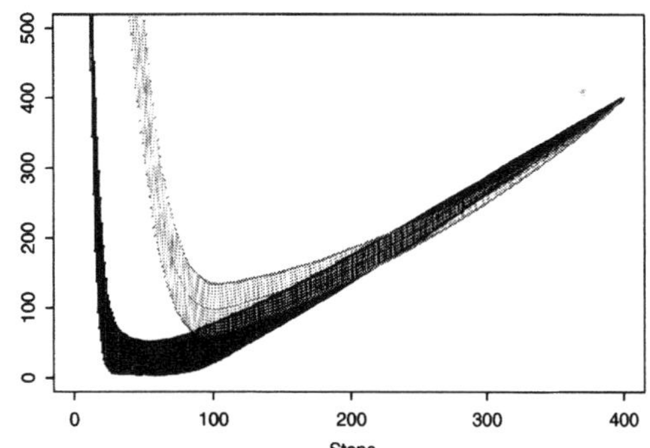

FIG. 3. *C_p values from simulations where* $\rho = 0$ (left) *and* $\rho = 0.9$ (right): *bottom curves are from SVS; top curves are from LARS. The lines seen on each curve are the mean C_p values based on* 100 *simulations. Note how the minimum value for SVS is near zero in both cases. Also superimposed on each curve are error bars representing mean values plus or minus one standard deviation.*

We can use Lemma 1 of the paper to gain some insight into this. For this argument I will assume that m is fixed (the lemma is stated for $m = n$ but applies in general) and I will need to assume that $X_{n\times m}$ is a random orthogonal matrix, chosen so that its rows are exchangeable. To produce such an X, choose m values $\mathbf{e}_{i_1}, \dots, \mathbf{e}_{i_m}$ without replacement from $\{\mathbf{e}_1, \dots, \mathbf{e}_n\}$, where $\mathbf{e}_j$ is defined as in Section 4.1, and set $X = [\mathbf{e}_{i_1}, \dots, \mathbf{e}_{i_m}]$. It is easy to see that this ensures row-exchangeability. Hence, $\mu_1, \dots, \mu_n$ are exchangeable and, therefore, $Y_i = \mu_i + \varepsilon_i$ are exchangeable since ε_i are i.i.d. I will assume, as in (4.1), that ε_i are independent $\mathrm{N}(0, \sigma^2)$ variables.

For simplicity take $\sigma^2 = \overline{\sigma}^2 = 1$. Let V_j, for $j = 0, \dots, m-1$, denote the $(j+1)$st largest value from the set of values $\{|Y_{i_1}|, \dots, |Y_{i_m}|\}$. Let k_0 denote the true dimension, that is, the number of nonzero coordinates of the true $\boldsymbol{\beta}$, and suppose that k is some dimension larger than k_0 such that $1 \le k_0 < k \le m \le n$. Notice that $V_k \le V_{k_0}$, and thus, by Lemma 1 and (4.10),

$$\begin{aligned} C_p(\widehat{\boldsymbol{\mu}}_k) - C_p(\widehat{\boldsymbol{\mu}}_{k_0}) &= (V_k^2 - V_{k_0}^2) \sum_{j=1}^m \mathbb{1}\{|Y_{i_j}| > V_{k_0}\} + V_k^2 \sum_{j=1}^m \mathbb{1}\{V_k < |Y_{i_j}| \le V_{k_0}\} \\ &\quad - \sum_{j=1}^m Y_{i_j}^2 \mathbb{1}\{V_k < |Y_{i_j}| \le V_{k_0}\} + 2(k - k_0) \\ &\le -\Delta_k B_k + 2(k - k_0), \end{aligned}$$

where $\Delta_k = V_{k_0}^2 - V_k^2 \ge 0$ and $B_k = \sum_{j=1}^m \mathbb{1}\{|Y_{i_j}| > V_{k_0}\}$. Observe that by exchangeability B_k is a Binomial$(m, k_0/m)$ random variable. It is a little messy to work out the distribution for Δ_k explicitly. However, it is not hard to see that Δ_k can be reasonably large with high probability. Now if $k_0 > k - k_0$ and k_0 is large, then B_k, which has a mean of k_0, will become the dominant term in $\Delta_k B_k$

and $\Delta_k B_k$ will become larger than $2(k - k_0)$ with high probability. This suggests, at least in this setting, that C_p will overfit if the dimension of the problem is high. In this case there will be too much improvement in the residual sums of squares when moving from k_0 to k because of the nonvanishing difference between the squared order statistics $V_{k_0}^2$ and V_k^2.

4. Summary. The use of C_p seems to encourage large models in LARS, especially in high-dimensional orthogonal problems, and can adversely affect variable selection performance. It can also be unreliable when used with stepwise regression. The use of C_p with SVS, however, seems better motivated due to the benefits of model averaging, which mitigates the selection bias effect. This suggests that C_p can be used effectively if model uncertainty is accounted for. This might be one remedy. Another remedy would be simply to use a different model selection criteria when using LARS.

REFERENCES

BREIMAN, L. (1992). The little bootstrap and other methods for dimensionality selection in regression: X-fixed prediction error. *J. Amer. Statist. Assoc.* **87** 738–754.

GEORGE, E. I. and MCCULLOCH, R. E. (1993). Variable selection via Gibbs sampling. *J. Amer. Statist. Assoc.* **88** 881–889.

ISHWARAN, H. and RAO, J. S. (2000). Bayesian nonparametric MCMC for large variable selection problems. Unpublished manuscript.

ISHWARAN, H. and RAO, J. S. (2003). Detecting differentially expressed genes in microarrays using Bayesian model selection. *J. Amer. Statist. Assoc.* **98** 438–455.

MALLOWS, C. (1973). Some comments on C_p. *Technometrics* **15** 661–675.

MITCHELL, T. J. and BEAUCHAMP, J. J. (1988). Bayesian variable selection in linear regression (with discussion). *J. Amer. Statist. Assoc.* **83** 1023–1036.

SHAO, J. (1993). Linear model selection by cross-validation. *J. Amer. Statist. Assoc.* **88** 486–494.

DEPARTMENT OF BIOSTATISTICS/WB4
CLEVELAND CLINIC FOUNDATION
9500 EUCLID AVENUE
CLEVELAND, OHIO 44195
USA
E-MAIL: ishwaran@bio.ri.ccf.org

DISCUSSION

BY KEITH KNIGHT

University of Toronto

First, I congratulate the authors for a truly stimulating paper. The paper resolves a number of important questions but, at the same time, raises many others. I would like to focus my comments to two specific points.

1. The similarity of Stagewise and LARS fitting to the Lasso suggests that the estimates produced by Stagewise and LARS fitting may minimize an objective function that is similar to the appropriate Lasso objective function. It is not at all (at least to me) obvious how this might work though. I note, though, that the construction of such an objective function may be easier than it seems. For example, in the case of bagging [Breiman (1996)] or subagging [Bühlmann and Yu (2002)], an "implied" objective function can be constructed. Suppose that $\widehat{\theta}_1, \dots, \widehat{\theta}_m$ are estimates (e.g., computed from subsamples or bootstrap samples) that minimize, respectively, objective functions $Z_1, \dots, Z_m$ and define

$$\widehat{\theta} = g(\widehat{\theta}_1, \dots, \widehat{\theta}_m);$$

then $\widehat{\theta}$ minimizes the objective function

$$Z(t) = \inf\{Z_1(t_1) + \cdots + Z_m(t_m) : g(t_1, \dots, t_m) = t\}.$$

(Thanks to Gib Bassett for pointing this out to me.) A similar construction for stagewise fitting (or LARS in general) could facilitate the analysis of the statistical properties of the estimators obtained via these algorithms.

2. When I first started experimenting with the Lasso, I was impressed by its robustness to small changes in its tuning parameter relative to more classical stepwise subset selection methods such as Forward Selection and Backward Elimination. (This is well illustrated by Figure 5; at its best, Forward Selection is comparable to LARS, Stagewise and the Lasso but the performance of Forward Selection is highly dependent on the model size.) Upon reflection, I realized that there was a simple explanation for this robustness. Specifically, the strict convexity in $\boldsymbol{\beta}$ for each t in the Lasso objective function (1.5) together with the continuity (in the appropriate sense) in t of these objective functions implies that the Lasso solutions $\widehat{\boldsymbol{\beta}}(t)$ are continuous in t; this continuity breaks down for nonconvex objective functions. Of course, the same can be said of other penalized least squares estimates whose penalty is convex. What seems to make the Lasso special is (i) its ability to produce exact 0 estimates and (ii) the "fact" that its bias seems to be more controllable than it is for other methods (e.g., ridge regression, which naturally overshrinks large effects) in the sense that for a fixed tuning parameter the bias is bounded by a constant that depends on the design but not the true parameter values. At the same time, though, it is perhaps unfair to compare stepwise methods to the Lasso, LARS or Stagewise fitting since the space of models considered by the latter methods seems to be "nicer" than it is for the former and (perhaps more important) since the underlying motivation for using Forward Selection is typically not prediction. For example, bagged Forward Selection might perform as well as the other methods in many situations.

REFERENCES

BREIMAN, L. (1996). Bagging predictors. *Machine Learning* **24** 123–140.
BÜHLMANN, P. and YU, B. (2002). Analyzing bagging. *Ann. Statist.* **30** 927–961.

DEPARTMENT OF STATISTICS
UNIVERSITY OF TORONTO
100 ST. GEORGE ST.
TORONTO, ONTARIO M5S 3G3
CANADA
E-MAIL: keith@utstat.toronto.edu

DISCUSSION

BY JEAN-MICHEL LOUBES AND PASCAL MASSART

Université Paris-Sud

The issue of model selection has drawn the attention of both applied and theoretical statisticians for a long time. Indeed, there has been an enormous range of contribution in model selection proposals, including work by Akaike (1973), Mallows (1973), Foster and George (1994), Birgé and Massart (2001a) and Abramovich, Benjamini, Donoho and Johnstone (2000). Over the last decade, modern computer-driven methods have been developed such as All Subsets, Forward Selection, Forward Stagewise or Lasso. Such methods are useful in the setting of the standard linear model, where we observe noisy data and wish to predict the response variable using only a few covariates, since they provide automatically linear models that fit the data. The procedure described in this paper is, on the one hand, numerically very efficient and, on the other hand, very general, since, with slight modifications, it enables us to recover the estimates given by the Lasso and Stagewise.

1. Estimation procedure. The "LARS" method is based on a recursive procedure selecting, at each step, the covariates having largest absolute correlation with the response y. In the case of an orthogonal design, the estimates can then be viewed as an l^1-penalized estimator. Consider the linear regression model where we observe y with some random noise ε, with orthogonal design assumptions:

$$y = X\beta + \varepsilon.$$

Using the soft-thresholding form of the estimator, we can write it, equivalently, as the minimum of an ordinary least squares and an l^1 penalty over the coefficients of the regression. As a matter of fact, at step $k = 1, \dots, m$, the estimators $\hat{\beta}^k = X^{-1}\hat{\mu}^k$ are given by

$$\hat{\mu}^k = \arg\min_{\mu \in \mathbb{R}^n} \big(\|Y - \mu\|_n^2 + 2\lambda_n^2(k)\|\mu\|_1 \big).$$

There is a trade-off between the two terms, balanced by the smoothing decreasing sequence $\lambda_n^2(k)$. The more stress is laid on the penalty, the more parsimonious the representation will be. The choice of the l^1 penalty enables us to keep the largest coefficients, while the smallest ones shrink toward zero in a soft-thresholding scheme. This point of view is investigated in the Lasso algorithm as well as in studying the false discovery rate (FDR).

So, choosing these weights in an optimal way determines the form of the estimator as well as its asymptotic behavior. In the case of the algorithmic procedure, the suggested level is the $(k+1)$-order statistic:

$$\lambda_n^2(k) = |y|_{(k+1)}.$$

As a result, it seems possible to study the asymptotic behavior of the LARS estimates under some conditions on the coefficients of β. For instance, if there exists a roughness parameter $\rho \in [0, 2]$, such that $\sum_{j=1}^m |\beta_j|^\rho \leq M$, metric entropy theory results lead to an upper bound for the mean square error $\|\hat{\beta} - \beta\|^2$. Here we refer to the results obtained in Loubes and van de Geer (2002). Consistency should be followed by the asymptotic distribution, as is done for the Lasso in Knight and Fu (2000).

The interest for such an investigation is double: first, it gives some insight into the properties of such estimators. Second, it suggests an approach for choosing the threshold λ_n^2 which can justify the empirical cross-validation method, developed later in the paper. Moreover, the asymptotic distributions of the estimators are needed for inference.

Other choices of penalty and loss functions can also be considered. First, for $\gamma \in (0, 1]$, consider

$$J_\gamma(\beta) = \sum_{j=1}^m (X\beta)_j^\gamma.$$

If $\gamma < 1$, the penalty is not convex anymore, but there exist algorithms to solve the minimization problem. Constraints on the l^γ norm of the coefficients are equivalent to lacunarity assumptions and may make estimation of sparse signals easier, which is often the case for high-dimensional data for instance.

Moreover, replacing the quadratic loss function with an l^1 loss gives rise to a robust estimator, the penalized absolute deviation of the form

$$\tilde{\mu}^k = \arg \min_{\mu \in \mathbb{R}^n} \left(\|Y - \mu\|_{n,1} + 2\lambda_n^2(k)\|\mu\|_1 \right).$$

Hence, it is possible to get rid of the problem of variance estimation for the model with these estimates whose asymptotic behavior can be derived from Loubes and van de Geer (2002), in the regression framework.

Finally, a penalty over both the number of coefficients and the smoothness of the coefficients can be used to study, from a theoretical point of view, the asymptotics of the estimate. Such a penalty is analogous to complexity penalties studied in van de Geer (2001):

$$\mu^\star = \arg \min_{\mu \in \mathbb{R}^n, k \in [1,m]} \left(\|Y - \mu\|_n^2 + 2\lambda_n^2(k)\|\mu\|_1 + \text{pen}(k)\right).$$

2. Mallows' C_p. We now discuss the crucial issue of selecting the number k of influential variables. To make this discussion clear, let us first assume the variance σ^2 of the regression errors to be known. Interestingly the penalized criterion which is proposed by the authors is exactly equivalent to Mallows' C_p when the design is orthogonal (this is indeed the meaning of their Theorem 3). More precisely, using the same notation as in the paper, let us focus on the following situation which illustrates what happens in the orthogonal case where LARS is equivalent to the Lasso. One observes some random vector y in $\mathbb{R}^n$, with expectation μ and covariance matrix $\sigma^2 I_n$. The variable selection problem that we want to solve here is to determine which components of y are influential. According to Lemma 1, given k, the kth LARS estimate $\widehat{\mu}_k$ of μ can be explicitly computed as a soft-thresholding estimator. Indeed, considering the order statistics of the absolute values of the data denoted by

$$|y|_{(1)} \geq |y|_{(2)} \geq \cdots \geq |y|_{(n)}$$

and defining the soft threshold function $\eta(\cdot, t)$ with level $t \geq 0$ as

$$\eta(x, t) = x \mathbb{1}_{|x|>t}\left(1 - \frac{t}{|x|}\right),$$

one has

$$\widehat{\mu}_{k,i} = \eta\big(y_i, |y|_{(k+1)}\big).$$

To select k, the authors propose to minimize the C_p criterion

$$C_p(\widehat{\mu}_k) = \|y - \widehat{\mu}_k\|^2 - n\sigma^2 + 2k\sigma^2. \tag{1}$$

Our purpose is to analyze this proposal with the help of the results on penalized model selection criteria proved in Birgé and Massart (2001a, b). In these papers some oracle type inequalities are proved for selection procedures among some arbitrary collection of projection estimators on linear models when the regression errors are Gaussian. In particular one can apply them to the variable subset selection problem above, assuming the random vector y to be Gaussian. If one decides to penalize in the same way the subsets of variables with the same cardinality, then the penalized criteria studied in Birgé and Massart (2001a, b) take the form

$$C_p'(\widetilde{\mu}_k) = \|y - \widetilde{\mu}_k\|^2 - n\sigma^2 + \text{pen}(k), \tag{2}$$

where pen(k) is some penalty to be chosen and $\tilde{\mu}_k$ denotes the hard threshold estimator with components

$$\tilde{\mu}_{k,i} = \eta'(y_i, |y|_{(k+1)}),$$

where

$$\eta'(x,t) = x\mathbb{1}_{|x|>t}.$$

The essence of the results proved by Birgé and Massart (2001a, b) in this case is the following. Their analysis covers penalties of the form

$$\text{pen}(k) = 2k\sigma^2 C\left(\log\left(\frac{n}{k}\right) + C'\right)$$

[note that the FDR penalty proposed in Abramovich, Benjamini, Donoho and Johnstone (2000) corresponds to the case $C = 1$]. It is proved in Birgé and Massart (2001a) that if the penalty pen(k) is heavy enough (i.e., $C > 1$ and C' is an adequate absolute constant), then the model selection procedure works in the sense that, up to a constant, the selected estimator $\tilde{\mu}_{\tilde{k}}$ performs as well as the best estimator among the collection $\{\tilde{\mu}_k, 1 \leq k \leq n\}$ in terms of quadratic risk. On the contrary, it is proved in Birgé and Massart (2001b) that if $C < 1$, then at least asymptotically, whatever C', the model selection does not work, in the sense that, even if $\mu = 0$, the procedure will systematically choose large values of k, leading to a suboptimal order for the quadratic risk of the selected estimator $\tilde{\mu}_{\tilde{k}}$. So, to summarize, some $2k\sigma^2 \log(n/k)$ term should be present in the penalty, in order to make the model selection criterion (2) work. In particular, the choice pen(k) $= 2k\sigma^2$ is not appropriate, which means that Mallows' C_p does not work in this context. At first glance, these results seem to indicate that some problems could occur with the use of the Mallows' C_p criterion (1). Fortunately, however, this is not at all the case because a very interesting phenomenon occurs, due to the soft-thresholding effect. As a matter of fact, if we compare the residual sums of squares of the soft threshold estimator $\hat{\mu}_k$ and the hard threshold estimator $\tilde{\mu}_k$, we easily get

$$\|y - \hat{\mu}_k\|^2 - \|y - \tilde{\mu}_k\|^2 = \sum_{i=1}^{n} |y|^2_{(k+1)} \mathbb{1}_{|y_i| > |y|_{(k+1)}} = k|y|^2_{(k+1)}$$

so that the "soft" C_p criterion (1) can be interpreted as a "hard" criterion (2) with random penalty

$$\text{pen}(k) = k|y|^2_{(k+1)} + 2k\sigma^2. \tag{3}$$

Of course this kind of penalty escapes *stricto sensu* to the analysis of Birgé and Massart (2001a, b) as described above since the penalty is not deterministic. However, it is quite easy to realize that, in this penalty, $|y|^2_{(k+1)}$ plays the role of the apparently "missing" logarithmic factor $2\sigma^2 \log(n/k)$. Indeed, let us consider the pure noise situation where $\mu = 0$ to keep the calculations as simple as possible. Then, if we consider the order statistics of a sample $U_1, \dots, U_n$ of the uniform distribution on [0, 1]

$$U_{(1)} \le U_{(2)} \le \cdots \le U_{(n)},$$

taking care of the fact that these statistics are taken according to the usual increasing order while the order statistics on the data are taken according to the reverse order, $\sigma^{-2}|y|^2_{(k+1)}$ has the same distribution as

$$Q^{-1}(U_{(k+1)}),$$

where Q denotes the tail function of the chi-square distribution with 1 degree of freedom. Now using the double approximations $Q^{-1}(u) \sim 2\log(|u|)$ as u goes to 0 and $U_{(k+1)} \approx (k+1)/n$ (which at least means that, given k, $nU_{(k+1)}$ tends to $k+1$ almost surely as n goes to infinity but can also be expressed with much more precise probability bounds) we derive that $|y|^2_{(k+1)} \approx 2\sigma^2 \log(n/(k+1))$. The conclusion is that it is possible to interpret the "soft" C_p criterion (1) as a randomly penalized "hard" criterion (2). The random part of the penalty $k|y|^2_{(k+1)}$ cleverly plays the role of the unavoidable logarithmic term $2\sigma^2 k \log(n/k)$, allowing the hope that the usual $2k\sigma^2$ term will be heavy enough to make the selection procedure work as we believe it does. A very interesting feature of the penalty (3) is that its random part depends neither on the scale parameter σ^2 nor on the tail of the errors. This means that one could think to adapt the data-driven strategy proposed in Birgé and Massart (2001b) to choose the penalty without knowing the scale parameter to this context, even if the errors are not Gaussian. This would lead to the following heuristics. For large values of k, one can expect the quantity $-\|y - \widehat{\mu}_k\|^2$ to behave as an affine function of k with slope $\alpha(n)\sigma^2$. If one is able to compute $\alpha(n)$, either theoretically or numerically (our guess is that it varies slowly with n and that it is close to 1.5), then one can just estimate the slope (for instance by making a regression of $-\|y - \widehat{\mu}_k\|^2$ with respect to k for large enough values of k) and plug the resulting estimate of σ^2 into (1). Of course, some more efforts would be required to complete this analysis and provide rigorous oracle inequalities in the spirit of those given in Birgé and Massart (2001a, b) or Abramovich, Benjamini, Donoho and Johnstone (2000) and also some simulations to check whether our proposal to estimate σ^2 is valid or not.

Our purpose here was just to mention some possible explorations starting from the present paper that we have found very stimulating. It seems to us that it solves practical questions of crucial interest and raises very interesting theoretical questions: consistency of LARS estimator; efficiency of Mallows' C_p in this context; use of random penalties in model selection for more general frameworks.

REFERENCES

ABRAMOVICH, F., BENJAMINI, Y., DONOHO, D. and JOHNSTONE, I. (2000). Adapting to unknown sparsity by controlling the false discovery rate. Technical Report 2000–19, Dept. Statistics, Stanford Univ.

AKAIKE, H. (1973). Maximum likelihood identification of Gaussian autoregressive moving average models. *Biometrika* **60** 255–265.

BIRGÉ, L. and MASSART, P. (2001a). Gaussian model selection. *J. Eur. Math. Soc.* **3** 203–268.

BIRGÉ, L. and MASSART, P. (2001b). A generalized C_p criterion for Gaussian model selection. Technical Report 647, Univ. Paris 6 & 7.

FOSTER, D. and GEORGE, E. (1994). The risk inflation criterion for multiple regression. *Ann. Statist.* **22** 1947–1975.

KNIGHT, K. and FU, B. (2000). Asymptotics for Lasso-type estimators. *Ann. Statist.* **28** 1356–1378.

LOUBES, J.-M. and VAN DE GEER, S. (2002). Adaptive estimation with soft thresholding penalties. *Statist. Neerlandica* **56** 453–478.

MALLOWS, C. (1973). Some comments on C_p. *Technometrics* **15** 661–675.

VAN DE GEER, S. (2001). Least squares estimation with complexity penalties. *Math. Methods Statist.* **10** 355–374.

CNRS AND LABORATOIRE DE MATHÉMATIQUES
UMR 8628
EQUIPE DE PROBABILITÉS, STATISTIQUE ET MODÉLISATION
UNIVERSITÉ PARIS-SUD, BÂT. 425
91405 ORDAY CEDEX
FRANCE
E-MAIL: Jean-Michel.Loubes@math.u-psud.fr

LABORATOIRE DE MATHÉMATIQUES
UMR 8628
EQUIPE DE PROBABILITÉS, STATISTIQUE ET MODÉLISATION
UNIVERSITÉ PARIS-SUD, BÂT. 425
91405 ORDAY CEDEX
FRANCE
E-MAIL: Pascal.Massart@math.u-psud.fr

DISCUSSION

BY DAVID MADIGAN AND GREG RIDGEWAY

Rutgers University and Avaya Labs Research, and RAND

Algorithms for simultaneous shrinkage and selection in regression and classification provide attractive solutions to knotty old statistical challenges. Nevertheless, as far as we can tell, Tibshirani's Lasso algorithm has had little impact on statistical practice. Two particular reasons for this may be the relative inefficiency of the original Lasso algorithm and the relative complexity of more recent Lasso algorithms [e.g., Osborne, Presnell and Turlach (2000)]. Efron, Hastie, Johnstone and Tibshirani have provided an efficient, simple algorithm for the Lasso as well as algorithms for stagewise regression and the new least angle regression. As such this paper is an important contribution to statistical computing.

1. Predictive performance. The authors say little about predictive performance issues. In our work, however, the relative out-of-sample predictive performance of LARS, Lasso and Forward Stagewise (and variants thereof) takes

TABLE 1
Stagewise, LARS and Lasso mean square error predictive performance, comparing cross-validation with C_p

Diabetes			Boston			Servo		
	CV	C_p		**CV**	C_p		**CV**	C_p
Stagewise	3083	3082	Stagewise	25.7	25.8	Stagewise	1.33	1.32
LARS	3080	3083	LARS	25.5	25.4	LARS	1.33	1.30
Lasso	3083	3082	Lasso	25.8	25.7	Lasso	1.34	1.31

center stage. Interesting connections exist between boosting and stagewise algorithms so predictive comparisons with boosting are also of interest.

The authors present a simple C_p statistic for LARS. In practice, a cross-validation (CV) type approach for selecting the degree of shrinkage, while computationally more expensive, may lead to better predictions. We considered this using the LARS software. Here we report results for the authors' diabetes data, the Boston housing data and the Servo data from the UCI Machine Learning Repository. Specifically, we held out 10% of the data and chose the shrinkage level using either C_p or nine-fold CV using 90% of the data. Then we estimated mean square error (MSE) on the 10% hold-out sample. Table 1 shows the results for main-effects models.

Table 1 exhibits two particular characteristics. First, as expected, Stagewise, LARS and Lasso perform similarly. Second, C_p performs as well as cross-validation; if this holds up more generally, larger-scale applications will want to use C_p to select the degree of shrinkage.

Table 2 presents a reanalysis of the same three datasets but now considering

TABLE 2
Predictive performance of competing methods: LM is a main-effects linear model with least squares fitting; LARS is least angle regression with main effects and CV shrinkage selection; LARS two-way C_p is least angle regression with main effects and all two-way interactions, shrinkage selection via C_p; GBM additive and GBM two-way use least squares boosting, the former using main effects only, the latter using main effects and all two-way interactions; MSE is mean square error on a 10% holdout sample; MAD is mean absolute deviation

	Diabetes		**Boston**		**Servo**	
	MSE	**MAD**	**MSE**	**MAD**	**MSE**	**MAD**
LM	3000	44.2	23.8	3.40	1.28	0.91
LARS	3087	45.4	24.7	3.53	1.33	0.95
LARS two-way C_p	3090	45.1	14.2	2.58	0.93	0.60
GBM additive	3198	46.7	16.5	2.75	0.90	0.65
GBM two-way	3185	46.8	14.1	2.52	0.80	0.60

five different models: least squares; LARS using cross-validation to select the coefficients; LARS using C_p to select the coefficients and allowing for two-way interactions; least squares boosting fitting only main effects; least squares boosting allowing for two-way interactions. Again we used the authors' LARS software and, for the boosting results, the gbm package in R [Ridgeway (2003)]. We evaluated all the models using the same cross-validation group assignments.

A plain linear model provides the best out-of-sample predictive performance for the diabetes dataset. By contrast, the Boston housing and Servo data exhibit more complex structure and models incorporating higher-order structure do a better job.

While no general conclusions can emerge from such a limited analysis, LARS seems to be competitive with these particular alternatives. We note, however, that for the Boston housing and Servo datasets Breiman (2001) reports substantially better predictive performance using random forests.

2. Extensions to generalized linear models. The minimal computational complexity of LARS derives largely from the squared error loss function. Applying LARS-type strategies to models with nonlinear loss functions will require some form of approximation. Here we consider LARS-type algorithms for logistic regression.

Consider the logistic log-likelihood for a regression function $f(\mathbf{x})$ which will be linear in $\mathbf{x}$:

$$\ell(f) = \sum_{i=1}^{N} y_i f(\mathbf{x}_i) - \log(1 + \exp(f(\mathbf{x}_i))). \tag{1}$$

We can initialize $f(\mathbf{x}) = \log(\bar{y}/(1 - \bar{y}))$. For some α we wish to find the covariate x_j that offers the greatest improvement in the logistic log-likelihood, $\ell(f(\mathbf{x}) + \mathbf{x}_j^t \alpha)$. To find this $\mathbf{x}_j$ we can compute the directional derivative for each j and choose the maximum,

$$j^* = \arg\max_j \left| \frac{d}{d\alpha} \ell(f(\mathbf{x}) + \mathbf{x}_j^t \alpha) \right|_{\alpha=0} \tag{2}$$

$$= \arg\max_j \left| \mathbf{x}_j^t \left(\mathbf{y} - \frac{1}{1 + \exp(-f(\mathbf{x}))} \right) \right|. \tag{3}$$

Note that as with LARS this is the covariate that is most highly correlated with the residuals. The selected covariate is the first member of the active set, A. For α small enough (3) implies

$$(s_{j^*} \mathbf{x}_{j^*} - s_j \mathbf{x}_j)^t \left(y - \frac{1}{1 + \exp(-f(\mathbf{x}) - \mathbf{x}_{j^*}^t \alpha)} \right) \geq 0 \tag{4}$$

for all $j \in A^C$, where s_j indicates the sign of the correlation as in the LARS development. Choosing α to have the largest magnitude while maintaining the constraint in (4) involves a nonlinear optimization. However, linearizing (4) yields a fairly simple approximate solution. If x_2 is the variable with the second largest correlation with the residual, then

$$\hat{\alpha} = \frac{(s_{j^*}\mathbf{x}_{j^*} - s_2\mathbf{x}_2)^t(y - p(\mathbf{x}))}{(s_{j^*}\mathbf{x}_{j^*} - s_2\mathbf{x}_2)^t(p(\mathbf{x})(1 - p(\mathbf{x}))\mathbf{x}_{j^*})}. \tag{5}$$

The algorithm may need to iterate (5) to obtain the exact optimal $\hat{\alpha}$. Similar logic yields an algorithm for the full solution.

We simulated $N = 1000$ observations with 10 independent normal covariates $\mathbf{x}_i \sim N_{10}(\mathbf{0}, \mathbf{I})$ with outcomes $Y_i \sim \text{Bern}(1/(1 + \exp(-\mathbf{x}_i^t\beta)))$, where $\beta \sim N_{10}(0, 1)$. Figure 1 shows a comparison of the coefficient estimates using Forward Stagewise and the Least Angle method of estimating coefficients, the final estimates arriving at the MLE. While the paper presents LARS for squared error problems, the Least Angle approach seems applicable to a wider family of models. However, an out-of-sample evaluation of predictive performance is essential to assess the utility of such a modeling strategy.

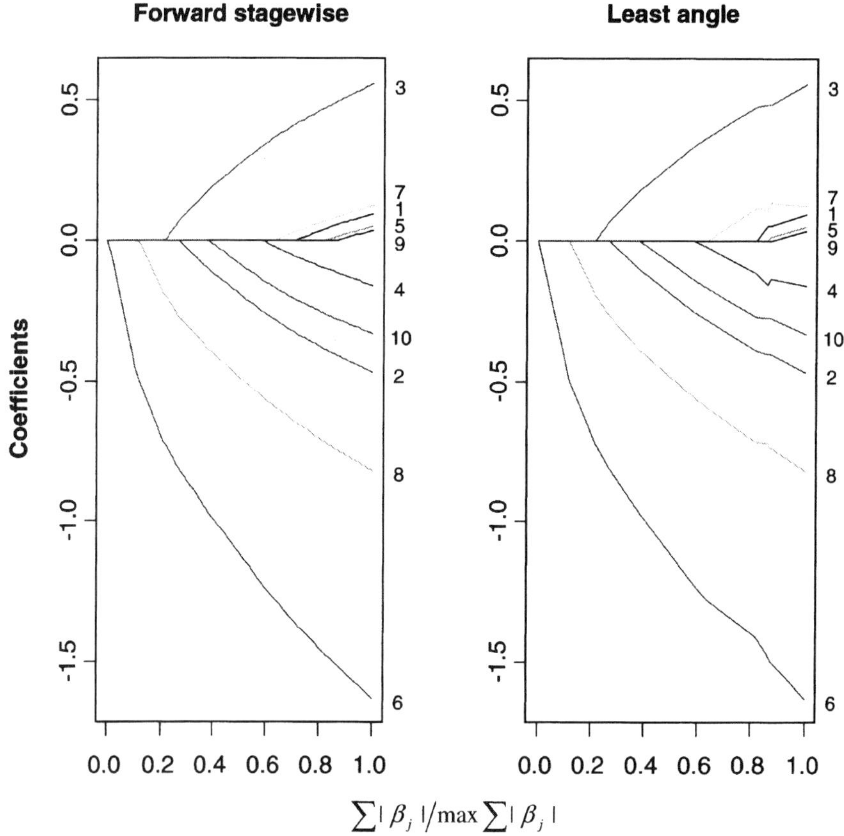

FIG. 1. *Comparison of coefficient estimation for Forward Stagewise and Least Angle Logistic Regression.*

Specifically for the Lasso, one alternative strategy for logistic regression is to use a quadratic approximation for the log-likelihood. Consider the Bayesian version of Lasso with hyperparameter γ (i.e., the penalized rather than constrained version of Lasso):

$$\begin{aligned}
&\log f(\boldsymbol{\beta}|y_1,\dots,y_n)\\
&\quad\propto \sum_{i=1}^{n} \log\bigl(y_i \Lambda(\mathbf{x}_i\boldsymbol{\beta}) + (1-y_i)(1-\Lambda(\mathbf{x}_i\boldsymbol{\beta}))\bigr) + d\log\Bigl(\frac{\gamma^{1/2}}{2}\Bigr) - \gamma^{1/2}\sum_{i=1}^{d}|\beta_i|\\
&\quad\approx \Biggl(\sum_{i=1}^{n} a_i(\mathbf{x}_i\boldsymbol{\beta})^2 + b_i(\mathbf{x}_i\boldsymbol{\beta}) + c_i\Biggr) + d\log\Bigl(\frac{\gamma^{1/2}}{2}\Bigr) - \gamma^{1/2}\sum_{i=1}^{d}|\beta_i|,
\end{aligned}$$

where Λ denotes the logistic link, d is the dimension of $\boldsymbol{\beta}$ and a_i, b_i and c_i are Taylor coefficients. Fu's elegant coordinatewise "Shooting algorithm" [Fu (1998)], can optimize this target starting from either the least squares solution or from zero. In our experience the shooting algorithm converges rapidly.

REFERENCES

BREIMAN, L. (2001). Random forests. Available at ftp://ftp.stat.berkeley.edu/pub/users/breiman/randomforest2001.pdf.

FU, W. J. (1998). Penalized regressions: The Bridge versus the Lasso. *J. Comput. Graph. Statist.* **7** 397–416.

OSBORNE, M. R., PRESNELL, B. and TURLACH, B. A. (2000). A new approach to variable selection in least squares problems. *IMA J. Numer. Anal.* **20** 389–403.

RIDGEWAY, G. (2003). GBM 0.7-2 package manual. Available at http://cran.r-project.org/doc/packages/gbm.pdf.

AVAYA LABS RESEARCH
AND
DEPARTMENT OF STATISTICS
RUTGERS UNIVERSITY
PISCATAWAY, NEW JERSEY 08855
USA
E-MAIL: madigan@stat.rutgers.edu

RAND STATISTICS GROUP
SANTA MONICA, CALIFORNIA 90407-2138
USA
E-MAIL: gregr@rand.org

DISCUSSION

BY SAHARON ROSSET AND JI ZHU

IBM T. J. Watson Research Center and Stanford University

1. Introduction. We congratulate the authors on their excellent work. The paper combines elegant theory and useful practical results in an intriguing manner. The LAR–Lasso–boosting relationship opens the door for new insights on existing methods' underlying statistical mechanisms and for the development of new and promising methodology. Two issues in particular have captured our attention, as their implications go beyond the squared error loss case presented in this paper, into wider statistical domains: robust fitting, classification, machine learning and more. We concentrate our discussion on these two results and their extensions.

2. Piecewise linear regularized solution paths. The first issue is the piecewise linear solution paths to regularized optimization problems. As the discussion paper shows, the path of optimal solutions to the "Lasso" regularized optimization problem

$$\hat{\beta}(\lambda) = \arg\min_{\beta} \|y - X\beta\|_2^2 + \lambda\|\beta\|_1 \tag{1}$$

is piecewise linear as a function of λ; that is, there exist $\infty > \lambda_0 > \lambda_1 > \cdots > \lambda_m = 0$ such that $\forall\, \lambda \geq 0$, with $\lambda_k \geq \lambda \geq \lambda_{k+1}$, we have

$$\hat{\beta}(\lambda) = \hat{\beta}(\lambda_k) - (\lambda - \lambda_k)\gamma_k.$$

In the discussion paper's terms, γ_k is the "LAR" direction for the kth step of the LAR–Lasso algorithm.

This property allows the LAR–Lasso algorithm to generate the whole path of Lasso solutions, $\hat{\beta}(\lambda)$, for "practically" the cost of one least squares calculation on the data (this is exactly the case for LAR but not for LAR–Lasso, which may be significantly more computationally intensive on some data sets). The important practical consequence is that it is not necessary to select the regularization parameter λ in advance, and it is now computationally feasible to optimize it based on cross-validation (or approximate C_p, as presented in the discussion paper).

The question we ask is: what makes the solution paths piecewise linear? Is it the use of squared error loss? Or the Lasso penalty? The answer is that both play an important role. However, the family of (loss, penalty) pairs which facilitates piecewise linear solution paths turns out to contain many other interesting and useful optimization problems.

We now briefly review our results, presented in detail in Rosset and Zhu (2004). Consider the general regularized optimization problem

$$\hat{\beta}(\lambda) = \arg\min_{\beta} \sum_i L(y_i, \mathbf{x}_i^t\beta) + \lambda J(\beta), \tag{2}$$

where we only assume that the loss L and the penalty J are both convex functions of β for any sample $\{\mathbf{x}_i^t, y_i\}_{i=1}^n$. For our discussion, the data sample is assumed fixed, and so we will use the notation $L(\beta)$, where the dependence on the data is implicit.

Notice that piecewise linearity is equivalent to requiring that

$$\frac{\partial\hat{\beta}(\lambda)}{\partial\lambda} \in \mathcal{R}^p$$

is piecewise constant as a function of λ. If L, J are twice differentiable functions of β, then it is easy to derive that

$$\frac{\partial \hat{\beta}(\lambda)}{\partial \lambda} = -(\nabla^2 L(\hat{\beta}(\lambda)) + \lambda \nabla^2 J(\hat{\beta}(\lambda)))^{-1} \nabla J(\hat{\beta}(\lambda)). \tag{3}$$

With a little more work we extend this result to "almost twice differentiable" loss and penalty functions (i.e., twice differentiable everywhere except at a finite number of points), which leads us to the following *sufficient conditions for piecewise linear* $\hat{\beta}(\lambda)$:

1. $\nabla^2 L(\hat{\beta}(\lambda))$ is piecewise constant as a function of λ. This condition is met if L is a piecewise-quadratic function of β. This class includes the squared error loss of the Lasso, but also absolute loss and combinations of the two, such as Huber's loss.
2. $\nabla J(\hat{\beta}(\lambda))$ is piecewise constant as a function of λ. This condition is met if J is a piecewise-linear function of β. This class includes the l_1 penalty of the Lasso, but also the l_∞ norm penalty.

2.1. *Examples.* Our first example is the "Huberized" Lasso; that is, we use the loss

$$L(y, \mathbf{x}\beta) = \begin{cases} (y - \mathbf{x}^t\beta)^2, & \text{if } |y - \mathbf{x}^t\beta| \le \delta, \\ \delta^2 + 2\delta(|y - \mathbf{x}\beta| - \delta), & \text{otherwise,} \end{cases} \tag{4}$$

with the Lasso penalty. This loss is more robust than squared error loss against outliers and long-tailed residual distributions, while still allowing us to calculate the whole path of regularized solutions efficiently.

To illustrate the importance of robustness in addition to regularization, consider the following simple simulated example: take $n = 100$ observations and $p = 80$ predictors, where all x_{ij} are i.i.d. $N(0, 1)$ and the true model is

$$y_i = 10 \cdot x_{i1} + \varepsilon_i, \tag{5}$$

$$\varepsilon_i \overset{\text{i.i.d.}}{\sim} 0.9 \cdot N(0, 1) + 0.1 \cdot N(0, 100). \tag{6}$$

So the normality of residuals, implicitly assumed by using squared error loss, is violated.

Figure 1 shows the optimal coefficient paths $\hat{\beta}(\lambda)$ for the Lasso (right) and "Huberized" Lasso, using $\delta = 1$ (left). We observe that the Lasso fails in identifying the correct model $E(Y|x) = 10x_1$ while the robust loss identifies it almost exactly, *if we choose the appropriate regularization parameter.*

As a second example, consider a classification scenario where the loss we use depends on the margin $y\mathbf{x}^t\beta$ rather than on the residual. In particular, consider the 1-*norm* support vector machine regularized optimization problem, popular in the

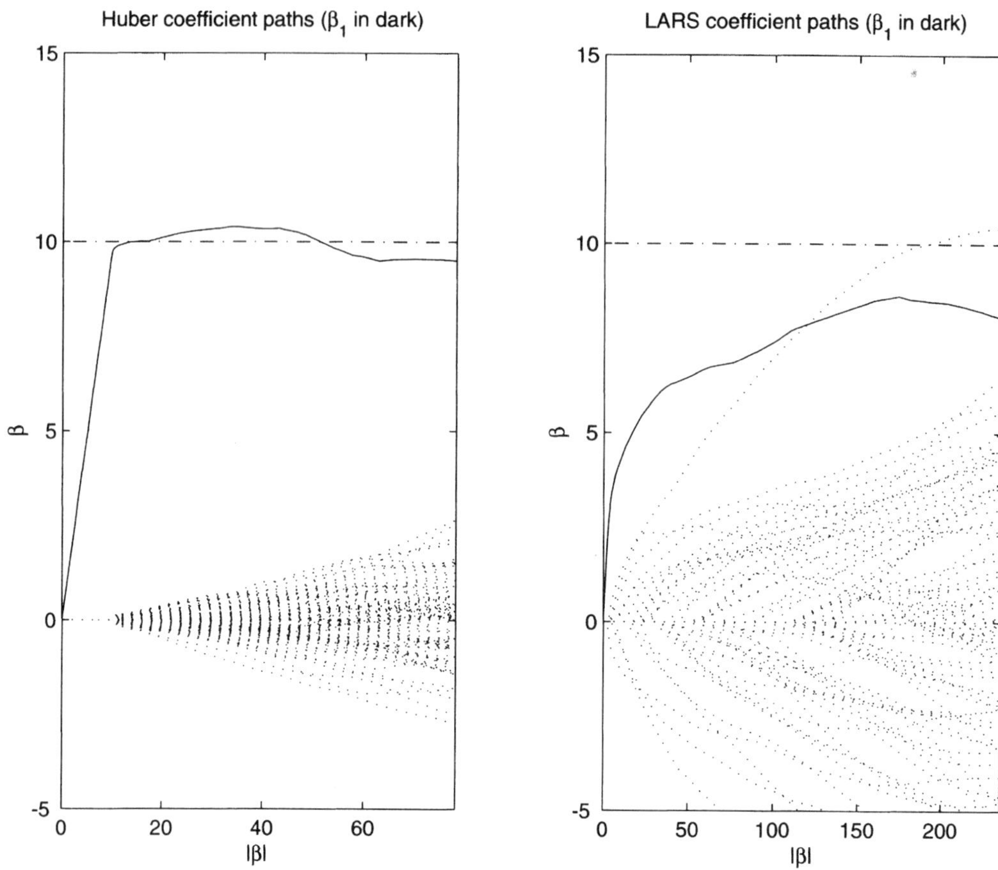

FIG. 1. *Coefficient paths for Huberized Lasso* (left) *and Lasso* (right) *for data example*: $\hat{\beta}_1(\lambda)$ *is the full line; the true model is* $E(Y|x) = 10x_1$.

machine learning community. It consists of minimizing the "hinge loss" with a Lasso penalty:

$$L(y, \mathbf{x}^t\beta) = \begin{cases} (1 - y\mathbf{x}^t\beta), & \text{if } y\mathbf{x}^t\beta \le 1, \\ 0, & \text{otherwise.} \end{cases} \tag{7}$$

This problem obeys our conditions for piecewise linearity, and so we can generate all regularized solutions for this fitting problem efficiently. This is particularly advantageous in high-dimensional machine learning problems, where regularization is critical, and it is usually not clear in advance what a good regularization parameter would be. A detailed discussion of the computational and methodological aspects of this example appears in Zhu, Rosset, Hastie, and Tibshirani (2004).

3. Relationship between "boosting" algorithms and l_1-regularized fitting. The discussion paper establishes the close relationship between ε-stagewise linear regression and the Lasso. Figure 1 in that paper illustrates the near-equivalence in the solution paths generated by the two methods, and Theorem 2 formally states a related result. It should be noted, however, that their theorem falls short of proving the global relation between the methods, which the examples suggest.

In Rosset, Zhu and Hastie (2003) we demonstrate that this relation between the path of l_1-regularized optimal solutions [which we have denoted above by $\hat{\beta}(\lambda)$] and the path of "generalized" ε-stagewise (AKA boosting) solutions extends beyond squared error loss and in fact applies to any convex differentiable loss.

More concretely, consider the following generic gradient-based "ε-boosting" algorithm [we follow Friedman (2001) and Mason, Baxter, Bartlett and Frean (2000) in this view of boosting], which iteratively builds the solution path $\beta^{(t)}$:

ALGORITHM 1 (Generic gradient-based boosting algorithm).

1. Set $\beta^{(0)} = 0$.
2. For $t = 1 : T$,

    (a) Let $j_t = \arg\max_j |\frac{\partial \sum_i L(y_i, \mathbf{x}_i^t \beta^{(t-1)})}{\partial \beta_j^{(t-1)}}|$.

    (b) Set $\beta_{j_t}^{(t)} = \beta_{j_t}^{(t-1)} - \varepsilon \operatorname{sign}(\frac{\partial \sum_i L(y_i, \mathbf{x}_i^t \beta^{(t-1)})}{\partial \beta_{j_t}^{(t-1)}})$ and $\beta_k^{(t)} = \beta_k^{(t-1)}$, $k \neq j_t$.

This is a coordinate descent algorithm, which reduces to forward stagewise, as defined in the discussion paper, if we take the loss to be squared error loss: $L(y_i, \mathbf{x}_i^t \beta^{(t-1)}) = (y_i - \mathbf{x}_i^t \beta^{(t-1)})^2$. If we take the loss to be the exponential loss,

$$L(y_i, \mathbf{x}_i^t \beta^{(t-1)}) = \exp(-y_i \mathbf{x}_i^t \beta^{(t-1)}),$$

we get a variant of AdaBoost [Freund and Schapire (1997)]—the original and most famous boosting algorithm.

Figure 2 illustrates the equivalence between Algorithm 1 and the optimal solution path for a simple logistic regression example, using five variables from the "spam" dataset. We can see that there is a perfect equivalence between the regularized solution path (left) and the "boosting" solution path (right).

In Rosset, Zhu and Hastie (2003) we formally state this equivalence, with the required conditions, as a conjecture. We also generalize the weaker result, proven by the discussion paper for the case of squared error loss, to any convex differentiable loss.

This result is interesting in the boosting context because it facilitates a view of boosting as approximate and implicit regularized optimization. The situations in which boosting is employed in practice are ones where explicitly solving regularized optimization problems is not practical (usually very high-dimensional predictor spaces). The approximate regularized optimization view which emerges from our results allows us to better understand boosting and its great empirical success [Breiman (1999)]. It also allows us to derive approximate convergence results for boosting.

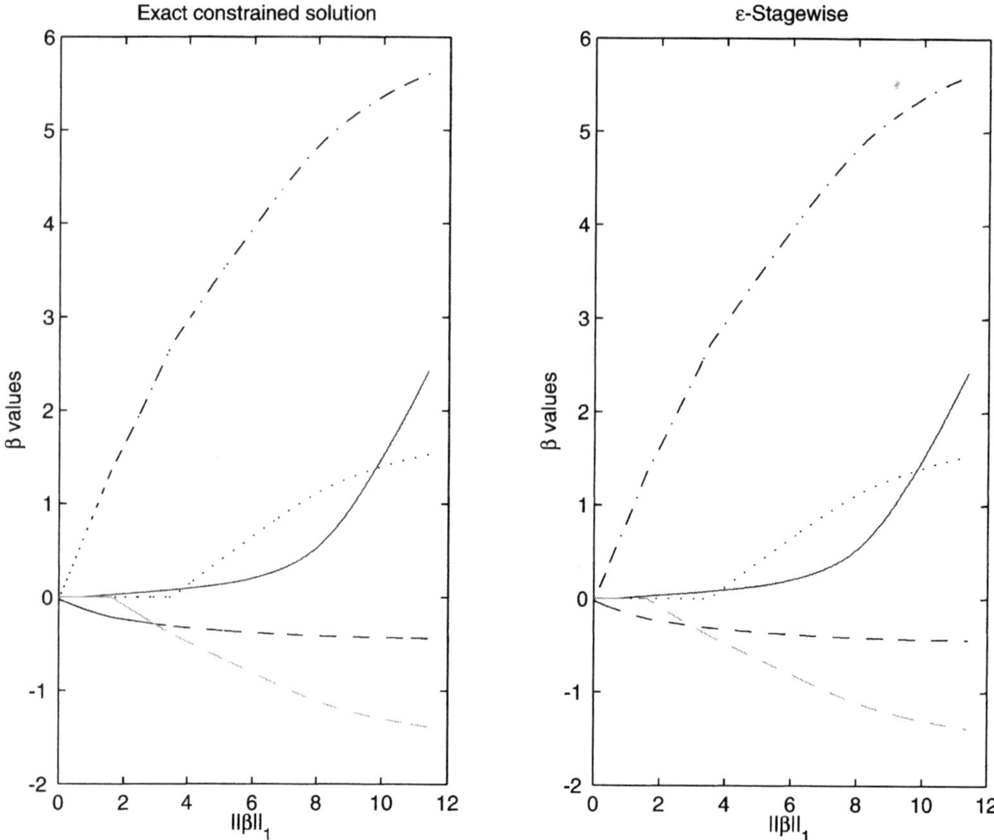

FIG. 2. *Exact coefficient paths* (left) *for l_1-constrained logistic regression and boosting coefficient paths* (right) *with binomial log-likelihood loss on five variables from the "spam" dataset. The boosting path was generated using $\varepsilon = 0.003$ and 7000 iterations.*

4. Conclusion. The computational and theoretical results of the discussion paper shed new light on variable selection and regularization methods for linear regression. However, we believe that variants of these results are useful and applicable beyond that domain. We hope that the two extensions that we have presented convey this message successfully.

Acknowledgment. We thank Giles Hooker for useful comments.

REFERENCES

BREIMAN, L. (1999). Prediction games and arcing algorithms. *Neural Computation* **11** 1493–1517.

FREUND, Y. and SCHAPIRE, R. E. (1997). A decision-theoretic generalization of on-line learning and an application to boosting. *J. Comput. System Sci.* **55** 119–139.

FRIEDMAN, J. H. (2001). Greedy function approximation: A gradient boosting machine. *Ann. Statist.* **29** 1189–1232.

MASON, L., BAXTER, J., BARTLETT, P. and FREAN, M. (2000). Boosting algorithms as gradient descent. In *Advances in Neural Information Processing Systems* **12** 512–518. MIT Press, Cambridge, MA.

ROSSET, S. and ZHU, J. (2004). Piecewise linear regularized solution paths. *Advances in Neural Information Processing Systems* **16**. To appear.

ROSSET, S., ZHU, J. and HASTIE, T. (2003). Boosting as a regularized path to a maximum margin classifier. Technical report, Dept. Statistics, Stanford Univ.

ZHU, J., ROSSET, S., HASTIE, T. and TIBSHIRANI, R. (2004). 1-norm support vector machines. *Neural Information Processing Systems* **16**. To appear.

IBM T. J. WATSON RESEARCH CENTER
P.O. BOX 218
YORKTOWN HEIGHTS, NEW YORK 10598
USA
E-MAIL: srosset@us.ibm.com

DEPARTMENT OF STATISTICS
UNIVERSITY OF MICHIGAN
550 EAST UNIVERSITY
ANN ARBOR, MICHIGAN 48109-1092
USA
E-MAIL: jizhu@umich.edu

DISCUSSION

BY ROBERT A. STINE

University of Pennsylvania

I have enjoyed reading the work of each of these authors over the years, so it is a real pleasure to have this opportunity to contribute to the discussion of this collaboration. The geometry of LARS furnishes an elegant bridge between the Lasso and Stagewise regression, methods that I would not have suspected to be so related. Toward my own interests, LARS offers a rather different way to construct a regression model by gradually blending predictors rather than using a predictor all at once. I feel that the problem of "automatic feature generation" (proposing predictors to consider in a model) is a current challenge in building regression models that can compete with those from computer science, and LARS suggests a new approach to this task. In the examples of Efron, Hastie, Johnstone and Tibshirani (EHJT) (particularly that summarized in their Figure 5), LARS produces models with smaller predictive error than the old workhorse, stepwise regression. Furthermore, as an added bonus, the code supplied by the authors runs faster for me than the `step` routine for stepwise regression supplied with *R*, the generic version of S-PLUS that I use.

My discussion focuses on the use of C_p to choose the number of predictors. The bootstrap simulations in EHJT show that LARS reaches higher levels of "proportion explained" than stepwise regression. Furthermore, the goodness-of-fit obtained by LARS remains high over a wide range of models, in sharp contrast to the narrow peak produced by stepwise selection. Because the cost of overfitting with LARS appears less severe than with stepwise, LARS would seem to have a clear advantage in this respect. Even if we do overfit, the fit of LARS degrades only slightly. The issue becomes learning how much LARS overfits, particularly in situations with many potential predictors (m as large as or larger than n).

To investigate the model-selection aspects of LARS further, I compared LARS to stepwise regression using a "reversed" five-fold cross-validation. The cross-validation is reversed in the sense that I estimate the models on one fold (88 observations) and then predict the other four. This division with more set aside for validation than used in estimation offers a better comparison of models. For example, Shao (1993) shows that one needs to let the proportion used for validation grow large in order to get cross validation to find the right model. This reversed design also adds a further benefit of making the variable selection harder. The quadratic model fitted to the diabetes data in EHJT selects from $m = 64$ predictors using a sample of $n = 442$ cases, or about 7 cases per predictor. Reversed cross-validation is closer to a typical data-mining problem. With only one fold of 88 observations to train the model, observation noise obscures subtle predictors. Also, only a few degrees of freedom remain to estimate the error variance $\overline{\sigma}^2$ that appears in C_p [equation (4.5)]. Because I also wanted to see what happens when $m > n$, I repeated the cross-validation with 5 additional possible predictors added to the 10 in the diabetes data. These 5 spurious predictors were simulated i.i.d. Gaussian noise; one can think of them as extraneous predictors that one might encounter when working with an energetic, creative colleague who suggests many ideas to explore. With these 15 base predictors, the search must consider $m = 15 + \binom{15}{2} + 14 = 134$ possible predictors.

Here are a few details of the cross-validation. To obtain the stepwise results, I ran forward stepwise using the hard threshold $2 \log m$, which is also known as the risk inflation criterion (RIC) [Donoho and Johnstone (1994) and Foster and George (1994)]. One begins with the most significant predictor. If the squared t-statistic for this predictor, say $t^2_{(1)}$, is less than the threshold $2 \log m$, then the search stops, leaving us with the "null" model that consists of just an intercept. If instead $t^2_{(1)} \geq 2 \log m$, the associated predictor, say $X_{(1)}$, joins the model and the search moves on to the next predictor. The second predictor $X_{(2)}$ joins the model if $t^2_{(2)} \geq 2 \log m$; otherwise the search stops with the one-predictor model. The search continues until reaching a predictor whose t-statistic fails this test, $t^2_{(q+1)} < 2 \log m$, leaving a model with q predictors. To obtain the results for LARS, I picked the order of the fit by minimizing C_p. Unlike the forward, sequential stepwise search, LARS globally searches a collection of models up to some large size, seeking the model which has the smallest C_p. I set the maximum model size to 50 (for $m = 64$) or 64 (for $m = 134$). In either case, the model is chosen from the collection of linear and quadratic effects in the 10 or 15 basic predictors. Neither search enforces the principle of marginality; an interaction can join the model without adding main effects.

I repeated the five-fold cross validation 20 times, each time randomly partitioning the 442 cases into 5 folds. This produces 100 stepwise and LARS fits. For each

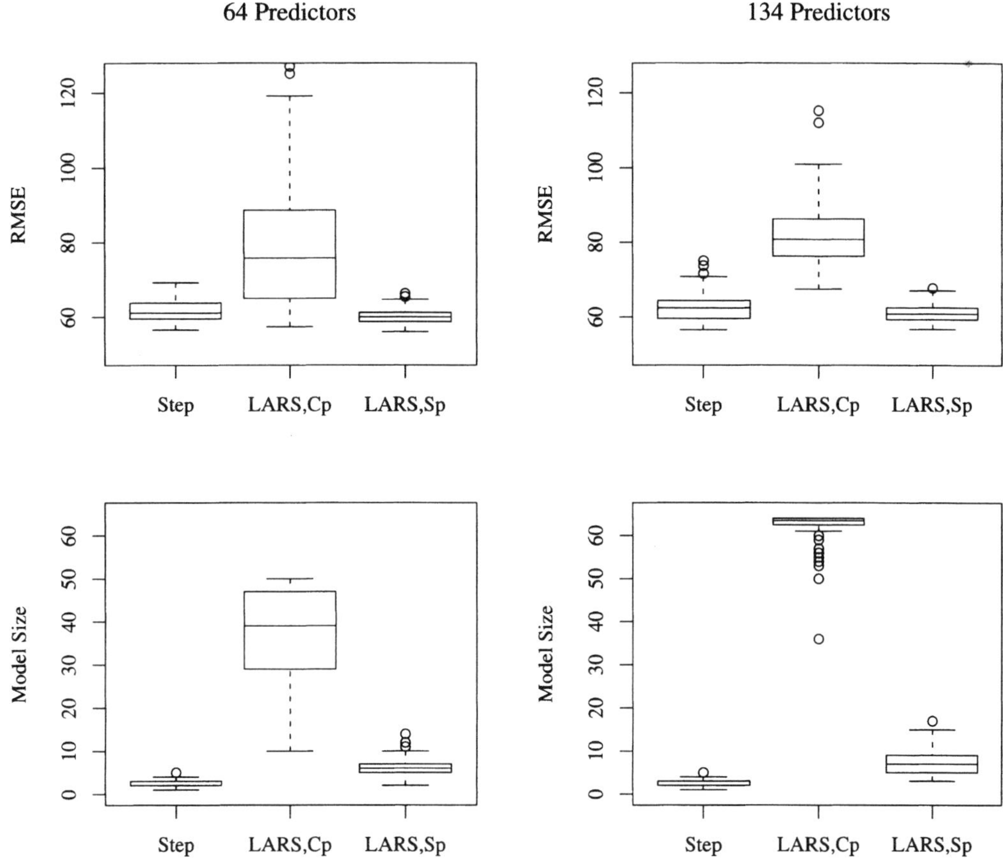

FIG. 1. *Five-fold cross-validation of the prediction error and size of stepwise regression and LARS when fitting models to a collection of* 64 (left) *or* 134 *predictors* (right). *LARS fits chosen by* C_p *overfit and have larger RMSE than stepwise; with* C_p *replaced by the alternative criterion* S_p *defined in* (3), *the LARS fits become more parsimonious with smaller RMSE. The random splitting into estimation and validation samples was repeated* 20 *times, for a total of* 100 *stepwise and LARS fits.*

of these, I computed the square root of the out-of-sample mean square error (MSE) when the model fit on one fold was used to predict the held-back 354 $[= 4(88) + 2]$ observations. I also saved the size q for each fit.

Figure 1 summarizes the results of the cross-validation. The comparison boxplots on the left compare the square root of the MSE (top) and selected model order (bottom) of stepwise to LARS when picking from $m = 64$ candidate predictors; those on the right summarize what happens with $m = 134$. When choosing from among 64 predictors, the median size of a LARS model identified by C_p is 39. The median stepwise model has but 2 predictors. (I will describe the S_p criterion further below.) The effects of overfitting on the prediction error of LARS are clear: LARS has higher RMSE than stepwise. The median RMSE for stepwise is near 62. For LARS, the median RMSE is larger, about 78. Although the predictive accuracy of LARS declines more slowly than that of stepwise when it overfits (imagine the fit of stepwise with 39 predictors), LARS overfits by enough in this case that it predicts worse than the far more parsimonious stepwise models. With more predictors ($m = 134$), the boxplots on the right of Figure 1 show that C_p tends to pick the largest possible model—here a model with 64 predictors.

Why does LARS overfit? As usual with variable selection in regression, it is simpler to try to understand when thinking about the utopian orthogonal regression with known σ^2. Assume, as in Lemma 1 of EHJT, that the predictors X_j are the columns of an identity matrix, $X_j = e_j = (0, \dots, 0, 1_j, 0, \dots, 0)$. Assume also that we know $\sigma^2 = 1$ and use it in place of the troublesome $\overline{\sigma}^2$ in C_p, so that for this discussion

$$C_p = \mathrm{RSS}(p) - n + 2p. \tag{1}$$

To define $\mathrm{RSS}(p)$ in this context, denote the ordered values of the response as

$$Y_{(1)}^2 > Y_{(2)}^2 > \cdots > Y_{(n)}^2.$$

The soft thresholding summarized in Lemma 1 of EHJT implies that the residual sum-of-squares of LARS with q predictors is

$$\mathrm{RSS}(q) = (q+1)Y_{(q+1)}^2 + \sum_{j=q+2}^{n} Y_{(j)}^2.$$

Consequently, the drop in C_p when going from the model with q to the model with $q+1$ predictors is

$$C_q - C_{q+1} = (q+1)\,d_q - 2,$$

with

$$d_q = Y_{(q+1)}^2 - Y_{(q+2)}^2;$$

C_p adds X_{q+1} to the model if $C_q - C_{q+1} > 0$.

This use of C_p works well for the orthogonal "null" model, but overfits when the model includes much signal. Figure 2 shows a graph of the mean and standard deviation of $\mathrm{RSS}(q) - \mathrm{RSS}(0) + 2q$ for an orthogonal model with $n = m = 100$ and $Y_i \overset{\text{i.i.d.}}{\sim} N(0,1)$. I subtracted $\mathrm{RSS}(0)$ rather than n to reduce the variation in the simulation. Figure 3 gives a rather different impression. The simulation is identical except that the data have some signal. Now, $EY_i = 3$ for $i = 1, \dots, 5$. The remaining 95 observations are $N(0,1)$. The "true" model has only 5 nonzero components, but the minimal expected C_p falls near 20.

This stylized example suggests an explanation for the overfitting—as well as motivates a way to avoid some of it. Consider the change in RSS for a null model when adding the sixth predictor. For this step, $\mathrm{RSS}(5) - \mathrm{RSS}(6) = 6(Y_{(6)}^2 - Y_{(7)}^2)$. Even though we multiply the difference between the squares by 6, adjacent order statistics become closer when removed from the extremes, and C_p tends to increase

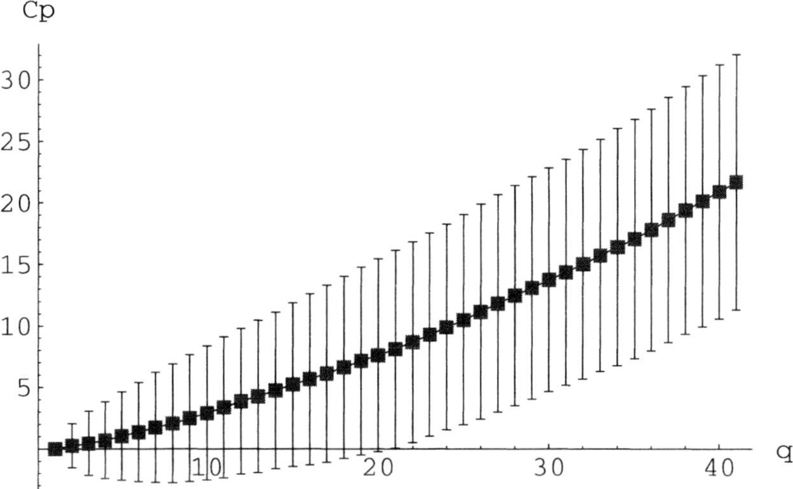

FIG. 2. *A simulation of C_p for LARS applied to orthogonal, normal data with no signal correctly identifies the null model. These results are from a simulation with* 1000 *replications, each consisting of a sample of* 100 *i.i.d. standard normal observations. Error bars indicate* ± 1 *standard deviation.*

as shown in Figure 2. The situation changes when signal is present. First, the five observations with mean 3 are likely to be the first five ordered observations. So, their spacing is likely to be larger because their order is determined by a sample of five normals; C_p adds these. When reaching the noise, the difference $Y_{(6)}^2 - Y_{(7)}^2$ now behaves like the difference between the *first two* squared order statistics in an

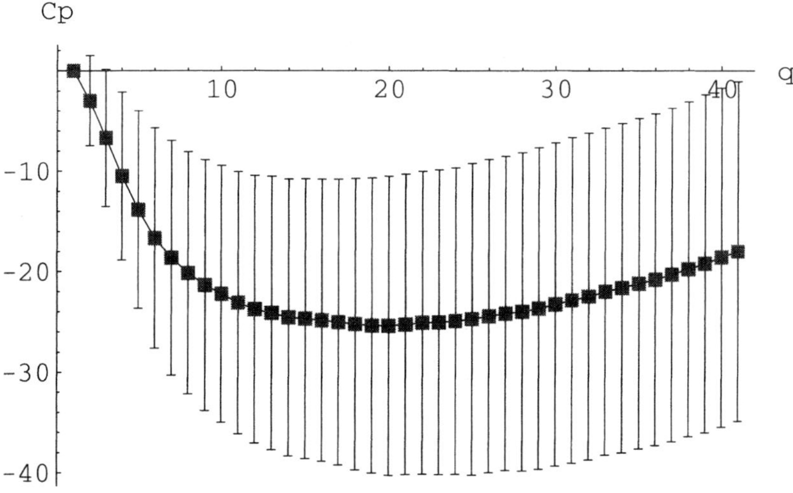

FIG. 3. *A simulation of C_p for LARS applied to orthogonal, normal data with signal present overfits. Results are from a simulation with* 1000 *replications, each consisting of* 5 *observations with mean* 3 *combined with a sample of* 95 *i.i.d. standard normal observations. Error bars indicate* ± 1 *standard deviation.*

i.i.d. sample of 95 standard normals. Consequently, this comparison involves the gap between the most extreme order statistics rather than those from within the sample, and as a result C_p drops to indicate a larger model.

This explanation of the overfitting suggests a simple alternative to C_p that leads to smaller LARS models. The idea is to compare the decreasing residual sum of squares $\mathrm{RSS}(q)$ to what is expected under a model that has fitted some signal *and* some noise. Since overfitting seems to have relatively benign effects on LARS, one does not want to take the hard-thresholding approach; my colleague Dean Foster suggested that the criterion might do better by assuming that some of the predictors already in the model are really noise. The criterion S_p suggested here adopts this notion. The form of S_p relies on approximations for normal order statistics commonly used in variable selection, particularly adaptive methods [Benjamini and Hochberg (1995) and Foster and Stine (1996)]. These approximate the size of the jth normal order statistic in a sample of n with $\sqrt{2\log(n/j)}$. To motivate the form of the S_p criterion, I return to the orthogonal situation and consider what happens when deciding whether to increase the size of the model from q to $q+1$ predictors. If I know that k of the already included q predictors represent signal and the rest of the predictors are noise, then $d_q = Y^2_{(q+1)} - Y^2_{(q+2)}$ is about

$$2\log\frac{m-k}{q+1-k} - 2\log\frac{m-k}{q+2-k}. \tag{2}$$

Since I do not know k, I will just set $k = q/2$ (i.e., assume that half of those already in the model are noise) and approximate d_q as

$$\delta(q) = 2\log\frac{q/2+2}{q/2+1}.$$

[Define $\delta(0) = 2\log 2$ and $\delta(1) = 2\log 3/2$.] This approximation suggests choosing the model that minimizes

$$S_q = \mathrm{RSS}(q) + \hat{\sigma}^2\sum_{j=1}^{q} j\delta(j), \tag{3}$$

where $\hat{\sigma}^2$ represents an "honest" estimate of σ^2 that avoids selection bias. The S_p criterion, like C_p, penalizes the residual sum-of-squares, but uses a different penalty.

The results for LARS with this criterion define the third set of boxplots in Figure 1. To avoid selection bias in the estimate of σ^2, I used a two-step procedure. First, fit a forward stepwise regression using hard thresholding. Second, use the estimated error variance from this stepwise fit as $\hat{\sigma}^2$ in S_p and proceed with LARS. Because hard thresholding avoids overfitting in the stepwise regression, the resulting estimator $\hat{\sigma}^2$ is probably conservative—but this is just what is needed when modeling data with an excess of possible predictors. If the variance estimate from the largest LARS model is used instead, the S_p criterion also overfits (though not so much as C_p). Returning to Figure 1, the combination of LARS with S_p obtains the smallest typical MSE with both $m = 64$ and 134 predictors. In either case, LARS includes more predictors than the parsimonious stepwise fits obtained by hard thresholding.

These results lead to more questions. What are the risk properties of the LARS predictor chosen by C_p or S_p? How is it that the number of possible predictors m does not appear in either criterion? This definition of S_p simply supposes half of the included predictors are noise; why half? What is a better way to set k in (2)? Finally, that the combination of LARS with either C_p or S_p has less MSE than stepwise when predicting diabetes is hardly convincing that such a pairing would do well in other applications. Statistics would be well served by having a repository of test problems comparable to those held at UC Irvine for judging machine learning algorithms [Blake and Merz (1998)].

REFERENCES

BENJAMINI, Y. and HOCHBERG, Y. (1995). Controlling the false discovery rate: A practical and powerful approach to multiple testing. *J. Roy. Statist. Soc. Ser. B* **57** 289–300.

BLAKE, C. and MERZ, C. (1998). UCI repository of machine learning databases. Technical report, School Information and Computer Science, Univ. California, Irvine. Available at www.ics.uci.edu/~mlearn/MLRepository.html.

DONOHO, D. L. and JOHNSTONE, I. M. (1994). Ideal spatial adaptation by wavelet shrinkage. *Biometrika* **81** 425–455.

FOSTER, D. P. and GEORGE, E. I. (1994). The risk inflation criterion for multiple regression. *Ann. Statist.* **22** 1947–1975.

FOSTER, D. P. and STINE, R. A. (1996). Variable selection via information theory. Technical Report Discussion Paper 1180, Center for Mathematical Studies in Economics and Management Science, Northwestern Univ.

SHAO, J. (1993). Linear model selection by cross-validation. *J. Amer. Statist. Assoc.* **88** 486–494.

DEPARTMENT OF STATISTICS
THE WHARTON SCHOOL
UNIVERSITY OF PENNSYLVANIA
PHILADELPHIA, PENNSYLVANIA 19104-6340
USA
E-MAIL: stine@wharton.upenn.edu

DISCUSSION

BY BERWIN A. TURLACH

University of Western Australia

I would like to begin by congratulating the authors (referred to below as EHJT) for their interesting paper in which they propose a new variable selection method

(LARS) for building linear models and show how their new method relates to other methods that have been proposed recently. I found the paper to be very stimulating and found the additional insight that it provides about the Lasso technique to be of particular interest.

My comments center around the question of how we can select linear models that conform with the marginality principle [Nelder (1977, 1994) and McCullagh and Nelder (1989)]; that is, the response surface is invariant under scaling and translation of the explanatory variables in the model. Recently one of my interests was to explore whether the Lasso technique or the nonnegative garrote [Breiman (1995)] could be modified such that it incorporates the marginality principle. However, it does not seem to be a trivial matter to change the criteria that these techniques minimize in such a way that the marginality principle is incorporated in a satisfactory manner.

On the other hand, it seems to be straightforward to modify the LARS technique to incorporate this principle. In their paper, EHJT address this issue somewhat in passing when they suggest toward the end of Section 3 that one first fit main effects only and interactions in a second step to control the order in which variables are allowed to enter the model. However, such a two-step procedure may have a somewhat less than optimal behavior as the following, admittedly artificial, example shows.

Assume we have a vector of explanatory variables $X = (X_1, X_2, \dots, X_{10})$ where the components are independent of each other and X_i, $i = 1, \dots, 10$, follows a uniform distribution on $[0, 1]$. Take as model

$$Y = (X_1 - 0.5)^2 + X_2 + X_3 + X_4 + X_5 + \varepsilon, \tag{1}$$

where ε has mean zero, has standard deviation 0.05 and is independent of X.

It is not difficult to verify that in this model X_1 and Y are uncorrelated. Moreover, since the X_i's are independent, X_1 is also uncorrelated with any residual vector coming from a linear model formed only by explanatory variables selected from $\{X_2, \dots, X_{10}\}$.

Thus, if one fits a main effects only model, one would expect that the LARS algorithm has problems identifying that X_1 should be part of the model. That this is indeed the case is shown in Figure 1. The picture presents the result of the LARS analysis for a typical data set generated from model (1); the sample size was $n = 500$. Note that, unlike Figure 3 in EHJT, Figure 1 (and similar figures below) has been produced using the standardized explanatory variables and no back-transformation to the original scale was done.

For this realization, the variables are selected in the sequence X_2, X_5, X_4, X_3, X_6, X_{10}, X_7, X_8, X_9 and, finally, X_1. Thus, as expected, the LARS algorithm has problems identifying X_1 as part of the model. To further verify this, 1000 different data sets, each of size $n = 500$, were simulated from model (1) and a LARS analysis performed on each of them. For each of the 10 explanatory variables the

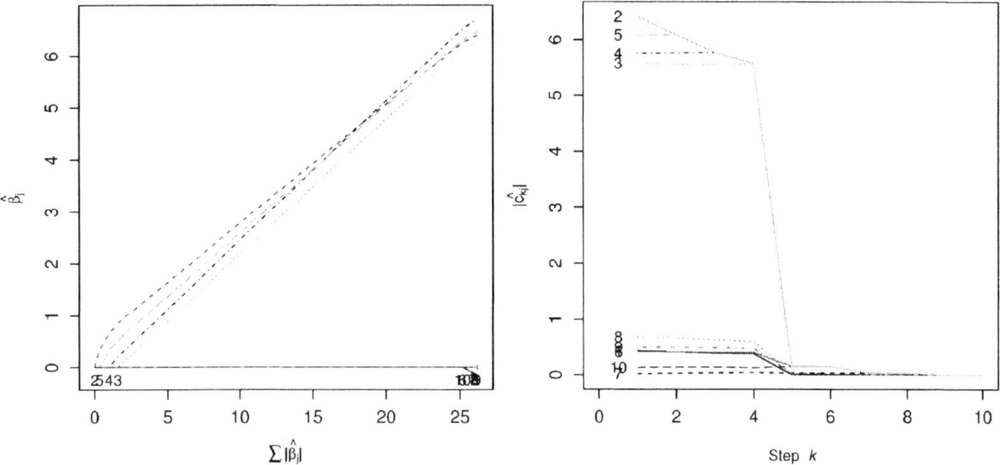

FIG. 1. *LARS analysis of simulated data with main terms only*: (left) *estimates of regression coefficients* $\hat{\beta}_j$, $j = 1, \ldots, 10$, *plotted versus* $\sum |\hat{\beta}_j|$; (right) *absolute current correlations as functions of LARS step.*

step at which it was selected to enter the model was recorded. Figure 2 shows for each of the variables the (percentage) histogram of these data.

It is clear that the LARS algorithm has no problems identifying that $X_2, \ldots, X_5$ should be in the model. These variables are all selected in the first four steps and, not surprisingly given the model, with more or less equal probability at any of these

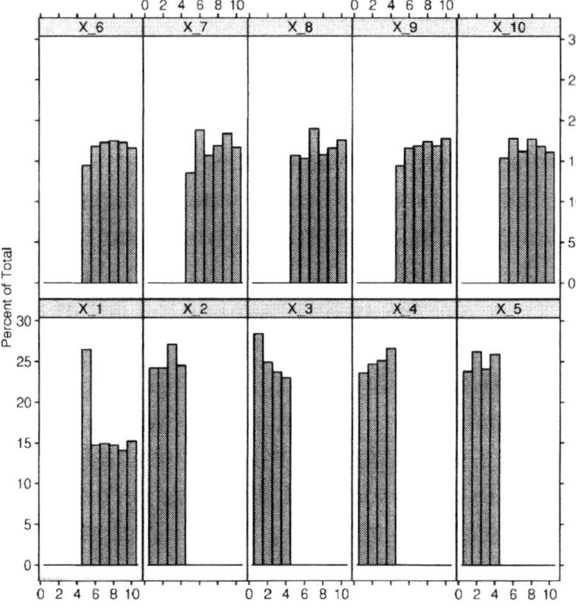

FIG. 2. *Percentage histogram of step at which each variable is selected based on* 1000 *replications: results shown for LARS analysis using main terms only.*

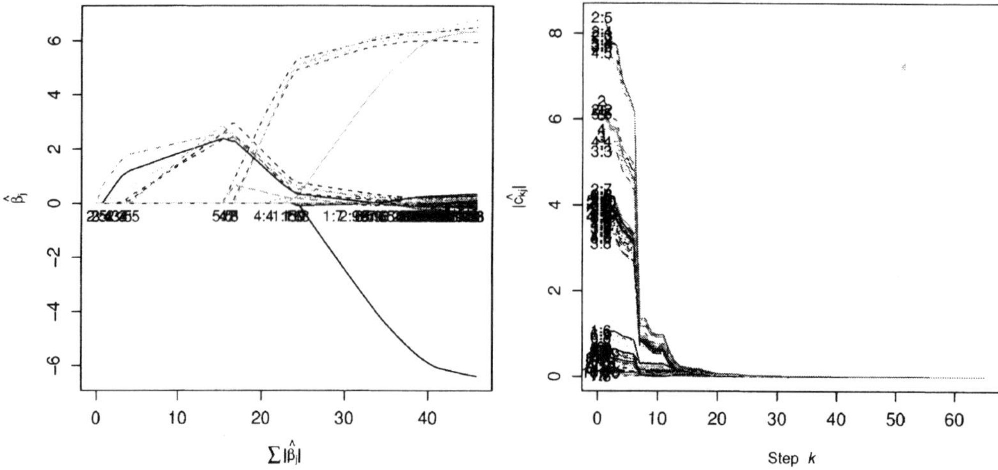

FIG. 3. *LARS analysis of simulated data with main terms and interaction terms*: (left) *estimates of regression coefficients* $\hat{\beta}_j$, $j = 1, \dots, 65$, *plotted versus* $\sum|\hat{\beta}_j|$; (right) *absolute current correlations as functions of LARS step.*

steps. X_1 has a chance of approximately 25% of being selected as the fifth variable, otherwise it may enter the model at step 6, 7, 8, 9 or 10 (each with probability roughly 15%). Finally, each of the variables X_6 to X_{10} seems to be selected with equal probability anywhere between step 5 and step 10.

This example shows that a main effects first LARS analysis followed by a check for interaction terms would not work in such cases. In most cases the LARS analysis would miss X_1 as fifth variable and even in the cases where it was selected at step 5 it would probably be deemed to be unimportant and excluded from further analysis.

How does LARS perform if one uses from the beginning all 10 main effects and all 55 interaction terms? Figure 3 shows the LARS analysis for the same data used to produce Figure 1 but this time the design matrix was augmented to contain all main effects and all interactions. The order in which the variables enter the model is $X_{2:5} = X_2 \times X_5$, $X_{2:4}$, $X_{3:4}$, $X_{2:3}$, $X_{3:5}$, $X_{4:5}$, $X_{5:5} = X_5^2$, X_4, X_3, X_2, X_5, $X_{4:4}$, $X_{1:1}$, $X_{1:6}$, $X_{1:9}$, $X_1, \dots$. In this example the last of the six terms that are actually in model (1) was selected by the LARS algorithm in step 16.

Using the same 1000 samples of size $n = 500$ as above and performing a LARS analysis on them using a design matrix with all main and interaction terms shows a surprising result. Again, for each replication the step at which a variable was selected into the model by LARS was recorded and Figure 4 shows for each variable histograms for these data. To avoid cluttering, the histograms in Figure 4 were truncated to [1, 20]; the complete histograms are shown on the left in Figure 7.

The most striking feature of these histograms is that the six interaction terms $X_{i:j}$, $i, j \in \{2, 3, 4, 5\}$, $i < j$, were always selected first. In no replication was any

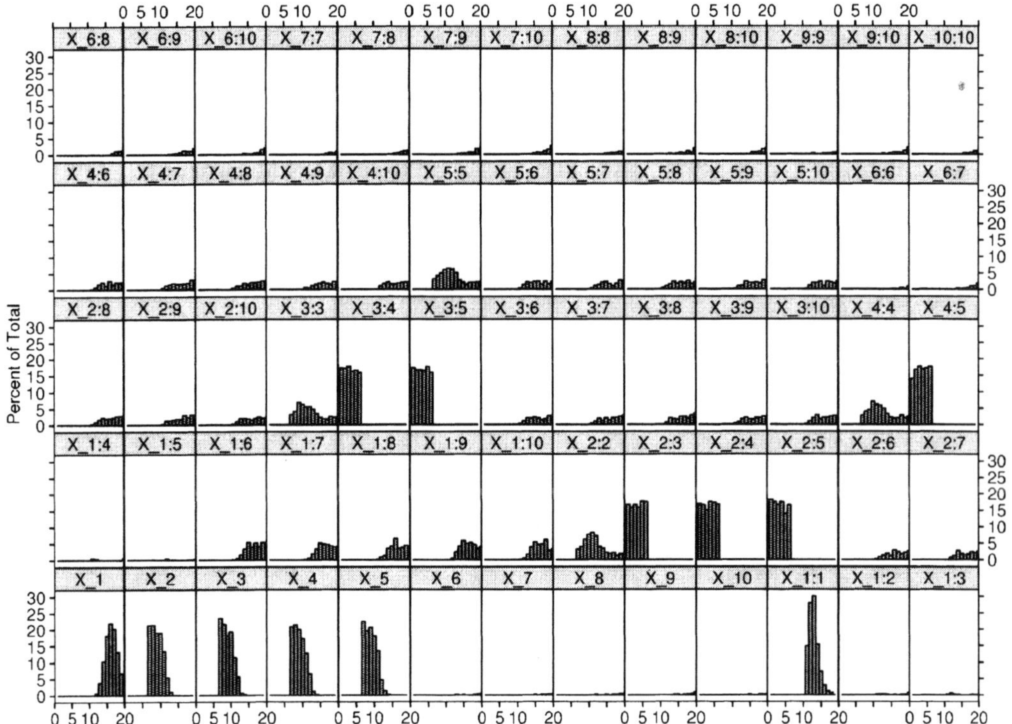

FIG. 4. *Percentage histogram of step at which each variable is selected based on* 1000 *replications: results shown for variables selected in the first* 20 *steps of a LARS analysis using main and interaction terms.*

of these terms selected after step 6 and no other variable was ever selected in the first six steps. That one of these terms is selected as the first term is not surprising as these variables have the highest correlation with the response variable Y. It can be shown that the covariance of these interaction terms with Y is by a factor $\sqrt{12/7} \approx 1.3$ larger than the covariance between X_i and Y for $i = 2, \dots, 5$. But that these six interaction terms dominate the early variable selection steps in such a manner came as a bit as a surprise.

After selecting these six interaction terms, the LARS algorithm then seems to select mostly X_2, X_3, X_4 and X_5, followed soon by $X_{1:1}$ and X_1. However, especially the latter one seems to be selected rather late and other terms may be selected earlier. Other remarkable features in Figure 4 are the peaks in histograms of $X_{i:i}$ for $i = 2, 3, 4, 5$; each of these terms seems to have a fair chance of being selected before the corresponding main term and before $X_{1:1}$ and X_1.

One of the problems seems to be the large number of interaction terms that the LARS algorithm selects without putting the corresponding main terms into the model too. This behavior violates the marginality principle. Also, for this model, one would expect that ensuring that for each higher-order term the corresponding lower-order terms are in the model too would alleviate the problem that the six interaction terms $X_{i:j}$, $i, j \in \{2, 3, 4, 5\}$, $i < j$, are always selected first.

I give an alternative description of the LARS algorithm first before I show how it can be modified to incorporate the marginality principle. This description is based on the discussion in EHJT and shown in Algorithm 1.

ALGORITHM 1 (An alternative description of the LARS algorithm).

1. Set $\hat{\boldsymbol{\mu}}_0 = \mathbf{0}$ and $k = 0$.
2. **repeat**
3. Calculate $\hat{\mathbf{c}} = X'(\mathbf{y} - \hat{\boldsymbol{\mu}}_k)$ and set $\hat{C} = \max_j\{|\hat{c}_j|\}$.
4. Let $\mathcal{A} = \{j : |\hat{c}_j| = \hat{C}\}$.
5. Set $X_{\mathcal{A}} = (\cdots \mathbf{x}_j \cdots)_{j\in\mathcal{A}}$ for calculating $\bar{\mathbf{y}}_{k+1} = (X'_{\mathcal{A}} X_{\mathcal{A}})^{-1} X'_{\mathcal{A}}\mathbf{y}$ and $\mathbf{a} = X'_{\mathcal{A}}(\bar{\mathbf{y}}_{k+1} - \hat{\boldsymbol{\mu}}_k)$.
6. Set

$$\hat{\boldsymbol{\mu}}_{k+1} = \hat{\boldsymbol{\mu}}_k + \hat{\gamma}(\bar{\mathbf{y}}_{k+1} - \hat{\boldsymbol{\mu}}_k),$$

where, if $\mathcal{A}^c \neq \varnothing$,

$$\hat{\gamma} = \min_{j\in\mathcal{A}^c}{}^{+}\left\{\frac{\hat{C} - \hat{c}_j}{\hat{C} - a_j}, \frac{\hat{C} + \hat{c}_j}{\hat{C} + a_j}\right\},$$

otherwise set $\hat{\gamma} = 1$.
7. $k \leftarrow k + 1$.
8. **until** $\mathcal{A}^c = \varnothing$.

We start with an estimated response $\hat{\boldsymbol{\mu}}_0 = \mathbf{0}$ and then iterate until all variables have been selected. In each iteration, we first determine (up to a constant factor) the correlation between all variables and the current residual vector. All variables whose absolute correlation with the residual vector equals the maximal achievable absolute correlation are chosen to be in the model and we calculate the least squares regression response, say $\bar{\mathbf{y}}_{k+1}$, using these variables. Then we move from our current estimated response $\hat{\boldsymbol{\mu}}_k$ toward $\bar{\mathbf{y}}_{k+1}$ until a new variable would enter the model.

Using this description of the LARS algorithm, it seems obvious how to modify the algorithm such that it respects the marginality principle. Assume that for each column i of the design matrix we set $d_{ij} = 1$ if column j should be in the model whenever column i is in the model and zero otherwise; here $j \neq i$ takes values in $\{1, \dots, m\}$, where m is the number of columns of the design matrix. For example, abusing this notation slightly, for model (1) we might set $d_{1:1,1} = 1$ and all other $d_{1:1,j} = 0$; or $d_{1:2,1} = 1$, $d_{1:2,2} = 1$ and all other $d_{1:2,j}$ equal to zero.

Having defined such a dependency structure between the columns of the design matrix, the obvious modification of the LARS algorithm is that when adding, say, column i to the selected columns one also adds all those columns for which $d_{ij} = 1$. This modification is described in Algorithm 2.

ALGORITHM 2 (The modified LARS algorithm).

1. Set $\hat{\boldsymbol{\mu}}_0 = \mathbf{0}$ and $k = 0$.
2. **repeat**
3. Calculate $\hat{\mathbf{c}} = X'(\mathbf{y} - \hat{\boldsymbol{\mu}}_k)$ and set $\hat{C} = \max_j\{|\hat{c}_j|\}$.
4. Let $\mathcal{A}_0 = \{j : |\hat{c}_j| = \hat{C}\}$, $\mathcal{A}_1 = \{j : d_{ij} \neq 0, i \in \mathcal{A}_0\}$ and $\mathcal{A} = \mathcal{A}_0 \cup \mathcal{A}_1$.
5. Set $X_{\mathcal{A}} = (\cdots \mathbf{x}_j \cdots)_{j \in \mathcal{A}}$ for calculating $\bar{\mathbf{y}}_{k+1} = (X'_{\mathcal{A}} X_{\mathcal{A}})^{-1} X'_{\mathcal{A}} \mathbf{y}$ and $\mathbf{a} = X'_{\mathcal{A}}(\bar{\mathbf{y}}_{k+1} - \hat{\boldsymbol{\mu}}_k)$.
6. Set
$$\hat{\boldsymbol{\mu}}_{k+1} = \hat{\boldsymbol{\mu}}_k + \hat{\gamma}(\bar{\mathbf{y}}_{k+1} - \hat{\boldsymbol{\mu}}_k),$$
where, if $\mathcal{A}^c \neq \varnothing$,
$$\hat{\gamma} = \min_{j \in \mathcal{A}^c}{}^{+}\left\{\frac{\hat{C} - \hat{c}_j}{\hat{C} - a_j}, \frac{\hat{C} + \hat{c}_j}{\hat{C} + a_j}\right\},$$
otherwise set $\hat{\gamma} = 1$.
7. $k \leftarrow k + 1$.
8. **until** $\mathcal{A}^c = \varnothing$.

Note that compared with the original Algorithm 1 only the fourth line changes. Furthermore, for all $i \in \mathcal{A}$ it is obvious that for $0 \leq \gamma \leq 1$ we have

$$|\hat{c}_i(\gamma)| = (1 - \gamma)|\hat{c}_i|, \tag{2}$$

where $\hat{\mathbf{c}}(\gamma) = X'(\mathbf{y} - \hat{\boldsymbol{\mu}}(\gamma))$ and $\hat{\boldsymbol{\mu}}(\gamma) = \hat{\boldsymbol{\mu}}_k + \gamma(\bar{\mathbf{y}}_{k+1} - \hat{\boldsymbol{\mu}}_k)$.

Note that, by definition, the value of $|\hat{c}_j|$ is the same for all $j \in \mathcal{A}_0$. Hence, the functions (2) for those variables are identical, namely $(1 - \gamma)\hat{C}$, and for all $j \in \mathcal{A}_1$ the corresponding functions $|\hat{c}_j(\gamma)|$ will intersect $(1 - \gamma)\hat{C}$ at $\gamma = 1$. This explains why in line 6 we only have to check for the first intersection between $(1 - \gamma)\hat{C}$ and $|\hat{c}_j(\gamma)|$ for $j \in \mathcal{A}^c$.

It also follows from (2) that, for all $j \in \mathcal{A}_0$, we have

$$\mathbf{x}'_j(\bar{\mathbf{y}}_{k+1} - \hat{\boldsymbol{\mu}}_k) = \operatorname{sign}(\hat{c}_j)\hat{C}.$$

Thus, for those variables that are in $\mathcal{A}_0$ we move in line 6 of the modified algorithm in a direction that has a similar geometric interpretation as the direction along which we move in the original LARS algorithm. Namely that for each $j \in \mathcal{A}_0$ the angle between the direction in which we move and $\operatorname{sign}(\hat{c}_j)\mathbf{x}_j$ is the same and this angle is less than 90°.

Figure 5 shows the result of the modified LARS analysis for the same data used above. Putting variables that enter the model simultaneously into brackets, the order in which the variables enter the model is $(X_{2:5}, X_2, X_5)$, $(X_{3:4}, X_3, X_4)$, $X_{2:5}$, $X_{2:3}$, $(X_{1:1}, X_1), \ldots$. That is, the modified LARS algorithm selects in this case in five steps a model with 10 terms, 6 of which are the terms that are indeed in model (1).

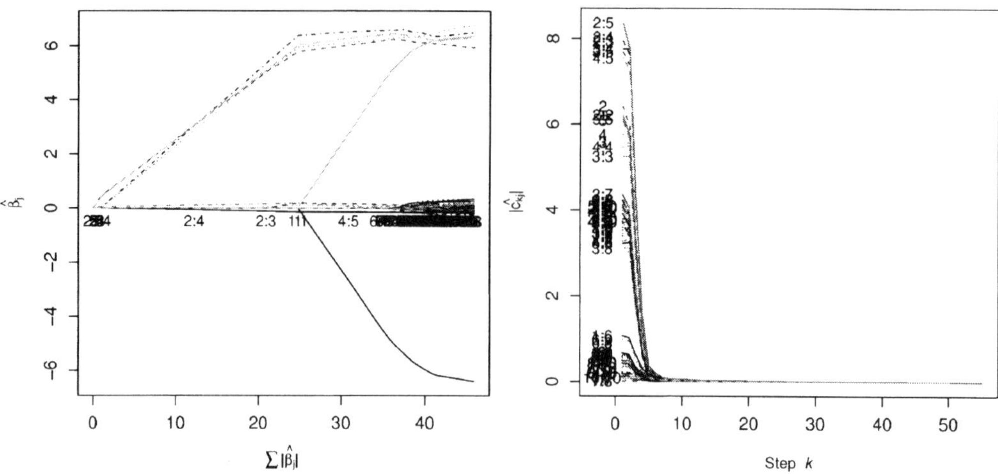

FIG. 5. *Modified LARS analysis of simulated data with main terms and interaction terms*: (left) *estimates of regression coefficients* $\hat{\beta}_j$, $j = 1, \ldots, 65$, *plotted versus* $\sum |\hat{\beta}_j|$; (right) *absolute current correlations as functions of* $k = \#\mathcal{A}^c$.

Using the same 1000 samples of size $n = 500$ as above and performing a modified LARS analysis on them using a design matrix with all main and interaction terms also shows markedly improved results. Again, for each replication the step at which a variable was selected into the model was recorded and Figure 6 shows for each variable histograms for these data. To avoid cluttering, the histograms in Figure 6 were truncated to $[1, 20]$; the complete histograms are shown on the right in Figure 7.

From Figure 6 it can be seen that now the variables X_2, X_3, X_4 and X_5 are all selected within the first three iterations of the modified LARS algorithm. Also $X_{1:1}$ and X_1 are picked up consistently and early. Compared with Figure 4 there are marked differences in the distribution of when the variable is selected for the interaction terms $X_{i:j}$, $i, j \in \{2, 3, 4, 5\}$, $i \leq j$, and the main terms X_i, $i = 6, \ldots, 10$. The latter can be explained by the fact that the algorithm now enforces the marginality principle. Thus, it seems that this modification does improve the performance of the LARS algorithm for model (1). Hopefully it would do so also for other models.

In conclusion, I offer two further remarks and a question. First, note that the modified LARS algorithm may also be used to incorporate factor variables with more than two levels. In such a situation, I would suggest that indicator variables for *all* levels are included in the initial design matrix; but this would be done mainly to easily calculate all the correlations. The dependencies d_{ij} would be set up such that if one indicator variable is selected, then all enter the model. However, to avoid redundancies one would only put all but one of these columns into the matrix $X_{\mathcal{A}}$. This would also avoid that $X_{\mathcal{A}}$ would eventually become singular if more than one explanatory variable is a factor variable.

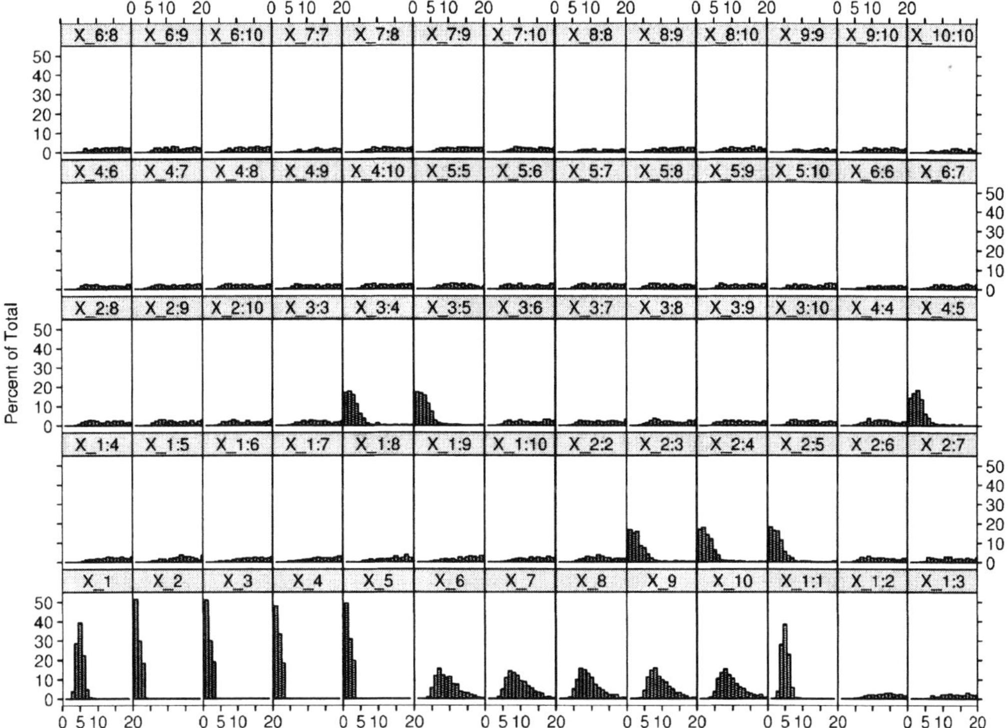

FIG. 6. *Percentage histogram of step at which each variable is selected based on* 1000 *replications: results shown for variables selected in the first* 20 *steps of a modified LARS analysis using main and interaction terms.*

Second, given the insight between the LARS algorithm and the Lasso algorithm described by EHJT, namely the sign constraint (3.1), it now seems also possible to modify the Lasso algorithm to incorporate the marginality principle by incorporating the sign constraint into Algorithm 2. However, whenever a variable

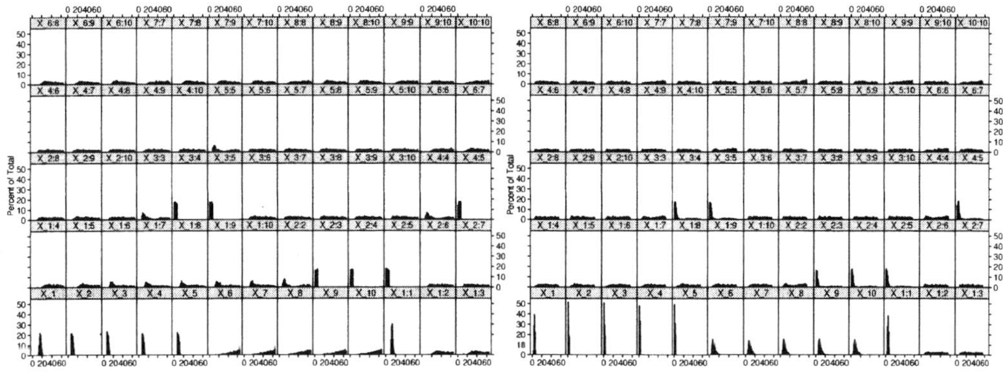

FIG. 7. *Percentage histogram of step at which each variable is selected based on* 1000 *replications:* (left) *LARS analysis*; (right) *modified LARS analysis.*

would be dropped from the set $\mathcal{A}_0$ due to violating the sign constraint there might also be variables dropped from $\mathcal{A}_1$. For the latter variables these might introduce discontinuities in the traces of the corresponding parameter estimates, a feature that does not seem to be desirable. Perhaps a better modification of the Lasso algorithm that incorporates the marginality principle can still be found?

Finally, the behavior of the LARS algorithm for model (1) when all main terms and interaction terms are used surprised me a bit. This behavior seems to raise a fundamental question, namely, although we try to build a linear model and, as we teach our students, correlation "measures the direction and strength of the linear relationship between two quantitative variables" [Moore and McCabe (1999)], one has to wonder whether selecting variables using correlation as a criterion is a sound principle? Or should we modify the algorithms to use another criterion?

REFERENCES

BREIMAN, L. (1995). Better subset regression using the nonnegative garrote. *Technometrics* **37** 373–384.

MCCULLAGH, P. and NELDER, J. A. (1989). *Generalized Linear Models*, 2nd ed. Chapman and Hall, London.

MOORE, D. S. and MCCABE, G. P. (1999). *Introduction to the Practice of Statistics*, 3rd ed. Freeman, New York.

NELDER, J. A. (1977). A reformulation of linear models (with discussion). *J. Roy. Statist. Soc. Ser. A* **140** 48–76.

NELDER, J. A. (1994). The statistics of linear models: Back to basics. *Statist. Comput.* **4** 221–234.

SCHOOL OF MATHEMATICS AND STATISTICS
UNIVERSITY OF WESTERN AUSTRALIA
35 STIRLING HIGHWAY
CRAWLEY WA 6009
AUSTRALIA
E-MAIL: berwin@maths.uwa.edu.au

DISCUSSION

BY SANFORD WEISBERG[1]

University of Minnesota

Most of this article concerns the uses of LARS and the two related methods in the age-old, "somewhat notorious," problem of "[a]utomatic model-building algorithms..." for linear regression. In the following, I will confine my comments to this notorious problem and to the use of LARS and its relatives to solve it.

[1]Supported by NSF Grant DMS-01-03983.

1. The implicit assumption. Suppose the response is y, and we collect the m predictors into a vector x, the realized data into an $n \times m$ matrix X and the response is the n-vector Y. If P is the projection onto the column space of $(1, X)$, then LARS, like ordinary least squares (OLS), assumes that, for the purposes of model building, Y can be replaced by $\hat{Y} = PY$ without loss of information. In large samples, this is equivalent to the assumption that the conditional distributions $F(y|x)$ can be written as

$$F(y|x) = F(y|x'\beta) \tag{1}$$

for some unknown vector β. Efron, Hastie, Johnstone and Tibshirani use this assumption in the definition of the LARS algorithm and in estimating residual variance by $\hat{\sigma}^2 = \|(I - P)Y\|^2/(n - m - 1)$. For LARS to be reasonable, we need to have some assurance that this particular assumption holds or that it is relatively benign. If this assumption is not benign, then LARS like OLS is unlikely to produce useful results.

A more general alternative to (1) is

$$F(y|x) = F(y|x'B), \tag{2}$$

where B is an $m \times d$ rank d matrix. The smallest value of d for which (2) holds is called the structural dimension of the regression problem [Cook (1998)]. An obvious precursor to fitting linear regression is deciding on the structural dimension, not proceeding as if $d = 1$. For the diabetes data used in the article, the R package `dr` [Weisberg (2002)] can be used to estimate d using any of several methods, including sliced inverse regression [Li (1991)]. For these data, fitting these methods suggests that (1) is appropriate.

Expanding x to include functionally related terms is another way to provide a large enough model that (1) holds. Efron, Hastie, Johnstone and Tibshirani illustrate this in the diabetes example in which they expand the 10 predictors to 65 including all quadratics and interactions. This alternative does not include (2) as a special case, as it includes a few models of various dimensions, and this seems to be much more complex than (2).

Another consequence of assumption (1) is the reliance of LARS, and of OLS, on correlations. The correlation measures the degree of linear association between two variables particularly for normally distributed or at least elliptically contoured variables. This requires not only linearity in the conditional distributions of y given subsets of the predictors, but also linearity in the conditional distributions of $a'x$ given $b'x$ for all a and b [see, e.g., Cook and Weisberg (1999a)]. When the variables are not linearly related, bizarre results can follow; see Cook and Weisberg (1999b) for examples. Any method that replaces Y by PY cannot be sensitive to nonlinearity in the conditional distributions.

Methods based on PY alone may be strongly influenced by outliers and high leverage cases. As a simple example of this, consider the formula for C_p given by

Efron, Hastie, Johnstone and Tibshirani:

$$C_p(\hat{\mu}) = \frac{\|Y - \hat{\mu}\|^2}{\sigma^2} - n + 2\sum_{i=1}^{n} \frac{\text{cov}(\hat{\mu}_i, y_i)}{\sigma^2}. \tag{3}$$

Estimating σ^2 by $\hat{\sigma}^2 = \|(I - P)Y\|^2/(n - m - 1)$, and adapting Weisberg (1981), (3) can be rewritten as a sum of n terms, the ith term given by

$$C_{pi}(\hat{\mu}) = \frac{(\hat{y}_i - \hat{\mu}_i)^2}{\hat{\sigma}^2} + \frac{\text{cov}(\hat{\mu}_i, y_i)}{\hat{\sigma}^2} - \left(\frac{h_i - \text{cov}(\hat{\mu}_i, y_i)}{\hat{\sigma}^2}\right),$$

where $\hat{y}_i$ is the ith element of PY and h_i is the ith leverage, a diagonal element of P. From the simulation reported in the article, a reasonable approximation to the covariance term is $\hat{\sigma}^2 u_i$, where u_i is the ith diagonal of the projection matrix on the columns of $(1, X)$ with nonzero coefficients at the current step of the algorithm. We then get

$$C_{pi}(\hat{\mu}) = \frac{(\hat{y}_i - \hat{\mu}_i)^2}{\hat{\sigma}^2} + u_i - (h_i - u_i),$$

which is the same as the formula given in Weisberg (1981) for OLS except that $\hat{\mu}_i$ is computed from LARS rather than from a projection. The point here is that the value of $C_{pi}(\hat{\mu})$ depends on the agreement between $\hat{\mu}_i$ and $\hat{y}_i$, on the leverage in the subset model and on the difference in the leverage between the full and subset models. Neither of these latter two terms has much to do with the problem of interest, which is the study of the conditional distribution of y given x, but they are determined by the predictors only.

2. Selecting variables. Suppose that we can write $x = (x_a, x_u)$ for some decomposition of x into two pieces, in which x_a represents the "active" predictors and x_u the unimportant or inactive predictors. The variable selection problem is to find the smallest possible x_a so that

$$F(y|x) = F(y|x_a) \tag{4}$$

thereby identifying the active predictors. Standard subset selection methods attack this problem by first assuming that (1) holds, and then fitting models with different choices for x_a, possibly all possible choices or a particular subset of them, and then using some sort of inferential method or criterion to decide if (4) holds, or more precisely if

$$F(y|x) = F(y|\gamma' x_a)$$

holds for some γ. Efron, Hastie, Johnstone and Tibshirani criticize the standard methods as being too greedy: once we put a variable, say, $x^* \in x_a$, then any predictor that is highly correlated with x^* will never be included. LARS, on the other hand, permits highly correlated predictors to be used.

LARS or any other methods based on correlations cannot be much better at finding x_a than are the standard methods. As a simple example of what can go wrong, I modified the diabetes data in the article by adding nine new predictors, created by multiplying each of the original predictors excluding the sex indicator by 2.2, and then rounding to the nearest integer. These rounded predictors are clearly less relevant than are the original predictors, since they are the original predictors with noise added by the rounding. We would hope that none of these would be among the active predictors.

Using the S-PLUS functions kindly provided by Efron, Hastie, Johnstone and Tibshirani, the LARS procedure applied to the original data selects a seven-predictor model, including, in order, BMI, S5, BP, S3, SEX, S6 and S1. LARS applied to the data augmented with the nine inferior predictors selects an eight-predictor model, including, in order, BMI, S5, rBP, rS3, BP, SEX, S6 and S1, where the prefix "r" indicates a rounded variable rather than the variable itself. LARS not only selects two of the inferior rounded variables, but it selects both BP and its rounded version rBP, effectively claiming that the rounding is informative with respect to the response.

Inclusion and exclusion of elements in x_a depends on the marginal distribution of x as much as on the conditional distribution of $y|x$. For example, suppose that the diabetes data were a random sample from a population. The variables S3 and S4 have a large sample correlation, and LARS selects one of them, S3, as an active variable. Suppose a therapy were available that could modify S4 without changing the value of S3, so in the future S3 and S4 would be nearly uncorrelated. Although this would arguably not change the distribution of $y|x$, it would certainly change the marginal distribution of x, and this could easily change the set of active predictors selected by LARS or any other method that starts with correlations.

A characteristic that LARS shares with the usual methodology for subset selection is that the results are invariant under rescaling of any individual predictor, but not invariant under reparameterization of functionally related predictors. In the article, the authors create more predictors by first rescaling predictors to have zero mean and common standard deviation, and then adding all possible cross-products and quadratics to the existing predictors. For this expanded definition of the predictors, LARS selects a 15 variable model, including 6 main-effects, 6 two-factor interactions and 3 quadratics. If we add quadratics and interactions first and then rescale, LARS picks an 8 variable model with 2 main-effects, 6 two-factor interactions, and only 3 variables in common with the model selected by scaling first. If we define the quadratics and interactions to be orthogonal to the main-effects, we again get a different result. The lack of invariance with regard to definition of functionally related predictors can be partly solved by considering the functionally related variables simultaneously rather than sequentially. This seems to be self-defeating, at least for the purpose of subset selection.

3. Summary. Long-standing problems often gain notoriety because solution of them is of wide interest and at the same time illusive. Automatic model building in linear regression is one such problem. My main point is that neither LARS nor, as near as I can tell, any other *automatic* method has any hope of solving this problem because automatic procedures by their very nature do not consider the context of the problem at hand. I cannot see any solution to this problem that is divorced from context. Most of the ideas in this discussion are not new, but I think they bear repeating when trying to understand LARS methodology in the context of linear regression. Similar comments can be found in Efron (2001) and elsewhere.

REFERENCES

COOK, R. D. (1998). *Regression Graphics*. Wiley, New York.

COOK, R. D. and WEISBERG, S. (1999a). *Applied Regression Including Computing and Graphics*. Wiley, New York.

COOK, R. D. and WEISBERG, S. (1999b). Graphs in statistical analysis: Is the medium the message? *Amer. Statist.* **53** 29–37.

EFRON, B. (2001). Discussion of "Statistical modeling: The two cultures," by L. Breiman. *Statist. Sci.* **16** 218–219.

LI, K. C. (1991). Sliced inverse regression for dimension reduction (with discussion). *J. Amer. Statist. Assoc.* **86** 316–342.

WEISBERG, S. (1981). A statistic for allocating C_p to individual cases. *Technometrics* **23** 27–31.

WEISBERG, S. (2002). Dimension reduction regression in R. *J. Statistical Software* **7**. (On-line journal available at www.jstatsoft.org. The software is available from cran.r-project.org.)

SCHOOL OF STATISTICS
UNIVERSITY OF MINNESOTA
1994 BUFORD AVENUE
ST. PAUL, MINNESOTA 55108
USA
E-MAIL: sandy@stat.umn.edu

REJOINDER

BY BRADLEY EFRON, TREVOR HASTIE, IAIN JOHNSTONE AND ROBERT TIBSHIRANI

The original goal of this project was to explain the striking similarities between models produced by the Lasso and Forward Stagewise algorithms, as exemplified by Figure 1. LARS, the Least Angle Regression algorithm, provided the explanation and proved attractive in its own right, its simple structure permitting theoretical insight into all three methods. In what follows "LAR" will refer to the basic, unmodified form of Least Angle Regression developed in Section 2, while "LARS" is the more general version giving LAR, Lasso, Forward

Stagewise and other variants as in Section 3.4. Here is a summary of the principal properties developed in the paper:

1. LAR builds a regression model in piecewise linear forward steps, accruing explanatory variables one at a time; each step is taken along the equiangular direction between the current set of explanators. The step size is less greedy than classical forward stepwise regression, smoothly blending in new variables rather than adding them discontinuously.
2. Simple modifications of the LAR procedure produce all Lasso and Forward Stagewise solutions, allowing their efficient computation and showing that these methods also follow piecewise linear equiangular paths. The Forward Stagewise connection suggests that LARS-type methods may also be useful in more general "boosting" applications.
3. The LARS algorithm is computationally efficient; calculating the full set of LARS models requires the same order of computation as ordinary least squares.
4. A k-step LAR fit uses approximately k degrees of freedom, in the sense of added prediction error (4.5). This approximation is exact in the case of orthogonal predictors and is generally quite accurate. It permits C_p-type stopping rules that do not require auxiliary bootstrap or cross-validation computations.
5. For orthogonal designs, LARS models amount to a succession of soft thresholding estimates, (4.17).

All of this is rather technical in nature, showing how one might efficiently carry out a program of automatic model-building ("machine learning"). Such programs seem increasingly necessary in a scientific world awash in huge data sets having hundreds or even thousands of available explanatory variables.

What this paper, strikingly, does not do is justify any of the three algorithms as providing *good* estimators in some decision-theoretic sense. A few hints appear, as in the simulation study of Section 3.3, but mainly we are relying on recent literature to say that LARS methods are at least reasonable algorithms and that it is worthwhile understanding their properties. Model selection, the great underdeveloped region of classical statistics, deserves careful theoretical examination but that does not happen here. We are not as pessimistic as Sandy Weisberg about the potential of automatic model selection, but agree that it requires critical examination as well as (over) enthusiastic algorithm building.

The LARS algorithm in any of its forms produces a one-dimensional path of prediction vectors going from the origin to the full least-squares solution. (Figures 1 and 3 display the paths for the diabetes data.) In the LAR case we can label the predictors $\widehat{\mu}(k)$, where k is identified with both the number of steps and the degrees of freedom. What the figures do not show is when to stop the model-building process and report $\widehat{\mu}$ back to the investigator. The examples in our paper rather casually used stopping rules based on minimization of the C_p error prediction formula.

Robert Stine and Hemant Ishwaran raise some reasonable doubts about C_p minimization as an effective stopping rule. For any one value of k, C_p is an unbiased estimator of prediction error, so in a crude sense C_p minimization is trying to be an unbiased estimator of the optimal stopping point k_{opt}. As such it is bound to overestimate k_{opt} in a large percentage of the cases, perhaps near 100% if k_{opt} is near zero.

We can try to improve C_p by increasing the *df* multiplier "2" in (4.5). Suppose we change 2 to some value *mult*. In standard normal-theory model building situations, for instance choosing between linear, quadratic, cubic, ... regression models, the *mult* rule will prefer model $k+1$ to model k if the relevant t-statistic exceeds $\sqrt{mult}$ in absolute value (here we are assuming σ^2 known); $mult = 2$ amounts to using a rejection rule with $\alpha = 16\%$. Stine's interesting S_p method chooses *mult* closer to 4, $\alpha = 5\%$.

This works fine for Stine's examples, where k_{opt} is indeed close to zero. We tried it on the simulation example of Section 3.3. Increasing *mult* from 2 to 4 decreased the average selected step size from 31 to 15.5, but with a small increase in actual squared estimation error. Perhaps this can be taken as support for Ishwaran's point that since LARS estimates have a broad plateau of good behavior, one can often get by with much smaller models than suggested by C_p minimization. Of course no one example is conclusive in an area as multifaceted as model selection, and perhaps no 50 examples either. A more powerful theory of model selection is sorely needed, but until it comes along we will have to make do with simulations, examples and bits and pieces of theory of the type presented here.

Bayesian analysis of prediction problems tends to favor *much* bigger choices of *mult*. In particular the Bayesian information criterion (BIC) uses $mult =$ log(sample size). This choice has favorable consistency properties, selecting the correct model with probability 1 as the sample size goes to infinity. However, it can easily select too-small models in nonasymptotic situations.

Jean-Michel Loubes and Pascal Massart provide two interpretations using penalized estimation criteria in the orthogonal regression setting. The first uses the link between soft thresholding and ℓ_1 penalties to motivate entropy methods for asymptotic analysis. The second is a striking perspective on the use of C_p with LARS. Their analysis suggests that our usual intuition about C_p, derived from selecting among projection estimates of different ranks, may be misleading in studying a nonlinear method like LARS that combines thresholding and shrinkage. They rewrite the LARS-C_p expression (4.5) in terms of a penalized criterion for selecting among orthogonal projections. Viewed in this unusual way (for the estimator to be used is *not* a projection!), they argue that *mult* in fact behaves like $\log(n/k)$ rather than 2 (in the case of a k-dimensional projection). It is indeed remarkable that this same model-dependent value of *mult*, which has emerged in several recent studies [Foster and Stine (1997), George and Foster (2000), Abramovich, Benjamini, Donoho and Johnstone (2000) and Birgé and Massart (2001)], should also appear as relevant for the analysis of LARS. We look forward to the further extension of the Birgé–Massart approach to handling these nondeterministic penalties.

Cross-validation is a nearly unbiased estimator of prediction error and as such will perform similarly to C_p (with $mult = 2$). The differences between the two methods concern generality, efficiency and computational ease. Cross-validation, and nonparametric bootstrap methods such as the 632+ rule, can be applied to almost any prediction problem. C_p is more specialized, but when it does apply it gives more efficient estimates of prediction error [Efron (2004)] at almost no computational cost. It applies here to LAR, at least when $m < n$, as in David Madigan and Greg Ridgeway's example.

We agree with Madigan and Ridgeway that our new LARS algorithm may provide a boost for the Lasso, making it more useful and attractive for data analysts. Their suggested extension of LARS to generalized linear models is interesting. In logistic regression, the L_1-constrained solution is not piecewise linear and hence the pathwise optimization is more difficult. Madigan and Ridgeway also compare LAR and Lasso to least squares boosting for prediction accuracy on three real examples, with no one method prevailing.

Saharon Rosset and Ji Zhu characterize a class of problems for which the coefficient paths, like those in this paper, are piecewise linear. This is a useful advance, as demonstrated with their robust version of the Lasso, and the ℓ_1-regularized Support Vector Machine. The former addresses some of the robustness concerns of Weisberg. They also report on their work that strengthens the connections between ε-boosting and ℓ_1-regularized function fitting.

Berwin Turlach's example with uniform predictors surprised us as well. It turns out that 10-fold cross-validation selects the model with $|\beta_1| \approx 45$ in his Figure 3 (left panel), and by then the correct variables are active and the interactions have died down. However, the same problem with 10 times the noise variance does not recover in a similar way. For this example, if the X_j are uniform on $[-\frac{1}{2}, \frac{1}{2}]$ rather than $[0, 1]$, the problem goes away, strongly suggesting that proper centering of predictors (in this case the interactions, since the original variables are automatically centered by the algorithm) is important for LARS.

Turlach also suggests an interesting proposal for enforcing marginality, the hierarchical relationship between the main effects and interactions. In his notation, marginality says that $\beta_{i:j}$ can be nonzero only if β_i and β_j are nonzero. An alternative approach, more in the "continuous spirit" of the Lasso, would be to include constraints

$$|\beta_{i:j}| \le \min\{|\beta_i|, |\beta_j|\}.$$

This implies marginality but is stronger. These constraints are linear and, according to Rosset and Zhu above, a LARS-type algorithm should be available for its estimation. Leblanc and Tibshirani (1998) used constraints like these for shrinking classification and regression trees.

As Turlach suggests, there are various ways to restate the LAR algorithm, including the following nonalgebraic purely statistical statement in terms of repeated fitting of the residual vector $\mathbf{r}$:

1. Start with $\mathbf{r} = \mathbf{y}$ and $\widehat{\beta}_j = 0 \ \forall j$.
2. Find the predictor $\mathbf{x}_j$ most correlated with $\mathbf{r}$.
3. Increase $\widehat{\beta}_j$ in the direction of the sign of $\text{corr}(\mathbf{r}, \mathbf{x}_j)$ until some other competitor $\mathbf{x}_k$ has as much correlation with the current residual as does $\mathbf{x}_j$.
4. Update $\mathbf{r}$, and move $(\widehat{\beta}_j, \widehat{\beta}_k)$ in the joint least squares direction for the regression of $\mathbf{r}$ on $(\mathbf{x}_j, \mathbf{x}_k)$ until some other competitor $\mathbf{x}_\ell$ has as much correlation with the current residual.
5. Continue in this way until all predictors have been entered. Stop when $\text{corr}(\mathbf{r}, \mathbf{x}_j) = 0 \ \forall j$, that is, the OLS solution.

Traditional forward stagewise would have completed the least-squares step at each stage; here it would go only a fraction of the way, until the next competitor joins in.

Keith Knight asks whether Forward Stagewise and LAR have implicit criteria that they are optimizing. In unpublished work with Trevor Hastie, Jonathan Taylor and Guenther Walther, we have made progress on that question. It can be shown that the Forward Stagewise procedure does a sequential minimization of the residual sum of squares, subject to

$$\sum_j \left| \int_0^t \beta_j'(s)\, ds \right| \leq t.$$

This quantity is the total L_1 arc-length of the coefficient curve $\beta(t)$. If each component $\beta_j(t)$ is monotone nondecreasing or nonincreasing, then L_1 arc-length equals the L_1-norm $\sum_j |\beta_j|$. Otherwise, they are different and L_1 arc-length discourages sign changes in the derivative. That is why the Forward Stagewise solutions tend to have long flat plateaus. We are less sure of the criterion for LAR, but currently believe that it uses a constraint of the form $\sum_j |\int_0^k \beta_j(s)\, ds| \leq A$.

Sandy Weisberg, as a ranking expert on the careful analysis of regression problems, has legitimate grounds for distrusting automatic methods. Only foolhardy statisticians dare to ignore a problem's context. (For instance it helps to know that diabetes progression behaves differently after menopause, implying strong age–sex interactions.) Nevertheless even for a "small" problem like the diabetes investigation there is a limit to how much context the investigator can provide. After that one is drawn to the use of automatic methods, even if the "automatic" part is not encapsulated in a single computer package.

In actual practice, or at least in good actual practice, there is a cycle of activity between the investigator, the statistician and the computer. For a multivariable prediction problem like the diabetes example, LARS-type programs are a good first step toward a solution, but hopefully not the last step. The statistician examines the output critically, as did several of our commentators, discussing the results with the investigator, who may at this point suggest adding or removing explanatory variables, and so on, and so on.

Fully automatic regression algorithms have one notable advantage: they permit an honest evaluation of estimation error. For instance the C_p-selected LAR quadratic model estimates that a patient one standard deviation above average on BMI has an increased response expectation of 23.8 points. The bootstrap analysis (3.16) provided a standard error of 3.48 for this estimate. Bootstrapping, jackknifing and cross-validation require us to repeat the original estimation procedure for different data sets, which is easier to do if you know what the original procedure actually was.

Our thanks go to the discussants for their thoughtful remarks, and to the Editors for the formidable job of organizing this discussion.

REFERENCES

ABRAMOVICH, F., BENJAMINI, Y., DONOHO, D. and JOHNSTONE, I. (2000). Adapting to unknown sparsity by controlling the false discovery rate. Technical Report 2000-19, Dept. Statistics, Stanford Univ.

BIRGÉ, L. and MASSART, P. (2001). Gaussian model selection. *J. Eur. Math. Soc.* **3** 203–268.

EFRON, B. (2004). The estimation of prediction error: Covariance penalties and cross-validation. *J. Amer. Statist. Assoc.* To appear.

FOSTER, D. and STINE, R. (1997). An information theoretic comparison of model selection criteria. Technical report, Dept. Statistics, Univ. Pennsylvania.

GEORGE, E. I. and FOSTER, D. P. (2000). Calibration and empirical Bayes variable selection. *Biometrika* **87** 731–747.

LEBLANC, M. and TIBSHIRANI, R. (1998). Monotone shrinkage of trees. *J. Comput. Graph. Statist.* **7** 417–433.

DEPARTMENT OF STATISTICS
STANFORD UNIVERSITY
SEQUOIA HALL
STANFORD, CALIFORNIA 94305-4065
USA
E-MAIL: brad@stat.stanford.edu

21

Large-Scale Simultaneous Hypothesis Testing: The Choice of a Null Hypothesis

Introduction by Michael A. Newton
University of Wisconsin, Madison

It is an honor and great pleasure to contribute to the present volume.

A theme connecting this paper to Professor Efron's earlier and subsequent work is the concept of local false discovery rate, defined here as an upper bound on the posterior probability that the null hypothesis in question is true, given the value of a relevant test statistic, and in the context of a two-point mixture model for the value of many such hypotheses (12). Elsewhere in his writings, indeed elsewhere in the paper (p. 101), Professor Efron defines the local fdr as this posterior probability itself, as opposed to an upper bound obtained by taking the prior mass equal to 1, and I find this concept slightly more useful. When I first read about local fdr in Efron et al. (2001), I found it rather odd that the concept required a special name. I realized that local fdr was simply a posterior probability, and such objects, as varied as they might be in distinct applications of statistics, exemplified the unity of an important statistical approach. Why give them a new name? Perhaps I was affected by having attended so many Valencia meetings! In so far as these values are rather specific posterior probabilities, obtained semi-parametrically and empirically, I guess one can justify a special name, but I still find it comforting to know that they are just posterior probabilities, as would be computed in a certain hierarchical probability model with certain methods of estimation in mind. They do have a rather neat duality property, in contrast to other test-specific test statistics like p-values, which is that a small local fdr value both identifies the corresponding null hypothesis as probably untrue and also measures the error rate of this identification (Newton et al. 2004, 2006).

A premise of this paper is that one task of high-dimensional hypothesis testing is to separate interesting from uninteresting cases. In contrast to my terminological distaste for local fdr, I am a fan of the new dichotomy, which helps to clarify the distinctive nature of the new inference problem as compared to the problem solved by significance testing. The first thing many statisticians would do to address the high-dimensional testing problem is to apply traditional significance testing separately to each dimension, and then to accommodate the multiplicity afterwards. Working statisticians have engrained notions of significance, but interesting – when are test results interesting? Having new terminology is liberating and potentially useful. Efron's paper provides no mathematical definition of interesting, but we can infer that non-null hypotheses are the truly interesting ones, and, importantly, a given application ought not have too many interesting cases: otherwise the cases found might not in themselves each be so interesting!

The empirical finding on which the paper is based is the observation that a collection of test statistics may not distribute themselves nicely as a simple mixture of mostly null values combined with a small fraction of rather extreme, interesting ones. Actual marginal distributions can be rather odd-looking when considered relative to this ideal, and the use of simplistic mixture models may not reveal the most interesting cases for further analysis. A very useful contribution of the paper is that it investigates reasons why marginal distributions might behave badly: (1) unmeasured confounders that variously affect the different tests, (2) within-test dependence across the samples, and (3) the possibility of a high incidence of mildly interesting cases. I would add that a high degree of among-test dependence can also distort the marginal empirical distribution. All these potential problems ought to be in the analyst's mind during data analysis; but before Efron's publication, there was no clear strategy to handle them.

In his characteristically-authoritative style, Professor Efron addresses the problems by questioning the central pillar of the statistical analysis: the null hypothesis. The empirical null is a fascinating idea introduced to handle apparent defects in the empirical distribution of test statistics. It is a new concept and one that would not have been possible before high-dimensional testing. And it feels like one of Captain Kirk's nifty solutions to an impossible situation, developed with similar style and ease! I find the solution to be quite reasonable in the case of confounders and dependence, but I am rather less convinced in case (3), when the null is simply not true in a large fraction of cases. One practical problem with the method (e.g., p. 101) is that interesting findings can disappear, leaving us little to report. Another problem is that we must simultaneously hold that p_0 is both small and large, which is a little disconcerting. Further, there may be detrimental effects on downstream calculations caused by converting the mildly interesting cases to uninteresting ones according to the empirical null. Often the cases are themselves organized according to exogenous information, such as in gene-set analysis, and it may be that the most interesting sets of cases are those that contain many mildly interesting ones. For instance, Sengupta et al. (2006) found that an unusually high fraction of immune-response genes were down-regulated in cancer cells that expressed high levels of certain genes from the infecting Epstein–Barr virus, though much of the down-regulation was very modest. Something about the biology by which the virus eludes the host defenses caused the extensive but modest shift, though it might not have been detected had data been compared to an empirical null. Professor Efron's screening model, in which only a small fraction of cases can be interesting, may be too rigid in certain applications.

Efron's 2004 work challenges basic notions of statistical inference and helps us to develop new terminology and methodology for high-dimensional testing. As we are required to incorporate more information into our statistical analyses, these methods and concepts surely will be developed further. And Professor Efron himself is leading the way (see Efron 2007). I expect we will have to allow that interestingness is a relative quantity affected by contextual information; surely it will be with great interest that we delve more deeply into the matter!

References

Efron, B. (2007). Simultaneous inference: When should hypothesis testing problems be combined. Unpublished Technical Report, Department of Statistics, Stanford University.

Efron, B., Tibshirani, R., Storey, J. D. and Tusher, V. (2001). Empirical Bayes analysis of a microarray experiment *Journal of the American Statistical Association* **96**, 1151–1160.

Newton, M. A., Noueiry, A., Sarkar, D. and Ahlquist, P. (2004). Detecting differential gene expression with a semiparametric hierarchical mixture method. *Biostatistics* **5**, 155–176.

Newton, M. A., Wang, P. and Kendziorski, C. (2006). Hierarchical mixture models for expression profiles. In *Bayesian inference for gene expression and proteomics*. Edited by K. Ahn Do, P. M. Mueller and M. Vannucci. Cambridge University Press. London.

Sengupta, S., den Boon, J. A., Chen, I. H., Newton, M. A., Dahl, D. B., Chen, M., Cheng, Y. J., Westra, W. H., Chen, C. J., Hildesheim, A., Sugden, B. and Ahlquist, P. (2006). Genome-wide expression profiling reveals Epstein Barr virus-associated inhibition of MHC class I expression in nasopharyngeal carcinoma. *Cancer Research* **66**, 7999–8006.

Large-Scale Simultaneous Hypothesis Testing: The Choice of a Null Hypothesis

Bradley EFRON

Current scientific techniques in genomics and image processing routinely produce hypothesis testing problems with hundreds or thousands of cases to consider simultaneously. This poses new difficulties for the statistician, but also opens new opportunities. In particular, it allows empirical estimation of an appropriate null hypothesis. The empirical null may be considerably more dispersed than the usual theoretical null distribution that would be used for any one case considered separately. An empirical Bayes analysis plan for this situation is developed, using a local version of the false discovery rate to examine the inference issues. Two genomics problems are used as examples to show the importance of correctly choosing the null hypothesis.

KEY WORDS: Empirical Bayes; Empirical null hypothesis; Local false discovery rate; Microarray analysis; Unobserved covariates.

1. INTRODUCTION

Until recently, "simultaneous inference" meant considering two or five or perhaps even 10 hypothesis tests at the same time, as in Miller's classic text (Miller 1981). Rapid progress in technology, particularly in genomics and imaging, has vastly upped the ante for simultaneous inference problems. Now 500 or 5,000 or even 50,000 tests may need to be evaluated simultaneously, raising new problems for the statistician, but also opening new analytic opportunities. This article explores choosing an appropriate null hypothesis in large-scale testing situations, and how this choice affects well-known inference methods, such as the false discovery rate (FDR).

Simultaneous hypothesis testing begins with a collection of null hypotheses,

$$H_1, H_2, \ldots, H_N; \tag{1}$$

corresponding test statistics, possibly not independent,

$$Y_1, Y_2, \ldots, Y_N; \tag{2}$$

and their p values, $P_1, P_2, \ldots, P_N$, with P_i measuring how strongly y_i, the observed value of Y_i, contradicts H_i; for instance, $P_i = \Pr_{H_i}\{|Y_i| > |y_i|\}$. "Large-scale" means that N is a big number, say at least $N > 100$.

It is convenient, although not necessary, to work with *z-values* instead of the Y_i's or P_i's,

$$z_i = \Phi^{-1}(P_i), \qquad i = 1, 2, \ldots, N, \tag{3}$$

with Φ indicating the standard normal *cumulative distribution function* (cdf), for example, $\Phi^{-1}(.95) = 1.645$. If H_i is exactly true, then z_i will have a standard normal distribution,

$$z_i | H_i \sim \mathrm{N}(0, 1). \tag{4}$$

I call (4) the *theoretical null hypothesis*.

Our motivating example concerns a study of 1,391 patients with human immunodeficiency virus (HIV) infection, investigating which of 6 protease inhibitor (PI) drugs cause mutations at which of 74 sites on the viral genome. Each patient provided a vector of predictors,

$$\mathbf{x} = (x_1, x_2, \ldots, x_6). \tag{5}$$

with $x_j = 1$ or 0 indicating whether or not the patient used PI_j, $1 \le \sum_1^6 x_j \le 6$; and a vector of responses,

$$\mathbf{v} = (v_1, v_2, \ldots, v_{74}), \tag{6}$$

$v_k = 1$ or 0 indicating whether or not a mutation occurred at site k. Remark A of Section 7 describes the study in more detail.

For each of the 74 genomic sites, a separate logistic regression analysis was run using all 1,391 cases, with that site's mutation indicators as responses and the PI indicators as predictors. Together these yielded $444 = 6 \times 74$ z-values, one for testing each null hypothesis that drug j does not cause mutations at site k, $j = 1, 2, \ldots, 6$ and $k = 1, 2, \ldots, 74$. The z-values were based on the usual approximation

$$z_i = y_i / se_i, \qquad i = 1, 2, \ldots, 444, \tag{7}$$

[using a single subscript i in place of (j, k)] where y_i is the maximum likelihood estimate (MLE) of the logistic regression coefficient and se_i is its approximate large-sample standard error.

Figure 1 shows a histogram of the 444 z-values, with negative z_i's indicating greater mutational effects. The smooth curve, $f(z)$, is a natural spline with 7 df, fit to the histogram counts by Poisson regression. It emphasizes the *central peak* near $z = 0$, presumably the large majority of uninteresting drug–site combinations that have negligible mutation effects. Near its center, the peak is well described by a normal density with mean $-.35$ and standard deviation 1.20, which will be called the *empirical null hypothesis*,

$$z_i | H_i \sim \mathrm{N}(-.35, 1.20^2). \tag{8}$$

Section 3 describes the estimation methodology for (8), with a brief discussion of the normality assumption in Remark D of Section 7.

The difference between the theoretical null N(0, 1) and empirical null $\mathrm{N}(-.35, 1.20^2)$ may not seem worrisome here, but it will be shown that it substantially affects any simultaneous inference procedure. More dramatic example is given in Section 6, for a microarray analysis in which going from the theoretical to empirical null totally negates any findings of significance. Situations going in the reverse direction can also occur.

Bradley Efron is Professor, Department of Statistics, Stanford University, Stanford, CA 94305 (E-mail: *brad@stat.stanford.edu*). The author thanks Robert Shafer, David Katzenstein, and Rami Kantor for bringing the drug mutation data to his attention, and Robert Tibshirani for several helpful discussions.

Journal of the American Statistical Association
March 2004, Vol. 99, No. 465, Theory and Methods
DOI 10.1198/016214504000000089

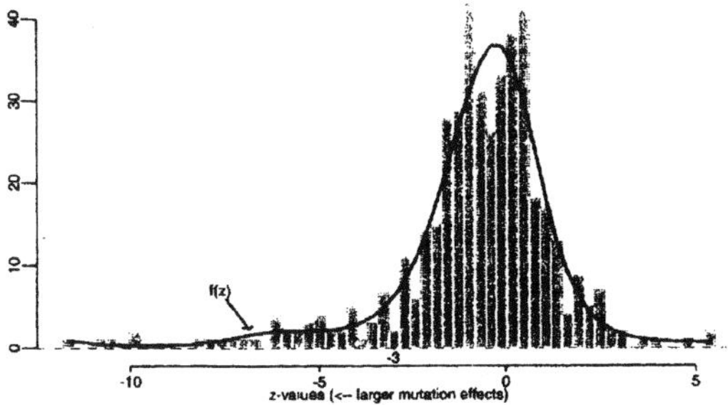

Figure 1. Histogram of 444 z-Values From the Drug Mutation Analysis. The smooth curve f(z) is a natural spline fit to histogram counts. The central peak near $z = 0$ is approximately $N(-.35, 1.20^2)$, the "empirical null hypothesis." Simultaneous hypothesis tests for the 444 cases depend critically on the choice between the empirical or theoretical N(0, 1) null.

In classic situations involving only a single hypothesis test, one must, out of necessity, use the theoretical null hypothesis, $z \sim N(0, 1)$. The main point of this article is that large-scale testing situations permit empirical estimation of the null distribution. Sections 3–5 explore reasons why the empirical and theoretical null might differ, and which might be preferable in different situations.

There are scientific as well as statistical differences between small-scale and large-scale hypothesis testing situations. A single hypothesis test is most often run with the expectation and hope of rejecting the null, "with 80% power" in a typical clinical trial. Nobody wants to reject 80% of $N = 5{,}000$ null hypotheses. The usual point of large-scale testing is to identify a small percentage of interesting cases that deserve further investigation. Although we are not exactly looking for a needle in a haystack, we do not want the whole haystack either. An important assumption of what follows is that the proportion of interesting cases is small, perhaps 1% or 5% of N, but not more than 10%. This is made explicit in Section 2, in the description of the local false discovery rate as an analytic tool for large-scale testing. There are situations in which the 10% limit is irrelevant (e.g., in constructing prediction models), but these lie outside our purpose here.

The terminology "Interesting/Uninteresting" used in this article in preference to "Significant/Nonsignificant" is discussed near the end of Section 5. We conclude in Sections 7 and 8 with remarks, including most of the technical details, and a summary.

2. THE LOCAL FALSE DISCOVERY RATE

It is convenient to discuss large-scale testing problems in terms of the *local false discovery rate* (fdr), an empirical Bayes version of Benjamini and Hochberg's (1995) methodology focusing on densities rather than tail areas (see Efron et al. 2001; Efron and Tibshirani 2002; Storey 2002, 2003).

We begin with a simple Bayes model. Suppose that each of the N z-values falls into one of two classes, "Uninteresting" or "Interesting," corresponding to whether or not z_i is generated according to the null hypothesis, with prior probabilities p_0 and $p_1 = 1 - p_0$ for the classes. Assume that z_i has density either $f_0(z)$ or $f_1(z)$, depending on its class,

$$p_0 = \Pr\{\text{Uninteresting}\}, \quad f_0(z) \text{ density if Uninteresting (Null)},$$
$$p_1 = \Pr\{\text{Interesting}\}, \quad f_1(z) \text{ density if Interesting (Nonnull)}. \tag{9}$$

The smooth curve in Figure 1 estimates the *mixture density*, $f(z)$,

$$f(z) = p_0 f_0(z) + p_1 f_1(z). \tag{10}$$

According to Bayes theorem, the a posteriori probability of being in the Uninteresting class given z is

$$\Pr\{\text{Uninteresting}|z\} = p_0 f_0(z)/f(z). \tag{11}$$

Here I define the fdr as

$$\text{fdr}(z) \equiv f_0(z)/f(z), \tag{12}$$

ignoring the factor p_0 in (11), so fdr(z) is an upper bound on Pr{Uninteresting|z}. In fact, p_0 can be roughly estimated (see Remark B in Sec. 7), but I am assuming that p_0 is near 1, say $p_0 \geq .90$, so fdr(z) is not a flagrant overestimator.

The fdr provides a useful methodology for identifying Interesting cases in a situation like that of Figure 1: (1) estimate $f(z)$ from the observed ensemble of z-values, for example, by the natural spline fit to the histogram counts; (2) assign a null density $f_0(z)$; (3) calculate $\text{fdr}(z) = f_0(z)/f(z)$; and (4) report as Interesting those cases with fdr(z_i) less than some threshold value, perhaps $\text{fdr}(z_i) \leq .10$. Remark B discusses the close connection between this algorithm and Benjamini and Hochberg's (1995) method.

This article concerns the choice of $f_0(z)$, the null hypothesis density. In the drug mutation example, it is crucial to determine whether f_0 is taken to be the theoretical null, N(0, 1), or the empirical null, $N(-.35, 1.20^2)$. This is illustrated in Figure 2, a close-up view of Figure 1 focusing on the bin containing $z = -3$. The expected number of the 444 z_i values falling into this bin is 6.37 for $f(z)$, and either .62 or 3.90 as $f_0(z)$

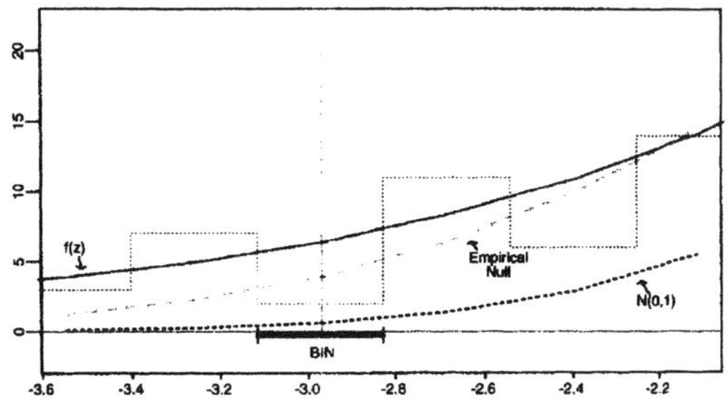

Figure 2. Close-Up View of the Bin Containing $z = -3$ in Figure 1. Expected numbers in the bin are 6.37 for f(z), .62 for $f_0 = N(0, 1)$, and 3.90 for $f_0 = N(.35, 1.20^2)$, the empirical null. Corresponding estimates of fdr(−3) are .097 for N(0, 1) versus .612 for $N(-.35, 1.20^2)$. Should we report the cases in this bin as Interesting?

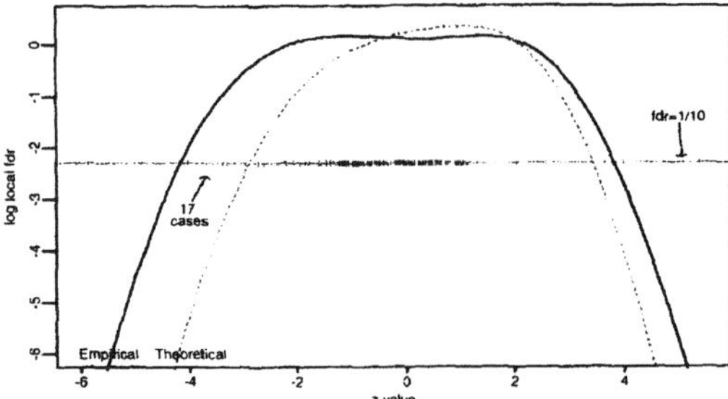

Figure 3. Comparison of Estimates of log fdr(z) for the Drug Mutation Data. The empirical null estimate (——) declines more slowly than the theoretical null estimate (······). Dashes indicate the 444 z-values. A total of 17 cases on left have fdr(z) < 1/10 for theoretical but > 1/10 for empirical.

is N(0, 1) or $N(-.35, 1.20^2)$. Thus $\mathrm{fdr}(z) = f_0(z)/f(z)$ at $z = -3$ is estimated to be either

$$\mathrm{fdr}(-3) = \begin{cases} .097 & \text{using the theoretical null N(0, 1)} \\ \text{or} & \\ .612 & \text{using the empirical null } N(-.35, 1.20^2). \end{cases} \tag{13}$$

In this bin, changing from the theoretical null to the empirical null changes the inferences from Interesting to definitely Uninteresting.

Figure 3 compares the two estimates of log fdr(z) over most of the z scale. As shown, 18 of the 444 z-values have $\mathrm{fdr}(z) < .10$ for $f = N(0, 1)$ but $> .10$ for $f_0 = N(-.35, 1.20^2)$, with 17 of these at the left end of the scale. All told, the empirical null yields only two-thirds as many cases with fdr < .10 as the theoretical null (35 versus 53).

3. ESTIMATING THE EMPIRICAL NULL DISTRIBUTION

The empirical null distribution for the drug mutation data is estimated in two steps: (1) Fit the curve $f(z)$ shown in Figure 1 to the histogram counts by Poisson regression, and (2) Obtain the center and half-width of the central peak, say δ_0 and σ_0, from $f(z)$,

$$\delta_0 = \arg\max\{f(z)\} \quad \text{and} \quad \sigma_0 = \left[-\frac{d^2}{dz^2}\log f(z)\right]_{\delta_0}^{-\frac{1}{2}}, \tag{14}$$

yielding $(\delta_0, \sigma_0) = (-.35, 1.20)$. Details are given in Remark D (Sec. 7), which briefly discusses the possibility of a nonnormal empirical null distribution.

More direct estimation methods for f_0 seem possible; for example, estimating δ_0 by the median of the z-values. Suppose, however, that 10% of the z-values came from the nonnull distribution and that all of these were located at the far left end of Figure 1. Then the median of all the z's would be the 4/9 quartile of the actual null distribution, not its median, yielding a badly biased estimate of δ_0. Similar comments apply to estimating σ_0 (see Remark D). Method (14) does not require preliminary estimates of the proportion p_0 in the null population of (9), a considerable practical advantage.

How accurate are the estimates (−.35, 1.20)? The usual standard error approximations for a Poisson regression fit are not appropriate here, because the z_i's are not independent of each other. A nonparametric bootstrap analysis was performed instead, with the 1,391 80-dimensional vectors $(\mathbf{x}, \mathbf{v})$ [(5) and (6)], as the resampling units. This yielded .09 and .08 for the bootstrap standard errors of δ_0 and σ_0, that is,

$$(\delta_0, \sigma_0) = (-.35, 1.20) \pm (.09, .08). \tag{15}$$

It seems quite unlikely that estimation error alone accounts for the difference between the empirical null and the theoretical values $(\delta_0, \sigma_0) = (0, 1)$. (Note that this type of bootstrap analysis, which requires independent sampling units, is not applicable to the microarray example of Sec. 6, where correlations among the genes are present.)

The next two sections concern other possible causes for empirical/theoretical differences, diagnostics for these causes, and their interpretations. This list is not exhaustive, and in fact the microarray example of Section 6 demonstrates another form of pathology.

4. PERMUTATION TESTS AND UNOBSERVED COVARIATES

The theoretical N(0, 1) null hypothesis (4) is usually based on asymptotic approximations like those for the logistic regression coefficients in the drug mutation study. Permutation methods can be used to avoid these approximations, perhaps in the hope that an improved theoretical null will more closely match the empirical.

This was not the case for the drug mutation data, for which permutation testing was implemented by randomly pairing the 1,391 predictor vectors $\mathbf{x}$, (5), with the 1,391 response vectors $\mathbf{v}$, (6), and recalculating the 444 z-values. This whole process was repeated independently 20 times, yielding a total of 20 × 444 permutation z's. Their distribution was well approximated by a $N(0, .965^2)$ density (the "permutation null"), except for a prominent spike near $z = .3$. In this case, the permutation-improved theoretical null differs more, rather than less, from the empirical null $N(-.35, 1.20^2)$.

Permutation methods are popular in the microarray literature as a way of avoiding assumptions and approximations (see Efron, Tibshirani, Storey, and Tusher 2001; Dudoit, Shaffer, and Boldrick 2003), *but they do not automatically resolve the question of an appropriate null hypothesis.* This can be seen in the following hypothetical example, which is a stylized version of the two-sample microarray testing problem discussed in Section 6. The data, x_{ij}, come from N simultaneous two-sample experiments, each comparing $2n$ subjects,

$$x_{ij} \begin{cases} \text{Controls}, & j = 1, 2, \ldots, n \\ \text{Treatments}, & j = n+1, n+2, \ldots, 2n \end{cases} \quad (i = 1, \ldots, N). \tag{16}$$

The ith test statistic, Y_i, is the usual two-sample t statistic, comparing Treatments versus Controls for the ith experiment.

Suppose that, unknown to the statistician, the data were actually generated from

$$x_{ij} = u_{ij} + \frac{I_j}{2}\beta_i \qquad \begin{cases} u_{ij} \sim N(0, 1) \\ \beta_i \sim N(0, \sigma_\beta^2). \end{cases} \tag{17}$$

with the u_{ij} and β_i mutually independent and

$$I_j = \begin{cases} -1, & j = 1, 2, \ldots, n \\ +1, & j = n+1, \ldots, 2n \end{cases} \tag{18}$$

Then it is easy to show that the statistics Y_i follow a dilated t distribution with $2n - 2$ df,

$$Y_i \sim \left(1 + \frac{n}{2}\sigma_\beta^2\right)^{\frac{1}{2}} \cdot t_{2n-2}, \tag{19}$$

whereas the permutation distribution, permuting Treatments and Controls within each experiment, has nearly a standard t_{2n-2} null distribution. So, for example, if $\sigma_\beta^2 = 2/n$, then the empirical density of the Y_i's will be $\sqrt{2}$ times as wide as the permutation null.

The quantity β_i in (17) and (18) produces the only consistent differences between Treatments and Controls in experiment i. If β_i is a dependable feature of the ith experiment, and would appear again with the same value in a replication of the study, then the permutation null t_{2n-2} is a reasonable basis for inference. With n large and $\sigma_\beta^2 = 2/n$, this results in $\text{fdr}(y_i) < .10$ for the most extreme 2% of the observed t statistics, favoring those with the largest values of $|\beta_i|$.

Suppose, however, that β_i is not inherent to experiment i, but rather is a purely random effect that would have a different value and perhaps a different sign if the study were repeated; that is, β_i is part of the noise and not part of the signal. In this case, the appropriate choice is the empirical null (19). The equivalent of Figure 1 would be *all* central peak, with no interesting outliers, and no cases with small values of $\text{fdr}(y_i)$. This is appropriate, because now there is no real Treatment effect.

In this latter context β_i acts as an *unobserved covariate*, a quantity that the statistician would use to correct the Treatment–Control comparison if it were observable. Unobserved covariates are ubiquitous in observational studies. There are several obvious ones in the drug mutation study, including personal patient characteristics, such as age and gender, previous use of AZT and other non-PI drugs, years since infection, geographic location, and so on.

The effect of important unobserved covariates is to dilate the null hypothesis density $f_0(z)$, as happens in (19). Unobserved covariates will also dilate the Interesting density $f_1(z)$ in (9), and the mixture density $f(z)$, (10). However, an empirical fitting method for estimating $f(z)$, such as the spline fit in Figure 1, automatically includes any dilation effects. In estimating $\text{fdr}(z) = f_0(z)/f(z)$, it is important to also allow for dilation of the numerator f_0. *This is a strong argument for preferring the empirical null hypothesis in observational studies.*

5. A STRUCTURAL MODEL FOR THE z-VALUES

The Bayesian specifications (9) underlying the fdr results have the advantage of not requiring a structural model for the z-values; in particular, it is not necessary to motivate, or even describe, the nonnull density $f_1(z)$. There is, however, a simple structural model that can help elucidate the Interesting–Uninteresting distinction in (9).

The structural model assumes that z_i, the ith z-value, is normally distributed around a "true value" μ_i, its expectation,

$$z_i \sim N(\mu_i, 1) \quad \text{for } i = 1, 2, \ldots, N, \tag{20}$$

with μ_i having some prior distribution $g(\mu)$,

$$\mu_i \sim g(\mu) \quad \text{for } i = 1, 2, \ldots, N. \tag{21}$$

Structure (20) is often a good approximation (see Efron 1988, sec. 4), and in fact proved reasonably accurate in the bootstrap experiment yielding (15). Together, (20) and (21) say that the mixture density $f(z)$, (10), is a convolution of $g(\mu)$ with the standard normal density $\varphi(z)$,

$$f(z) = \int_{-\infty}^{\infty} \varphi(z - \mu) g(\mu)\, d\mu \tag{22}$$

[with the understanding that $g(\mu)$ may include discrete probability atoms].

As a first application of the structural model, suppose that we insist that $g(\mu)$ put probability p_0 on $\mu = 0$,

$$\Pr_g\{\mu = 0\} = p_0, \tag{23}$$

for some fixed value of p_0 between 0 and 1. This amounts to the original Bayes model (9) with $p_0 = \Pr\{\text{Uninteresting}\}$, $f_0(z)$ the theoretical null hypothesis $N(0, 1)$, and

$$f_1(z) = \int_{\mu \neq 0} \varphi(z - \mu) g(\mu)\, d\mu \Big/ (1 - p_0). \tag{24}$$

In the context of this article, p_0 should be .90 or greater.

For any $f(z)$ of the convolution form (22), let (δ_g, σ_g) be the center and width parameters (δ_0, σ_0) defined by (14). Figure 4 answers the following question: For a given choice of p_0 in constraint (23), what are the maximum possible values of $|\delta_g|$ and of σ_g,

$$\delta_{\max} = \max\{|\delta_g| | p_0\} \quad \text{and} \quad \sigma_{\max} = \max\{\sigma_g | p_0\}? \tag{25}$$

Three curves appear for $\sigma_{\max}$, for the general case just described, for the case where the nonzero component of $g(\mu)$ is required to be symmetric around 0, and for the case where it is also required to be normal. Here only the general case will be discussed. Remark F (Sec. 7) discusses the solution of (25), which turns out to have a simple "single-point" form.

The notable feature of Figure 4 is that for $p_0 \geq .90$, my preferred realm for large-scale hypothesis testing, $(\delta_{\max}, \sigma_{\max})$ must be quite near the theoretical null values (0, 1),

$$\delta_{\max} \leq .07 \quad \text{and} \quad \sigma_{\max} \leq 1.04. \tag{26}$$

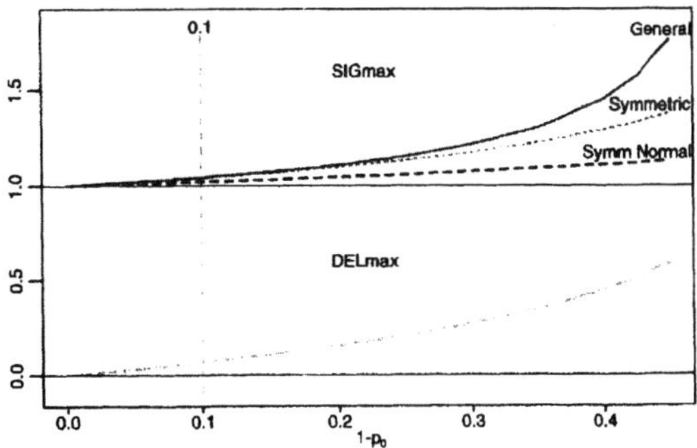

Figure 4. Maximum Possible Values of the Center and Width Parameters (δ_0, σ_0), (14), When the Structural Model (20)–(22) Is Constrained to Put Probability p_0 on $\mu = 0$. For $1 - p_0 \leq .10$, the maxima are not much greater than the theoretical null values (0, 1), as shown in Table 1.

Table 1. Value of σ_{max} and δ_{max} as a Function of $1 - p_0$ (23)

$1 - p_0$:	.05	**.10**	.20	.30	(Drug mutation)
σ_{max}:	1.02	**1.04**	1.11	1.22	(1.20)
δ_{max}:	.03	**.07**	.15	.27	(−.35)

Table 1 shows $(\delta_{max}, \sigma_{max})$ for various choices of p_0. It shows that the "Interesting" probability $1 - p_0$ would have to be nearly .30, very large by the standards of large-scale testing, to obtain the observed drug mutation values $(\delta_0, \sigma_0) = (-.35, 1.20)$. The inference is that Uninteresting effects, such as the unobserved covariates of Section 4, are dilating the null hypothesis.

The main point here is that the measures (14) of center and width are quite robust to the arrangement of Interesting values μ_i, as long as the Interesting percentage does not exceed 10%. If (δ_0, σ_0) for the central peak is much different than (0, 1), as it is in Figure 1, then using the theoretical null is bound to result in identifying an uncomfortably large percentage of supposedly Interesting cases.

We can pursue this last point for the drug mutation data by removing constraint (23). Figure 5 shows an unconstrained estimate of $g(\mu)$. For computational simplicity, $g(\mu)$ was assumed to be discrete, with at most $J = 8$ support points $\mu_1, \mu_2, \ldots, \mu_J$, so that (22) becomes

$$f(z) = \sum_{j=1}^{J} \pi_j \varphi(z - \mu_j), \tag{27}$$

π_j being the probability g puts on μ_j, with $\pi_j \geq 0$ and $\sum \pi_j = 1$. A nonlinear minimization program was employed to find the best-fitting curve of form (27) to the histogram counts in Figure 1, using Poisson deviance as the fitting criterion. The vertical bars in Figure 5 are located at the resulting eight values μ_j, with the bar's height proportional to π_j. For example, the little bar at far left represents an atom of probability $\pi_1 = .015$ at $\mu_1 = -10.9$. The resulting $f(z)$ estimate, (26), closely resembles the natural spline fit of Figure 1. Table 2 shows all eight (π_j, μ_j) pairs.

Suppose for a moment that the estimated $g(\mu)$ is exactly correct, so 1.5% of the 444 cases have their μ_i's equal to −10.9, 1.3%, to −7.0, and so on, and that an oracle has told us the eight (π_j, μ_j) values. Given an observed z_i, we can now calculate Pr{Uninteresting|z}, (11), exactly, *once the scientist specifies the definition of Uninteresting versus Interesting*. It seems obvious that the 60.8% at $\mu_j = 0$ are Uninteresting, and that the 10.6% at $\mu_j = -10.9, -7.0, -4.9$, and 6.1 deserve Interesting status. However, the status of the 28.6% at $\mu_j = -1.8$, −1.1, and 2.4 is less clear.

If the 28.6% are deemed Interesting, then this leaves only the 60.8% at $\mu_j = 0$ as Uninteresting. In terms of the Bayes

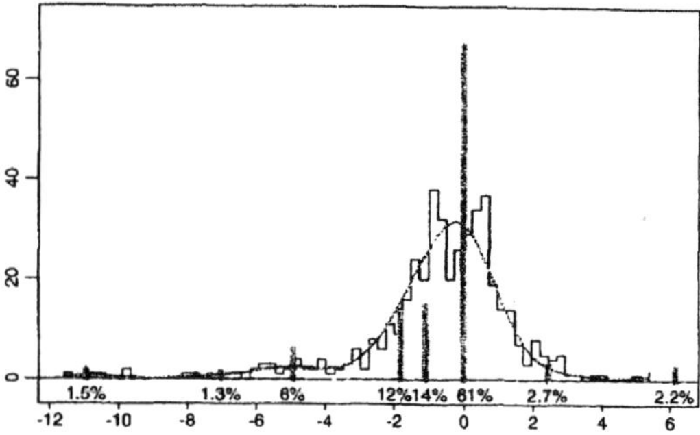

Figure 5. Best-Fit Discrete Mixing Function $g(\mu)$, (21), for Drug Mutation Data. The bars are located at support points μ_j, the heights are proportional to weights π_j, and the tall bar at $\mu_j = 0$ has weight $\pi_j = .61$. Solid curve is a best-fit estimate $f(z) = \sum \pi_j \varphi(z - \mu_j)$; it closely matches the natural spline fit from Figure 1 (- - - -).

model (9), this yields $p_0 = .608$ and $f_0(z) \sim N(0, 1)$, the theoretical null. About 174 of the 444 cases will be identified as Interesting, too many for a typical screening exercise. Shifting the 28.6% to the Uninteresting classification increases p_0 to .608 + .286 = .894, a more manageable value, and changes $f_0(z)$ to the version of (27) supported on the four Uninteresting μ_j's,

$$f_0(z) = \sum_{j=4}^{7} \pi_j \varphi(z - \mu_j) \Big/ \sum_{j=4}^{7} \pi_j. \tag{28}$$

This is approximately $N(-.34, 1.19^2)$, almost the same as the empirical null (8).

In other words, the definition of "Interesting" determines the relevant choice of the null hypothesis f_0. If the goal is to keep the proportion of Interesting cases manageably small, then $f_0(z)$ must grow wider than N(0, 1).

Use of the term "Interesting" rather than "Significant" reflects a difference in intent between large-scale and classical testing. In the hypothetical context of Figure 5 and Table 2, all of the 39.2% of the cases with nonzero μ_i's would eventually be declared as "significantly different from 0" if the sample size of patients was vastly increased. Section 4 suggests that minor deviations from N(0, 1) might arise from scientifically uninteresting causes, such as unobserved covariates. However, even if a modestly nonzero μ_i is genuine in some sense, it may still be Uninteresting when viewed in comparison with an ensemble of more dramatic possibilities. Nonsignificant implies Uninteresting, but not conversely.

6. A MICROARRAY EXAMPLE

Microarrays have become a prime source of large-scale simultaneous testing problems. Figure 6 relates to a well-known

Table 2. Weights π_j and Locations μ_j for the Eight-Point Best-Fit Estimate $g(\mu)$ of Figure 8

	−Interesting−			?	?	Uninteresting	?	Interesting
$100 \cdot \pi_j$	1.5%	1.3%	5.6%	12.3%	13.6%	60.8%	2.7%	2.2%
μ_j	**−10.9**	**−7.0**	**−4.9**	**−1.8**	**−1.1**	**0**	**2.4**	**6.1**

NOTE: Which locations deemed Interesting versus Uninteresting determines the choice between the theoretical or empirical null hypothesis. (Numerical results accurate to one decimal place.)

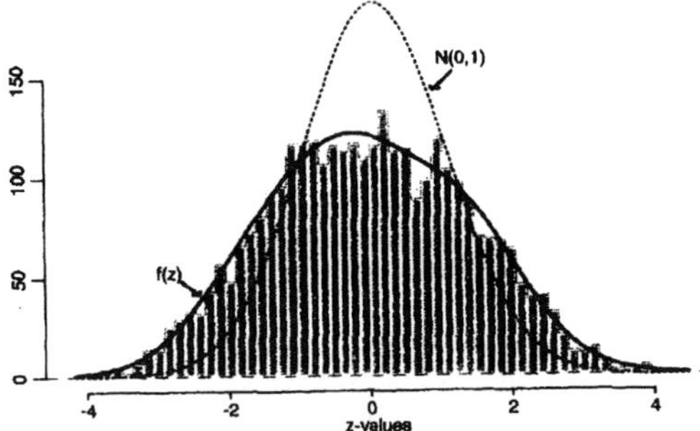

Figure 6. Histogram of $N = 3{,}226$ z-Values From the Breast Cancer Study. The theoretical N(0, 1) null is much narrower than the central peak, which has $(\delta_0, \sigma_0) = (-.02, 1.58)$. In this case the central peak seems to include the entire histogram.

microarray experiment concerning differences between two types of genetic mutations causing increased breast cancer risk, BRCA1 and BRCA2 (see Hedenfalk, Duggen, and Chen 2001; Efron and Tibshirani 2002; Efron 2003).

The experiment included 15 breast cancer patients, 7 with BRCA1 and 8 with BRCA2. Each women's tumor was analyzed on a separate microarray, each microarray reporting on the same set of $N = 3{,}226$ genes. For each gene, the two-sample t statistic y_i comparing the seven BRCA1 responses with the eight BRCA2's was computed. The y_i's were then converted to z-values,

$$z_i = \Phi^{-1} F_{13}(y_i), \tag{29}$$

where F_{13} is the cdf of a standard t distribution with 13 df. Figure 6 displays the histogram of the 3,226 z-values.

The central peak is wider here than in Figure 1, with center-width estimates $(\delta_0, \sigma_0) = (-.02, 1.58)$. More importantly, the histogram seems to be *all* central peak, with no interesting outliers such as those seen at the left of Figure 1. This was reflected in the local fdr calculations; using the theoretical N(0, 1) null yielded 35 genes with $\text{fdr}(z_i) < .1$, those with $|z_i| > 3.35$; using the empirical $N(-.02, 1.58^2)$ null, no genes at all had fdr $< .1$—or, for that matter, fdr $< .9$, the histogram in fact being a little short-tailed compared with $N(-.02, 1.58^2)$.

There is ample reason to distrust the theoretical null in this case. The microarray experiment, for all its impressive technology, is still an observational study, with a wide range of unobserved covariates possibly distorting the BRCA1–BRCA2 comparison.

Another reason for doubt can be found in the data itself. The fdr methodology does not require independence of the y_i's or z_i's across genes. However, it does require that the 15 measurements for *each* gene be independent across the microarrays. Otherwise, the two-sample t statistic y_i will not have an t_{13} null distribution, not even approximately.

Unfortunately the experimental methodology used in the breast cancer study seems to have induced substantial correlations among the various microarrays. In particular, as discussed in Remark G, the first four microarrays in the BRCA2 groups were mutually correlated, and likewise the last four. Correlations reduce the effective sample size for a two-sample t statistic, just the type of effect that would induce overdispersion in (29).

This does not say that there are no BRCA1–BRCA2 differences, only that it is dangerous to compare the t statistics with a standard t_{13} null distribution, even if simultaneous inference is accounted for.

7. REMARKS

Remark A (Drug mutation study). The data base for the drug mutation study (Wu et al. 2002), included 2,497 patients having HIV subtype B, of whom 1,391 had received at least one of six popular protease inhibitor (PI) drugs: amprenavir, indinavir, lopinavir, nelfinavir, ritonavir, or saquinavir. Among the 1,391, the mean number of PI drugs taken was 2.05 per patient. Amino acid sequences were obtained at all 99 positions on the HIV protease gene, and mutations from wild-type recorded; 25 positions showed 3 or fewer mutations among the 1,391 patients, deemed too few for analysis, leaving 74 positions for the investigation here. Each of the 74 individual logistic regressions included an intercept term as well as the six PI main effects, but no other covariates.

Remark B (The local false discovery rate). The local fdr, (11) or (12), is closely related to Benjamini and Hochberg's (1995) "tail-area" FDR, as discussed by Efron et al. (2001), Storey (2002), and Efron and Tibshirani (2002). Substituting cdf's F_0 and F for the densities f_0 and f, Bayes's theorem gives a tail-area version of (11),

$$\begin{aligned}\Pr\{\text{Uninteresting}|z \le z_0\} &= p_0 F_0(z_0)/F(z_0)\\ &\equiv \text{FDR}(z_0).\end{aligned} \tag{30}$$

Here $\text{FDR}(z_0)$ turns out to be the conditional expectation of $\text{fdr}(z) \equiv p_0 f_0(z)/f(z)$ given $z \le z_0$,

$$\text{FDR}(z_0) = \int_{-\infty}^{z_0} \text{fdr}(z) f(z)\,dz \Big/ \int_{-\infty}^{z_0} f(z)\,dz. \tag{31}$$

Benjamini and Hochberg worked in a frequentist framework, but their FDR control rule can be stated in empirical Bayes terms. Given F_0, which they usually took to be what has been called here the theoretical null, they estimate $\text{FDR}(z_0)$ by

$$\widehat{\text{FDR}}(z_0) = p_0 F_0(z)/\widehat{F}(z_0), \tag{32}$$

where $\widehat{F}$ is the empirical cdf of the z_i's. For a desired control level α, say $\alpha = .05$, define

$$z_0 = \arg\max_z \{\widehat{\text{FDR}}(z) \le \alpha\}; \tag{33}$$

then rejecting all cases with $z_i \le z_0$ gives an expected (frequentist) rate of false discoveries no greater than α.

With z_0 as in (33), relation (31) (applied to the estimated versions of FDR, fdr, and f) says that the weighted average of $\text{fdr}(z_i)$ for the cases rejected by the FDR level-α rule is itself α. As an example, take $\alpha = .05$ and f_0 equal the theoretical N(0, 1) null. Applying the FDR control rule to the negative side of Figure 1's drug mutation data rejects the null hypothesis for the 56 cases having $z_i \le -2.61$; the corresponding 56 values of $\text{fdr}(z_i)$ have weighted average $\alpha = .05$. They vary from nearly 0 at the far left to .19 at the boundary value $z = -2.61$, justifying the term "local"; z_i's near the boundary are more likely to be false discoveries than the overall .05 rate suggests.

Our concern with a correct choice of null hypothesis applies to FDR just as well as to fdr. In the microarray study, FDR with $F_0 = N(0, 1)$ gives 24 significant genes at $\alpha = .05$, whereas $F_0 = N(-.02, 1.58^2)$ gives none. In fact, any simultaneous testing procedure, the popular Westfall–Young method (Westfall and Young, 1993), for example, will depend on a correct assessment of p values for the individual cases, that is, on the choice of F_0.

Remark C [Estimating $f(z)$]. The Poisson regression method used in Figure 1 to estimate the mixture density $f(z)$, (10), originates in an idea of Lindsey described by Efron and Tibshirani (1996, sec. 2). The range of the sample $z_1, z_2, \ldots, z_N$ is partitioned into K equal intervals, with interval k having midpoint x_k and containing count s_k of the N z-values; the expectation λ_k of s_k is nearly proportional to $f_k \equiv f(x_k)$, and if the z_i's are independent, then the counts approximate independent Poisson variates,

$$s_k \overset{\text{ind}}{\sim} \text{Poi}(\lambda_k) \quad \text{and} \quad \lambda_k = cf_k, \qquad k = 1, 2, \ldots, K, \tag{34}$$

where c is a constant depending on N and the interval length.

Lindsey's method is to estimate the λ_k's with a Poisson regression, which because of (34) amounts to estimating a scaled version of the f_k's; in other words, estimating $f(z)$. K equals 60 in Figure 1, with the regression model being a natural spline with 7 df, roughly equivalent to a sixth degree polynomial fit in z.

Poisson regression based on (34) is almost fully efficient for estimating $f(z)$ if the z_i's are independent. Here one does not expect independence, but we still have the expectation of s_k proportional to f_k. The Poisson regression method will still tend to unbiasedly estimate $f(z)$, assuming the regression model is sufficiently flexible, though we may lose estimating efficiency.

I also used the bootstrap analysis that gave the standard errors in (15) to check (34). This turned out to be surprisingly accurate for the drug mutation data. If it had not, then I might have used the bootstrap estimate of covariance for the s_k's to motivate a more efficient estimation procedure, though this is unlikely to be important for large values of N. In any case bootstrap analyses as in (15) will provide legitimate standard errors for the Poisson regression whether or not (34) is valid.

Remark D (Estimating the empirical null distribution). The main tactic of this article is to estimate the null distribution $f_0(x)$ in (9) from the central peak in the z-values' histogram. Assuming normality for f_0 gives

$$\log f(z) \doteq -\frac{1}{2}\left(\frac{z-\delta_0}{\sigma_0}\right)^2 + \text{constant} \tag{35}$$

for z near 0, so that δ_0 and σ_0 can be estimated by differentiating $\log f(z)$ as in (14). The constant depends on N and p_0, but the constant has no effect on the derivatives in (14).

Directly differentiating the spline estimate of $\log f(z)$ can give an overly variable estimate of σ_0. One more smoothing step was used here, fitting a quadratic curve $a_0 + a_1 x_k + a_2 x_k^2$ by ordinary least squares to the estimated values $\log f_k$, for x_k within 1.5 units of the maximum δ_0, yielding $\sigma_0 = [-2a_2]^{-\frac{1}{2}}$ as in (14). This procedure gave the small bootstrap standard error estimate in (15).

None of this methodology is crucial, although it is important that the estimates δ_0 and σ_0 relate directly to $f_0(z)$ and are not much affected by the nonnull distribution $f_1(z)$ in (9). As an example of what can go wrong, suppose that one tries to estimate σ_0 by a "robust" scale measure, such as (84th quantile minus 16th quantile)/2. This gives $\sigma_0 = 1.47$ for the drug mutation data, reflecting long tails due to the Interesting cases in Figure 1. Similar difficulties arise using the central slope of a qq plot. Basically, a density estimate of the central peak is required, and then some assessment of its center and width.

More ambitiously, one might try extending the estimation of $f_0(z)$ to third moments, permitting a skew null distribution. Expression (35) could be generalized to

$$-\log f(z) \doteq c_0 + c_1 z + c_2 z^2/2 + c_3 z^3/6, \tag{36}$$

now requiring three derivates to estimate the coefficients rather than the two of (14). This is an unexplored path, and in particular Table 1 has not been extended to include skewness bounds.

Familiarity was the only reason for using z-values instead of t-values in Figures 1 and 6.

Remark E (Estimating p_0). One can obtain reasonable upper bounds for p_0 in (9) from estimates of

$$\pi(c) \equiv \Pr_f\{z_i \in \delta_0 \pm c\sigma_0\}. \tag{37}$$

Supposing that $f_0(z) = N(\delta_0, \sigma_0^2)$, define

$$G_0(c) = 2\Phi(c) - 1 \quad \text{and} \quad G_1(c) = \int_{\delta_0 - c\sigma_0}^{\delta_0 + c\sigma_0} f_1(z)\,dz, \tag{38}$$

the probabilities that $z_i \in \delta_0 \pm c\sigma_0$ under f_0 and f_1. Then

$$p_0 = \frac{\pi(c) - G_1(c)}{G_0(c) - G_1(c)} \le \frac{\pi(c)}{G_0(c)}, \tag{39}$$

the inequality following from the assumption that $G_1(c) \le G_0(c)$; that is, the f_1 density is more dispersed than f_0.

This leads to the estimated upper bound for p_0,

$$\widehat{p}_0 = \frac{\widehat{\pi}(c)}{G_0(c)}, \quad \text{with } \widehat{\pi}(c) = \#\{z_i \in \delta_0 \pm c\sigma_0\}/N. \tag{40}$$

In particular, if it is assumed that $G_1(c) = 0$—in other words, that the Interesting z_i's always fall outside $\delta_0 \pm c\sigma_0$—then $\widehat{p}_0 = \widehat{\pi}(c)/G_0(c)$ is unbiased. (This is the same estimate suggested in remark F of Efron et al. 2001 and Storey 2002.) Choosing $(\delta_0, \sigma_0) = (-.35, 1.20)$ and $c = 1.5$ gave $\widehat{p}_0 = .88$ for the drug mutation data, with bootstrap standard error .024.

Remark F [Single-point solutions for $(\delta_{\max}, \sigma_{\max})$]. The distributions $g(\mu)$ providing $(\delta_{\max}, \sigma_{\max})$ in (25), as graphed in Figure 4, have their nonzero components supported at a single point μ_1. For example, $g(\mu)$ for the entry giving $\sigma_{\max} = 1.04$ in Table 1 puts probability .90 at $\mu = 0$ and .10 at $\mu_1 = 1.47$. Single-point optimality was proved for three of the four cases in Figure 4, and verified by numerical maximization for the "General" case. Here is the proof for the $\sigma_{\max}$ "Symmetric" case; the other two proofs are similar.

Consider symmetric distributions putting probability p_0 on $\mu = 0$ and probabilities p_j on symmetric pairs $(-\mu_j, \mu_j)$, $j = 1, 2, \ldots, J$, so (22) becomes

$$f(z) = p_0\varphi(z) + \sum_{j=1}^{J} p_j[\varphi(z - \mu_j) + \varphi(z + \mu_j)]/2. \quad (41)$$

Defining $c_0 = p_0/(1 - p_0)$, $r_j = p_j/p_0$, and $r_+ = \sum_1^J r_j = 1/c_0$, σ_0 in (14) can be expressed as

$$\sigma_0 = (1 - Q)^{-\frac{1}{2}}, \quad \text{where } Q = \frac{\sum_1^J r_j \mu_j^2 e^{-\mu_j^2/2}}{c_0 r_+ + \sum_1^J r_j e^{-\mu_j^2/2}}. \quad (42)$$

Here $\delta_0 = 0$, which is true by symmetry assuming that $p_0 \geq 1/2$. Then σ_{max} in (25) can be found by maximizing Q.

It will be shown that with p_0 (and c_0) and $\mu_1, \mu_2, \ldots, \mu_J$ held fixed in (41), Q is maximized by a choice of $p_1, p_2, \ldots, p_J$ having $J - 1$ zero values; this is a stronger version of the single-point result. Because Q is homogeneous in $\mathbf{r} = (r_1, r_2, \ldots, r_J)$ in (42), the unconstrained maximization of $Q(\mathbf{r})$, subject only to $r_j \geq 0$ for $j = 1, 2, \ldots, J$, can be considered.

Differentiation gives

$$\partial Q/\partial r_j = \frac{1}{\text{den}}\left[\mu_j^2 e^{-\mu_j^2/2} - Q \cdot \left(c_0 + e^{-\mu_j^2/2}\right)\right], \quad (43)$$

with "den" the denominator of Q. At a maximizing point $\mathbf{r}$, we must have

$$\frac{\partial Q(\mathbf{r})}{\partial r_j} \leq 0 \quad \text{with equality if } r_j > 0. \quad (44)$$

Defining $R_j = \mu_j^2/(1 + c_0 e^{\mu_j^2/2})$, (43) and (44) yield

$$Q(\mathbf{r}) \geq R_j \quad \text{with equality if } r_j > 0. \quad (45)$$

Because $Q(\mathbf{r})$ is the maximum, this says that r_j, and p_j can be nonzero only if j maximizes R_j. In case of ties, one of the maximizing j's can be arbitrarily chosen.

All of this shows that only $J = 1$ need be considered in (41). The global maximized value of r_0 in (41) is $\sigma_{max} = (1 - R_{max})^{-\frac{1}{2}}$, where

$$R_{max} = \max_{\mu_1}\{\mu_1^2/(1 + c_0 e^{\mu_1^2/2})\}. \quad (46)$$

The maximizing argument μ_1 ranges from 1.43 for $p_0 = .95$ to 1.51 for $p_0 = .70$. The corresponding result for δ_{max} is simpler, $\mu_1 = \delta_{max} + 1$.

Remark G (Microarray correlation in the breast cancer study). It is easy to spot an unwanted correlation structure among the eight BRCA2 microarrays. Let $\mathbf{X}$ be the $3{,}226 \times 8$ matrix of BRCA2 data, with the columns of $\mathbf{X}$ standardized to have mean 0 and variance 1. A "de-gened" matrix $\widetilde{\mathbf{X}}$ was formed by subtracting row-wise averages from each element of $\mathbf{X}$,

$$\widetilde{x}_{ij} = x_{ij} - \sum_{k=1}^{8} x_{ik}/8. \quad (47)$$

Table 3 shows the 8×8 correlation matrix of $\widetilde{\mathbf{X}}$. With genuine gene effects subtracted out, the correlations should vary around $-1/7 = -.14$ if the columns of $\mathbf{X}$ are independent. Instead, the columns are correlated in blocks of four, with the off-diagonal blocks too negative and the on-diagonal blocks too positive.

Table 3. Correlation Matrix for the BRCA2 Data With Row-Wise Means Subtracted off (46), Indicating Positive Correlations Within the Two Blocks of Four

	1	2	3	4	5	6	7	8
1	1.00	.02	.02	.23	−.36	−.35	−.39	−.34
2	.02	1.00	.10	−.08	−.30	−.30	−.23	−.33
3	.02	.10	1.00	−.17	−.21	−.26	−.31	−.27
4	.23	−.08	−.17	1.00	−.30	−.23	−.27	−.32
5	−.36	−.30	−.21	−.30	1.00	−.02	.11	.22
6	−.35	−.30	−.26	−.23	−.02	1.00	.15	.13
7	−.39	−.23	−.31	−.27	.11	.15	1.00	.07
8	−.34	−.33	−.27	−.32	.22	.13	.07	1.00

Remark H (Scaling properties). The associate editor pointed out that the combination of empirical null hypotheses with false discovery rate methodology "scales up" nicely, in terms of both the number of tests and the amount of information per test. Consider the structural model (20), (21) with $g(\mu)$ a mixture of 99% $\mu \sim N(0, .01)$ and 1% of $\mu = 5$. For N the number of tests large enough, methods like Bonferroni bounds that control the family-wise error rate will eventually accept all N null hypotheses; fdr methods, using either the empirical or theoretical null, will correctly identify most of the Interesting 1%.

Suppose now that the amount of information per test increases by a factor of n, so that each $\mu_i \to \sqrt{n}\,\mu_i$ in (21). Using the theoretical N(0, 1) null makes fdr reject all N cases for n sufficiently large, whereas the empirical null continues to identify only the Interesting 1%. In this context, the fdr/empirical combination avoids the standard criticism of hypotheses testing, that rejection becomes certain for large sample sizes.

8. SUMMARY

Large-scale simultaneous hypothesis testing, where the number of cases exceeds, say 100, permits the empirical estimation of a null hypothesis distribution. The empirical null may be wider (more dispersed) than the theoretical null distribution that would ordinarily be used for a single hypothesis test. The choice between empirical and theoretical nulls can greatly influence which cases are identified as "Significant" or "Interesting," as opposed to "Null" or "Uninteresting," this being true no matter which simultaneous hypothesis testing method is used.

This article presents an analysis plan for large-scale testing situations:

- A density fitting technique is used to estimate the null hypothesis distribution f_0, (Fig. 1 and Sec. 3).
- The local false discovery rate (fdr), an empirical Bayes version of standard FDR theory, provides inferences for the N cases (Fig. 3 and Sec. 2).

There are many possible reasons for overdispersion of the empirical null distribution that would lead to the empirical null being preferred for simultaneous testing including:

- Unobserved covariates in an observational study, (Sec. 4)
- Hidden correlations (Sec. 6)
- A large proportion of genuine but uninterestingly small effects (Fig. 5).

Large-scale testing differs in scientific intent from an individual hypothesis test. The latter is most often designed to reject the null hypothesis with high probability. Large-scale testing is usually more of a screening operation, intended to identify a *small* percentage of Interesting cases, assumed to be on the order of 10% or less in this article. The empirical null hypothesis methodology is designed to be accurate under this constraint (Fig. 4). More traditional estimation methods, involving permutations or quantiles, give incorrect f_0 estimates (Sec. 4 and Remark D).

[Received June 2003. Revised August 2003.]

REFERENCES

Benjamini, Y., and Hochberg, Y. (1995), "Controlling the False Discovery Rate: A Practical and Powerful Approach to Multiple Testing," *Journal of the Royal Statistical Society*, Ser. B, 57, 289–300.

Dudoit, S., Shaffer J., and Boldrick J. (2003), "Multiple Hypothesis Testing in Microarray Experiments," *Statistical Science*, 18, 71–103.

Efron, B. (1988), "Three Examples of Computer-Intensive Statistical Inference," *Sankhyā*, 50, 338–362.

——— (2003), "Robbins, Empirical Bayes, and Microarrays," *The Annals of Statistics*, 31, 366–378.

Efron, B., and Tibshirani, R. (1996), "Using Specially Designed Exponential Families for Density Estimation," *The Annals of Statistics*, 24, 2431–2461.

——— (2002), "Empirical Bayes Methods and False Discovery Rates for Microarrays," *Genetic Epidemiology*, 23, 70–86.

Efron, B., Tibshirani, R., Storey, J., and Tusher, V. (2001), "Empirical Bayes Analysis of a Microarray Experiment," *Journal of the American Statistical Association*, 96, 1151–1160.

Hedenfalk, I., Duggen, D., Chen, Y. et al. (2001), "Gene Expression Profiles in Hereditary Breast Cancer," *New England Journal of Medicine*, 344, 539–548.

Miller, R. (1981), *Simultaneous Statistical Inference* (2nd ed.), New York: Springer-Verlag.

Storey, J. (2002), "A Direct Approach to False Discovery Rates," *Journal of the Royal Statistical Society*, Ser. B, 64, 479–498.

——— (2003), "The Positive False Discovery Rate: A Bayesian Interpretation and the q-Value," *The Annals of Statistics*, 31, to appear.

Westfall, P., and Young, S. (1993), *Resampling-Based Multiple Testing: Examples and Methods for p-Value Adjustments*, New York: Wiley.

Wu, T., Schiffer, C., Shafer, R. et al. (2003), "Mutation Patterns and Structural Correlates in Human Immunodeficiency Virus Type 1 Protease Following Different Protease Inhibitor Treatments," *Journal of Virology*, 77, 4836–4847.

Bradley Efron

But What Do Statisticians *Do?*

At a recent birthday party a friend produced a fake photo of me sitting next to Fidel Castro, with the caption "Brad tries to look interested as Fidel explains that statistics was his worst class in college." There is a depressing frequency to this sort of conversation, for me and I suspect for most of you, too. What is it about statistics that elicits confessions of incompetence? It's hard to imagine "I was a complete dunce in English" as common party chatter, but statistics seems to be fair game.

These thoughts came back to me recently when I was interviewed for an ongoing column in the *Stanford Magazine* called "What do they do?" The nice (and smart) interviewer told me that this was a fairly new feature designed to explain "unusual" campus professions to the Stanford readership. The previous interviewee was the campus gardener, whom I expect had an easier time conveying his duties than I. He could point (authoritatively—the campus looks great) to trees and flowerbeds and pristine paths, things that most people know and treasure. What does a statistician point to?

Well, I tried my hardest. An ongoing consulting job at the medical school seemed promising. It involved 20,000 genes and 88 microarrays in a massive experiment concerning the genetic basis of atherosclerosis. I hauled out graphs and charts and computer displays, trying to convey the dangers of answering "which genes are important?" in a context with 20,000 candidates. I think, or maybe just hope, that some feeling for the collaboration between scientists and statisticians came across. The fact is that it was a complicated story, and my interviewer would need advanced reporting skills not to get lost in a maze of microarrays before getting to the statistics part.

Which brings us to the nub of the "what do statisticians *do*" problem. My research collaborators are trying to cure heart disease. I'm trying to help them understand the data they've collected in trying to cure heart disease. Statisticians work at least two levels away from nature (three levels when we write *JASA* papers intended to help other statisticians help scientists understand nature). In its essence our work is more abstract than most of science, and science itself is scary enough for most people.

Another way to say this is that statistics is an information science, in fact, the first information science, and the most developed. In the first four decades of the 20th century, an enormously ambitious and successful intellectual effort produced a theory of inference that cuts across individual scientific disciplines. It was by no means obvious that such a theory should exist—why should a method like regression apply to economics as well as astronomy, geology, and medicine? It does exist, much to the benefit of science, as I was trying to get across with the microarray story. We don't need to construct a special theory for each new area of application, though context usually determines the specific details. Modern statisticians have combined 20th century theory with 21st century computing power to build a general inference machine of impressive force. This machine has become the principal mode of scientific inference in literally dozens of fields, and the list keeps growing.

This makes it more surprising how little many scientists seem to know about statistics. Unless they have had occasion to need our services, they are likely to think of statistics as a computer program, something like TurboTax, or maybe a dimly remembered formula for the standard error. My rule of thumb has been that scientific colleagues can do quite well with probability models but are not to be trusted with statistical inference. We are the only ones trained in the arduous kind of reverse logic required to go from data, back to an unseen model.

A legitimate answer to the *what-do-statisticians-do* question is "everything." In the past few months I've heard statistics department seminars touching on medicine, biology, sociology, geology (image analysis), particle physics, and chemistry. Statistics departments tend to annoy university and business administrators by not fitting neatly into the organization chart, sprawling across division and school lines. One of the real charms of statistics is the opportunity to peek into everyone else's science business. In an age of specialization, we might be the last remaining scientific generalists.

Of course "I'm a generalist" doesn't get one very far at the cocktail party. Most people seem to know, or think they know, what biologists or physicists or economists do for a living. Even computer scientists have a movie-made persona these days, the geeky guy eating pizza in front of multiple screens, while he hacks into secret web sites. Statisticians don't have a signature piece of equipment to fall back upon. Our profession is defined by its ideas and methodology, rather than specific subject matter or technology, which is noble enough but difficult to explain.

"How do statisticians *think*?" is a better question than what do we do. I still remember how difficult it was for me, an undergraduate mathematician, to understand why something like ANOVA made sense in analyzing real scientific problems. NFL cornerbacks learn the difficult skill of running backwards; we have to learn to think backwards, from the specific instance to the general rule. Along the way we also learn proper respect for the power of randomness to confuse the human mind. Statisticians are good at spotting patterns hidden in random noise, and even better at not being fooled by apparent patterns.

There are a few of us, not me, I'm sorry to say, who do have the gift of conveying statistical ideas to the general public. (Fred Mosteller comes to mind.) Telling people what statisticians do is an important job for us, and also for society as a whole, which could use a good deal more clear statistical thinking about crucial issues like environmental policy, terrorism, and preventive medicine. Those few journalists who can write clearly about statistical topics, Gina Kolata for one, are our prime allies in the war against ignorance. To this end, the ASA has instituted an Excellence in Statistical Reporting Award (ERISA). The selection committee, Don Berry, Betz Halloran, and John Rolph, are looking for the first winner among "members of the communications media who best display a commitment to statistics and to advancing the role of the media in the science of statistics in public life." You can find out more on the ASA web site, *www.amstat.org*, and even nominate your own favorite journalist.

I hope that one of my guest columnists will have something helpful to say about dealing with that awkward pause at the cocktail party, just after you've answered, "I'm a statistician." I can report my only success along this line, which occurred at a university gathering for winners of a minor award. Here is the conversation in full:

Nice Lady: "And what do you do?"

Me: "I'm a statistician"

Lady (after a pause): "Uh... what did you win for?"

Me: "I invented the mean."

She left perfectly happy.

Bradley Efron

Index

Springer Series in Statistics

(continued from p. ii)

Le/Zidek: Statistical Analysis of Environmental Space-Time Processes
Liese/Miescke: Statistical Decision Theory: Estimation, Testing, Selection
Liu: Monte Carlo Strategies in Scientific Computing
Manski: Partial Identification of Probability Distributions
Mielke/Berry: Permutation Methods: A Distance Function Approach, 2nd edition
Molenberghs/Verbeke: Models for Discrete Longitudinal Data
Morris/Tibshirani: The Science of Bradley Efron: Selected Papers
Mukerjee/Wu: A Modern Theory of Factorial Designs
Nelsen: An Introduction to Copulas, 2nd edition
Pan/Fang: Growth Curve Models and Statistical Diagnostics
Politis/Romano/Wolf: Subsampling
Ramsay/Silverman: Applied Functional Data Analysis: Methods and Case Studies
Ramsay/Silverman: Functional Data Analysis, 2nd edition
Reinsel: Elements of Multivariate Time Series Analysis, 2nd edition
Rosenbaum: Observational Studies, 2nd edition
Rosenblatt: Gaussian and Non-Gaussian Linear Time Series and Random Fields
Särndal/Swensson/Wretman: Model Assisted Survey Sampling
Santner/Williams/Notz: The Design and Analysis of Computer Experiments
Schervish: Theory of Statistics
Shaked/Shanthikumar: Stochastic Orders
Shao/Tu: The Jackknife and Bootstrap
Simonoff: Smoothing Methods in Statistics
Song: Correlated Data Analysis: Modeling, Analytics, and Applications
Sprott: Statistical Inference in Science
Stein: Interpolation of Spatial Data: Some Theory for Kriging
Taniguchi/Kakizawa: Asymptotic Theory for Statistical Inference for Time Series
Tanner: Tools for Statistical Inference: Methods for the Exploration of Posterior Distributions and Likelihood Functions, 3rd edition
Tillé: Sampling Algorithms
Tsaitis: Semiparametric Theory and Missing Data
van der Laan/Robins: Unified Methods for Censored Longitudinal Data and Causality
van der Vaart/Wellner: Weak Convergence and Empirical Processes: With Applications to Statistics
Verbeke/Molenberghs: Linear Mixed Models for Longitudinal Data
Weerahandi: Exact Statistical Methods for Data Analysis